Probability and Statistics
for Engineering and the Sciences

Probability and Statistics
for Engineering and the Sciences

JAY L. DEVORE

California Polytechnic State University

San Luis Obispo

Brooks/Cole Publishing Company   Monterey, California

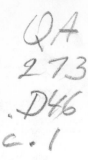

To Carol, Allie, and Teri

Brooks/Cole Publishing Company
A Division of Wadsworth, Inc.

Printed in the United States of America

10 9 8 7 6 5 4 3 2 1

Library of Congress Cataloging in Publication Data

Devore, Jay L.
   Probability and statistics for engineering
and the sciences.

   Bibliography: p.
   Includes index.
   1. Probabilities.   2. Mathematical
statistics.   I. Title.
QA273.D46   519.5   81-21744
ISBN 0-8185-0514-1   AACR2

Acquisition Editor: Craig Barth
Production Coordinator: Cece Munson
Book Design: Albert Burkhardt
Production: Greg Hubit Bookworks, Larkspur, California
Typesetting: Interactive Composition Corporation, Pleasant Hill, California

Preface

The use of probability models and statistical methods for analyzing data has become common practice in virtually all scientific disciplines. This book attempts to provide a comprehensive introduction to those models and methods most likely to be encountered and used by students in their careers in engineering and the natural sciences. Although the examples and exercises have been designed with scientists and engineers in mind, most of the methods covered are basic to statistical analyses in many other disciplines, so that students of business and the social sciences will also profit from reading the book.

Students in a statistics course designed to serve other majors may be initially skeptical of the value and relevance of the subject matter, but my experience is that students *can* be turned on to statistics by the use of good examples and exercises which blend their everyday experiences with their scientific interests. Consequently, I have worked hard to find examples of real, rather than artificial, data—data that someone thought was worth collecting and analyzing. Many of the methods presented, especially in the later chapters on statistical inference, are illustrated by analyzing data taken from a published source, and many of the exercises also involve working with such data. Sometimes the reader may be unfamiliar with the context of a particular problem (as indeed I often was), but I have found that students are more attracted by real problems with a somewhat strange context than by patently artificial problems in a familiar setting.

The exposition is relatively modest in terms of mathematical development. Substantial use of the calculus is made only in Chapter 4 and parts of Chapters 5 and 6. In particular, with the exception of an occasional remark or aside, calculus appears in the inference part of the book only in the second section of Chapter 6. A background in matrix algebra is a prerequisite only for Section 5 of the chapter on multiple regression, and no such background is required for the discussion of multiple regression models and computer analysis of such data given in Section 4 of that chapter. Thus almost all the exposition should be accessible to those whose mathematical background includes one semester or two quarters of differential and integral calculus.

Although the book's mathematical level should give most science and engineering students little difficulty, working toward an understanding of the concepts and gaining an appreciation for the logical development of the methodology may sometimes require substantial effort. To help students gain such an understanding and appreciation, I have provided numerous exercises ranging in difficulty from many that involve routine application of text material to some that ask the reader to extend concepts discussed in the text to somewhat new situations. Most of the more conceptually oriented exercises in Chapters 1–9, where the basic methodology is presented, appear in supplementary exercise sets at the end of each chapter (mixed in with more straightforward questions). I have not provided supplementary exercises for Chapters 10–15, where the more advanced methods appear, so each end-of-section exercise set contains a few more challenging problems as well as many relatively straightforward ones. There are many more exercises than most instructors would want to assign during any particular course, but I recommend that students be required to work a substantial number of them; in a problem-solving discipline, active involvement of this sort is the surest way to identify and close the gaps in understanding that inevitably arise.

There is enough material in the book for a full-year (30-week) course, so in courses of shorter duration a selection of topics will be necessary. Because goals, backgrounds, and abilities of students—and instructors' tastes—vary widely, I hesitate to make specific recommendations regarding coverage of topics. Experience in teaching a two-quarter sequence at Cal Poly (three lectures per week and no quiz or discussion sections) may provide helpful guidelines.

First-quarter coverage includes Chapter 1, most of Chapters 2 and 3, the first three sections of Chapter 4 (with little time spent on continuous families of distributions other than the normal), the last two sections of Chapter 5 (joint distributions and expected values, presented in the first two sections, are deemphasized), Section 1 of Chapter 6, and the first one or two sections of Chapters 7 and 9. In the second quarter we cover the remainder of Chapters 7 and 9, Chapter 8, selected material from Chapters 10 and 11 (typically Section 1 and parts of Sections 2 and 3 of Chapter 10, and brief mention of multifactor analyses), Chapter 12, selected portions from Sections 1 through 4 of Chapter 13, and material from the first three sections of Chapter 14. I have had success in introducing hypothesis testing right after a discussion of the binomial distribution (before any normal theory), so I have included such a section in Chapter 3, but this section can easily be skipped and the subject postponed until the course reaches Chapter 7. There always seems to be too little time in lectures to discuss all the topics that we statisticians think ought to be discussed. I hope that my presentation of material is readable enough so that students can be asked to read on their own selected portions of the text which have not or will not be covered in lecture.

Acknowledgments

I gratefully acknowledge the numerous suggestions and constructive criticisms provided by the following reviewers: Lyle Broemeling, Oklahoma State University; Maurice C. Bryson, Los Alamos National Laboratory; Louis J. Cote, Purdue University, Jonathan Cryer, University of Iowa; Arthur Dayton, Kansas State University; Janice DuBien, Western Michigan University; John Groves, California State Polytechnic University; William Lesso, University of Texas, Austin; Frederick Morgan, Clemson University; Larry Ringer, Texas A&M University; Paul Shaman, University of Pennsylvania; Martyn Smith, Michigan Technical University; and Richard Van Nostrand, Eastman Kodak Company. Typing chores were admirably handled by Lynda Alamo, Carol Devore, Fran Fairbrother, and Meri Kay Gurnee. Finally, I very much appreciate the editorial and production services rendered by Craig Barth, Greg Hubit, Joan Marsh, Cece Munson, Mike Needham, and the staff at Brooks/Cole.

Jay L. Devore

Contents

Probability and Statistics
for Engineering and the Sciences

CHAPTER 1

Introduction and Descriptive Statistics

1.1 An Overview of Probability and Statistics

In our own work, through conversations with others, and through contact with media of various sorts (books, television, newspapers, and the like), we are continually being confronted with collections of facts or **data.** Statistics is the branch of scientific inquiry which provides methods for organizing and summarizing data, and for using information in the data to draw various conclusions.

Frequently we wish to acquire information or draw some conclusion about an entire **population** consisting of all individuals or objects of a particular type. The population of interest might consist of all radial tires manufactured by a particular company during the previous calendar year, or it might be the collection of all individuals who had been inoculated with a particular flu vaccine, or it might consist of all U.S. colleges and universities. In this last case, we might be interested specifically in the number of students enrolled at each school. If so, rather than think of the schools as the population members, we may speak of the **numerical population** in which each population member is an enrollment figure such as 1536 or 21,311. In the inoculation example, we might wish to focus on whether or not individuals had subsequently exhibited a certain condition (a rash, dizziness, and so on). We might then visualize the population as consisting of Y's (for yes, the condition was present) and N's (no, the condition was not present). This is an example of a **dichotomous** (two-valued) population. In general, we will define the population to reflect our particular interests at the time of the investigation.

The data at our disposal frequently consists of a portion or subset of the population; any such subset is called a **sample.** If the population is all U.S. colleges and universities, one sample would be {Stanford University, University of Washington, Oberlin College, California Institute of Technology, Iowa State University}. If the population comprises all college and university enrollment figures, a sample might consist of {13,043, 35,234, 2756, 21,831}.

The objectives of organization and summarization of data have been pursued for hundreds of years. The part of statistics that deals with methods for performing these operations is called **descriptive statistics.** Descriptive methods can be used either when we have a list of all population members (a **census**), or when the data consists of a sample.

When the data is a sample and the objective is to go beyond the sample to draw conclusions about the population based on sample information, methods from **inferential statistics** are used. With a few isolated exceptions, the development of inferential statistics has occurred only since the early 1900s, making it of much more recent vintage than descriptive statistics. Yet much of the interest and activity in statistics today, particularly as it relates to scientific activity and experimentation, concerns inferences rather than just description. The psychologist who tries her behavior modification technique on a sample of obese individuals would like to infer something about what the technique would do if applied to all such individuals. The engineer who accumulates data on a sample of computer systems will ultimately wish to draw conclusions about all such systems. The medical team that develops a new vaccine for a disease threatening a particular population is interested in what would happen if the vaccine were administered to all people in the population. The marketing expert may test a product in a few "representative" areas; from the resulting information he will draw conclusions about what would happen if the product were made available to all potential purchasers.

The main focus of this book is on presenting and illustrating methods of inferential statistics which are useful in scientific work. The three important types of inferential procedures—point estimation, hypothesis testing, and estimation by confidence intervals—are introduced in Chapters 6–9, and then used in more complicated settings in Chapters 10–15 (a brief introduction to hypothesis testing appears earlier, in Section 3.5). The remainder of this chapter presents methods from descriptive statistics which are most used in the development of inference.

Chapters 2–5 present material from the discipline of probability. This material ultimately forms a bridge between the descriptive and inferential techniques and leads to a better understanding of how inferential procedures are developed and used, how statistical conclusions can be translated into everyday language and interpreted, and when and where pitfalls may occur in applying the methods. Probability and statistics both deal with questions involving populations and samples, but do so in an "inverse manner" to one another.

In a probability problem, properties of the population under study are assumed known (in a numerical population, for example, some specified distribution of the population values may be assumed), and questions regarding a sample taken from the population are posed and answered. In a statistics problem, characteristics of a sample are available to the experimenter, and this information enables the experimenter to draw conclusions about the population. The relationship between the two disciplines can be summarized by saying that probability reasons from the population to the sample (deductive reasoning), while statistics reasons from the sample to the population (inductive reasoning). This is illustrated in Figure 1.1.

Before we can understand what a particular sample can tell us about the population, we should first understand the uncertainty associated with taking a

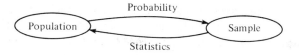

Figure 1.1   The relationship between probability and inferential statistics

sample from a given population. This is why we study probability before statistics. The following two examples pursue the distinction between the two disciplines and the types of questions asked within each.

Example 1.1   A new type of gasoline pump nozzle has been developed in order to minimize the emission of pollutants into the atmosphere during pumping. One potential design defect is that "topping off" the tanks may cause gasoline to be sucked back into the underground storage tank. (This phenomenon was described in an article which appeared in a May 1981 issue of the *Los Angeles Times.*) Suppose that 10,000 nozzles manufactured by a particular company are currently in use at Southern California gas stations. In probability, we might assume that 10% of them have the defect described above (an assumption about the population of nozzles) and ask, "How likely is it that a sample of 25 nozzles will include at least five which are defective?" or, "How many defective nozzles can we expect in a sample of 25?" On the other hand, in statistics we might find that there were five defectives in a sample of 25 nozzles (sample information) and then ask, "Does this strongly indicate that at least 10% of all nozzles currently in use are defective?" The last question concerns the population as a whole.

Example 1.2   The following measurements (reported in *Science* 167, pp. 277–279) of the ratio of the mass of the earth to that of the moon were obtained during several different spacecraft flights: 81.3001 (*Mariner 2*), 81.3015 (*Mariner 4*), 81.3006 (*Mariner 5*), 81.3011 (*Mariner 6*), 81.2997 (*Mariner 7*), 81.3005 (*Pioneer 6*), and 81.3021 (*Pioneer 7*). These numbers differ from one another (and presumably from the true ratio) because of measurement error. In probability, we might assume that the distribution of all possible measurements is bell shaped and centered at the true value of 81.3035, and then ask, "How likely is it that all seven measurements actually made result in observed ratios which are less than the true value?" In statistics, having been given the above seven measurements, we might ask, "Does this sample information conclusively demonstrate that the ratio is something other than 81.3035?" or, "How confident can we be that the true ratio is in the interval (81.2998, 81.3018)?"

In Example 1.1, the population is a well defined concrete or existing one—all nozzles presently in use. In Example 1.2, however, while the sample consists of seven observations, the population does not actually exist; instead, the population here is one consisting of all possible measurements that might be made under similar experimental conditions. Such a population is referred to as a **conceptual** or **hypothetical population.** There are a number of problems in which we fit questions into the framework of inferential statistics by conceptualizing a population. As another example, imagine that a sample of five catalytic converters with a new design is experimentally manufactured and tested. Then the inferences made will refer to the conceptual population consisting of all converters of this type which could be manufactured.

Many newcomers to the study of statistics are unaware of the broad potential application of probability and statistical methods. To remedy this, a number of statisticians contributed short, nontechnical essays to a book of readings edited by Judith Tanur, entitled *Statistics: A Guide to the Unknown* (Holden-Day, 1978). These essays are about areas of application, rather than particular methods of analysis, and serve as an excellent supplement to the problem-oriented textbook.

Exercises / Section 1.1

1. List a sample of size four from each of the following populations.
   a. The population of all daily newspapers published in the United States.
   b. The population of all U.S. corporations.
   c. The population of all companies which manufacture hand-held calculators.
   d. The population of all radio stations.

2. For each of the following hypothetical populations, list a plausible sample of size four.
   a. The population of all distances which might result when you throw a football.
   b. The population of page lengths of books which you might read during the next year.

   c. The population of all possible earthquake strength measurements (Richter scale) which might be recorded in California during the next year.
   d. The population consisting of all pH measurements made on soil samples from a particular region.

3. List three different examples of concrete populations and three different examples of hypothetical populations.

4. For one each of your concrete and your conceptual examples (exercise 3), give an example of a probability question and an example of an inferential statistics question.

1.2 Pictorial and Tabular Methods in Descriptive Statistics

Descriptive statistics can be divided into two general subject areas. In this section we will discuss the first of these areas—representing a data set using visual techniques. In Sections 1.3 and 1.4, we will develop some numerical summary measures for data sets. Many visual techniques may already be familiar to you: frequency tables, tally sheets, histograms, pie charts, bar graphs, scatter diagrams, and the like. Here we focus on a selected few of these techniques which are most useful and relevant to probability and inferential statistics.

Notation

Some general notation will make it easier to apply our methods and formulas to a wide variety of practical problems. The number of observations in a single data set will often be denoted by n, so that $n = 4$ for the sample of universities {Stanford, Iowa State, Wyoming, Rochester} and also for the sample of pH measurements {6.3,

6.2, 5.9, 6.5}. If two data sets are simultaneously under consideration, either m and n or n_1 and n_2 may be used to denote the numbers of observations. Thus if {29.7, 31.6, 30.9} and {28.7, 29.5, 29.4, 30.3} are thermal efficiency measurements for two different types of diesel engines, then $m = 3$ and $n = 4$.

Given a data set consisting of n observations, the observations themselves will be represented by subscripting a selected letter. We will frequently represent the observations by $x_1, x_2, x_3, \ldots, x_n$ (though any other letter could be used in place of x). The subscript bears no relation to the magnitude of a particular observation, so that x_1 will not in general be the smallest observation in the set, nor will x_n typically be the largest. In many applications x_1 will be the first observation gathered by the experimenter, x_2 the second, and so on. The ith observation in the data set will be denoted by x_i.

Stem and Leaf Displays

Suppose we have a data set $x_1, x_2, \ldots, x_n$ for which each x_i consists of at least two digits. A quick way to obtain an informative visual representation of the data set is to construct a stem and leaf display. To do this, split each x_i into two parts: a stem, consisting of one or more of the leading digits, and a leaf, which consists of the remaining digits. Thus if the data set consists of exam scores between 0 and 100, then we would split the score 83 into the stem 8 and the leaf 3. If the data set consists of automobile gas mileages, each recorded to a tenth of a mile per gallon, lying between 7.1 and 47.8, then a reasonable choice of stems would be 0, 1, 2, 3, and 4; 32.6 would then have stem 3 and leaf 2.6, and 7.1 would have stem 0 and leaf 7.1. In general, the stems should be chosen so that there are relatively few stems compared with the number of observations—between five and 20 stems is usually desirable. For the gas mileages, choosing stems 7, 8, . . ., 47 and each leaf as the digit to the right of the decimal point (so 32.6 has stem 32 and leaf .6) results in too many stems unless the data set is huge.

Once the set of stems has been established, the stem values are listed out along the left-hand margin of the page, and beside each stem all leaves corresponding to data values are listed out in the order in which they are encountered as we proceed through the set.

Example 1.3

The following data on motor octane ratings for various gasoline blends is taken from an article in *Technometrics* (vol. 19, p. 425), a journal devoted to applications of statistics in the physical sciences and engineering:

> 88.5, 87.7, 83.4, 86.7, 87.5, 91.5, 88.6, 100.3, 95.6, 93.3, 94.7, 91.1, 91.0,
> 94.2, 87.8, 89.9, 88.3, 87.6, 84.3, 86.7, 88.2, 90.8, 88.3, 98.8, 94.2, 92.7,
> 93.2, 91.0, 90.3, 93.4, 88.5, 90.1, 89.2, 88.3, 85.3, 87.9, 88.6, 90.9, 89.0,
> 96.1, 93.3, 91.8, 92.3, 90.4, 90.1, 93.0, 88.7, 89.9, 89.8, 89.6, 87.4, 88.4,
> 88.9, 91.2, 89.3, 94.4, 92.7, 91.8, 91.6, 90.4, 91.1, 92.6, 89.8, 90.6, 91.1,
> 90.4, 89.3, 89.7, 90.3, 91.6, 90.5, 93.7, 92.7, 92.2, 92.2, 91.2, 91.0, 92.2,
> 90.0, 90.7.

Since the smallest observation is 83.4 and the largest is 100.3, we choose as stem values the numbers 83, 84, . . ., 100. The resulting stem and leaf display is given

```
 83 ‖ .4
 84 ‖ .3
 85 ‖ .3
 86 ‖ .7, .7
 87 ‖ .7, .5, .8, .6, .9, .4
 88 ‖ .5, .6, .3, .2, .3, .5, .3, .6, .7, .4, .9
 89 ‖ .9, .2, .0, .9, .8, .6, .3, .8, .3, .7
 90 ‖ .8, .3, .1, .9, .4, .1, .4, .6, .4, .3, .5, .0, .7 ⎫ mode
 91 ‖ .5, .1, .0, .0, .8, .2, .8, .6, .1, .1, .6, .2, .0 ⎭
 92 ‖ .7, .3, .7, .6, .7, .2, .2, .2
 93 ‖ .3, .2, .4, .3, .0, .7
 94 ‖ .7, .2, .2, .4
 95 ‖ .6
 96 ‖ .1
 97 ‖
 98 ‖ .8
 99 ‖
100 ‖ .3
```

Figure 1.2 Stem and leaf display for octane data

in Figure 1.2, which shows immediately that most of the octane ratings lie between 86 and 95, and that the middle value (which divides the data set into two sets of equal size) is someplace between 90 and 92. It also demonstrates that the octane ratings are distributed in a roughly symmetric fashion about the middle value. The display thus allows us to extract some of the salient features of the data set.

Example 1.4

As a second example of a stem and leaf display, Figure 1.3 presents in this format a random sample of yardages of golf courses which have been designated by *Golf Magazine* as among the most challenging in the United States. Among the sample of 40 yardages, the shortest course is 6433 yards long while the longest is 7280 yards. The yardages appear to be distributed in a roughly uniform fashion over the range of values in the sample. Notice that a stem choice here of either a single digit

Figure 1.3 Stem and leaf display for golf course yardage

```
64 ‖ 35, 64, 33, 70
65 ‖ 26, 27, 06, 83
66 ‖ 05, 94, 14
67 ‖ 90, 70, 00, 98, 70, 45, 13
68 ‖ 90, 70, 73, 50
69 ‖ 00, 27, 36, 04
70 ‖ 51, 05, 11, 40, 50, 22
71 ‖ 31, 69, 68, 05, 13, 65
72 ‖ 80, 09
```

(6 or 7) or three digits (643, . . ., 728) would yield an uninformative display, the first because of too few stems and the latter because of too many.

The book *Exploratory Data Analysis* listed in the chapter bibliography provides many interesting examples, variations on, and extensions of stem and leaf displays, as well as a wealth of information on other new descriptive techniques.

Frequency Distributions for Quantitative Data

A frequency distribution provides an even more compact summary of a data set than does a stem and leaf display. Rather than retaining the entire data set in the display, a frequency distribution essentially provides a count of only the number of observations associated with each stem choice. The first step in constructing a frequency distribution is to divide the relevant measurement axis into a collection of disjoint (nonoverlapping) intervals such that each observation in the data set is contained in one of these intervals. If, for example, the data set consists of heights of professional basketball players and all heights are between 70 in. and 86 in., one choice of intervals would be 70 in.–under 71 in., 71 in.–under 72 in., . . ., 85 in.–86 in. Each of the resulting intervals is called a **class interval,** or simply a class.

Once the class intervals have been chosen, they are listed along the left-hand margin, and then beside each interval a tally mark is recorded for every number falling in the interval as it is encountered in the data set. Finally, the number of tally marks beside each interval is recorded in a column marked "frequency," and the numbers in that column, along with the class intervals, make up the frequency distribution.

Example 1.5   The following data consists of lifetimes (in thousands of cycles) of 101 rectangular strips of aluminum subjected to repeated alternating stress at 21,000 psi, 18 cycles per second. The information is taken from an article in the *Journal of the American Statistical Association* (1958, p. 159).

1293, 1567, 1222, 1199, 1782, 1055, 797, 1420, 1016, 2100, 930, 1505, 1310, 1115, 1881, 1238, 370, 1522, 1419, 1262, 1763, 1020, 1102, 1730, 1893, 1594, 716, 1475, 1540, 1203, 988, 2268, 1485, 1792, 1330, 886, 1602, 2023, 2440, 1102, 990, 1390, 1502, 1270, 1140, 1910, 1000, 1450, 1608, 2130, 1290, 706, 1313, 1578, 1478, 1258, 1010, 1018, 1820, 1530, 1420, 2215, 1269, 746, 1134, 1522, 1235, 960, 1768, 1315, 844, 1452, 1940, 1252, 1781, 1108, 785, 1200, 1416, 886, 1750, 1923, 1355, 1085, 858, 1674, 1200, 1890, 1513, 1945, 1120, 1604, 1750, 1481, 855, 1868, 1560, 1300, 1630, 1895, 1642

Since the smallest observation is 370 and the largest is 2440, a reasonable choice of class intervals (in that it produces neither two few nor too many intervals) is obtained by starting the first interval at 350 and letting each interval have length 200. The resulting frequency distribution is given in Table 1.1. The frequency distribution indicates a small number of relatively small and relatively large lifetimes, with the distribution of values in the middle intervals being relatively flat.

Table 1.1 *A Frequency Distribution for the Lifetimes of Example 1.5*

Class interval	Tally	Frequency = f_i	Relative freq. = f_i/n
1. 350–under 550	I	1	.010
2. 550–under 750	III	3	.030
3. 750–under 950	N̄ III	8	.079
4. 950–under 1150	N̄ N̄ N̄ II	17	.168
5. 1150–under 1350	N̄ N̄ N̄ IIII	19	.188
6. 1350–under 1550	N̄ N̄ N̄ IIII	19	.188
7. 1550–under 1750	N̄ N̄ I	11	.109
8. 1750–under 1950	N̄ N̄ N̄ II	17	.168
9. 1950–under 2150	III	3	.030
10. 2150–under 2350	II	2	.020
11. 2350–under 2550	I	1	.010
		$n = 101$	1.000

There are no hard and fast rules for selecting class intervals for a frequency distribution. A distribution constructed from either very few or a great many intervals will not typically be very informative; most distributions are based on between five and 20 intervals. It is not even necessary that all intervals have the same length; often there are very few observations on either end of the distribution, so that the end intervals are chosen to be longer than the middle intervals. Thus in Example 1.5 we could have combined the two smallest intervals, 350–550 and 550–750, into one larger interval, and done the same for the two highest intervals.

We can obtain the **relative frequency distribution** from the frequency distribution by dividing each number in the frequency column by the number of observations n in the data set. The relative frequency distribution thus gives the *proportion* of the data set falling into each of the class intervals. In Example 1.5 each frequency is divided by $n = 101$, and the resulting relative frequencies lie between 0 and 1 and (unless there is round-off error) add to 1. That is, if f_i denotes the frequency for the ith interval, then the ith relative frequency is f_i/n and $\Sigma f_i/n = 1$. If the data set constitutes a sample from a population and the sample size is large, the relative frequency f_i/n will tend to be close to the proportion of the entire population falling in the ith interval. More will be said about this subsequently.

Histograms

A pictorial representation of a frequency distribution can be obtained by constructing a histogram. First draw a horizontal line to represent the measurement axis, and then mark the boundaries of adjacent class intervals on the axis. Now above each class interval draw a rectangle whose area is proportional to the frequency of that interval. Thus if interval i contains twice as many observations as interval j, the rectangle above interval i would have twice the area of the rectangle above interval j. Figure 1.4 pictures the histogram corresponding to the frequency distribution of Table 1.2. Often relative frequencies (or frequencies) are included either just above each rectangle or inside each rectangle.

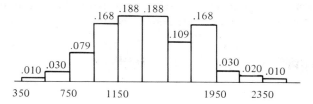

Figure 1.4 A histogram for the frequency distribution of Table 1.1

A histogram is easiest to construct and gain information from if the class intervals are of equal length, for then we need only construct rectangles whose heights, rather than areas, are proportional to the frequencies. It is much easier to look at two rectangles with the same base length and judge that one is twice as high as the other than it is to compare areas of two rectangles which have different bases and heights.

A vertical axis to the left of the histogram can be used instead of entering each relative frequency on the histogram. One possibility is to mark relative frequencies on this axis. Another frequently used method is to mark the axis so that the area of each rectangle is its relative frequency. For this to be the case, the height of a rectangle should equal (relative frequency)/(length of base). This results in the sum of the areas of all rectangles being 1. In Figure 1.5, the histogram of Figure 1.4 has been redrawn. Each base length is 200, so the height of the highest rectangle is $.188/200 = .00094$, and so on.

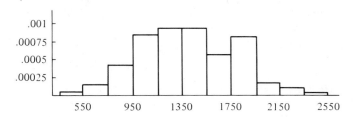

Figure 1.5 Histogram with area of each rectangle = relative frequency

Qualitative Data

Both a frequency distribution and a histogram can be constructed when the data set is qualitative (categorical) in nature. In some cases there will be a natural ordering of classes—for example, freshmen, sophomores, juniors, seniors, graduate students—while in other cases the order will be arbitrary—for example, Catholic, Jewish, Protestant, and the like. With such categorical data the intervals above which rectangles are constructed should have equal length.

Example 1.6 Each member of a sample of 120 individuals owning motorcycles was asked for the name of manufacturer of his or her bike. The frequency distribution and histogram for the resulting data are given in Table 1.2.

Table 1.2 *Frequency Distribution and Histogram for Motorcycle Data*

	Frequency distribution		Histogram
Manufacturer	Frequency	Relative frequency	
1. Honda	41	.34	
2. Yamaha	27	.23	
3. Kawasaki	20	.17	
4. Suzuki	18	.15	
5. Harley-Davidson	3	.03	
6. Other	11	.09	
	120	1.01	

Multivariate Data

The techniques presented so far have been exclusively for situations in which each observation in a data set has been either a single number or a single category. Often, however, the data is multivariate in nature. That is, if we obtain a sample of individuals or objects, and on each one we make two or more measurements, then each "observation" would consist of several measurements on one individual or object. The sample is bivariate if each observation consists of two measurements, so that the data set is represented as $(x_1, y_1), \ldots, (x_n, y_n)$. For example, x might refer to engine size and y to miles per gallon, or x might refer to brand of calculator owned and y to academic major. In Chapters 12–14 we shall analyze multivariate data sets of this sort, so we will postpone a detailed discussion until that time.

Exercises / Section 1.2

1. The following data set refers to the tonnage (in thousands of tons) for a sample of large oil tankers. These numbers were taken from the 1976 *International Petroleum Encyclopedia*.

   229, 232, 239, 232, 259, 361, 220, 260, 231, 229, 249, 254, 257, 214, 237, 253, 274, 230, 223, 253, 195, 269, 231, 268, 189, 290, 218, 313, 220, 270, 277, 375, 222, 290, 231, 258, 227, 269, 220, 224

   a. Construct a stem and leaf display in which the stems are 18, 19, 20, ...
   b. Use the display of (a) to construct a frequency distribution consisting of eight class intervals of equal length, with the first being 175–under 200.
   c. Draw the histogram corresponding to the distribution of (b).

2. The following observations are values of the cost of living index for various cities of the world (New York City = 100), as given in the *1978 World Almanac* (p. 602), .published by the Newspaper Enterprise Association.

   129, 136, 96, 88, 112, 91, 80, 94, 80, 99, 79, 115, 77, 142, 90, 72, 121, 77, 131, 88, 98, 67, 75, 92, 67, 132, 119, 81, 101, 127, 93, 129, 68, 133, 84, 77, 112, 82, 117, 76, 132, 99, 138, 84, 101, 124, 84, 123, 84, 83, 97, 95, 84, 75, 58, 89, 103, 79, 89, 82, 73, 100, 115, 68, 135, 111, 87, 114, 91, 70, 79, 81, 95, 87, 87, 84, 100, 93, 87, 85, 97, 102, 78, 101, 98, 89, 97, 130, 109, 99, 96, 68, 118, 79, 93, 109

a. Construct a stem and leaf display for this data set. Does the distribution of cost of living index appear to be bell-shaped or uniform over the set of recorded values?

b. Construct a frequency and relative frequency distribution for the data.

c. Construct a histogram for the distribution of (b).

3. The following data set constitutes measurements made on the steam rate (lb/hr) of a distillation tower ("A Self-Descaling Distillation Tower," *Chem. Eng. Prog.*, 1968, pp. 79–84).*

1170, 1350, 1640, 1800, 1800, 1260, 1440, 1730, 1710, 1350, 1440, 1710, 1530, 1800, 1530, 1170, 1440, 1350, 1260, 1530, 1350, 1440, 1170, 1350, 1170, 1620, 1800, 1170, 1440, 1800, 1260, 1170, 1260, 1710, 1710, 1530, 1350, 1530, 1440, 1530, 1170, 1350, 1620, 1495, 1440, 1260, 1540, 1170, 1170, 1440

a. Construct a stem and leaf plot of the data using the leading two digits as stems (so that 1160 has stem 11 and leaf 60).

b. Construct a relative frequency distribution and histogram for the data using class intervals of length 100 with the first interval beginning at 1100.

4. Each of the numbers in the accompanying data set is the length of service in years of a U.S. Supreme Court Justice who has terminated his court service:

5, 1, 20, 8, 6, 9, 1, 13, 15, 4, 31, 34, 30, 16, 18, 33, 22, 20, 2, 32, 14, 32, 28, 4, 28, 15, 19, 27, 5, 23, 6, 8, 23, 18, 28, 14, 34, 8, 10, 21, 9, 14, 34, 6, 7, 20, 11, 5, 21, 20, 15, 10, 2, 16, 13, 26, 29, 19, 3, 4, 5, 26, 5, 10, 10, 26, 22, 5, 8, 15, 16, 7, 16, 11, 15, 6, 34, 19, 23, 36, 9, 5, 1, 12, 6, 13, 7, 18, 7, 16, 16, 5, 3, 4

*Throughout this book, we will consistently cite articles from areas other than statistics to show the wide practical applications statistics has. In these cases because the data drawn from the articles is often not pertinent to the main thrust of the articles' authors, we have chosen not to cite these authors by name.

a. Construct a frequency and relative frequency distribution for this data set using class intervals 0–under 5, 5–under 10, . . . , 35–under 40.

b. Draw a histogram corresponding to the distributions of (a) so that area = relative frequency.

c. Regarding the above data set as a sample from the conceptual population of all possible choices for Supreme Court Justices, does it appear as though a randomly selected Justice can be expected to serve 10 years or more? Justify your answer.

5. An article in *Environmental Concentration and Toxicology* ("Trace Metals in Sea Scallops," vol. 19, pp. 326–1334) reported the amount of cadmium in sea scallops observed at a number of different stations in North Atlantic waters. The observed values follow:

5.1, 14.4, 14.7, 10.8, 6.5, 5.7, 7.7, 14.1, 9.5, 3.7, 8.9, 7.9, 7.9, 4.5, 10.1, 5.0, 9.6, 5.5, 5.1, 11.4, 8.0, 12.1, 7.5, 8.5, 13.1, 6.4, 18.0, 27.0, 18.9, 10.8, 13.1, 8.4, 16.9, 2.7, 9.6, 4.5, 12.4, 5.5, 12.7, 17.1

a. Construct a frequency and relative frequency distribution for the data set using 0.0–under 4.0 as the first class interval.

b. Draw a histogram corresponding to the distributions of (a) so that area = relative frequency.

6. If in Example 1.5 the relative frequencies are computed to only two decimal places, do they add to 1?

7. For quantitative data, the **cumulative frequency** and cumulative relative frequency for a particular class interval are the sum of frequencies and relative frequencies respectively for that interval and all intervals lying below it. If, for example, there are four intervals with frequencies 9, 16, 13, and 12, then the cumulative frequencies are 9, 25, 38, and 50 and the cumulative relative frequencies are .18, .50, .76, and 1.00. Compute the cumulative frequencies and cumulative relative frequencies for the data of Exercise 5.

1.3 Measures of Location

Having briefly studied tabular and pictorial methods for organizing and summarizing data, in this section and the next we will focus on numerical summary measures for a given data set. That is, from the data we try to extract several summarizing numbers, numbers which might serve to characterize the data set and convey some of its salient features. Our primary concern will be with numerical data, though some comments regarding categorical data appear at the end of the section.

Suppose, then, that our data set is of the form $x_1, x_2, \ldots, x_n$, where each x_i is a number. What features of such a set of numbers are of most interest and deserve emphasis? One important characteristic of a set of numbers is its location, and in particular its center. This section presents methods for describing the location of a data set, while in Section 1.4 we will turn to methods for measuring the variability in a set of numbers.

The Mean

For a given set of numbers $x_1, x_2, \ldots, x_n$, the most familiar and useful measure of the center is the mean, or arithmetic average of the set. Because we will almost always think of the x_i' s as constituting a sample, we will often refer to the arithmetic average as the **sample mean** and denote it by $\bar{x}$.

Definition: The **sample mean** $\bar{x}$ of a set of numbers $x_1, x_2, \ldots, x_n$ is given by

$$\bar{x} = \frac{x_1 + x_2 + \cdots + x_n}{n} = \frac{\sum\limits_{i=1}^{n} x_i}{n}$$

Example 1.7   Recall the data set $x_1 = 81.3001$, $x_2 = 81.3015$, $x_3 = 81.3006$, $x_4 = 81.3011$, $x_5 = 81.2997$, $x_6 = 81.3005$, $x_7 = 81.3021$ of Example 1.2, where each x_i is an estimate of the ratio of earth mass to moon mass obtained from a different spacecraft mission. The sample mean is

$$\bar{x} = \frac{\sum\limits_{i=1}^{7} x_i}{7} = \frac{569.1056}{7} = 81.3008$$

There is a physical interpretation of $\bar{x}$ that illustrates how it measures location. If we draw a horizontal measurement axis and think of hanging a one pound weight at each of the observed data values, then a fulcrum placed at $\bar{x}$ will cause the system of weights to balance, while placing the fulcrum at any other point will cause a tilt. This is illustrated in Figure 1.6 for the spacecraft data.

In Example 1.7 the sample size $n = 7$ divided the sample total 569.1056 evenly to produce an $\bar{x}$ with the same number of decimal places as each x_i. If, though, $\Sigma x_i = 569.1058$, then $\bar{x}$ is a nonterminating decimal. Because the sample mean is really more precise than any single x_i, we will usually report one more decimal place for $\bar{x}$ than for the x_i's themselves. With $\Sigma x_i = 569.1058$, $\bar{x} = 81.30083$.

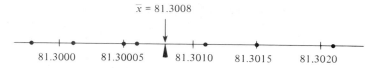

$\bar{x} = 81.3008$

81.3000   81.30005   81.3010   81.3015   81.3020

Figure 1.6   The sample mean as the balance point for a system of weights

Just as $\bar{x}$ represents the average value of the observations in the sample, we can think of calculating the average of all values in the population. This average is called the **population mean**, and is denoted by the Greek letter μ (mu). When there are N values in the population (a finite population), represent them by $y_1, y_2, \ldots, y_N$; then $\mu = \left(\sum\limits_{i=1}^{N} y_i \right) / N$.

Example 1.8   Exercise 1.6 listed the length of service of each U.S. Supreme Court Justice whose service terminated before 1978. Regarding these $N = 94$ numbers as constituting the population of interest, the population mean is

$$\mu = \frac{5 + 1 + 20 + \cdots + 3 + 4}{94} = \frac{1406}{94} = 15.0$$

If a sample of size $n = 5$ is selected and the result is $x_1 = 8$, $x_2 = 23$, $x_3 = 16$, $x_4 = 7$, and $x_5 = 18$, then the sample mean is $\bar{x} = 14.4$. If we did not know μ, we could estimate it as 14.4, the value of the sample mean.

In Chapters 3 and 4 we will discuss models for infinite populations, so we will postpone until then a general definition of μ. Just as $\bar{x}$ is an interesting and important measure of sample location, μ is an interesting and important (often the most important) characteristic of a population. In the chapters on statistical inference, we will present methods based on the sample mean for drawing conclusions about a population mean. For example, we might use the sample mean $\bar{x} = 81.3008$ computed in Example 1.7 as a point estimate (a single number which is our "best" guess) of $\mu =$ the true earth/moon mass ratio.

The sample mean does possess one property which renders it a somewhat unsatisfactory measure of location for some data sets: The computed value of $\bar{x}$ can be greatly influenced by the presence of just one observation which lies very far to one side or the other of the other values.

Example 1.9   Suppose that we randomly select five recordings of classical music from the *Schwann Record Catalog* (which lists all current recordings of classical and non-classical music), and determine the listening time for each. If the data values are (rounded to the nearest minute) $x_1 = 37$, $x_2 = 46$, $x_3 = 40$, $x_4 = 57$, and $x_5 = 50$, then $\bar{x} = 46.0$ minutes. However, if the fifth recording selected is not a Tschaikovsky symphony but instead a Wagner opera, so that $x_5 = 200$ (and seems much longer), then $\bar{x} = 76$. Since most of the data values are considerably smaller than

76, many would feel that $\bar{x}$ here is not a reliable measure of location. Of course, this effect would be even more pronounced if x_5 were, for example, 400 (so $\bar{x} = 116$) or 1000.

A sample of incomes often produces a few such outlying values (those lucky few who earn astronomical incomes), and the use of average income as a measure of location will often be misleading. Such examples suggest that we look for a measure which is less sensitive to outlying values than $\bar{x}$, and we shall momentarily propose one. However, while $\bar{x}$ does have this potential defect, it is still the most widely used measure, largely because there are many populations for which an extreme outlier in the sample would be highly unlikely. When sampling from such a population (a normal or bell-shaped population being the most important example), the sample mean will be stable and quite representative of the sample.

The Median

The word "median" is synonomous with "middle," and the sample median is indeed the middle value when the observations are ordered from smallest to largest in magnitude. When the observations are denoted by $x_1, \ldots, x_n$, we will use the symbol $\tilde{x}$ to represent the sample median.

Definition: Given the sample $x_1, x_2, \ldots, x_n$, rearrange the observations in increasing order (from most negative to most positive). Then the **sample median** is given by

$$\tilde{x} = \begin{cases} \text{single middle value in the ordered list if } n \text{ is odd;} \\ \text{the average of the two middle values in the ordered list if } n \text{ is even} \end{cases}$$

Example 1.10   The following data refers to active repair times (hours) for an airborne communication receiver.

1.1, 4.0, .5, 5.4, 2.0, .5, .8, 9.0, 5.0, 3.3, .3, .7, 2.2, 22.0, 4.0, 2.7, 1.0, 3.0, 1.0, 1.5, 24.5, 1.5, 3.3, 2.5, 1.0, .8, 1.5, .6, 10.3, 1.3, .7, 2.0, 4.7, 3.0, .8, 1.0, 8.8, .6, 4.5, 7.0, .7, 5.4, 1.5, .2, 7.5, .5

Because $n = 46$ is even, $\tilde{x}$ will be the average of the two middle values in the ordered list. Ordering $x_1, \ldots, x_{46}$ gives

.2, .3, .5, .5, .5, .5, .6, .6, .7, .7, .7, .8, .8, .8, 1.0, 1.0, 1.0, 1.0, 1.1, 1.3, 1.5, 1.5, 1.5, 1.5, 2.0, 2.0, 2.2, 2.5, 2.7, 3.0, 3.0, 3.3, 3.3, 4.0, 4.0, 4.5, 4.7, 5.0, 5.4, 5.4, 7.0, 7.5, 8.8, 9.0, 10.3, 22.0, 24.5

The two middle values are 1.5 and 1.5, so that $\tilde{x}$ is 1.5. Notice that if we had not listed all replications of data values, $\tilde{x}$ would have been much larger, since these replications occur for values of small magnitude but not values of large magnitude.

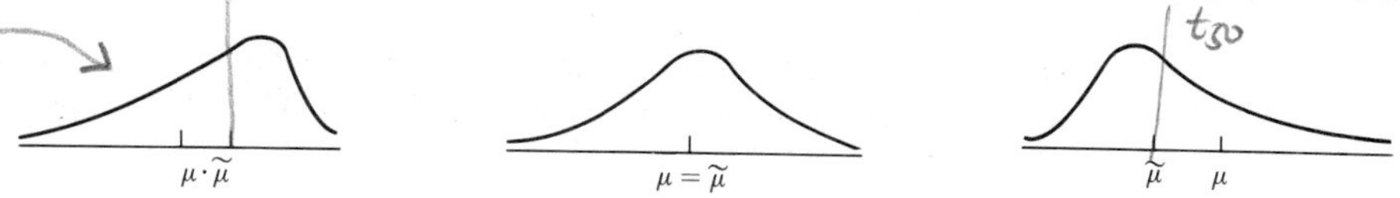

t_{50}

$\mu < \tilde{\mu}$

t_{50}

$\mu \cdot \tilde{\mu}$ (a) Negative skew

$\mu = \tilde{\mu}$ (b) Symmetric

$\tilde{\mu} \quad \mu$ (c) Positive skew

Figure 1.7 Three different shapes for a population distribution

$\mu > \tilde{\mu}$

The data set in Example 1.10 illustrates an important property of $\tilde{x}$ in contrast to $\bar{x}$; the sample median is very insensitive to a number of extremely small or extremely large data values. If, for example, we increased the two largest x_i's from 22.0 and 24.5 to 122.0 and 424.5, $\tilde{x}$ would be unaffected. Thus in the treatment of outlying data values, $\bar{x}$ and $\tilde{x}$ are at opposite ends of a spectrum: $\bar{x}$ is sensitive to just one such value, while $\tilde{x}$ is insensitive to a large number of outlying values.

Because the large values in the sample of Example 1.10 affect $\bar{x}$ more than $\tilde{x}$, $\tilde{x} < \bar{x}$ for that data. While $\bar{x}$ and $\tilde{x}$ both provide a measure for the center of a data set, they will not in general be equal, since they focus on different aspects of the sample.

Analagous to $\tilde{x}$ as the middle value in the sample, there is a middle value in the population, the **population median,** denoted by $\tilde{\mu}$. As with $\bar{x}$ and μ, we can think of using the sample median $\tilde{x}$ to make an inference about $\tilde{\mu}$. In Example 1.10 we might use $\tilde{x} = 1.5$ as an estimate of the median number of hours necessary to repair all such communication receivers when they fail. A median is often used to describe income or salary data (because it is not greatly influenced by a few large salaries), so if the median salary for a sample of engineers were $\tilde{x} = \$26,416$, we might use this as a basis for concluding that the median salary for all engineers exceeds $25,000.

The population mean μ and median $\tilde{\mu}$ will not generally be equal to one another. If the population distribution is either positively or negatively skewed, as pictured in Figure 1.7, then $\mu \neq \tilde{\mu}$. When this is the case, in making inferences we must first decide which of the two population characteristics is of most interest and then proceed accordingly.

Other Measures of Location: Quartiles, Percentiles, and Trimmed Means

The median (population or sample) divides the data set into two parts of equal size. To obtain finer measures of location, we could divide the data into more than two such parts. Roughly speaking, quartiles divide the data set into four equal parts, with the observations above the third quartile constituting the upper quarter of the data set, the second quartile being identical to the median, and the first quartile separating the lower quarter from the upper three-quarters. Similarly, a data set (sample or population) can be divided into 100 equal parts using percentiles; the ninty-ninth percentile separates off the highest 1% from the bottom 99%, and so on. Unless the number of observations is a multiple of 100, care must be exercised in obtaining percentiles. We will use percentiles in Chapter 4 in connection with certain models for infinite populations, so we shall postpone discussion until that point.

As emphasized above, the sample mean and sample median are influenced by outlying values in a very different manner: the mean greatly and the median not at all. Since extreme behavior of either type might not always be desirable, we briefly consider alternative measures which are neither as sensitive as $\bar{x}$ nor as insensitive as $\tilde{x}$. To motivate these alternatives, note that $\bar{x}$ and $\tilde{x}$ are at opposite extremes of the same "family" of measures. After ordering the data, $\tilde{x}$ is computed by throwing away as many values on either end as one can without eliminating everything (leaving just one or two middle values) and averaging what is left, while to compute $\bar{x}$ one throws away nothing before averaging. To paraphrase, the mean involves trimming 0% from either end of the sample, while for the median the maximum possible amount is trimmed from either end. A **trimmed mean** is a compromise between $\bar{x}$ and $\tilde{x}$. A 10% trimmed mean, for example, would be computed by eliminating the smallest 10% and the largest 10% of the sample and then averaging what is left over.

Example 1.11   Consider the following 20 observations, ordered from smallest to largest, each one representing the lifetime in hours of a certain type of incandescent lamp.

612, 623, 666, 744, 883, 898, 964, 970, 983, 1003, 1016, 1022, 1029, 1058, 1085, 1088, 1122, 1135, 1197, 1201

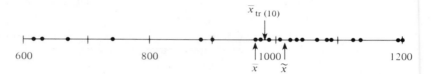

Figure 1.8

The average of all 20 observations is $\bar{x} = 965.0$, while $\tilde{x} = 1009.5$. The 10% trimmed mean is obtained by deleting the smallest two observations (612 and 623) and the largest two (1197 and 1201), and then averaging the remaining 16 to obtain $\bar{x}_{tr(10)} = 979.1$. The effect of trimming here is to produce a "central value" which is somewhat above the mean ($\bar{x}$ is pulled down by a few small lifetimes) and yet considerably below the median. Similarly, the 20% trimmed mean averages the middle 12 values to obtain $\bar{x}_{tr(20)} = 999.9$, even closer to the median.

Generally speaking, using a trimmed mean with a moderate trimming proportion (10 or 20%) will yield a measure which is neither as sensitive to outliers as the mean (since any small number of outliers will be deleted before averaging) nor as insensitive as the median. For this reason trimmed means have merited increasing attention from statisticians both for descriptive and inferential purposes. More will be said about trimmed means when point estimation is discussed in Chapter 6. As a final point if the trimming proportion is denoted by α and $n\alpha$ is not an integer, then it is not obvious how the $100\alpha\%$ trimmed mean should be computed. For example, if $\alpha = .10$ (10%) and $n = 22$, then $n\alpha = (22)(.10) = 2.2$, and we cannot trim 2.2 observations from each end of the ordered sample. In this case, the 10% trimmed mean would be obtained by first trimming two observations from each end and calculating $\bar{x}_{tr}$, then trimming three and calculating $\bar{x}_{tr}$, and finally interpolating between the two values to obtain $\bar{x}_{tr(10)}$.

Categorical Data and Sample Proportions

When the data is categorical, a frequency distribution or relative frequency distribution provides an effective tabular summary of the data. The natural numerical summary quantities in this situation are the individual frequencies and the relative frequencies. For example, if a survey of individuals who own stereo receivers is undertaken to study brand preference, then each individual in the sample would identify the brand of receiver that he or she owned, from which we could count the number owning Sony, Marantz, Pioneer, and so on. Consider sampling a dichotomous population—one which consists of only two categories (such as voted or did not vote in the last election, does or does not own a stereo receiver, and so on). If we let x denote the number in the sample falling in category one, then the number in category two is $n - x$. The relative frequency or *sample proportion* in category one is x/n and the sample proportion in category two is $1 - x/n$. To see that x/n *can be regarded as a sample mean,* identify a category one response by a 1 and a category two response by a 0. A sample of size $n = 10$ might then yield the responses 1, 1, 0, 1, 1, 1, 0, 0, 1, 1. The sample mean for this numerical sample is (since number of 1's $= x = 7$)

$$\frac{x_1 + \cdots + x_n}{n} = \frac{1 + 1 + 0 + \cdots + 1 + 1}{10} = \frac{7}{10} = \frac{x}{n} = \frac{\text{sample}}{\text{proportion}}$$

This result can be generalized and summarized as follows: *If in a categorical data situation we focus attention on a particular category and code the sample results so that a 1 is recorded for an individual in the category and a 0 for an individual not in the category, then the sample proportion of individuals in the category is the sample mean of the sequence of 1's and 0's.* Thus a sample mean can be used to summarize the results of a categorical sample. These remarks also apply to situations in which categories are defined by grouping values in a numerical sample or population (for example, we might be interested in knowing whether or not individuals have owned their present automobile for at least five years, rather than studying the exact length of ownership).

Analogous to the sample proportion x/n of individuals falling in a particular category, let p represent the proportion of individuals in the entire population falling in the category. As with x/n, p is a quantity between 0 and 1. While x/n is a sample characteristic, p is a characteristic of the population; the relationship between the two parallels the relationship between $\tilde{x}$ and $\tilde{\mu}$ and between $\bar{x}$ and μ. In particular, we shall subsequently use x/n to make inferences about p. If, for example, a sample of 100 car owners reveals that 22 owned their car at least five years, then we might use $22/100 = .22$ as a point estimate of the proportion of all owners who have owned their car at least five years. We will study the properties of x/n as an estimator of p and see how x/n can be used to answer other inferential questions. With k categories ($k > 2$) we can use the k sample proportions $x_1/n, \ldots, x_k/n$ to answer questions about the population proportions $p_1, \ldots, p_k$.

Exercises / Section 1.3

1. The nine measurements which follow are temperature determinations at various locations beneath the surface of Lake Ontario, as reported in *Limnology and Oceanography* (vol. 22, pp. 158–159):

   4.45, 3.91, 3.86, 3.93, 3.94, 3.90, 3.80, 3.73, 3.69

   a. Compute the sample mean of these data values.
   b. Compute the sample median of these data values.

2. Bacterial mutants of the bacterium *E. Coli* that are resistant to the virus T_1 are called *Ton^r* variants. A recent paper in *Genetics* reported the following results on the number of *Ton^r* bacteria found in 10 different bulk cultures when a particular experimental culture method was used:

   14, 15, 13, 21, 15, 14, 26, 16, 20, 13

   a. Compute the sample mean number of bacteria per culture.
   b. Compute the sample median number of bacteria per culture.
   c. Compute the 10% trimmed mean for the data set.

3. The 20-year average annual costs for $50,000 5-year-renewable term insurance policies (men, age 35) for ten companies are

   214, 352, 379, 344, 337, 347, 338, 300, 300, and 361

   a. Compute the sample mean cost.
   b. Compute the sample median.
   c. Compute the 10% trimmed mean.

4. Compute the sample median, 10% trimmed mean, 25% trimmed mean, and sample mean for the data of Exercise 5, Section 1.2.

5. In an attempt to study the effect of choice of postage stamps on response rate in a mail survey, W. E. Hensley (*Public Opinion Quarterly,* vol. 38, pp. 280–283), reported the following data on number of mailings n_i and number of returns x_i both when inside and outside stamps were dissimilar ($i = 1$) and similar ($i = 2$): $n_1 = 354$, $x_1 = 217$, $n_2 = 176$, $x_2 = 89$.

   a. Compute the sample proportion of returns both for dissimilar stamps and similar stamps.
   b. The sample proportions of (a) can be viewed as estimates of true return proportions p_1 and p_2 for hypothetical populations. If there is actually no difference in response rate due to the types of stamps, then $p_1 = p_2$, and we have a sample of size $n = 354 + 176 = 530$ from a single hypothetical population, with $x = 306$ returns. Compute the sample proportion for this pooled (combined) sample.

6. In Exercise 3, obtain the sample proportion of insurance companies which charge at least $350 for such policies. What is the sample proportion of companies which charge between $300 and $350 inclusive?

7. **a.** If a constant c is added to each x_i in the sample, yielding $y_i = x_i + c$, how do the sample mean and median of the y_i's relate to the mean and median of the x_i's?
   b. If each x_i is multiplied by a constant c, yielding $y_i = cx_i$, answer the question of part (a).

1.4   Measures of Variability

No single measure of location can give a complete summarization of a data set. Consider the *x* data set 20, 100, 0, 60, 70 and the *y* data set 60, 20, 80, 60, 30. Since $\bar{x} = \bar{y} = 50$ and $\tilde{x} = \tilde{y} = 60$, the two standard measures of location by themselves do not distinguish between the two sets. Yet the plots of these two data sets in Figure 1.9 show that the observations in the *y* data set cluster more closely about their center than do the observations from the *x* data set. That is, there is more variability or dispersion in the *x*'s than in the *y*'s.

    One simple measure of variability is the **sample range,** defined as the difference between the largest observation and the smallest observation—that is, sample

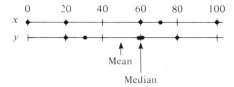

Figure 1.9   Data sets with the same center and differing variability

range $=$ max(x_i) $-$ min(x_i). We might then say that a small (large) range indicates little (great) variability. A defect of the sample range, though, is that it depends only on the two most extreme observations and disregards the positions of the middle observations. For example, the two samples 0, 5, 5, 5, 10 and 0, 1, 5, 9, 10 both have the same sample range, yet in the first sample there is variability only in the two extreme values, while the second sample has more variability in its middle values. We would like a measure which somehow depends on all observations rather than just a few.

Deviations from the Mean

The quantity $x_i - \bar{x}$ is called the deviation of the ith observation from the mean, or just the *ith deviation.* A positive deviation indicates an observation to the right of $\bar{x}$ on the measurement axis, while a negative deviation indicates an observation to the left of $\bar{x}$. If all the deviations $x_1 - \bar{x}$, . . ., $x_n - \bar{x}$ are small in absolute magnitude, then all x_i's are close to $\bar{x}$ and thus to one another, suggesting a relatively small amount of variability in the sample. On the other hand, if some of the $(x_i - \bar{x})$'s are large in absolute magnitude, then some of the x_i's lie far from $\bar{x}$, suggesting large variation. A simple way of combining the n deviations into a single quantity is to average them (sum them and divide by n). However, since some deviations will be negative and others positive, adding results in cancellation.

Proposition: $\displaystyle\sum_{i=1}^{n}$ (ith deviation) $= \displaystyle\sum_{i=1}^{n} (x_i - \bar{x}) = 0$

so the average deviation from the mean is always 0.

Proof: The verification uses several standard rules of summation and the fact that $\bar{x}$ is a constant in the summation:

$$\sum (x_i - \bar{x}) = \sum x_i - \sum \bar{x} = \sum x_i - n\bar{x} = \sum x_i - n\left(\frac{1}{n} \sum x_i\right) = 0$$

To obtain an informative measure of variability, we need to change the deviations to nonnegative quantities before combining. One possibility is to average the absolute deviations $|x_1 - \bar{x}|$, . . ., $|x_n - \bar{x}|$. Because this leads to a multitude of theoretical difficulties, consider instead the squared deviations $(x_1 - \bar{x})^2$, . . ., $(x_n - \bar{x})^2$. We might now use the average of the squared deviations. There is, however, a technical reason (to be discussed shortly) for dividing the sum of squared deviations not by n but by $n - 1$.

The Sample Variance

Definition: The **sample variance** of the set $x_1, \ldots, x_n$ of numerical observations, denoted by s^2, is given by

$$s^2 = \frac{\sum_{i=1}^{n} (x_i - \overline{x})^2}{n - 1}$$

The divisor of s^2 is $n - 1$, which is smaller than n; the effect of dividing by $n - 1$ is to obtain a number somewhat larger than the average squared deviation.

Example 1.12   The following seven measurements are of cerebral metabolic rate for glucose in a sample of adult rhesus monkeys (mg per 100 g min^{-1}) using the technique of quantitative emission tomography (cf. *Science*, vol. 199, pp. 986–987).

Table 1.3

Observations	Deviations $x_i - \overline{x}$	Squared deviations
$x_1 = 4.51$	−.701	.491
$x_2 = 4.59$	−.621	.386
$x_3 = 4.90$	−.311	.097
$x_4 = 4.93$	−.281	.079
$x_5 = 6.80$	1.589	2.525
$x_6 = 5.08$	−.131	.017
$x_7 = 5.67$	.459	.211
$\overline{x} = 5.211$	sum = .003	sum = 3.806

Effects of rounding account for the sum of deviations not being exactly 0. Adding the entries in the "squared deviations" column gives the numerator of s^2, so that $s^2 = 3.806/(n - 1) = 3.806/6 = .634$.

Admittedly s^2 does not have as much intuitive appeal as our measures of location had. Variability is a somewhat less familiar characteristic than is location, so it is less obvious which measures might be useful. In particular, we are not in a position to say that $s^2 = .634$ indicates a large or a small amount of variability. All that you should believe at this point is that if "eyeballing" two different samples suggests that the first clearly has less variability than the second, then s^2 for the first sample should be smaller than s^2 for the second.

Whatever the units of measurement of the observations (ft, lb, °F, and so on), the variance s^2 will be expressed in squared units (ft^2, lb^2, °F^2, and the like). A measure of variability expressed in the original units of measurement is obtained simply by taking the (positive) square root of s^2.

> **Definition:** The **sample standard deviation,** denoted by s, is the positive square root of the sample variance.

In Example 1.12, $s^2 = .634$, so the standard deviation is $s = \sqrt{.634} = .796$ mg per 100 g min^{-1}. In inferential problems s will typically be more useful than s^2, but the latter must be computed first.

Motivation for s^2

In order to explain why s^2 rather than the average squared deviation is used to measure variability, note first that whereas s^2 measures sample variability, there is a measure of variability in the population called the population variance. We shall use σ^2 (the square of the lowercase Greek letter sigma) to denote the population variance and σ to denote the population standard deviation (the square root of σ^2).

When the population is finite and consists of values $y_1, y_2, \ldots, y_N$, $\sigma^2 = \sum_{i=1}^{N} \times (y_i - \mu)^2/N$, which is the average of all squared deviations from the population mean (for the population, the divisor is N and not $N - 1$). More general definitions of σ^2 appear in Chapters 3 and 4.

Just as $\bar{x}$ will be used to make inferences about the population mean μ, we should define the sample variance so that it can be used to make inferences about σ^2. Now note that σ^2 involves squared deviations about the population mean μ. If we actually knew the value of μ, then we could define the sample variance as the average squared deviation of the sample x_i's about μ. However, the value of μ is almost never known, so the sum of squared deviations about $\bar{x}$ must be used. But *the x_i's tend to be closer to their average $\bar{x}$ than to the population average μ, so to compensate for this the divisor $n - 1$ is used rather than n.* Said another way, if we used a divisor n in the sample variance, then the resulting quantity would tend to underestimate σ^2 (produce estimated values which are too small on the average), while dividing by the slightly smaller $n - 1$ corrects this underestimating.

It is customary to refer to s^2 as being based on $n - 1$ **"degrees of freedom."** This terminology results from the fact that while s^2 is based on the n quantities $x_1 - \bar{x}, x_2 - \bar{x}, \ldots, x_n - \bar{x}$, these sum to 0, so specifying the values of any $n - 1$ of the quantities determines the remaining one. For example, if $n = 4$ and $x_1 - \bar{x} = 8$, $x_2 - \bar{x} = -6$, and $x_4 - \bar{x} = -4$, then automatically $x_3 - \bar{x} = 2$, so only three of the four $x_i - \bar{x}$'s are freely determined (3 degrees of freedom).

The Computation of s^2

With n observations the computation of s^2 from the definition involves n subtractions and n squaring operations. If $\bar{x}$ is not an integer, the subtractions to obtain the deviations can be tedious, and the deviations themselves may be unpleasant to square. There is an equivalent expression for s^2 which yields a more efficient method for computing it.

Proposition:

$$s^2 = \frac{\sum_{i=1}^{n} (x_i - \bar{x})^2}{n - 1} = \frac{\sum_{i=1}^{n} x_i^2 - \left(\sum_{i=1}^{n} x_i\right)^2 / n}{n - 1}$$

Proof: Because $\bar{x} = \sum x_i/n$, $n\bar{x}^2 = \left(\sum x_i\right)^2 / n$. Then

$$\sum_{i=1}^{n} (x_i - \bar{x})^2 = \sum_{i=1}^{n} (x_i^2 - 2\bar{x}\, x_i + \bar{x}^2) = \sum_{i=1}^{n} x_i^2 - 2\bar{x} \sum_{i=1}^{n} x_i + \sum_{i=1}^{n} (\bar{x})^2$$

$$= \sum_{i=1}^{n} x_i^2 - 2\bar{x} \cdot n\bar{x} + n(\bar{x})^2 = \sum_{i=1}^{n} x_i^2 - n(\bar{x})^2$$

$$= \sum_{i=1}^{n} x_i^2 - \frac{\left(\sum_{i=1}^{n} x_i\right)^2}{n}$$

To use this method for computing s^2, square each x_i (before subtraction), then add the squares, and subtract $(\sum x_i)^2/n$ from $\sum x_i^2$. While this involves squaring $n + 1$ numbers ($\sum x_i$ in addition to each x_i), only one subtraction is necessary. We shall refer to this formula for computing s^2 as the "shortcut method for s^2." The shortcut for s involves computing s^2 using the shortcut and then taking the square root.

Example 1.12
(continued)

x_i:	4.51	4.59	4.90	4.93	6.80	5.08	5.67
x_i^2:	20.340	21.068	24.010	24.305	46.240	25.806	32.149

The quantity $\sum x_i^2$ is computed (by summing the numbers in the x_i^2 row) as 193.918, and $(\sum x_i)^2/n = (36.48)^2/7 = 190.113$, so $s^2 = (193.918 - 190.113)/6 = 3.805/6 = .634$ and $s = .796$.

The shortcut method can yield values of s^2 and s which differ from the values computed using the definitions. These differences are due to effects of rounding and will not be important in most problems. In order to minimize the effects of rounding when using the shortcut formula, particularly when there is little variability in the data, intermediate calculations should be done using several more significant digits than are to be retained in the final answer. Because the numerator of s^2 is the sum of nonnegative quantities (squared deviations), s^2 is guaranteed to be nonnegative. Yet if the shortcut method is used, particularly with data having little variability, a slight numerical error can result in a negative numerator ($\sum x_i^2$ smaller than $(\sum x_i)^2/n$). If your value of s^2 is negative you have made a computational error.

There are several other properties of s^2 which can sometimes be used to increase computational efficiency. These are summarized in the following proposition.

> **Proposition:** Let $x_1, x_2, \ldots, x_n$ be a sample and c be any nonzero constant.
>
> 1. If $y_1 = x_1 + c$, $y_2 = x_2 + c$, $\ldots$, $y_n = x_n + c$, then $s_y^2 = s_x^2$ and
> 2. If $y_1 = cx_1, \ldots, y_n = cx_n$, then $s_y^2 = c^2 s_x^2$, $s_y = |c| s_x$,
>
> where s_x^2 is the sample variance of the x's and s_y^2 is the sample variance of the y's.

In words, (1) says that if a constant c is added to (or subtracted from) each data value, the variance is unchanged. This is intuitive, since adding or subtracting c shifts the location of the data set but leaves distances between data values unchanged. According to (2), multiplication of each x_i by c results in s^2 being multiplied by a factor c^2. These properties can be proved by noting in (1) that $\bar{y} = \bar{x} + c$ and in (2) that $\bar{y} = c\bar{x}$.

Example 1.13 Recall the spacecraft data (Example 1.2) $x_1 = 81.3001$, $x_2 = 81.3015$, $x_3 = 81.3006$, $x_4 = 81.3011$, $x_5 = 81.2997$, $x_6 = 81.3005$, and $x_7 = 81.3021$. If we subtract the smallest value 81.2997 from all observations, the resulting values are .0004, .0018, .0009, .0014, .0090, .0008, and .0024; the variance of this set is the same as that of the original data and is easier to compute. Now if we multiply each value by 1000, the variance of the resulting set is $(1000)^2$ times the original variance. The new data set is 4, 18, 9, 14, 0, 8, and 24 with mean 11.0 and sum of squares 1257, so the variance is $[1257 - (77)^2/7]/6 = 68.33$. The variance of the original data set is therefore $68.33/(1000)^2 = .00006833$.

Exercises / Section 1.4

1. The amount of leaf protein (mg/g fresh weight) in soybean plants of a particular variety was determined for a sample of six plants, yielding the following data:

11.7, 16.1, 14.0, 6.1, 5.1, 4.9

(*Science*, vol. 199, p. 974)
  a. Compute the sample range.
  b. Compute the sample variance s^2 from the definition (that is, by first computing deviations, then squaring them, and so on).
  c. Compute the sample standard deviation.

  d. Compute s^2 using the shortcut method.

2. A sample of eight resistors of a certain type resulted in the sample resistances (ohms) $x_1 = 40$, $x_2 = 43$, $x_3 = 39$, $x_4 = 35$, $x_5 = 37$, $x_6 = 43$, $x_7 = 46$, $x_8 = 37$.
  a. Compute s^2 and s directly from the definitions.
  b. Compute s^2 and s using the shortcut formula.
  c. Subtract 35 from each x_i and then compute s^2.
  d. If the resistances were 400, 430, 390, 350, 370, 430, 460, and 370, how would you use the results of (a), (b), or (c) to compute s^2 and s?

3. The following observations are weight gains (mg/cm^3) of Ti-Cr alloy samples due to oxidation when exposed to CO_2 for one hour at $1000°$ C: 6.4, 5.9, 6.1, 5.8, 6.6, 6.0.
   a. Using $\bar{x}$ computed to two decimal places, compute s^2 directly from the definition, and then compute s.
   b. Repeat (a) with $\bar{x}$ rounded to three decimal places.
   c. Compute s^2 using the shortcut formula.

4. Compute s^2 and s for the bacteria data from Exercise 2, section 1.3.

5. Compute s^2 and s for the temperature data from Exercise 1, section 1.3, by first subtracting the number 3.70 off each data value. Is there anything special about the choice of 3.70, or would other choices work just as well?

6. Compute s^2 and s for the insurance policy data of Exercise 3, Section 1.3, by first subtracting a convenient number off each of the 10 data values.

7. In Exercise 5 above, after subtracting 3.70 off each data value, multiply each resulting number by 100 and then compute s^2 for the resulting set of numbers. How can you now obtain s^2 for the original data set?

8. The standard deviation and variance are both measures of variability which depend on the units of measurement, with s expressed in the original units of measurement. The **coefficient of variation,** defined by $c.v. = s/\bar{x}$, is a dimensionless quantity which measures the amount of variability relative to the value of the mean. Compute the value of $c.v.$ for the leaf protein data of Exercise 1 and the weight gain data of Exercise 3. Which data set has more relative variation?

Supplementary Exercises / Chapter 1

1. The mode of a numerical data set is that value which occurs most frequently in the set. (a) What is the mode for the Supreme Court service length data from Exercise 1.2.4? (b) Repeat (a) for the cost of living data from Exercise 1.2.2. (c) For a categorical sample, how would you define the modal category?

2. The task completion time in a learning experiment was recorded both on the first trial (x) and the tenth trial (y) for six individuals:

   individual i: 1 2 3 4 5 6
   x_i: 23.7 36.9 25.5 30.2 28.0 34.8
   y_i: 15.0 25.4 21.0 22.3 25.2 28.8

   a. Compute the decrease $d_i = x_i - y_i$ for each individual, and then compute the sample average decrease.
   b. Compute the sample average time on the first trial, the sample average time on the tenth trial, and then the difference in the two sample averages.
   c. Show in general using rules of summation that the average of the differences $d_i = x_i - y_i$ equals the difference $\bar{x} - \bar{y}$ of the averages.

3. Fifteen air samples from a certain region were obtained and for each one the carbon monoxide concentration was determined. The results were (ppm)

   9.3, 10.7, 8.5, 9.6, 12.2, 15.6, 9.2, 10.5, 9.0, 13.2, 11.0, 8.8, 13.7, 12.1, 9.8

   Using the interpolation method suggested in section 1.3, compute the 10% trimmed mean.

4. a. For what value of c is the quantity $\sum (x_i - c)^2$ minimized? *Hint:* Take the derivative with respect to c, set equal to 0, and solve.
   b. Using the result of (a), which of the two quantities $\sum (x_i - \bar{x})^2$ and $\sum (x_i - \mu)^2$ will be smaller than the other (assuming that $\bar{x} \neq \mu$)?

5. a. Let a and b be constants and let $y_i = ax_i + b$ for $i = 1, 2, \ldots, n$. What is the relationship between $\bar{x}$ and $\bar{y}$ and between s_x^2 and s_y^2?
   b. A sample of temperatures for initiating a certain chemical reaction yielded a sample average (°C) of 87.3 and a sample standard deviation of 1.04. What are the sample average and standard deviation measured in °F? *Hint:* $F = \frac{9}{5}C + 32$.

Bibliography

Anderson, Theodore, and Sclove, Stanley, *An Introduction to the Statistical Analysis of Data,* Houghton Mifflin, Boston, 1978. Contains four good chapters on descriptive methods.

Hoaglin, David and Velleman, Paul, *Applications, Basics, and Computing of Exploratory Data Analysis,* Duxbury, Boston, 1980. A good discussion of some basic exploratory methods; it's easier to read than the book by Tukey, though less comprehensive.

Moore, David, *Statistics: Concepts and Controversies,* W. H. Freeman, San Francisco, 1979. An extremely readable and entertaining paperback which contains an intuitive discussion of problems connected with sampling and designed experiments.

Freedman, David, Pisani, Robert, and Purves, Roger, *Statistics,* W. W. Norton, New York, 1978. An excellent very nonmathematical survey of basic statistical reasoning and methodology.

Tanur, Judith (ed.), *Statistics: A Guide to the Unknown* (2nd ed.), Holden-Day, San Francisco, 1978. Contains many short nontechnical articles describing various applications of statistics.

Tukey, John, *Exploratory Data Analysis,* Addison-Wesley, Reading, Mass., 1977. Introduces and illustrates a broad range of newly developed descriptive and "data-snooping" methods.

CHAPTER 2

Probability

Introduction

The term **probability** refers to the study of randomness and uncertainty. In any situation in which one of a number of possible outcomes may occur, the theory of probability provides methods for quantifying the chances or likelihoods associated with the various outcomes. The language of probability is constantly used in an informal manner in both written and spoken contexts. Examples include such statements as "It is quite likely that the Dow-Jones average will increase by the end of the year," "There is a 50-50 chance that the incumbent will seek reelection," "There will probably be at least one section of that course offered next year," "The odds favor a quick settlement of the strike," and "It is expected that at least 20,000 concert tickets will be sold." In this chapter we introduce some elementary probability concepts, indicate how probabilities can be interpreted, and show how the rules of probability can be applied to compute the probabilities of many interesting events. The methodology of probability will then permit us to express in precise language such informal statements as those given above.

The study of probability as a branch of mathematics goes back over 300 years, where it had its genesis in connection with questions involving games of chance. Many books are devoted exclusively to probability, but our objective here is to cover only that part of the subject which has the most direct bearing on problems of statistical inference.

2.1 Sample Spaces and Events

An **experiment** is any action or process which generates observations. Although the word "experiment" generally suggests a planned or carefully controlled laboratory-testing situation, we use it here in a much wider sense. Thus, experiments which may be of interest include tossing a coin once or several times, selecting a card or cards from a deck, weighing a loaf of bread, ascertaining the commuting time from home to work on a particular morning, obtaining blood types from a group of individuals, or measuring the compressive strengths of different steel beams.

The Sample Space of an Experiment

> **Definition:** The **sample space** of an experiment, denoted by $\mathcal{S}$, is the set of all possible outcomes of that experiment.

Example 2.1   The simplest experiment to which probability applies is one with two possible outcomes. One such experiment consists of examining a single fuse to see whether or not it is defective. The sample space for this experiment can be abbreviated as $\mathcal{S} = \{N, D\}$, where N represents not defective, D represents defective, and the braces are used to enclose the elements of a set. Another such experiment would involve tossing a thumbtack and noting whether it landed point up or point down, with sample space $\mathcal{S} = \{U, D\}$, and yet another would consist of observing the sex of the next child born at the local hospital, with $\mathcal{S} = \{B, G\}$.

Example 2.2   If we examine three fuses in sequence and note the result of each examination, then an outcome for the entire experiment is any sequence of N's and D's of length 3, so

$$\mathcal{S} = \{NNN, NND, NDN, NDD, DNN, DND, DDN, DDD\}$$

If we had tossed a thumbtack three times, the sample space would be obtained by replacing N by U in $\mathcal{S}$ above, with a similar notational change yielding the sample space for the experiment in which the sexes of three newborn children are observed.

Example 2.3   If a six-sided die is tossed once and the outcome is the number on the upturned face, then $\mathcal{S} = \{1, 2, 3, 4, 5, 6\}$. If both a red die and a green die are tossed together, then a particular outcome specifies both the number on the red die and the number on the green die. An outcome consists of a *pair* of numbers, such as (1, 1) or (3, 2), and the sample space consists of the 36 outcomes listed in the accompanying table.

		Green Die					
		1	2	3	4	5	6
	1	(1, 1)	(1, 2)	(1, 3)	(1, 4)	(1, 5)	(1, 6)
	2	(2, 1)	(2, 2)	(2, 3)	(2, 4)	(2, 5)	(2, 6)
Red Die	3	(3, 1)	(3, 2)	(3, 3)	(3, 4)	(3, 5)	(3, 6)
	4	(4, 1)	(4, 2)	(4, 3)	(4, 4)	(4, 5)	(4, 6)
	5	(5, 1)	(5, 2)	(5, 3)	(5, 4)	(5, 5)	(5, 6)
	6	(6, 1)	(6, 2)	(6, 3)	(6, 4)	(6, 5)	(6, 6)

Note that the outcome (2, 1) is different from (1, 2) and that both are included in the sample space; a similar comment applies to any other pair for which the first toss and the second toss yield different numbers.

Example 2.4

If a new type-D flashlight battery has a voltage which is outside certain limits, that battery is characterized as a failure (F); if the battery has a voltage within the prescribed limits, it is a success (S). Suppose that an experiment consists of testing each battery as it comes off an assembly line until we first observe a success. Although it may not be very likely, a possible outcome of this experiment is that the first 10 (or 100 or 1000 or . . .) are F's and the next one is an S. That is, for any positive integer n, we may have to examine n batteries before seeing the first S. The sample space is $\mathscr{S} = \{S, FS, FFS, FFFS, . . .\}$, which contains an infinite number of possible outcomes. The same abbreviated form of the sample space is appropriate for an experiment in which, starting at a specified time, the sex of each newborn infant is recorded until the birth of a female is observed.

Events

In our study of probability, we will be interested not only in the individual outcomes of $\mathscr{S}$ but also in any collection of outcomes from $\mathscr{S}$.

> **Definition:** An **event** is any collection (subset) of outcomes contained in the sample space $\mathscr{S}$. An event is said to be **simple** if it consists of exactly one outcome and **compound** if it consists of more than one outcome.

Example 2.2 (continued)

If we examine three fuses, there are eight outcomes in $\mathscr{S}$, so there are eight simple events; these simple events include $A_1 = \{NNN\}$, $A_2 = \{NND\}$, and $A_3 = \{NDN\}$. Compound events include

$B = \{NNN, DDD\}$ = the event that either none or all three of the fuses are defective,

$C = \{NNN, NND, NDN, DNN\}$ = the event that at most one defective fuse is observed, and

$E = \{NND, NDN, DNN\}$ = the event that exactly one among the three fuses is defective.

Example 2.3 (continued)

When two dice are tossed, there are 36 possible outcomes, so there are 36 simple events: $E_1 = \{(1, 1)\}$, $E_2 = \{(1, 2)\}$, . . ., $E_{36} = \{(6, 6)\}$. Examples of compound events are

$A = \{(1, 1), (2, 2), (3, 3), (4, 4), (5, 5), (6, 6)\}$ = the event that both tosses show the same face,

$B = \{(1, 3), (2, 2), (3, 1)\}$ = the event that the sum of the numbers resulting from the two tosses is four, and

$C = \{(6, 6), (6, 5), (5, 6)\}$ = the event that the sum of the two numbers is at least 11.

Example 2.4
(continued)

The sample space for the battery examination experiment contains an infinite number of outcomes, so there are an infinite number of simple events. Compound events include

$A = \{S, FS, FFS\}$ = the event that at most three batteries are examined, and

$E = \{FS, FFFS, FFFFFS, \ldots\}$ = the event that an even number of batteries are examined.

Some Relations from Set Theory

An event is nothing but a set, so that relationships and results from elementary set theory can be used to study events. The following concepts from set theory will be used to construct new events from given events:

Definition: 1. The **union** of two events A and B denoted by $A \cup B$ and read "A *or* B," is the event consisting of all outcomes which are *either in A or in B or in both events* (so that the union includes outcomes for which both A and B occur as well as outcomes for which exactly one occurs).
2. The **intersection** of two events A and B, denoted by $A \cap B$ and read "A *and* B," is the event consisting of all outcomes which are in *both A and B*.
3. The **complement** of an event A, denoted by A', is the set of all outcomes in $\mathcal{S}$ which are not contained in A.

Example 2.3
(continued)

For the experiment consisting of a single die toss, let

$A = \{1, 2, 3\} = \{$outcome is $\leq 3\}$

$B = \{1, 2, 5, 6\}$

$C = \{1, 3, 5\} = \{$outcome is odd$\}$

Then

$A \cup B = \{1, 2, 3, 5, 6\}$, $A \cup C = \{1, 2, 3, 5\}$, $A \cap B = \{1, 2\}$

$A \cap C = \{1, 3\}$, $A' = \{4, 5, 6\}$, and $C' = \{2, 4, 6\}$

Example 2.4
(continued)

In the battery experiment, define *A*, *B*, and *C* by

$$A = \{S, FS, FFS\}$$
$$B = \{S, FFS, FFFFS\}$$

and

$$C = \{FS, FFFS, FFFFFS, \ldots\}$$

Then

$$A \cup B = \{S, FS, FFS, FFFFS\}$$
$$A \cap B = \{S, FFS\}$$
$$A' = \{FFFS, FFFFS, FFFFFS, \ldots\}$$

and

$$C' = \{S, FFS, FFFFS, \ldots\} = \{\text{an odd number of batteries are examined}\}$$

Sometimes *A* and *B* have no outcomes in common, so that the intersection of *A* and *B* contains no outcomes.

> **Definition:** When *A* and *B* have no outcomes in common, they are said to be **mutually exclusive** or **disjoint** events.

Example 2.5

A small city has three automobile dealerships: a GM dealer selling Chevrolets, Pontiacs, and Buicks; a Ford dealer selling Fords and Mercurys; and a Chrysler dealer selling Plymouths and Chryslers. If an experiment consists of observing the brand of the next car sold, then the events *A* = {Chevrolet, Pontiac, Buick} and *B* = {Ford, Mercury} are mutually exclusive, since the next car sold cannot be both a GM product and a Ford product.

The operations of union and intersection can be extended to more than two events. For any three events *A*, *B*, and *C*, the event $A \cup B \cup C$ is the set of outcomes contained in at least one of the three events, while $A \cap B \cap C$ is the set of outcomes contained in all three events. Given events $A_1, A_2, A_3, \ldots$, these events are said to be mutually exclusive (or pairwise disjoint) if no two events have any outcomes in common.

A pictorial representation of events and manipulations with events is obtained by using Venn diagrams. To construct a Venn diagram, draw a rectangle whose interior will represent the sample space $\mathscr{S}$. Then any event *A* is represented as the interior of a closed curve (often a circle) contained in $\mathscr{S}$. Examples of Venn diagrams appear in Figure 2.1.

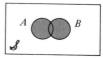

(a)   Venn diagram of
events A and B

(b) Shaded region
is $A \cap B$

(c) Shaded region
is $A \cup B$

(d) Shaded region
is A'

(e)   Mutually exclusive
events

2.1   Venn diagrams illustrated

Exercises / Section 2.1

1. A family which owns two cars is selected, and for both the older car and the newer car we note whether the car was manufactured in America, Europe, or Japan.
   a. What are the possible outcomes of this experiment?
   b. What outcomes are contained in the event that one car is American and the other is foreign?
   c. What outcomes are contained in the event that at least one of the two cars is foreign? What is the complement of this event? Is either of these two events a simple event?

2. Each of a sample of four life insurance companies is classified as a mutual (M) or nonmutual (N) company.
   a. What are the 16 outcomes in $\mathscr{S}$?
   b. What outcomes are in the event that exactly three of the companies are of type M?
   c. What outcomes are in the event that all four companies are of the same type?
   d. What outcomes are in the event that at most one of the four companies is of type M?
   e. What is the union of the events in (c) and (d), and what is the intersection of these two events?
   f. What are the union and intersection of the two events in (b) and (c)?

3. A family consisting of three persons—A, B, and C—belongs to a medical clinic which always has a doctor at each of stations 1, 2, and 3. During a cer-

tain week each member of the family visits the clinic once and is assigned at random to a station. The experiment consists of recording the station number for each member. One outcome is (1, 2, 1) for A to station 1, B to station 2, and C to station 1.
   a. List the 27 outcomes in the sample space.
   b. List all outcomes in the event that all three members go to the same station.
   c. List all outcomes in the event that all members go to different stations.
   d. List all outcomes in the event that no one goes to station 2.

4. A small town has both an Arco gas station and a Conoco gas station. Each station sells its own brand of motor oil and also Pennzoil and Quaker State. An experiment consists of recording both the station and brand of oil chosen by someone who has stopped to purchase a single can of oil.
   a. List in abbreviated fashion the sample space of this experiment.
   b. Is the event $P = \{$the chosen brand is Pennzoil$\}$ a simple event? Why or why not?
   c. Is the event $A = \{$the chosen brand is Arco$\}$ simple? Why or why not?
   d. List the outcomes in the events $A \cup P$, A', and P'.
   e. Are the events A and P mutually exclusive? Why or why not?

f. Suppose that the type of gasoline purchased (R = regular, U = unleaded, E = premium) is also recorded. What is the new sample space?

5. To demonstrate my skill at clairvoyance, I shuffle a deck consisting of three cards marked 1, 2, and 3 and place the deck face down on a table. I then guess at the number on the first card and turn it over. If my guess is correct, the experiment terminates. If my first guess is incorrect, I guess again and then turn over the second card. At this point I know what the third card is, so at most the first two guesses and cards need be reported. An outcome is described by a sequence of either two or four numbers in which a guess and card alternate.

a. List all outcomes in the event that the experiment terminates after just one guess.

b. List all outcomes in the event that the two guesses are both incorrect (I will only guess a number which is possible).

6. Jack Twain has just visited a friend at location A below. He wishes to get to location I. He will travel only along indicated straight-line paths and will only go either north (up) or east (to the right). At each intersection at which there is a choice, he will toss a coin to decide in which of the two possible directions he should go. An outcome for Jack's experiment consists of specifying the corners visited on his trip.

a. List all outcomes in the sample space.

b. List all outcomes for which three coin tosses are necessary.

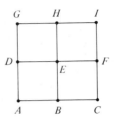

7. Use Venn diagrams to verify the following two relationships for any events A and B (these are called De Morgan's laws):

a. $(A \cup B)' = A' \cap B'$ **b.** $(A \cap B)' = A' \cup B'$

8. a. In Example 2.5, identify three events which are mutually exclusive.

b. Suppose that there is no outcome common to all three of the events A, B, and C. Are these three events necessarily mutually exclusive? If your answer is yes, explain why, while if your answer is no, give a counterexample using the experiment of Example 2.5.

2.2 Axioms, Interpretations, and Properties of Probability

Given an experiment and a sample space $\mathscr{S}$, *the objective of probability is to assign to each event A a number P(A), called the probability of the event A, which will give a precise measure of the chance that A will occur.* To ensure that the probability assignments will be consistent with our intuitive notions of probability, all assignments should satisfy the following axioms (basic properties) of probability.

Axiom 1: For any event A, $P(A) \geq 0$.

Axiom 2: $P(\mathscr{S}) = 1$.

Axiom 3: a. If $A_1, A_2, \ldots, A_n$ is a finite collection of mutually exclusive events, then

$$P(A_1 \cup A_2 \cup \ldots \cup A_n) = \sum_{i=1}^{n} P(A_i)$$

b. If $A_1, A_2, A_3, \ldots$ is an infinite collection of mutually exclusive events, then

$$P(A_1 \cup A_2 \cup A_3 \cup \ldots) = \sum_{i=1}^{\infty} P(A_i)$$

Axiom 1 reflects the intuitive notion that the chance of A occurring should be at least 0, so that negative probabilities are not allowed. The sample space is by definition an event which must occur when the experiment is performed ($\mathcal{S}$ contains all possible outcomes), so Axiom 2 says that the maximum possible probability of one is assigned to $\mathcal{S}$. The third axiom formalizes the idea that if we wish the probability that at least one of a number of events will occur and no two of the events can occur simultaneously, then the chance of at least one occurring is the sum of the chances of the individual events.

Example 2.6
In the experiment in which a single coin is tossed, the sample space is $\mathcal{S} = \{H, T\}$. The axioms specify $P(\mathcal{S}) = 1$, so to complete the probability assignment, it remains only to determine $P(H)$ and $P(T)$. Since H and T are disjoint events and $H \cup T = \mathcal{S}$, Axiom 3 implies that $1 = P(\mathcal{S}) = P(H) + P(T)$, So $P(T) = 1 - P(H)$. Thus the only freedom allowed by the axioms in this experiment is the probability asigned to H. One possible assignment of probabilities is $P(H) = .5$, $P(T) = .5$, while another possible assignment is $P(H) = .75, P(T) = .25$. In fact, letting p represent any fixed number between 0 and 1, $P(H) = p$ and $P(T) = 1 - p$ is an assignment which is consistent with the axioms.

Example 2.7
Consider the experiment in Example 2.4, in which batteries coming off an assembly line are tested one by one until one having a voltage within prescribed limits is found. The simple events are $E_1 = \{S\}$, $E_2 = \{FS\}$, $E_3 = \{FFS\}$, $E_4 = \{FFFS\}$, ... Suppose that the probability of any particular battery being satisfactory is .99. Then it can be shown that $P(E_1) = .99$, $P(E_2) = (.01)(.99)$, $P(E_3) = (.01)^2(.99)$, ... is an assignment of probabilities to the simple events which satisfies the axioms. In particular, because the E_i's are disjoint and $\mathcal{S} = E_1 \cup E_2 \cup E_3 \cup \ldots$, we must have $1 = P(\mathcal{S}) = P(E_1) + P(E_2) + P(E_3) + \cdots = .99[1 + .01 + (.01)^2 + (.01)^3 + \cdots]$. The validity of this equality is a consequence of a mathematical result concerning the sum of a geometric series.

However, another legitimate (according to the axioms) probability assignment of the same "geometric" type is obtained by replacing .99 by any other number p between 0 and 1 (and .01 by $1 - p$). Thus there are an infinite number of legitimate probability assignments of this type.

Interpreting Probability

Examples 2.6 and 2.7 show that the axioms do not completely determine an assignment of probabilities to events. The axioms only serve to rule out assignments inconsistent with our intuitive notions of probability. In the coin-tossing experiment of Example 2.6, two particular assignments were suggested. The appropriate or correct assignment depends on the manner in which the experiment is carried out and also on one's interpretation of probability. The interpretation which is most frequently used and most easily understood is based on the notion of relative frequencies.

Consider an experiment which can be repeatedly performed in an identical and independent fashion, and let A be an event consisting of a fixed set of outcomes of the experiment. Simple examples of such repeatable experiments include the coin-tossing and die-tossing experiments previously discussed. If the experiment is performed n times, on some of the replications the event A will occur (the outcome will be in the set A), and on others, A will not occur. Let $n(A)$ denote the number of replications on which A does occur. Then the ratio $n(A)/n$ is called the *relative frequency* of occurrence of the event A in the sequence of n replications. Empirical evidence, based on the results of many of these sequences of repeatable experiments, indicates that as n grows large, the relative frequency $n(A)/n$ stabilizes, as pictured in Figure 2.2. That is, as n gets arbitrarily large, the relative frequency approaches a limiting number which we refer to as the *limiting relative frequency* of the event A. The objective interpretation of probability identifies this limiting relative frequency with $P(A)$.

If probabilities are assigned to events in accordance with their limiting relative frequencies, then we can interpret a statement such as "The probability of that coin landing with the head facing up when it is tossed is .5" to mean that in a large number of such tosses, a head will appear on approximately half the tosses and a tail on the other half.

This relative frequency interpretation of probability is referred to as an objective interpretation because it rests on a property of the experiment rather than any particular individual concerned with the experiment. For example, two different observers of a sequence of coin tosses should both use the same probability assignments since the observers have nothing to do with limiting relative frequency. In practice, this interpretation is not as objective as it might seem, since the limiting relative frequency of an event will not be known. Thus, we will have to assign probabilities based on our beliefs about the limiting relative frequency of events under study. Fortunately, there are many experiments for which there will be a consensus with respect to probability assignments. When we speak of a fair coin, we shall mean $P(H) = P(T) = .5$, and a fair die is one for which limiting relative frequencies of the six outcomes are all $\frac{1}{6}$, suggesting probability assignments $P(\{1\}) = \cdots = P(\{6\}) = \frac{1}{6}$.

Because the objective interpretation of probability is based on the notion of limiting frequency, its applicability is limited to experimental situations which are repeatable. Yet the language of probability is often used in connection with situations that are inherently unrepeatable. Examples include: "The chances are good for

Figure 2.2   Stabilization of relative frequency

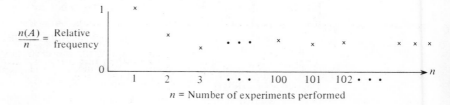

n = Number of experiments performed

a peace agreement"; "It is quite likely that our company will be awarded the contract"; and "Because their best quarterback is injured, I expect them to score no more than 10 points against us." In such situations we would like, as before, to assign numerical probabilities to various outcomes and events (for example, the probability is .9 that we will get the contract). We must therefore adopt an alternative interpretation of these probabilities. Because different observers may have different prior information and opinions concerning such experimental situations, probability assignments may now differ from individual to individual. Interpretations in such situations are thus referred to as subjective. The book by Winkler listed in the chapter references gives a very readable survey of several subjective interpretations.

Properties of Probability

> **Proposition:** For any event A, $P(A) = 1 - P(A')$.

Proof: In Axiom 3a, let $n = 2$, $A_1 = A$, and $A_2 = A'$. Since by definition of A', $A \cup A' = \mathcal{S}$ while A and A' are disjoint, $1 = P(\mathcal{S}) = P(A \cup A') = P(A) + P(A')$, from which the desired result follows.

This proposition is surprisingly useful, since there are many situations in which one of $P(A)$ or $P(A')$ is more easily computed than is the other.

Example 2.8   Of all Americans, 85% have an agglutinating factor in their blood which classifies them as Rh positive, while 15% lack the factor, so are classified as Rh negative. Suppose that a doctor wishes to do an analysis of blood from a newborn Rh negative infant, so he examines blood-typing results from a sequence of newborn infants until he finds an Rh negative infant. What is the probability that at least two infants must be typed before the search is terminated?

This experiment has the same structure as the battery experiment of Example 2.4. The event that at least two typings are necessary is the union of an infinite number of simple events (two typings, three typings, . . .). However, let

$B = \{$the first infant examined is Rh negative$\}$

Since 15% of all infants will be Rh negative, $P(B) = .15$. Then

$B' = \{$the first infant is not Rh negative$\}$

$= \{$at least two infants must be examined$\}$

with $P(B') = 1 - P(B) = 1 - .15 = .85$ as the desired probability.

In general, the above proposition is useful when the event of interest can be expressed as "at least . . .," since then the complement, "less than . . ." is often easier to work with (in some problems, "more than . . ." is easier to deal with than "at most . . . .").

> **Proposition:** If A and B are mutually exclusive, then $P(A \cap B) = 0$.

Proof: Because $A \cap B$ contains no outcomes, $(A \cap B)' = \mathscr{S}$. Thus $1 = P[(A \cap B)'] = 1 - P(A \cap B)$.

When events A and B are mutually exclusive, Axiom 3 gives $P(A \cup B) = P(A) + P(B)$. When A and B are not mutually exclusive, the proba-bility of the union is obtained from the following result.

> **Proposition:** For any two events A and B,
>
> $$P(A \cup B) = P(A) + P(B) - P(A \cap B)$$

Notice that the proposition is valid even if A and B are mutually exclusive, since then $P(A \cap B) = 0$. The key idea is that, in adding $P(A)$ and $P(B)$, the intersection $A \cap B$ is actually counted twice, so $P(A \cap B)$ must be subtracted out.

Proof: Note first that $A \cup B = A \cup (B \cap A')$, as illustrated in Figure 2.3.
Since A and $(B \cap A')$ are mutually exclusive, $P(A \cup B) = P(A) + P(B \cap A')$. But $B = (B \cap A) \cup (B \cap A')$ (the union of that part of B in A and that part of B not in A), with $(B \cap A)$ and $(B \cap A')$ mutually exclusive, so that $P(B) = P(B \cap A) + P(B \cap A')$. This all gives

$$P(A \cup B) = P(A) + P(B \cap A') = P(A) + [P(B) - P(A \cap B)]$$
$$= P(A) + P(B) - P(A \cap B)$$

Figure 2.3 $A \cup B = A \cup (B \cap A')$

Example 2.9

In a certain residential suburb, 60% of all households subscribe to the metropolitan newspaper published in a nearby city, 80% subscribe to the local afternoon paper, and 50% of all households subscribe to both papers. If a household is selected at random, what is the probability that it subscribes to (a) at least one of the two newspapers? and (b) exactly one of the two newspapers?

With $A = \{$subscribes to the metropolitan paper$\}$ and $B = \{$subscribes to the local paper$\}$, the given information implies that $P(A) = .6$, $P(B) = .8$, and $P(A \cap B) = .5$. The above proposition then applies to give

P (subscribes to at least one of the two newspapers) $=$

$$P(A \cup B) = P(A) + P(B) - P(A \cap B) = .6 + .8 - .5 = .9$$

Because {exactly one} and {both} are mutually exclusive with {exactly one} ∪ {both} = {at least one}

$$P(\text{exactly one}) + P(\text{both}) = P(\text{at least one}) = .9$$

giving

$$P(\text{exactly one}) = .9 - P(\text{both}) = .9 - .5 = .4.$$

The probability of a union of more than two events can be computed analogously. For three events A, B, and C, the result is

$$P(A \cup B \cup C) = P(A) + P(B) + P(C) - P(A \cap B) - P(A \cap C) - P(B \cap C)$$
$$+ P(A \cap B \cap C)$$

This can be seen by examining a Venn diagram of $A \cup B \cup C$:

Figure 2.4

When $P(A)$, $P(B)$, and $P(C)$ are added, certain intersections are counted twice, so must be subtracted out, but this results in $P(A \cap B \cap C)$ being subtracted once too often.

Determining Probabilities Systematically

When the number of possible outcomes (simple events) is large, there will be many compound events. A simple way to determine probabilities for these events that avoids violating the axioms and derived properties is to first determine probabilities $P(E_i)$ for all simple events. These should satisfy $P(E_i) \geq 0$ and $\sum_{\text{all } i} P(E_i) = 1$. Then the probability of any compound event A is computed by adding together the $P(E_i)$'s for all E_i's in A:

$$P(A) = \sum_{\text{all } E_i\text{'s in } A} P(E_i)$$

Example 2.10 Denote the six elementary events {1}, . . ., {6} associated with tossing a six-sided die once by E_1, . . ., E_6. If the die is constructed so that any of the three even outcomes is twice as likely to occur as any of the three odd outcomes, then an appropriate assignment of probabilities to elementary events is $P(E_1) = P(E_3) = P(E_5) = \frac{1}{9}$, $P(E_2) = P(E_4) = P(E_6) = \frac{2}{9}$. Then for the event $A = \{$outcome is even$\} = E_2 \cup E_4 \cup E_6$, $P(A) = P(E_2) + P(E_4) + P(E_6) = \frac{6}{9} = \frac{2}{3}$; for $B = \{$outcome $\leq 3\} = E_1 \cup E_2 \cup E_3$, $P(B) = \frac{1}{9} + \frac{2}{9} + \frac{1}{9} = \frac{4}{9}$.

Equally Likely Outcomes

In many experiments consisting of N outcomes, it is reasonable to assign equal probabilities to all N simple events. These include such obvious examples as tossing a fair coin or fair die once or twice (or any fixed number of times), or selecting one or several cards from a well-shuffled deck of 52. With $p = P(E_i)$ for every i,

$$1 = \sum_{i=1}^{N} P(E_i) = \sum_{i=1}^{N} p = p \cdot N; \quad \text{so } p = \frac{1}{N}$$

That is, if there are N possible outcomes, then the probability assigned to each is $1/N$.

Now consider an event A, with $N(A)$ denoting the number of outcomes contained in A. Then

$$P(A) = \sum_{E_i \text{ in } A} P(E_i) = \sum_{E_i \text{ in } A} \frac{1}{N} = \frac{N(A)}{N}$$

Once we have counted the number of outcomes N in the sample space, to compute the probability of any event we must count the number of outcomes contained in that event and take the ratio of the two numbers. Thus when outcomes are equally likely, computing probabilities reduces to counting.

Example 2.11   When two dice are rolled separately as in Example 2.3, there are 36 possible outcomes ($N = 36$). If the dice are both fair, all 36 outcomes are equally likely, so $P(E_i) = \frac{1}{36}$. Then the event $A = \{$sum of two numbers $= 7\}$ consists of the six outcomes (1, 6), (2, 5), (3, 4), (4, 3), (5, 2), and (6, 1), so

$$P(A) = \frac{N(A)}{N} = \frac{6}{36} = \frac{1}{6}$$

Exercises / Section 2.2

1. A family that owns two automobiles is selected at random. Let $A_1 = \{$the older car is American$\}$ and $A_2 = \{$the newer car is American$\}$. If $P(A_1) = .7$, $P(A_2) = .5$, and $P(A_1 \cap A_2) = .4$, compute
   a. $P(A_1 \cup A_2)$ (the probability that at least one car is American)
   b. the probability that neither car is American
   c. the probability that exactly one of the two cars is American

2. An engineering construction firm presently has bids out on three projects. Let $A_i = \{$awarded project $i\}$ for $i = 1, 2, 3$, and suppose that $P(A_1) = .22$, $P(A_2) = .25$, $P(A_3) = .28$, $P(A_1 \cap A_2) = .11$, $P(A_1 \cap A_3) = .05$, $P(A_2 \cap A_3) = .07$, $P(A_1 \cap A_2 \cap A_3) = .01$. Express in words each of the following events and compute the probability of each event.

a. $A_1 \cup A_2$
b. $A_1' \cap A_2'$ *Hint:* $(A_1 \cup A_2)' = A_1' \cap A_2'$
c. $A_1 \cup A_2 \cup A_3$ **d.** $A_1' \cap A_2' \cap A_3'$
e. $A_1' \cap A_2' \cap A_3$ **f.** $(A_1' \cap A_2') \cup A_3$

3. A utility company offers a lifeline rate to any household whose electricty usage falls below 240 kWh during a particular month. Let A denote the event that a randomly selected household in a certain community does not exceed the lifeline usage during January, and let B be the analogous event for the month of July (A and B refer to the same household). Suppose that $P(A) = .8$, $P(B) = .7$, and $P(A \cup B) = .9$.
a. Compute $P(A \cap B)$.
b. Compute the probability that the lifeline usage amount is exceeded in exactly one of the two months. Describe this event in terms of A and B.

4. In the experiment in which two fair dice (one red and one green) are rolled, let $S_2 = \{$the sum of the two resulting numbers is divisible by 2$\}$, $S_3 = \{$the sum is divisible by 3$\}$, $S_4 = \{$the sum is divisible by 4$\}$, and $Q = \{$the square root of the sum is an integer$\}$.
a. Compute $P(S_2)$, $P(S_3)$, $P(S_4)$, and $P(Q)$, justifying each computation by identifying any axiom, rule, or property of probability used.
b. Compute $P(S_4 \cup Q)$ first by listing outcomes in $S_4 \cup Q$ and then by using the results of (a) along with an appropriate axiom or property.
c. Compute $P(S_2 \cap S_3)$ by listing outcomes in $S_2 \cap S_3$.
d. Compute $P(S_2 \cup S_3)$ by using the results of (a) and (c).
e. Compute $P(S_2 \cap S_4)$. Have you already computed this probability?
f. Compute $P(S_2 \cup S_3 \cup S_4)$.
g. Compute $P(S_2 \cup S_3)$ without listing outcomes. *Hint:* See Exercise 7, Section 2.1.

5. In a school machine shop, 60% of all machine breakdowns occur on lathes and 15% on drills. Let $A = \{$the next machine breakdown is a lathe$\}$, and $B = \{$the next machine breakdown is a drill$\}$ (so that A and B are mutually exclusive). With $P(A) = .60$ and $P(B) = .15$, calculate
a. $P(A')$ **b.** $P(A \cup B)$ **c.** $P(A' \cap B')$

6. A state park has two campgrounds. Let A denote the event that there is still space in the first campground at 5:00 P.M. on any given day, and define

B analogously for the second campground. If $P(A) = .3$, $P(B) = .2$, and $P(A \cap B) = .1$, what is the probability that a camper who shows up at precisely 5:00 P.M.
a. will be able to camp in at least one of the campgrounds
b. will be able to camp in neither of the campgrounds
c. will be able to camp in exactly one of the two campgrounds

7. If when I visit the local library, the probability that someone is reading the current issue of *Sports Illustrated* is .4, the probability that someone is reading *Time* is .3, and the probability that at least one of these two magazines is being read by someone is .5, what is the probability that
a. both of the magazines are being read
b. neither of the two is being read
c. exactly one is being read

8. Use the axioms to show that if one event A is contained in another event B (A is a subset of B), then $P(A) \leq P(B)$. *Hint:* For such A and B, $B = A \cup (B \cap A')$, draw a Venn diagram.

9. If a standard deck of 52 cards (suits spades, hearts, diamonds, and clubs; with ace, 2, . . ., 10, jack, queen, and king in each suit) is well mixed before a single card is drawn, all 52 outcomes are equally likely. Define events by $A = \{$heart$\}$, $B = \{$black card$\}$, $C = \{$jack, queen, or king$\}$, and $D = \{$selected card is at most a five$\}$, and use elementary counting along with the axioms and properties to compute the following probabilities.
a. $P(A)$, $P(B)$, $P(C)$, and $P(D)$
b. $P(A \cup B)$ and $P(C \cup D)$
c. $P(A \cup C)$ and $P(B \cup D)$
d. $P(A \cup B \cup C \cup D)$

10. I have a cassette deck in my car along with 25 tapes, of which 15 consist of classical music. If while driving I reach out and begin selecting tapes at random, what is the probability that I must select at least two before finding one with classical music on it?

11. The three most popular options on a certain type of new car are automatic transmission (A), power steering (B), and a radio (C). If 70% of all purchasers request A, 75% request B, 80% request C, 80% request A or B, 85% request A or C, 90% request B or C, and 95% request A or B or C, compute the probabilities of the following events. *Note*: "A or B" could mean that both were requested. *Hint*: Try drawing a Venn diagram and labeling all regions.

a. The next purchaser will select at least one of the three options ($A \cup B \cup C$).
b. The next purchaser will select none of the three options.
c. The next purchaser will select only a radio.
d. The next purchaser will select exactly one of the three options.

12. An academic department with five faculty members—Anderson, Box, Cox, Cramer, and Fisher—must select two of its members to serve on a personnel review committee. Because the work will be time consuming, no one is anxious to serve, so it is decided that the representative will be selected by putting five slips of paper in a box, mixing them, and selecting two.

a. What is the probability that Anderson and Box are both selected? *Hint*: There are 20 equally likely outcomes for this experiment.
b. What is the probability that at least one of the two members whose name begins with C is selected?
c. If the five faculty members have taught for 3, 6, 7, 10, and 14 years, respectively, at the university, what is the probability that the two chosen representatives have at least 15 years' teaching experience at the university?

13. In Exercise 3 of Section 2.1, suppose that any incoming individual is equally likely to be assigned to any of the three stations irrespective of where other individuals have been assigned.

a. What is the probability that all three family members are assigned to the same station?
b. What is the probability that at most two family members are assigned to the same station?
c. What is the probability that every family member is assigned to a different station?

2.3 Counting Techniques

When the various outcomes of an experiment are equally likely (the same probability is assigned to each simple event), then the task of computing probabilities reduces to counting. In particular, if N is the number of outcomes in a sample space and $N(A)$ is the number of outcomes contained in an event A, then

$$P(A) = \frac{N(A)}{N} \tag{2.1}$$

If a list of the outcomes is available or easy to construct and N is small, then the numerator and denominator of (1) can be obtained without the benefit of any general counting principles.

There are, however, many experiments for which the effort involved in constructing such a list is prohibitive because N is quite large. By exploiting some general counting rules, it is possible to compute probabilities of the form (2.1) without a listing of outcomes. These rules are also useful in many problems involving outcomes which are not equally likely. Several of the rules developed here will be used in studying probability distributions in the next chapter.

The Product Rule for Ordered Pairs

Our first counting rule applies to any situation in which a set (event) consists of ordered pairs of objects, and we wish to count the number of such pairs. By an ordered pair we mean that, if O_1 and O_2 are objects, then the pair (O_1, O_2) is different from the pair (O_2, O_1). For example, in tossing a die twice, each outcome consists of a pair of numbers referring to the outcomes on the two successive tosses, and (1, 2) and (2, 1) are different outcomes.

> **Proposition:** If the first element or object of an ordered pair can be selected in n_1 ways, and for each of these n_1 ways the second element of the pair can be selected in n_2 ways, then the number of pairs is $n_1 n_2$.

Example 2.12   A student wishes to commute first to a junior college for two years and then to a state college campus. Within commuting range there are four junior colleges and three state college campuses. How many choices of junior college and state college are available to her? Numbering junior colleges by 1, 2, 3, 4 and state colleges by a, b, c, choices are $(1, a)$, $(1, b)$, . . ., $(4, c)$, a total of 12 choices. With $n_1 = 4$ and $n_2 = 3$, $N = n_1 n_2 = 12$ without a list.

Example 2.13   A homeowner doing some remodeling requires the services of both a plumbing contractor and an electrical contractor. If there are 12 plumbing contractors and nine electrical contractors available in the area, in how many ways can the contractors be chosen? If we denote the plumbers by P_1, . . ., P_{12} and the electricians by Q_1, . . ., Q_9, then we wish the number of pairs of the form (P_i, Q_j). With $n_1 = 12$ and $n_2 = 9$, the product rule yields $N = (12)(9) = 108$ possible ways of choosing the two types of contractors.

In both Examples 2.12 and 2.13, the choice of the second element of the pair did not depend on which first element was chosen or occurred. As long as there are the same number of choices of the second element for each first element, the product rule is valid even when the set of possible second elements depends on the first element.

Example 2.14   A family has just moved to a new city and requires the services of both an obstetrician and a pediatrician. There are two easily accessible medical clinics, each having two obstetricians and three pediatricians. The family will obtain maximum health insurance benefits by joining a clinic and selecting both doctors from that clinic. In how many ways can this be done? Denote the obstetricians by $O_1, O_2, O_3,$ and O_4 and the pediatricians by P_1, . . ., P_6. Then we wish the number of pairs (O_i, P_j) for which O_i and P_j are associated with the same clinic. Because there are four obstetricians, $n_1 = 4$, and for each there are three choices of pediatrician, so $n_2 = 3$. Applying the product rule gives $N = n_1 n_2 = 12$ possible choices.

Tree Diagrams

In problems in which the product rule can be applied, a configuration called a tree diagram can be used to represent pictorially all the possibilities. The tree diagram

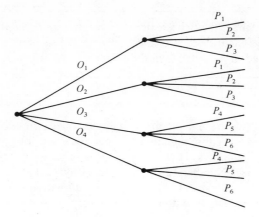

Figure 2.5   Tree diagram for Example 2.14

associated with Example 2.14 appears in Figure 2.5. Starting from a point on the left side of the diagram, for each possible first element of a pair a straight-line segment emanates rightward. Each of these lines is referred to as a first-generation branch. Now for any given first-generation branch we construct another line segment emanating from the tip of the branch for each possible choice of a second element of the pair. Each such line segment is a second-generation branch. Because there are four obstetricians, there are four first-generation branches, and three pediatricians for each obstetrician yields three second-generation branches emanating from each first-generation branch.

In the general case there are n_1 first-generation branches, and for each first-generation branch there are n_2 second-generation branches. The total number of second-generation branches is, therefore, $n_1 n_2$. Since each second-generation branch corresponds to exactly one possible pair (choosing a first element and then a second puts us at the end of exactly one second-generation branch), there are $n_1 n_2$ pairs, so the product rule is verified.

The construction of a tree diagram does not depend on having the same number of second-generation branches emanating from each first generation branch. If the second clinic had four pediatricians then there would be only three branches emanating from two of the first-generation branches and four emanating from each of the other two first-generation branches. A tree diagram can thus be used to represent pictorially experiments other than those to which the product rule applies.

A More General Product Rule

If a six-sided die is tossed five times in succession rather than just twice, then each possible outcome is an ordered collection of five numbers such as (1, 3, 1, 2, 4) or (6, 5, 2, 2, 2). We shall call an ordered collection of k objects a **k-tuple** (so a pair is a 2-tuple and a triple is a 3-tuple). Each outcome of the die-tossing experiment is then a 5-tuple.

> **Product rule for k-tuples:** Suppose that a set consists of ordered collections of k elements (k-tuples) and that there are n_1 possible choices for the first element; for each choice of first element there are n_2 possible choices of the second element; . . .; for each possible choice of the first $k - 1$ elements, there are n_k choices of the kth element. Then there are $n_1 n_2 \ldots n_k$ possible k-tuples.

This more general rule can also be illustrated by a tree diagram; simply construct a more elaborate diagram by adding third-generation branches emanating from the tip of each second generation, then fourth-generation branches, and so on, until finally kth-generation branches are added.

Example 2.13 (continued)

Suppose that the home remodeling job involves first purchasing several kitchen appliances. They will all be purchased from the same dealer, and there are five dealers in the area. With the dealers denoted by $D_1, \ldots, D_5$ there are $N = n_1 n_2 n_3 = (5)(12)(9) = 540$ 3-tuples of the form (D_i, P_j, Q_k), so there are 540 ways to choose first an appliance dealer, then a plumbing contractor, and finally an electrical contractor.

Example 2.14 (continued)

If each clinic has both three specialists in internal medicine and two general surgeons, there are $n_1 n_2 n_3 n_4 = (4)(3)(3)(2) = 72$ ways to select one doctor of each type such that all doctors practice at the same clinic.

Permutations

So far the successive elements of a k-tuple were selected from entirely different sets (for example, appliance dealers, then plumbers, and finally electricians). In several tosses of a die, the set from which successive elements are chosen is always $\{1, 2, 3, 4, 5, 6\}$, but the choices are made "with replacement," so that the same element can appear more than once. We now consider a fixed set consisting of n distinct elements and suppose that a k-tuple is formed by selecting successively from this set *without replacement*, so that an element can appear in at most one of the k positions.

> **Definition:** Any ordered sequence of k objects taken from a set of n distinct objects is called a **permutation** of size k of the objects. The number of permutations of size k which can be constructed from the n objects is denoted by $P_{k, n}$.

The number of permutations of size k is obtained immediately from the general product rule. The first element can be chosen in n ways, for each of these n ways the second element can be chosen in $n - 1$ ways, and so on; finally, for each way of choosing the first $k - 1$ elements, there are $n - (k - 1) = n - k + 1$ ways of choosing the kth element, so

$$P_{k,n} = n(n - 1)(n - 2) \cdots (n - k + 2)(n - k + 1)$$

Example 2.15   There are eight teaching assistants available for grading papers in a particular course. The first exam consists of four questions, and the professor wishes to select a different assistant to grade each question (only one assistant per question). In how many ways can assistants be chosen to grade the exam? Here $n =$ the number of assistants $= 8$ and $k =$ the number of questions $= 4$. The number of different grading assignments is then $P_{4,8} = (8)(7)(6)(5) = 1680$.

The use of factorial notation allows $P_{k,n}$ to be expressed more compactly.

> **Definition:** For any positive integer m, $m!$ is read "m factorial" and is defined by $m! = m(m - 1) \cdots (2)(1)$. Also, $0! = 1$.

Using factorial notation yields

$$P_{k,n} = n(n - 1) \cdots (n - k + 1)$$

$$= \frac{n(n - 1) \cdots (n - k + 1)(n - k)(n - k - 1) \cdots (2)(1)}{(n - k)(n - k - 1) \cdots (2)(1)}$$

which becomes

$$P_{k,n} = \frac{n!}{(n - k)!}$$

For example, $P_{3,9} = 9!/(9 - 3)! = 9!/6! = 9 \cdot 8 \cdot 7 \cdot 6!/6! = 9 \cdot 8 \cdot 7$. Note also that because $0! = 1$, $P_{n,n} = n!/(n - n)! = n!/0! = n!/1 = n!$, as it should.

Combinations

There are many counting problems in which one is given a set of n distinct objects and wishes to count the number of unordered subsets of size k. For example, in bridge it is only the 13 cards in a hand and not the order in which they are dealt that is important; in the formation of a committee, the order in which committee members are listed is frequently unimportant.

> **Definition:** Given a set of n distinct objects, any unordered subset of size k of the objects is called a **combination**. The number of combinations of size k which can be formed from n distinct objects will be denoted by $\binom{n}{k}$. (This notation is more common in probability than $C_{k,n}$ which would be analogous to notation for permutations).

The number of combinations of size k from a particular set is smaller than the number of permutations since, when order is disregarded, a number of permutations correspond to the same combination. Consider, for example, the set $\{A, B, C, D, E\}$ consisting of 5 elements. We know that there are $5!/(5 - 3)! = 60$ permutations of size 3. There are 6 permutations of size 3 consisting of the elements A, B, and C since these 3 can be ordered $3 \cdot 2 \cdot 1 = 3! = 6$ ways: (A, B, C), (A, C, B), (B, A, C), (B, C, A), (C, A, B), and (C, B, A). These 6 permutations are equivalent to the single combination $\{A, B, C\}$. Similarly for any other combination of size 3, there are 3! permutations, each obtained by ordering the 3 objects. Thus,

$$60 = P_{3,5} = \binom{5}{3} \cdot 3!; \quad \text{so} \quad \binom{5}{3} = \frac{60}{3!} = 10$$

These 10 combinations are

$\{A, B, C\}, \{A, B, D\}, \{A, B, E\}, \{A, C, D\}, \{A, C, E\}, \{A, D, E\},$
$\{B, C, D\}, \{B, C, E\}, \{B, D, E\}, \{C, D, E\}$

When there are n distinct objects, any permutation of size k is obtained by ordering the k unordered objects of a combination in $k!$ ways, so the number of permutations is the product of $k!$ and the number of combinations. This gives

$$\binom{n}{k} = \frac{P_{k,n}}{k!} = \frac{n!}{k!\,(n - k)!}$$

Notice that $\binom{n}{n} = 1$ and $\binom{n}{0} = 1$ since there is only one way to choose a set of (all) n elements or of no elements, and $\binom{n}{1} = n$ since there are n subsets of size 1.

Example 2.16 A bridge hand consists of any 13 cards selected from a 52-card deck without regard to order. There are $\binom{52}{13} = \frac{52!}{13!39!}$ different bridge hands, which works out to approximately 635 billion. Since there are 13 cards in each suit, the number of hands consisting entirely of clubs and spades is $\binom{26}{13} = \frac{26!}{13!\,13!} = 10,400,597$. Since all hands are equally likely, letting A = {you are dealt a hand consisting entirely of spades and clubs},

$$P(A) = \frac{N(A)}{N} = \frac{\binom{26}{13}}{\binom{52}{13}} = .0000164$$

Since there are $\binom{4}{2}$ = 6 combinations consisting of two suits, of which spades and clubs is one pair, if we let B = {a bridge hand consists of exactly two suits}, then

$$P(B) = \frac{N(B)}{N} = \frac{6\binom{26}{13}}{\binom{52}{13}} = .0000984$$

That is, a hand consisting entirely of cards from exactly two of the four suits will occur roughly once in every 10,000 hands. If you play bridge only once a month, it is likely that you will never be dealt such a hand!

Example 2.17 A rental car service facility has 10 foreign cars and 15 domestic cars waiting to be serviced on a particular Saturday morning. Because there are so few mechanics working on Saturday, only 6 can be serviced. If the 6 are chosen at random, what is the probability that 3 of the cars selected are foreign and the other 3 domestic?

Let D_3 = {exactly 3 of the 6 cars chosen are foreign}. Assuming that any particular set of 6 cars is as likely to be chosen as is any other set of 6, we have equally likely outcomes, so $P(D_3) = N(D_3)/N$, where N is the number of ways of choosing 6 cars from the 25 and $N(D_3)$ is the number of ways of choosing 3 domestic cars and 3 foreign cars. We have immediately that $N = \binom{25}{6}$. To obtain $N(D_3)$, think of first choosing 3 of the 15 domestic cars and then 3 of the foreign cars. There are $\binom{15}{3}$ ways of choosing the 3 domestic cars, and there are $\binom{10}{3}$ ways of choosing the 3 foreign cars; $N(D_3)$ is now the product of these two numbers (visualize a tree diagram—we are really using a product rule argument here), so

$$P(D_3) = \frac{N(D_3)}{N} = \frac{\binom{15}{3}\binom{10}{3}}{\binom{25}{6}} = \frac{\frac{15!}{3!\,12!} \cdot \frac{10!}{3!\,7!}}{\frac{25!}{6!\,19!}} = .3083$$

Similarly, the probability that at least 3 domestic cars are selected is

$$P(D_3 \cup D_4 \cup D_5 \cup D_6) = P(D_3) + P(D_4) + P(D_5) + P(D_6)$$

$$= \frac{\binom{15}{3}\binom{10}{3}}{\binom{25}{6}} + \frac{\binom{15}{4}\binom{10}{2}}{\binom{25}{6}} + \frac{\binom{15}{5}\binom{10}{1}}{\binom{25}{6}} + \frac{\binom{15}{6}\binom{10}{0}}{\binom{25}{6}} = .8530$$

Exercises / Section 2.3

1. The Student Engineers Council at a certain college has one student representative from each of the five engineering majors (civil, electrical, industrial, materials, and mechanical).

   a. In how many ways can both a council president and a vice president be selected?

   b. In how many ways can a president, a vice president, and a secretary be selected?

   c. In how many ways can two members be selected for the President's Council?

2. A real estate agent is showing homes to a prospective buyer. There are 10 homes in the desired price range listed in the area. The buyer only has time to visit three of them.

   a. In how many ways could the three homes be chosen if the order of visiting is considered?

   b. In how many ways could the three homes be chosen if the order is unimportant?

   c. If four of the homes are new and six have previously been occupied, and if the three homes to visit are randomly chosen, what is the probability that all three are new? (The same answer results whether or not order is considered.)

3. a. Beethoven wrote 9 symphonies and Mozart wrote 27 piano concertos. If a university radio station announcer wishes to play first a Beethoven symphony and then a Mozart concerto, in how many ways can this be done?

   b. The station manager has decided that on each successive night (seven days per week), a Beethoven symphony will be played, followed by a Mozart piano concerto, followed by a Schubert string quartet (of which there are 15). For roughly how many years could this policy be continued before exactly the same program would have to be repeated?

4. During the fall quarter of an academic year, Statistics 1A will be offered at 12 different times, while during the winter quarter Statistics 1B will be offered at 10 different times.

   a. If a student wishes to take both 1A and 1B during the fall and winter quarters, respectively, how many different time choices for the two courses are possible?

   b. If each section of each course can accommodate up to 35 students, what is the maximum

number of students that could be enrolled in these courses?

   c. If six of the 1A sections are given in the morning and seven of the 1B sections are given in the morning, what is the probability that a student will end up with two morning sections if the student selects one of the time choices of (a) in a completely random fashion?

   d. If Professor I. N. Coherent is scheduled to teach three sections of 1A and three sections of 1B, what is the probability that a student who randomly selects sections will end up having Professor Coherent for both quarters? for neither quarter?

5. A city is serviced by a cable television company which provides programs on seven different channels for its subscribers. From 5:30 to 6:00 P.M. on weekdays three channels show news, from 6:00 to 6:30 five channels show news, from 6:30 to 7:00 four channels show news, and from 7:00 to 7:30 two channels show news.

   a. If a viewing sequence consists of four half-hour programs between 5:30 and 7:30, how many viewing sequences are there?

   b. If a viewing sequence is chosen at random from the set of all possible sequences, what is the probability that all four programs selected are news programs?

   c. Choosing a sequence as in (b), what is the probability that at least two programs are news programs?

   d. Choosing a sequence as in (b), what is the probability that a news program is selected both at 6:00 P.M. and at 6:30 P.M.?

6. A chain of high-fidelity stores is offering a special price on a complete set of components (receiver, turntable, speakers, cassette deck). A purchaser is offered a choice of manufacturer for each component:

   Receiver: Kenwood, Pioneer, Sansui, Sony, Sherwood

   Turntable: BSR, Dual, Sony, Technics

   Speakers: AR, KLH, JBL

   Cassette Deck: Advent, Sony, Teac, Technics

A switchboard display in the store allows a customer to hook together any selection of components (consisting of one of each type). Use the product rules to answer the following questions.

a. In how many ways can one component of each type be selected?

b. In how many ways can components be selected if both the receiver and the turntable are to be Sony?

c. In how many ways can components be selected if none is to be Sony?

d. In how many ways can a selection be made if at least one Sony component is to be included?

e. If someone flips switches on the selection in a completely random fashion, what is the probability that the system selected contains at least one Sony component? Exactly one Sony component?

7. Shortly after being put into service, some buses manufactured by a certain company have developed cracks on the underside of the main frame. Suppose that a particular city has 20 of these buses, and cracks have actually appeared in eight of them.

a. How many ways are there to select a sample of five buses from the 20 for a thorough inspection?

b. In how many ways can a sample of five buses contain exactly five with visible cracks?

c. If a sample of five buses is chosen at random, what is the probability that at least four of the five have visible cracks?

8. A consumer group is concerned about the possibility that there is systematic underweighing in the meat department of a local supermarket. A representative is sent to purchase five 1-lb packages of ground meat. If there are 20 such packages on display and eight of them are actually underweight, what is the probability that at least three of the five packages purchased are underweight?

9. A mathematics professor wishes to schedule an ap-

pointment with each of her eight teaching assistants, four male and four female, to discuss her calculus course. Suppose that all possible orderings of appointments are equally likely to be selected.

a. What is the probability that at least one female assistant is among the first three that she meets with?

b. What is the probability that after the first five appointments she has met with all female assistants?

c. Suppose that the professor has the same eight assistants the following semester and again schedules appointments without regard to the ordering during the first semester. What is the probability that the orderings of appointments are different?

10. Three married couples have purchased theatre tickets and are seated in a row consisting of just six seats. If they take their seats in a completely random fashion (random order), what is the probability that Jim and Paula (husband and wife) sit in the two seats on the far left? What is the probability that Jim and Paula end up sitting next to one another? What is the probability that at least one of the wives ends up sitting next to her husband?

11. In five-card poker, a straight consists of five cards with adjacent denominations (for example, 9 of clubs, 10 of hearts, jack of hearts, queen of spades, and king of clubs). Assuming that aces can be high or low, if you are dealt a five-card hand, what is the probability that it will be a straight with high card 10? What is the probability that it will be a straight? What is the probability that it will be a straight flush (all cards in the same suit)?

12. Show that $\binom{n}{k} = \binom{n}{n-k}$. Give an interpretation involving subsets.

2.4 Conditional Probability

The probabilities assigned to various events depend upon what is known about the experimental situation when the assignment is made. Subsequent to the initial assignment, partial information about or relevant to the outcome of the experiment may become available, and this information may cause us to revise some of our probability assignments. For a particular event A, we have used $P(A)$ to represent the probability asigned to A; we now think of $P(A)$ as the original or unconditional probability of the event A.

In this section we examine how the information "an event B has occurred" affects the probability assigned to A. For example, A might refer to an individual having a particular disease in the presence of certain symptoms. If a blood test is performed on the individual and the result is negative (B = negative blood test), then the probability of having the disease will change (it should decrease, but not usually to zero, since blood tests are not infallible). We will use the notation $P(A|B)$ to represent the **conditional probability of A given that the event B has occurred.**

Example 2.18   Complex components are assembled in a plant which uses two different assembly lines, A and A'. Line A uses older equipment than A', so it is somewhat slower and less reliable. Suppose that on a given day line A has assembled 8 components, of which 2 have been identified as defective (B) and 6 as nondefective (B'), while A' has produced 1 defective and 9 nondefective components. This information is summarized in the accompanying table.

		Condition	
		B	B'
Line	A	2	6
	A'	1	9

Unaware of this information, the sales manager randomly selects 1 of these 18 components for a demonstration. Prior to the demonstration

$$P(\text{line } A \text{ component selected}) = P(A) = \frac{N(A)}{N} = \frac{8}{18} = .44$$

However, if the chosen component turns out to be defective, then the event B has occurred, so the component must have been 1 of the 3 in the B column of the table. Since these 3 components are equally likely among themselves after B has occurred

$$P(A \mid B) = \frac{2}{3} = \frac{\frac{2}{18}}{\frac{3}{18}} = \frac{P(A \cap B)}{P(B)} \qquad (2.2)$$

In (2.2) the conditional probability is expressed as a ratio of unconditional probabilities: The numerator is the probability of the intersection of the two events, while the denominator is the probability of the conditioning event B. A Venn diagram illuminates this relationship (Figure 2.6).

Given that B has occurred, the relevant sample space is no longer $\mathscr{S}$ but consists of outcomes in B; A has occurred if and only if one of the outcomes in the intersection occurred, so that the conditional probability of A given B is proportional to $P(A \cap B)$. The proportionality constant $1/P(B)$ is used to ensure that the probability $P(B \mid B)$ of the new sample space B equals one.

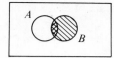

Figure 2.6

The Definition of Conditional Probability

Example 2.18 demonstrates that when outcomes are equally likely, computation of conditional probabilities can be based on intuition. When experiments are more complicated, though, intuition may fail us, so we want to have a general definition of conditional probability which will yield intuitive answers in simple problems. The Venn diagram and (2.2) suggest the appropriate definition.

> **Definition:** For any two events A and B with $P(B) > 0$, the **conditional probability of A given that B has occurred** is defined by
>
> $$P(A \mid B) = \frac{P(A \cap B)}{P(B)} \tag{2.3}$$

Example 2.19   In a BASIC programming course, students are allowed up to six runs (program submissions) to get each program to run correctly. Suppose that for a randomly chosen student on a certain assignment, $P(1 \text{ run}) = \frac{1}{21}$, $P(2 \text{ runs}) = \frac{2}{21}$, . . ., $P(6 \text{ runs}) = \frac{6}{21}$. Then with $A = \{\text{at most 4 runs}\}$ and $B = \{\text{at least 3 runs}\}$, $A \cap B = \{3 \text{ or } 4 \text{ runs}\}$. Thus $P(A) = \frac{10}{21} = .48$, $P(B) = \frac{18}{21} = .83$, and $P(A \cap B) = \frac{7}{21}$, so $P(A \mid B) = P(\text{at most 4 runs} \mid \text{at least 3 runs}) = P(A \cap B)/P(B) = \frac{7}{18} = .39$, while $P(B \mid A) = P(\text{at least 3 runs} \mid \text{at most 4 runs}) = P(A \cap B)/P(A) = \frac{7}{10} = .70$. Notice that $P(A \mid B) \neq P(A)$ and $P(B \mid A) \neq P(B)$.

Example 2.20   A news magazine publishes three columns entitled "Art" (A), "Books" (B), and "Cinema" (C). Reading habits of a randomly selected reader with respect to these columns are

Read regularly	A	B	C	$A \cap B$	$A \cap C$	$B \cap C$	$A \cap B \cap C$
Probability	.14	.23	.37	.08	.09	.13	.05

We thus have

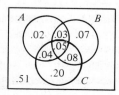

Figure 2.7

$$P(A \mid B) = \frac{P(A \cap B)}{P(B)} = \frac{.08}{.23} = .348$$

$$P(A \mid B \cup C) = \frac{P(A \cap (B \cup C))}{P(B \cup C)} = \frac{.04 + .05 + .03}{.47} = \frac{.12}{.47} = .255$$

$$P(A \mid \text{reads at least one}) = P(A \mid A \cup B \cup C) = \frac{P(A \cap (A \cup B \cup C))}{P(A \cup B \cup C)}$$

$$= \frac{P(A)}{P(A \cup B \cup C)} = \frac{.14}{.49} = .286$$

and

$$P(A \cup B \mid C) = \frac{P((A \cup B) \cap C)}{P(C)} = \frac{.04 + .05 + .08}{.37} = .459$$

The Multiplication Rule for $P(A \cap B)$

The definition of conditional probability yields the following result, obtained by multiplying both sides of (2.3) by $P(B)$.

> **The multiplication rule:** $P(A \cap B) = P(A \mid B) \cdot P(B)$

This rule is important because it is often the case that $P(A \cap B)$ is desired while both $P(B)$ and $P(A \mid B)$ can be specified from the problem description.

Example 2.21

Four individuals have responded to a request by a blood bank for blood donations. None of them has donated before, so their blood types are unknown. Suppose that only type A positive is desired, and that only one of the four actually has this type. If the potential donors are selected in random order for typing, what is the probability that at least three individuals must be typed to obtain the desired type?

Making the identification $B = \{\text{1st type not A+}\}$ and $A = \{\text{2nd type not A+}\}$, $P(B) = \frac{3}{4}$. Given that the first type is not A+, two of the three individuals left are not A+, so $P(A \mid B) = \frac{2}{3}$. The multiplication rule now gives

$$P(\text{at least three individuals are typed}) = P(A \cap B)$$

$$= P(A \mid B) \cdot P(B) = \frac{3}{4} \cdot \frac{2}{3} = \frac{6}{12} = .5$$

The multiplication rule is most useful when the experiment consists of several stages in succession. The conditioning event B then describes the outcome of the first stage and A the outcome of the second, so that $P(A \mid B)$—conditioning on what occurs first—will often be known. The rule is easily extended to experiments involving more than two stages. For example

$$P(A_1 \cap A_2 \cap A_3) = P(A_3 \mid A_1 \cap A_2) \cdot P(A_1 \cap A_2)$$

$$= P(A_3 \mid A_1 \cap A_2) \cdot P(A_2 \mid A_1) \cdot P(A_1) \qquad (2.4)$$

where A_1 occurs first, followed by A_2, and finally A_3.

Example 2.21
(continued)

$P(\text{third type is A}+) = P(\text{third is} \mid \text{first isn't} \cap \text{second isn't})$

$\cdot P(\text{second isn't} \mid \text{first isn't}) \cdot P(\text{first isn't})$

$$= \tfrac{1}{2} \cdot \tfrac{2}{3} \cdot \tfrac{3}{4} = \tfrac{1}{4} = .25$$

When the experiment of interest consists of a sequence of several stages, it is convenient to represent these with a tree diagram. Once we have an appropriate tree diagram, probabilities and conditional probabilities can be entered on the various branches; this will make repeated use of the multiplication rule quite straight-forward.

Example 2.22

Three different airlines (1, 2, and 3) operate night flights from Los Angeles to New York City. Experience has shown that 40% of airline 1's flights are late in takeoff, 50% of airline 2's flights are late in takeoff, and that 70% of airline 3's flights are late taking off. On a particular night, I select in a completely random fashion one of the three airlines and fly on its night flight.

   a. What is the probability that I select airline 1 and I'm late taking off?
   b. What is the probability that I'm late taking off?
   c. Given that I was late in takeoff, what is the probability that I selected airline 1?

The first stage of the problem involves selecting one of the three airlines on which to fly. If we let A_1, A_2, and A_3 be events representing the selection of airlines 1, 2, and 3, then $P(A_1) = P(A_2) = P(A_3) = \tfrac{1}{3}$. Once an airline is selected, the second stage involves observing whether or not the airplane chosen is late taking off. With $B = \{\text{late in takeoff}\}$ and $B' = \{\text{not late}\}$, the information about lateness implies $P(B \mid A_1) = .4$, $P(B \mid A_2) = .5$, and $P(B \mid A_3) = .7$.

The tree diagram representing this experiment appears in Figure 2.8. The three initial branches correspond to different choices of airline, and there are two second-generation branches—one for "late takeoff" and one for "not late"—emanating from each of the three. On the ith initial branch we record $P(A_i)$, the probability of initially selecting that branch. On each second-generation branch the appropriate conditional probability $P(B \mid A_i)$ or $P(B' \mid A_i)$ appears—that is, the chance of going out on the particular second-generation branch given that the ith initial branch was chosen.

To the right of each second-generation branch corresponding to a late takeoff is the product of probabilities on the branches leading out to that point. This is just the multiplication rule in action. For example, the answer to question (a) is

$$P(A_1 \cap B) = P(A_1) \cdot P(B \mid A_1) = \tfrac{1}{3} \cdot \tfrac{4}{10} = \tfrac{4}{30} = .133$$

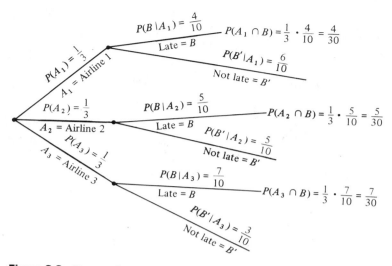

Figure 2.8   The tree diagram for the night flight problem

To answer (b), note that a late takeoff can result from taking any one of the three airlines, so

$$P(B) = P(A_1 \cap B) + P(A_2 \cap B) + P(A_3 \cap B) = \tfrac{4}{30} + \tfrac{5}{30} + \tfrac{7}{30} = \tfrac{8}{15} = .533$$

Each $P(A_i \cap B)$ appears to the right of its associated second-generation branch. To obtain a probability such as $P(B)$ from a tree diagram like Figure 2.8, we need only identify second-generation branches for which B occurs and add the probabilities appearing on their right. The answer to (c) now follows immediately, since

$$P(A_1 \mid B) = \frac{P(A_1 \cap B)}{P(B)} = \frac{\tfrac{4}{30}}{\tfrac{8}{15}} = \frac{1}{4} = .25$$

Notice that the initial or *prior* probability of selecting airline 1 was $\tfrac{1}{3}$, whereas once we are informed that the selected airplane was late in taking off, the *posterior* (after new information) probability of airline 1 is $\tfrac{1}{4}$, which is less than the prior probability. This is because airline 1 has a somewhat better prompt takeoff record than does either airline 2 or 3.

Bayes' Theorem

The computation of a posterior probability $P(A_k \mid B)$ from given prior probabilities $P(A_i)$ and conditional probabilities $P(B \mid A_i)$ occupies a central position in elementary probability. The general rule for such computations, which is really just a simple application of the multiplication rule, goes back to Reverend Thomas Bayes, who lived in the eighteenth century. To state it we first need another result. Recall events $A_1, \ldots, A_n$ are *mutually exclusive* if no two have any common outcomes. The events are *exhaustive* if one A_i must occur, so that $A_1 \cup \cdots \cup A_n = \mathcal{S}$.

The law of total probability: Let $A_1, \ldots, A_n$ be mutually exclusive and exhaustive events. Then for any other event B,

$$P(B) = \sum_{i=1}^{n} P(B \mid A_i)\, P(A_i) \qquad (2.5)$$

Proof: Because the A_i's are mutually exclusive and exhaustive, if B occurs it must be in conjunction with exactly one of the A_i's. That is, $B = (A_1 \text{ and } B)$ or $\cdots$ or $(A_n \text{ and } B) = (A_1 \cap B) \cup \cdots \cup (A_n \cap B)$, where the $(A_i \cap B)$'s are exclusive. This "partitioning of B" is illustrated in Figure 2.9. Thus

$$P(B) = \sum_{i=1}^{n} P(A_i \cap B) = \sum_{i=1}^{n} P(B \mid A_i)\, P(A_i) \quad \text{as desired.}$$

An example of the use of (2.5) appeared in answering question (b) of Example 2.22, where $A_1 = \{\text{airline 1}\}$, $A_2 = \{\text{airline 2}\}$, $A_3 = \{\text{airline 3}\}$, and $B = \{\text{late}\}$.

Bayes' theorem: Let $A_1, A_2, \ldots, A_n$ be a collection of n mutually exclusive and exhaustive events with $P(A_i) > 0$ for $i = 1, \ldots, n$. Then for any other event B for which $P(B) > 0$

$$P(A_k \mid B) = \frac{P(A_k \cap B)}{P(B)} = \frac{P(B \mid A_k)\, P(A_k)}{\sum_{i=1}^{n} P(B \mid A_i) \cdot P(A_i)} \qquad k = 1, \ldots, n \qquad (2.6)$$

The transition from the second to the third expression in (2.6) rests on using the multiplication rule in the numerator and the law of total probability in the denominator.

While the right-hand side expression in (2.6) enables one to compute posterior probabilities by substituting prior and conditional probabilities in the appropriate places, the proliferation of events and subscripts can often intimidate a newcomer to probability. Example 2.22 shows that the posterior probabilities can be computed from a tree diagram by using the middle expression in (2.6) without ever referring explicitly to Bayes' theorem.

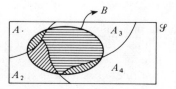

Figure 2.9   Partition of B by exclusive and exhaustive A_i's

Example 2.23   *Incidence of a rare disease.*   Only one in 1000 adults is afflicted with a rare disease for which a diagnostic test has been developed. The test is such that, when an individual actually has the disease, a positive result will occur 99% of the time, while an individual without the disease will show a positive test result only 2% of the time. If a randomly selected individual is tested and the result is positive, what is the probability that the individual has the disease?

To use Bayes' theorem, let A_1 = {individual has the disease}, A_2 = {individual does not have the disease}, and B = {positive test result}. Then $P(A_1) = .001$, $P(A_2) = .999$, $P(B \mid A_1) = .99$, and $P(B \mid A_2) = .02$. The tree diagram for this problem appears in Figure 2.10.

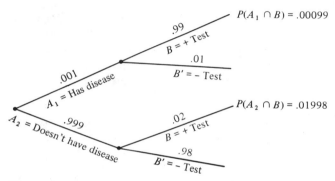

Figure 2.10   Tree diagram for the rare disease problem

Next to each branch corresponding to a positive test result, the multiplication rule yields the recorded probabilities. Therefore $P(B) = .00099 + .01998 = .02097$, from which we have

$$P(A_1 \mid B) = \frac{P(A_1 \cap B)}{P(B)} = \frac{.00099}{.02097} = .047$$

This result seems counter-intuitive; the diagnostic test appears so accurate we expect someone with a positive test result to be highly likely to have the disease, whereas the computed conditional probability is only .047. The reason for this seemingly paradoxical result is that, because the disease is rare and the test only moderately reliable, most positive test results arise from errors rather than from diseased individuals. The probability of having the disease has increased by a multiplicative factor of 47 (from prior .001 to posterior .047); but to get a further increase in the posterior probability, a diagnostic test with much smaller error rates is needed. If the disease were not so rare (for example, 25% incidence in the population), then the error rates for the present test would provide good diagnoses.

Exercises / Section 2.4

1. A study of the relationship among adult drivers between income level (L = low, M = medium, and H = high) and preference for one of the "big three" automobile manufacturers (denoted by A, B, and C here) yielded the accompanying table of joint probabilities.

		Income		
	L	*M*	*H*	
A	.10	.13	.02	.25
Preference B	.20	.12	.08	.40
C	.10	.15	.10	.35
	.40	.40	.20	

This table shows, for example, that P(low income and prefer A) = $P(L \cap A)$ = .10, P(low income) = $P(L)$ = .40, and P(prefer A) = $P(A)$ = .25. Use this table to compute the following conditional probabilities.

a. $P(B \mid H)$ **b.** $P(M \mid C)$ **c.** $P(A' \mid M)$
d. $P(M \mid A')$ **e.** $P(M \mid B \cup C)$
f. $P(L \cup M \mid C)$
g. What is the probability that a randomly selected adult driver will either prefer A or have a high income?

2. A mathematics professor is teaching both a morning and an afternoon section of introductory calculus. Let A = {the profesor gives a bad morning lecture} and B = {the professor gives a bad afternoon lecture}. If $P(A)$ = .3, $P(B)$ = .2, and $P(A \cap B)$ = .1, calculate the following probabilities (a Venn diagram might help):

a. $P(B \mid A)$ **b.** $P(B' \mid A)$ **c.** $P(B \mid A')$
d. $P(B' \mid A')$
e. If at the conclusion of the afternoon class, the professor is heard to mutter "what a rotten lecture," what is the probability that the morning lecture was also bad?

3. When a roulette wheel is spun once, there are 38 possible outcomes—18 red, 18 black, and two green (if the outcome is green, the house wins everything). If a wheel is spun twice, all (38)(38) outcomes are equally likely. If you are told that in two spins, at least one resulted in a green outcome, what is the probability that both outcomes were green?

4. Refer to Exercise 1 of Section 2.2, where A_1 = {older car is American}, A_2 = {newer car is American}, $P(A_1)$ = .7, $P(A_2)$ = .5, and $P(A_1 \cap A_2)$ = .4. Compute the following probabilities:

a. $P(A_2 \mid A_1)$ **b.** $P(A_1 \mid A_2)$ **c.** $P(A_2' \mid A_1')$
d. $P(A_1 \cap A_2 \mid$ at least one car is American)

5. In Exercise 2 of Section 2.2, A_i = {awarded project i} for i = 1, 2, 3. Use the probabilities given there to compute the following probabilities:

a. $P(A_2 \mid A_1)$ **b.** $P(A_2 \cap A_3 \mid A_1)$
c. $P(A_2 \cup A_3 \mid A_1)$
d. $P(A_1 \cap A_2 \cap A_3 \mid A_1 \cup A_2 \cup A_3)$. Express in words the event whose probability you have calculated.

6. In Exercise 10 of Section 2.3, six people (three married couples) chose seats at random in a row consisting of six seats.
a. Use the multiplication rule to compute the probability that Jim and Paula sit together on the far left (event A) and that John and Mary Lou (husband and wife) sit together in the middle (event B).
b. Given that John and Mary Lou sit together in the middle, what is the probability that the two other husbands sit next to their wives?
c. Given that John and Mary Lou sit together, what is the probability that all husbands sit next to their wives?

7. For any events A and B with $P(B) > 0$, show that $P(A \mid B) + P(A' \mid B) = 1$.

8. At a certain gas station 40% of the customers request regular gas (A_1), 35% request unleaded gas (A_2), and 25% request premium gas (A_3). Of those customers requesting regular gas, only 30% fill their tanks (event B). Of those customers requesting unleaded gas, 60% fill their tanks, while of those requesting premium, 50% fill their tanks.
a. What is the probability that the next customer will request unleaded gas and fill his tank ($A_2 \cap B$)?
b. What is the probability that the next customer fills his tank?
c. If the next customer fills his tank, what is the probability that he requested regular gas? Unleaded gas? Premium gas?

9. A company employing 10,000 workers offers deluxe medical coverage, standard medical coverage, and economy medical coverage. Of the employees 30% have deluxe coverage, 60% have standard coverage, and 10% have economy coverage. From past experience, the probability that an employee with deluxe coverage will submit no claims during the next year is .1, the probability of an employee with standard coverage submitting no claim is .4, and the probability of an employee with economy coverage submitting no claim is .7. If an employee is selected at random, draw an appropriate tree diagram and answer the following questions.
   a. What is the probability that the selected employee has standard coverage and will submit no claims?
   b. What is the probability that the selected employee will submit no claims?
   c. If the selected employee submits no claims during the next year, what is the probability that the employee had standard coverage? Economy coverage?

10. For customers purchasing a full set of tires at a particular tire store, consider the events

   A = {tires purchased were made in the United States}
   B = {purchaser has tires balanced immediately}
   C = {purchaser requests front end alignment}

   along with A', B', and C'. Assume the following unconditional and conditional probabilities:

   $P(A) = .75$, $P(B \mid A) = .9$, $P(B \mid A') = .8$,
   $P(C \mid A \cap B) = .8$, $P(C \mid A \cap B') = .6$,
   $P(C \mid A' \cap B) = .7$, $P(C \mid A' \cap B') = .3$

   a. Construct a tree diagram consisting of first-, second-, and third-generation branches, and place an event label and appropriate probability next to each branch.
   b. Compute $P(A \cap B \cap C)$.
   c. Compute $P(B \cap C)$.
   d. Compute $P(C)$.
   e. Compute $P(A \mid B \cap C)$, the probability of a purchase of U.S. tires given that both balancing and an alignment were requested.

11. In Example 2.23 suppose that the incidence rate for the disease is 1 in 25 rather than 1 in 1000. What then is the probability of a positive test re-sult? Given that the test result is positive, what is the probability that the individual has the disease? Given a negative test result, what is the probability that the individual does not have the disease?

12. At a large university, in the never-ending quest for a satisfactory textbook, the Statistics Department has tried a different text during each of the last three quarters. During the fall quarter 500 students used the text by Professor Mean, during the winter quarter 300 students used the text by Professor Median, and during the spring quarter 200 students used the text by Professor Mode. A survey at the end of each quarter showed that 200 students were satisfied with Mean's book, 150 were satisfied with Median's book, and 160 were satisfied with Mode's book. If a student who took statistics during one of these quarters is selected at random and admits to having been satisfied with the text, is the student most likely to have used the book by Mean, Median, or Mode? Who is the least likely author? *Hint*: Draw a tree diagram or use Bayes' theorem.

13. A friend who works in a big city owns two cars, one small and one large. Three-quarters of the time he drives the small car to work, and one-quarter of the time he takes the large car. If he takes the small car, he usually has little trouble parking, and so is at work on time with probability .9. If he takes the large car, he is on time to work with probability .6. Given that he was on time on a particular morning, what is the probability that he drove the small car?

14. In Exercise 2.4.8, consider the following additional information on credit card usage:
   70% of all regular fill-up customers use a credit card,
   50% of all regular non-fill-up customers use a credit card,
   60% of all unleaded fill-up customers use a credit card,
   50% of all unleaded non-fill-up customers use a credit card,
   50% of all premium fill-up customers use a credit card,
   and 40% of all premium non-fill-up customers use a credit card.
Compute the probability of each of the following events for the next customer to arrive (a tree diagram might help).

a. {unleaded and fill-up and credit card}
b. {premium and non-fill-up and credit card}
c. {premium and fill-up and credit card}
d. {premium and credit card}

e. {fill-up and credit card}
f. {credit card}
g. If the next customer uses a credit card, what is the probability that he requested premium?

2.5   Independence

The definition of conditional probability enables us to revise the probability $P(A)$ originally assigned to A when we are subsequently informed that another event, B, has occurred; the new probability of A is $P(A \mid B)$. In our examples, it was frequently the case that $P(A \mid B)$ was unequal to the unconditional probability $P(A)$, indicating that the information "B has occurred" resulted in a change in the chance of A occurring. There are other situations, though, in which the chance that A will occur or has occurred is not affected by knowledge that B has occurred, so that $P(A \mid B) = P(A)$. It is then natural to think of A and B as independent events, meaning that the occurrence or nonoccurrence of one event has no bearing on the chance that the other will occur.

Definition: Two events A and B are **independent** if $P(A \mid B) = P(A)$ and are **dependent** otherwise.

The definition of independence might seem "unsymmetric," since we do not demand that $P(B \mid A) = P(B)$ also. However, using the definition of conditional probability and the multiplication rule,

$$P(B \mid A) = \frac{P(A \cap B)}{P(A)} = \frac{P(A \mid B)\, P(B)}{P(A)} \tag{2.7}$$

The right-hand side of (2.7) is $P(B)$ if and only if $P(A \mid B) = P(A)$ (independence), so the equality in the definition implies the other equality (and vice versa).

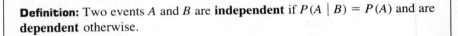

Example 2.24   Consider tossing a fair six-sided die once and define events $A = \{2, 4, 6\}$, $B = \{1, 2, 3\}$, and $C = \{1, 2, 3, 4\}$. We then have $P(A) = \frac{1}{2}$, $P(A \mid B) = \frac{1}{3}$, and $P(A \mid C) = \frac{1}{2}$. That is, events A and B are dependent while events A and C are independent. Intuitively, if such a die is tossed and we are informed that the outcome was either 1, 2, 3, or 4 (C has occurred), then the probability that A occurred is $\frac{1}{2}$ as it originally was, since two of the four relevant outcomes are even and the outcomes are still equally likely.

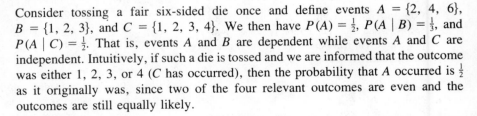

Example 2.25   Let A and B be any two mutually exclusive events with $P(A) > 0$. For example, for a randomly chosen automobile, let $A = \{$the car has four cylinders$\}$ and $B = \{$the car has six cylinders$\}$. Since the events are mutually exclusive, if B occurs, then A

cannot possibly have occurred, so $P(A \mid B) = 0 \neq P(A)$. The message here is that *if two events are mutually exclusive, they cannot be independent.* When A and B are mutually exclusive, the information that A occurred says something about B (it cannot have occurred), so independence is precluded.

P(A ∩ B) When Events Are Independent

There are a number of experimental situations in which, rather than having to verify the independence of A and B by using the definition, we wish to build the independence of A and B into our probability assignments. For example, if a card is selected from a deck of 52, then replaced and the deck reshuffled before a second card is drawn, then it is reasonable to assume that any event A defined with respect to the first card selected (red card, picture card, five, and so on) is independent of an event defined with respect to the second card. The next proposition tells us how to compute $P(A \cap B)$ when the events are independent.

Proposition: A and B are independent if and only if

$$P(A \cap B) = P(A) \cdot P(B) \tag{2.8}$$

To paraphrase the proposition, A and B are independent events iff* the probability that they both occur ($A \cap B$) is the product of the two individual probabilities. The verification is as follows:

$$P(A \cap B) = P(A \mid B) \cdot P(B) = P(A) \cdot P(B) \tag{2.9}$$

where the second equality in (2.9) is valid iff A and B are independent. Because of the equivalence of independence with (2.8), the latter can be used as a definition of independence.

Example 2.26 Two different record companies, X and Y, both produce classical music recordings. Label X is a "budget" label, and 5% of X's new records exhibit a significant degree of warpage. Label Y is manufactured under more stringent quality-control conditions (and sold at a higher price) than X, so only 2% of its new pressings are warped. If you purchase one label X recording and one label Y recording at your local record store, what is the probability that both records are warped?

To answer this question, define events A and B by $A = \{$X record is warped$\}$ and $B = \{$Y record is warped$\}$. Then $P(A) = .05$, $P(B) = .02$, and we wish $P(A \cap B)$. Because labels X and Y have no relationship to one another, it is reasonable to assume that A and B are independent events, whence $P(A \cap B) = P(A) \cdot P(B) = .001$.

---

* iff is an abbreviation for "if and only if."

Example 2.27 A minicomputer salesperson makes either one or two sales contacts on any given day, with probabilities .6 and .4, respectively. If only one contact is made, the probability is .2 that a sale will result and .8 that no sale will result. If two contacts are made, the two customers will make their purchase decisions independently of one another, each purchasing with probability .2 and not purchasing with probability .8. What is the probability that the salesperson has made two sales at the end of the day? To begin,

$$P(\text{two sales}) = P(\text{two contacts and both purchase})$$

$$= P(\text{both purchase} \mid \text{two contacts}) \cdot P(\text{two contacts})$$

Since, given that two contacts are made, the two purchase decisions are independent, $P(\text{both purchase} \mid \text{two contacts}) = (.2)(.2) = .04$, while $P(\text{two contacts}) = .4$. This gives $P(\text{two sales}) = (.04)(.4) = .016$.

Independence of More Than Two Events

The notion of independence of two events can be extended to collections of more than two events. While it is possible to extend the definition for two independent events by working in terms of conditional and unconditional probabilities, it is more direct and less cumbersome to proceed along the lines of the last proposition.

> **Definition:** Events $A_1, \ldots, A_n$ are **mutually independent** if for every k, $k = 2, 3, \ldots, n$, and every subset of indices $i_1, i_2, \ldots, i_k$, $P(A_{i_1} \cap A_{i_2} \cap \cdots \cap A_{i_k}) = P(A_{i_1}) \cdot P(A_{i_2}) \cdots P(A_{i_k})$

Paraphrasing the definition, the events are mutually independent if the probability of the intersection of any subset of the n events is equal to the product of the individual probabilities. When $P(A_i \cap A_j) = P(A_i) \cdot P(A_j)$ for every pair i, j with $i \neq j$, the events are said to be **pairwise independent**. Intuitively, pairwise independence means that occurrence of any single one of the A_i's does not affect the chance that any other single one might occur. The next example shows that mutual independence is a more general notion than pairwise independence.

Example 2.28 The winner of a contest will be rewarded with one or more of three different prizes in the following manner. Four slips of paper marked 1, 2, 3, and (1, 2, 3), respectively, will be placed in a box. After the box is shaken, one slip will be withdrawn, and the contest winner will receive the prize (or prizes) whose number (numbers) is (are) on the slip. Let A_i ($i = 1, 2, 3$) denote the event that the ith prize is awarded. Notice that the ith prize will be awarded if either the slip with the number i on it is selected or if the slip with (1, 2, 3) on it is selected. Assigning probability $\frac{1}{4}$ to each of the four slips then gives $P(A_1) = P(A_2) = P(A_3) = \frac{1}{2}$; $P(A_1 \cap A_2) = P(\text{win both prizes 1 and 2}) = P[(1, 2, 3) \text{ is selected}] = \frac{1}{4}$; and similarly $P(A_1 \cap A_3) = P(A_2 \cap A_3) = \frac{1}{4}$. Since $P(A_i \cap A_j) = P(A_i) \cdot P(A_j)$ for all i, j with $i \neq j$, the three events are pairwise independent. However, $P(A_1 \cap A_2 \cap A_3) = P[(1, 2, 3) \text{ selected}] = \frac{1}{4} \neq P(A_1) \cdot P(A_2) \cdot P(A_3) = \frac{1}{8}$, so the events are not mutually independent. That is, while the unconditional probability of winning prize 1 is $\frac{1}{2}$, if we are told that A_2 and A_3 both occurred, it must have been the case that the slip marked (1, 2, 3) was selected, so that the conditional probability of A_1 is 1. Given only that A_2 occurred, both the slip marked 2 and the one marked (1, 2, 3) are possible, so the conditional probability equals the unconditional probability. To summarize, mutual independence implies pairwise independence, but the reverse implication is not correct.

As was the case with two events, we frequently specify at the outset of a problem the independence of certain events. The definition can then be used to calculate the probability of an intersection.

Example 2.29   A system consists of four components as illustrated in Figure 2.11. The entire system will work if either the 1–2 subsystem works or if the 3–4 subsystem works (since the two subsystems are connected in parallel). Since the two components in each subsystem are connected in series, a subsystem will work only if both its components work. If components work or fail independently of one another, and if each works with probability .9, what is the probability that the entire system will work (the system reliability coefficient)? Letting A_i ($i = 1, 2, 3, 4$) be the event that the ith component works, the A_i's are mutually independent. The event that the 1–2 subsystem works is $A_1 \cap A_2$, and similarly $A_3 \cap A_4$ denotes the event that the 3–4 subsystem works. The event that the system works is $(A_1 \cap A_2) \cup (A_3 \cap A_4)$, so

$$P[(A_1 \cap A_2) \cup (A_3 \cap A_4)] = P(A_1 \cap A_2) + P(A_3 \cap A_4)$$
$$- P[(A_1 \cap A_2) \cap (A_3 \cap A_4)]$$
$$= P(A_1) \cdot P(A_2) + P(A_3) \cdot P(A_4)$$
$$- P(A_1) \cdot P(A_2) \cdot P(A_3) \cdot P(A_4)$$
$$= (.9)(.9) + (.9)(.9) - (.9)(.9)(.9)(.9) = .9639$$

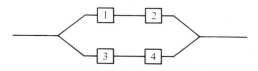

Figure 2.11

Letting $x = P(A_i)$ for $i = 1, 2, 3, 4$, what value of x would yield a system reliability of .99? Proceeding analogously, $P(\text{system works}) = x^2 + x^2 - x^4 = .99$ or $y^2 - 2y + .99 = 0$ where $y = x^2$. Solving this quadratic gives $y = .9$, so

$x = \sqrt{.9} \doteq .95$ (recall that $\doteq$ means approximately equal). To achieve a system reliability of .99 without introducing more components or changing the system configuration, the reliability of each component would have to be increased from .9 to .95.

Our last example shows how elementary probability techniques can be used to derive a famous result from genetics called the Hardy-Weinberg Law. This law states that under conditions of random mating (described below), the genetic makeup of a population (with respect to a particular gene) will reach equilibrium after one generation of mating.

Example 2.30   Consider a single gene which can be in one of two states: dominant, denoted by A, and recessive, denoted by a. Each individual in a large population carries two such genes, so the possible gene combinations are AA (homozygous dominant), Aa (heterozygous), and aa (homozygous recessive). Suppose that in the first generation of a large population, the proportions of individuals having these three genotypes are p_1, p_2, and p_3, respectively (for both males and females). In random mating, both a male and a female are selected at random and each contributes independently with equal probability one of its two genes to the offspring. The probability that the male contributes an A gene to the offspring is

$$p = P(A \text{ transmitted}) = P[AA \text{ or } (Aa \text{ and gene } A \text{ is transmitted})]$$
$$= P(AA) + P(A \text{ transmitted} \mid Aa) \cdot P(Aa)$$
$$= p_1 + \tfrac{1}{2}p_2$$

Thus $P(a \text{ transmitted}) = 1 - p = \tfrac{1}{2}p_2 + p_3$ (since $p_1 + p_2 + p_3 = 1$). Notice that these probabilities are the same for both the male and the female contribution. Thus, in the second generation, the genotype proportions are $p^2 = (p_1 + \tfrac{1}{2}p_2)^2$ for AA, $2p(1-p) = 2(p_1 + \tfrac{1}{2}p_2)(\tfrac{1}{2}p_2 + p_3)$ for Aa, and $(1-p)^2 = (\tfrac{1}{2}p_2 + p_3)^2$ for aa. Repeating the same argument for the third generation using these second-generation proportions, the probability that the second-generation male (or female) contributes an A is $p^2 + \tfrac{1}{2}[2p(1-p)] = p$ which is the same as for first-generation transmission. Therefore, the third-generation genotypes (and those of all succeeding generations) will be represented in the same proportions as in the second generation, so an equilibrium distribution of genotypes is attained after only one generation. This is exactly the Hardy-Weinberg Law, which was first proved by the mathematician G. H. Hardy in a 1908 letter published in *Science*.

Exercises / Section 2.5

1. If $P(A_1) = .7$, $P(A_2) = .5$, and $P(A_1 \cap A_2) = .4$ as in Exercise 1, Section 2.2, show that A_1 and A_2 are dependent first by using the definition of independence and then by verifying that the multiplication property does not hold.

2. In Exercise 2 of Section 2.2, is any A_i independent of any other A_j? Answer using the multiplication property for independent events.

3. An executive has both a morning and an afternoon meeting on a particular day. Let $A = \{$late to the morning meeting$\}$ and $B = \{$late to the afternoon meeting$\}$.
 a. If $P(A) = .4$, $P(B) = .5$, and $P(A \cap B) = .25$, are A and B independent events?
 b. If A and B are independent events with $P(A) = .4$ and $P(B) = .5$, what is the probability that the executive is on time to both meetings? To exactly one meeting?

4. The probability that a grader will make a marking error on any particular question of a multiple choice exam is .1. If there are n questions and questions are marked independently, what is the probability that no errors are made? What is the probability that at least one error is made? If the probability of a marking error is p rather than .1, answer these two questions.

5. Referring back to Example 2.27, calculate the probability that the salesperson has made exactly one sale at the end of the day. *Hint*: One sale can be made with either one contact or with two contacts; if there are two contacts, the first could result in a purchase and the second in no purchase or vice versa. A tree diagram might be helpful.

6. Consider the system of components connected as in the accompanying picture. Components 1 and 2 are connected in parallel, so that subsystem works iff either 1 or 2 works; since 3 and 4 are connected in series, that subsystem works iff both 3 and 4 work. If components work independently of one another and P(component works) = .9, calculate P(system works).

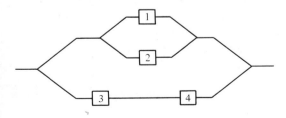

7. Suppose that with probability $\frac{2}{3}$, a randomly selected vehicle will pass inspection at a headlight inspection station. Assuming that successive vehi-

cles pass or fail independently of one another, calculate the following probabilities.
 a. P(all of the next three vehicles inspected pass).
 b. P(at least one of the next three inspected fails).
 c. P(exactly one of the next three inspected passes). *Hint*: With $A = \{$exactly one passes$\}$, $A = E_1 \cup E_2 \cup E_3$ where $E_i = \{$ith passes and the other two don't$\}$.
 d. P(at most one of the next three vehicles inspected passes).
 e. Given that at least one of the next three vehicles passes inspection, what is the probability that all three pass (a conditional probability)?

8. A quality-control inspector is inspecting newly produced items for faults. The inspector searches an item for faults in a series of independent fixations, each of a fixed duration. Given that a flaw is actually present, let p denote the probability that the flaw is detected during any one fixation (this model is discussed in "Human Performance in Sampling Inspection," *Human Factors*, 1979, pp. 99–105).
 a. Assuming that an item has a flaw, what is the probability that it is detected by the end of the second fixation (once a flaw has been detected, the sequence of fixations terminates). *Hint*: $\{$flaw detected in at most 2 fixations$\}$ = $\{$flaw detected on the first fixation$\}$ $\cup$ $\{$flaw undetected on the first and detected on the second$\}$.
 b. Give an expression for the probability that a flaw will be detected by the end of the nth fixation.
 c. If when a flaw has not been detected in three fixations, the item is passed, what is the probability that a flawed item will pass inspection?
 d. Suppose that 10% of all items contain a flaw [P(randomly chosen item is flawed) = .1]. With the assumption of (c), what is the probability that a randomly chosen item will pass inspection (it will automatically pass if it is not flawed, but could also pass if it is flawed)?
 e. Given that an item has passed inspection (no flaws in three fixations), what is the probability that it is actually flawed? Calculate for $p = .5$.

9. a. A lumber company has just taken delivery on a lot of 10,000 2 × 4 boards. Suppose that 20% of these boards (2000) are actually too green to be used in first-quality construction. Two boards are selected at random, one after the other. Let A = {the first board is green} and B = {the second board is green}. Compute $P(A)$, $P(B)$, and $P(A \cap B)$ (a tree diagram might help). Are A and B independent?

b. With A and B independent and $P(A) = P(B) = .2$, what is $P(A \cap B)$? How much difference is there between this answer and $P(A \cap B)$ in (a)? For purposes of calculating $P(A \cap B)$, can we assume that A and B of (a) are independent to obtain essentially the correct probability?

c. Suppose that the lot consists of ten boards, of which two are green. Does the assumption of independence now yield approximately the correct answer for $P(A \cap B)$? What is the critical difference between the situation here and that of (a)? When do you think that an independence assumption would be valid in obtaining an approximately correct answer to $P(A \cap B)$?

10. Answer Question 6, above, for the system in the accompanying picture. How would this probability change if another component is added in parallel to component 7?

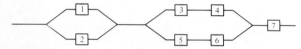

11. Professor Stan der Deviation can take one of two routes on his way home from work. On the first route, there are four railroad crossings. The probability that he will be stopped by a train at any particular one of the crossings is .1, and trains operate independently at the four crossings. The other route is longer but there are only two crossings, independent of one another with the same stoppage probability for each as on the first route. On a particular day, Professor Deviation has a meeting scheduled at home for a certain time. Whichever route he takes, he calculates that he will be late if he is stopped by trains at at least half the crossings encountered.

a. Which route should he take to minimize the probability of being late to the meeting?

b. If he tosses a fair coin to decide on a route and he is late, what is the probability that he took the four-crossing route?

12. In Exercise 6 of section 2.1, suppose that Jack Twain is at location A on the diagram and that his sister Jill is at location I. At exactly the same instant, Jack and Jill decide to leave their respective locations and journey toward one another (Jack traveling north and east only and Jill traveling south and west). If they walk at exactly the same pace and toss a fair coin independently at each corner location at which there is a choice of direction, what is the probability that the Twains meet at G? What is the probability that the Twains do not meet?

13. a. Suppose that identical tags are placed on both the left ear and the right ear of a fox. The fox is then let loose for a period of time. Consider the two events C_1 = {left ear tag is lost} and C_2 = {right ear tag is lost}. Let $\pi = P(C_1) = P(C_2)$, and assume that C_1 and C_2 are independent events. Derive an expression (involving π) for the probability that exactly one tag is lost given that at most one is lost ("Ear Tag Loss in Red Foxes," *J. Wildlife Mgmt.*, 1976, pp. 164–167). *Hint*: Draw a tree diagram in which the two initial branches refer to whether or not the left ear tag was lost.

b. If each of n foxes is given two ear tags and n_2 of these foxes have lost both tags at the end of the time period, π can be estimated by $\sqrt{n_2/n}$. The difficulty in practice is that some foxes captured without tags at the end of the period may not have been given tags initially, so n_2 is not observable. As an alternative approach, let m_1 be the number of foxes caught with one tag remaining and m_2 be the number captured with two tags. Show how m_1 and m_2 can be used along with the result of (a) to estimate π.

14. If A and B are independent, show that A' and B' are also. *Hint*: $A' \cap B' = (A \cup B)'$. Use this and the addition rule.

Supplementary Exercises / Chapter 2

1. A small manufacturing company will start operating a night shift. There are 20 machinists employed by the company.

a. If a night crew consists of three machinists,

how many different crews are possible?

b. If the machinists are ranked 1, 2, . . ., 20 in order of competence, how many of these crews would not have the best machinist?

c. How many of the crews would have at least one of the best 10 machinists?

d. If one of these crews is selected at random to work on a particular night, what is the probability that the best machinist will not work that night?

2. A claims officer at a Social Security office will have time to examine six claims during a particular day. There are 10 claims on her desk, of which four concern disability and six concern old-age benefits. If claims are selected at random from the 10, what is the probability that all disability claims will have been examined by the end of the day? What is the probability that by the end of the day, only one of the two types of claims will remain on her desk?

3. A personnel manager is to interview four candidates for a job. These are ranked 1, 2, 3, and 4 in order of preference, and will be interviewed in random order. However, at the conclusion of each interview, the manager will know only how the current candidate compares to those previously interviewed. For example, the interview order 3, 4, 1, 2 generates no information after the first interview, shows that the second candidate is worse than the first, and that the third is better than the first two. However, the order 3, 4, 2, 1 would generate the same information after each of the first three interviews. The manager wants to hire the best candidate, but must make an irrevocable hire–no hire decision after each interview. Consider the strategy: automatically reject the first s candidates, and then hire either the first subsequent candidate who is best among those already interviewed or the last candidate.

For example, with $s = 2$ the order 3, 4, 1, 2 would result in the best being hired, while 3, 1, 2, 4 would not. Of the four possible s values (0, 1, 2, and 3), which one maximizes P(best is hired)? *Hint:* Write out the 24 equally likely interviewing orderings; $s = 0$ means that the first candidate is automatically hired.

4. Let A be the event that a certain book on fluid mechanics is presently checked out of the university library, and let B be the event that another book on the same subject is checked out.

a. If $P(A) = .5$, $P(B) = .4$, and $P(A \cup B) = .65$, compute $P(A \cap B)$.

b. Using the probabilities of (a), compute P(exactly one of the two books is checked out). *Hint:* Venn diagram.

c. If $P(A \cup B) = .7$, $P(A \cap B) = .2$, and P (only the first book is checked out) $= .4$, compute $P(A)$ and $P(B)$. *Hint:* Venn diagram.

d. If $P(A \cup B) = .7$, $P(A \cap B) = .2$, and P (exactly one book is checked out) $= .5$, can $P(A)$ and $P(B)$ be determined?

5. Suppose that I fly to New York on airline X and return to Los Angeles on airline Y. Let $A = \{$X loses my baggage$\}$ and $B = \{$Y loses my baggage$\}$. If A and B are independent events with $P(A) > P(B)$, $P(A \cap B) = .0002$, and $P(A \cup B) = .03$, determine $P(A)$ and $P(B)$.

6. Individual A has a circle of five close friends (B, C, D, E, and F). A has heard a certain rumor from outside the circle and has invited the five friends to a party in order to circulate the rumor. To begin, A selects one of the five at random and tells the rumor to the chosen individual. That individual then selects at random one of the four remaining individuals and repeats the rumor. Continuing, a new individual is selected from those not already having heard the rumor by the individual who has just heard it, until everyone has been told. What is the probability that the rumor is repeated in the order B, C, D, E, and F?

7. a. In Question 6, what is the probability that F is the third person at the party to be told the rumor?

b. What is the probability that F is the last person to hear the rumor?

8. Referring to Question 6, if at each stage the person who currently "has" the rumor does not know who has already heard it, so selects the next recipient at random from all five possible individuals, what is the probability that F has still not heard the rumor after it has been told 10 times at the party?

9. An automobile insurance company classifies each driver as a good risk (A_1), a medium risk (A_2), or a poor risk (A_3). Of those currently insured, 30% are good risks, 50% are medium risks, and 20% are poor risks. In any given year the probability that a driver will have at least one accident is .1 for a good risk, .3 for a medium risk, and .5 for a poor risk. If a randomly selected driver insured by this company has an accident during the next year, what is the probability that the driver was actually a good risk? A medium risk?

10. In Exercise 9, suppose that each insuree has a three-year policy. If years are independent and a randomly selected driver reports no accidents during the three years, what is the probability that the driver was actually a good risk?

11. Each contestant on a quiz show is asked to specify one of six possible categories from which questions will be asked. Suppose that P (contestant requests category i) = $\frac{1}{6}$ and that successive contestants choose their categories independently of one another. If there are three contestants on each show and all three contestants on a particular show select different categories, what is the probability that exactly one has selected category 1?

12. One method used to distinguish between granitic (G) and basaltic (B) rocks is to examine a portion of the infrared spectrum of the sun's energy reflected from the rock surface. Let R_1, R_2, and R_3 denote measured spectrum intensities at three different wavelengths; typically, for granite $R_1 < R_2 < R_3$ while for basalt $R_3 < R_1 < R_2$. When measurements are made remotely (using aircraft), various orderings of the R_i's may arise whether the rock is basalt or granite.

Flights over regions of known composition have yielded the following information.

	$R_1 < R_2 < R_3$	$R_1 < R_3 < R_2$	$R_3 < R_1 < R_2$
Granite	60%	25%	15%
Basalt	10%	20%	70%

Suppose that for a randomly selected rock in a certain region, P(granite) = .25 and P(basalt) = .75.

a. Show that P (granite $\mid R_1 < R_2 < R_3$) > P (basalt $\mid R_1 < R_2 < R_3$). If measurements yielded $R_1 < R_2 < R_3$, would you classify the rock as granite or basalt?

b. If measurements yielded $R_1 < R_3 < R_2$, how would you classify the rock? Same question for $R_3 < R_1 < R_2$.

c. Using the classification rule indicated in (a) and

(b), when selecting a rock from this region, what is the probability of an erroneous classification? *Hint:* Either G could be classified as B or B as G, and $P(B)$ and $P(G)$ are known.

d. If P(granite) = p rather than .25, are there values of p (other than 1) for which one would always classify a rock as granite?

13. A subject is allowed a sequence of glimpses to detect a target. Let G_i = {the target is detected on the ith glimpse}, with $p_i = P(G_i)$. Suppose that the G_i's are independent events, and write an expression for the probability that the target has been detected by the end of the nth glimpse. *Note:* This model is discussed in "Predicting Aircraft Detectability," *Human Factors* (1979, pp. 277–91).

14. In a Little League baseball game, team A's pitcher throws a strike 50% of the time and a ball 50% of the time, successive pitches are independent of one another, and the pitcher never hits a batter. Knowing this, team B's manager has instructed the first batter not to swing at anything. Calculate the probablity that

a. the batter walks on the fourth pitch
b. the batter walks on the sixth pitch (so two of the first five must be strikes), using a counting argument or constructing a tree diagram
c. the batter walks
d. What is the probability that the first batter up scores while no one is out (assuming that each batter pursues a no-swing strategy)?

15. Four engineers, A, B, C, and D, have been scheduled for job interviews at 10 A.M. on Friday, January 13, at Random Sampling, Inc. The personnel manager has scheduled the four for interview rooms 1, 2, 3, and 4. However, the manager's secretary does not know this, so assigns them in a completely random fashion (what else!) to the four rooms.

a. What is the probablity that all four end up in the correct rooms?
b. What is the probablity that none of the four ends up in the correct room?

Bibliography   Derman, Cyrus, Glaser, Leon, and Olkin, Ingram, *Probability Models and Applications*, Macmillan, New York, 1980. A comprehensive introduction to probability, written at a slightly higher mathematical level than this text but containing many good examples.

Larsen, Richard, and Marx, Morris, *Introduction to Mathematical Statistics,* Prentice-Hall, Englewood Cliffs, N.J., 1980. Includes a nice concise exposition of probability written at a relatively modest mathematical level.

Mosteller, Frederick, Rourke, Robert, and Thomas, George, *Probability with Statistical Applications* (2nd ed.), Addison-Wesley, Reading, Mass., 1970. A very good pre-calculus introduction to probability, with many entertaining examples; especially good on counting rules and their application.

Winkler, Robert, *Introduction to Bayesian Inference and Decision,* Holt, Rinehart & Winston, New York, 1972. A very good introduction to subjective probability.

CHAPTER 3

Discrete Random Variables and Probability Distributions

Introduction

Whether an experiment yields qualitative or quantitative outcomes, methods of statistical analysis require that we focus on certain numerical aspects of the data (such as a sample proportion x/n, mean $\bar{x}$, or standard deviation s). The concept of a random variable allows us to pass from the experimental outcomes themselves to a numerical function of the outcomes. There are two fundamentally different types of random variables—discrete random variables and continuous random variables. In this chapter we examine the basic properties and discuss the most important examples of discrete variables, and we will study continuous variables in Chapter 4.

3.1 Random Variables

In any experiment there are numerous characteristics which can be observed or measured, but in most cases an experimenter will focus on some specific aspect or aspects of a sample. For example, in a study of commuting patterns in a metropolitan area, each individual in a sample might be asked about commuting distance and the number of people commuting in the same vehicle, but not about IQ, income, family size, and other such characteristics. Alternatively, a researcher may test a sample of components and record only the number which have failed within 1000 hours, rather than recording the individual failure times.

In general, each outcome of an experiment can be associated with a number by specifying a rule of association (for example, the number among the sample of 10 components which fail to last 1000 hours, or the total weight of baggage for a sample of 25 airline passengers). Such a rule of association is called a **random variable**—a variable because different numerical values are possible, and random because the observed value depends on which of the possible experimental outcomes results.

> **Definition:** For a given sample space $\mathcal{S}$ of some experiment, a **random variable** is any rule which associates a number with each outcome in $\mathcal{S}$.

$$X = 3x + 2$$

Figure 3.1 A random variable

We will often use the abbreviation r.v. in place of random variable. As is customary in probability and statistics, capital letters such as X or Y will be used to denote random variables. The notation $X(s) = x$ means that x is the number associated with the outcome s by the random variable X, so x is called the value of the variable associated with s.

Example 3.1

When a student attempts to log on to a computer time-sharing system, either all ports could be busy (F), in which case the student will fail to obtain access, or else there will be at least one port free (S), in which case the student will be successful in accessing the system. With $\mathcal{S} = \{S, F\}$, define a random variable X by

$$X(S) = 1, \quad X(F) = 0$$

The r.v. X indicates whether (1) or not (0) the student can log on.

In Example 3.1, the r.v. X was specified by explicitly listing each element of $\mathcal{S}$ and the associated number. If $\mathcal{S}$ contains more than a few outcomes, such a listing is tedious, but can frequently be avoided.

Example 3.2

Consider the experiment in which an automobile with California license plates is selected, and define a random variable Y by

$$Y = \begin{cases} 1 & \text{if the selected automobile uses unleaded gas} \\ 0 & \text{if the selected automobile does not use unleaded gas} \end{cases}$$

For example, if California license UZF002 uses unleaded gas, then $Y(\text{UZF002}) = 1$, while $Y(\text{AXJ375}) = 0$ tells us that the car with license AXJ375 does not use unleaded gas. A word description of this sort is more economical than a complete listing, so wherever possible we shall use such a description.

In Examples 3.1 and 3.2, the only possible values of the random variable were 0 and 1. Such a random variable arises frequently enough to be given a special name, after the individual who first studied it.

> **Definition:** Any random variable whose only possible values are 0 and 1 is called a **Bernoulli random variable.**

We will often want to define and study several different random variables from the same sample space.

Example 3.3

In the experiment in which a red and a green die are each rolled once (Example 2.3), there are 36 outcomes. Define random variables X, Y, and U by

X = sum of the two resulting numbers,

Y = difference between the number on the red die and the number on the green die, and

U = the maximum of the two resulting numbers.

If this experiment is performed and $s = (2, 3)$ results, then $X((2, 3)) = 2 + 3 = 5$, so we say that the observed value of X was $x = 5$. Similarly, the observed value of Y would be $y = 2 - 3 = -1$, and the observed value of U would be $u = \max(2, 3) = 3$.

An Infinite Set of Possible X Values

Each of the random variables of Examples 3.1–3.3 can assume only a finite number of possible values. This need not be the case.

Example 3.4

In Example 2.4 we considered the experiment in which batteries coming off an assembly line were examined until a good one (S) was obtained. The sample space was $\mathcal{S} = \{S, FS, FFS, \ldots\}$. Define a r.v. X by

X = the number of batteries examined before the experiment terminates

Then $X(S) = 1, X(FS) = 2, X(FFS) = 3, \ldots, X(FFFFFFS) = 7$, and so on. Any positive integer is a possible values of X, so the set of possible values is infinite.

Example 3.5

Suppose that in some random fashion, a location (latitude and longitude) in the continental United States is selected, and define a r.v. Y by

Y = the height above sea level at the selected location

For example, if the selected location were (39° 50′ N, 98° 35′ W), then we might have $Y((39° 50′ N, 98° 35′ W)) = 1748.26$ ft. The largest possible value of Y is 14,494 (Mt. Whitney) and the smallest possible value is -282 (Death Valley). The set of all possible values of X is the set of all numbers in the interval between -282 and 14,494—that is,

$$\{x: x \text{ is a number}, -282 \le x \le 14{,}494\}$$

and there are an infinite number of numbers in this interval.

Discrete Random Variables

Although both X of Example 3.4 and Y of Example 3.5 can assume any one of an infinite number of possible values, the two infinite sets are actually quite different from one another. The variable X is more closely related to the finite valued variables of Examples 3.1–3.3 than to Y above.

> **Definition:** A set is **discrete** either if it consists of a finite number of elements, or if its elements can be listed so that there is a first element, a second element, a third element, and so on, in the list.

A discrete set which is infinite is sometimes called countably infinite, since according to the definition we can count $(1, 2, 3, \ldots)$ its elements. Most of the infinite discrete sets that we shall encounter will consist either of the nonnegative integers $(\{0, 1, 2, \ldots\})$ or some subset of these integers. The set of possible values of the r.v. X in Example 3.4 is $D = \{1, 2, 3, \ldots\}$, which is clearly discrete.

That there are infinite sets which are not discrete is a profound and important result from pure mathematics. If D is any interval of real numbers, such as $\{x: -282 \le x \le 14{,}494\}$ of Example 3.5, then it can be shown that there is no way to list (or count) the elements of this set. More will be said about this at the outset of Chapter 4.

> **Definition:** A random variable is said to be **discrete** if its set of possible values is a discrete set.

Example 3.6

All random variables of Examples 3.1–3.4 are discrete. As another example, suppose that we select married couples at random and do a blood test on each person until we find a husband and wife who both have the same Rh factor. With $X =$ the number of blood tests to be performed, possible values of X are $D = \{2, 4, 6, 8, \ldots\}$. Since the possible values have been listed in sequence, D is a discrete set, so X is a discrete random variable.

To study basic properties of discrete random variables, only the tools of discrete mathematics—summation and differences—are required, while the study of continuous variables requires the continuous mathematics of the calculus—integrals and derivatives. This is the reason for making the distinction between the two types of variables.

Exercises / Section 3.1

1. Three automobiles are selected at random, and each is categorized as having a diesel (S) or nondiesel (F) engine (so outcomes are *SSS, SSF,* and so on). If X = the number of cars among the three which have a diesel engine, list each outcome in $\mathscr{S}$ and its associated X value.

2. Give three examples of Bernoulli random variables (other than those in the text).

3. Using the experiment in Example 3.3, define two more random variables and list the possible values of each.

4. Let X = the number of nonzero digits in a randomly selected zip code. What are the possible values of X? Give three possible outcomes and their associated X values.

5. If the sample space $\mathscr{S}$ is an infinite set, does this necessarily imply that any r.v. X defined from $\mathscr{S}$ will have an infinite set of possible values? If yes, say why. If no, give an example.

6. Starting at a fixed time, each car entering an intersection is observed to see whether it turns left (L), right (R), or goes straight ahead (A). The experiment terminates as soon as a car is observed to turn left. Let X = the number of cars observed. What are possible X values? List five outcomes and their associated X values.

7. For each random variable defined below, describe the set of possible values for the variable and state whether or not the variable is discrete.
  a. X = the number of unbroken eggs in a randomly chosen standard egg carton.
  b. Y = the number of students on a class list for a particular course who are absent on the first day of classes.
  c. U = the number of times that a duffer has to swing at a golf ball before hitting it.
  d. X = the length of a randomly selected rattlesnake.
  e. Z = the amount of royalties earned from the sale of a first edition of 10,000 textbooks.
  f. Y = the pH of a randomly chosen soil sample.

g. X = the tension (p.s.i.) at which a randomly selected tennis racket has been strung.

h. X = the total number of coin tosses required for three individuals to obtain a match (*HHH* or *TTT*).

8. Consider an experiment in which an individual named Claudius is located at the point 0 in the diagram below.

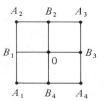

Using an appropriate randomization device (such as a tetrahedral die, one having four sides), Claudius first moves to one of the four locations B_1, B_2, B_3, B_4. Once at one of these locations, another randomization device is used to decide whether Claudius next returns to 0 or next visits one of the two other adjacent points. The experiment then continues in this fashion; after each move, another move to one of the (new) adjacent points is determined by tossing an appropriate die or coin.
  a. Let X = the number of moves that Claudius makes before first returning to 0. What are possible values of X? Is X discrete or continuous?
  b. If moves are allowed also along the diagonal paths connecting 0 to A_1, A_2, A_3, and A_4, respectively, answer the questions in part (a).

9. If first an ordinary six-sided die is tosed, after which a four-sided die (tetrahedral, with faces marked 1, 2, 3, and 4) is tossed, list the possible values for the following random variables:
  a. T = the total of the two numerical outcomes.
  b. X = the absolute value of the difference between the two outcomes.
  c. Y = the number of tosses for which the outcome is even.
  d. Z = the number of tosses for which the outcome is six.

3.2   Probability Distributions for Discrete Random Variables

When probabilities are assigned to various outcomes in $\mathscr{S}$, these in turn determine probabilities associated with values of any particular random variable X. The **proba-**

bility distribution of X says how the total probability of 1 is distributed among (allocated to) the various possible X values.

Example 3.7 To decide who has the first turn at a computer terminal, A and B will each put out one or two fingers on the count of three, with A going first if the total number of fingers is an even number. Assuming that each puts out one or two fingers with equal probability independently of the number put out by the other person, the four equally likely outcomes are (1, 1), (1, 2), (2, 1), and (2, 2). Let X denote the total number of fingers put out by A and B; the possible values of X are 2, 3, and 4. Then

$$P(X = 2) = P(\text{outcome is } (1, 1)) = \tfrac{1}{4}$$

$$P(X = 3) = P(\text{outcome is } (1, 2) \text{ or } (2,1)) = \tfrac{1}{2}$$

and

$$P(X = 4) = P(\text{outcome is } (2, 2)) = \tfrac{1}{4}$$

The values of X and their probabilities collectively specify the **probability distribution** or **probability mass function** of X.   *pᵈf*

Definition: The **probability distribution** or **probability mass function** (p.m.f.) of a discrete random variable is defined for every number x by
$$p(x) = P(X = x) = P(\text{all } s \in \mathcal{S}: X(s) = x).*$$

In words, for every possible value x of the random variable, the p.m.f. specifies the probability of observing that value when the experiment is performed. The conditions $p(x) \geq 0$ and $\sum\limits_{\text{all possible } x} p(x) = 1$ are required of any p.m.f.

Example 3.8 Suppose we go to a large tire store and observe whether the next customer to purchase tires purchases a radial or a bias-ply tire. Let

$$X = \begin{cases} 1 & \text{if the customer purchases a radial tire} \\ 0 & \text{if the customer purchases a bias-ply tire} \end{cases}$$

If 60% of all customers purchase radials, then the p.m.f. for X is

$$p(0) = P(X = 0) = P(\text{next customer purchases a bias-ply tire}) = .4$$

$$p(1) = P(X = 1) = P(\text{next customer purchases a radial tire}) = .6$$

$$p(x) = P(X = x) = 0 \text{ for } x \neq 0 \text{ or } 1$$

An equivalent description is

*$P(X = x)$ is read "the probability that the r.v. X assumes the value x." For example, $P(X = 2)$ denotes the probability that the resulting X value is 2.

$$p(x) = \begin{cases} .4 & \text{if } x = 0, \\ .6 & \text{if } x = 1, \\ 0 & \text{if } x \neq 0 \text{ or } 1. \end{cases}$$

A picture of this p.m.f., called a line graph, appears in Figure 3.2.

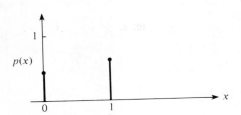

Figure 3.2   A line graph for a p.m.f.

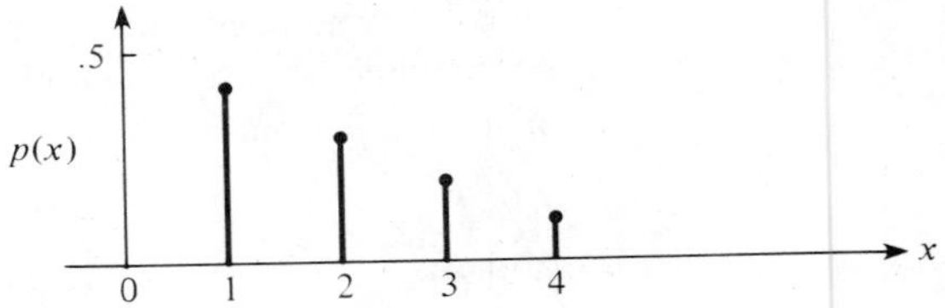

Figure 3.3   The line graph for the p.m.f. in Example 3.8

Example 3.9   Consider a group of five potential blood donors—A, B, C, D, and E—of whom only A and B have type O+ blood. Five blood samples, one from each individual, will be typed in random order until an O+ individual is identified. Let the r.v. Y = the number of typings necessary to identify an O+ individual. Then the p.m.f. of Y is

$$p(1) = P(Y = 1) = P(\text{A or B typed first}) = \tfrac{2}{5} = .4$$

$$p(2) = P(Y = 2) = P(\text{C, D, or E first, and then A or B}) = (\tfrac{3}{5})(\tfrac{2}{4}) = .3$$

$$p(3) = P(Y = 3) = P(\text{C, D, or E first and second, and then A or B})$$

$$= (\tfrac{3}{5})(\tfrac{2}{4})(\tfrac{2}{3}) = .2$$

$$p(4) = P(Y = 4) = P(\text{C, D, and E all done first}) = (\tfrac{3}{5})(\tfrac{2}{4})(\tfrac{1}{3}) = .1$$

$$p(y) = 0 \quad \text{if } y \neq 1, 2, 3, 4$$

The p.m.f. can be presented nicely in tabular form:

y	1	2	3	4
$p(y)$	.4	.3	.2	.1

where any y value not listed receives zero probability.

The name "probability mass function" is suggested by a model used in physics for a system of "point masses." In this model masses are distributed at various locations x along a one-dimensional axis. Our p.m.f. describes how the total probability mass of 1 is distributed at various points along the axis of possible values of the random variable (where, and how much mass at each x).

Another useful pictorial representation of a p.m.f., called a **probability histo-**

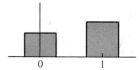 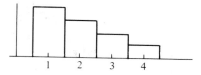

Figure 3.4 Probability histograms for Examples 3.7 and 3.8

gram, is similar to histograms discussed in Chapter 1. Above each y with $p(y) > 0$, construct a rectangle centered at y. The height of each rectangle is proportional to $p(y)$, and the base is the same for all rectangles. When possible values are equally spaced, the base is frequently chosen as the distance between successive y values (though it could be smaller). Figure 3.4 pictures several probability histograms.

A Parameter of a Probability Distribution

In Example 3.8 we were able to obtain $p(x)$ because we had specific information about the probabilities of various outcomes (radial or bias ply). Even without this specific information, we can write the general form of the p.m.f. Suppose that

$$P(\text{the next customer purchases a radial tire}) = \alpha$$

and

$$P(\text{the next customer purchases a bias-ply tire}) = 1 - \alpha$$

where $0 < \alpha < 1$ but α is otherwise unspecified. Then the p.m.f. can be written

$$p(0) = P(X = 0) = P(\text{bias ply}) = 1 - \alpha$$
$$p(1) = P(X = 1) = P(\text{radial}) = \alpha$$
$$p(x) = 0 \quad \text{for} \quad x \neq 0 \text{ or } 1$$

Because the p.m.f. depends on the particular value of α, we often write $p(x; \alpha)$ rather than just $p(x)$:

$$p(x; \alpha) = \begin{cases} 1 - \alpha & \text{if} \quad x = 0 \\ \alpha & \text{if} \quad x = 1 \\ 0 & \text{otherwise} \end{cases} \tag{3.1}$$

Then each choice of α in (3.1) yields a different p.m.f. If the two events {radial} and {bias ply} are equally likely, then $\alpha = .5$.

Definition: Suppose that $p(x)$ depends on a quantity which can be assigned any one of a number of possible values, with each different value determining a different probability distribution. Such a quantity is called a **parameter** of the distribution. The collection of all probability distributions for different values of the parameter is called a **family** of probability distributions.

The quantity α in (3.1) is a parameter. Each different number α between 0 and 1 determines a different member of a family of distributions; two such members are

$$p(x; .6) = \begin{cases} .4 & \text{if } x = 0 \\ .6 & \text{if } x = 1 \\ 0 & \text{otherwise} \end{cases} \text{ and } p(x; .5) = \begin{cases} .5 & \text{if } x = 0 \\ .5 & \text{if } x = 1 \\ 0 & \text{otherwise} \end{cases}$$

Every probability distribution for a Bernoulli r.v. has the form (3.1), so (3.1) is called the family of Bernoulli distributions.

Example 3.10   Starting at a fixed time, we observe the sex of each newborn child at a certain hospital until a boy (B) is born. Let $p = P(B)$, assume that successive births are independent, and define the r.v. X by X = number of births observed. Then

$$p(1) = P(X = 1) = P(B) = p$$

$$p(2) = P(X = 2) = P(GB) = P(G) \cdot P(B) = (1 - p)p$$

and

$$p(3) = P(X = 3) = P(GGB) = P(G) \cdot P(G) \cdot P(B) = (1 - p)^2 p$$

Continuing in this way, a general formula emerges:

$$p(x) = \begin{cases} (1 - p)^{x-1}p & x = 1, 2, 3, \cdots \\ 0 & \text{otherwise} \end{cases} \tag{3.2}$$

The quantity p in (3.2) represents a number between 0 and 1 and is a parameter of the probability distribution. In the sex example, $p = .51$ might be appropriate, but if we were looking for the first child with Rh positive blood, then we might let $p = .85$.

The Cumulative Distribution Function

For some fixed value x, we often wish to compute the probability that the observed X will be at most x. For example, the p.m.f. in Example 3.7 was

$$p(x) = \begin{cases} \frac{1}{4} & \text{for } x = 2 \text{ or } 4 \\ \frac{1}{2} & \text{for } x = 3 \\ 0 & \text{otherwise} \end{cases}$$

The probability that X is at most 3 is then $P(X \le 3) = \frac{1}{4} + \frac{1}{2} = \frac{3}{4}$. Similarly, $P(X \le 2) = \frac{1}{4}$ and $P(X \le 4) = 1$. Also, $P(X \le \frac{2}{3}) = 0$, $P(X \le 2.5) = \frac{1}{4}$, $P(X \le 4.7) = 1$, and so on. Evaluating $P(X \le x)$ for every number x yields the **cumulative distribution function** of X—"cumulative" because we accumulate (add) probabilities of all X values which are at most x.

> **Definition:** The **cumulative distribution function** (c.d.f.) $F(x)$ of a discrete random variable X with p.m.f. $p(x)$ is defined for every number x by
>
> $$F(x) = P(X \le x) = \sum_{y:y \le x} p(y)$$

For any number x, $F(x)$ is the probability that the observed value of X will be at most x.

Example 3.9
(continued)

The p.m.f. of Y for Example 3.9 was

y	1	2	3	4
$p(y)$	.4	.3	.2	.1

We first determine $F(y)$ for each value in the set $\{1, 2, 3, 4\}$ of possible values:

$$F(1) = P(Y \le 1) = P(Y = 1) = p(1) = .4$$
$$F(2) = P(Y \le 2) = P(Y = 1 \text{ or } 2) = p(1) + p(2) = .7$$
$$F(3) = P(Y \le 3) = P(Y = 1 \text{ or } 2 \text{ or } 3) = p(1) + p(2) + p(3) = .9$$
$$F(4) = P(Y \le 4) = P(Y = 1 \text{ or } 2 \text{ or } 3 \text{ or } 4) = 1$$

Now for any other number y, $F(y)$ will equal the value of F at the closest possible value of Y to the left of y. For example, $F(2.7) = P(Y \le 2.7) = P(Y \le 2) = .7$, and $F(3.999) = F(3) = .9$. The c.d.f. is thus

$$F(y) = \begin{cases} 0 & \text{if } y < 1 \\ .4 & \text{if } 1 \le y < 2 \\ .7 & \text{if } 2 \le y < 3 \\ .9 & \text{if } 3 \le y < 4 \\ 1 & \text{if } 4 \le y \end{cases} \qquad (3.3)$$

A graph of $F(y)$ appears in Figure 3.5.

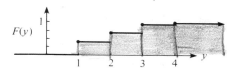

Figure 3.5   A graph of the c.d.f. of Example 3.9

For X a discrete random variable, the graph of $F(x)$ will have a jump at every possible value of X and will be flat between possible values. Such a graph is called a **step function.**

Example 3.10
(continued)

In Example 3.10 any positive integer was a possible X value, and the p.m.f. was

$$p(x) = \begin{cases} (1 - p)^{x-1}p & x = 1, 2, 3, \ldots \\ 0 & \text{otherwise} \end{cases}$$

For any positive integer x,

$$F(x) = \sum_{y \leq x} p(y) = \sum_{y=1}^{x} (1 - p)^{y-1}p = p \sum_{y=0}^{x-1} (1 - p)^y \tag{3.4}$$

To evaluate this sum, we use the fact that the partial sum of a geometric series is

$$\sum_{y=0}^{k} a^y = \frac{1 - a^{k+1}}{1 - a}$$

Using this in (3.4) with $a = 1 - p$ and $k = x - 1$ gives

$$F(x) = p \cdot \frac{1 - (1 - p)^x}{1 - (1 - p)} = 1 - (1 - p)^x, \ x \text{ a positive integer}$$

Since F is constant in between positive integers,

$$F(x) = \begin{cases} 0 & x < 1 \\ 1 - (1 - p)^{[x]} & x \geq 1 \end{cases} \tag{3.5}$$

where $[x]$ is the largest integer $\leq x$ (such as $[2.7] = 2$). For example, if $p = .51$ as in the birth example, then the probability of having to examine at most five births to see the first boy is $F(5) = 1 - (.49)^5 = 1 - .0282 = .9718$, while $F(10) \doteq 1.0000$.

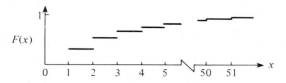

Figure 3.6   A graph of $F(x)$ for Example 3.9

In our examples, the c.d.f. has been constructed from the p.m.f., but it is possible to reverse this procedure and obtain the p.m.f. from the c.d.f. If the possible X values are, in increasing order, $x_1, x_2, x_3, \ldots$, then $p(x_i) = F(x_i) - F(x_{i-1})$, which is just the amount by which F jumps at x_i. More generally, the probability of any interval of X values equals the difference between the c.d.f. at two different values.

> **Proposition:** For any two numbers a and b with $a \leq b$,
>
> $$P(a \leq X \leq b) = F(b) - F(a-)$$
>
> where "$a-$" represents the largest possible X value which is strictly less than a. In particular, if the only possible values are integers and if a and b are integers,
>
> $$P(a \leq X \leq b) = P(X = a \text{ or } a + 1 \text{ or } \cdots \text{ or } b) = F(b) - F(a-1)$$

Example 3.11 Let X = the number of days of sick leave taken by a randomly selected employee of a large company during 1980. If the maximum number of allowable sick days per year is 14, possible values of X are 0, 1, . . ., 14. With $F(0) = .58$, $F(1) = .72$, $F(3) = .81$, $F(4) = .88$, and $F(5) = .94$,

$$P(2 \leq X \leq 5) = P(X = 2, 3, 4, \text{ or } 5) = F(5) - F(1) = .22$$

and

$$P(X = 3) = F(3) - F(2) = .09$$

The reason for subtracting $F(a-)$ rather than $F(a)$ is that we want to include $P(X = a)$; $F(b) - F(a)$ gives $P(a < X \leq b)$. This proposition will be used extensively when computing binomial and Poisson probabilities in Sections 3.4 and 3.7.

Another View of p.m.f.'s

It is often helpful to think of a p.m.f. as specifying a mathematical model for a discrete population.

Example 3.12 Consider selecting at random a student who is among the 15,000 registered for the current term at Mega University. Let X = the number of courses for which the selected student is registered, and suppose that X has p.m.f.

x	1	2	3	4	5	6	7
$p(x)$	.01	.03	.13	.25	.39	.17	.02

One way to view this situation is to think of the population as consisting of 15,000 individuals, each having his or her own X value; the proportion with each X value is given by $p(x)$ above. An alternative viewpoint is to forget about the students and think of the population itself as consisting of the X values: there are some 1's in the population, some 2's, . . ., and finally some 7's. The population then consists of the numbers 1, 2, . . ., 7 (so is discrete), and $p(x)$ gives a model for the distribution of population values.

Once we have such a mathematical model for a population, we will use it to compute values of population characteristics (such as the mean μ) and make inferences about such characteristics.

Exercises / Section 3.2

1. An automobile service facility specializing in engine tuneups knows that 25% of all tuneups are done on four-cylinder automobiles, 40% on six-cylinder automobiles, and 35% on eight-cylinder automobiles. Let X = the number of cylinders on the next car to be tuned.
  a. What is the probability mass function of X?
  b. Draw both a line graph and a probability histogram for the p.m.f. of (a).
  c. Compute the cumulative distribution function of X, and then graph it.

2. Let X = the number of tires on a randomly selected automobile which are underinflated.
  a. Which of the following three $p(x)$ functions is a legitimate p.m.f. for X, and why are the other two not allowed?

x	0	1	2	3	4
$p(x)$	.3	.2	.1	.05	.05
$p(x)$	.4	.1	.1	.1	.3
$p(x)$	.4	.1	.2	.1	.3

  b. For the legitimate p.m.f. of (a), compute $P(2 \le X \le 4)$, $P(X \le 2)$, and $P(X \ne 0)$.
  c. If $p(x) = c \cdot (5 - x)$ for $x = 0, 1, \ldots, 4$, what is the value of c? *Hint:* $\sum_{x=0}^{4} p(x) = 1$.

3. A contractor is required by a county planning department to submit one, two, three, four, or five forms (depending on the nature of the project) in applying for a building permit. Let Y = the number of forms required of the next applicant. The probability that y forms are required is known to be proportional to y—that is, $p(y) = ky$ for $y = 1, \ldots, 5$.
  a. What is the value of k? *Hint:* $\sum_{y=1}^{5} p(y) = 1$.
  b. What is the probability that at most three forms are required?
  c. What is the probability that between two and four forms (inclusive) are required?

  d. Could $p(y) = y^2/50$ for $y = 1, \ldots, 5$ be the p.m.f. of Y?

4. Two fair six-sided dice are tossed independently. Let M = the maximum of the two tosses.
  a. What is the p.m.f. of M? *Hint:* First determine $p(1)$, then $p(2)$, and so on.
  b. Compute the c.d.f. of M and graph it.

5. In Example 3.9 suppose that there are only four potential blood donors, of whom only one has type 0+ blood. Compute the p.m.f. of Y.

6. A library subscribes to two different weekly news magazines, each of which is supposed to arrive in Wednesday's mail. In actuality, each one may arrive on Wednesday, Thursday, Friday, or Saturday. Suppose that the two arrive independently of one another, and that for each one $P(\text{Wed.}) = .4$, $P(\text{Thurs.}) = .3$, $P(\text{Fri.}) = .2$, and $P(\text{Sat.}) = .1$. Let Y = the number of days beyond Wednesday that it takes for both magazines to arrive (so possible Y values are 0, 1, 2, or 3) Compute the p.m.f. of Y.

7. A small town situated on a main highway has two gas stations, A and B. Station A sells regular, unleaded, and premium gas for 144.9, 147.6, and 149.9 cents per gallon, respectively, while B sells no regular, but sells unleaded and premium for 145.9 and 149.9 cents per gallon, respectively. Of the cars which stop at Station A, 50% buy regular, 30% buy unleaded, and 20% buy premium. Of the cars stopping at B, 60% buy unleaded and 40% buy premium. Suppose that A gets 60% of the cars which stop for gas in this town, while 40% go to B. Let V = the price per gallon paid by the next car which stops for gas in this town.
  a. Compute the p.m.f. of V. Then draw a probability histogram of the p.m.f.
  b. Compute and graph the c.d.f. of V.

8. An insurance company offers its policyholders a number of different premium payment options. For a randomly selected policyholder, let X = the number of months between successive payments. The c.d.f. of X is

$$F(x) = \begin{cases} 0 & \text{if } x < 1 \\ .3 & \text{if } 1 \le x < 3 \\ .4 & \text{if } 3 \le x < 4 \\ .45 & \text{if } 4 \le x < 6 \\ .60 & \text{if } 6 \le x < 12 \\ 1 & \text{if } 12 \le x \end{cases}$$

a. What is the p.m.f. of X?

b. Using just the c.d.f., compute $P(3 \le X \le 6)$ and $P(4 \le X)$.

9. In Example 3.10 let Y = the number of girls born before the experiment terminates. With $p = P(B)$ and $1 - p = P(G)$, what is the p.m.f. of Y? *Hint:* First list the possible values of Y, starting with the smallest, and proceed until you see a general formula.

10. In Example 3.10 suppose that the experiment does not terminate until two boys are born, and let Y = the number of births examined before the experiment terminates. Then possible Y values are 2, 3, 4, ... Again assuming independent births and $P(B) = p$, derive the p.m.f. of Y. *Hint:* To have $Y = y$, the yth birth must be a boy, and one of the first $y - 1$ must also be a boy, while the others are girls; there are $y - 1$ different ways for this latter event to occur, so first write down the probability of each such way. It might help to select, for example, y = 5 and list all corresponding outcomes.

11. In the proposition which uses the c.d.f. to compute $P(a \le X \le b)$,
  a. What is the result if $a = b$?
  b. What is the result if b is replaced by ∞?

12. Alvie Singer lives at 0 in the diagram below, and has four friends who live at A, B, C, and D. One day Alvie decides to go visiting, so he tosses a fair coin twice to decide which of the four to visit. Once at a friend's house, he will either return home or else proceed to one of the two adjacent houses (such as 0, A, or C when at B), with each of the three possibilities having probability $\frac{1}{3}$. In this way Alvie continues to visit friends until he returns home.

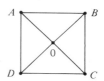

  a. Let X = the number of times that Alvie visits a friend. Derive the p.m.f. of X.
  b. Let Y = the number of straight-line segments that Alvie traverses (including those leading to and from 0). What is the p.m.f. of Y?
  c. Suppose that female friends live at A and C and male friends at B and D. If Z = the number of visits to female friends, what is the p.m.f. of Z?

3.3 Expected Values of Discrete Random Variables

In Example 3.12 we considered a university having 15,000 students and let X = the number of courses for which a randomly selected student is registered. The probability mass function of X appears below. Since $p(1) = .01$, we know that $(.01) \cdot (15,000) = 150$ of the students are registered for one course, and similarly for the other x values.

x	1	2	3	4	5	6	7	
$p(x)$	.01	.03	.13	.25	.39	.17	.02	(3.6)
number registered	150	450	1950	3750	5850	2550	300	

To compute the average number of courses per student, or the average value of X in the population, we should compute the total number of courses and divide by the total number of students. Since each of 150 students is taking one course, these 150 contribute 150 courses to the total. Similarly, 450 students contribute 2(450) courses, and so on. The population average value of X is then

$$\frac{1(150) + 2(450) + 3(1950) + \cdots + 7(300)}{15,000} = 4.57 \tag{3.7}$$

Since $150/15,000 = .01 = p(1)$, $450/15,000 = .03 = p(2)$, and so on, an alternative expression for (3.7) is

$$1 \cdot p(1) + 2 \cdot p(2) + \cdots + 7 \cdot p(7) \tag{3.8}$$

Expression (3.8) shows that to compute the population average value of X, we need only the possible values of X along with their probabilities (proportions). In particular the population size is irrelevant as long as the probability mass function is given by (3.6). The average or mean value of X is then a weighted average of the possible values 1, . . ., 7, where the weights are the probabilities of those values.

The Expected Value of *X*

> **Definition:** Let X be a discrete random variable with set of possible values D and probability mass function $p(x)$. The **expected value** or **mean value** of X, denoted by $E(X)$ or μ_X, is
>
> $$E(X) = \mu_X = \sum_{x \in D} x \cdot p(x)$$

When it is clear to which X the expected value refers, μ rather than μ_X is often used.

Example 3.13   For the p.m.f. in (3.6)

$$\begin{aligned}
\mu &= 1 \cdot p(1) + 2 \cdot p(2) + \cdots + 7 \cdot p(7) \\
&= (1)(.01) + 2(.03) + \cdots + (7)(.02) \\
&= .01 + .06 + .39 + 1.00 + 1.95 + 1.02 + .14 = 4.57
\end{aligned}$$

If we think of the population as consisting of the X values 1, 2, . . ., 7, then $\mu = 4.57$ is the population mean. In the sequel we shall often refer to μ as the population mean rather than the mean of X in the population.

In the above example, the expected value μ was 4.57, which is not a possible value of X. The word "expected" should be interpreted with caution, since one would not expect to see an X value of 4.57 when a single student is selected.

Example 3.14 Just after birth, each newborn child is rated on a scale called the Apgar scale. The possible ratings are 0, 1, . . ., 10, with the child's rating determined by color, muscle tone, respiratory effort, heartbeat, and reflex irritability (the best possible score is 10). Let X be the Apgar score of a randomly selected child born at a certain hospital during the next year, and suppose that the p.m.f. of X is

x	0	1	2	3	4	5	6	7	8	9	10
$p(x)$	.002	.001	.002	.005	.02	.04	.18	.37	.25	.12	.01

Then the mean value of X is

$$E(X) = \mu = 0(.002) + 1(.001) + 2(.002) + \cdots + 8(.25) + 9(.12)$$
$$+ 10(.01) = 7.16$$

Again μ is not a possible value of the variable X. Also, because the variable refers to a future child, there is no concrete existing population to which μ refers. Instead, we think of the p.m.f. as a model for a conceptual population consisting of the values 0, 1, 2, . . ., 10. The mean value of this conceptual population is then $\mu = 7.16$.

Example 3.15 Let X be a Bernoulli random variable with p.m.f.

$$p(X=x) = p(x) = \begin{cases} 1 - p & x = 0 \\ p & x = 1 \\ 0 & x \neq 0, 1 \end{cases}$$

Then $E(X) = 0 \cdot p(0) + 1 \cdot p(1) = 0(1 - p) + 1(p) = p$. That is, the expected value of X is just the probability that X takes on the value 1. If we conceptualize a population consisting of 0's in proportion $1 - p$ and 1's in proportion p, then the population average is $\mu = p$.

Example 3.16 The general form for the p.m.f. of X = number of children born up to and including the first boy is

$$p(x) = \begin{cases} p(1 - p)^{x-1} & x = 1, 2, 3, \ldots \\ 0 & \text{otherwise} \end{cases}$$

From the definition,

$$E(X) = \sum_D x \cdot p(x) = \sum_{x=1}^{\infty} xp(1 - p)^{x-1} = p \sum_{x=1}^{\infty} \left[\frac{d}{dp}(1 - p)^x \right] \tag{3.9}$$

If we exchange the order of taking the derivative and the summation, the sum is that of a geometric series. After computing it, the derivative is taken, and the final result is $E(X) = 1/p$. If p is near 1, we expect to see a boy very soon, while if p is near 0, we expect many births before the first boy. For $p = .5$, $E(X) = 2$.

There is another frequently used interpretation of μ. Consider the p.m.f.

$$p(x) = \begin{cases} (.5) \cdot (.5)^{x-1} & \text{if } x = 1, 2, 3, \ldots \\ 0 & \text{otherwise} \end{cases}$$

This is the p.m.f. of X = the number of tosses of a fair coin necessary to obtain the first H (a special case of Example 3.16). Suppose that we observe a value x from this p.m.f. (toss a coin until an H appears), then observe independently another value (keep tossing), then another, and so on. If after observing a very large number of x values, we average them, the resulting sample average will be very near to $\mu = 2$. That is, μ can be interpreted as the long-run average observed value of X when the experiment is performed repeatedly.

Example 3.17   Let X have p.m.f.

$$p(x) = \begin{cases} \dfrac{k}{x^2} & x = 1, 2, 3, \ldots \\ 0 & \text{otherwise} \end{cases}$$

where k is chosen so that $\sum_{x=1}^{\infty} (k/x^2) = 1$ (in a mathematics course on infinite series, it is shown that $\sum_{x=1}^{\infty} (1/x^2) < \infty$, which implies that such a k exists, but its exact value need not concern us). The expected value of X is

$$\mu = E(X) = \sum_{x=1}^{\infty} x \cdot \frac{k}{x^2} = k \sum_{x=1}^{\infty} \frac{1}{x} \qquad (3.10)$$

The sum on the right of (3.10) is the famous harmonic series of mathematics, and can be shown to equal ∞. $E(X)$ is not finite here because $p(x)$ does not decrease sufficiently fast as x increases; statisticians say that the probability distribution of X has "a heavy tail." If a sequence of X values is chosen using this distribution, the sample average will not settle down to some finite number but will tend to grow without bound.

Statisticians use the phrase "heavy tails" in connection with any distribution having a large amount of probability far from μ (so heavy tails does not require $\mu = \infty$). Such heavy tails make it difficult to make inferences about μ.

The Expected Value of a Function

Often we will be interested in the expected value of some function $h(X)$ rather than X itself.

Example 3.18   Suppose that a bookstore purchases 10 copies of a book at $6.00 each, to sell at $12.00 with the understanding that at the end of a three-month period, any unsold copies can be redeemed for $2.00. If X = the number of copies purchased, then net revenue = $h(X) = 12X + 2(10 - X) - 60 = 10X - 40$.

An easy way of computing the expected value of $h(X)$ is suggested by Example 3.19.

Example 3.19   Let X be the outcome when a fair six-sided die is tossed once, and let $h(X) = X^2$. The p.m.f. of X is

x	1	2	3	4	5	6
$p(x)$	$\frac{1}{6}$	$\frac{1}{6}$	$\frac{1}{6}$	$\frac{1}{6}$	$\frac{1}{6}$	$\frac{1}{6}$

If we let $Y = X^2$, then Y is a random variable with set of possible values $D^* = \{1, 4, 9, 16, 25, 36\}$ and the p.m.f. of Y is

y	1	4	9	16	25	36
$p(y)$	$\frac{1}{6}$	$\frac{1}{6}$	$\frac{1}{6}$	$\frac{1}{6}$	$\frac{1}{6}$	$\frac{1}{6}$

Having obtained the p.m.f. of Y, its expected value (that is, the expected value of X^2) is

$$E(Y) = \sum_{D^*} y \cdot p(y) = (1) \cdot \tfrac{1}{6} + (4) \cdot \tfrac{1}{6} + \cdots + (25) \cdot \tfrac{1}{6} + (36) \cdot \tfrac{1}{6} = \tfrac{91}{6}$$

$$\text{(3.11)}$$

$$= (1)^2 \cdot \tfrac{1}{6} + (2)^2 \cdot \tfrac{1}{6} + \cdots + (5)^2 \cdot \tfrac{1}{6} + (6)^2 \cdot \tfrac{1}{6} = \sum_{D} x^2 \cdot p(x)$$

According to (3.11) it was not necessary to compute the p.m.f. of $Y = X^2$ to obtain $E(Y)$; instead the desired expected value is a weighted average of the possible $h(x)$ (rather than x) values.

> **Proposition:** If the random variable X has set of possible values D and p.m.f. $p(x)$, then the expected value of any function $h(X)$, denoted by $E[h(X)]$ or $\mu_{h(X)}$, is computed by
>
> $$E[h(X)] = \sum_{D} h(x) \cdot p(x)$$

According to this proposition, $E[h(X)]$ is computed in the same way that $E(X)$ itself is, except that $h(x)$ is substituted in place of x.

Example 3.20   A small drugstore orders four copies of a news magazine for its magazine rack each week. Let X = demand for the magazine, with p.m.f.

x	1	2	3	4	5	6
$p(x)$	$\frac{1}{15}$	$\frac{2}{15}$	$\frac{3}{15}$	$\frac{4}{15}$	$\frac{3}{15}$	$\frac{2}{15}$

The number of copies left at the end of the week is $h(X) = \text{maximum}(4 - X, 0)$, so the expected number left is

$$E[h(x)] = \sum_D h(x) \cdot p(x) = \sum_{x=1}^{6} \max(4 - x, 0) \cdot p(x)$$

$$= 3 \cdot \tfrac{1}{15} + 2 \cdot \tfrac{2}{15} + 1 \cdot \tfrac{3}{15} + 0 \cdot \tfrac{4}{15} + 0 \cdot \tfrac{3}{15} + 0 \cdot \tfrac{2}{15} = \tfrac{10}{15}$$

Rules of Expected Value

The $h(X)$ function of interest is quite frequently a linear function $aX + b$. In this case $E[h(X)]$ is easily computed from $E(X)$.

> **Proposition:** $E(aX + b) = a \cdot E(X) + b$ (or, using alternative notation, $\mu_{aX+b} = a \cdot \mu_x + b$).

To paraphrase, the expected value of a linear function equals the linear function of the expected value $E(X)$. If $E(X) = 10$ and the linear function is $2X - 3$, then $E(2X - 3) = (2) \cdot (10) - 3 = 17$.

Proof:

$$E(aX + b) = \sum_D (ax + b) \cdot p(x) = a \sum_D x \cdot p(x) + b \sum_D p(x)$$

$$= aE(X) + b$$

Example 3.18
(continued)   With $X = $ the number of copies purchased, net revenue $= h(X) = 10X - 40$. If $E(X) = 7.5$, then expected net revenue is $E(10X - 40) = 10(7.5) - 40 = 35$.

Two special cases of the proposition yield two important rules of expected value.

> **1.** For any constant a, $E(aX) = a \cdot E(X)$ (take $b = 0$)
> **2.** For any constant b, $E(X + b) = E(X) + b$ (take $a = 1$)   (3.12)

Multiplication of X by a constant a changes the units of measurement (from dollars to cents, where $a = 100$, inches to cm, where $a = 2.54$, and so on). Rule (1) says that the expected value in the new units equals the expected value in the old units multiplied by the conversion factor a. Similarly, if a constant b is added to each possible value of X, then the expected value will be shifted by that same constant amount.

The Variance of X

The expected value of X measures where the probability distribution is centered. Using the physical analogy of placing point mass $p(x)$ at the value x on a one-dimensional axis, if the axis were then supported by a fulcrum placed at μ, there would be no tendency for the axis to tilt. This is illustrated for two different distributions in Figure 3.7.

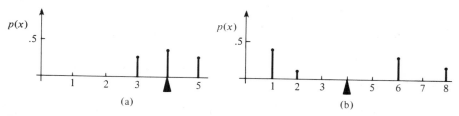

Figure 3.7 Two different probability distributions with $\mu = 4$

Although both distributions pictured in Figure 3.7 have the same center μ, the distribution of (*b*) has greater spread or variability or dispersion than does that of (*a*). We will use the variance of X to measure the amount of variability in (the distribution of) X, just as s^2 was used in Chapter 1 to measure variability in a sample.

Definition: Let X have probability function $p(x)$ and expected value μ. Then **the variance of X,** denoted by Var(X) or σ_X^2, or just σ^2, is

$$\text{Var}(X) = \sum_D (x - \mu)^2 \cdot p(x) = E[(x - \mu)^2]$$

The **standard deviation** of X is

$$\sigma_X = \sqrt{\sigma_X^2}$$

The quantity $h(X) = (X - \mu)^2$ is the squared deviation of X from its mean, and σ^2 is the expected squared deviation. If most of the probability distribution is close to μ, then σ^2 will be relatively small, while if there are x values far from μ which have large $p(x)$, then σ^2 will be quite large.

Example 3.21 If X has p.m.f.

x	3	4	5
$p(x)$	.3	.4	.3

(as in Figure 3.7a) with mean $\mu = 4$, then

$$\text{Var}(X) = \sigma^2 = \sum_{x=3}^{5} (x - 4)^2 \cdot p(x)$$

$$= (3 - 4)^2 \cdot (.3) + (4 - 4)^2 \cdot (.4) + (5 - 4)^2 \cdot (.3) = .6$$

The standard deviation of X is $\sigma = \sqrt{.6} = .77$.

If X has the p.m.f. of Figure 3.7b,

x	1	2	6	8
$p(x)$	.4	.1	.3	.2

then

$$\sigma^2 = (1 - 4)^2 \cdot (.4) + (2 - 4)^2 \cdot (.1) + (6 - 4)^2 \cdot (.3) + (8 - 4)^2 \cdot (.2)$$

$$= 8.4$$

and

$$\sigma = 2.90$$

When the p.m.f. $p(x)$ specifies a mathematical model for the distribution of population values, both σ^2 and σ measure the spread of values in the population; σ^2 is the population variance and σ is the population standard deviation.

A Shortcut Formula for σ^2

The number of arithmetic operations necessary to compute σ^2 can be reduced by using an alternative computing formula.

> **Proposition:** $\text{Var}(X) = \sigma^2 = \left[\sum_{D} x^2 \cdot p(x)\right] - \mu^2 = E(X^2) - [E(X)]^2$

In using this formula, $E(X^2)$ is computed first without any subtraction; then $E(X)$ is computed, squared, and subtracted (once) from $E(X^2)$.

Example 3.22   If X is the outcome of a single roll of a fair die, then

$$E(X^2) = \sum_{D} x^2 \cdot p(x) = \sum_{x=1}^{6} x^2 \cdot \tfrac{1}{6} = \tfrac{91}{6}$$

and

$$E(X) = \mu = \sum_{D} x \cdot p(x) = \sum_{x=1}^{6} x \cdot \tfrac{1}{6} = \tfrac{7}{2}$$

The variance is then

$$\sigma^2 = \tfrac{91}{6} - \left(\tfrac{7}{2}\right)^2 = \tfrac{91}{6} - \tfrac{49}{4} = \tfrac{70}{24} = 2.92,$$

and $\sigma = \sqrt{2.92} = 1.71$.

Proof of the shortcut formula: Expand $(x - \mu)^2$ in the definition of σ^2 to obtain $x^2 - 2\mu x + \mu^2$, and then carry Σ through to each of the three terms:

$$\sigma^2 = \sum_D x^2 \cdot p(x) - 2\mu \cdot \sum_D x \cdot p(x) + \mu^2 \sum_D p(x)$$

$$= E(X^2) - 2\mu \cdot \mu + \mu^2 = E(X^2) - \mu^2$$

Rules of Variance

The variance of $h(X)$ is the expected value of the squared difference between $h(X)$ and its expected value:

$$\text{Var}[h(X)] = \sigma^2_{h(X)} = \sum_D \{h(x) - E[h(X)]\}^2 \cdot p(x) \tag{3.13}$$

When $h(X)$ is a linear function, $\text{Var}[h(X)]$ is easily related to $\text{Var}(X)$.

> **Proposition:** $\text{Var}(aX + b) = \sigma^2_{aX+b} = a^2 \cdot \sigma^2_X$, and $\sigma_{aX+b} = |a| \cdot \sigma_X$

This result says that the addition of the constant b does not affect the variance, which is intuitive, since the addition of b changes the location (mean value) but not the spread of values. In particular,

> 1. $\sigma^2_{aX} = a^2 \cdot \sigma^2_X$, $\sigma_{aX} = |a| \cdot \sigma_X$
> 2. $\sigma^2_{X+b} = \sigma^2_X$ (3.14)

The reason for the absolute value in σ_{aX} is that a may be negative, while a standard deviation cannot be negative; a^2 results when a is brought outside the term being squared in (3.13).

Example 3.18 (continued)

The revenue from selling the book was $10X - 40$, where X denoted the number of copies purchased. If $\sigma^2_X = 7$, then $\text{Var}(10X - 40) = \text{Var}(10X) = 100 \cdot \text{Var}(X) = 700$, and the standard deviation of revenue is $\sqrt{700} = 26.46$.

Exercises / Section 3.3

1. The probability mass function for X = the number of cars owned by a randomly selected family in Suburbanville is

x	0	1	2	3	4
$p(x)$	.08	.15	.45	.27	.05

   a. Compute $E(X)$.
   b. Compute Var(X) directly from the definition.
   c. Compute the standard deviation of X.
   d. Compute Var(X) using the shortcut formula.

2. An instructor in a technical writing class has asked that a certain report be turned in the following week, adding the restriction that any report exceeding four pages will not be accepted. Let Y = the number of pages in a randomly chosen student's report, and suppose that X has p.m.f.

y	1	2	3	4
$p(y)$	.01	.19	.35	.45

   a. Compute $E(Y)$.
   b. Suppose that the instructor spends $\sqrt{Y}$ minutes grading a paper consisting of Y pages. What is the expected amount of time $[E(\sqrt{Y})]$ spent grading a randomly selected paper?
   c. Compute the variance and standard deviation of Y.

3. An appliance dealer sells three different models of upright freezers having 13.5, 15.9, and 19.1 cubic feet of storage space respectively. Let X = the amount of storage space purchased by the next customer to buy a freezer. Suppose that X has p.m.f.

x	13.5	15.9	19.1
$p(x)$	.2	.5	.3

   a. Compute $E(X)$, $E(X^2)$, and Var(X).
   b. If the price of a freezer having capacity X cubic feet is $25X - 8.5$, what is the expected price paid by the next customer to buy a freezer?
   c. What is the variance of the price $25X - 8.5$ paid by the next customer?
   d. Suppose that while the rated capacity of a freezer is X, the actual capacity is $h(X) = X - .01X^2$. What is the expected actual capacity of the freezer purchased by the next customer?

4. Let X be a Bernoulli r.v. with p.m.f. as in Example 3.15.
   a. Compute $E(X^2)$.
   b. Show that Var$(X) = p(1 - p)$.
   c. Compute $E(X^{79})$.

5. Suppose that the number of plants of a particular type which are found in a rectangular region

(called a quadrat by ecologists) in a certain geographical area is a r.v. X with p.m.f.

$$p(x) = \begin{cases} c/x^3 & x = 1, 2, 3, \ldots \\ 0 & \text{otherwise} \end{cases}$$

Is $E(X)$ finite? Justify your answer (this is another distribution which statisticians would call heavy-tailed).

6. In Example 3.20 suppose that the store owner actually pays $\$.25$ for each copy of the news magazine, and the price to customers is $\$1.00$. If magazines left at the end of the week have no salvage value, is it better to order three or four copies of the magazine? *Hint:* For both three and four copies purchased, express net revenue as a function of demand X, and then compute the expected revenue.

7. You are offered the choice of playing one of two games. In game A, you roll a fair die once and receive the amount X = the resulting number (in dollars). In game B, you roll a fair die twice and receive Y = the maximum of the two outcomes. If the entry fee for game A is $\$3.00$ and for game B is $\$3.50$, which game should you play?

8. The n candidates for a job have been ranked 1, 2, 3, . . ., n. Let X = the rank of a randomly selected candidate, so that X has p.m.f.

$$p(x) = \begin{cases} 1/n & x = 1, 2, \ldots, n \\ 0 & \text{otherwise} \end{cases}$$

(this is called the discrete uniform distribution). Compute $E(X)$ and Var(X) using the shortcut formula. *Hint:* The sum of the first n positive integers is $n(n + 1)/2$, while the sum of their squares is $n(n + 1)(2n + 1)/6$.

9. Let X = the outcome when a fair die is rolled once. If before the die is rolled you are offered either $(1/3.5)$ dollars or $h(X) = 1/X$ dollars, would you accept the guaranteed amount or would you gamble? *Note:* It is not generally true that $1/E(X) = E(1/X)$.

10. A chemical supply company currently has in stock 100 lbs. of a certain chemical, which it sells to customers in 5-lb. lots. Let X = the number of lots ordered by a randomly chosen customer, and suppose that X has p.m.f.

x	1	2	3	4
$p(x)$	.2	.4	.3	.1

Compute $E(X)$ and Var(X). Then compute the expected number of pounds left after the next customer's order is shipped, and the variance of the number of pounds left. *Hint:* The number of pounds left is a linear function of X.

11. a. Draw a line graph of the p.m.f. of X in Example 3.22. Then determine the p.m.f. of $-X$ and draw its line graph. From these two pictures, what can you say about Var(X) and Var($-X$)?

b. Use the proposition involving Var($aX + b$) to establish a general relationship between Var(X) and Var($-X$).

12. Use the definition in Expression (3.13) to prove that Var($aX + b$) $= a^2 \cdot \sigma_X^2$. *Hint:* With $h(X) = aX + b$, $E[h(X)] = a\mu + b$ where $\mu = E(X)$.

13. Suppose that $E(X) = 5$ and $E[X(X - 1)] = 27.5$.
  a. What is $E(X^2)$? *Hint:* $E[X(X - 1)] = E[X^2 - X] = E(X^2) - E(X)$.
  b. What is Var(X)?
  c. What is the general relationship between $E(X)$, $E[X(X - 1)]$, and Var(X)?

14. Write a general rule for $E(X - c)$ where c is a constant. What happens when you let $c = \mu$, the expected value of X?

3.4 The Binomial Probability Distribution

There are many experiments which conform either exactly or approximately to the following list of requirements:

1. The experiment consists of a sequence of n trials, where n is fixed in advance of the experiment.
2. The trials are identical, and each trial can result in one of the same two possible outcomes, which we denote by success (S) or failure (F).
3. The trials are independent, so that the outcome on any particular trial does not influence the outcome on any other trial.
4. The probability of success is constant from trial to trial; we denote this probability by p.

> **Definition:** An experiment for which conditions 1–4 are satisfied is called a **binomial** experiment.

Example 3.23 The same coin is tossed successively and independently n times. We arbitrarily use S to denote the outcome H (heads) and F to denote the outcome T (tails). Then this experiment satisfies 1–4.

While many experiments involve a sequence of independent trials, there are often more than two possible outcomes on any one trial. A binomial experiment can then be created by dividing the possible outcomes into two groups.

Example 3.24 The color of pea seeds is determined by a single genetic locus. If the two alleles at this locus are AA or Aa (the genotype), then the pea will be yellow (the phenotype), and if the allele is aa, the pea will be green. Suppose we pair off 20 Aa seeds and cross the two seeds in each of the 10 pairs to obtain 10 new genotypes. Call each new genotype a success S if it is aa and a failure otherwise. Then with this identification of S and F, the experiment is binomial with $n = 10$ and $p = P(\text{aa}$ genotype). If each member of the pair is equally likely to contribute a or A, $p = P(\text{a}) \cdot P(\text{a}) = (\frac{1}{2})(\frac{1}{2}) = \frac{1}{4}$.

Example 3.25   Suppose that a certain city has 50 licensed restaurants, of which 15 currently have at least one serious health code violation while the other 35 have no serious violations. There are five inspectors, each of whom will inspect one restaurant during the coming week. The name of each restaurant is written on a different slip of paper, and after the slips are thoroughly mixed, each inspector in turn draws one of the slips *without replacement*. Label the ith trial as a success if the ith restaurant selected $(i = 1, \ldots, 5)$ has no serious violations. Then

$$P(S \text{ on first trial}) = \tfrac{35}{50} = .70$$

and

$$\begin{aligned}
P(S \text{ on second trial}) &= P(SS) + P(FS) = P(\text{second } S \mid \text{first } S)\,P(\text{first } S) \\
&\quad + P(\text{second } S \mid \text{first } F)\,P(\text{first } F) \\
&= \tfrac{34}{49} \cdot \tfrac{35}{50} + \tfrac{35}{49} \cdot \tfrac{15}{50} = \tfrac{35}{50}[\tfrac{34}{49} + \tfrac{15}{49}] = \tfrac{35}{50} = .70
\end{aligned}$$

Similarly, it can be shown that $P(S \text{ on } i\text{th trial}) = .70$ for $i = 3, 4, 5$. However,

$$P(S \text{ on fifth trial} \mid SSSS) = \tfrac{31}{46} = .67 \neq .70 = P(S \text{ on fifth trial})$$

so the experiment is not binomial because the trials are not independent. In general, if sampling is without replacement, the experiment will not yield independent trials. If each slip had been replaced after being drawn, then trials would be independent, but this might result in the same restaurant being inspected by more than one inspector.

Example 3.26   Suppose that a certain state has 500,000 registered automobiles, of which 400,000 were manufactured domestically. A sample of 10 cars is chosen without replacement. The ith trial is labeled S if the ith car chosen is a domestic model. Although this situation would seem identical to that of Example 3.25, the important difference is that the size of the population being sampled is very large relative to the sample size. In this case

$$P(S \text{ on } 2 \mid S \text{ on } 1) = \frac{399,999}{499,999} = .80000$$

and

$$P(S \text{ on } 10 \mid S \text{ on first } 9) = \frac{399,991}{499,991} = .799996 \doteq .80000$$

These calculations indicate that while the trials are not exactly independent, the conditional probabilities differ so slightly from one another that for practical purposes the trials can be regarded as independent with constant $P(S) = .8$. Thus to a very good approximation, the experiment is binomial with $n = 10$ and $p = .8$.

We will use the following rule of thumb in deciding whether a "without replacement" experiment can be treated as a binomial experiment.

> **Rule:** Suppose that each trial of an experiment can result in S or F, but the sampling is without replacement from a population of size N. If the sample size (number of trials) n is at most 5% of the population size, the experiment can be analyzed as though it is exactly a binomial experiment.

By "analyzed," we mean that probabilities based on the binomial experiment assumptions will be quite close to the actual "without replacement" probabilities, which are typically more difficult to calculate. In Example 3.25 $n/N = 5/50 = .1 > .05$, so the binomial experiment is not a good approximation, but in Example 3.26 $n/N = 10/500,000 < .05$.

The Binomial Random Variable and Distribution

In most binomial experiments it is the total number of S's, rather than knowledge of exactly which trials yielded S's, that is of interest.

> **Definition:** Given a binomial experiment consisting of n trials, the **binomial random variable** X associated with this experiment is defined as
>
> $X =$ the number of S's among the n trials

Suppose, for example, that $n = 3$. Then there are eight possible outcomes for the experiment:

$$SSS, \ SSF, \ SFS, \ SFF, \ FSS, \ FSF, \ FFS, \ FFF$$

From the definition of X, $X(SSF) = 2$, $X(SFF) = 1$, and so on. Possible values for X in an n trial experiment are $x = 0, 1, 2, \ldots, n$. We will often write $X \sim \text{Bin}(n, p)$ to indicate that X is a binomial random variable based on n trials with success probability p.

> **Notation:** Because the p.m.f. of a binomial r.v. X depends on the two parameters n and p, we denote the p.m.f. by $b(x; n, p)$.

To gain insight into $b(x; n, p)$ for general n, consider the case $n = 4$ for which each outcome, its probability, and corresponding x value are listed in Table 3.1. For example,

$$P(SSFS) = P(S) \cdot P(S) \cdot P(F) \cdot P(S) \quad \text{(independent trials)}$$

$$= p \cdot p \cdot (1 - p) \cdot p \quad \text{(constant } P(S)\text{)}$$

$$= p^3 \cdot (1 - p)$$

Table 3.1 *Outcomes and Probabilities for a Binomial Experiment with Four Trials*

Outcome	Observed x	Probability	Outcome	Observed x	Probability
SSSS	4	p^4	FSSS	3	$p^3(1 - p)$
SSSF	3	$p^3(1 - p)$	FSSF	2	$p^2(1 - p)^2$
SSFS	3	$p^3(1 - p)$	FSFS	2	$p^2(1 - p)^2$
SSFF	2	$p^2(1 - p)^2$	FSFF	1	$p(1 - p)^3$
SFSS	3	$p^3(1 - p)$	FFSS	2	$p^2(1 - p)^2$
SFSF	2	$p^2(1 - p)^2$	FFSF	1	$p(1 - p)^3$
SFFS	2	$p^2(1 - p)^2$	FFFS	1	$p(1 - p)^3$
SFFF	1	$p(1 - p)^3$	FFFF	0	$(1 - p)^4$

In this special case we wish $b(x; 4, p)$ for $x = 0, 1, 2, 3,$ and 4. For $b(3; 4, p)$, we identify which of the 16 outcomes yields an x value of 3 and add up the probabilities associated with each such outcome:

$$b(3; 4, p) = P(FSSS) + P(SFSS) + P(SSFS) + P(SSSF) = 4p^3(1 - p)$$

Since the number of outcomes with $x = 3$ is four, and each of the four has probability $p^3(1 - p)$ (the order of S's and F's is not important, but only the number of S's),

$$b(3; 4, p) = \left\{ \begin{array}{l} \text{number of outcomes} \\ \text{with } X = 3 \end{array} \right\} \cdot \left\{ \begin{array}{l} \text{probability of any particular} \\ \text{outcome with } X = 3 \end{array} \right\}$$

Similarly, $b(2; 4, p) = 6p^2(1 - p)^2$, which is also the product of the number of outcomes with $X = 2$ and the probability of any such outcome.

In general,

$$b(x; n, p) = \left\{ \begin{array}{l} \text{number of sequences of} \\ \text{length } n \text{ consisting of } x \text{ } S\text{'s} \end{array} \right\} \cdot \left\{ \begin{array}{l} \text{probability of any} \\ \text{particular such sequence} \end{array} \right\}$$

Since the ordering of S's and F's is not important, the second factor above is $p^x(1 - p)^{n-x}$ (for example, the first x trials resulting in S and the last $n - x$ in F). The first factor is the number of ways of choosing x of the n trials to be S's—that is, the number of combinations of size x which can be constructed from n distinct objects (trials here).

> **Theorem:**
>
> $$b(x; n, p) = \begin{cases} \binom{n}{x} p^x (1 - p)^{n-x} & x = 0, 1, 2, \ldots, n \\ 0 & \text{otherwise} \end{cases}$$

Example 3.27   Each of six randomly selected beer drinkers is given a glass containing beer S and a glass containing F. The glasses are identical in appearance except for a code on the bottom to identify the beer. Suppose that there is actually no tendency among beer drinkers to prefer one beer to the other. Then $p = P$(a selected individual prefers S) $= .5$, so with $X =$ the number among the six who prefer S, $X \sim$ Bin(6, .5).

Thus

$$P(X = 3) = b(3; 6, .5) = \binom{6}{3} (.5)^3 (.5)^3 = 20 \, (.5)^6 = .313$$

The probability that at least three prefer S is

$$P(3 \le X) = \sum_{x=3}^{6} b(x; 6, .5) = \sum_{x=3}^{6} \binom{6}{x} (.5)^x (.5)^{6-x} = .656$$

while the probability that at most one prefers S is

$$P(X \le 1) = \sum_{x=0}^{1} b(x; 6, .5) = .109$$

Using Binomial Tables

Even for a relatively small value of n, the computation of binomial probabilities can be tedious. Appendix Table A.1 tabulates for $n = 5, 10, 15, 20, 25$ and selected values of p, the cumulative distribution function $F(x) = P(X \le x)$. Various other probabilities can then be calculated using the proposition on c.d.f.'s from Section 3.2.

> **Notation:** For $X \sim$ Bin(n, p), the cumulative distribution function will be denoted by
>
> $$P(X \le x) = B(x; n, p) = \sum_{y=0}^{x} b(y; n, p), \quad x = 0, 1, \ldots, n$$

Example 3.28   A multiple-choice test consists of 15 questions, and for each question there are five possible answers. If for each question an answer is selected in a completely random fashion, what is the probability that (a) at most eight of the questions are answered correctly? (b) Exactly eight? (c) At least eight? (d) Between four and seven inclusive?

The binomial r.v. here is X = the number of correctly answered questions, with $n = 15$ and $p = \frac{1}{5} = .2$. We wish first $P(X \leq 8) = \sum_{x=0}^{8} b(x; 15, .2) = B(8; 15, .2)$, which is the $x = 8$ entry (row $x = 8$) in the $p = .2$ column of the $n = 15$ table; the tabulated probability is $B(8; 15, .2) = .999$. Similarly $P(X = 8) = P(X \leq 8) - P(X \leq 7) = B(8; 15, .2) - B(7; 15, .2) = .999 - .996 = .003$, while $P(X \geq 8) = 1 - P(X \leq 7) = 1 - B(7; 15, .2) = 1 - .996 = .004$. Finally, for question (d), $P(4 \leq X \leq 7) = P(X = 4, 5, 6, \text{or } 7) = P(X \leq 7) - P(X \leq 3) = B(7; 15, .2) - B(3; 15, .2) = .996 - .648 = .348$.

Example 3.29   Cars coming to a dead-end intersection can turn either left or right. Suppose that successive cars choose a turning direction independently of one another and that $P(\text{left turn}) = .7$. Among the next 15 cars, what is the probability that at least 10 turn left? Among the next 15 cars, what is the probability that at least 10 turn in the same direction?

With X = the number among the next 15 cars which turn left, $X \sim \text{Bin}(15, .7)$, so

$$P(\text{at least 10 left turns}) = P(10 \leq X) = 1 - P(X \leq 9)$$

$$= 1 - B(9; 15, .7) = 1 - .278$$

$$= .722$$

where .278 is the $n = 15$ binomial table entry in the $p = .7$ column and $x = 9$ row. Similarly

$$P\left(\begin{array}{c}\text{at least 10 turns in} \\ \text{the same direction}\end{array}\right) = P\left(\begin{array}{c}\text{at least 10 left turns or} \\ \text{at least 10 right turns}\end{array}\right)$$

$$= P(10 \leq X \text{ or } X \leq 5) = P(X \leq 5) + P(10 \leq X)$$

$$= B(5; 15, .7) + 1 - B(9; 15, .7)$$

$$= .004 + .722 = .726$$

Note that a table entry of 0 signifies only that a probability is 0 to three significant digits, for all entries in the table are actually positive. Much more extensive tables of binomial probabilities have been published. In Chapter 5 we show how quickly and accurately to approximate binomial probabilities when n is large.

The Mean and Variance of X

For $n = 1$, the binomial distribution becomes the Bernoulli distribution with p.m.f.

$$p(x) = \begin{cases} 1 - p & x = 0 \\ p & x = 1 \\ 0 & \text{otherwise} \end{cases}$$

From Example 3.15 the mean value of a Bernoulli variable is $\mu = p$, so the expected number of S's on any single trial is p. Since a binomial experiment consists of n trials, intuition suggests that for $X \sim \text{Bin}(n, p)$, $E(X) = np$, the product of the number of trials and the probability of success on a single trial. The expression for Var (X) is not so intuitive.

Proposition: If $X \sim \text{Bin}(n, p)$, then $E(X) = np$, Var $(X) = np(1 - p) = npq$, and $\sigma_X = \sqrt{npq}$ (where $q = 1 - p$).

Proof of $E(X) = np$:

$$E(X) = \sum_{x=0}^{n} x \cdot \binom{n}{x} p^x (1 - p)^{n-x} = \sum_{x=1}^{n} x \cdot \frac{n!}{x! \, (n - x)!} p^x (1 - p)^{n-x}$$

$$= \sum_{x=1}^{n} \frac{n!}{(x - 1)! \, (n - x)!} p^x (1 - p)^{n-x}$$

$$= np \sum_{x=1}^{n} \frac{(n - 1)!}{(x - 1)! \, (n - x)!} p^{x-1} (1 - p)^{n-x}$$

$$= np \sum_{y=0}^{n-1} \frac{(n - 1)!}{y! \, (n - 1 - y)!} p^y (1 - p)^{n-1-y} \quad (y \text{ replaces } x - 1)$$

$$= np \left\{ \sum_{y=0}^{n-1} \binom{n - 1}{y} p^y (1 - p)^{n-1-y} \right\}$$

The expression in braces is the sum over all possible values $y = 0, 1, \ldots, n - 1$ of a binomial p.m.f. based on $n - 1$ trials, so equals 1, leaving only np as desired. The proof of the result for variance is sketched in Exercise 3.4.11.

Example 3.30   If 75% of all consultations handled by student consultants at a computing center involve programs with syntax errors, and X is the number of programs with syntax errors in 10 randomly chosen consultations, then $X \sim \text{Bin}(10, .75)$. Thus $E(X) = np = (10)(.75) = 7.5$, Var$(X) = npq = 10(.75)(.25) = 1.875$, and $\sigma = \sqrt{1.875}$. Again, even though X can take on only integer values, $E(X)$ need not be an integer. If we perform a large number of independent binomial experiments, each with $n = 10$ trials and $p = .75$, then the average number of S's per experiment will be close to 7.5.

Exercises / Section 3.4*

1. Compute the following binomial probabilities directly from the formula for $b(x; n, p)$.

   a. $b(3; 8, .6)$ **b.** $b(5; 8, .6)$

   c. $P(3 \le X \le 5)$ when $n = 8$ and $p = .6$

   d. $P(1 \le X)$ when $n = 12$ and $p = .1$

   *"Between a and b inclusive" is equivalent to $(a \le X \le b)$.

2. Use Appendix Table A.1 to obtain the following probabilities:

    a. $B(4; 10, .3)$ **b.** $b(4; 10, .3)$ **c.** $b(6; 10, .7)$
    d. $P(2 \leq X \leq 4)$ when $X \sim$ Bin $(10, .3)$
    e. $P(2 \leq X)$ when $X \sim$ Bin $(10, .3)$
    f. $P(X \leq 1)$ when $X \sim$ Bin $(10, .7)$
    g. $P(2 < X < 6)$ when $X \sim$ Bin $(10, .3)$

3. A large supermarket stocks both national brands of coffee and its own house brand. Consider a single randomly selected customer purchasing coffee and let success = {the customer purchases a national brand}. Assume that $p = .75$ and that customers make coffee purchase decisions independently of one another. Then with $X =$ the number among 20 randomly selected coffee purchasers who select a national brand, $X \sim$ Bin $(20, .75)$.

    a. Compute $P(X \leq 15)$
    b. Compute $P(X = 15)$
    c. Compute the probability that at most 17 of the 20 customers inclusive select a national brand.
    d. Compute the probability that at least 12 customers select a national brand.
    e. Compute the probability that between 14 and 18 customers inclusive select a national brand.
    f. Compute the probability that fewer than half the customers select a national brand.
    g. To increase the sales of its own brand, the store instituted a promotional campaign. After the campaign, if fewer than half of 20 randomly selected purchases had selected the national brand, would this be strong evidence that p was now less than .75?
    h. With $X \sim$ Bin $(20, .75)$, compute $E(X)$ and Var (X). Why is Var (X) here larger than Example 3.30?

4. Suppose that only 20% of all drivers come to a complete stop at an intersection having flashing red lights in all directions when no other cars are visible. What is the probability that, of 20 randomly chosen drivers coming to an intersection under these conditions,

    a. at most 5 will come to a complete stop
    b. exactly 5 will come to a complete stop
    c. at least 5 will come to a complete stop
    d. How many of the next 20 drivers do you expect to come to a complete stop?

5. If 90% of all students taking a beginning computer programming course fail to get their first program to run on first submission, what is the probability that among 15 randomly chosen such students,

    a. at least 12 fail on first submission
    b. between 10 and 13 inclusive fail on first submission
    c. at most 2 get their program to run properly on first submission

6. A particular type of tennis racket comes strung with either nylon or gut strings. The local tennis shop currently has 10 such rackets strung with nylon and 10 strung with gut. If 60% of all customers who request this racket choose nylon strings, what is the probability that

    a. more than 10 of the next 15 customers request nylon
    b. all of the next 15 customers who ask for this racket can be given their choice of strings without the store having to reorder

7. A manufacturer of flashlight batteries wishes to control the quality of its product by rejecting any lot in which the proportion of batteries having unacceptable voltage appears to be too high. To this end, out of each large lot (10,000 batteries), 25 will be selected and tested. If at least five of these generate an unacceptable voltage, the entire lot will be rejected. What is the probability that a lot will be rejected if

    a. 5% of the batteries in the lot have unacceptable voltage
    b. 10% of the batteries in the lost have unacceptable voltage
    c. 20% of the batteries in the lot have unacceptable voltage
    d. What would happen to the above probabilities if the critical rejection number were increased from five to six?

8. A student who is trying to write a paper for a course has a choice of two topics, A and B. If topic A is chosen, the student will order two books through interlibrary loan, while if topic B is chosen, the student will order four books. The student feels that a good paper necessitates receiving and using at least half the books ordered for either topic chosen. If the probability that a book ordered through interlibrary loan actually arrives in time is .9 and books arrive independently of one another, which topic should the student choose to maximize the probability of writing a good paper? What if the arrival probability is only .5 instead of .9?

9. a. For fixed n, are there values of $p\,(0 \le p \le 1)$ for which Var $(X) = 0$? Explain why this is so.

b. For what value of p is Var (X) maximized? *Hint:* Either graph Var (X) as a function of p or else take a derivative.

10. a. Show that $b(x;\, n,\, 1 - p) = b(n - x;\, n,\, p)$.

b. Show that $B(x;\, n,\, 1 - p) = 1 - B(n - x - 1;\, n,\, p)$. *Hint:* At most x S's is equivalent to at least $n - x$ F's.

c. What do (a) and (b) imply about the necessity of including values of p greater than .5 in Appendix Table A.1?

11. a. Proceeding in a manner similar to the proof of $E(X) = np$, show that $E[X(X - 1)] = np^2(1 - p)$. *Hint:* The first two terms of the expected value sum are zero, and then

$x(x - 1)$ cancels part of $\binom{n}{x}$.

b. Using the fact that $E[X(X - 1)] = E(X^2) - E(X)$ and the shortcut formula Var $(X) = E(X^2) - [E(X)]^2$, show that Var $(X) = np(1 - p)$.

12. Customers at a gas station select either leaded regular (A), leaded premium (B), or unleaded fuel (C). Assume that successive customers make independent choices, with $P(A) = .3$, $P(B) = .2$, and $P(C) = .5$.

a. Among the next 100 customers, what are the mean and variance of the number who select leaded regular fuel? Explain your reasoning.

b. Answer (a) for the number among the 100 who select a leaded fuel.

3.5 Hypothesis Testing Using the Binomial Distribution

Methods of statistical inference are often used to draw a conclusion about the true value of a single population characteristic, such as a population mean μ or population proportion p. If p refers to the proportion of individuals in a population who possess a particular trait (smoke, drive a car, voted in the last election, have a college degree, and so on) and the sample size n is small relative to the population size, then the number X in the sample who possess the trait has approximately a binomial distribution with parameters n and p. Thus the population characteristic p is also a parameter of the binomial probability distribution.

In a hypothesis-testing problem, one is presented with two claims about the true value of a parameter, of which exactly one must be true. Based on experimental evidence—a sample from the population or distribution of interest—one wishes to decide which of the two contrasting claims is correct.

Example 3.31 It has been reported (see "Behavioral Treatment of Obesity," *J. Amer. Med. Assn.* 237, pp. 2829–2831, 1977) that only 25% of obese patients treated in medical settings lose 9 kg or more by the end of the treatment period. Suppose that a group of psychologists has developed a behavior modification procedure for inducing weight loss in obese individuals. They believe that the use of this procedure will result in more than 25% of all obese patients having weight losses of at least 9 kg by the end of a standard treatment period (where obesity is a result of overeating rather than any accompanying medical condition). Let p denote the proportion of (overeating induced) obese individuals who would lose at least 9 kg when treated by this behavior modification technique. Alternatively, if a success S denotes an obese individual who loses at least 9 kg when treated by this technique, then for a randomly selected obese individual, $p = P(S)$.

One claim of interest here is that the new technique is no more successful than what has been tried before, so that $p \leq .25$. The alternative claim is that $p > .25$, which says that the new technique is an improvement on other treatments.

The Null and Alternative Hypotheses

One of the two claims in a hypothesis-testing problem is called the **null hypothesis,** which we denote by H_0, and the other is called the **alternative hypothesis** and is denoted by H_a. The word null means "of no value, effect, or consequence," which suggests that H_0 should be identified with the hypothesis of no change (from current opinion), no difference, no improvement, no departure (from current beliefs), and so on. In Example 3.31 the no-improvement claim is that $p \leq .25$.

The decision as to which claim becomes H_0 and which becomes H_a is important, since the two hypotheses will not be treated symmetrically. This has its analogue in a judicial system. In legal proceedings against a defendent, one claim is that the defendent is innocent and the other is that the defendent is guilty. In the United States, innocence is the favored claim and the burden of proof is on those who want to show guilt (in some other systems the reverse is the case). Similarly, for us the null hypothesis H_0 will be the favored claim. The burden of proof will rest with H_a in the sense that we shall continue to believe in H_0 unless the experimental evidence strongly contradicts it. In scientific investigations H_0 is often the "status quo" claim, stating that previously accepted theory remains valid, while H_a is the "research" hypothesis which contradicts or extends in a new manner the accepted theory.

For the moment we replace in Example 3.31 the claim $p \leq .25$ by the simpler claim that $p = .25$, so that the null and alternative hypotheses become

$$H_0 : p = .25 \quad \text{versus} \quad H_a : p > .25 \tag{3.15}$$

We shall see that if this new H_0 is rejected in favor of H_a, the original H_0 would also be rejected. In the sequel a null hypothesis will usually be phrased as an equality ($=$), while the alternative will specify one of three inequalities $>$, $<$, or $\neq$.

Decision Rules or Test Procedures

To decide between H_0 and H_a, we must obtain a sample from the relevant population. In Example 3.31 suppose that we select at random 20 obese individuals and use the new technique on each one. The random variable

X = the number among the 20 subjects who have lost at least 9 kg by the end of the treatment period

is then used to reach a decision. This is done by dividing the possible values of X into two sets. One set is called the **rejection region** or **critical region;** if the observed value of X is in the rejection region, then H_0 is rejected in favor of H_a. The other set consists of the complement of the rejection region, and if the observed X lies in this set then H_0 is not rejected. The r.v. X is called the **test statistic** because it is the function of the sample used to make a decision.

The rejection region consists of X values which provide strong evidence in support of H_a as opposed to H_0. In Example 3.31 a large value of X would be consistent with $p > .25$, so an appropriate rejection region would consist of large X values. One possible rejection region is

$$R_1 = \{8, 9, \ldots, 20\}$$

If we used this region and the observed value of X was $x = 11$, then H_0 would be rejected in favor of the claim that $p > .25$ and the behavior modification technique would be declared an improvement on past techniques. Before using R_1, we should justify the inclusion of $x = 8$ and noninclusion of smaller values.

The null hypothesis in (3.15) is of the general form $H_0 : p = p_0$; p_0 is called the **null value** of the parameter. The alternative specifies values of p which lie entirely to one side of (above) p_0, so H_a is called a **one-sided** hypothesis. Another one-sided hypothesis would be $H_a : p < .7$, while an example of a **two-sided** hypothesis is $H_a : p \neq .5$. The form of the rejection region will depend on whether H_a is one-sided above, one-sided below, or two-sided. In general a rejection region is **upper-tailed** if it consists only of large X values, **lower-tailed** if it consists only of small X values, and **two-tailed** if it contains both small and large X values. The rejection region R_1, as well as any other region appropriate for Example 3.31, is upper-tailed.

Choosing a Test Procedure: Errors in Hypothesis Testing

Should we use region R_1 or some other upper-tailed region to decide between H_0 and H_a in Example 3.31? To answer this, we must first recognize that use of any reasonable rejection region may lead to an erroneous conclusion. If R_1 is used and $p = .25$, it is possible that $X = 10$ will be observed, while if $p = .5$, we might still observe $X = 6$. There are two different errors which must be considered.

Definition: A **type I error** consists of rejecting the null hypothesis H_0 when it is true. A **type II error** arises if H_0 is not rejected when it is false.

Figure 3.8 illustrates the possibilities for Example 3.31.

		True State of Nature	
		H_0 True ($p = .25$)	H_0 False ($p > .25$)
Decision	Accept H_0	No error	Type II error
	Reject H_0	Type I error	No error

Figure 3.8 Errors in hypothesis testing

In our judicial system H_0 states that the defendant is innocent, so a type I error results when an innocent person is convicted and a type II error results when a guilty person is set free.

A test procedure will be good if both its type I and type II error probabilities are small, so that an error is unlikely to be made when the procedure is used. Consider the procedure specified by region R_1 and let P (type I error) be denoted by α. Since for R_1 "rejecting H_0" is equivalent to $X = 8, 9, \ldots,$ or 20, while "H_0 true" is equivalent to $X \sim \text{Bin}(20, .25)$,

$$\alpha = P(\text{rejecting } H_0 \text{ when } H_0 \text{ is true})$$

$$= P[X = 8, 9, \ldots, \text{ or } 20 \quad \text{when} \quad X \sim \text{Bin}(20, .25)]$$

$$= 1 - B(7; 20, .25) = 1 - .898 = .102$$

Similarly,

$$P(\text{type II error}) = P(\text{not rejecting } H_0 \text{ when } H_0 \text{ is false})$$

$$= P(X \le 7 \text{ when } X \sim \text{Bin}(20, p) \text{ with } p > .25)$$

Because $X \le 7$ is less likely when p is large than when p is small, the desired probability depends on which $p > .25$ is used. Let $\beta(p)$ denote the probability of a type II error for the alternative value p. Then

$$\beta(.5) = P(X \le 7 \text{ when } X \sim \text{Bin}(20, .5)) = B(7; 20, .5) = .132$$

For each different $p > .25$, a different $\beta(p)$ is obtained; $\beta(p)$ is tabulated below for selected values of p (using R_1):

p	.3	.4	.5	.6	.7	.8
$\beta(p)$	.772	.416	.132	.021	.001	.000

To interpret these probabilities, think of being faced with exactly the same testing problem (same H_0, H_a, and n) over and over and using R_1 each time. Then $\alpha = .102$ implies that when H_0 is true, we will end up incorrectly rejecting it approximately 10% of the time. Since $\beta(.5) = .132$, even when $p = .5$ so that the null hypothesis is false, roughly 13% of the time such a departure from H_0 will not be detected.

Searching for a Better Procedure

Although for R_1 the probability of a type II error is reasonably small when p is far from the null value .25, both α and $\beta(p)$ for p near .25 are rather large. Perhaps there is a better choice of rejection region. Consider the two regions

$$R_2 = \{9, 10, \ldots, 20\} \quad \text{and} \quad R_3 = \{10, 11, \ldots, 20\}$$

If either R_2 or R_3 is used in place of R_1, then more convincing evidence against H_0 is required in order to reject it. Intuitively it is less likely that H_0 will be rejected whether H_0 is true or false. For region R_3,

$$\alpha = P[\text{rejecting } H_0 \text{ when } H_0 \text{ is true})$$

$$= P[X \ge 10 \text{ when } X \sim \text{Bin}(20, .25)] = 1 - B(9; 20, .25) = .014$$

and

$$\beta(.5) = P[\text{not rejecting } H_0 \text{ when } X \sim \text{Bin}(20, .5)] = B(9; 20, .5) = .412$$

The following table contains summary information on R_1, R_2, and R_3.

	α	$\beta(p)$ for these values of p					
		.3	.4	.5	.6	.7	.8
R_1	.102	.772	.416	.132	.021	.001	.000
R_2	.041	.887	.596	.252	.057	.005	.000
R_3	.014	.952	.755	.412	.128	.017	.001

From the table we see that for these three regions, a smaller α can be achieved only at the expense of a larger $\beta(p)$ for each p!

Proposition: Let R be any fixed rejection region. Then any other rejection region which is contained in R will have a smaller value of α than R but a larger value of $\beta(p)$ for every alternative value of p.

This result follows because α is computed by adding probabilities of values in the rejection region, so a smaller region yields a smaller α. But $\beta(p)$ is computed by summing probabilities of values in the complement of R, so a smaller R corresponds to a larger R' and a larger $\beta(p)$.

The Level of Significance and Best Tests

The above proposition says that as long as the sample size n is fixed, there is no region R of the appropriate form which will simultaneously make both α and $\beta(p)$ small. In choosing a region, we must compromise between small α and small β.

Definition: A test procedure is said to have **level of significance** α, or to be a **level α test** if $P(\text{type I error}) \leq \alpha$.

For example, the test procedure which uses region R_2 is a level .05 test, but the test which uses region R_1 is not. The idea now is to specify the maximum type I error probability that can be tolerated, and then to consider using only tests which have this level of significance. If our maximum tolerable type I error probability in Example 3.31 is .05, then R_2 and R_3 both specify level .05 tests (.041 $\leq$ .05 and .014 $\leq$ .05), but R_1 does not (.102 $>$ .05). By considering only level α tests, we ensure that the type I error probability is controlled at a small value. Since a type I error often involves rejecting the status quo in favor of a new theory, a serious error, controlling the probability of this happening makes good sense, though it may result in a large β. The traditional levels of significance are .01, .05, and .1, but a smaller level of significance may be desirable if a type I error is very serious.

> **General principle for test selection:** Choose the level of significance α as the maximum tolerable type I error probability. Then among all level α tests, the **best level α test** is the one which minimizes $\beta(p)$ for every alternative value of p.

For level of significance .05, the rejection region R_2 specifies the best test in Example 3.31. Regions smaller than R_2 (but still upper-tailed) are also level .05 but have larger β, while larger regions are not level .05. Because the test statistic X has a discrete probability distribution, we cannot find a region with P (type I error) $= .05$, so we look for a region with error probability as close to .05 as possible.

If we are willing to take a larger sample (more expensive and time consuming), then both α and β can be decreased. For $n = 25$, the region $\{10, 11, \ldots, 25\}$ has $\alpha = .030$ and $\beta(p)$ uniformly smaller than the region R_2. This region specifies a best level .05 (upper-tailed) test for sample size 25.

Example 3.32 A department of mathematics has decided to change its calculus text for the following year and has narrowed the choice to two different texts, A and B. To select the winner, the faculty has decided on the unorthodox approach of obtaining student feedback. If a student consensus shows a strong preference for one or the other, that book will be chosen without any further consideration (so that next year's students will know who to blame). If no clear consensus develops, then the faculty members will make the final choice. To obtain information, 15 students who are finishing first-year calculus are selected at random, asked (and paid) to read material in several matched sections of each text. Then each student is asked for a preference. If 12 of the 15 students prefer A, does this indicate a degree of preference strong enough to make faculty action unnecessary?

Let

$p = P$ (a randomly selected student prefers A)

$X =$ number among the 15 students who prefer A

Then no preference is equivalent to $p = .5$, while $p \neq .5$ indicates a preference for one of the two books. The null and alternative hypotheses are

$$H_0 : p = .5 \quad \text{versus} \quad H_a : p \neq .5$$

The alternative is two-sided. Using X as the test statistic, either a small or a large observed value provides evidence against H_0, suggesting a two-tailed rejection region. For the region

$$R = \{0, 1, 2, 3, 12, 13, 14, 15\}$$

$$\alpha = P(X \leq 3 \text{ or } \geq 12 \text{ when } X \sim \text{Bin}(15, .5))$$

$$= B(3; 15, .5) + 1 - B(11; 15, .5) = .035$$

Adding 4 and 11 to the region yields $\alpha > .05$, so R specifies the best level .05 test. Since the observed value of X is $x = 12$, which is in R, H_0 is rejected at level of significance .05 and text A is chosen without any further ado.

Because $p_0 = .5$ in Example 3.32, the rejection set should contain x if and only if it contains $15 - x$ (0 and 15, 1 and 14, and so on), because both values provide the same amount of evidence for or against H_0. If the alternative is $H_a : p \neq p_0$ with p_0 something other than .5, the form of a good region may not be obvious. For example, if $p_0 = .3$ then $R = \{0, 9, 10, 11, \ldots, 15\}$ may be best. These issues will be discussed in Chapter 7.

The Effect of a More Realistic H_0

Regions R_1, R_2, and R_3 are all plausible for testing the original $H_0 : p \leq .25$ against $H_a : p > .25$, since only large X values make us believe strongly in H_a. Only the error probability calculations are affected by the more complicated H_0; instead of a single type I error probability, there will be a different one for each p, resulting in $\alpha(p)$ for every $p \leq .25$. However, the largest $\alpha(p)$ (maximum type I error probability) will be for $p = p_0 = .25$, the null value. *By controlling the type I error probability for the simpler H_0, we also control it for the more complicated H_0.* The value $p = p_0$ corresponds to the "worst case."

The Power of a Test

There is a way of combining $\alpha(p)$ and $\beta(p)$ into a single function, called the power function of a test.

> **Definition:** The **power function** of a test for hypotheses about p, denoted by $\pi(p)$, is given by
> $$\pi(p) = P(\text{rejecting } H_0 \text{ when the true proportion is } p), \quad 0 \leq p \leq 1.$$

For $H_0 : p \leq .25$ versus $H_a : p > .25$ we don't want to reject H_0 when p is $\leq .25$, so the power should be small for such values. For a value $p > .25$, we want large power, since then H_0 should be rejected. From the definitions of $\alpha(p)$ and $\beta(p)$, it follows that

$$\pi(p) = \begin{cases} \alpha(p) & \text{if } p \leq .25 \ (H_0 \text{ true}) \\ 1 - \beta(p) & \text{if } p > .25 \ (H_0 \text{ false}) \end{cases}$$

Therefore small β corresponds to large power for alternative values, which is what we want. The power function for the test with region R_2 is sketched in Figure 3.9.

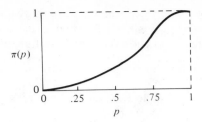

Figure 3.9 The graph of a power function

Exercises / Section 3.5

1. Let p = the probability that an ambulance rescue unit can respond to an emergency call in a community within 10 minutes. Based on $n = 15$ independent trials, it is desired to test $H_0 : p = .7$ versus $H_a : p < .7$. Let X = the number of responses among the 15 which are within 10 minutes, so $X \sim$ Bin $(15, p)$.
   a. Compute α for the rejection region $R = \{0, 1, \ldots, 6, 7\}$.
   b. Compute β for the region of (a) when $p = .5$ and when $p = .3$.
   c. If $x = 9$ and the region of (a) is used, what would you conclude? What type of error might you have made in reaching this conclusion? Explain.

2. To ascertain which of two mixtures of asphalt will result in less road damage under certain conditions, 20 road sections are chosen and each is paved partly with type-A and partly with type-B asphalt. After a period of time each section is examined and the asphalt type resulting in less damage is determined. Let X = the number among the 20 for which type A was better. With $p = P$ (type A is better), $X \sim$ Bin $(20, p)$. We wish to test $H_0 : p = .5$ versus $H_a : p \neq .5$.
   a. For the rejection region $R = \{0, 1, \ldots, 5, 15, 16, \ldots, 20\}$, if $x = 16$ what would you conclude? What type of error might you have made in reaching this conclusion?
   b. Compute α for the region of (a).
   c. Compute β for the region of (a) when $p = .3$ and when $p = .7$.
   d. Without doing any computations, explain how you would change R to obtain a test procedure with a smaller value of α (and say why this is

the case). How would values of β compare for the two regions?

3. (Based on an article in the May 1978 issue of *Consumer Reports*.) To decide whether tennis enthusiasts strongly prefer gut strings to nylon strings in their tennis rackets, each of 20 randomly selected enthusiasts is given two tennis rackets which are identical except in that one is strung with nylon and the other with gut. After several weeks of playing alternately with each racket, each of the enthusiasts will be asked to state a preference for one of the two rackets. Let p = the true proportion of enthusiasts who prefer gut strings, and X = the number of enthusiasts in the sample who prefer gut. Because gut is more expensive and breaks sooner than nylon, the null hypothesis is that at most 50% of all enthusiasts prefer gut; we simplify this to $H_0 : p = .5$, planning to reject H_0 only if experimental evidence strongly favors gut strings.
   a. Which of the rejection regions $\{15, 16, 17, 18, 19, 20\}$, $\{0, 1, 2, 3, 4, 5\}$, or $\{0, 1, 2, 3, 17, 18, 19, 20\}$ is most appropriate, and why are the other two not appropriate?
   b. What is the probability of a type I error for the chosen region of (a)? Does the region specify a level .05 test? It is the best level .05 test?
   c. If 60% of all enthusiasts prefer gut, calculate the probability of a type II error using the appropriate region from (a). Repeat if 80% of all enthusiasts prefer gut.
   d. Calculate the power of the test for each alternative in (c), and sketch the power function of the test.
   e. If 13 out of the 20 players prefer gut, should

H_0 be rejected using a .10 level of significance?

4. An ordinance requiring that a smoke detector be installed in all previously constructed houses has been in effect in a particular city for one year. The fire department is concerned that many houses remain without detectors. Let p = the true proportion of such houses having detectors, and suppose that a random sample of 25 homes is inspected. If the sample strongly indicates that fewer than 80% of all houses have a detector, the fire department will campaign for a mandatory inspection program. gram. Because of the costliness of the program, the department prefers not to call for such inspections unless sample evidence strongly argues for their necessity. Let the null hypothesis be $H_0 : p = .8$ (actually $p \geq .8$) and X = the number of houses in the sample having a detector.

a. Calculate the probability of a type I error for the rejection region $\{0, 1, 2, \ldots, 14, 15\}$. Does this region prescribe a best level .01 test? If not, what region does?

b. Calculate the probability of a type II error when $p = .7$ using the region $\{0, 1, \ldots, 14, 15\}$. Repeat for $p = .5$ and $p = .3$.

c. If the simple H_0 is replaced by the more realistic $H_0 : p \geq .8$, calculate the probability of a type I error when $p = .9$ and when $p = .95$ using the region $\{0, 1, \ldots, 14, 15\}$.

d. Use the calculations of parts (b) and (c) to sketch the power function of the test for $H_0 : p \geq .8$ versus $H_a : p < .8$.

e. If 15 of the 25 houses inspected have detectors and the desired level of significance is .01, should the department campaign for mandatory inspections? What type of error might have been made?

5. A manufacturer of plumbing fixtures has developed a new type of washerless faucet. Let $p = P$ (a randomly selected faucet of this type will develop a leak within two years under normal use). The manufacturer has decided to proceed with production unless it can be determined that p is too large; the borderline acceptable value of p is specified as .10. The manufacturer decides to subject n of these faucets to accelerated testing (approximating two years of normal use). With X = the number among the n faucets which leak before the test concludes, production will commence unless the observed X is too large. It is de-

cided that if $p = .10$, the probability of not proceeding should be at most .10, while if $p = .30$ the probability of proceeding should be at most .10. Can $n = 10$ be used? $n = 20$? $n = 25$? What is the appropriate rejection region for the chosen n, and what are the actual error probabilities when this region is used?

6. A company which sells classical music recordings by mail is trying to decide which of two new recordings of Beethoven's Ninth Symphony to add to its catalogue. If both recordings are equally appealing to subscribers, then both should be offered, while if one is clearly preferred to the other, then only the preferred one should be offered. The relevant hypotheses are $H_0 : p = .5$ versus $H_a : p \neq .5$, where p = the proportion of subscribers who prefer recording A to B. Suppose that 10 subscribers are randomly selected, and each is asked to listen to both recordings and express a preference. Let X = the number who prefer recording A.

a. Compute α for the rejection region $\{0, 1, 2, 8, 9, 10\}$. Is this a level .10 test? What region specifies the best level .10 test?

b. Compute β if $p = .4$ using the best level .10 test. Repeat for $p = .6$ and $p = .8$.

c. If $X = 9$ is observed and a level .10 test is desired, should both recordings be offered? What type of error might have been made?

7. The April 1978 issue of *Consumer Reports* reported on an experiment to discover whether there was a difference between domestic Lowenbrau ($2.50 per six-pack) and Miller High Life ($1.80) which would be discernible to the average beer drinker. Their sample size was 24, but here we suppose that $n = 25$ individuals were gathered together. Each individual was given three glasses of beer to taste, two from the same bottle, and asked to identify which contained the different beer. If 12 of the 25 correctly identified the different beer, would you conclude that there was a discernible difference (the actual result was 11/24)? *Hint:* The parameter of interest is $p = P$ (the different beer is correctly identified). First state H_0 and H_a, then specify a reasonable level of signficance and find the best rejection region.

8. After a period of apprenticeship, an organization gives an exam which must be passed to be eligible for membership. Let $p = P$ (randomly chosen apprentice passes). The organization wishes an exam which most but not all should be able to pass, so decides that $p = .90$ is desirable. For a particular exam the relevant hypotheses are $H_0 : p = .90$ versus $H_a : p \neq .90$. Suppose that 10 people take the exam, and let $X =$ the number who pass.

a. Does the lower-tailed region $\{0, 1, \ldots, 5\}$

specify a level .01 test?

b. Show that even though H_a is two-sided, no two-tailed test is a level .01 test.

c. Sketch the graph of the power function of the test in (a). Sketch the graph of the desired power function for this situation.

3.6 The Hypergeometric and Negative Binomial Distributions

The hypergeometric and negative binomial distributions are both closely related to the binomial distribution. Whereas the binomial distribution is the approximate probability model for sampling without replacement from a finite dichotomous (S–F) population, the hypergeometric distribution is the exact probability model for the number of S's in the sample. The binomial random variable X is the number of S's when the number of trials n is fixed, whereas the negative binomial distribution arises from fixing the number of S's and letting the number of trials be random.

The Hypergeometric Distribution

The assumptions leading to the hypergeometric distribution are:

1. The population or set to be sampled consists of N individuals, objects, or elements (a *finite* population).
2. Each individual can be characterized as a success (S) or a failure (F), and there are M successes in the population.
3. A sample of n individuals is drawn in such a way that each subset of size n is equally likely to be chosen.

The random variable of interest is $X =$ the number of S's in the sample. The probability distribution of X depends on the parameters n, M, and N, so we wish to obtain $P(X = x) = h(x; n, M, N)$.

Example 3.33   An undergraduate library has 20 copies of a certain introductory economics text, of which eight are first printings and 12 are second printings (containing corrections of some minor errors which appeared in the first printing). The course instructor has requested that five copies be put on two-hour reserve. If the copies are selected in a completely random fashion, so that every subset of size five has the same probability of being selected, what is the probability that x ($x = 0, 1, 2, 3, 4,$ or 5) of those selected are second printings?

In this example the population size is $N = 20$, the sample size is $n = 5$, and the number of S's (second printing = S) and F's in the population are $M = 12$ and $N - M = 8$, respectively. Consider the value $x = 2$. Because all outcomes (each one consisting of five particular books) are equally likely,

$$P(X = 2) = h(2; 5, 12, 20) = \frac{\text{number of outcomes having } X = 2}{\text{number of possible outcomes}}$$

The number of possible outcomes in the experiment is the number of ways of selecting five from the 20 objects without regard to order—that is, $\binom{20}{5}$. To count the number of outcomes having $X = 2$, note that there are $\binom{12}{2}$ ways of selecting two of the second printings, and for each such way there are $\binom{8}{3}$ ways of selecting the three first printings to fill out the sample. This gives, using the product rule for counting, $\binom{12}{2}\binom{8}{3}$ outcomes with $X = 2$, so that

$$h(2; 5, 12, 20) = \frac{\binom{12}{2}\binom{8}{3}}{\binom{20}{5}} = \frac{77}{323} = .238$$

Proceeding in the same way in the general case gives the following result.

> **Proposition:** If X is the number of S's in a completely random sample of size n drawn from a population consisting of M S's and $N - M$ F's, then the probability distribution of X, called the **hypergeometric distribution,** is given by
>
> $$P(X = x) = h(x; n, M, N) = \frac{\binom{M}{x}\binom{N - M}{n - x}}{\binom{N}{n}} \tag{3.16}$$
>
> $$x = 0, 1, \ldots, \min(n, M)$$

For the problem of Example 3.33 $n = 5$, $M = 12$, and $N = 20$, so $h(x; 5, 12, 20)$ for $x = 0, 1, 2, 3, 4, 5$ can be obtained by substituting these numbers into (3.16).

Example 3.34 Five individuals from an animal population thought to be near extinction in a certain region have been caught, tagged, and released to mix into the population. After they have had an opportunity to mix in, a random sample of 10 of these animals is selected. Let $X =$ the number of tagged animals in the second sample. If there are actually 25 animals of this type in the region, what is the probability that (a) $X = 2$? (b) $X \leq 2$?

The parameter values are $n = 10$, $M = 5$ (five tagged animals in the population), and $N = 25$, so

$$h(x; 10, 5, 25) = \frac{\binom{5}{x}\binom{20}{5-x}}{\binom{25}{5}} \qquad x = 0, 1, 2, 3, 4, 5$$

For (a), $P(X = 2) = h(2; 10, 5, 25) = \dfrac{\binom{5}{2}\binom{20}{3}}{\binom{25}{5}} = .292$

For (b), $P(X \le 2) = P(X = 0, 1, \text{ or } 2) = \displaystyle\sum_{x=0}^{2} h(x; 10, 5, 25)$

$$= .215 + .456 + .292 = .963$$

Comprehensive tables of the hypergeometric distribution are available, but because the distribution has three parameters, these tables require much more space than tables for the binomial distribution.

The Mean and Variance of X

Proposition: The mean and variance of the hypergeometric random variable X having p.m.f. $h(x; n, M, N)$ are

$$E(X) = n \cdot \frac{M}{N} \quad \text{and} \quad \text{Var}(X) = \left(\frac{N-n}{N-1}\right) \cdot n \cdot \frac{M}{N} \cdot \left(1 - \frac{M}{N}\right)$$

The derivations of $E(X)$ and $\text{Var}(X)$ parallel those for the binomial r.v., but the algebra is somewhat messy. The ratio M/N is the proportion of S's in the population. If we replace M/N by p in $E(X)$ and $\text{Var}(X)$, we get

$$E(X) = np$$

$$\text{Var}(X) = \left(\frac{N-n}{N-1}\right) \cdot np(1-p) \tag{3.17}$$

Expression (3.17) shows that the means of the binomial and hypergeometric random variables are equal, while the variances of the two r.v.'s differ by the factor $(N-n)/(N-1)$, often called the **finite population correction factor.** This factor is less than one, so the hypergeometric variable has smaller variance than does the binomial r.v. The correction factor can be written $(1 - n/N)/(1 - 1/N)$, which is approximately 1 when n is small relative to N.

Example 3.34
(continued)

In the animal-tagging example, $n = 10$, $M = 5$, and $N = 25$, so $p = \frac{5}{25} = .2$ and

$$E(X) = 10(.2) = 2$$

$$\text{Var}(X) = \frac{15}{24}(10)(.2)(.8) = (.625)(1.6) = 1$$

If the sampling was carried out with replacement, $\text{Var}(X) = 1.6$. Suppose that the population size N is not actually known, so the value x is observed and we wish to estimate N. It is reasonable to equate the observed sample proportion of S's x/n with the population proportion M/N, giving the estimate

$$\hat{N} = \frac{M \cdot n}{x}$$

If $M = 100$, $n = 40$, and $x = 16$, then $\hat{N} = 250$.

Approximating Hypergeometric Probabilities

Our general rule of thumb in Section 3.4 stated that if sampling was without replacement but n/N was at most .05, then the binomial distribution could be used to compute approximate probabilities involving the number of S's in the sample. A more precise statement is as follows: Let the population size N and number of population S's M get large with the ratio M/N remaining fixed at p. Then $h(x; n, M, N)$ approaches $b(x; n, p)$, so for n/N small, the two are approximately equal provided that p is not too near either 0 or 1. This is the rationale for our rule of thumb.

The Negative Binomial Distribution

The negative binomial random variable and distribution are based on an experiment satisfying the following conditions:

1. The experiment consists of a sequence of independent trials.
2. Each trial can result in either a success (*S*) or a failure (*F*).
3. The probability of success is constant from trial to trial, so $P(S$ on trial $i) = p$ for $i = 1, 2, 3 \ldots$.
4. The experiment continues (trials are performed) until a total of r successes have been observed, where r is a specified positive integer.

The random variable of interest is $X =$ the number of failures which precede the rth success; X is called a **negative binomial** random variable because, in contrast to the binomial r.v., the number of successes is fixed and the number of trials is random.

Possible values of X are $0, 1, 2, \ldots$. Let $nb(x; r, p)$ denote the p.m.f. of X. The event $\{X = x\}$ is equivalent to $\{r - 1\ S$'s in the first $x + r - 1$ trials and an S on the $(x + r)$th trial$\}$ (if, for example, $r = 5$ and $x = 10$, then there must be four S's in the first 14 trials and the trial 15 must be a S). Since trials are independent,

$$nb(x; r, p) = P(X = x)$$

$$= P(r - 1 \text{ S's on the first } x + r - 1 \text{ trials}) \cdot P(S) \qquad (3.18)$$

The first probability on the far right of (3.18) is the binomial probability

$$\binom{x + r - 1}{r - 1} p^{r-1} (1 - p)^x, \text{ while } P(S) = p.$$

Proposition: The p.m.f. of the negative binomial r.v. X with parameters $r = $ number of S's and $p = P(S)$ is

$$nb(x; r, p) = \binom{x + r - 1}{r - 1} p^r (1 - p)^x \quad x = 0, 1, 2, \dots$$

Example 3.35   A pediatrician wishes to recruit five couples, each of whom is expecting their first child, to participate in a new natural childbirth regimen. Let $p = P$ (a randomly selected couple agrees to participate). If $p = .2$, what is the probability that 15 couples must be asked before five are found who agree to participate? That is, with $S = \{$agrees to participate$\}$, what is the probability that 10 F's occur before the fifth S? Substituting $r = 5$, $p = .2$, and $x = 10$ into $nb(x; r, p)$ gives

$$nb(10; 5, .2) = \binom{14}{4}(.2)^5(.8)^{10} = .034$$

The probability that at most 10 F's are observed (at most 15 couples are asked) is

$$P(X \le 10) = \sum_{x=0}^{10} nb(x; 5, .2) = (.2)^5 \sum_{x=0}^{10} \binom{x + 4}{4}(.8)^x = .164$$

In some sources the negative binomial r.v. is taken to be the number of trials $X + r$ rather than the number of failures.

In the special case $r = 1$, the p.m.f. is

$$nb(x; 1, p) = (1 - p)^x p \quad x = 0, 1, 2, \dots \qquad (3.19)$$

In Example 3.9 we derived the p.m.f. for the number of trials necessary to obtain the first S, and the p.m.f. there is quite similar to (3.19). Both $X = $ number of F's and $Y = $ number of trials ($= 1 + X$) are referred to in the literature as **geometric random variables**, and the p.m.f. (3.19) is called the **geometric distribution.**

The Mean and Variance of a Negative Binomial Variable

In Example 3.16 the expected number of trials until the first S was shown to be $1/p$, so that the expected number of F's until the first S is $(1/p) - 1 = (1 - p)/p$. Intuitively, we would expect to see $r \cdot (1 - p)/p$ F's before the rth S. The variance of X is not so easy to motivate or derive, so we state the result without proof.

> **Proposition:** If X is a negative binomial random variable with p.m.f. $nb(x; r, p)$, then
>
> $$E(X) = \frac{r(1-p)}{p} \quad \text{and} \quad \text{Var}(X) = \frac{r(1-p)}{p^2}$$

Generalizing the Negative Binomial Distribution

The coefficient $\binom{x+r-1}{r-1}$ in $nb(x; r, p)$ can be expressed in a different form:

$$\binom{x+r-1}{r-1} = \frac{(x+r-1)!}{(r-1)!\,x!} = \frac{(x+r-1)(x+r-2)\ldots(x+r-x)}{x!}$$

$$= k(r, x) \tag{3.20}$$

The quantity $k(r, x)$ in (3.20) now makes sense even when r is not an integer. For example, if $r = \frac{1}{3}$ and $x = 4$,

$$k(4, \tfrac{1}{3}) = \frac{(4+\frac{1}{3}-1)(4+\frac{1}{3}-2)(4+\frac{1}{3}-3)(4+\frac{1}{3}-4)}{4!}$$

$$= \frac{\frac{10}{3} \cdot \frac{7}{3} \cdot \frac{4}{3} \cdot \frac{1}{3}}{24} = \frac{35}{243}$$

For any positive number, define a function $nb(x; r, p)$ by

$$nb(x; r, p) = k(r, x) \cdot p^r (1-p)^x \quad x = 0, 1, 2, \ldots \tag{3.21}$$

When r is not an integer, it is not possible to give an intuitive derivation of (3.21) as a p.m.f., but $nb(x; r, p) > 0$ and the sum over $x = 0, 1, \ldots$ can be shown to equal 1, so this function does specify a p.m.f. which is also referred to as the negative binomial distribution. It has been found to fit observed data quite well in biological, economic, medical, and psychological applications. For example, Bliss and Fisher (*Biometrics*, 1953) reported data on the number of Salicornia stricta plants growing in a salt marsh in each of 98 different quadrats (rectangular sampling regions). The negative binomial distribution with $r = 2.69$ and $p = .28$ yields probabilities $nb(0; 2.69, .28)$, $nb(1; 2.69, .28)$, . . . which are quite close to the observed proportion of quadrats containing 0, 1, . . . plants. In the 1960 *Journal of Geology*, Griffiths obtained good agreement between observed zircon counts in different rock samples and $nb(x; .239, .186)$. In Chapter 14 we will present methods for determining whether a particular probability distribution or type of distribution provides a plausible model for such count data.

Exercises / Section 3.6

1. A shipment of 15 concrete cylinders has been received by a contractor, five for a small project and the other ten for a larger project. Suppose that six of the 15 have a crushing strength which is below

the specified minimum. If the five for the smaller project are randomly selected from the 15 and X = the number among the five which have a below minimum crushing strength, then X has a hypergeometric distribution with parameters $n = 5$, $M = 6$, and $N = 15$.
   a. Compute $P(X = 2)$
   b. Compute $P(X \leq 2)$
   c. Compute $P(X \geq 1)$
   d. Compute $E(X)$ and Var(X).

2. Each of 12 refrigerators of a certain type has been returned to a distributor because of the presence of a high-pitched oscillating noise when the refrigerator is running. Suppose that four of these 12 have defective compressors and the other eight have less serious problems. If they are examined in random order, let X = the number among the first six examined which have a defective compressor.
   a. Compute $P(X = 1)$
   b. Compute $P(X \geq 4)$
   c. Compute $P(1 \leq X \leq 3)$

3. Professor Samuel P. Ling, a new faculty member in the Statistics Department at Hypergeometric State University, is to be assigned five student advisees. The department has 15 advisees to assign, of whom 10 are honor students, and will assign them in such a way that any five of the 15 are equally likely to be assigned to Professor Ling. Let X = the number of honor students assigned to Professor Ling. Write the probability mass function of X.
   a. What is the probability that all five of Professor Ling's advisees are honor students?
   b. What is the probability that exactly three of the five are honor students?
   c. Compute $E(X)$ and Var(X).

4. A tennis instructor has 12 cans of Wilson tennis balls and 15 cans of Penn tennis balls. If there are 20 students in the class and after pairing off, each of the 10 pairs is given a can chosen at random from those available, write down the p.m.f. of the random variable X = the number of cans of Wilson balls distributed to the class. What is the probability that five cans of each type of ball are used by the class?

5. A geologist has collected 10 specimens of basaltic rock and 10 specimens of granite. If the geologist instructs a laboratory assistant to randomly select

15 of the specimens for analysis, what is the p.m.f. of the number of basalt specimens selected for analysis? What is the probability that all specimens of one of the two types of rock are selected for analysis?

6. A personnel director interviewing 11 senior engineers for four job openings has scheduled six interviews for the first day and five for the second day of interviewing. Assume that the candidates are interviewed in random order.
   a. What is the probability that x of the top four candidates are interviewed on the first day?
   b. How many of the top four candidates can be expected to be interviewed on the first day?

7. Twenty pairs of individuals playing in a bridge tournament have been seeded 1, . . ., 20. In the first part of the tournament, the 20 are randomly divided into 10 east–west pairs and 10 north–south pairs.
   a. What is the probability that x of the top 10 pairs end up playing east–west?
   b. What is the probability that all of the top five pairs end up playing the same direction?
   c. If there are $2n$ pairs, what is the p.m.f. of X = the number among the top n pairs who end up playing east–west, and what are $E(X)$ and Var(X)?

8. A second-stage smog alert has been called in a certain area of Los Angeles County in which there are 50 industrial firms. An inspector will visit 10 randomly selected firms to check for violations of regulations.
   a. If 15 of the firms are actually violating at least one regulation, what is the p.m.f. of the number of firms visited by the inspector which are in violation of at least one regulation?
   b. If there are 500 firms in the area, of which 150 are in violation, approximate the p.m.f. of (a) by a simpler p.m.f.
   c. For X = the number among the 10 visited which are in violation, compute $E(X)$ and Var(X) both for the exact p.m.f. and the approximating p.m.f. in (b).

9. Suppose that $p = P$ (male birth) = .5. A couple wishes to have exactly two female children in their family. They will have children until this condition is fulfilled.
   a. What is the probability that the family has x male children?

b. What is the probability that the family has four children?

c. What is the probability that the family has at most four children?

d. How many male children would you expect this family to have? How many children would you expect this family to have?

10. A family decides to have children until it has three children of the same sex. Assuming $P(B) = P(G) = .5$, what is the p.m.f. of $X =$ the number of children in the family?

11. Three brothers and their wives decide to have children until each family has two female children. What is the p.m.f. of $X =$ the total number of male children born to the brothers? What is $E(X)$, and how does it compare to the expected number of male children born to each brother?

12. Individual A has a red die and B has a green die (both fair). If they each roll until they obtain five "doubles" (1-1, . . ., 6-6), what is the p.m.f. of $X =$ the total number of times a die is rolled? What are $E(X)$ and $Var(X)$?

3.7 The Poisson Probability Distribution

The binomial, hypergeometric, and negative binomial distributions were all derived by starting with an experiment consisting of trials or draws and applying the laws of probability to various outcomes of the experiment. There is no simple experiment on which the Poisson distribution is based, though we will shortly describe how it can be obtained by certain limiting operations.

Definition: A random variable X is said to have a **Poisson distribution** if the probability mass function of X is

$$p(x; \lambda) = \frac{e^{-\lambda} \lambda^x}{x!} \quad x = 0, 1, 2, \ldots$$

for some $\lambda > 0$.

The value of λ is frequently a rate per unit time or per unit area. The letter e in $p(x; \lambda)$ represents the base of the natural logarithm system, and its numerical value is approximately 2.71828. Because λ must be positive, $p(x; \lambda) > 0$ for all possible x values. The fact that $\sum_{x=0}^{\infty} p(x; \lambda) = 1$ is a consequence of the Maclaurin infinite series expansion of e^λ, which appears in most calculus texts:

$$e^\lambda = 1 + \lambda + \frac{\lambda^2}{2!} + \frac{\lambda^3}{3!} + \cdots = \sum_{x=0}^{\infty} \frac{\lambda^x}{x!} \tag{3.22}$$

If the two extreme terms in (3.22) are multiplied by $e^{-\lambda}$, and then $e^{-\lambda}$ is placed inside the summation, the result is

$$1 = \sum_{x=0}^{\infty} e^{-\lambda} \frac{\lambda^x}{x!}$$

which shows that $p(x; \lambda)$ fulfills the second condition necessary for specifying a p.m.f.

Example 3.36 Let X = the number of trees in a certain forest found to be suffering from a rare disease, and suppose that X has a Poisson distribution with $\lambda = 5$. Then the probability that exactly two trees have the disease is

$$P(X = 2) = \frac{e^{-5}(5)^2}{2!} = .084$$

while the probability that at most two suffer from the disease is

$$P(X \leq 2) = \sum_{x=0}^{2} \frac{e^{-5}(5)^x}{x!} = e^{-5}\left(1 + 5 + \frac{25}{2}\right) = .125$$

The Poisson Distribution as a Limit

The rationale for using the Poisson distribution in many situations similar to that in Example 3.36 is provided by the following proposition.

> **Proposition:** Suppose that in the binomial p.m.f. $b(x; n, p)$, we let $n \to \infty$ and $p \to 0$ in such a way that np remains fixed at value $\lambda > 0$. Then $b(x; n, p) \to p(x; \lambda)$.

According to this proposition, *in any binomial experiment in which n is large and p is small, $b(x; n, p) \doteq p(x; \lambda)$ where λ = np*. As a rule of thumb, this approximation can safely be applied if $n \geq 100$, $p \leq .01$, and $np \leq 20$.

Proof:

$$b(x; n, p) = \frac{n!}{x!\,(n-x)!}p^x(1-p)^{n-x} = \frac{n!}{x!\,(n-x)!}\left(\frac{\lambda}{n}\right)^x\left(1-\frac{\lambda}{n}\right)^{n-x}$$

$$= \frac{n(n-1)\cdots(n-x+1)}{n^x}\cdot\left(1-\frac{\lambda}{n}\right)^{-x}\cdot\left(1-\frac{\lambda}{n}\right)^{n}\cdot\frac{\lambda^x}{x!}$$

$$= \left[1\left(1-\frac{1}{n}\right)\cdots\left(1-\frac{x-1}{n}\right)\left(1-\frac{\lambda}{n}\right)^{-x}\right]$$

$$\cdot\left(1-\frac{\lambda}{n}\right)^{n}\cdot\frac{\lambda^x}{x!} \tag{3.23}$$

Each of the terms inside the square brackets in (3.23) approaches 1 as $n \to \infty$. Using the result that as $n \to \infty$, $(1 + c/n)^n \to e^c$, which is frequently presented in introductory calculus, the last expression in (3.23) approaches $e^{-\lambda}\cdot\lambda^x/x!$ as $n \to \infty$.

Example 3.37   If a publisher of nontechnical books takes great pains to ensure that its books are free of typographical errors, so that the probability of any given page containing at least one such error is .005 and errors are independent from page to page, what is the probability that one of its 400-page novels will contain exactly one page with errors? At most three pages with errors?

With S denoting a page containing at least one error and F an error-free page, the number of pages X containing at least one error is a binomial r.v. with $n = 400$ and $p = .005$, so $np = 2$. We wish

$$P(X = 1) = b(1; 400, .05) \doteq p(1; 2) = \frac{e^{-2}(2)^1}{1!} = .271$$

Similarly,

$$P(X \le 3) \doteq \sum_{x=0}^{3} p(x; 2) = \sum_{x=0}^{3} e^{-2} \frac{2^x}{x!} = .135 + .271 + .271 + .180 = .857$$

Appendix Table A.2 exhibits the cumulative distribution function $F(x; \lambda)$ for $\lambda = .1, .2, \ldots, 1, 2, \ldots, 10, 15,$ and 20. For example, if $\lambda = 2$, then $P(X \le 3) = F(3; 2) = .857$ as in Example 3.37, while $P(X = 3) = F(3; 2) - F(2; 2) = .180$.

The Mean and Variance of X

Since $b(x; n, p) \to p(x; \lambda)$ as $n \to \infty$, $p \to 0$, $np = \lambda$, the mean and variance of a binomial variable should approach those of a Poisson variable. These limits are $np = \lambda$ and $np(1 - p) \to \lambda$.

> **Proposition:** If X has a Poisson distribution with parameter λ, then $E(X) = \text{Var}(X) = \lambda$.

These results can also be derived directly from the definitions of mean and variance (see Exercise 3.7.8).

Example 3.36 (continued)   Both the expected number of diseased trees and the variance of the number of diseased trees equal five.

The Poisson Process

Another important application of the Poisson distribution arises in connection with the occurrence of events of a particular type over time. As an example, suppose that starting from a time point which we label $t = 0$, we are interested in counting the number of radioactive pulses recorded by a Geiger counter. We make the following assumptions about the way in which pulses occur:

1. There exists a parameter $\lambda > 0$ such that for any short time interval of length Δt, the probability that exactly one pulse is received is $\lambda \cdot \Delta t + o(\Delta t)$.*

2. The probability of more than one pulse being received during Δt is $o(\Delta t)$ [which, along with (1), implies that the probability of no pulses during Δt is $1 - \lambda \cdot \Delta t - o(\Delta t)$].

3. The number of pulses received during the time interval Δt is independent of the number received prior to this time interval.

Informally, Assumption 1 says that for a short interval of time, the probability of receiving a single pulse is approximately proportional to the length of the time interval, where λ is the constant of proportionality. Now let $P_k(t)$ denote the probability that k pulses have been received by the counter during any particular time interval of length t.

Proposition: $P_k(t) = e^{-\lambda t} \cdot (\lambda t)^k / k!$, so that the number of pulses during a time interval of length t is a Poisson random variable with parameter λt. The expected number of pulses during any such time interval is then λt, so the expected number during a unit interval of time is λ.

Example 3.38   Suppose that pulses arrive at the counter at an average rate of six per minute, so that $\lambda = 6$. To find the probability that in a 1/2-minute interval at least one pulse is received, note that the number of pulses in such an interval has a Poisson distribution with parameter $\lambda t = 6(.5) = 3$ (.5 min is used because λ is expressed as a rate per minute). Then with $X =$ the number of pulses received in the 30-sec interval,

$$P(1 \leq X) = 1 - P(X = 0) = 1 - \frac{e^{-3}(3)^0}{0!} = .950$$

If in (1), (2), and (3) we replace "pulse" by "event," then the number of events occurring during a fixed time interval of length t has a Poisson distribution with parameter λt. Any process which has this distribution is called a **Poisson process,** and λ is called the rate of the process. Other examples of situations giving rise to a Poisson process include monitoring the status of a computer system over time, with breakdowns constituting the events of interest, recording the number of accidents in an industrial facility over time, answering calls at a telephone switchboard, and observing the number of cosmic-ray showers from a particular observatory over time.

---

* A quantity is $o(\Delta t)$ (read "little o of delta t") if as Δt approaches 0, so does $o(\Delta t)/\Delta t$. That is, $o(\Delta t)$ is even more negligible than Δt itself. The quantity $(\Delta t)^2$ has this property, but $\sin(\Delta t)$ does not.

Instead of observing events over time, consider observing events of some type which occur in a two- or three-dimensional region. For example, we might select on a map a certain region R of a forest, go to that region and count the number of trees. Each tree would represent an event occurring at a particular point in space. Under assumptions similar to (1)–(3), it can be shown that the number of events occurring in a region R has a Poisson distribution with parameter $\lambda \cdot a(R)$, where $a(R)$ is the area of R. The quantity λ is the expected number of events per unit area or volume.

Exercises / Section 3.7

1. Let X have a Poisson distribution with parameter $\lambda = 5$, and use Appendix Table A.2 to compute the following probabilities.
   a. $P(X \leq 8)$ b. $P(X = 8)$ c. $P(9 \leq X)$
   d. $P(5 \leq X \leq 8)$ e. $P(5 < X < 8)$

2. Suppose that the number of tornadoes X observed in a particular region during a one-year period has a Poisson distribution with $\lambda = 8$.
   a. Compute $P(X \leq 5)$
   b. Compute $P(6 \leq X \leq 9)$
   c. Compute $P(10 \leq X)$
   d. How many tornadoes can be expected to be observed during the one-year period, and what is the standard deviation of the number of observed tornadoes?

3. Let X = the number of automobiles of a particular year and model which will at some time in the future suffer a catastrophic failure in the steering mechanism, causing complete loss of control at high speed. Suppose that X has a Poisson distribution with parameter $\lambda = 10$.
   a. What is the probability that at most 10 cars suffer such a failure?
   b. What is the probability that between 10 and 15 (inclusive) cars will suffer such a failure?
   c. What are $E(X)$ and σ_X?

4. A library employee shelves 1000 books on a particular day. If the probability that any particular book is misshelved is .001 and books are shelved independently of one another, what is the approximate probability distribution for the number of books misshelved on that day? Use this distribution to calculate the probability that
   a. exactly three books are misshelved on that day
   b. at least one book is misshelved on that day

5. The number of parking tickets issued per hour by a meter reader for parking meter violations has a

Poisson distribution with parameter $\lambda = 5$ (so that tickets are given out at the rate of five per hour).
   a. What is the probability that exactly five tickets are given out during a particular hour?
   b. What is the probability that at least three tickets are given out during a particular hour?
   c. How many tickets do you expect to be given during a 45-min period?

6. The number of requests for assistance received by a towing service is a Poisson process with rate $\lambda = 4$ per hour.
   a. Compute the probability that exactly 10 requests are received during a particular two-hour period.
   b. If the operators of the towing service take a 30-minute break for lunch, what is the probability that they do not miss any calls for assistance?
   c. How many calls would you expect during their break?

7. Suppose that the performance of the librarian (Exercise 4, above) in shelving books is independent on any one day of what happens on any other day. What is the probability that the librarian misshelves exactly one book during a five-day work week?

8. Let X have a Poisson distribution with parameter λ.
   a. Compute $E(X)$ directly from the definition of expected value. *Hint:* The first term in the sum equals zero, and then x can be canceled. Now factor out λ, and show that what is left sums to one.
   b. Compute $E[X(X - 1)]$, then $E(X^2)$, and finally $\text{Var}(X)$.

9. Suppose that the number of trees in any region of a forest has a Poisson distribution with $\lambda =$ the

expected number of trees per acre = 80.

a. What is the probability that in a certain one-quarter-acre plot, there will be at most 16 trees?

b. If the forest covers 85,000 acres, what is the expected number of trees in the forest?

c. Suppose that you select a point in the forest and construct a circle of radius .1 miles. Let X = the number of trees within that circular region. What is the p.m.f. of X? *Hint:* 1 sq. mile = 640 acres.

10. Automobiles arrive at a vehicle equipment inspection station according to a Poisson process with rate $\lambda = 10$ per hour. Suppose that with probability .5, an arriving vehicle will have no equipment violations.

a. What is the probability that exactly 10 arrive during the hour and all 10 have no violations?

b. For any fixed $y \geq 10$, what is the probability that y arrive during the hour, of which 10 have no violations?

c. What is the probability that 10 "no violation" cars arrive during the next hour? *Hint:* Sum the probabilities in (b) from $y = 10$ to ∞.

11. a. In a Poisson process, what has to happen in both the time interval $(0, t)$ and the interval $(t, t + \Delta t)$ in order that there be no events occurring in the entire interval $(0, t + \Delta t)$? Use this and assumptions (1)–(3) to write a relationship between $P_0(t + \Delta t)$ and $P_0(t)$.

b. Use the result of (a) to write an expression for $P_0(t + \Delta t) - P_0(t)$. Then divide by Δt and let $\Delta t \to 0$ to obtain an equation involving $d/dt\, P_0(t)$, the derivative of $P_0(t)$ with respect to t.

c. Verify that $P_0(t) = e^{-\lambda t}$ satisfies the equation of (b).

d. It can be shown in a manner similar to (a) and (b) that the $P_k(t)$'s must satisfy the system of differential equations

$$\frac{d}{dt} P_k(t) = \lambda P_{k-1}(t) - \lambda P_k(t)$$

$$k = 1, 2, 3, \ldots$$

Verify that $P_k(t) = e^{-\lambda t} (\lambda t)^k/k!$ satisfies the system. (This is actually the only solution).

Supplementary Exercises / Chapter 3

1. Consider a deck consisting of seven cards, marked 1, 2, . . ., 7. Three of these cards are selected at random. Define a r.v. W by W = the sum of the resulting numbers, and compute the p.m.f. of W. Then compute μ and σ^2. *Hint:* Consider outcomes as unordered, so that (1, 3, 7) and (3, 1, 7) are not different outcomes. Then there are 35 outcomes, and they can be listed (this type of r.v. actually arises in connection with a hypothesis test called Wilcoxon's rank sum test, in which there is an x sample and a y sample, and W is the sum of the ranks of the x's in the combined sample).

2. After shuffling a deck of 52 cards, a dealer deals out five. Let X = the number of suits represented in the five-card hand.

a. Show that the p.m.f. of X is

x	1	2	3	4
$p(x)$	.002	.146	.588	.264

Hint: $p(1) = 4\, P(\text{all are spades})$, $p(2) = 6\, P(\text{at}$

least one spade $\cap$ at least one heart), and $p(4) = 4\, P(2$ spades $\cap$ one of each other suit).

b. Compute μ, σ^2, and σ.

3. The negative binomial random variable X was defined as the number of F's preceding the rth S. Let Y = the number of trials necessary to obtain the rth S. In the same manner in which the p.m.f. of X was derived, derive the p.m.f. of Y.

4. Of all customers purchasing automatic garage door openers, 75% purchase a chain-driven model. Let X = the number among the next 15 purchasers who select the chain-driven model.

a. What is the p.m.f. of X?

b. Compute $P(X \leq 12)$.

c. Compute $P(9 \leq X \leq 12)$

d. Compute μ and σ^2.

e. If the store currently has in stock 10 of the chain-driven models and eight shaft-driven models, what is the probability that the requests of these 15 customers can all be met from existing stock?

5. Of the people passing through an airport metal detector, 1% activate it; let X = the number among a randomly selected group of 50 who activate the detector.
   a. What is the (approximate) p.m.f. of X?
   b. Compute $P(X = 2)$.   c. Compute $P(2 \le X)$.

6. An educational consulting firm is trying to decide whether high school students who have never before used a hand-held calculator can solve a certain type of problem more easily with a calculator which uses reverse Polish notation or one which doesn't use this notation. A sample of 25 students is selected and allowed to practice on both calculators. Then each student is asked to work one problem on the reverse Polish calculator and a similar problem on the other. Let $p = P(S)$ where S denotes a student who worked the problem more quickly using reverse Polish notation than without, and let X = number of S's.
   a. If $p = .5$, what is $P(7 \le X \le 18)$?
   b. If $p = .8$, what is $P(7 \le X \le 18)$?
   c. If the claim that $p = .5$ is to be rejected when either $X \le 7$ or $X \ge 18$, what is the probability of rejecting the claim when it is actually correct?
   d. If the decision to reject the claim $p = .5$ is made as in (c), what is the probability that the claim is not rejected when $p = .6$? When $p = .8$?
   e. What decision rule would you choose for rejecting the claim $p = .5$ if you wanted the probability in (c) to be at most .01?

7. Of all repair work done on TV sets by a certain store, 80% is done on sets which are no longer under warranty (so if $S = \{$not under warranty$\}$, $P(S) = .8$).
   a. Among 20 sets brought in for repair work, what is the expected number not under warranty?
   b. Among these 20 sets, what is the expected number under warranty?
   c. Among these 20 sets, what is the probability that at least 75% are not under warranty?
   d. Suppose that there are 12 sets now in the shop of which four are under warranty. If the 12 were brought in in completely random order and are serviced in the same order, what is the

p.m.f. of X = the number of sets under warranty among the first five serviced? What are $E(X)$ and $\text{Var}(X)$?

8. A trial has just resulted in a hung jury because eight of the jury were in favor of a guilty verdict while the other four were for acquittal. If the jurors leave the jury room in random order and the first four leaving the room are each accosted by a reporter in quest of an interview, what is the p.m.f. of X = the number of jurors favoring acquittal among those interviewed? How many of those favoring acquittal do you expect to be interviewed?

9. A telephone company employs five information operators who receive requests for information independently of one another, each according to a Poisson process with rate $\lambda = 2$ per minute.
   a. What is the probability that during a given one-minute period, the first operator receives no requests?
   b. What is the probability that during a given one-minute period, exactly four of the five operators receive no requests?
   c. Write an expression for the probability that during a given one-minute period, all of the operators receive exactly the same number of requests.

10. Grasshoppers are distributed at random in a large field according to a Poisson distribution with parameter $\lambda = 2$ per square yard. How large should the radius R of a circular sampling region be taken so that the probability of finding at least one in the region equals .99?

11. A newsstand has ordered five copies of a certain issue of a photography magazine. Let X = the number of individuals who come in to purchase this magazine. If X has a Poisson distribution with parameter $\lambda = 4$, what is the expected number of copies which are sold?

12. Individuals A and B begin to play a sequence of chess games. Let $S = \{$A wins a game$\}$, and suppose that outcomes of successive games are independent with $P(S) = p$ and $P(F) = 1 - p$ (they never draw). They will play until one of them wins 10 games. Let X = the number of games played (with possible values 10, 11, . . ., 19).
   a. For $x = 10, 11, . . ., 19$, obtain an expression for $p(x) = P(X = x)$.

b. If a draw is possible, with $p = P(S)$, $q = P(F)$, $1 - p - q = P$ (draw), what are the possible values of X? What is $P(20 \leq X)$? *Hint:* $P(20 \leq X) = 1 - P(X < 20)$.

13. Two drugs are being considered for treatment of the same disease. Let $p_1 = P(S$ using drug A) and $p_2 = P(S$ using drug B). Patients will be paired off and then one patient from each pair will be treated with drug A and the other with drug B. This clinical trial experiment will continue until one of the drugs first accumulates 10 S's. Let $X =$ the number of pairs of patients treated.

a. What is $P(X = 10)$?

b. What is the probability that both drugs achieve 10 S's on the same pair? *Hint:* Both can achieve 10 S's on, say, the fifteenth pair if each of them has nine S's on the first 14 and a S on the fifteenth.

c. Derive an expression for $F(x) = P(X \leq x)$ $x = 10, 11, \ldots$.

14. The negative binomial p.m.f. is given by

$$nb(x; r, p) = k(r, x) \cdot p^r (1 - p)^x$$

$$x = 0, 1, 2, \ldots$$

Let $X =$ the number of observed plant species in a region in which there is known to be at least one species. The p.m.f. of X is often taken to be the zero-truncated negative binomial p.m.f.

$$nbt(x; r, p) = c \cdot k(r, x) \cdot p^r (1 - p)^x$$

$$x = 1, 2, 3, \ldots$$

a. What value of c makes this a p.m.f.? *Hint:* $\sum_{x=0}^{\infty} nb(x; r, p) = 1$.

b. What is the probability that at least two species are observed when $r = \frac{5}{2}$ and $p = .5$?

15. Define a function $p(x; \lambda, \mu)$ by

$p(x; \lambda, \mu)$

$$= \begin{cases} \dfrac{1}{2} e^{-\lambda} \dfrac{\lambda^x}{x!} + \dfrac{1}{2} e^{-\mu} \dfrac{\mu^x}{x!} & x = 0, 1, 2, \ldots \\ 0 & \text{otherwise} \end{cases}$$

a. Show that $p(x; \lambda, \mu)$ satisfies the two conditions necessary for specifying a p.m.f. *Note:* If a firm employs two typists, one of whom makes typographical errors at the rate of λ per page and the other at rate μ per page, and they each do half the firm's typing, then $p(x; \lambda, \mu)$ is the p.m.f. of $X =$ the number of errors on a randomly chosen page.

b. If the first typist (rate λ) types 60% of all pages, what is the p.m.f. of X of part (a)?

c. What is $E(X)$ for $p(x; \lambda, \mu)$ given by the expression just above?

d. What is σ^2 for $p(x; \lambda, \mu)$ given by that expression?

16. The *mode* of a discrete r.v. X with p.m.f. $p(x)$ is that value x^* for which $p(x)$ is largest (the most probable x value).

a. Let $X \sim \text{Bin}(n, p)$. By considering the ratio $b(x + 1; n, p)/b(x; n, p)$, show that $b(x; n, p)$ increases with x as long as $x < np - (1 - p)$. Conclude that the mode x^* is the integer satisfying $(n + 1)p - 1 \leq x^* \leq (n + 1)p$.

b. Show that if x has a Poisson distribution with parameter λ, the mode is the largest integer $< \lambda$. If λ is an integer, show that both $\lambda - 1$ and λ are modes.

17. A computer disk storage device has 10 concentric tracks, numbered 1, 2, . . ., 10 from outermost to innermost, and a single access arm. Let $p_i =$ the probability that any particular requests for data will take the arm to track i ($i = 1, \ldots, 10$). Assume that the tracks accessed in successive seeks are independent, and let $X =$ the number of tracks over which the access arm passes during two successive requests (excluding the track which the arm has just left, so possible X values are $x = 0$, 1, . . ., 9.) Compute the p.m.f. of X. *Hint:* P (the arm is now on track i and $X = j$) $= P(X = j \mid$ arm now on $i) \cdot p_i$. After writing the conditional probability in terms of $p_1, \ldots, p_{10}$, by the law of total probability, the desired probability is obtained by summing over i.

18. If X is a hypergeometric r.v., show directly from the definition that $E(X) = nM/N$ (consider only the case $n < M$). *Hint:* Factor nM/N out of the sum for $E(X)$, and show that the terms inside the sum are of the form $h(y; n - 1, M, N - 1)$ where $y = x - 1$.

Bibliography   Derman, Cyrus, Glaser, Leon, and Olkin, Ingram, *Probability Models and Applications,* Macmillan, New York, 1980. Contains an in-depth discussion of both general properties of discrete and continuous distributions and results for specific distributions.

Johnson, Norman and Kotz, Samuel, *Distributions in Statistics: Discrete Distributions,* Houghton Mifflin, Boston, 1969. An encyclopedia of information on discrete distributions.

Ross, Sheldon, *Introduction to Applied Probability Models* (2nd ed.), Academic Press, New York, 1980. A good source of material on the Poisson process and generalizations, and a nice introduction to other topics in applied probability.

CHAPTER 4

Continuous Random Variables and Probability Distributions

Introduction

As mentioned at the beginning of Chapter 3, not all random variables are discrete. In this chapter we study the second general type of random variable that arises in many applied problems. Sections 4.1 and 4.2 present the basic definitions and properties of continuous random variables and their probability distributions. In Section 4.3 we study in detail the normal random variable and distribution, unquestionably the most important and useful in probability and statistics. Sections 4.4 and 4.5 discuss some other continuous distributions which are often used in applied work.

4.1 Continuous Random Variables and Probability Density Functions

A discrete random variable is one whose possible values either constitute a finite set or else can be listed in an infinite sequence (a list in which there is a first element, a second element, and so on). A random variable whose set of possible values is an entire interval of numbers is not discrete.

Continuous Random Variables

> **Definition:** A random variable X is said to be **continuous** if its set of possible values is an entire interval of numbers—that is, if for some $A < B$, any number x between A and B is possible.

Example 4.1 If in the study of the ecology of a lake, we make depth measurements at randomly chosen locations, then X = the depth at such a location is a continuous random variable. Here A is the minimum depth in the region being sampled and B is the maximum depth.

Example 4.2   If a chemical compound is randomly selected and its pH X is determined, then X is a continuous random variable, because any pH value between 0 and 14 is possible. If more is known about the compound selected for analysis, then the set of possible values might be a subinterval of $[0, 14]$ such as $5.5 \leq x \leq 6.5$, but X would still be continuous.

If the measurement scale of X can be subdivided to any extent desired, then the variable is continuous; if it cannot, the variable is discrete. For example, if the variable is height or length, then it can be measured in kilometers, meters, centi-meters, millimeters, and so on, so the variable is continuous. If, however, $X =$ the billing amount on a randomly chosen monthly natural gas statement, then the smallest unit of measurement is pennies, so any value of X is a multiple of $\$.01$ and X is a discrete variable.

One might argue that although in principle variables such as height, weight, and temperature are continuous, in practice the limitations of our measuring instruments restrict us to a discrete (though sometimes very finely subdivided) world. However, continuous models often approximate real-world situations very well, and con-tinuous mathematics (the calculus) is frequently easier to work with than mathe-matics of discrete variables and distributions.

Probability Distributions for Continuous Variables

Suppose that the variable of interest X is the depth of a lake at a randomly chosen point on the surface. Let $M =$ the maximum depth (in meters), so that any number in the interval $[0, M]$ is a possible value of X. If we "discretize" X by measuring depth to the nearest meter, then possible values are nonnegative integers $\leq M$. The resulting discrete distribution of depth can be pictured using a probability histogram. If we draw the histogram so that the area of the rectangle above any possible integer k is the proportion of the lake whose depth is (to the nearest meter) k, then the total area of all rectangles is one. A possible histogram appears in Figure 4.1a.

If depth is measured to the nearest centimeter and the same measurement axis as in Figure 4.1a is used, each rectangle in the resulting probability histogram is much narrower, though the total area of all rectangles is still one. A possible

Figure 4.1   (*a*) Probability histogram of depth measured to the nearest meter; (*b*) probability histogram of depth measured to the nearest centimeter; (*c*) a limit of a sequence of discrete histograms

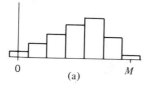
(a)

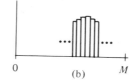
(b)

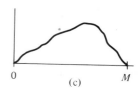
(c)

histogram is pictured in Figure 4.1b; it has a much smoother appearance than the histogram in Figure 4.1a. If we continue in this way to measure depth more and more finely, the resulting sequence of histograms approaches a smooth curve such as is pictured in Figure 4.1c. Because for each histogram the total area of all rectangles equals one, the total area under the smooth curve is also one. The probability that the depth at a randomly chosen point is between a and b is just the area under the smooth curve between a and b. It is exactly a smooth curve of the type pictured in Figure 4.1c which specifies a continuous probability distribution.

Definition: Let X be a continuous random variable. Then a **probability distribution** or **probability density function** (p.d.f.) of X is a function $f(x)$ such that for any two numbers a and b with $a \leq b$,

$$P(a \leq X \leq b) = \int_a^b f(x)\, dx$$

That is, the probability that X takes on a value in the interval $[a, b]$ is the area under the graph of the density function.

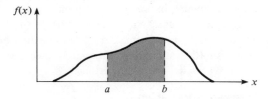

Figure 4.2 $P(a \leq X \leq b)$ = the area under the graph of $f(x)$ between a and b

In order that $f(x)$ be a legitimate p.d.f., it must satisfy the two conditions

    1. $f(x) \geq 0$ for all x
    2. $\int_{-\infty}^{\infty} f(x)\, dx$ = area under the entire graph of $f(x)$ = 1

Example 4.3

Suppose that I take a bus to work, and that every five minutes a bus arrives at my stop. Because of variation in the time that I leave my house, I don't always arrive at the bus stop at the same time, so my waiting time X for the next bus is a continuous random variable. The set of possible values of X is the interval $[0, 5]$. One possible probability density function for X is

$$f(x) = \begin{cases} \frac{1}{5} & 0 \leq x \leq 5 \\ 0 & \text{otherwise} \end{cases}$$

The p.d.f. $f(x)$ is graphed in Figure 4.3. Clearly $f(x) \geq 0$, and the total area under the graph is $5 \cdot (\frac{1}{5}) = 1$. The probability that I wait between one and three minutes is

$$P(1 \leq X \leq 3) = \int_1^3 f(x)\, dx = \int_1^3 \frac{1}{5}\, dx = \frac{x}{5}\Big|_{x=1}^{x=3} = \frac{2}{5}$$

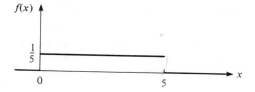

Figure 4.3   The p.d.f. of Example 4.3

Similarly, $P(2 \le X \le 4) = \frac{2}{5}$. The probability that I wait at least four minutes is

$$P(4 \le X) = \int_4^\infty f(x)\,dx = \int_4^5 \frac{1}{5}\,dx + \int_5^\infty 0\,dx = \frac{x}{5}\Big|_{x=4}^{x=5} = \frac{1}{5}$$

Because whenever $0 \le a \le b \le 5$ in Example 4.1, the probability $P(a \le X \le b)$ depends only on the length $b - a$ of the interval, X is said to have a uniform distribution.

> **Definition:** A continuous r.v. X is said to have a **uniform distribution** on the interval $[A, B]$ if the p.d.f. of X is
>
> $$f(x; A, B) = \begin{cases} \dfrac{1}{B - A} & A \le x \le B \\ 0 & \text{otherwise} \end{cases}$$

The graph of any uniform p.d.f. looks like the graph in Figure 4.3 except that the graph is positive above the interval $[A, B]$ rather than $[0, 5]$.

In the discrete case a probability mass function tells us how little "blobs" of probability mass of various magnitudes are distributed along the measurement axis. In the continuous case, probability density is "smeared" in a continuous fashion along the interval of possible values. When density is smeared uniformly over the interval, a uniform p.d.f., as in Figure 4.3, results.

When X is a discrete r.v., each possible value was assigned positive probability, but this is not the case when X is continuous.

> **Proposition:** If X is a continuous r.v., then for any number c, $P(X = c) = 0$. Furthermore, for any two numbers a and b with $a < b$,
>
> $$P(a \le X \le b) = P(a < X \le b)$$
> $$= P(a \le X < b) = P(a < X < b) \qquad (4.1)$$

In words the probability assigned to any particular value is zero, and the probability of an interval does not depend on whether or not either of its end points is included. These properties follow from the fact that the area under the graph of $f(x)$ and above the single value c is zero, and that the area under the graph above an interval is unaffected by exclusion or inclusion of the end points of the interval. If X is discrete and there is positive probability mass at both $X = a$ and $X = b$, then all four probabilities in (4.1) will differ from one another.

The fact that a continuous distribution assigns probability zero to each value has a physical analogue. Consider a solid circular rod with cross-sectional area = 1 in.[2] Place the rod alongside a measurement axis and suppose that the density of the rod at any point x is given by the value $f(x)$ of a density function. Then if the rod is sliced at points a and b and this segment is removed, the amount of mass removed is $\int_a^b f(x)\,dx$, while if the rod is sliced just at the point c, no mass is removed. Mass is assigned to interval segments of the rod but not to individual points.

Example 4.4   "Time headway" in traffic flow is the elapsed time between the time that one car finishes passing a fixed point and the instant that the next car begins to pass that point. Let X = the time headway for two randomly chosen consecutive cars on a freeway during a period of heavy flow. The following p.d.f. of X is essentially the one suggested in "The Statistical Properties of Freeway Traffic," *Transportation Research*, vol. 11, pp. 221–228:

$$f(x) = \begin{cases} .15e^{-.15(x-.5)} & x \geq .5 \\ 0 & \text{otherwise} \end{cases}$$

The graph of $f(x)$ appears in Figure 4.4; there is no density associated with headway times less than .5, and headway density decreases rapidly (exponentially fast) as x increases from .5. Clearly $f(x) \geq 0$; to show that $\int_{-\infty}^{\infty} f(x)\,dx = 1$ we use the calculus result $\int_a^{\infty} e^{-kx}\,dx = (1/k)\,e^{-k \cdot a}$

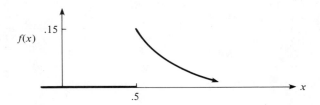

Figure 4.4   The p.d.f. of headway time in Example 4.4

Then

$$\int_{-\infty}^{\infty} f(x)\,dx = \int_{.5}^{\infty} .15e^{-.15(x-.5)}\,dx = .15e^{.075} \int_{.5}^{\infty} e^{-.15x}\,dx$$

$$= .15e^{.075} \cdot \frac{1}{.15} e^{-(.15)(.5)} = 1$$

The probability that headway time is at most five seconds is

$$P(X \le 5) = \int_{-\infty}^{5} f(x) \, dx = \int_{.5}^{5} .15e^{-.15(x-.5)} \, dx$$

$$= .15e^{.075} \int_{.5}^{5} e^{-.15x} \, dx = .15e^{.075} \cdot -\frac{1}{.15} e^{-.15x} \Big|_{x=.5}^{x=5}$$

$$= e^{.075}[-e^{-.75} + e^{-.075}] = .928[-.472 + .928] = .423$$

$$= P(\text{less than 5 seconds}) = P(X < 5)$$

Unlike discrete distributions such as the binomial, hypergeometric, and negative binomial, the distribution of any given continuous random variable cannot usually be derived using simple probabilistic arguments. Instead, one must make a judicious choice of p.d.f. based on prior knowledge and available data. Fortunately there are some general families of p.d.f.'s which have been found to fit well in a wide variety of experimental situations; several of these are discussed later in the chapter.

Just as in the discrete case, it often helpful to think of the population of interest as consisting of X values rather than individuals or objects. The probability density function is then a model for the distribution of values in this numerical population, and from this model various population characteristics (such as the mean) can be calculated.

Exercises / Section 4.1

1. Let X denote the amount of time for which a book on two-hour reserve at a college library is checked out by a randomly selected student, and suppose that X has density function

$$f(x) = \begin{cases} \frac{1}{2}x & 0 \le x \le 2 \\ 0 & \text{otherwise} \end{cases}$$

  a. Calculate $P(X \le 1)$
  b. Calculate $P(.5 \le X \le 1.5)$
  c. Calculate $P(1.5 < X)$

2. Suppose that the reaction temperature X (in °C) in a certain chemical process has a uniform distribution with $A = -5$ and $B = 5$.
  a. Compute $P(X < 0)$
  b. Compute $P(-2 < X < 2)$
  c. Compute $P(-2 \le X \le 3)$
  d. For k satisfying $-5 < k < k + 4 < 5$, compute $P(k < X < k + 4)$

3. Suppose that the distance X between a point target and a shot aimed at the point in a coin-operated target game is a continuous random variable with p.d.f.

$$f(x) = \begin{cases} \frac{3}{4}(1 - x^2) & -1 \le x \le 1 \\ 0 & \text{otherwise} \end{cases}$$

  a. Sketch the graph of $f(x)$.
  b. Compute $P(X > 0)$
  c. Compute $P(-.5 < X < .5)$
  d. Compute $P(X < -.25$ or $X > .25)$

4. A college professor never finishes his lecture before the bell rings to end the period, and always finishes his lectures within one minute after the bell rings. Let $X =$ the time which elapses between the bell and the end of the lecture, and suppose that the p.d.f. of X is

$$f(x) = \begin{cases} kx^2 & 0 \le x \le 1 \\ 0 & \text{otherwise} \end{cases}$$

a. Find the value of k. *Hint:* Total area under the graph of $f(x)$ is 1.

b. What is the probability that the lecture ends within $1/2$ minute of the bell ringing?

c. What is the probability that the lecture continues beyond the bell for between 15 and 30 seconds?

d. What is the probability that the lecture continues for at least 40 seconds beyond the bell?

5. The actual tracking weight of a stereo cartridge which is set to track at 3 gm on a particular changer can be regarded as a continuous r.v. X with p.d.f.

$$f(x) = \begin{cases} k[1 - (x - 3)^2] & 2 \le x \le 4 \\ 0 & \text{otherwise} \end{cases}$$

a. Sketch the graph of $f(x)$.

b. Find the value of k.

c. What is the probability that the actual tracking weight is greater than the prescribed weight?

d. What is the probability that the actual weight is within .25 gm of the prescribed weight?

e. What is the probability that the actual weight differs from the prescribed weight by more than .5 gm?

6. At a well-known restaurant the preparation time for a house speciality dish is listed on the menu as 30 minutes. Assume the actual preparation time X has a uniform distribution with $A = 25$ and $B = 35$.

a. Write the p.d.f. of X and sketch its graph.

b. What is the probability that actual preparation time exceeds 33 minutes?

c. What is the probability that actual preparation time is within two minutes of the time advertised?

d. For any a such that $25 < a < a + 2 < 35$, what is the probability that actual preparation time is between a and $a + 2$ minutes?

7. In commuting to work I must first get on a bus near my house and then transfer to a second bus. If the waiting time at each stop has a uniform distribution with $A = 0$ and $B = 5$, then it can be shown that my total waiting time Y has the p.d.f.

$$f(y) = \begin{cases} \frac{1}{25}y & 0 \le y < 5 \\ \frac{2}{5} - \frac{1}{25}y & 5 \le y \le 10 \\ 0 & y < 0 \ \text{ or } \ y > 10 \end{cases}$$

a. Sketch a graph of the p.d.f. of Y.

b. Verify that $\int_{-\infty}^{\infty} f(y)\, dy = 1$.

c. What is the probability that total waiting time is at most three minutes?

d. What is the probability that total waiting time is at most eight minutes?

e. What is the probability that total waiting time is between three and eight minutes?

f. What is the probability that total waiting time is either less than two minutes or more than six minutes?

8. Consider again the p.d.f. of $X = $ time headway given in Example 4.4.

a. What is the probability that time headway is at most six seconds? (If you don't have a calculator with an exponential function, just express the answer in terms of e.)

b. What is the probability that time headway is more than six seconds? At least six seconds?

c. What is the probability that time headway is between five and six seconds?

9. A family of p.d.f.'s which has been used to approximate the distribution of income, city population size, and size of firms is the Pareto family. The family has two parameters, k and θ, both > 0, and the p.d.f. is

$$f(x; k, \theta) = \begin{cases} \dfrac{k \cdot \theta^k}{x^{k+1}} & x \ge \theta \\ 0 & x < \theta \end{cases}$$

a. Sketch the graph of $f(x; k, \theta)$.

b. Verify that the total area under the graph equals 1.

c. If the r.v. X has p.d.f. $f(x; k, \theta)$, for any fixed $b > \theta$ obtain an expression for $P(X \le b)$.

d. For $\theta < a < b$, obtain an expression for $P(a \le X \le b)$.

e. A truncated Pareto density function of the form

$$f(x; k, \theta_1, \theta_2) = \begin{cases} c \cdot \left(\dfrac{\theta_1}{x}\right)^{k+1} & \theta_1 \le x \le \theta_2 \\ 0 & \text{otherwise} \end{cases}$$

has been used to model the distribution of oil field size. What value of the constant c (in terms of k, θ_1, and θ_2) makes this a p.d.f.?

4.2 Cumulative Distribution Functions and Expected Values for Continuous Random Variables

The Cumulative Distribution Function for a Continuous Variable

> **Definition:** The **cumulative distribution function** (c.d.f.) $F(x)$ for a continuous r.v. X is defined for every number x by
>
> $$F(x) = P(X \le x) = \int_{-\infty}^{x} f(y) \, dy$$

For X discrete, $F(x)$ is the sum of the p.m.f. over values $\le x$, whereas here summation is replaced by integration and the p.m.f. by the p.d.f.

Example 4.5 Let X have a uniform distribution on $[A, B]$. The density function appears in Figure 4.5. For $x < A$, $F(x) = 0$, since there is no area under the graph of the density function to the left of such an x. For $x \ge B$, $F(x) = 1$, since all the area is accumulated to the left of such an x. Finally, for $A \le x \le B$,

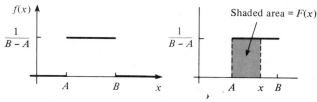

Figure 4.5 The p.d.f. for a uniform distribution

$$F(x) = \int_{-\infty}^{x} f(y) \, dy = \int_{A}^{x} \frac{1}{B - A} \, dy = \frac{1}{B - A} \cdot y \Big|_{y=A}^{y=x} = \frac{x - A}{B - A}$$

The entire c.d.f. is

$$F(x) = \begin{cases} 0 & x < A \\ \dfrac{x - A}{B - A} & A \le x < B \\ 1 & x > B \end{cases}$$

The graph of this c.d.f. appears in Figure 4.6.

Figure 4.6 The c.d.f. for a uniform distribution

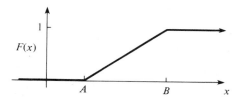

Using *F(x)* to Compute Probabilities

The importance of the c.d.f. here, just as for discrete random variables, is that probabilities of various intervals can be computed from a formula for or table of $F(x)$.

Proposition: Let X be a continuous random variable with p.d.f. $f(x)$ and c.d.f. $F(x)$. Then for any two numbers a, b with $a < b$,

$$P(a \leq X \leq b) = F(b) - F(a)$$

Figure 4.7 illustrates this proposition; the desired probability is the shaded area under the density curve between a and b and equals the difference between the two shaded cumulative areas.

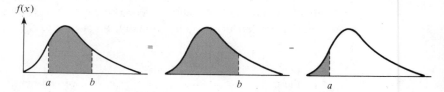

Figure 4.7 Computing $P(a \leq X \leq b)$ from cumulative probabilities

---

Example 4.5
(continued)

If X has a uniform distribution with $A = 0$ and $B = 5$, as in the waiting-time example of Section 4.1, then

$$F(x) = \begin{cases} 0 & x < 0 \\ \dfrac{x}{5} & 0 \leq x < 5 \\ 1 & x \geq 5 \end{cases}$$

Then $P(2 \leq X \leq 4) = F(4) - F(2) = \frac{4}{5} - \frac{2}{5} = \frac{2}{5}$.

In this example $P(2 \leq X \leq 4)$ could just as easily have been obtained by integrating $f(x)$, but when the integration must be done numerically (as with the normal and gamma distributions to be encountered shortly), tabulating $F(x)$ enables probabilities to be computed quickly.

Obtaining *f(x)* from *F(x)*

For X discrete, the p.m.f. is obtained from the c.d.f. by taking the difference between two $F(x)$ values. The continuous analogue of a difference is a derivative. The following result is a consequence of the Fundamental Theorem of Calculus.

> **Proposition:** If X is a continuous r.v. with p.d.f. $f(x)$ and c.d.f. $F(x)$, then at every x at which the derivative $F'(x)$ exists, $F'(x) = f(x)$.

Example 4.5
(continued)

When X has a uniform distribution, $F(x)$ is differentiable except at $x = A$ and $x = B$, where the graph of $F(x)$ has sharp corners. Since $F(x) = 0$ for $x < A$ and $F(x) = 1$ for $x > B$, $F'(x) = 0 = f(x)$ for such x. For $A < x < B$,

$$F'(x) = \frac{d}{dx}\left(\frac{x - A}{B - A}\right) = \frac{1}{B - A} = f(x)$$

Percentiles of a Continuous Distribution

When we say that individual A's score was at the eighty-fifth percentile of the population, we mean that 85% of all population scores were below A's while 15% were above A's score.

> **Definition:** Let p be a number between 0 and 1. The **$(100p)$th percentile** of the distribution of a continuous r.v. X, denoted by x_p, is defined by
>
> $$p = F(x_p) = \int_{-\infty}^{x_p} f(y)\, dy \qquad (4.2)$$

According to (4.2) x_p is that value on the measurement axis such that $100p\%$ of the area under the graph of $f(x)$ lies to the left of x_p and $100(1 - p)\%$ lies to the right. This is illustrated in Figure 4.8.

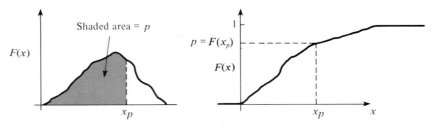

Figure 4.8   The $(100p)$th percentile of a continuous distribution

Example 4.6

The distribution of the amount of gravel (in tons) sold by a particular construction supply company in a given week is a continuous r.v. X with p.d.f.

$$f(x) = \begin{cases} \dfrac{3}{2}(1 - x^2) & 0 \le x \le 1 \\ 0 & \text{otherwise} \end{cases}$$

The cumulative distribution function of sales is then, for $0 < x < 1$,

$$F(x) = \int_0^x \frac{3}{2}(1 - y^2) \, dy = \frac{3}{2}\left(y - \frac{y^3}{3}\right)\Bigg|_{y=0}^{y=x} = \frac{3}{2}\left(x - \frac{x^3}{3}\right)$$

The graphs of both $f(x)$ and $F(x)$ appear in Figure 4.9. The (100p)th percentile of this distribution satisfies the equation

$$p = F(x_p) = \frac{3}{2}\left(x_p - \frac{x_p^3}{3}\right) \quad \text{or} \quad x_p^3 - 3x_p + 2p = 0$$

For the fiftieth percentile, $p = .5$ and the equation to be solved is $x^3 - 3x + 1 = 0$; the solution is $x = x_{.5} = .347$. If the distribution remains the same from week to week, then in the long run 50% of all weeks will result in sales of less than .347 tons and 50% in more than .347 tons.

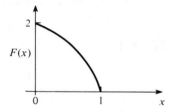 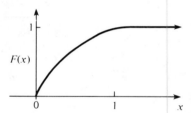

Figure 4.9   The p.d.f. and c.d.f. for Example 4.6

> **Definition:** The **median** of a continuous distribution, denoted by $\tilde{\mu}$, is the fiftieth percentile, so $\tilde{\mu}$ satisfies $.5 = F(\tilde{\mu})$. That is, half the area under the density curve is to the left of $\tilde{\mu}$ and half is to the right of $\tilde{\mu}$.

Example 4.7   A continuous distribution whose p.d.f. is **symmetric**—which means that the graph of the p.d.f. to the left of some point is a mirror image of the graph to the right of

Figure 4.10   Medians of symmetric distributions

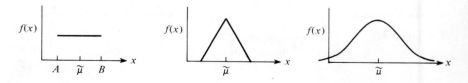

that point—has median $\widetilde{\mu}$ equal to the point of symmetry, since half the area under the curve lies to either side of this point. Figure 4.10 gives several examples. The amount of error in a measurement of a physical quantity is often assumed to have a symmetric distribution.

Expected Values for Continuous Random Variables

For a discrete r.v. X, $E(X)$ was obtained by summing $x \cdot p(x)$ over possible X values. Here we replace summation by integration and the p.m.f. by the p.d.f. to get a continuous weighted average.

Definition: The **expected** or **mean value** of a continuous r.v. X with p.d.f. $f(x)$ is

$$\mu_X = E(X) = \int_{-\infty}^{\infty} x \cdot f(x)\, dx$$

Example 4.6 (continued)

The p.d.f. of weekly gravel sales X was

$$f(x) = \begin{cases} \frac{3}{2}(1 - x^2) & 0 \le x \le 1 \\ 0 & \text{otherwise} \end{cases}$$

so

$$E(X) = \int_{-\infty}^{\infty} x \cdot f(x)\, dx = \int_{0}^{1} x \cdot \tfrac{3}{2}(1 - x^2)\, dx$$

$$= \frac{3}{2} \int_{0}^{1} (x - x^3)\, dx = \frac{3}{2} \left(\frac{x^2}{2} - \frac{x^4}{4} \right) \Big|_{x=0}^{x=1} = \frac{3}{8}$$

When the p.d.f. $f(x)$ specifies a model for the distribution of values in a numerical population, then μ is the population mean, which is the most frequently used measure of population location or center.

Often we wish to compute the expected value of some function $h(X)$ of the r.v. X. If we think of $h(X)$ as a new r.v. Y, techniques from mathematical statistics can be used to derive the p.d.f. of Y, and $E(Y)$ can be computed from the definition. Fortunately, as in the discrete case, there is an easier way to compute $E[h(X)]$.

Proposition: If X is a continuous r.v. with p.d.f. $f(x)$ and $h(X)$ is any function of X, then

$$E[h(X)] = \mu_{h(X)} = \int_{-\infty}^{\infty} h(x) \cdot f(x)\, dx$$

Example 4.8

Two species are competing in a region for control of a limited amount of a certain resource. Let X = the proportion of the resource controlled by species 1, and suppose that X has p.d.f.

$$f(x) = \begin{cases} 1 & 0 \le x \le 1 \\ 0 & \text{otherwise} \end{cases}$$

which is a uniform distribution on $[0, 1]$ (in her book *Ecological Diversity*, E. C. Pielou calls this the "broken stick" model for resource allocation, since it is analogous to breaking a stick at a randomly chosen point). Then the species which controls the majority of this resource controls the amount

$$h(X) = \max(X, 1 - X) = \begin{cases} 1 - X & \text{if} \quad 0 \le X < \tfrac{1}{2} \\ X & \text{if} \quad \tfrac{1}{2} \le X \le 1 \end{cases}$$

The expected amount controlled by the species having majority control is then

$$E[h(X)] = \int_{-\infty}^{\infty} \max(x, 1 - x) \cdot f(x)\, dx = \int_{0}^{1} \max(x, 1 - x) \cdot 1\, dx$$

$$= \int_{0}^{1/2} (1 - x) \cdot 1\, dx + \int_{1/2}^{1} x \cdot 1\, dx = \tfrac{3}{4}$$

For $h(X)$ a linear function, $E[h(X)] = E(aX + b) = aE(X) + b$.

The Variance of a Continuous Random Variable

> **Definition:** The **variance** of a continuous r.v. X with p.d.f. $f(x)$ and mean value μ is
>
> $$\sigma_X^2 = \text{Var}(X) = \int_{-\infty}^{\infty} (x - \mu)^2 \cdot f(x)\, dx = E[(X - \mu)^2]$$
>
> The **standard deviation** of X is $\sigma_X = \sqrt{\text{Var}(X)}$.

As in the discrete case, σ_X^2 is the expected or average squared deviation about the mean μ, and gives a measure of how much spread there is in the distribution or population of x values. The easiest way to compute σ^2 is to again use a shortcut formula.

> **Proposition:** $\text{Var}(X) = E(X^2) - [E(X)]^2$

Example 4.6 (continued)

For X = weekly gravel sales, we computed $E(X) = \tfrac{3}{8}$. Since

$$E(X^2) = \int_{-\infty}^{\infty} x^2 \cdot f(x)\, dx = \int_0^1 x^2 \cdot \tfrac{3}{2}(1 - x^2)\, dx = \int_0^1 \tfrac{3}{2}(x^2 - x^4)\, dx = \tfrac{1}{5}$$

$$\text{Var}(X) = \tfrac{1}{5} - (\tfrac{3}{8})^2 = \tfrac{19}{320} = .059 \quad \text{and} \quad \sigma_X = .244$$

When $h(X)$ is a linear function and $\text{Var}(X) = \sigma^2$, $\text{Var}[h(X)] =$
$\text{Var}(aX + b) = a^2 \cdot \sigma^2$ and $\sigma_{aX+b} = |a| \cdot \sigma$.

Exercises / Section 4.2

1. The cumulative distribution function of checkout duration X as described in Exercise 1, Section 4.1, is

$$F(x) = \begin{cases} 0 & x < 0 \\ \dfrac{x^2}{4} & 0 \le x < 2 \\ 1 & 2 \le x \end{cases}$$

Use this to compute
a. $P(X \le 1)$ **b.** $P(.5 \le X \le 1)$ **c.** $P(X > .5)$
d. the median checkout duration $\tilde{\mu}$ [solve $.5 = F(\tilde{\mu})$]
e. Compute $F'(x)$ to obtain the density function $f(x)$.

2. The cumulative distribution function for $X(=$ shot-to-target distance) of Exercise 3, Section 4.1, is

$$F(x) = \begin{cases} 0 & x < -1 \\ \dfrac{1}{2} + \dfrac{3}{4}\left(x - \dfrac{x^3}{3}\right) & -1 \le x < 1 \\ 1 & 1 \le x \end{cases}$$

a. Compute $P(X < 0)$
b. Compute $P(-.5 < X < .5)$
c. Compute $P(.25 < X)$
d. Verify that $f(x)$ is as given in Exercise 3, Section 4.1, by obtaining $F'(x)$.
e. Verify that $\tilde{\mu} = 0$.

3. Let X denote checkout time duration with p.d.f. given in Exercise 1, Section 4.1
a. Compute $E(X)$.   **b.** Compute $\text{Var}(X)$ and σ_X.
c. If the borrower is charged an amount $h(X) = X^2$ when checkout duration is X, compute the expected charge $E[h(X)]$.

4. Suppose that the p.d.f. of weekly gravel sales X (in tons) is

$$f(x) = \begin{cases} 2(1 - x) & 0 \le x \le 1 \\ 0 & \text{otherwise} \end{cases}$$

a. Obtain the c.d.f. of X and graph it.
b. What is $P(X \le .5)$ [that is, $F(.5)$]?
c. Using (a), what is $P(.25 < X \le .5)$? $P(.25 \le X \le .5)$?
d. What is the seventy-fifth percentile of the sales distribution?
e. What is the median $\tilde{\mu}$ of the sales distribution?
f. Compute $E(X)$, $\text{Var}(X)$, and σ_X.

5. Answer (a)–(f) above for the distribution of $X =$ lecture time past the bell given in Exercise 4, Section 4.1.

6. Consider the p.d.f. of $X =$ actual tracking weight given in Exercise 5, Section 4.1.
a. Obtain and graph the c.d.f. of X.
b. From the graph of $f(x)$, what is $\tilde{\mu}$?
c. Compute $E(X)$ and $\text{Var}(X)$.

7. Let X have a uniform distribution on the interval $[A, B]$.
a. Obtain an expression for the $(100\, p)$th percentile.
b. Compute $E(X)$, $\text{Var}(X)$, and σ_X.
c. For n a positive integer, compute $E(X^n)$.

8. Consider the p.d.f. for total waiting time Y for two buses

$$f(y) = \begin{cases} \dfrac{1}{25} y & 0 \le y < 5 \\ \dfrac{2}{5} - \dfrac{1}{25} y & 5 \le y \le 10 \\ 0 & \text{otherwise} \end{cases}$$

introduced in Exercise 7, of Section 4.1
a. Compute and sketch the c.d.f. of Y. *Hint:* Consider separately $0 \le y < 5$ and $5 \le y \le 10$ in

computing $F(y)$. A picture of the p.d.f. should be helpful.

b. Obtain an expression for the $(100p)$th percentile. *Hint:* Consider separately $0 < p < .5$ and $.5 < p < 1$.

c. Compute $E(Y)$ and $Var(Y)$. How do these compare with the expected waiting time and variance for a single bus when the time is uniformly distributed on $[0, 5]$?

9. An ecologist wishes to mark off a circular sampling region having radius 10 meters. However, the radius of the resulting region is actually a random variable R with p.d.f.

$$f(r) = \begin{cases} \frac{3}{4}[1 - (10 - r)^2] & 9 \le r \le 11 \\ 0 & \text{otherwise} \end{cases}$$

What is the expected area of the resulting circular region?

10. The weekly demand for propane gas (in thousands of gallons) from a particular facility is a r.v. X with p.d.f.

$$f(x) = \begin{cases} 2\left(1 - \dfrac{1}{x^2}\right) & 1 \le x \le 2 \\ 0 & \text{otherwise} \end{cases}$$

a. Compute the c.d.f. of X.

b. Obtain an expression for the $(100p)$th percentile. What is the value of $\tilde{\mu}$?

c. Compute $E(X)$ and $Var(X)$.

d. If 1.5 thousand gallons is in stock at the beginning of the week and no new supply is due in during the week, how much of the 1.5 thousand gallons is expected to be left at the end of the week?

11. If the temperature of a compound at a particular time during a chemical reaction is a r.v. with mean value 120 °C and standard deviation 2 °C, what are the mean temperature and standard deviation measured in °F? *Hint:* °F $= \frac{9}{5}$ °C $+ 32$.

12. Let X have the Pareto p.d.f.

$$f(x; k, \theta) = \begin{cases} \dfrac{k \cdot \theta^k}{x^{k+1}} & x \ge \theta \\ 0 & x < \theta \end{cases}$$

introduced in Exercise 9, Section 4.1.

a. If $k > 1$, compute $E(X)$.

b. What can you say about $E(X)$ if $k = 1$?

c. If $k > 2$, show that $Var(X) = k\theta^2(k - 1)^{-2}(k - 2)^{-1}$.

d. If $k = 2$, what can you say about $Var(X)$?

e. What conditions on k are necessary to ensure that $E(X^n)$ is finite?

4.3   The Normal Distribution

Definition: A continuous random variable X is said to have a **normal distribution with parameters μ and σ (or μ and σ^2)**, where $-\infty < \mu < \infty$ and $0 < \sigma$, if the probability density function of X is

$$f(x; \mu, \sigma) = \frac{1}{\sqrt{2\pi}\,\sigma} e^{-(x-u)^2/2\sigma^2} \qquad -\infty < x < \infty \tag{4.3}$$

Again e denotes the base of the natural logarithm system and equals approximately 2.71828, while π represents the familiar mathematical constant with approximate value 3.14159. The statement that X is normally distributed with parameters μ and σ^2 is often abbreviated $X \sim N(\mu, \sigma^2)$.

Clearly $f(x; \mu, \sigma) \ge 0$ for any number x, but techniques from multivariate calculus must be used to show that $\int_{-\infty}^{\infty} f(x; \mu, \sigma)\, dx = 1$. It can be shown that $E(X) = \mu$ and $Var(X) = \sigma^2$, so the parameters are the mean and standard deviation of X. Figure 4.11 presents graphs of $f(x; \mu, \sigma)$ for several different (μ, σ) pairs. Each graph is symmetric about μ and bell-shaped, so the center of the bell (point of symmetry) is both the mean of the distribution and the median. Large values of σ

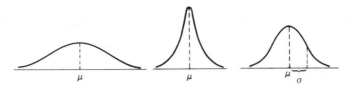

Figure 4.11 Graphs of normal p.d.f.'s.

yield graphs which are quite spread out about μ, whereas small values of σ yield graphs with a high peak above μ and most of the area under the graph quite close to μ. Thus a large σ implies that a value of X far from μ may well be observed, while such a value is quite unlikely when σ is small.

The normal distribution is the most important one in all of probability and statistics. Many numerical populations have distributions which can be fit very closely by a normal curve with appropriate values of μ and σ. Examples include heights, weights and other physical characteristics (the famous 1903 *Biometrika* paper "On the Laws of Inheritance in Man" discussed many examples of this sort), measurement errors in scientific experiments, anthropometric measurements on fossils, reaction times in psychological experiments, measurements of intelligence and aptitude, scores on various tests, and numerous economic measures and indicators. Even when the underlying distribution is discrete, the normal curve often gives an excellent approximation. In addition, even when individual variables themselves are not normally distributed, sums and averages of the variables will under suitable conditions have approximately a normal distribution; this is the content of the Central Limit Theorem discussed in the next chapter.

The Standard Normal Distribution

To compute $P(a \leq X \leq b)$ when X is a normal r.v. with parameters μ and σ, we must evaluate

$$\int_a^b \frac{1}{\sqrt{2\pi}\,\sigma} e^{-(x-u)^2/2\sigma^2}\,dx \tag{4.4}$$

None of the standard integration techniques can be used to evaluate (4.4). Instead, for $\mu = 0$ and $\sigma = 1$, (4.4) has been numerically evaluated and tabulated for certain values of a and b. Using this table, probabilities can also be computed for any other values of μ and σ under consideration.

Definition: The normal distribution with parameter values $\mu = 0$ and $\sigma = 1$ is called a **standard normal distribution.** A random variable which has a standard normal distribution is called a **standard normal random variable,** and will be denoted by Z. The p.d.f. of Z is

$$f(z;\,0,\,1) = \frac{1}{\sqrt{2\pi}} e^{-z^2/2} \qquad -\infty < z < \infty$$

The cumulative distribution function of Z is $P(Z \leq z) = \int_{-\infty}^z f(y;\,0,\,1)\,dy$, which we shall denote by $\Phi(z)$.

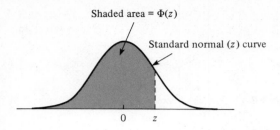

Figure 4.12   Standard normal curve areas tabulated in Appendix Table A.3

The standard normal distribution does not frequently serve as a model for a naturally arising population. Instead it is a reference distribution from which information about other normal distributions can be obtained. Appendix Table 4.3 gives $\Phi(z) = P(Z \le z)$, the area under the graph of the standard normal p.d.f. to the left of z, for $z = -3.49, -3.48, \ldots, 3.48, 3.49$. From this table various other probabilities involving Z can be calculated.

Example 4.9   Compute the following probabilities:

(a) $P(Z \le 1.25)$ (b) $P(Z > 1.25)$ (c) $P(Z \le -1.25)$
(d) $P(-.38 \le Z \le 1.25)$

Solutions

1. $P(Z \le 1.25) = \Phi(1.25)$, a probability which is tabulated in Appendix Table A.3 at the intersection of the row marked 1.2 and the column marked .05. The number there is .8944, so $P(Z \le 1.25) = .8944$. (See Figure 4.13.)

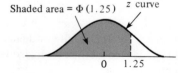

Figure 4.13

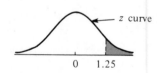

Figure 4.14

2. $P(Z > 1.25) = 1 - P(Z \le 1.25) = 1 - \Phi(1.25)$, the area under the standard normal curve to the right of 1.25 (an upper-tail area). Since $\Phi(1.25) = .8944$, $P(Z > 1.25) = .1056$. Since Z is a continuous r.v., $P(Z \ge 1.25)$ also equals .1056. (See Figure 4.14.)

3. $P(Z \le -1.25) = \Phi(-1.25)$, a lower-tail area. Directly from Table A.3, $\Phi(-1.25) = .1056$. By symmetry of the normal curve, this is the same answer as in (b).

4. $P(-.38 \le Z \le 1.25)$ is the area under the standard normal curve above the

interval whose left endpoint is $-.38$ and whose right endpoint is 1.25. From Section 4.2 if X is a continuous r.v. with c.d.f. $F(x)$, then $P(a \leq X \leq b) = F(b) - F(a)$. This gives $P(-.38 \leq Z \leq 1.25) = \Phi(1.25) - \Phi(-.38) = .8944 - .3520 = .5424$. (See Figure 4.15.)

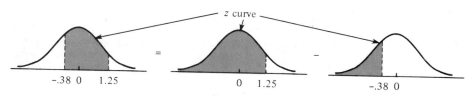

Figure 4.15

Percentiles of the Standard Normal Distribution

For any p between 0 and 1, Appendix Table 4.3 can be used to obtain the $(100p)$th percentile of the standard normal distribution.

Example 4.10   The ninety-ninth percentile of the standard normal distribution is that point on the axis such that the area under the curve to the left of the point is $.9900$. Now Appendix Table A.3 gives for fixed z the area under the standard normal curve to the left of z, whereas here we have the area and want the value of z. This is the "inverse" problem to "$P(Z \leq z) = ?$" so the table is used in an inverse fashion: Find in the middle of the table $.9900$, and then the row and column in which it lies identify the ninety-ninth percentile z. Here $.9901$ lies in the row marked 2.3 and column marked $.03$, so the ninety-ninth percentile is (approximately) $z = 2.33$. By symmetry, the first percentile is the negative of the ninety-ninth percentile, so equals -2.33 (1% lies below the first and above the ninety-ninth).

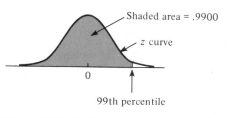

Figure 4.16   Finding the ninety-ninth percentile

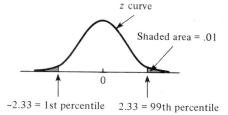

Figure 4.17   The relationship between the first and ninety-ninth percentiles

In general, the $(100p)$th percentile is identified by the row and column of Appendix Table A.3 in which the entry p is found (for example, the sixty-seventh percentile is obtained by finding .6700 in the body of the table, which gives $z = 0.44$). If p does not appear, the number closest to it is often used, though linear interpolation gives a more accurate answer.

Example 4.11 To find the ninety-fifth percentile, we look for .9500 inside the table. While .9500 does not appear, both .9495 and .9505 do, corresponding to $z = 1.64$ and 1.65, respectively. Since .9500 is halfway between the two probabilities which do appear, we shall use 1.645 as the ninety-fifth percentile and -1.645 as the fifth percentile.

z_α Notation

In statistical inference we shall need the values on the measurement axis which cut off or capture certain small tail areas under the standard normal curve.

> **Notation:** z_α will be used to denote the value on the measurement axis for which α of the area under the standard normal curve lies above z_α.

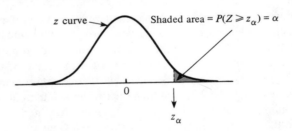

Figure 4.18 z_α notation illustrated

Since α of the area under the standard normal curve lies to the right of z_α, $1 - \alpha$ of the area lies to the left of z_α. Thus z_α *is the $100(1 - \alpha)$th percentile of the standard normal distribution*. Further, by symmetry the area under the standard normal curve to the left of $-z_\alpha$ is also α. The z_α's are usually referred to as **critical values**. Table 4.1 lists the most important standard normal percentiles and z_α values.

Table 4.1 *Standard Normal Percentiles and Critical Values*

Percentile	90	95	97.5	99	99.5	99.9	99.95
α	.1	.05	.025	.01	.005	.001	.0005
$z_\alpha = 100(1 - \alpha)$th percentile	1.28	1.645	1.96	2.33	2.58	3.08	3.27

Example 4.12 $z_{.05}$ is the $100(1 - .05)$th = ninety-fifth percentile of the standard normal distribution, so $z_{.05} = 1.645$. The area under the standard normal curve to the left of $-z_{.05}$ is also .05.

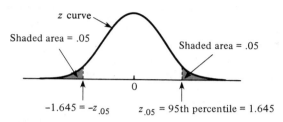

Figure 4.19 Finding $z_{.05}$

Nonstandard Normal Distributions

When $X \sim N(\mu, \sigma^2)$, probabilities involving X are computed by "standardizing." The **standardized variable** is $(X - \mu)/\sigma$. Subtracting μ shifts the mean from μ to zero, and then dividing by σ scales the variable so that the standard deviation is one rather than σ.

> **Proposition:** If X has a normal distribution with mean μ and standard deviation σ, then
>
> $$Z = \frac{X - \mu}{\sigma}$$
>
> is a standard normal random variable.

The key idea of the proposition is that by standardizing, any probability involving X can be expressed as a probability involving a standard normal r.v. Z, so that Appendix Table A.3 can be used. In particular,

$$P(X \le x) = P[Z \le (x - \mu)/\sigma] = \Phi[(x - \mu)/\sigma]$$

as illustrated in Figure 4.20. The proposition can be proved by writing the c.d.f. of $Z = (X - \mu)/\sigma$ as $P(Z \le z) = P(X \le \sigma z + \mu) = \int_{-\infty}^{\sigma z + \mu} f(x; \mu, \sigma) \, dx$. Using a result from calculus, this integral can be differentiated with respect to z to yield the desired p.d.f. $f(z; 0, 1)$.

Figure 4.20 Equality of nonstandard and standard normal curve areas

Example 4.13 The breakdown voltage X of a randomly chosen diode of a particular type is known to be normally distributed with $\mu = 40$ volts and $\sigma = 1.5$ volts. What is the probability that the breakdown voltage will be between 39 and 42 volts?

The desired probability is $P(39 \leq X \leq 42)$. To compute it we express the event $39 \leq X \leq 42$ in equivalent form by standardizing:

$$39 \leq X \leq 42 \quad \text{iff} \quad \frac{39 - 40}{1.5} \leq \frac{X - 40}{1.5} \leq \frac{42 - 40}{1.5}$$

that is, $-.67 \leq Z \leq 1.33$. Thus

$$P(39 \leq X \leq 42) = P(-.67 \leq Z \leq 1.33) = \Phi(1.33) - \Phi(-.67)$$

$$= .9082 - .2514 = .6568$$

Similarly,

$$P(40 \leq X \leq 43) = P\left(\frac{40 - 40}{1.5} \leq \frac{X - 40}{1.5} \leq \frac{43 - 40}{1.5}\right)$$

$$= P(0 \leq Z \leq 2) = .4772$$

Example 4.14 Tissue respiration rate in diaphragms of rats (microliters per mg dry wt per hr) under normal temperature conditions is normally distributed with $\mu = 2.03$ and $\sigma = .44$. To compute the probability that a randomly chosen rat has rate X which is at least 2.5, we again standardize:

$$P(X \geq 2.5) = P\left(\frac{X - 2.03}{.44} \geq \frac{2.5 - 2.03}{.44}\right)$$

$$= P(Z \geq 1.07) = 1 - \Phi(1.07) = .1423$$

The probability that X falls outside the interval (1.59, 2.47) is

$$P(X \leq 1.59 \text{ or } \geq 2.47) = P(X \leq 1.59) + P(X \geq 2.47)$$

$$= P\left(\frac{X - 2.03}{.44} \leq \frac{1.59 - 2.03}{.44}\right) + P\left(\frac{X - 2.03}{.44} \geq \frac{2.47 - 2.03}{.44}\right)$$

$$= P(Z \leq -1) + P(Z \geq 1) = .1587 + .1587 = .3174$$

The difference $X - \mu$ gives the distance between X and its mean, so $Z = (X - \mu)/\sigma$ is this distance measured as a number of standard deviations. For example, if $\mu = 40$ and $\sigma = 1.5$ as in Example 4.13, then $x = 43$ corresponds to $z = (43 - 40)/1.5 = 2$, so the value 43 is two standard deviations above the mean. The value $x = 38.5$ is one standard deviation below the mean since $z = (38.5 - 40)/1.5 = -1$. Thus the probability that any normal r.v. falls within, say, one standard deviation of its mean is $P(-1 \leq Z \leq 1) = .6826$, independent of μ and σ. Similarly the probability of an observed value within two standard deviations of μ is $P(-2 \leq Z \leq 2) = .9544$, and within 3σ of μ is $P(-3 \leq Z \leq 3) = .9974$.

> Approximately 68% of the values in any normal population lie within one standard deviation of the mean, approximately 95% lie within two standard deviations of μ, and approximately 99.7% lie within three standard deviations of μ.

It is indeed rare to observe a value from a normal population which is much further than two standard deviations from μ. These results will be important in testing hypotheses about μ.

Percentiles of an Arbitrary Normal Distribution

The $(100p)$th percentile of a normal distribution with mean μ and standard deviation σ is easily related to the $(100p)$th percentile of the standard normal distribution.

> **Proposition:** $\begin{array}{l}(100p)\text{th percentile} \\ \text{for normal } (\mu, \sigma)\end{array} = \mu + \left[\begin{array}{l}(100p)\text{th for} \\ \text{standard normal}\end{array}\right] \cdot \sigma$

Another way of saying this is that if z is the desired percentile for the standard normal distribution, then the desired percentile for the normal (μ, σ) distribution is z standard deviations from μ.

Proof: $P(X \leq \mu + [(100p)\text{th for standard normal}] \cdot \sigma) = P[(X - \mu)/\sigma \leq (100p)\text{th for standard normal}] = P[Z \leq (100p)\text{th for standard normal}] = p$ by definition of the $(100p)$th percentile.

Example 4.15   A famous restaurant does not accept reservations, but knows that on Saturday night a customer's waiting time X for a table is normally distributed with $\mu = 20$ and $\sigma = 3.876$. The restaurant wishes to advertise that anyone not seated within t minutes will be given a free drink. What value of t will ensure that one-half of 1% of all Saturday evening customers will be given free drinks? The desired value of t satisfies $P(X \geq t) = .005$, which means that t is the 99.5th percentile of the normal distribution with $\mu = 20$ and $\sigma = 3.876$. Since the 99.5th percentile of the standard normal distribution is 2.58, $t = 20 + (2.58)(3.876) = 30$ min. (See Figure 4.21.)

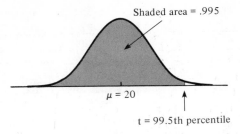

Figure 4.21   The waiting time distribution for Example 4.15

The Normal Distribution and Discrete Populations

The normal distribution is often used as an approximation to the distribution of values in a discrete population. In such situations extra care must be taken to ensure that probabilities are computed in an accurate manner.

Example 4.16   IQ in a particular population (as measured by a standard test) is known to be approximately normally distributed with $\mu = 100$ and $\sigma = 15$. What is the probability that a randomly selected individual has an IQ of at least 125? Letting $X =$ the IQ of a randomly chosen person, we wish $P(X \geq 125)$. The temptation here is to standardize $X \geq 125$ immediately as in previous examples. However, the IQ population is actually discrete, since IQ's are integer valued, so the normal curve is an approximation to a discrete probability histogram as pictured in Figure 4.22.

The rectangles of the histogram are *centered* at integers, so IQ's of at least 125 correspond to rectangles beginning at 124.5, as shaded in Figure 4.22. Thus we really want $P(X \geq 124.5)$, which can now be standardized to yield $P(Z \geq 1.63) = .0516$. If we had standardized $X \geq 125$, we would have obtained $P(Z \geq 1.67) = .0475$. The difference is not great, but the answer .0516 is more accurate. Similarly, $P(X = 125)$ would be approximated by the area between 124.5 and 125.5, since the area under the normal curve above the single value 125 is zero.

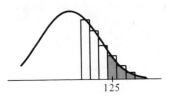

Figure 4.22   A normal approximation to a discrete distribution

The correction for discreteness of the underlying distribution in Example 4.16 is often called a **continuity correction.** It will be used again in the next chapter when approximating binomial probabilities.

Is the Distribution Normal?

A number of important inferential procedures are based on the assumption that the distribution being sampled is (at least approximately) normal. In Chapter 14 we describe several "goodness-of-fit" procedures for deciding whether a normality assumption is reasonable. For now an informal check can be based on the fact that approximately 68%, 95%, and 99.7% of a normal sample should lie within 1, 2, and 3 standard deviations, respectively, of the mean.

Let the sample be denoted by $x_1, \ldots, x_n$. The probability that a single observation falls within $(\mu - \sigma, \mu + \sigma)$ is approximately $p = .68$, so we expect $np = .68n$ of the x_i's to be in this interval. How much of a discrepancy from this expected number would be surprising? The standard deviation of the number falling inside this interval is $\sqrt{npq} = .47\sqrt{n}$, so three standard deviations is $1.41\sqrt{n}$. Our rule of thumb is that if the observed number in the interval differs from $.68n$ by more than $1.41\sqrt{n}$, the assumption of normality is suspect. Similar results apply to the intervals $(\mu - 2\sigma, \mu + 2\sigma)$ with $p = .95$ and $(\mu - 3\sigma, \mu + 3\sigma)$ with $p = .997$.

Since μ and σ are usually unknown, when n is large we replace them by the sample mean $\bar{x}$ and sample standard deviation s. This leads to the following empirical rule.

Empirical rule for checking normality: If either

1. $|\text{number in } (\bar{x} - s, \bar{x} + s) - .68n| \geq 1.41\sqrt{n}$
2. $|\text{number in } (\bar{x} - 2s, \bar{x} + 2s) - .95n| \geq .654\sqrt{n}$

or

3. $|\text{number in } (\bar{x} - 3s, \bar{x} + 3s) - .997n| \geq .164\sqrt{n}$

the assumption of normality is of doubtful validity.

Example 4.17

In Example 1.3 we presented $n = 80$ observations on octane rating. The sample mean and standard deviation are $\bar{x} = 90.64$ and $s = 2.80$, so the three intervals are $(87.84, 93.44)$, $(85.04, 96.24)$, and $(82.24, 99.04)$, containing 61, 75, and 79 observations, respectively. The inequalities 1, 2, and 3 become $|61-54.4| \geq 12.61$, $|75-76| \geq 5.85$, and $|79-79.8| \geq 1.47$. Since none of the inequalities is satisfied, an assumption of normality is plausible.

When the x_i's themselves do not appear to have come from a normal population, sometimes a transformation applied to the x_i's results in a data set which is more nearly normal. For example, the transformations $\sqrt{x}$ and $\log(x)$ make larger values smaller, which counteracts the presence of a few outlying values. The book *Data Analysis and Regression* (in the Chapter 6 bibliography) discusses such transformations.

Another informal method for checking normality is to plot the data on a special type of paper called normal probability paper. This paper is constructed so that a sample from a normal population should give roughly a straight line plot. A good description of such plots is contained in the book *Fitting Equations to Data* (listed in the Chapter 13 bibliography).

Exercises / Section 4.3

1. Let Z be a standard normal random variable and calculate the following probabilities, drawing pictures wherever appropriate:
   a. $P(0 \leq Z \leq 2.17)$ b. $P(0 \leq Z \leq 1)$
   c. $P(-2.50 \leq Z \leq 0)$
   d. $P(-2.50 \leq Z \leq 2.50)$
   e. $P(Z \leq 1.37)$ f. $P(-1.75 \leq Z)$
   g. $P(-1.50 \leq Z \leq 2.00)$
   h. $P(1.37 \leq Z \leq 2.50)$ i. $P(1.50 \leq Z)$
   j. $P(|Z| \leq 2.50)$

2. In each case below, determine the value of the constant c which makes the probability statement correct.
   a. $\Phi(c) = .9838$ b. $P(0 \leq Z \leq c) = .291$
   c. $P(c \leq Z) = .121$
   d. $P(-c \leq Z \leq c) = .668$
   e. $P(c \leq |Z|) = .016$

3. Find the following percentiles for the standard normal distribution. Interpolate where appropriate.
   a. ninety-first   b. ninth   c. seventy-fifth
   d. twenty-fifth   e. sixth

4. Determine z_α for
   a. $\alpha = .0055$ b. $\alpha = .09$ c. $\alpha = .663$

5. If X is a normal random variable with mean 80 and standard deviation 10, compute the following probabilities by standardizing:
   a. $P(X \leq 100)$ b. $P(X \leq 80)$
   c. $P(65 \leq X \leq 100)$ d. $P(70 \leq X)$
   e. $P(85 \leq X \leq 95)$ f. $P(|X - 80| \leq 10)$

6. Suppose that the force acting on a column which helps to support a building is normally distributed with mean 15.0 kips and standard deviation 1.25 kips.
   a. What is the probability that the force is at most 17 kips?
   b. What is the probability that the force is between 12 and 17 kips?
   c. What is the probability that the force differs from 15.0 kips by at most two standard deviations?

7. Assume that development time for a particular type of photographic printing paper when it is exposed to a light source for five seconds is normally distributed with a mean of 25 seconds and a standard deviation of 1.3 seconds.
   a. What is the probability that a particular print will require more than 26.5 seconds to develop?
   b. What is the probability that development time is at least 23 seconds?
   c. What is the probability that development time differs from the expected time by more than 2.5 seconds?

8. A particular type of gasoline tank for a compact car is designed to hold 15 gallons. Suppose that the actual capacity X of a randomly chosen tank of this type is normally distributed with mean 15 and standard deviation .2 gallons.
   a. What is the probability that a randomly selected tank will hold at most 14.8 gallons?
   b. What is the probability that a randomly selected tank will hold between 14.7 and 15.1 gallons?
   c. If the car on which a randomly chosen tank is mounted gets exactly 25 m.p.g., what is the probability that the car can travel 370 miles without refueling?

9. a. If $X \sim N(\mu, \sigma^2)$, show that $E(X) = \mu$. *Hint:* After setting up the integral expression for $E(X)$, let $y = x - \mu$. Then the integral can be split into two parts, one of which equals zero.
   b. If $X \sim N(\mu, \sigma^2)$, show that Var $(X) = \sigma^2$. *Hint:* After writing the integral expression for $E[(X - \mu)^2]$, make the change of variable $y = (x - \mu)/\sigma$. Then integrate by parts using $dv = ye^{-y^2/2}dy$.

10. a. If X has a normal distribution with parameters μ and σ, show that $Y = aX + b$ (a linear function of X) also has a normal distribution. What are the parameters of the distribution of Y [that is, $E(Y)$ and Var (Y)]? *Hint:* Write the c.d.f. of

Y, $P(Y \leq y)$, as an integral involving the p.d.f. of X, and then differentiate with respect to y to get the p.d.f. of Y.

b. If when measured in °C, temperature is normally distributed with mean 115 and standard deviation 2, what can be said about the distribution of temperature measured in °F?

11. a. If a normal distribution has $\mu = 25$ and $\sigma = 5$, what is the ninety-first percentile of the distribution?

b. What is the sixth percentile of the distribution of (a)?

c. If bearing diameter is normally distributed with mean .25 in. and standard deviation .002 in., below what value will 95% of all bearings have diameters?

d. As in (c), above what value will 10% of all bearings have diameters?

12. Suppose that the pH of soil samples taken from a certain geographic region is normally distributed with mean pH 6.00 and standard deviation .10. If the pH of a randomly selected soil sample from this region is determined,

a. What is the probability that the resulting pH is between 5.90 and 6.15?

b. What is the probability that the resulting pH exceeds 6.10?

c. What is the probability that the resulting pH is at most 5.95?

d. What value will be exceeded by only 5% of all such pH's?

13. The Rockwell hardness of a metal is determined by impressing a hardened point into the surface of the metal and then measuring the depth of penetration of the point. Suppose that the Rockwell hardness of a particular alloy is normally distributed with mean 70 and standard deviation 3 (assume that Rockwell hardness is measured on a continuous scale).

a. If a specimen is acceptable only if its hardness is between 65 and 75, what is the probability that a randomly chosen specimen has an acceptable hardness?

b. If the acceptable range of hardness was $(70 - c, 70 + c)$, for what value of c would 95% of all specimens have acceptable hardness?

c. If the acceptable range is as in (a) and the hardness of each of 10 randomly selected specimens

is independently determined, what is the expected number of acceptable specimens among the 10?

d. What is the probability that at most eight of 10 independently selected specimens have a hardness of less than 73.84? *Hint:* Y = the number among the 10 specimens with hardness less than 73.84 is a binomial variable; what is p?

14. Suppose that Appendix Table A.3 contained $\Phi(z)$ only for $z \geq 0$. Explain how you could still compute

a. $P(-1.72 \leq Z \leq -.55)$

b. $P(-1.72 \leq Z \leq .55)$

Is it necessary to table $\Phi(z)$ for z negative? What property of the standard normal curve justifies your answer?

15. The distribution of time taken to read through once and answer questions on a multiple-choice exam is known to be normal with mean 65 min and standard deviation 15.2 min. How long should the exam last if the examiners want to ensure that 95% of all those taking the exam have at least 10 min to check back over their work after going through the exam once? *Hint:* First answer the question with 0 min replacing 10 min.

16. If the diameter of bearings produced by a certain machine has a normal distribution with $\sigma = .002$ in., what is the probability that a randomly chosen bearing will have a diameter which differs from the mean diameter by at least .005 in.?

17. Let X represent the number of pages of text in a randomly chosen mathematics Ph.D. thesis. While X can assume only positive integer values, suppose that it is approximately normally distributed with expected value 90 and standard deviation 15.

a. What is the probability that a randomly chosen thesis contains at most 100 pages (using the continuity correction)?

b. What is the probability that a randomly chosen thesis contains between 80 and 110 pages (using the continuity correction)?

18. Is it generally true that if X has a distribution which belongs to a certain family, then the standardized version of X also belongs to that family (as is the case for X normal)? *Hint:* Take X binomial.

19. In Example 1.5 101 observations on lifetime of aluminum strips subjected to stress are given. The sample mean and standard deviation are $\bar{x} =$ 1400.9 and 389.4, respectively. Does the empir-

ical rule indicate that a normal distribution for life-times is a reasonable assumption?

4.4 **The Gamma Distribution and Its Relatives**

The graph of any normal p.d.f. is bell-shaped and thus symmetric. There are many practical situations in which the variable of interest to the experimenter might have a skewed distribution. A family of p.d.f.'s that yields a wide variety of skewed distributional shapes is the gamma family. To define the family of gamma distributions, we first need to introduce a function which plays an important role in many branches of mathematics.

> **Definition:** For $\alpha > 0$, the **gamma function** $\Gamma(\alpha)$ is defined by
>
> $$\Gamma(\alpha) = \int_0^\infty x^{\alpha-1} e^{-x} \, dx \tag{4.5}$$

The most important properties of the gamma function are:

1. For any $\alpha > 1$, $\Gamma(\alpha) = (\alpha - 1) \cdot \Gamma(\alpha - 1)$
2. For any positive integer n, $\Gamma(n) = (n - 1)!$
3. $\Gamma(\frac{1}{2}) = \sqrt{\pi}$

By (4.5) if we let

$$f(x; \alpha) = \begin{cases} \dfrac{x^{\alpha-1} e^{-x}}{\Gamma(\alpha)} & x \geq 0 \\ 0 & \text{otherwise} \end{cases} \tag{4.6}$$

then $f(x; \alpha) \geq 0$ and $\int_0^\infty f(x; \alpha) \, dx = \dfrac{\Gamma(\alpha)}{\Gamma(\alpha)} = 1$, so $f(x; \alpha)$ satisfies the two basic properties of a p.d.f.

The Family of Gamma Distributions

> **Definition:** A continuous random variable X will be said to have a **gamma distribution** if the p.d.f. of X is
>
> $$f(x; \alpha, \beta) = \begin{cases} \dfrac{1}{\beta^\alpha \Gamma(\alpha)} x^{\alpha-1} e^{-x/\beta} & x \geq 0 \\ 0 & \text{otherwise} \end{cases} \tag{4.7}$$
>
> where the parameters α and β satisfy $\alpha > 0$, $\beta > 0$. The **standard gamma distribution** has $\beta = 1$, so the p.d.f. of a standard gamma r.v. is given by (4.6).

Figure 4.23 illustrates the graphs of the gamma p.d.f. (4.7) for several (α, β) pairs, while Figure 4.24 presents graphs of the standard gamma p.d.f. For the

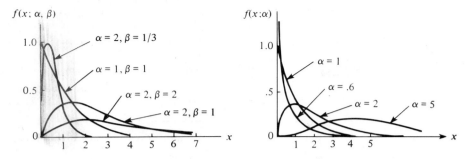

Figure 4.23 Gamma density functions   **Figure 4.24** Standard gamma density functions

standard p.d.f., when $\alpha \leq 1$, $f(x; \alpha)$ is strictly decreasing as x increases from 0; when $\alpha > 1$, $f(x; \alpha)$ rises from 0 at $x = 0$ to a maximum and then decreases. The parameter β in (4.7) is called the scale parameter because values other than 1 either stretch or compress the p.d.f. in the x direction.

The Mean and Variance of X

> **Proposition:** The mean and variance of a random variable X having the gamma distribution (4.7) are
>
> $$E(X) = \mu = \alpha\beta, \quad \text{Var}(X) = \sigma^2 = \alpha\beta^2$$

$E(X)$ and $E(X^2)$ can be obtained from a reasonably straightforward integration, and then $\text{Var}(X) = E(X^2) - [E(X)]^2$.

Computing Probabilities from the Gamma Distribution

When X is a standard gamma r.v., the c.d.f. of X,

$$F(x; \alpha) = \int_0^x \frac{y^{\alpha-1}e^{-y}}{\Gamma(\alpha)} \, dy \quad x > 0 \tag{4.8}$$

is called the **incomplete gamma function** [sometimes the incomplete gamma function refers to (4.8) without the denominator $\Gamma(\alpha)$ in the integrand]. There are extensive tables of $F(x; \alpha)$ available; in Appendix Table A.4 we present a small table for $\alpha = 1, 2, \ldots, 10$ and $x = 1, 2, \ldots, 15$.

Example 4.18 Suppose that the reaction time X of a randomly selected individual to a certain stimulus has a standard gamma distribution with $\alpha = 2$ seconds. Since

$$P(a \leq X \leq b) = F(b) - F(a) \text{ when } X \text{ is continuous,}$$

$$P(3 \leq X \leq 5) = F(5; 2) - F(3; 2) = .960 - .801 = .159$$

The probability that the reaction time is more than four seconds is

$$P(X > 4) = 1 - P(X \le 4) = 1 - F(4; 2) = 1 - .908 = .092$$

The incomplete gamma function can also be used to compute probabilities involving nonstandard gamma distributions.

Proposition: Let X have a gamma distribution with parameters α and β. Then for any $x > 0$, the c.d.f. of X is given by

$$P(X \le x) = F\left(\frac{x}{\beta}; \alpha\right)$$

where $F(\cdot; \alpha)$ is the incomplete gamma function.

Example 4.19   Suppose that the survival time X in weeks of a randomly selected male mouse exposed to 240 rads of gamma radiation has a gamma distribution with $\alpha = 8$ and $\beta = 15$ (data in *Survival Distributions: Reliability Applications in the Biomedical Services* by A. J. Gross and V. Clark, suggests $\alpha \doteq 8.5$ and $\beta \doteq 13.3$). The expected survival time is $E(X) = (8)(15) = 120$ weeks, while Var $(X) = (8)(15)^2 = 1800$ and $\sigma_X = \sqrt{1800} = 42.43$ weeks. The probability that a mouse survives between 60 and 120 weeks is

$$P(60 \le X \le 120) = P(X \le 120) - P(X \le 60)$$
$$= F(120/15; 8) - F(60/15; 8)$$
$$= F(8; 8) - F(4; 8) = .547 - .051 = .496$$

The probability that a mouse survives at least 30 weeks is

$$P(X \ge 30) = 1 - P(X < 30) = 1 - P(X \le 30)$$
$$= 1 - F(30/15; 8) = .999$$

The Exponential Distribution

Definition: X is said to have an exponential distribution if the p.d.f. of X is

$$f(x; \lambda) = \begin{cases} \lambda e^{-\lambda x} & x \ge 0 \\ 0 & \text{otherwise} \end{cases} \quad \text{where} \quad \lambda > 0 \qquad (4.9)$$

The exponential p.d.f. is a special case of the general gamma p.d.f. (4.7) in which $\alpha = 1$ and β has been replaced by $1/\lambda$ (some authors use the form $1/\beta\, e^{-x/\beta}$). The mean and variance of X are then

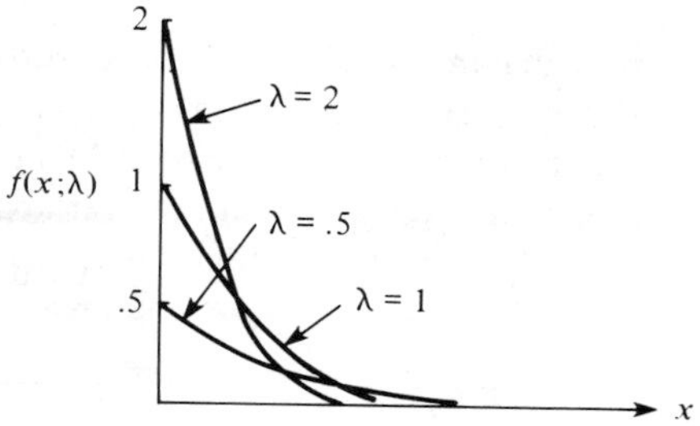

Figure 4.25 Exponential density functions

$$\mu = \alpha\beta = \frac{1}{\lambda}, \qquad \sigma^2 = \alpha\beta^2 = \frac{1}{\lambda^2}$$

Both the mean and standard deviation of the exponential distribution equal $1/\lambda$. Graphs of several exponential p.d.f.'s appear in Figure 4.25.

Unlike the general gamma p.d.f, the exponential p.d.f. can be easily integrated. In particular, the c.d.f. of X is

$$F(x; \lambda) = \begin{cases} 0 & x < 0 \\ 1 - e^{-\lambda x} & x \geq 0 \end{cases}$$

Example 4.20 Suppose that the response time X at a certain on-line computer terminal (the elapsed time between the end of a user's inquiry and the beginning of the system's response to that inquiry) has an exponential distribution with expected response time equal to 5 seconds. Then $E(X) = 1/\lambda = 5$, so $\lambda = .2$. The probability that the response time is at most 10 seconds is

$$P(X \leq 10) = F(10; .2) = 1 - e^{-(.2)(10)} = 1 - e^{-2} = 1 - .135 = .865$$

The probability that response time is between 5 and 10 seconds is

$$P(5 \leq X \leq 10) = F(10; 2) - F(5; .2) = (1 - e^{-2}) - (1 - e^{-1}) = .233$$

Applications of the Exponential Distribution

The exponential distribution is frequently used as a model for the distribution of times between the occurrence of successive events such as customers arriving at a service facility or calls coming in to a switchboard. The reason for this is that the exponential distribution is closely related to the Poisson process discussed in Chapter 3.

> **Proposition:** Suppose that the number of events occurring in any time interval of length t has a Poisson distribution with parameter λt (where λ, the rate of the event process, is the expected number of events occurring in one unit of time) and that numbers of occurrences in nonoverlapping intervals are independent of one another. Then the distribution of elapsed time between the occurrence of two successive events is exponential with parameter λ.

While a complete proof is beyond the scope of the text, the result is easily verified for the time X_1 until the first event occurs:

$$P(X_1 \leq t) = 1 - P(X_1 > t) = 1 - P\,[\text{no events in } (0, t)]$$

$$= 1 - \frac{e^{-\lambda t} \cdot (\lambda t)^0}{0!} = 1 - e^{-\lambda t}$$

which is exactly the c.d.f. of the exponential distribution.

Example 4.21 Suppose that calls are received at a 24-hour "suicide hotline" according to a Poisson process with rate $\lambda = .5$ calls per day. Then the number of days X between successive calls has an exponential distribution with parameter $\lambda = .5$, so the probability that more than two days elapse between calls is

$$P(X > 2) = 1 - P(X \leq 2) = 1 - F(2; .5) = e^{-(.5)(2)} = .368$$

Another important application of the exponential distribution is to model the distribution of component lifetime. A partial reason for the popularity of such applications is the **"memoryless" property** of the exponential distribution. Suppose that component lifetime is exponentially distributed with parameter λ. After putting the component into service, we leave for a period of t_0 hours and then return to find the component still working; what now is the probability that it lasts at least an additional t hours? In symbols we wish $P(X \geq t + t_0 \mid X \geq t_0)$. By the definition of conditional probability,

$$P(X \geq t + t_0 \mid X \geq t_0) = \frac{P\,[(X \geq t + t_0) \cap (X \geq t_0)]}{P(X \geq t_0)}$$

But the event $X \geq t_0$ in the numerator is redundant, since both events can occur if and only if $X \geq t + t_0$. Therefore

$$P(X \geq t + t_0 \mid X \geq t_0) = \frac{P(X \geq t + t_0)}{P(X \geq t_0)} = \frac{1 - F(t + t_0; \lambda)}{1 - F(t_0; \lambda)} = e^{-\lambda t}$$

This conditional probability is identical to the original probability $P(X \geq t)$ that the component lasted t hours. Thus *the distribution of additional lifetime is exactly the same as the original distribution of lifetime,* so at each point in time the component shows no effect of wear. A way to paraphrase this is to say that the distribution of remaining lifetime is independent of current age.

While the memoryless property can be justified at least approximately in many applied problems, in other situations components deteriorate with age or occasionally improve with age (at least up to a certain point). More general lifetime models are then furnished by the gamma, Weibull, or lognormal distributions (the latter two are discussed in the next section).

The Chi-Squared Distribution

> **Definition:** Let ν be a positive integer. Then a r.v. X is said to have a **chi-squared distribution** with parameter ν if the p.d.f. of X is the gamma density (4.7) with $\alpha = \nu/2$ and $\beta = 2$. The p.d.f. of a chi-squared r.v. is thus
>
> $$f(x; \nu) = \begin{cases} \dfrac{1}{2^{\nu/2}\,\Gamma(\nu/2)} x^{(\nu/2)-1} e^{-x/2} & x \geq 0 \\ 0 & x < 0 \end{cases}$$
>
> The parameter ν is called the **number of degrees of freedom** (d.f.) of X. The symbol χ^2 is often used in place of "chi-squared."

The chi-squared distribution is important because it is the basis for a number of procedures in statistical inference. The reason for this is that chi-squared distributions are intimately related to normal distributions (see Exercise 4.4.12). We will discuss the chi-squared distribution in more detail in the chapters on inference.

Exercises / Section 4.4

1. Evaluate
   a. $\Gamma(6)$
   b. $\Gamma(5/2)$
   c. $F(4; 5)$ (the incomplete gamma function)
   d. $F(5; 4)$
   e. $F(0; 4)$

2. Let X have a standard gamma distribution with $\alpha = 7$. Evaluate
   a. $P(X \leq 5)$
   b. $P(X < 5)$
   c. $P(X > 8)$
   d. $P(3 \leq X \leq 8)$
   e. $P(3 < X < 8)$
   f. $P(X < 4 \ \text{ or } \ X > 6)$

3. Suppose that the time (in hours) taken by a homeowner to mow his lawn is a r.v. X having a gamma distribution with parameters $\alpha = 2$ and $\beta = \frac{1}{2}$.
   a. What is the probability that it takes at most 1 hour to mow the lawn?
   b. What is the probability that it takes at least 2 hours to mow the lawn?

c. What is the probability that it takes between .5 and 1.5 hours to mow the lawn?

4. Suppose that the time spent by a randomly selected student who uses a terminal connected to a local timesharing computer facility has a gamma distribution with mean 20 min and variance 80 min^2.
   a. What are the values of α and β?
   b. What is the probability that a student uses the terminal for at most 24 minutes?
   c. What is the probability that a student spends between 20 and 40 minutes using the terminal?

5. Suppose that when a transistor of a certain type is subjected to an accelerated life test, the lifetime X (in weeks) has a gamma distribution with mean 24 weeks and standard deviation 12 weeks.
   a. What is the probability that a transistor will last

between 12 and 24 weeks?

b. What is the probability that a transistor will last at most 24 weeks? Is the median of the lifetime distribution less than 24? Why or why not?

c. What is the ninety-ninth percentile of the lifetime distribution?

d. Suppose that the test will actually be terminated after t weeks. What value of t is such that only one-half of 1% of all transistors would still be operating at termination?

6. Let X = the time between two successive arrivals at the drive-up window of a local bank. If X has an exponential distribution with $\lambda = 1$ (which is identical to a standard gamma distribution with $\alpha = 1$), compute

a. the expected time between two successive arrivals

b. the standard deviation of the time between successive arrivals

c. $P(X \le 4)$ **d.** $P(2 \le X \le 5)$

7. The time X (seconds) that it takes a librarian to locate a card in a file of records on checked-out books has an exponential distribution with expected time = 20 seconds. Express the following in terms of e, and obtain numerical answers if your calculator has an e^x button.

a. $P(X \le 30)$ **b.** $P(X \ge 20)$ **c.** $P(20 \le X \le 30)$

d. For what value of t is $P(X \le t) = .5$? (t is the fiftieth percentile of the distribution)

8. If X has an exponential distribution with parameter λ, derive a general expression for the $(100p)$th percentile of the distribution. Then specialize to obtain the median.

9. Show that $\int_0^\infty f(x; \alpha, \beta)\, dx = 1$ for the gamma p.d.f. of (4.7). *Hint:* Let $y = x/\beta$ to obtain an integrand like that of (4.6).

10. The special case of the gamma distribution in which α is a positive integer n is called an Erlang distribution. If we replace β by $1/\lambda$ in (4.7) the Erlang p.d.f. is

$$f(x; \lambda, n) = \begin{cases} \dfrac{\lambda(\lambda x)^{n-1} e^{-\lambda x}}{(n-1)!} & x \ge 0 \\ 0 & x < 0 \end{cases}$$

It can be shown that if the times between successive events are independent, each with an exponential distribution with parameter λ, then the

total time X which elapses before all of the next n events occur has p.d.f. $f(x; \lambda, n)$.

a. What is the expected value of X? If the time (min) between arrivals of successive customers is exponentially distributed with $\lambda = .5$, how much time can be expected to elapse before the tenth customer arrives?

b. If customer interarrival time is exponentially distributed with $\lambda = .5$, what is the probability that the tenth customer (after the one which has just arrived) will arrive within the next 30 minutes?

c. The event $\{X \le t\}$ occurs iff at least n events occur in the next t units of time. Use the fact that the number of events occurring in an interval of length t has a Poisson distribution with parameter λt to write down an expression (involving Poisson probabilities) for the c.d.f. $F(t; \lambda, n)) = P(X \le t)$ of the Erlang distribution.

11. A system consists of five identical components connected in series as shown:

As soon as one component fails the entire system will fail. Suppose that each component has a lifetime which is exponentially distributed with $\lambda = .01$, and that components fail independently of one another. Define events $A_i = \{i$th component lasts at least t hours$\}$, $i = 1, \ldots, 5$, so that the A_i's are independent events. Let X = the time at which the system fails = the shortest (minimum) lifetime among the five components.

a. The event $\{X \ge t\}$ is equivalent to what event involving $A_1, \ldots, A_5$?

b. Using the independence of the A_i's, compute $P(X \ge t)$. Then obtain $F(t) = P(X \le t)$ and the p.d.f. of X. What type of distribution does X have?

c. Suppose that there are n components, each having exponential lifetime with parameter λ. What type of distribution does X have?

12. a. The event $\{X^2 \le y\}$ is equivalent to what event involving X itself?

b. If X has a standard normal distribution, use (a) to write the integral which equals $P(X^2 \le y)$. Then differentiate this with respect to y to obtain the p.d.f. of X^2 [the square of a $N(0, 1)$

variable]. Finally, show that X^2 has a chi-squared distribution with $\nu = 1$ degrees of freedom.

$$Hint: \quad \frac{d}{dy}\left\{\int_{a(y)}^{b(y)} f(x)\, dx\right\}$$

$$= f[b(y)] \cdot b'(y) - f[a(y)] \cdot a'(y)$$

4.5   Other Continuous Distributions

The normal, gamma (including exponential), and uniform families of distributions provide a wide variety of probability models for continuous variables, but there are many practical situations in which no member of these families fits a set of observed data very well. Statisticians and other investigators have developed other families of distributions which are often appropriate in practice.

The Weibull Distribution

The family of Weibull distributions was introduced by the Swedish physicist Waloddi Weibull in 1939; his 1951 paper "A Statistical Distribution Function of Wide Applicability" (*J. Applied Mechanics,* vol. 18, pp. 293–297) discusses a number of applications.

Definition: A random variable X is said to have a **Weibull distribution** with parameters α and β ($\alpha > 0$, $\beta > 0$) if the p.d.f. of X is

$$f(x; \alpha, \beta) = \begin{cases} \dfrac{\alpha}{\beta^\alpha} x^{\alpha-1} e^{-(x/\beta)^\alpha} & x \geq 0 \\[2mm] 0 & x < 0 \end{cases} \tag{4.11}$$

While in some situations there are theoretical justifications for the appropriateness of the Weibull distribution, in many applications $f(x; \alpha, \beta)$ simply provides a good fit to observed data for particular values of α and β. When $\alpha = 1$, the p.d.f. reduces to the exponential distribution (with $\lambda = 1/\beta$), so the exponential distribution is a special case of both the gamma and Weibull distributions. However, there are gamma distributions which are not Weibull distributions and vice versa, so one family is not a subset of the other. By varying α and β a number of different distributional shapes can be obtained, as illustrated in Figure 4.26. β is a scale parameter, so different values stretch or compress the graph in the x direction.

Integrating to obtain $E(X)$ and $E(X^2)$ yields

$$\mu = \beta\Gamma\left(1 + \frac{1}{\alpha}\right), \quad \sigma^2 = \beta^2\left\{\Gamma\left(1 + \frac{2}{\alpha}\right) - \left[\Gamma\left(1 + \frac{1}{\alpha}\right)\right]^2\right\}$$

The computation of μ and σ^2 thus necessitates using a table of the gamma function.

The integration $\int_0^x f(y; \alpha, \beta)\, dy$ is easily carried out to obtain the cumulative distribution function of X.

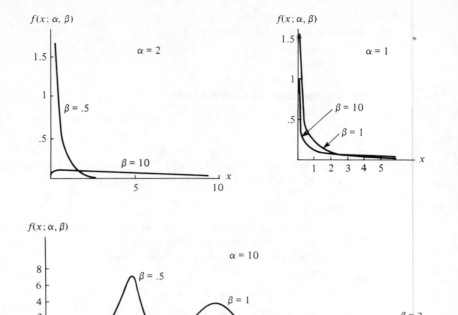

Figure 4.26   Graphs of Weibull p.d.f.'s

Proposition: The c.d.f. of a Weibull r.v. having parameters α and β is

$$F(x;\alpha,\beta) = \begin{cases} 0 & x < 0 \\ 1 - e^{-(x/\beta)^{\alpha}} & x \geq 0 \end{cases} \qquad (4.12)$$

Example 4.22   Let X = the ultimate tensile strength (ksi) at $-200\,°F$ of a type of steel which exhibits "cold brittleness" at low temperatures. Suppose that X has a Weibull distribution with parameters $\alpha = 20$ and $\beta = 100$. Then

$$P(X \leq 105) = F(105;20,100) = 1 - e^{-(105/100)^{20}} = 1 - .070 = .930$$

and

$$P(98 \leq X \leq 102) = F(102;20,100) - F(98;20,100)$$
$$= e^{-(.98)^{20}} - e^{-(1.02)^{20}} = .513 - .226 = .287$$

Frequently in practical situations a Weibull model may be reasonable except that the smallest possible X value may be some value γ not assumed to be zero (this would also apply to a gamma model). The quantity γ can then be regarded as a third parameter of the distribution, which is what Weibull did in his original work. For, say, $\gamma = 3$, all curves in Figure 4.26 would be shifted three units to the right. This is equivalent to saying that $X - \gamma$ has the p.d.f. (4.11), so that the c.d.f. of X is obtained by replacing x in (4.12) by $x - \gamma$.

Example 4.23 Let $X =$ the corrosion weight loss for a small square magnesium alloy plate immersed for seven days in an inhibited aqueous 20% solution of $MgBr_2$. Suppose that the minimum possible weight loss is $\gamma = 3$ and that the excess $X - 3$ over this minimum has a Weibull distribution with $\alpha = 2$ and $\beta = 4$ (this example was considered in "Practical Applications of the Weibull Distribution," *Industrial Quality Control,* August 1964, pp. 71–78; values for α and β were taken to be 1.8 and 3.67, respectively, though a slightly different choice of parameters was used in the article). The c.d.f. of X is then

$$F(x; \alpha, \beta, \gamma) = F(x; 2, 4, 3) = \begin{cases} 0 & x < 3 \\ 1 - e^{-[(x-3)/4]^2} & x \geq 3 \end{cases}$$

Therefore

$$P(X > 3.5) = 1 - F(3.5; 2, 4, 3) = e^{-.0156} = .985$$

and

$$P(7 \leq X \leq 9) = 1 - e^{-2.25} - (1 \cdot e^{-1}) = .895 - .632 = .263$$

The Lognormal Distribution

Definition: A nonnegative r.v. X is said to have a **lognormal distribution** if the r.v. $Y = \ln(X)$ has a normal distribution. The resulting p.d.f. of a lognormal r.v. when $\ln(X)$ is normally distributed with parameters μ and σ is

$$f(x; \mu, \sigma) = \begin{cases} \dfrac{1}{\sqrt{2\pi}\,\sigma x} e^{-\frac{1}{2\sigma^2}[\ln(x) - \mu]^2} & x \geq 0 \\ 0 & x < 0 \end{cases}$$

Be careful here; μ and σ are not the mean and standard deviation of X but of $\ln(X)$. The mean and variance of X can be shown to be

$$E(X) = e^{\mu + \sigma^2/2}, \quad \mathrm{Var}(X) = e^{2\mu + \sigma^2} \cdot (e^{\sigma^2} - 1)$$

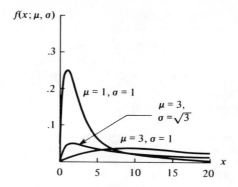

Figure 4.27 Graphs of the lognormal p.d.f.

In the next chapter we will present a theoretical justification for this distribution in connection with the Central Limit Theorem, but as with other distributions, the lognormal can be used as a model even in the absence of such justification. Figure 4.27 illustrates the graphs of the lognormal p.d.f.; although the normal curve is symmetric, a lognormal curve has a positive skew.

Because ln (X) has a normal distribution, the c.d.f. of X can be expressed in terms of the c.d.f. $\Phi(z)$ of a standard normal r.v. Z. For $x \geq 0$,

$$F(x; \mu, \sigma) = P(X \leq x) = P[\ln (X) \leq \ln(x)]$$

$$= P\left(Z \leq \frac{\ln(x) - \mu}{\sigma}\right) = \Phi\left(\frac{\ln(x) - \mu}{\sigma}\right) \qquad (4.13)$$

Example 4.24 Let X = the hourly median power (in decibels) of received radio signals transmitted between two cities. Because a decibel is a unit of measurement equal to the logarithm of the ratio I/I_0, where I is the intensity of the signal of interest and I_0 is the intensity of a standard signal or sound, it is reasonable to assume that X has a lognormal distribution ("Families of Distributions for Hourly Median Power and Instantaneous Power of Received Radio Signals," *J. Research National Bureau of Standards,* 1963, vol. 67D, pp. 753–762). If the parameter values are $\mu = 3.5$ and $\sigma = 1.2$, then

$$E(X) = e^{3.5 + .72} = 68.0, \quad \text{Var}(X) = e^{8.44}(e^{1.44} - 1) = 14,907.2$$

The probability that received power is between 50 and 250 db is

$$P(50 \le X \le 250) = F(250; 3.5, 1.2) - F(50; 3.5, 1.2)$$

$$= \Phi\left(\frac{\ln(250) - 3.5}{1.2}\right) - \Phi\left(\frac{\ln(50) - 3.5}{1.2}\right)$$

$$= \Phi(1.68) - \Phi(.41)$$

$$= .9535 - .6591 = .2944 \quad \text{(Appendix Table A.3)}$$

The probability that X does not exceed its mean is

$$P(X \le 68.0) = \Phi\left(\frac{\ln(68.0) - 3.5}{1.2}\right) = \Phi(.60) = .7257$$

If the distribution were symmetric, this probability would equal .5; it is much larger because of the positive skew (long upper tail) of the distribution, which pulls μ outward past the median.

The Beta Distribution

All families of continuous distributions discussed so far except for the uniform distribution have positive density over an infinite interval (though typically the density function decreases rapidly to zero beyond a few standard deviations from the mean). The beta distribution provides positive density only for X in an interval of finite length.

Definition: The random variable X is said to have a **standard beta distribution** with parameters α and β (both positive) if the p.d.f. of X is

$$f(x; \alpha, \beta) = \begin{cases} \dfrac{\Gamma(\alpha + \beta)}{\Gamma(\alpha) \cdot \Gamma(\beta)} x^{\alpha-1}(1 - x)^{\beta-1} & 0 \le x \le 1 \\ 0 & \text{otherwise} \end{cases}$$

A tricky integration is required to show that $\int_0^1 f(x; \alpha, \beta)\, dx = 1$. The c.d.f.

$$F(x; \alpha, \beta) = \int_0^x \frac{\Gamma(\alpha + \beta)}{\Gamma(\alpha) \cdot \Gamma(\beta)} y^{\alpha-1}(1 - y)^{\beta-1}\, dy$$

must be computed using numerical integration. It is called the incomplete beta function, and there are tables available (more cumbersome than the incomplete gamma function because of the extra parameter). The expected value and variance are

---

$$E(X) = \mu = \frac{\alpha}{\alpha + \beta}, \qquad \text{Var}(X) = \sigma^2 = \frac{\alpha\beta}{(\alpha + \beta)^2(\alpha + \beta + 1)}$$

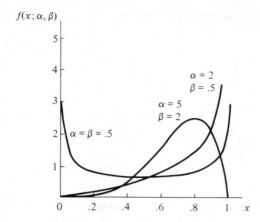

Figure 4.28   Graphs of standard beta p.d.f.'s

Figure 4.28 illustrates the beta p.d.f. for various choices of α and β. If both $\alpha > 1$ and $\beta > 1$, the p.d.f. is 0 at $x = 0$ and $x = 1$ and rises to a unique maximum between 0 and 1. If both $\alpha < 1$ and $\beta < 1$, the p.d.f. decreases from $+\infty$ at $x = 0$ and $x = 1$ to a unique minimum. If $\alpha > 1$ while $\beta < 1$ the p.d.f. increases from 0 to $+\infty$ as x increases, and the reverse occurs if $\alpha < 1$, $\beta > 1$. For $\alpha = \beta = 1$, X has a uniform distribution on [0, 1].

The standard beta p.d.f. is easily modified to obtain a model with positive density on [A, B]. A r.v. Y is said to have a beta distribution with parameters α, β, A, and B if $Y = A + (B - A)X$ where X has a standard beta distribution with parameters α and β. Since

$$P(Y \le y) = P\left(X \le \frac{y - A}{B - A}\right) = F\left(\frac{y - A}{B - A}; \alpha, \beta\right)$$

probabilities involving Y can be computed from a table of the incomplete beta function. Graphs of the p.d.f.'s of Y look like those of standard beta distributions, except that they have been shifted and then stretched or compressed to fit on the interval [A, B].

Example 4.25   Project managers often use a method labelled PERT—for program evaluation and review technique—to coordinate the various activities making up a large project (one successful application was in the construction of the Apollo spacecraft). A standard assumption in PERT analysis is that the time necessary to complete any particular activity once it has been started has a beta distribution with A = the optimistic time (if everything goes well) and B = the pessimistic time (if everything goes badly). Suppose that in constructing a single-family house, the time Y (in days) necessary for laying the foundation has a beta distribution with $A = 2$, $B = 5$, $\alpha = 2$, and $\beta = 3$. Then $E(X) = \alpha/(\alpha + \beta) = .4$, so $E(Y) = A + (B - A) \cdot E(X) = 3.2$. For these values of α and β, $f(x; \alpha, \beta)$ is a simple polynomial function, so tables of the incomplete beta function are unnecessary. The probability that it takes at most three days to lay the foundation is

$$P(Y \leq 3) = P\left(X \leq \frac{3-2}{5-2}\right) = P\left(X \leq \frac{1}{3}\right) = \int_0^{1/3} 6x(1-x)^2 \, dx = .204$$

The standard beta distribution is commonly used to model variation in the proportion or percentage of a quantity occurring in different samples, such as the proportion of a 24-hour day that an individual is asleep or the proportion of a certain element in a chemical compound.

Exercises / Section 4.5

1. The lifetime X (in hundreds of hours) of a certain type of vacuum tube has a Weibull distribution with parameters $\alpha = 2$ and $\beta = 3$.
  a. Compute $E(X)$ and Var (X), using $\sqrt{\pi} = 1.773$.
  b. Compute $P(X \leq 6)$.
  c. Compute $P(1.5 \leq X \leq 6)$.

2. Let $X =$ the time (in 10^{-1} weeks) from shipment of a defective product until the customer returns the product. Suppose that the minimum return time is $\gamma = 3.5$ and that the excess $X - 3.5$ over the minimum has a Weibull distribution with parameters $\alpha = 2$ and $\beta = 1.5$ (see the *Industrial Quality Control* article referenced in Example 4.23).
  a. What is the c.d.f. of X?
  b. What are the expected return time and variance of return time? *Hint:* First obtain $E(X - 3.5)$ and Var $(X - 3.5)$.
  c. Compute $P(X > 5)$.
  d. Compute $P(5 \leq X \leq 8)$.

3. Let X have a Weibull distribution with p.d.f. (4.11). Verify that $\mu = \beta \Gamma(1 + \alpha)$. *Hint:* In the integral for $E(X)$, make the change of variable $y = (x/\beta)^{\alpha}$, so that $x = \beta y^{1/\alpha}$.

4. a. In Exercise 1, what is the median lifetime of such tubes? *Hint:* Use expression (4.12).
  b. In Exercise 2, what is the median return time?
  c. If X has a Weibull distribution with c.d.f. (4.12), obtain a general expression for the $(100p)$th percentile of the distribution.
  d. In Exercise 2, the company wants to refuse to accept returns after t weeks. For what value of t will only 10% of all returns be refused?

5. In Example 4.24 in which X has a lognormal distribution with parameters $\mu = 3.5$ and $\sigma = 1.2$, compute
  a. $P(X \leq 100)$ **b.** $P(100 \leq X \leq 200)$

6. a. Use (4.13) to write a formula for the median $\tilde{\mu}$ of the lognormal distribution. What is the median for the power distribution of Example 4.24?
  b. Recalling that z_α is our notation for the $100(1 - \alpha)$ percentile of the standard normal distribution, write an expression for the $100(1 - \alpha)$ percentile of the lognormal distribution. In Example 4.24, above what value will median power be only 5% of the time?

7. A theoretical justification based on a certain material failure mechanism underlies the assumption that ductile strength X of a material has a lognormal distribution. Suppose that the parameters are $\mu = 5$ and $\sigma = .1$.
  a. Compute $E(X)$ and Var (X).
  b. Compute $P(X > 120)$.
  c. Compute $P(110 \leq X \leq 130)$.
  d. What is the value of median ductile strength?
  e. If 10 different samples of an alloy steel of this type were subjected to a strength test, how many would you expect to have strength at least 120?
  f. If the smallest 5% of strength values were unacceptable, what would the minimum acceptable strength be?

8. If Y has a beta distribution with parameters α, β, A, and B, what are $E(Y)$ and Var (Y)?

9. Suppose that the proportion X of surface area in a randomly selected quadrat which is covered by a certain plant has a standard beta distribution with $\alpha = 5$ and $\beta = 2$.
   a. Compute $E(X)$ and $\text{Var}(X)$.
   b. Compute $P(X \le .2)$.
   c. Compute $P(.2 \le X \le .4)$.
   d. What is the expected proportion of the sampling region which is not covered by the plant?

10. Let X have a standard beta density with parameters α and β.
   a. Verify the formula for $E(X)$ given in the section.
   b. Compute $E[(1 - X)^m]$. If X represents the proportion of a substance consisting of a particular

ingredient, what is the expected proportion which does not consist of this ingredient?
   c. What condition on α and β is necessary for the graph of $f(x; \alpha, \beta)$ to be symmetric?

11. Stress is applied to a 20-in. steel bar which is clamped in a fixed position at each end. Let $Y =$ the distance from the left end at which the bar snaps. Suppose that $Y/20$ has a standard beta distribution with $E(Y) = 10$ and $\text{Var}(Y) = \frac{100}{7}$.
   a. What are the parameters of the relevant standard beta distribution?
   b. Compute $P(8 \le Y \le 12)$.
   c. Compute the probability that the bar snaps more than two inches from where you expect it to.

Supplementary Exercises / Chapter 4

1. Let $X =$ the time that it takes a read/write head to locate a desired record on a computer disc memory device once the head has been positioned over the correct track. If the discs rotate once every 25 milliseconds, a reasonable assumption is that X is uniformly distributed on the interval $[0, 25]$.
   a. Compute $P(10 \le X \le 20)$.
   b. Compute $P(X \ge 10)$.
   c. Obtain the c.d.f. $F(X)$.
   d. Compute $E(X)$ and $\text{Var}(X)$.

2. A 12-in. bar which is clamped at both ends is to be subjected to an increasing amount of stress until it snaps. Let $Y =$ the distance from the left end at which the break occurs, and suppose that Y has p.d.f.

$$f(y) = \begin{cases} \left(\frac{1}{24}\right) y \left(1 - \frac{y}{12}\right) & 0 \le y \le 12 \\ 0 & \text{otherwise} \end{cases}$$

   a. Compute the c.d.f. of Y and graph it.
   b. Compute $P(Y \le 4)$, $P(Y > 6)$, and $P(4 \le Y \le 6)$.
   c. Compute $E(Y)$, $E(Y^2)$, and $\text{Var}(Y)$.
   d. Compute the probability that the break point occurs more than 2 in. from the expected break point.
   e. Compute the expected length of the shorter segment when the break occurs.

3. The completion time X for a certain task has cumulative distribution function

$$F(x) = \begin{cases} 0 & x < 0 \\ \dfrac{x^3}{3} & 0 \le x < 1 \\ 1 - \dfrac{1}{2}\left(\dfrac{7}{3} - x\right)\left(\dfrac{7}{4} - \dfrac{3}{4}x\right) & 1 \le x < \dfrac{7}{3} \\ 1 & x \ge \dfrac{7}{3} \end{cases}$$

   a. Obtain the p.d.f. $f(x)$ and sketch its graph.
   b. Compute $P(.5 \le X \le 2)$.
   c. Compute $E(X)$.

4. Suppose that the time X necessary to process an application for a license plate for a newly purchased automobile is normally distributed with $\mu = 6$ and $\sigma = 1.5$ min.
   a. What is the probability that it takes at least 9.50 min to process a single application?
   b. If 10 such applications are processed independently of one another, what is the probability that all take less than 9.5 min?
   c. If 1000 such applications are processed and $Y =$ the number among the 1000 which take at least 10.62 min (a "success"), what is $P(Y \le 1)$ (approximately)?

5. Suppose that a component has a lifetime X which

is exponentially distributed with parameter λ.

a. If the cost of operation per unit time is c, what is the expected cost of operating this component over its lifetime?

b. Instead of a constant cost rate c as in part (a), suppose that the cost rate is $c(1 - .5e^{ax})$, so that the cost per unit time is less than c when the component is new and gets more expensive as the component ages. Now compute the expected cost of operation over the lifetime of the component.

6. The *mode* of a continuous distribution is that value x^* which maximizes $f(x)$.

a. What is the mode of a normal distribution with parameters μ and σ?

b. Does the uniform distribution with parameters A and B have a single mode? Why or why not?

c. What is the mode of an exponential distribution with parameter λ? (Draw a picture here.)

d. If X has a gamma distribution with parameters α and β, and $\alpha > 1$, find the mode. *Hint:* $\ln[f(x)]$ will be maximized iff $f(x)$ is, and it may be simpler to take the derivative of $\ln[f(x)]$.

e. What is the mode of a chi-squared distribution having ν degrees of freedom?

7. The article "Error Distribution in Navigation" (*J. Institute of Navigation*, 1971, pp. 429–442) suggests that the frequency distribution of positive errors (magnitudes of errors) is well approximated by an exponential distribution. Let $X =$ the lateral position error (nautical miles), which can be either negative or positive, and suppose that p.d.f. of X is

$$f(x) = \begin{cases} (.1)e^{-.2|x|} & -\infty < x < \infty \\ 0 & \text{otherwise} \end{cases}$$

a. Sketch a graph of $f(x)$, and verify that $f(x)$ is a legitimate p.d.f. (show that it integrates to 1).

b. Obtain the c.d.f. of X and sketch it.

c. Compute $P(X \leq 0)$, $P(X \leq 2)$, $P(-1 \leq X \leq 2)$, and the probability that an error of more than two miles is made.

d. The magnitude of a navigation error is $Y = |X|$. Show that Y has an exponential distribution. *Hint:* $Y \leq y$ iff $-y \leq X \leq y$. Compute the probability of this and differentiate to obtain the p.d.f. of Y.

8. In some systems, a customer is allocated to one of two service facilities. If the service time for a customer served by facility i has an exponential distribution with parameter λ_i ($i = 1, 2$) and p is the proportion of all customers served by facility 1, then the p.d.f. of $X =$ the service time of a randomly selected customer is

$f(x; \lambda_1, \lambda_2, p)$

$$= \begin{cases} p\lambda_1 e^{-\lambda_1 x} + (1 - p)\lambda_2 e^{-\lambda_2 x} & x \geq 0 \\ 0 & \text{otherwise} \end{cases}$$

This is often called the hyperexponential distribution. As an example, many computer systems process some programs using both a fast in-core compiler (FORTRAN WATFIV) and a slower compiler.

a. Verify that $f(x; \lambda_1, \lambda_2, p)$ is indeed a p.d.f.

b. What is the c.d.f. $F(x; \lambda_1, \lambda_2, p)$?

c. If X has $f(x; \lambda_1, \lambda_2, p)$ as its p.d.f., what is $E(X)$?

d. Using the fact that $E(X^2) = 2/\lambda^2$ when X has an exponential distribution with parameter λ, compute $E(X^2)$ when X has p.d.f. $f(x; \lambda_1, \lambda_2, p)$. Then compute $\text{Var}(X)$.

e. The coefficient of variation of a random variable (or distribution) is $CV = \sigma/\mu$. What is CV for an exponential r.v.? What can you say about the value of CV when X has a hyperexponential distribution?

f. What is CV for an Erlang distribution with parameters λ and n as defined in Exercise 9, Section 4.4? *Note*: In applied work, the sample CV is used to decide which of the three distributions might be appropriate.

9. Suppose that a particular state allows individuals filing tax returns to itemize deductions only if the total of all itemized deductions is at least \$5,000.00, and let X (in thousands of dollars) be the total of itemized deductions on a randomly chosen form. Assume that X has the p.d.f.

$$f(x; \alpha) = \begin{cases} \dfrac{k}{x^\alpha} & x \geq 5 \\ 0 & \text{otherwise} \end{cases}$$

a. Find the value of k. What restriction on α is necessary?

b. What is the c.d.f. of X?

c. What is the expected total deduction on a randomly chosen form. What restriction on α is necessary for $E(X)$ to be finite?

d. Show that $\ln(X/5)$ has an exponential distribution with parameter α.

10. Let I_i be the input current to a transitor and I_0 be the output current. Then the current gain is proportional to $\ln(I_0/I_i)$. Suppose that the constant of proportionality is 1 (which amounts to choosing a particular unit of measurement), so that current gain $= X = \ln(I_0/I_i)$. Assume X is normally distributed with $\mu = 1$ and $\sigma = .05$.

a. What type of distribution does the ratio I_0/I_i have?

b. What is the probability that the output current is more than twice the input current?

c. What are the expected value and variance of the ratio of output to input current?

11. Let Z have a standard normal distribution, and define a new r.v. Y by $Y = \sigma Z + \mu$. Show that Y has a normal distribution with parameters μ and σ. *Hint:* $Y \leq y$ iff $Z \leq$? Use this to find the c.d.f. of Y and then differentiate it with respect to y.

12. a. Suppose that the lifetime X of a component, when measured in hours, has a gamma distribution with parameters α and β. Let $Y =$ the lifetime measured in minutes. Derive the p.d.f. of Y. *Hint:* $Y \leq y$ iff $X \leq y/60$. Use this to obtain the c.d.f. of Y and then differentiate to obtain the p.d.f.

b. If X has a gamma distribution with parameters α and β, what is the probability distribution of $Y = cX$?

13. In Exercises 11 and 12 above, as well as many other situations, one has the p.d.f. $f(x)$ of X and wishes the p.d.f. of $Y = h(X)$. Assume that $h(\cdot)$ is an invertible function, so that $y = h(x)$ can be solved for x to yield $x = k(y)$. Then it can be shown that the p.d.f. of Y is

$$g(y) = f[k(y)] \cdot |k'(y)|$$

a. If X has a uniform distribution with $A = 0$, $B = 1$, derive the p.d.f. of $Y = -\ln(X)$.

b. Work Exercise 11 using this result.

c. Work Exercise 12(b) using this result.

14. Let X denote the lifetime of a component, with $f(x)$ and $F(x)$ the p.d.f. and c.d.f. of X. The probability that the component fails in the interval $(x, x + \Delta x)$ is approximately $f(x) \cdot \Delta x$. The probability that it fails in $(x, x + \Delta x)$ given that it has lasted at least x is $f(x) \cdot \Delta x/[1 - F(x)]$. Dividing this by Δx produces the **failure rate function:**

$$r(x) = \frac{f(x)}{1 - F(x)}$$

An increasing failure rate function indicates that older components are increasingly likely to wear out, while a decreasing failure rate is evidence of increasing reliability with age. In practice a "bathtub-shaped" failure is often assumed.

a. If X is exponentially distributed, what is $r(x)$?

b. If X has a Weibull distribution with parameters α and β, what is $r(x)$? For what parameter values will $r(x)$ be increasing? For what parameter values will $r(x)$ decrease with x?

c. Since $r(x) = -d/dx \ln[1 - F(x)]$, $\ln[1 - F(x)] = -\int r(x)\,dx$. Suppose that

$$r(x) = \begin{cases} \alpha\left(1 - \dfrac{x}{\beta}\right) & 0 \leq x \leq \beta \\ 0 & \text{otherwise} \end{cases}$$

so that if a component lasts β hours, it will last forever (while seemingly unreasonable, this model can be used to study just "initial wearout"). What are the c.d.f. and p.d.f. of X?

15. Let U have a uniform distribution on the interval $[0, 1]$. Then observed values having this distribution can be obtained from a computer's random number generator. Let $X = -(1/\lambda)\ln(1 - U)$.

a. Show that X has an exponential distribution with parameter λ. *Hint:* The c.d.f. of X is $F(x) = P(X \leq x)$; $X \leq x$ is equivalent to $U \leq$?

b. How would you use (a) and a random number generator to obtain observed values from an exponential distribution with parameter $\lambda = 10$?

Bibliography

Bury, Karl, *Statistical Models in Applied Science,* John Wiley, New York, 1975. Somewhat difficult to read, but the author presents a number of good engineering and science examples involving continuous probability models.

Derman, Cyrus, Glaser, Leon, and Olkin, Ingram, *Probability Models and Applications,* Macmillan, New York, 1980. Good coverage of general properties and specific distributions.

Johnson, Normal and Kotz, Samuel, *Distributions in Statistics: Continuous Distributions 1 and 2,* Houghton Mifflin, Boston, 1970. These two volumes together present an exhaustive survey of various continuous distributions.

CHAPTER 5

Joint Probability Distributions and Random Samples

Introduction

In Chapters 3 and 4 we studied probability models for a single random variable. Many problems in probability and statistics lead to models involving several random variables simultaneously. In the present chapter we first discuss probability models for the joint behavior of several random variables, putting special emphasis on the case in which the variables are independent of one another. We then study expected values of functions of several random variables, including covariance and correlation as measures of the degree of association between two variables.

Many statistical procedures are based on linear functions of random variables, especially totals $X_1 + X_2 + \ldots + X_n$ and averages $(X_1 + X_2 + \ldots + X_n)/n$, so properties of such functions are presented next. The last section focuses on the relationship between linear functions and the normal distribution. The key result is the Central Limit Theorem (CLT), which provides the justification for many large sample statistical procedures.

5.1 Jointly Distributed Random Variables

There are many experimental situations in which more than one random variable will be of interest to an investigator. We shall first consider joint probability distributions for two discrete random variables, then for two continuous variables, and finally for more than two variables.

The Joint P.M.F. for Two Discrete Random Variables

> **Definition:** Let X and Y be two discrete random variables defined on the sample space $\mathscr{S}$ of an experiment. The **joint probability mass function** $p(x, y)$ is defined for each pair of numbers (x, y) by
>
> $$p(x, y) = P(X = x \text{ and } Y = y)$$
>
> For any set A consisting of pairs of (x, y) values, the probability $P[(X, Y) \in A]$ is obtained by summing the joint p.m.f. over pairs in A:
>
> $$P[(X, Y) \in A] = \sum_{(x, y) \in A} \sum [p(x, y)]$$

Example 5.1 A large insurance agency services a number of customers who have purchased both a homeowners policy and an automobile policy from the agency. For each type of policy, a deductible amount must be specified. For an automobile policy, the choices are $100 and $250, while for a homeowners policy the choices are 0, $100, and $200. Suppose that an individual with both types of policy is selected at random from the agency's files, and let X = the deductible amount on the auto policy and Y = the deductible amount on the homeowner policy. Possible (X, Y) pairs are then (100, 0), (100, 100), (100, 200), (250, 0), (250, 100), and (250, 200); the joint p.m.f. specifies the probability associated with each one of these pairs, with any other pair having probability zero. Suppose that the joint p.m.f. is given in the accompanying **joint probability table:**

	$p(x, y)$	0	*y* 100	200
x	100	.20	.10	.20
	250	.05	.15	.30

Then $p(100, 100) = P(X = 100 \text{ and } Y = 100) = P(\$100 \text{ deductible on both poli-cies}) = .10$. The probability $P(Y \geq 100)$ is computed by summing probabilities of all (x, y) pairs for which $y \geq 100$:

$$P(Y \geq 100) = p(100, 100) + p(250, 100) + p(100, 200)$$
$$+ p(250, 200) = .75$$

A function $p(x, y)$ can be used as a joint p.m.f. provided that $p(x, y) \geq 0$ for all x and y and $\sum_x \sum_y p(x, y) = 1$.

The p.m.f. of one of the variables alone is obtained by summing $p(x, y)$ over values of the other variable. The result is called a marginal p.m.f. because when the $p(x, y)$'s appear in a rectangular table, the sums are just marginal (row or column) totals.

> **Definition:** The **marginal probability mass functions** of X and of Y, denoted by $p_X(x)$ and $p_Y(y)$, respectively, are given by
>
> $$p_X(x) = \sum_y p(x, y) \quad p_Y(y) = \sum_x p(x, y)$$

Thus to obtain the marginal p.m.f. of X evaluated at, say, $x = 100$, the probabilities $p(100, y)$ are added over all possible y values. Doing this for each possible X value gives the marginal p.m.f. of X alone (without reference to Y). From the marginal p.m.f.'s, probabilities of events involving only X or only Y can be computed.

Example 5.1
(continued)

The possible X values are $x = 100$ and $x = 250$, so computing row totals in the joint probability table yields

$$p_X(100) = p(100, 0) + p(100, 100) + p(100, 200) = .50$$

and

$$p_X(250) = p(250, 0) + p(250, 100) + p(250, 200) = .50$$

The marginal p.m.f. of X is then

$$p_X(x) = \begin{cases} .5 & x = 100, 250 \\ 0 & \text{otherwise} \end{cases}$$

Similarly, the marginal p.m.f. of Y is obtained from column totals as

$$p_Y(y) = \begin{cases} .25 & y = 0, 100 \\ .50 & y = 200 \\ 0 & \text{otherwise} \end{cases}$$

so $P(Y \geq 100) = p_Y(100) + p_Y(200) = .75$ as before.

The Joint P.D.F. for Two Continuous Random Variables

> **Definition:** Let X and Y be continuous random variables. Then $f(x, y)$ is the **joint probability density function** for X and Y if for any two-dimensional set A
>
> $$P[(X, Y) \in A] = \int\int_A f(x, y) \, dx \, dy$$
>
> In particular, if A is the two-dimensional rectangle $\{(x, y): a \leq x \leq b, c \leq y \leq d\}$, then
>
> $$P[(X, Y) \in A] = P(a \leq X \leq b, c \leq Y \leq d) = \int_a^b \int_c^d f(x, y) \, dy \, dx$$

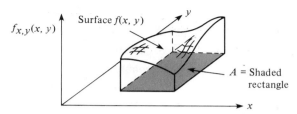

$f_{x,y}(x, y)$ Surface $f(x, y)$

A = Shaded rectangle

Figure 5.1 $P[(X, Y) \in A]$ = volume under density surface above A

As in the one-variable case, probabilities are computed by integrating a density function. For $f(x, y)$ to be a candidate for a joint p.d.f., it must satisfy $f(x, y) \geq 0$ and $\int_{-\infty}^{\infty} \int_{-\infty}^{\infty} f(x, y) \, dx \, dy = 1$. We can think of $f(x, y)$ as specifying a surface at height $f(x, y)$ above the point (x, y) in a three-dimensional coordinate system. Then $P[(X, Y) \in A]$ is the volume underneath this surface and above the region A, analogous to the area under a curve in the one-dimensional case. This is illustrated in Figure 5.1.

Example 5.2

A bank operates both a drive-up facility and a walk-in facility. On a randomly selected day, let X = the proportion of time that the drive-up facility is in use (at least one customer is being served or waiting to be served) and Y = the proportion of time that the walk-in facility is in use. Then the set of possible values for (X, Y) is the rectangle $D = \{(x, y): 0 \leq x \leq 1, 0 \leq y \leq 1\}$. Suppose that the joint p.d.f. of (X, Y) is given by

$$f(x, y) = \begin{cases} \frac{6}{5}(x + y^2) & 0 \leq x \leq 1, 0 \leq y \leq 1 \\ 0 & \text{otherwise} \end{cases}$$

To verify that this is a legitimate p.d.f., note that $f(x, y) \geq 0$ and

$$\int_{-\infty}^{\infty} \int_{-\infty}^{\infty} f(x, y) \, dx \, dy = \int_0^1 \int_0^1 \tfrac{6}{5}(x + y^2) \, dx \, dy = \int_0^1 \int_0^1 \tfrac{6}{5} x \, dx \, dy$$

$$+ \int_0^1 \int_0^1 \tfrac{6}{5} y^2 \, dx \, dy = \int_0^1 \tfrac{6}{5} x \, dx + \int_0^1 \tfrac{6}{5} y^2 \, dy = \tfrac{6}{10} + \tfrac{6}{15} = 1$$

The probability that neither facility is busy more than one-quarter of the time is

$$P(0 \leq X \leq \tfrac{1}{4}, 0 \leq Y \leq \tfrac{1}{4}) = \int_0^{1/4} \int_0^{1/4} \tfrac{6}{5}(x + y^2) \, dx \, dy$$

$$= \tfrac{6}{5} \int_0^{1/4} \int_0^{1/4} x \, dx \, dy + \tfrac{6}{5} \int_0^{1/4} \int_0^{1/4} y^2 \, dx \, dy$$

$$= \tfrac{6}{20} \cdot \frac{x^2}{2} \Big|_{x=0}^{x=1/4} + \tfrac{6}{20} \cdot \frac{y^3}{3} \Big|_{y=0}^{y=1/4} = \tfrac{7}{640} = .0109$$

As with joint p.m.f.'s, from the joint p.d.f. of X and Y each of the two marginal density functions can be computed.

Definition: The **marginal probability density functions** of X and Y, denoted by $f_X(x)$ and $f_Y(y)$, respectively, are given by

$$f_X(x) = \int_{-\infty}^{\infty} f(x, y) \, dy \quad \text{for} -\infty < x < \infty$$

$$f_Y(y) = \int_{-\infty}^{\infty} f(x, y) \, dx \quad \text{for} -\infty < y < \infty$$

Example 5.2 (continued)

The marginal p.d.f. of X, which gives the probability distribution of busy time for the drive-up facility without reference to the walk-in facility, is

$$f_X(x) = \int_{-\infty}^{\infty} f(x, y) \, dy = \int_0^1 \tfrac{6}{5}(x + y^2) \, dy = \tfrac{6}{5}x + \tfrac{2}{5}$$

for $0 \le x \le 1$ and 0 otherwise. The marginal p.d.f. of Y is

$$f_Y(y) = \begin{cases} \tfrac{6}{5}y^2 + \tfrac{3}{5} & 0 \le y \le 1 \\ 0 & \text{otherwise} \end{cases}$$

Then

$$P(\tfrac{1}{4} \le Y \le \tfrac{3}{4}) = \int_{1/4}^{3/4} f_Y(y) \, dy = \tfrac{37}{80} = .4625$$

In Example 5.2 the region of positive joint density was a rectangle, which made computation of the marginal p.d.f.'s relatively easy. Consider now an example in which the region of positive density is a more complicated figure.

Example 5.3

A nut company markets cans of deluxe mixed nuts containing almonds, cashews, and peanuts. Suppose that the net weight of each can is exactly one pound, but that the weight contribution of each type of nut is random. Because the three weights sum to one, a joint probability model for any two gives all necessary information about the weight of the third type. Let X = the weight of almonds in the can and Y = the weight of cashews. Then the region of positive density is $D = \{(x, y): 0 \le x \le 1, 0 \le y \le 1, x + y \le 1\}$, the shaded region pictured in Figure 5.2. Now let the joint p.d.f. for (X, Y) be

$$f(x, y) = \begin{cases} 24xy & 0 \le x \le 1, 0 \le y \le 1, x + y \le 1 \\ 0 & \text{otherwise} \end{cases}$$

For any fixed x, $f(x, y)$ increases with y, and for fixed y, $f(x, y)$ increases with x. This is appropriate since the word "deluxe" implies that most of the can should

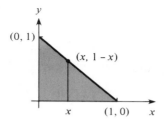

Figure 5.2 Region of positive density for Example 5.3

consist of almonds and cashews rather than peanuts, so that the density function should be large near the upper boundary and small near the origin. The surface determined by $f(x, y)$ slopes upward from zero as (x, y) moves away from either axis.

Clearly $f(x, y) \geq 0$. To verify the second condition on a joint p.d.f., recall that a double integral is computed as an iterated integral by holding one variable fixed (such as x as in Figure 5.2), integrating over values of the other variable lying along the straight line passing through the value of the fixed variable, and finally integrating over all possible values of the fixed variable. Thus

$$\int_{-\infty}^{\infty} \int_{-\infty}^{\infty} f(x, y) \, dy \, dx = \iint_{D} f(x, y) \, dy \, dx = \int_0^1 \left\{ \int_0^{1-x} 24xy \, dy \right\} dx$$

$$= \int_0^1 24x \left\{ \frac{y^2}{2} \Big|_{y=0}^{y=1-x} \right\} dx = \int_0^1 12x(1 - x)^2 \, dx = 1$$

To compute the probability that the two types of nuts together make up at most 50% of the can, let $A = \{(x, y): 0 \leq x \leq 1, 0 \leq y \leq 1, \text{ and } x + y \leq .5\}$. Then

$$P((X, Y) \in A) = \iint_{A} f(x, y) \, dx \, dy = \int_0^{.5} \int_0^{.5-x} 24xy \, dy \, dx = .125$$

The marginal p.d.f. for almonds is obtained by holding X fixed at x and integrating $f(x, y)$ along the vertical line through x:

Figure 5.3 Computing $P[(X, Y) \in A]$ for Example 5.3

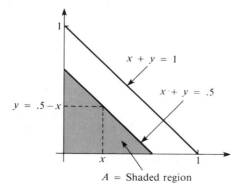

A = Shaded region

$$f_X(x) = \int_{-\infty}^{\infty} f(x, y) \, dy = \begin{cases} \int_0^{1-x} 24 \, xy \, dy = 12x(1-x)^2 & 0 \le x \le 1 \\ 0 & \text{otherwise} \end{cases}$$

By symmetry of $f(x, y)$ and the region D, the marginal p.d.f. of Y is obtained by replacing x and X in $f_X(x)$ by y and Y, respectively.

Independent Random Variables

In many situations information about the observed value of one of the two variables X and Y gives information about the value of the other variable. In Example 5.1 the marginal probability of X at $x = 250$ was .5, as was the probability that $X = 100$. If, however, we are told that the selected individual had $Y = 0$, then $X = 100$ is four times as likely as $X = 250$. Thus there is a dependence between the two variables.

In Chapter 2 we pointed out that one way of defining independence of two events was to say that A and B are independent if $P(A \cap B) = P(A) \cdot P(B)$. We now use an analogous definition for the independence of two random variables.

Definition: Two random variables X and Y are said to be **independent** if for every pair of x and y values,

$$p(x, y) = p_X(x) \cdot p_Y(y) \quad \text{when } X \text{ and } Y \text{ are discrete}$$

or   (5.1)

$$f(x, y) = f_X(x) \cdot f_Y(y) \quad \text{when } X \text{ and } Y \text{ are continuous}$$

If (5.1) is not satisfied for all (x, y), X and Y are said to be **dependent.**

The definition says that two variables are independent if their joint p.m.f. or p.d.f. is the product of the two marginal p.m.f.'s or p.d.f.'s.

Example 5.1 (continued)

$p(100, 100) = .10$, while $p_X(100) = .5$ and $p_Y(100) = .25$. Since $.10 \ne (.5)(.25)$, X and Y are not independent.

Example 5.3 (continued)

Because $f(x, y)$ has the form of a product, X and Y would appear to be independent. However, while $f_X(\frac{3}{4}) = f_Y(\frac{3}{4}) = (\frac{9}{16})^2$, $f(\frac{3}{4}, \frac{3}{4}) = 0 \ne (\frac{9}{16})^2 \cdot (\frac{9}{16})^2$, so the variables are not in fact independent. To be independent, $f(x, y)$ must have the form $g(x) \cdot h(y)$ and the region of positive density must be a rectangle whose sides are parallel to the coordinate axes.

Independence of two random variables is most useful when the description of the experiment under study tells us that X and Y have no effect on one another. Then once the marginal p.m.f.'s or p.d.f.'s have been specified, the joint p.m.f. or p.d.f. is simply the product of the two marginal functions.

Example 5.4   Suppose that the lifetimes of two components are independent of one another and that the first lifetime X_1 has an exponential distribution with parameter λ_1 while the second X_2 has an exponential distribution with parameter λ_2. Then the joint p.d.f. is

$$f(x_1, x_2) = f_{X_1}(x_1) \cdot f_{X_2}(x_2)$$

$$= \begin{cases} \lambda_1 e^{-\lambda_1 x_1} \cdot \lambda_2 e^{-\lambda_2 x_2} = \lambda_1 \lambda_2 e^{-\lambda_1 x_1 - \lambda_2 x_2} & x_1 > 0, x_2 > 0 \\ 0 & \text{otherwise} \end{cases}$$

More Than Two Random Variables

To model the joint behavior of more than two random variables, we extend the concept of a joint distribution of two variables.

Definition: If $X_1, X_2, \ldots, X_n$ are all discrete random variables, the joint p.m.f. of the variables is the function

$$p(x_1, x_2, \ldots, x_n) = P(X_1 = x_1, X_2 = x_2, \ldots, X_n = x_n)$$

If the variables are continuous, the joint p.d.f. of $X_1, \ldots, X_n$ is the function $f(x_1, x_2, \ldots, x_n)$ such that for any n intervals $[a_1, b_1], \ldots, [a_n, b_n]$,

$$P(a_1 \leq X_1 \leq b_1, \ldots, a_n \leq X_n \leq b_n)$$

$$= \int_{a_1}^{b_1} \cdots \int_{a_n}^{b_n} f(x_1, \ldots, x_n) \, dx_n \ldots dx_1$$

Example 5.5   In a binomial experiment each trial could result in one of only two possible outcomes. Consider now an experiment consisting of n independent and identical trials, in which each trial can result in any one of r possible outcomes. Let $p_i = P(\text{outcome } i \text{ on any particular trial})$, and define random variables by $X_i =$ the number of trials resulting in outcome i ($i = 1, \ldots, r$). Such an experiment is called a multinomial experiment, and the joint p.m.f. of $X_1, \ldots, X_r$ is called the multinomial distribution. By using a counting argument analogous to the one used in deriving the binomial distribution, the joint p.m.f. of $X_1, \ldots, X_r$ can be shown to be

$$p(x_1, \ldots, x_r) = \begin{cases} \dfrac{n!}{(x_1!)\,(x_2!)\,\ldots\,(x_r!)} p_1^{x_1} \cdots p_r^{x_r} & \begin{matrix} x_i = 0, 1, 2, \ldots \\ x_1 + \cdots + x_r = n \end{matrix} \text{ with} \\ 0 & \text{otherwise} \end{cases}$$

As an example, if the allele of each of 10 independently obtained pea sections is determined, and $p_1 = P(AA)$, $p_2 = P(Aa)$, $p_3 = P(aa)$, $X_1 =$ number of AA's, $X_2 =$ number of Aa's, $X_3 =$ number of aa's, then

$$p(x_1, x_2, x_3) = \frac{10!}{(x_1!)\,(x_2!)\,(x_3!)}p_1^{x_1}p_2^{x_2}p_3^{x_3} \qquad \begin{array}{l} x_i = 0, 1, \ldots \quad \text{and} \\ x_1 + x_2 + x_3 = 10 \end{array}$$

If

$$p_1 = p_3 = .25, p_2 = .5$$

then

$$P(X_1 = 2, X_2 = 5, X_3 = 3) = p(2, 5, 3)$$

$$= \frac{10!}{2!\,5!\,3!}(.25)^2(.5)^5(.25)^3 = .0769$$

Example 5.6

When a certain method is used to collect a fixed volume of rock samples in a region, there are four resulting rock types. Let X_1, X_2, and X_3 denote the proportion by volume of rock types 1, 2, and 3 in a randomly selected sample (the proportion of rock type 4 is $1 - X_1 - X_2 - X_3$, so a variable X_4 would be redundant). If the joint p.d.f. of X_1, X_2, X_3 is

$$f(x_1, x_2, x_3)$$

$$= \begin{cases} kx_1x_2(1 - x_3) & 0 \le x_1 \le 1, 0 \le x_2 \le 1, 0 \le x_3 \le 1, x_1 + x_2 + x_3 \le 1 \\ 0 & \text{otherwise} \end{cases}$$

then k is determined by

$$1 = \int_{-\infty}^{\infty}\int_{-\infty}^{\infty}\int_{-\infty}^{\infty} f(x_1, x_2, x_3)\, dx_3\, dx_2\, dx_1$$

$$= \int_0^1 \left\{ \int_0^{1-x_1}\left[\int_0^{1-x_1-x_2} kx_1x_2(1 - x_3)\, dx_3\right]dx_2\right\}dx_1$$

This iterated integral has value $k/144$, so $k = 144$. The probability that rocks of type 1 and 2 together account for at most 50% of the sample is

$$P(X_1 + X_2 \le .5) = \iiint\limits_{\left\{\substack{0 \le x_i \le 1 \ \text{for} \ i = 1, 2, 3 \\ x_1 + x_2 + x_3 \le 1, \ x_1 + x_2 \le .5}\right\}} f(x_1, x_2, x_3)\, dx_3\, dx_2\, dx_1$$

$$= \int_0^{.5} \left\{ \int_0^{.5-x_2}\left[\int_0^{1-x_1-x_2} 144x_1\, x_2(1 - x_3)\, dx_3\right]dx_2\right\}dx_1$$

$$= .6066$$

The notion of independence of more than two random variables is similar to the notion of independence of more than two events.

> **Definition:** The random variables $X_1, X_2, \ldots, X_n$ are said to be **independent** if for *every* subset $X_{i_1}, X_{i_2}, \ldots, X_{i_k}$ of the variables (each pair, each triple, and so on), the **joint p.m.f.** or **p.d.f.** of the subset is equal to the product of the marginal p.m.f.'s or p.d.f.'s.

Thus if the variables are independent with $n = 4$, then the joint p.m.f. or p.d.f. of any two variables is the product of the two marginals, and similarly for any three variables and all four variables together. Most importantly, once we are told that n variables are independent, then the joint p.m.f. or p.d.f. is the product of the n marginals.

Example 5.7

If $X_1, \ldots, X_n$ represent the lifetimes of n components, the components operate independently of one another, and each lifetime is exponentially distributed with parameter λ, then

$$f(x_1, x_2, \ldots, x_n) = (\lambda e^{-\lambda x_1}) \cdot (\lambda e^{-\lambda x_2}) \cdots (\lambda e^{-\lambda x_n})$$

$$= \begin{cases} \lambda^n e^{-\lambda \Sigma x_i} & x_1 \geq 0, \, x_2 \geq 0, \ldots, x_n \geq 0 \\ 0 & \text{otherwise} \end{cases}$$

If these n components constitute a system which will fail as soon as a single component fails, then the probability that the system lasts past time t is

$$P(X_1 \geq t, \ldots, X_n \geq t) = \int_t^\infty \cdots \int_t^\infty f(x_1, \ldots, x_n) \, dx_1, \ldots, dx_n$$

$$= \left(\int_t^\infty \lambda e^{-\lambda x_1} \, dx_1 \right) \cdots \left(\int_t^\infty \lambda e^{-\lambda x_n} \, dx_n \right)$$

$$= (e^{-\lambda t})^n = e^{-n\lambda t}$$

Therefore

$$P(\text{system lifetime} \leq t) = 1 - e^{-n\lambda t} \quad \text{for} \quad t \geq 0$$

which shows that system lifetime has an exponential distribution with parameter λn.

In many experimental situations to be considered in this book, independence is a reasonable assumption, so that specifying the joint distribution reduces to deciding on appropriate marginal distributions.

Exercises / Section 5.1

1. A service station has both self-service and full-service islands. On each island there is a single regular unleaded pump with two hoses. Let X denote the number of hoses being used on the self-service island at a particular time, and let Y denote the number of hoses on the full-service island in use at that time. The joint p.m.f. of X and Y appears in the accompanying tabulation.

$p(x, y)$		0	y 1	2
	0	.10	.04	.02
x	1	.08	.20	.06
	2	.06	.14	.30

a. What is $P(X = 1$ and $Y = 1)$?

b. Compute $P(X \leq 1$ and $Y \leq 1)$.

c. Give a word description of the event $\{X \neq 0$ and $Y \neq 0\}$, and compute the probability of this event.

d. Compute the marginal p.m.f. of X and of Y. Using $p_X(x)$, what is $P(X \leq 1)$?

e. Are X and Y independent ramdom variables? Explain.

2. When an automobile is stopped by a roving safety patrol, each tire is checked for tire wear and each headlight is checked to see whether it is properly aimed. Let X denote the number of headlights which need adjustment and let Y denote the number of defective tires.

a. If X and Y are independent with $p_X(0) = .5$, $p_X(1) = .3$, $p_X(2) = .2$, and $p_Y(0) = .6$, $p_Y(1) = .1$, $p_Y(2) = p_Y(3) = .05$, $p_Y(4) = .2$, display the joint p.m.f. of (X, Y) in a joint probability table.

b. Compute $P(X \leq 1$ and $Y \leq 1)$ from the joint probability table and verify that it equals $P(X \leq 1) \cdot P(Y \leq 1)$.

c. What is $P(X + Y = 0)$ (the probability of no violations)?

c. Compute $P(X + Y \leq 1)$.

3. A fair four-sided (tetrahedral) die is rolled twice independently; possible outcomes on each toss are 1, 2, 3, and 4.

a. Let X = the number of times that a one results, and Y = the number of times that a two results. What is the joint p.m.f. of X and Y?

b. Let X = the outcome of the first toss, and Y = the outcome of the second toss. What is the joint p.m.f. of X and Y?

4. A college bookstore carries three different editions of Shakespeare's *Hamlet*. At the beginning of the term it has on the shelf eight copies of publisher A's edition, 10 copies of publisher B's, and 12 copies of publisher C's. Six students have signed up for a course in which *Hamlet* will be read (this is a technical college). Each enters the bookstore and randomly selects a book. Let X = the number of publisher A's edition selected, Y = the number of publisher B's edition selected, and $p(x, y)$ denote the joint p.m.f. of X and Y.

a. What is $p(3, 2)$? *Hint:* Each sample of size six is equally likely to be selected. Therefore $p(3, 2)$ = (number of outcomes with $X = 3$ and $Y = 2$)/(total number of outcomes). Now use the product rule for counting to obtain the numerator and denominator.

b. Using the logic of (a), obtain $p(x, y)$ (This can be thought of as a multivariate hypergeometric distribution—sampling without replacement from a finite population consisting of more than two categories.)

5. Each front tire on a particular type of automobile is supposed to be filled to a pressure of 26 p.s.i. Suppose that the actual air pressure in each tire is a r.v.—X for the right tire and Y for the left tire, with joint p.d.f.

$$f(x, y) = \begin{cases} K(x^2 + y^2) & 20 \leq x \leq 30, \ 20 \leq y \leq 30 \\ 0 & \text{otherwise} \end{cases}$$

a. What is the value of K?

b. What is the probability that both tires are underfilled?

c. What is the probability that the difference in air pressure between the two tires is at most 2 p.s.i.?

d. Determine the (marginal) distribution of air pressure in the right tire alone.

e. Are X and Y independent random variables?

6. Annie and Alvie have agreed to meet between 5:00 P.M. and 6:00 P.M. for dinner at a local health food restaurant. Let X = Annie's arrival time and Y = Alvie's arrival time, and suppose that X and Y are independent with each uniformly distribuated on the interval [5, 6].

a. What is the joint p.d.f. of X and Y?

b. What is the probability that they both arrive between 5:15 and 5:45?

c. If the first one to arrive will wait only 10 minutes before leaving to eat elsewhere, what is the probability that they have dinner at the health food restaurant? *Hint:* The event of interest is $A = \{(x, y): |x - y| \leq \frac{1}{6}\}$.

7. A professor has just given one long technical paper to one typist and a somewhat shorter paper to

another typist. Let X = the number of typing errors on the first paper, and Y the number of typing errors on the second paper. Suppose that X has a Poisson distribution with parameter λ, Y has a Poisson distribution with parameter μ, and that X and Y are independent.

a. What is the joint p.m.f. of X and Y?

b. What is the probability that at most one error is made on both papers combined?

c. Obtain a general expression for the probability that the total number of errors in the two papers is m(where m is a nonnegative integer). *Hint:* $A = \{(x, y): x + y = m\} = \{(m, 0),$ $(m - 1, 1), \cdots, (1, m - 1), (0, m)\}$. Now sum the joint p.m.f. over $(x, y) \in A$ and use the binomial theorem, which says that for any a, b

$$\sum_{k=0}^{m} \binom{m}{k} a^k b^{m-k} = (a + b)^m$$

8. You have two lightbulbs for a particular lamp. Let X = the lifetime of the first bulb, and Y = the lifetime of the second bulb (both in 1000's of hours). Suppose that X and Y are independent and that each has an exponential distribution with parameter $\lambda = 1$.

a. What is the joint p.d.f. of X and Y?

b. What is the probability that each bulb lasts at most 1000 hours (that is, $X \leq 1$ and $Y \leq 1$)?

c. What is the probability that the total lifetime of the two bulbs is at most 2? *Hint:* Draw a picture of the region $A = \{(x, y): x \geq 0, y \geq 0,$ $x + y \leq 2\}$ before integrating.

d. What is the probability that the total lifetime is between 1 and 2?

9. Suppose that you have 10 lightbulbs, that the lifetime of each is independent of all the other lifetimes, and that each lifetime has an exponential distribution with parameter λ.

a. What is the probability that all 10 bulbs fail before time t?

b. What is the probability that exactly k of the 10 bulbs fail before time t?

c. Suppose that 9 of the bulbs have lifetimes which are exponentially distributed with parameter λ, and that the remaining bulb has a lifetime which is exponentially distributed with parameter μ (it's made by another manufacturer). What is the probability that exactly 5 of the 10 bulbs fail before time t?

10. An ecologist wishes to select a point inside a circular sampling region according to a uniform distribution (in practice this could be done by first selecting a direction and then a distance from the center in that direction). Let X = the x coordinate of the point selected, and Y = the y coordinate of the point selected. If the circle is centered at $(0, 0)$ and has radius R, then the joint p.d.f. of X and Y is

$$f(x, y) = \begin{cases} \dfrac{1}{\pi R^2} & x^2 + y^2 \leq R^2 \\ 0 & \text{otherwise} \end{cases}$$

a. What is the probability that the selected point is within $R/2$ of the center of the circular region? *Hint:* Draw a picture of the region of positive density D. Because $f(x, y)$ is constant on D, computing a probability reduces to computing an area.

b. What is the probability that both X and Y differ from 0 by at most $R/2$?

c. Answer (b) for $R/\sqrt{2}$ replacing $R/2$.

d. What is the marginal p.d.f. of X? Of Y? Are X and Y independent?

11. a. For $f(x_1, x_2, x_3)$ as given in Example 5.6, compute the **joint marginal density function of** X_1 and X_3 alone (by integrating over x_2).

b. What is the probability that rocks of type 1 and 3 together make up at most 50% of the sample? *Hint:* Use the result of (a).

c. Compute the marginal p.d.f. of X_3 alone. *Hint:* Use the result of (a).

5.2 Expected Values, Covariance, and Correlation

We previously saw that any function $h(X)$ of a single r.v. X is itself a random variable. However, to compute $E[h(X)]$, it was not necessary to obtain the probability distribution of $h(X)$; instead $E[h(X)]$ was computed as a weighted average of $h(x)$ values, where the weight function was the p.m.f. $p(x)$ or p.d.f. $f(x)$ of X. A similar result holds for a function $h(X, Y)$ of two jointly distributed random variables.

Proposition: Let X and Y be jointly distributed random variables with p.m.f. $p(x, y)$ or p.d.f. $f(x, y)$ according to whether the variables are discrete or continuous. Then the expected value of a function $h(X, Y)$, denoted by $E[h(X, Y)]$ or $\mu_{h(X,Y)}$, is given by

$$E[h(X, Y)] = \begin{cases} \sum_x \sum_y h(x, y) \cdot p(x, y) & \text{if } X \text{ and } Y \text{ are discrete} \\ \int_{-\infty}^{\infty} \int_{-\infty}^{\infty} h(x, y) \, dx \, dy & \text{if } X \text{ and } Y \text{ are continuous} \end{cases}$$

Example 5.8   Five friends have purchased tickets to a certain concert. If the tickets are for seats 1–5 in a particular row and the tickets are randomly distributed among the five, what is the expected number of seats separating any particular two of the five? Let X and Y denote the seat number of the first and second individuals, respectively. Possible (X, Y) pairs are $\{(1, 2), (1, 3), \ldots, (5, 4)\}$, and the joint p.m.f. of (X, Y) is

$$p(x, y) = \begin{cases} \dfrac{1}{20} & x = 1, \ldots, 5; y = 1, \ldots, 5; x \neq y \\ 0 & \text{otherwise} \end{cases}$$

The number of seats separating the two individuals is $h(X, Y) = |X - Y|$. The accompanying table gives $h(x, y)$ for each possible (x, y) pair.

$h(x, y)$	x 1	2	3	4	5
1	—	0	1	2	3
2	0	—	0	1	2
y 3	1	0	—	0	1
4	2	1	0	—	0
5	3	2	1	0	—

Thus

$$E[h(X, Y)] = \sum_{(x, y)} \sum h(x, y) \cdot p(x, y) = \sum_{\substack{x=1 \\ }}^{5} \sum_{\substack{y=1 \\ x \neq y}}^{5} |x - y| \cdot \frac{1}{20} = 1$$

Example 5.9   In Example 5.3 of the previous section, the joint p.d.f. of the amount X of almonds and amount Y of cashews in a one-pound can of nuts was

$$f(x, y) = \begin{cases} 24xy & 0 \le x \le 1, \, 0 \le y \le 1, \, x + y \le 1 \\ 0 & \text{otherwise} \end{cases}$$

If a pound of almonds costs the company 1.00, a pound of cashews costs 1.50, and a pound of peanuts costs $.50$, then the total cost of the contents of a can is

$$h(X, Y) = (1)X + (1.5)Y + (.5)(1 - X - Y) = .5 + .5X + Y$$

(since $1 - X - Y$ of the weight consists of peanuts). The expected total cost is

$$E[h(X, Y)] = \int_{-\infty}^{\infty} \int_{-\infty}^{\infty} h(x, y) \cdot f(x, y) \, dx \, dy$$

$$= \int_{0}^{1} \int_{0}^{1-x} (.5 + .5x + y) \cdot 24xy \, dy \, dx = \$1.10$$

The method of computing the expected value of a function $h(X_1, \ldots, X_n)$ of n random variables is similar to that for two random variables. If the X_i's are discrete, $E[h(X_1, \ldots, X_n)]$ is an n-dimensional sum, and if the X_i's are continuous, it is an n-dimensional integral.

Covariance

When two random variables X and Y are not independent, it is frequently of interest to measure how strongly they are related to one another.

Definition: The **covariance** between two random variables X and Y, denoted by $\text{Cov}(X, Y)$, is defined by

$$\text{Cov}(X, Y) = E[(X - \mu_X)(Y - \mu_Y)]$$

$$= \begin{cases} \sum_x \sum_y (x - \mu_X)(y - \mu_Y)p(x, y) & X, Y \text{ discrete} \\ \int_{-\infty}^{\infty} \int_{-\infty}^{\infty} (x - \mu_X)(y - \mu_Y)f(x, y) \, dx \, dy & X, Y \text{ continuous} \end{cases}$$

The rationale for the definition is as follows. Suppose that X and Y have a strong positive relationship to one another, by which we mean that large values of X tend to occur with large values of Y and small values of X with small values of Y. Then most of the probability mass or density will be associated with $(x - \mu_X)$ and $(y - \mu_Y)$ either both positive (both X and Y above their respective means) or both negative, so the product $(x - \mu_X)(y - \mu_Y)$ will tend to be positive. Thus for a strong positive relationship, $\text{Cov}(X, Y)$ should be quite positive. For a strong negative relationship, the signs of $(x - \mu_X)$ and $(y - \mu_Y)$ will tend to be opposite to one another, yielding a negative product. Thus for a strong negative relationship, $\text{Cov}(X, Y)$ should be quite negative. If X and Y are not strongly related, positive and negative products will tend to cancel one another, yielding a covariance near 0.

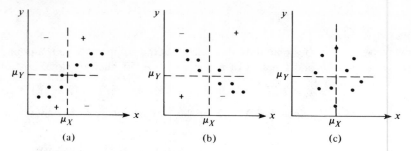

Figure 5.4 $p(x, y) = \frac{1}{10}$ for each of ten pairs corresponding to indicated points (a) positive covariance, (b) negative covariance, (c) covariance near zero

Figure 5.4 illustrates the different possibilities. The covariance depends *both* on the set of possible pairs and on the probabilities. In Figure 5.4 the probabilities could be changed without altering the set of possible pairs, and this could drastically change the value of Cov(X, Y).

Example 5.10 The joint and marginal p.m.f.'s for X = automobile policy deductible amount and Y = homeowner policy deductible amount in Example 5.1 were

$p(x, y)$	y 0	100	200
x 100	.20	.10	.20
250	.05	.15	.30

x	100	250
$p_X(x)$	.5	.5

y	0	100	200
$p_Y(y)$	.25	.25	.5

from which $\mu_X = \Sigma x\, p_X(x) = 175$ and $\mu_Y = 125$. Therefore

$$\text{Cov}(X, Y) = \sum\sum_{(x, y)} (x - 175)(y - 125)p(x, y)$$

$$= (100 - 175)(0 - 125)(.20) + \cdots$$

$$+ (250 - 175)(200 - 125)(.30) = 1875$$

The following shortcut formula for Cov(X, Y) simplifies the computations.

Proposition: $\text{Cov}(X, Y) = E(XY) - \mu_X \cdot \mu_Y$

According to this formula, no intermediate subtractions are necessary; only at the end of the computation is $\mu_X \cdot \mu_Y$ subtracted from $E(XY)$. The proof involves expanding $(X - \mu_X)(Y - \mu_Y)$ and then taking the expected value of each term separately. Note that $\text{Cov}(X, X) = E(X^2) - \mu_X^2 = \text{Var}(X)$.

Example 5.9
(continued)

The joint and marginal p.m.f.'s of X = amount of almonds and Y = amount of cashews were

$$f(x, y) = \begin{cases} 24xy & 0 \le x \le 1, 0 \le y \le 1, x + y \le 1 \\ 0 & \text{otherwise} \end{cases}$$

$$f_X(x) = \begin{cases} 12x(1 - x)^2 & 0 \le x \le 1 \\ 0 & \text{otherwise} \end{cases}$$

with $f_Y(y)$ obtained by replacing x by y in $f_X(x)$. It is easily verified that $\mu_X = \mu_Y = \frac{2}{5}$, and

$$E(XY) = \int_{-\infty}^{\infty} \int_{-\infty}^{\infty} xy \, f(x, y) \, dx \, dy = \int_0^1 \int_0^{1-x} xy \cdot 24xy \, dy \, dx$$

$$= 8 \int_0^1 x^2(1 - x)^3 \, dx = \frac{2}{15}$$

Thus Cov(X, Y) $= \frac{2}{15} - \left(\frac{2}{5}\right)\left(\frac{2}{5}\right) = \frac{2}{15} - \frac{4}{25} = -\frac{2}{75}$. A negative covariance is reasonable here, since more almonds in the can implies fewer cashews.

It would appear that the relationship in the insurance example is quite strong, since Cov(X, Y) = 1875, while Cov(X, Y) $= -\frac{2}{75}$ in the nut example would seem to imply quite a weak relationship. Unfortunately the covariance has a serious defect which makes it impossible to say whether a computed covariance is large or small. In the insurance example, suppose that we had measured deductible amount in cents rather than dollars. Then $100X$ would replace X, $100Y$ would replace Y, and the resulting covariance would be Cov($100X$, $100Y$) = (100)(100) Cov(X, Y) = 18,750,000. If, on the other hand, the deductible amount had been measured in 100's of dollars, the computed covariance would have been (.01)(.01)(1875) = .1875. *The defect of covariance is that its computed value depends critically on the units of measurement.* Ideally the choice of units should have no effect on a measure of strength of relationship. This is achieved by scaling the covariance.

Correlation

> **Definition:** The **correlation coefficient** of X and Y, denoted by Corr(X, Y), $\rho_{X,Y}$, or just ρ, is defined by
>
> $$\rho_{X,Y} = \frac{\text{Cov}(X, Y)}{\sigma_X \cdot \sigma_Y}$$

Example 5.10
(continued)

$E(X^2) = 36{,}250$, so $\sigma_X^2 = 36{,}250 - (175)^2 = 5625$ and $\sigma_X = 75$, while $E(Y^2) = 22{,}500$ so $\sigma_Y^2 = 6875$ and $\sigma_Y = 82.92$. This gives

$$\rho = \frac{1875}{(75)\,(82.92)} = .301$$

The following proposition shows that ρ remedies the defect of Cov(X, Y) and also suggests how to recognize the existence of a strong relationship.

Proposition:

1. If a and c are either both positive or both negative

   Corr($aX + b$, $cY + d$) = Corr(X, Y)

2. For any two r.v.'s X and Y, $-1 \le$ Corr(X, Y) ≤ 1.

Statement 1 says precisely that the correlation coefficient is not affected by a linear change in the units of measurement (if, say, X = temperature in °C, then ($9X/5 + 32$ = temperature in °F). According to statement 2 the strongest possible positive relationship is evidenced by $\rho = +1$, while the strongest possible negative relationship corresponds to $\rho = -1$. The proof of the first statement is sketched out in Exercise 5.2.11, while that of the second appears in a supplementary exercise at the end of the chapter. For descriptive purposes, the relationship will be described as strong if $|\rho| \ge .8$, moderate if $.5 < |\rho| < .8$, and weak if $|\rho| \le .5$.

If we think of $p(x, y)$ or $f(x, y)$ as prescribing a mathematical model for how the two numerical variables X and Y are distributed in some population (height and weight, verbal SAT score and quantitative SAT score, and the like), then ρ is a population characteristic or parameter which measures how strongly X and Y are related in the population. In Chapter 12 we will consider taking a sample of pairs $(x_1, y_1), \ldots, (x_n, y_n)$ from the population. The sample correlation coefficient r will then be defined and used to make inferences about ρ.

The correlation coefficient ρ is actually not a completely general measure of the strength of a relationship.

Proposition:

1. If X and Y are independent, then $\rho = 0$, but $\rho = 0$ does not imply independence.

2. $\rho = 1$ or -1 iff $Y = aX + b$ for some numbers a and b with $a \ne 0$.

This proposition says that ρ is a measure of the degree of **linear** relationship between X and Y, and only when the two variables are perfectly related in a linear manner will ρ be as positive or negative as it can be. A ρ less than 1 in absolute value indicates only that the relationship is not completely linear, but there may still be a very strong nonlinear relation. Also, $\rho = 0$ does not imply that X and Y are independent, but only that there is complete absence of a linear relationship. When $\rho = 0$, X and Y are said to be **uncorrelated**. Two variables could be uncorrelated yet highly dependent because there is a strong nonlinear relationship, so be careful not to conclude too much from knowing that $\rho = 0$.

Example 5.11 Let X and Y be discrete r.v.'s with joint p.m.f.

$$f(x, y) = \begin{cases} \frac{1}{4} & (x, y) = (-4, 1), (4, -1), (2, 2), (-2, -2) \\ 0 & \text{otherwise} \end{cases}$$

The points which receive positive probability mass are identified on the (x, y) coordinate system in Figure 5.5. It is evident from the figure that the value of X is completely determined by the value of Y and vice versa, so the two variables are completely dependent. However, by symmetry $\mu_X = \mu_Y = 0$ and $E(XY) = (-4)\frac{1}{4} + (-4)\frac{1}{4} + (4)\frac{1}{4} + (4)\frac{1}{4} = 0$, so $\text{Cov}(X, Y) = E(XY) - \mu_X \cdot \mu_Y = 0$ and thus $\rho_{X,Y} = 0$. Although there is perfect dependence, there is also complete absence of any linear relationship!

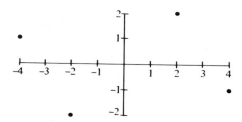

Figure 5.5

Exercises / Section 5.2

1. An instructor has given a short quiz consisting of two parts. For a randomly selected student, let X = the number of points earned on the first part, Y = the number of points earned on the second part, and suppose that the joint p.m.f. of X and Y is given in the accompanying tabulation.

$p(x, y)$		y		
	0	5	10	15
0	.02	.06	.02	.10
x 5	.04	.15	.20	.10
10	.01	.15	.14	.01

a. If the score recorded in the grade book is the total number of points earned on the two parts, what is the expected recorded score $E(X + Y)$?

b. If the maximum of the two scores is recorded, what is the expected recorded score?

2. Six individuals, including A and B, take seats around a circular table in a completely random fashion. Suppose that the seats are numbered 1, . . ., 6 and let $X = A$'s seat number and $Y = B$'s seat number. If A sends a written message around the table to B in the direction in which they are closest, how many individuals (including A and B) would you expect to handle the message?

3. A surveyor wishes to lay out a square region with each side having length L. However, because of measurement error, he instead lays out a rectangle in which the north/south sides both have length X and the east/west sides both have length Y. Suppose that X and Y are independent and that each is uniformly distributed on the interval $[L - A,$ $L + A]$ (where $0 < A < L$). What is the expected area of the resulting rectangle?

4. Annie and Alvie have agreed to meet for lunch between noon (0:00 P.M.) and 1:00 P.M. Denote Annie's arrival time by X, Alvie's by Y, and suppose that X and Y are independent with p.d.f.'s

$$f_X(x) = \begin{cases} 3x^2 & 0 \le x \le 1 \\ 0 & \text{otherwise} \end{cases}$$

$$f_Y(y) = \begin{cases} 2y & 0 \le y \le 1 \\ 0 & \text{otherwise} \end{cases}$$

What is the expected amount of time that the one who arrives first must wait for the other person? *Hint:* $h(X, Y) = |X - Y|$.

5. Show that if X and Y are independent random variables, then $E(XY) = E(X) \cdot E(Y)$.

6. Compute the correlation coefficient ρ for X and Y of Example 5.9 (we have already computed the covariance).

7. **a.** Compute the covariance for X and Y in Exercise 5.2.1.
   b. Compute ρ for X and Y in the same exercise.

8. **a.** Compute the covariance between X and Y in Exercise 5 of Section 5.1.
   b. Compute the correlation coefficient ρ for this X and Y.

9. Use the result of Exercise 5.2.5 to show that when X and Y are independent, Cov$(X, Y) =$ Corr$(X, Y) = 0$.

10. **a.** Recalling the definition of σ^2 for a single r.v. X, write a formula that would be appropriate for computing the variance of a function $h(X, Y)$ of two r.v.'s. *Hint:* Remember that variance is just a special expected value.
    b. Use this formula to compute the variance of the recorded score $h(X, Y) [= \max(X, Y)]$ in (b) of Exercise 5.2.1.

11. **a.** Use the rules of expected value to show that Cov$(aX + b, cY + d) = ac$ Cov(X, Y).
    b. Use (a) along with the rules of variance and standard deviation to show that Corr$(aX + b, cY + d) =$ Corr(X, Y) when a and c have the same sign.
    c. What happens if a and c have opposite signs?

12. Show that if $Y = aX + b (a \neq 0)$, then Corr$(X, Y) = +1$ or -1. Under what conditions will $\rho = +1$?

5.3   Sums and Averages of Random Variables

In many statistics problems the observations in a sample can be thought of as observed values of a sequence of random variables. In this section we let $X_1, X_2, \ldots, X_n$ be such a sequence and define and study several new random variables derived from the sequence.

> **Definition:** Let $X_1, X_2, \ldots, X_n$ be a collection of n random variables. These random variables are said to constitute a **random sample** of size n if (a) the X_i's are independent random variables, and (b) every X_i has the same probability distribution.

When (a) and (b) are satisfied, the X_i's are said to be **independent and identically distributed,** abbreviated **i.i.d.** If sampling is with replacement or from an infinite conceptual population, (a) and (b) are satisfied exactly. If the sample is taken without replacement from a finite population with the population size N much larger than the sample size n, (a) and (b) are approximately satisfied, so for all practical purposes the X_i's can be regarded as a random sample. In most problems we will proceed with our analysis by making the assumption that we have a random sample. A random sample will thus describe a collection of observations taken from the same population. The observed values of $X_1, X_2, \ldots, X_n$ will be denoted by $x_1, x_2, \ldots, x_n$.

The Distribution of the Sample Mean and Sample Total

> **Definition:** The random variables
> $$\bar{X} = \frac{1}{n} \sum_{i=1}^{n} X_i \quad \text{and} \quad T_o = \sum_{i=1}^{n} X_i$$
> are referred to as the **sample mean** and **sample total,** respectively, of the sample $X_1, \ldots, X_n$.

In Chapter 1 the sample mean was an average of a fixed set of numbers. Now we are thinking of the sample mean $\bar{X}$ before the X_i's are observed, so that the observed value $\bar{x}$ is not yet known. Information about the probability distribution of $\bar{X}$ will be used to specify inferential procedures in subsequent chapters.

Example 5.12 A large automobile service center charges $40, $45, and $50 for a tuneup of 4-, 6-, and 8-cylinder cars, respectively. If 20% of its tuneups are done on 4-cylinder cars, 30% on 6-cylinder cars, and 50% on 8-cylinder cars, then the probability distribution of revenue from a single randomly selected tuneup is given by

x	40	45	50
$p(x)$	.2	.3	.5

with $\mu = 46.5$, $\sigma^2 = 15.25$ $\qquad$ (5.2)

Suppose that on a particular day only two servicing jobs involve tuneups. Let $X_1 = $ the revenue from the first tuneup, $X_2 = $ the revenue from the second, and suppose that X_1 and X_2 are independent, each having probability distribution shown in expression (5.2) [so that X_1 and X_2 constitute a random sample from the distribution (5.2)]. Table 5.1 lists possible (x_1, x_2) pairs, the probability of each (computed

Table 5.1

x_1	x_2	$p(x_1, x_2)$	t	$\bar{x}$
40	40	.04	80	40
40	45	.06	85	42.5
40	50	.10	90	45
45	40	.06	85	42.5
45	45	.09	90	45
45	50	.15	95	47.5
50	40	.10	90	45
50	45	.15	95	47.5
50	50	.25	100	50

using (5.2) and the assumption of independence), and the resulting t (total) and $\bar{x}$ values. Now to obtain the probability distribution of $\bar{X}$, the sample average revenue per tuneup, we must consider each possible value $\bar{x}$ and compute its probability. For example, $\bar{x} = 45$ occurs three times in the table with probabilities .10, .09, and .10, so $P(\bar{X} = 45) = .10 + .09 + .10 = .29$. Proceeding in this manner, we obtain the p.m.f. $p_{\bar{X}}(\bar{x})$ of $\bar{X}$ and $p_{T_o}(t)$ of the sample total revenue T_o.

$\bar{x}$	40	42.5	45	47.5	50	
$p_{\bar{X}}(\bar{x})$	.04	.12	.29	.30	.25	(5.3)
t	80	85	90	95	100	
$p_{T_o}(t)$	.04	.12	.29	.30	.25	(5.4)

Figure 5.6 pictures a probability histogram both for the original distribution (5.2) and the $\bar{X}$ distribution (5.3). Figure 5.6 suggests first that the mean (expected value) of the $\bar{X}$ distribution is equal to the mean 46.5 of the original distribution, since both histograms appear to be centered at the same place. From (5.3)

$$\mu_{\bar{X}} = E(\bar{X}) = \sum \bar{x}\, p_{\bar{X}}(\bar{x}) = (40)\,(.04) + \cdots + (50)\,(.25) = 46.5 = \mu$$

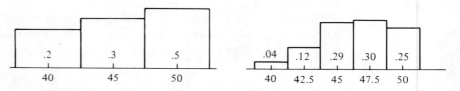

Figure 5.6   Probability histograms for the underlying distribution and $\bar{X}$ distribution in Example 5.12

Second, it appears that the $\bar{X}$ distribution has smaller spread (variability) than the original distribution, since probability mass has moved in toward the mean. Again from (5.3)

$$\sigma_{\bar{X}}^2 = \text{Var}(\bar{X}) = \sum \bar{x}^2 \cdot p_{\bar{X}}(\bar{x}) - \mu_{\bar{X}}^2$$

$$= (40)^2(.04) + \cdots + (50)^2(.25) - (46.5)^2$$

$$= 7.625 = \frac{(15.25)}{2} = \frac{\sigma^2}{2}$$

The variance of $\bar{X}$ is precisely half that of the original variance. Similar calculations with $p_{T_o}(t)$ yield $\mu_{T_o} = E(T_o) = 93 = 2\mu$ and $\sigma_{T_o}^2 = 30.5 = 2\sigma^2$, so both the expected value and variance of T_o are twice the mean and variance of the original distribution.

If four tuneups had been done on the day of interest, the sample average revenue $\bar{X}$ would be based on a random sample of four X_i's, each having distribution

(5.2). More calculation eventually yields the p.m.f. of $\overline{X}$ for $n = 4$ as

$\overline{x}$	40	41.25	42.5	43.75	45	46.25	47.5	48.75	50
$p_{\overline{X}}(\overline{x})$	.0016	.0096	.0376	.0936	.1761	.2340	.2350	.1500	.0625

From this, for $n = 4$, $\mu_{\overline{X}} = 46.50 = \mu$ and $\sigma_{\overline{X}}^2 = 3.8125 = \sigma^2/4$. A probability histogram of this p.m.f. appears in Figure 5.7.

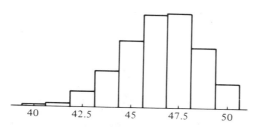

Figure 5.7 Probability histogram for $\overline{X}$ based on $n = 4$ in Example 5.12

Example 5.12 should suggest first of all that the computation of $p_{\overline{X}}(\overline{x})$ and $p_{T_o}(t)$ can be tedious. If the original distribution (5.2) had allowed for more than the three possible values 40, 45, and 50, then even for $n = 2$ the computations would have been more involved. The example should also suggest, however, that there are some general relationships between $E(\overline{X})$, $\mathrm{Var}(\overline{X})$, $E(T_o)$, $\mathrm{Var}(T_o)$, and the mean μ and variance σ^2 of the original distribution. Before developing these relationships, consider an example in which the random sample is drawn from a continuous distribution.

Example 5.13 The time that it takes to serve a customer at the cash register in a mini-market is a random variable having an exponential distribution with parameter λ. Suppose that X_1 and X_2 are service times for two different customers, assumed independent of each other. The c.d.f. of the total service time $T_o = X_1 + X_2$ for the two customers is, for $t \geq 0$,

$$F_{T_o}(t) = P(X_1 + X_2 \leq t) = \iint\limits_{\{(x_1, x_2):x_1 + x_2 \leq t\}} f(x_1, x_2) \, dx_1 \, dx_2$$

$$= \int_0^t \int_0^{t-x_1} \lambda e^{-\lambda x_1} \cdot \lambda e^{-\lambda x_2} \, dx_2 \, dx_1 = \int_0^t [\lambda e^{-\lambda x_1} - \lambda e^{-\lambda t}] \, dx_1$$

$$= 1 - e^{-\lambda t} - \lambda t e^{-\lambda t}$$

The p.d.f. of T_o is obtained by differentiating $F_{T_o}(t)$:

$$f_{T_o}(t) = \begin{cases} \lambda^2 t e^{-\lambda t} & t \geq 0 \\ 0 & t < 0 \end{cases} \tag{5.5}$$

The p.d.f. of $\bar{X} = T_o/2$ is obtained from the relation $\{\bar{X} \le \bar{x}\}$ iff $\{T_o \le 2\bar{x}\}$ as

$$f_{\bar{X}}(\bar{x}) = \begin{cases} 4\lambda^2 \bar{x}e^{-2\lambda\bar{x}} & \bar{x} \ge 0 \\ 0 & \bar{x} < 0 \end{cases} \tag{5.6}$$

The mean and variance of the underlying exponential distribution are $\mu = 1/\lambda$ and $\sigma^2 = 1/\lambda^2$. From (5.5) and (5.6) it can be verified that $E(\bar{X}) = 1/\lambda$, $\text{Var}(\bar{X}) = 1/2\lambda^2$, $E(T_o) = 2/\lambda$, and $\text{Var}(T_o) = 2/\lambda^2$. These results again suggest some general relationships between means and variances of $\bar{X}$, T_o, and the underlying distribution.

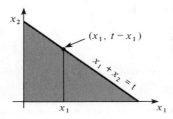

Figure 5.8   Region of integration to obtain c.d.f. of T_o in Example 5.13

Rules for Expected Value

Examples 5.12 and 5.13 demonstrate that the derivation of the distribution of $\bar{X}$ or T_o can be difficult (if, say, the integration is difficult) and time-consuming (though methods of mathematical statistics can sometimes be used to obtain the distributions). Given the individual p.m.f.'s or p.d.f.'s of the X_i's, and in particular $E(X_i)$ and $\text{Var}(X_i)$ for each i, what can we conclude about the distributions of $\bar{X}$ and T_o? To answer this, we first note that $\bar{X}$ and T_o are special cases of a general type of function of the X_i's.

> **Definition:** Given a collection of n random variables $X_1, \ldots, X_n$ and n numerical constants $a_1, \ldots, a_n$, the random variable
>
> $$Y = a_1X_1 + \cdots + a_nX_n = \sum_{i=1}^{n} a_iX_i \tag{5.7}$$
>
> is called a **linear combination** of the X_i's.

If we take $a_1 = \cdots = a_n = 1$ in (5.7), $Y = T_o$, while $a_1 = \cdots = a_n = 1/n$ produces $Y = \bar{X}$. Thus both T_o and $\bar{X}$ are special linear combinations.

Proposition: For any linear combination of $X_1, \ldots, X_n$, whether or not the X_i's are independent,

$$E(a_1X_1 + \cdots + a_nX_n) = a_1E(X_1) + \cdots + a_nE(X_n) \qquad (5.8)$$

To paraphrase (5.8), the expected value of a linear combination is the same linear combination of expected values.

Proof for $n = 2$: Assuming that X_1 and X_2 are continuous with joint p.d.f. $f(x_1, x_2)$,

$$E(a_1X_1 + a_2X_2) = \int_{-\infty}^{\infty} \int_{-\infty}^{\infty} (a_1x_1 + a_2x_2) f(x_1, x_2) \, dx_1 \, dx_2$$

$$= a_1 \int_{-\infty}^{\infty} \int_{-\infty}^{\infty} x_1 f(x_1, x_2) \, dx_2 \, dx_1$$

$$+ a_2 \int_{-\infty}^{\infty} \int_{-\infty}^{\infty} x_2 f(x_1, x_2) \, dx_1 \, dx_2$$

$$= a_1 \int_{-\infty}^{\infty} x_1 f_{X_1}(x_1) \, dx_1 + a_2 \int_{-\infty}^{\infty} x_2 f_{X_2}(x_2) \, dx_2$$

$$= a_1 E(X_1) + a_2 E(X_2)$$

The general case is proved in a similar manner.

Example 5.14 A bookstore stocks three different hardbound dictionaries selling for $10, $12, and $15, respectively. Let X_1, X_2, and X_3 refer to the number of $10, $12, and $15 dictionaries, respectively, sold during a given period. Then the total revenue from dictionary sales is $Y = 10X_1 + 12X_2 + 15X_3$. If $E(X_1) = 13$, $E(X_2) = 10$, and $E(X_3) = 6$, then

$$E(Y) = E(10X_1 + 12X_2 + 15X_3) = 10E(X_1) + 12E(X_2) + 15E(X_3) = 340$$

Corollary: When $E(X_i) = \mu$ for $i = 1, \ldots, n$ (as would be the case for a random sample),

$$E(\bar{X}) = \mu_{\bar{X}} = \mu \qquad (5.9)$$

and

$$E(T_o) = n\mu \qquad (5.10)$$

Expression (5.9) comes from letting $a_1 = \cdots = a_n = 1/n$ in (5.8), yielding $E(\bar{X})$ on the left and $(1/n) \mu + \cdots + (1/n) \mu = \mu$ on the right. Similarly (5.10) comes from $a_i = 1$ for all i in (5.8).

Example 5.15 In a notched tensile fatigue test on a titanium specimen, the expected number of cycles to first acoustic emission (used to indicate crack initiation) is 28,000. Let $X_1, \ldots, X_{10}$ be a random sample of size 10, where each X_i is the number of cycles on a different specimen. Then the expected average number of cycles per specimen until first emission is $E(\bar{X}) = \mu = 28,000$, while the expected total number of cycles on all specimens is $E(T_o) = 10\mu = 280,000$.

Rules for Variance

> **Proposition:** Let $X_1, \ldots, X_n$ be a collection of n random variables with $\text{Var}(X_i) = \sigma_i^2$.
>
> If $X_1, \ldots, X_n$ are independent, then
>
> $$\text{Var}(a_1 X_1 + \cdots + a_n X_n) = a_1^2 \sigma_1^2 + \cdots + a_n^2 \sigma_n^2$$
>
> $$= \sum_{i=1}^{n} a_i^2 \cdot \text{Var}(X_i) \qquad (5.11)$$
>
> For any $X_1, \ldots, X_n$,
>
> $$\text{Var}(a_1 X_1 + \cdots + a_n X_n) = \sum_{i=1}^{n} \sum_{j=1}^{n} a_i a_j \, \text{Cov}(X_i, X_j) \qquad (5.12)$$

The first part of this proposition is a special case of the second part; when the X_i's are independent, $\text{Cov}(X_i, X_j) = 0$ for $i \neq j$ and $= \text{Var}(X_i)$ for $i = j$, so (5.12) reduces to (5.11).

Proof for $n = 2$ (the general proof is similar):

With

$$Y = a_1 X_1 + a_2 X_2, \quad E(Y) = a_1 \mu_1 + a_2 \mu_2 \quad \text{where} \quad \mu_i = E(X_i).$$

Then

$$
\begin{aligned}
\text{Var}(Y) = E[Y - E(Y)^2] &= E\{[(a_1 X_1 + a_2 X_2) - (a_1 \mu_1 + a_2 \mu_2)]^2\} \\
&= E\{[(a_1 X_1 - a_1 \mu_1) + (a_2 X_2 - a_2 \mu_2)]^2\} \\
&= a_1^2 \, E[(X_1 - \mu_1)^2] + 2 a_1 a_2 \, E[(X_1 - \mu_1)(X_2 - \mu_2)] \qquad (5.13) \\
&\quad + a_2^2 \, E[(X_2 - \mu_2)^2] \\
&= a_1^2 \, \text{Var}(X_1) + 2 a_1 a_2 \, \text{Cov}(X_1, X_2) + a_2^2 \, \text{Var}(X_2)
\end{aligned}
$$

If X_1 and X_2 are independent, only the two variance terms at the end of (5.13) remain; if not, since $\text{Cov}(X_1, X_2) = \text{Cov}(X_2, X_1)$, (5.13) agrees with (5.12).

Corollary: Let $X_1, \ldots, X_n$ be a random sample (independent and identically distributed) from a distribution with variance σ^2 [so $\text{Var}(X_i) = \sigma^2$, $i = 1, \ldots, n$]. Then

$$\text{Var}(\overline{X}) = \sigma_{\overline{X}}^2 = \frac{\sigma^2}{n}, \qquad \sigma_{\overline{X}} = \frac{\sigma}{\sqrt{n}} \tag{5.14}$$

and

$$\text{Var}(T_o) = n\sigma^2, \qquad \sigma_{T_o} = \sqrt{n}\,\sigma \tag{5.15}$$

Result (5.14) is extremely important and useful. It says that if σ^2 is the variance of the distribution or population being sampled and n is the size of the random sample, then the variance of the sample mean is smaller than that of the original variance by the factor $1/n$. This result along with $E(\overline{X}) = \mu$ is illustrated in Figure 5.9.

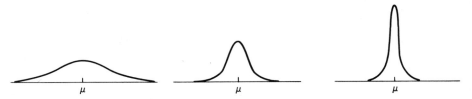

Figure 5.9 Distribution of the sample mean $\overline{X}$ for (a) $n = 1$ (original distribution), (b) $n = 10$, and (c) $n = 25$.

For every value of n, the distribution is centered at μ, the mean of the population being sampled, but as n increases, more and more probability is "squeezed in" close to μ. Intuitively this is quite plausible, since the average value in a large sample should be close to the population average.

The story is different for T_o. If μ is positive, $E(T_o) = n\mu$ increases with n, so the distribution of T_o is centered further and further to the right of zero, and since $\text{Var}(T_o) = n\sigma^2$ the distribution spreads out as n increases. This is because a sum is more variable than any single component in the sum.

**Example 5.15
(continued)**   If the standard deviation of the number of cycles to first acoustic emission is $\sigma = 5000$ cycles, then for a random sample of size $n = 25$, $\sigma_{\overline{X}} = 5000/\sqrt{25} = 1000$, $\sigma_{\overline{X}}^2 = 1{,}000{,}000$, while $\sigma_{T_o} = \sqrt{25}\,(5000) = 25{,}000$.

Another special linear combination is obtained by taking $n = 2$, $a_1 = 1$, and $a_2 = -1$, so that $Y = X_1 - X_2$. Then (5.11) gives

> **Corollary:** When X_1 and X_2 are independent, $\text{Var}(X_1 - X_2) = \sigma_1^2 + \sigma_2^2$.

That is, *the variance of a difference is the sum, not the difference, of the individual variances*. If we write $X_1 - X_2$ as $X_1 + (-X_2)$, this is plausible since $-X_2$ has exactly the same amount of variability as X_2 itself.

Example 5.16   My commute time X_1 to work has expected value 25 min and standard deviation 5 min, and my commute time X_2 home from work has expected value 20 min and standard deviation 4 min. The expected difference in the two times is then

$$E(X_1 - X_2) = E(X_1) - E(X_2) = 25 - 20 = 5 \text{ min}$$

Assuming that X_1 and X_2 are independent, the variance of the difference in times is

$$\text{Var}(X_1 - X_2) = \sigma_1^2 + \sigma_2^2 = (5)^2 + (4)^2 = 41$$

so

$$\sigma_{X_1 - X_2} = \sqrt{41} = 6.40 \text{ min}$$

These are also the variance and standard deviation of total commute time $T_o = X_1 + X_2$.

Exercises / Section 5.3

1. Let X_1 and X_2 be independent, each having the probability distribution

x	1	2	3	4
$p(x)$	.2	.3	.4	.1

a. List all possible (x_1, x_2) pairs, the probability of each one, and the corresponding values of $x_1 + x_2$ and $\bar{x}$. Then write the p.m.f. of $T_o = X_1 + X_2$ and $\bar{X}$.

b. Draw probability histograms of the distributions of X_1, T_o, and $\bar{X}$. What can you say about the centers of the histograms relative to one another? Which has the smallest spread and which has the largest?

c. Compute $E(X_1)$, and then use the rules of expected value to compute $E(T_o)$ and $E(\bar{X})$.

d. Compute $\text{Var}(X_1)$, and then use the rules of variance to obtain $\text{Var}(T_o)$, σ_{T_o}, $\text{Var}(\bar{X})$, and $\sigma_{\bar{X}}$.

2. A local clothing store carries three different brands of blue oxford cloth shirts, priced at \$12, \$15, and

\$20, respectively. Based on past experience, the probability distribution of the price X paid by a randomly selected customer purchasing such a shirt is known to be

x	12	15	20
$p(x)$	.5	.2	.3

a. Let X_1 and X_2 denote the prices paid by two different randomly selected individuals for such shirts, and suppose that X_1, X_2 constitute a random sample of size $n = 2$ from the distribution. Compute the p.m.f. of $T_o = X_1 + X_2$ and of $\bar{X}$.

b. Compute $\mu = E(X)$. Then compute $\mu_{\bar{X}}$ and μ_{T_o} from the distributions derived in (a), and verify that $\mu_{\bar{X}} = \mu$ and $\mu_{T_o} = 2\mu$.

c. Compute $\sigma^2 = \text{Var}(X)$. Then compute $\sigma_{\bar{X}}^2$ and $\sigma_{T_o}^2$ from the distributions derived in (a), and verify that $\sigma_{\bar{X}}^2 = \sigma^2/2$ and $\sigma_{T_o}^2 = 2\sigma^2$.

d. In the course of a week, 20 different customers purchase (independently of one another) one shirt according to the probability distribution. What is the expected total revenue from these 20 sales? What is the expected average revenue per sale?

e. As in (d), what is the variance of total revenue, what is the variance of average revenue, and what is the standard deviation of average revenue?

3. A friend of mine keeps a supply of yogurt in both his apartment refrigerator for breakfast and his office refrigerator for lunch. His favorite flavors are plain, vanilla, and strawberry, listed as containing 160, 200 and 240 calories, respectively. Suppose that he selects a container at random for breakfast and (independently) selects another for lunch, and let X_1 and X_2 denote the number of calories in the breakfast and lunch container, respectively.

a. Compute the p.m.f. of $T_o = X_1 + X_2$ and $\bar{X}$ if each refrigerator contains two containers of plain, two of vanilla, and one of strawberry yogurt.

b. Compute the p.m.f. of T_o if the apartment refrigerator is stocked as in (a), but the office refrigerator contains two plain, one vanilla, and two strawberry. Is X_1, X_2 a random sample in this case? Why or why not?

c. Use the p.m.f. derived in (b) to compute μ_{T_o} and $\sigma_{T_o}^2$. Then verify that $\mu_{T_o} = \mu_1 + \mu_2$, where $\mu_i = E(X_i)$, and that $\sigma_{T_o}^2 = \sigma_1^2 + \sigma_2^2$, where $\sigma_i^2 = \text{Var}(X_i)$.

d. Suppose that the refrigerators are stocked as in (b), but now after having selected at random for breakfast, my friend selects at random a different flavor for lunch. Are X_1 and X_2 still independent? Why or why not? Compute the p.m.f. of T_o. *Hint:* A tree diagram might help here.

e. In the situation described in (d), compute $\mu_{T_o} = E(X_1 + X_2)$ from the distribution that you have derived. Then compute $\mu_1 = E(X_1)$ and $\mu_2 = E(X_2)$, and verify that $\mu_{T_o} = \mu_1 + \mu_2$. *Hint:* You must be careful in writing down the distribution of X_2. It's best to use a tree diagram.

4. Let X_1 denote the time that you must wait before the bus that you take to work arrives, and let X_2 denote your waiting time for the bus which takes you home. Suppose that X_1 and X_2 are indepen-

dent, each distributed uniformly on [0, 5], and let $T = X_1 + X_2$ be your total waiting time.

a. Compute the c.d.f. $F_{T_o}(t)$ of T_o. Then differentiate it to obtain $f_{T_o}(t)$. *Hint:* The region of positive density is a square with side length $= 5$, and the joint density of (X_1, X_2) is $f(x_1, x_2) = 1/25$ on this square. In computing $F_{T_o}(t) = P(T_o \leq t)$, consider separately $t \leq 5$ and $t > 5$.

b. If X_1 and X_2 are independent with X_1 uniform on [0, 5] and X_2 uniform on [0, 10], compute $f_{T_o}(t)$.

5. Suppose that your waiting time for a bus in the morning is uniformly distributed on [0, 5], while waiting time in the evening is uniformly distributed on [0, 10] independent of morning waiting time.

a. If you take the bus each morning and evening for a week, what is your total expected waiting time? *Hint:* Define random variables $X_1, \ldots, X_{10}$ and use a rule of expected value.

b. What is the variance of your total waiting time?

c. What are the expected value and variance of the difference between morning and evening waiting time on a given day?

d. What are the expected value and variance of the difference between total morning waiting time and total evening waiting time for a particular week?

6. A store owner has purchased 20 lightbulbs of type A and 25 of type B. Let $X_1, \ldots, X_{20}$ represent the 20 type-A lifetimes, assumed independent with mean and standard deviation 1000 hours and 100 hours, respectively. Similarly, let $Y_1, \ldots, Y_{25}$ denote the type-B lifetimes, assumed independent with mean and standard deviation 1200 hours and 150 hours, respectively. Use appropriate rules of expected value and variance to obtain the expected value, variance, and standard deviation of the difference $\bar{X} - \bar{Y}$ between the sample average lifetimes for the two types of bulbs.

7. Although a lecture period at a certain university lasts exactly 50 min, the actual lecture time of a statistics instructor on any particular day is a random variable (what else!) with expected value 52 min and standard deviation 2 min. If times of different lectures are independent of one another and the instructor gives 25 lectures in a particular course, compute

a. the expected total lecture time, variance of total lecture time, and standard deviation of total lecture time in the course

b. the expected average lecture time per day, variance of average lecture time per day, and standard deviation of average lecture time per day for the 25-day course

8. In Sections 5.1 and 5.2 we considered the following joint p.d.f. for X = amount of almonds and Y = amount of cashews in a one-pound can.

$$f(x, y) = \begin{cases} \dfrac{1}{24} xy & x \geq 0, y \geq 0, x + y \leq 1 \\ 0 & \text{otherwise} \end{cases}$$

Compute the expected value of the total amount $X + Y$ of almonds and cashews in the can, and compute the variance of this total. *Hint:* The marginal p.d.f. of both X and Y as well as Cov (X, Y) were previously computed.

9. Three different roads feed into a particular freeway entrance. Suppose that during a fixed time period, the number of cars coming from each road onto the freeway is a random variable, with expected value and standard deviation as given.

	Road 1	Road 2	Road 3
Expected value	800	1000	600
Standard deviation	16	25	18

a. What is the expected total number of cars entering the freeway at this point during the period? *Hint:* Let X_i = the number from road i.

b. What is the variance of the total number of entering cars? Have you made any assumptions about the relationship between the number of cars on the different roads?

c. With X_i denoting the number of cars entering from road i during the period, suppose that Cov $(X_1, X_2) = 80$, Cov $(X_1, X_3) = 90$, and Cov $(X_2, X_3) = 100$ (so that the three streams of traffic are not independent). Compute the expected total number of entering cars and the standard deviation of the total.

10. Suppose we take a random sample of size n from a continuous distribution having median 0, so that the probability of any one observation being positive is .5. We now disregard the signs of the observations, rank them from smallest to largest in absolute value, and then let W = the sum of the ranks of the observations having positive signs. For example, if the observations are $-.3$, $+.7$, $+2.1$, and -2.5, then the ranks of positive observations are 2 and 3, so $W = 5$. In Chapter 15, W will be called Wilcoxon's signed rank statistic. W can be represented as follows:

$$W = 1 \cdot Y_1 + 2 \cdot Y_2 + 3 \cdot Y_3 + \cdots + n \cdot Y_n$$
$$= \sum_{i=1}^{n} i \cdot Y_i$$

where the Y_i's are independent Bernoulli random variables, each with p.m.f.

$$p_{Y_i}(y) = \begin{cases} .5 & y = 0 \\ .5 & y = 1 \end{cases}$$

(so $Y_i = 1$ corresponds to the observation with rank i being positive).

a. Compute $E(Y_i)$ and then $E(W)$ using the equation for W. *Hint:* The first n positive integers sum to $n(n + 1)/2$.

b. Compute Var (Y_i) and then Var (W). *Hint:* The sum of the squares of the first n positive integers is $n(n + 1)(2n + 1)/6$.

5.4 The Central Limit Theorem

In the previous section we gave several examples in which the probability distributions of the sample mean $\overline{X}$ and sample total T_o were computed. In most cases these distributions are quite difficult to obtain, though we did obtain information about the expected values and variances of these variables. There is, though, one very important special case in which the distribution of any linear combination of the X_i's is easily obtained.

Linear Combinations of Normal Random Variables

> **Proposition:** If $X_1, X_2, \ldots, X_n$ are independent normal random variables, then any linear combination
>
> $$Y = a_1X_1 + a_2X_2 + \cdots + a_nX_n$$
>
> also has a normal distribution with mean $\mu_Y = \Sigma \, a_i\mu_i$ and variance $\sigma_Y^2 = \Sigma \, a_i^2\sigma_i^2$. In particular if each X_i has the same distribution (a normal random sample) with mean μ and variance σ^2, then $\bar{X}$ is normally distributed with mean μ and variance σ^2/n, and T_o is normally distributed with mean $n\mu$ and variance $n\sigma^2$.

Figure 5.9 on p. 193 illustrates this result for $\bar{X}$. When the distribution being sampled is normal, then for any n the p.d.f. of $\bar{X}$ is also normal and centered at μ; the larger the sample size, the smaller the spread of the p.d.f. about μ.

A proof of the proposition for the linear combination $Y = a_1X_1 + a_2X_2$ is possible using the method in Example 5.13, but even here the details are messy. The general result is usually proved using a theoretical tool called a moment generating function. A good discussion of moment generating functions appears in both of the books listed in the chapter bibliography.

Once we know that a linear combination of interest is normally distributed, probabilities involving it are computed by standardizing exactly as before.

Example 5.17   The time that it takes a randomly selected rat of a certain subspecies to find its way through a maze is a normally distributed random variable with $\mu = 1.5$ min and $\sigma = .35$ min. Suppose that five rats are selected and let $X_1, \ldots, X_5$ denote their times in the maze. Assuming the X_i's to be a random sample from this normal distribution, what is the probability that the total time $T_o = X_1 + \cdots + X_5$ for the five is between 6 and 8 min? By the proposition, T_o has a normal distribution with $\mu_{T_o} = n\mu = 5(1.5) = 7.5$ and variance $\sigma_{T_o}^2 = n\sigma^2 = 5(.1225) = .6125$, so $\sigma_{T_o} = .783$. To standardize T_o, subtract μ_{T_o} and divide by σ_{T_o}:

$$P(6 \le T_o \le 8) = P\left(\frac{6 - 7.5}{.783} \le Z \le \frac{8 - 7.5}{.783}\right)$$

$$= P(-1.92 \le Z \le .64) = \Phi(.64) - \Phi(-1.92) = .7115$$

Example 5.18   Suppose that the lifetime of a particular brand of phonograph needle is normally distributed with $\mu = 1000$ hours and $\sigma = 150$ hours. If a high-fidelity store uses nine of these needles for demonstration purposes, what is the probability that the average lifetime $\bar{X}$ for the nine needles exceeds 1100 hours? Let $X_1, \ldots, X_9$ denote the nine lifetimes, assumed to be a random sample from this normal distribution. According to the proposition, $\bar{X}$ also has a normal distribution with $\mu_{\bar{X}} = \mu = 1000$

and $\sigma_{\bar{X}}^2 = \sigma^2/n = (150)^2/9$, so $\sigma_{\bar{X}} = \sigma/\sqrt{n} = 150/3 = 50$. The desired probability is now obtained by standardizing:

$$P(\bar{X} \geq 1100) = P\left(\frac{\bar{X} - 1000}{50} \geq \frac{1100 - 1000}{50}\right) = P(Z \geq 2) = .0228$$

Suppose that the actual observed value of $\bar{X}$ was $\bar{x} = 850$. Since the standard deviation of $\bar{X}$ is $\sigma/\sqrt{n} = 50$, 850 is three standard deviations (of $\bar{X}$) below the mean of 1000. The probability of observing $\bar{X}$ at least this far below μ is

$$P(\bar{X} \leq 850) = P(Z \leq -3) = \Phi(-3) = .0013$$

If the true value of μ is 1000, we have observed something very unlikely; we might begin to suspect that $\mu < 1000$.

Example 5.19   On Mondays, Wednesdays, and Fridays I drive to work by myself, while on Tuesdays and Thursdays I carpool with another individual by driving every other week. Assume that all driving times are independent of one another and normally distributed as follows: my house directly to work: $\mu = 30$ min, $\sigma = 8$ min; my house to carpooler's house: $\mu = 15$ min, $\sigma = 3$ min; carpooler's house to work: $\mu = 20$ min, $\sigma = 6$ min. What is the probability that my total driving time to work exceeds 200 min during a week in which I drive every day? To answer this, define random variables by $X_1, X_2, X_3 = $ M, W, F direct driving times; $X_4, X_5 = $ T, Th times to carpooler's house; $X_6, X_7 = $ T, Th times from carpooler's house to work.

The total driving time is $T_o = X_1 + \cdots + X_7$, and we wish $P(T_o \geq 200)$. To standardize, we compute

$$\mu_{T_o} = E(X_1 + \cdots + X_7) = \sum_{i=1}^{7} \mu_i$$

$$= 30 + 30 + 30 + 15 + 15 + 20 + 20 = 160$$

$$\sigma_{T_o}^2 = \text{Var}(X_1 + \cdots + X_7) = \sum_{i=1}^{7} \sigma_i^2$$

$$= 64 + 64 + 64 + 9 + 9 + 36 + 36 = 282$$

and

$$\sigma_{T_o} = \sqrt{282} = 16.79$$

Therefore, since T_o has a normal distribution

$$P(T_o \geq 200) = P\left(Z \geq \frac{200 - 160}{16.79}\right) = P(Z \geq 2.38) = .0087$$

The Central Limit Theorem

When the X_i's are normally distributed, so is $\bar{X}$ for every value of n. Figures 5.6 and 5.7 of the previous section suggest that even when the distribution of each X_i is highly nonnormal, averaging produces a distribution more symmetric and bell-shaped than the distribution being sampled. Notice that even when $n = 4$, probability has begun to pile up around μ in a bell-shaped fashion, though the original p.m.f. in Figure 5.6 is quite skewed. A reasonable conjecture is that if n is large, a suitable normal curve will approximate the actual distribution of $\bar{X}$. The formal statement of this result is the most important theorem of probability.

> **The Central Limit Theorem (C.L.T.):** Let $X_1, X_2, \ldots, X_n$ be a random sample from a distribution with mean μ and variance σ^2. Then if n is sufficiently large, $\bar{X}$ has approximately a normal distribution with $\mu_{\bar{X}} = \mu$ and $\sigma_{\bar{X}}^2 = \sigma^2/n$, and T_o also has approximately a normal distribution with $\mu_{T_o} = n\mu$, $\sigma_{T_o}^2 = n\sigma^2$. The larger the value of n, the better the approximation.

According to the C.L.T., when n is large and we wish to calculate, say, $P(a \leq \bar{X} \leq b)$, we need only "pretend" that $\bar{X}$ is normal, standardize it, and use the normal table. The resulting answer will be approximately correct. The exact answer could be obtained only by first finding the distribution of $\bar{X}$, so the C.L.T. provides a truly impressive shortcut. The proof of the theorem involves much advanced mathematics, so we proceed to examples.

Example 5.20 The nicotine content in a single cigarette of a particular brand is a random variable with mean $\mu = .8$ mg and standard deviation .1 mg. If an individual smokes five packs of these cigarettes per week, what is the probability that the total amount of nicotine consumed in a week is at least 82 mg?

Notice that there is no assumption of normality here, but five packs consist of $n = 100$ cigarettes. Letting $X_1, \ldots, X_{100}$ be the nicotine contents of the 100 cigarettes, the X_i's constitute a random sample from a distribution with $\mu = .8$ and $\sigma = .1$. Assuming that $n = 100$ is sufficiently large for the C.L.T. to apply, T_o is approximately normal with mean $(100)(.8) = 80$ and standard deviation $(.1) \cdot \sqrt{100} = 1$. Therefore $(T_o - 80)/1$ is *approximately* standard normal, so that

$$P(T_o \geq 82) \doteq P[Z \geq (82 - 80)/1] = P(Z \geq 2) = .0228$$

Thus the probability that at least 82 mg is consumed is approximately .0228.

Example 5.21 A well-groomed professor has his hair styled once a week at the local salon while classes are in session. His stylist is always the same person, and he believes his "styling time" on any particular visit to be a random variable with mean $\mu = 40$ min and standard deviation 4.2 min. If he visits the stylist 36 times during the academic year, what is the probability that the average styling time for the 36 visits will be between 38 and 41 min?

Again there is no mention of normality, but presuming that $n = 36$ is large enough for the C.L.T. to be invoked, the approximate probability can be obtained by standardizing $\bar{X}$. Since $\mu_{\bar{X}} = \mu = 40$ and $\sigma_{\bar{X}} = \sigma/\sqrt{n} = .7$,

$$P(38 \leq \bar{X} \leq 41) \doteq P\left(\frac{38 - 40}{.7} \leq Z \leq \frac{41 - 40}{.7}\right)$$

$$= P(-2.86 \leq Z \leq 1.43) = .9215$$

Understanding the C.L.T.

The Central Limit Theorem provides insight into why many random variables have probability distributions which are approximately normal. For example, the measurement error in a scientific experiment can be thought of as a sum of a number of underlying perturbations and errors of small magnitude.

Although the usefulness of the C.L.T. for inference will soon be apparent, the intuitive content of the result gives many beginning students difficulty. Again looking back to Figure 5.6, the probability histogram on the left is a picture of the distribution being sampled. It is discrete and quite skewed, so does not look at all like a normal distribution. The distribution of $\bar{X}$ for $n = 2$ starts to exhibit some symmetry, and this is even more pronounced for $n = 4$ in Figure 5.7. Figure 5.10 contains the probability distribution of $\bar{X}$ for $n = 8$, as well as a probability histogram for this distribution. With $\mu_{\bar{x}} = \mu = 46.5$ and $\sigma_{\bar{x}} = \sigma/\sqrt{n} = 3.905/\sqrt{8} = 1.38$, if we fit a normal curve with this mean and standard deviation through the

$\bar{x}$	40	40.625	41.25	41.875	42.5	43.125
$p(\bar{x})$	.0000	.0000	.0003	.0012	.0038	.0112
$\bar{x}$	43.75	44.375	45	45.625	46.25	46.875
$p(\bar{x})$	.0274	.0556	.0954	.1378	.1704	.1746
$\bar{x}$	47.5	48.125	48.75	49.375	50	
$p(\bar{x})$	.1474	.0998	.0519	.0188	.0039	

Figure 5.10a   Probability distribution of $\bar{X}$ for $n = 8$ when the original distribution is as in Example 5.12

Figure 5.10b   Probability histogram and normal approximation to the distribution of $\bar{X}$ in Figure 5.10a

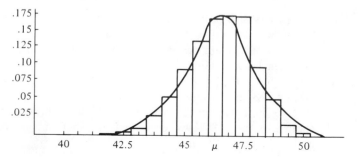

histogram of $\bar{X}$, the areas of rectangles in the probability histogram are reasonably well approximated by the normal curve areas, at least in the central part of the distribution. The picture for T_o is similar except that the horizontal scale is much more spread out, with T_o ranging between 320 ($\bar{x} = 40$) and 360 ($\bar{x} = 50$).

A better understanding of the C.L.T. can also be obtained by performing a simulation experiment of the following sort: Select a probability distribution and a reasonably large value of n, such as $n = 50$. Then obtain 50 observed x_i's from this distribution and compute the sample average $\bar{x}$ for the 50. To do this with the p.m.f.

x	40	45	50
$p(x)$	.2	.3	.5

that we have been working with, one could put two slips marked 40, three marked 45, and five marked 50 into a box and then sample with replacement. Alternatively, there are computer program packages such as MINITAB or GPSS which allow for such simulation. Next obtain another 50 x_i's and compute $\bar{x}$, and so on until 1000 $\bar{x}$'s, each an average of 50 observations, have been obtained. Now construct a sample histogram of the 1000 $\bar{x}$'s. This histogram will have the appearance of the true distribution of $\bar{X}$ *and* will be very close to bell-shaped. If $n = 100$ x_i's had been averaged to obtain each $\bar{x}$, the bell-shaped appearance would be even more pronounced, since a larger n improves the approximation.

A practical difficulty in applying the C.L.T. is in knowing when n is sufficiently large. The problem is that the accuracy of the approximation for a particular n depends on the shape of the original underlying distribution being sampled. If the underlying distribution is close to bell-shaped, then the approximation will be good even for a small n, while if it is far from bell-shaped then a large n will be required. We will use the following rule of thumb, which is frequently somewhat conservative.

Rule of thumb: If $n > 30$, the C.L.T. can be used.

The Normal Approximation to the Binomial Distribution

Recall that the binomial variable X is the number of successes in a binomial experiment, and X/n is the sample proportion of successes. Define n new random variables as follows:

$$X_1 = \begin{cases} 1 & \text{if trial 1 yields a success} \\ 0 & \text{if trial 1 yields a failure} \end{cases}$$

.

.

.

$$X_n = \begin{cases} 1 & \text{if trial } n \text{ yields a success} \\ 0 & \text{if trial } n \text{ yields a failure} \end{cases}$$

Because the trials are independent, the X_i's are independent of one another, and they all have the same probability distribution (each is a Bernoulli r.v. with success parameter p). Thus if n is sufficiently large, both the sum and the average of the X_i's have approximately normal distributions. But when the X_i's are summed, a 1 is added in for each S and a 0 for each F, so X = number of S's = $X_1 + \cdots + X_n$.

Proposition: Let X be a binomial r.v. based on n trials with $p = P(S)$, so that $\mu_X = np$ and $\sigma_X = \sqrt{npq}$. Then if n is sufficiently large, X has approximately a normal distribution with mean np and standard deviation $\sqrt{npq}$, and the sample proportion X/n has approximately a normal distribution with mean p and standard deviation $\sqrt{pq/n}$.

For p close to .5, the approximation will be good even for n as small as 15 or 20, but for p near 0 or 1, a much larger n is necessary.

Rule of thumb: If both $np \geq 5$ and $n(1 - p) \geq 5$, the normal approximation to the binomial distribution can be used.

Figure 5.11 illustrates how areas under a binomial probability histogram (that is, binomial probabilities) can be approximated by areas under the normal curve with $\mu = np$ and $\sigma = \sqrt{npq}$. To increase the accuracy of the approximation, the **"continuity correction"** can be used: The probability $P(a \leq X \leq b)$ is approximated by the area under the normal curve between $a - \frac{1}{2}$ and $b + \frac{1}{2}$. This is because the base of the rectangle representing $P(X = a)$ is centered at $x = a$ and extends from $a - \frac{1}{2}$ to $a + \frac{1}{2}$. This is indicated in Figure 5.11 for $P(7 \leq X \leq 11)$.

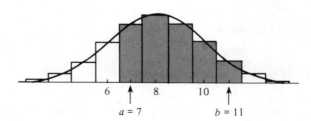

Figure 5.11   Normal approximation to binomial distribution when $n = 16$ and $p = .5$.

Example 5.22   A large supermarket chain has reported that 40% of all beer purchases are "light" beers (lower in calories than regular beer). What is the probability that between six and 10 of the next 20 beer purchases are of the light variety? If we identify a success S with a purchase of a light beer, then $n = 20$ and $p = .4$ for this binomial experiment. The mean and standard deviation of X = number of S's are $\mu = np = 8$ and $\sigma = \sqrt{npq} = 2.19$. Using the normal approximation and the continuity correction,

$$P(15 \leq X \leq 25) \doteq P(14.5 \leq \text{normal variable} \leq 25.5)$$

$$= P\left(\frac{5.5 - 8}{2.19} \leq Z \leq \frac{10.5 - 8}{2.19}\right) = P(-1.14 \leq Z \leq .68) = .6246$$

From Appendix Table A.1, the exact probability is $B(10; 20, .4) - B(5; 20, .4) = .6469$, which is close to the approximate answer.

Other Applications of the C.L.T.

Recall from Section 4.5 that a r.v. X has a lognormal distribution if $\ln(X)$ has a normal distribution.

Proposition: Let $X_1, X_2, \ldots, X_n$ be a random sample from a distribution for which only positive values are possible $[P(X_i > 0) = 1]$. Then if n is sufficiently large, the product $Y = X_1 X_2 \cdots X_n$ has approximately a lognormal distribution.

To verify this, note that

$$\ln(Y) = \ln(X_1) + \ln(X_2) + \cdots + \ln(X_n)$$

Since $\ln(Y)$ is a sum of independent and identically distributed r.v.'s [the $\ln(X_i)$'s], it is approximately normal when n is large, so Y itself has approximately a lognormal distribution.

As an example of the applicability of this result, Bury (*Statistical Models in Applied Science*, p. 590) argues that the damage process in plastic flow and crack propagation is a multiplicative process, so that variables such as percentage elongation and rupture strength have approximately lognormal distributions.

We have stated that a sum of independent normal variables is also normally distributed. A way to paraphrase this is to say that if $X_1, \ldots, X_n$ are members of the normal family of distributions, then so is ΣX_i. There are other situations in which ΣX_i has a distribution in the same family as each of the X_i's in the sum. Several of these are identified in the next proposition.

Proposition: Let $X_1, \ldots, X_n$ be independent random variables.

1. If X_i has a Poisson distribution with parameter λ_i, then $\sum_{i=1}^{n} X_i$ has a Poisson distribution with parameter $\lambda = \sum_{i=1}^{n} \lambda_i$.
2. If X_i has a gamma distribution with parameters α_i and β, then $\sum_{i=1}^{n} X_i$ has a gamma distribution with parameters $\alpha = \sum_{i=1}^{n} \alpha_i$ and β.

The statement of (1) can be "turned around" in the following manner: If X has a Poisson distribution with, say, $\lambda = 50$, then X is the sum of 50 Poisson variables, each having $\lambda = 1$. It can be shown that $n = 50$ is large enough to "overcome" the skewness of the Poisson distribution with $\lambda = 1$ (but not with, say, $\lambda = .01$), so that by the C.L.T. a Poisson variable with $\lambda = 50$ has approximately a normal distribution. Similar logic with other values of λ and with the gamma distribution yields

---

Proposition:

1. Let X have a Poisson distribution with parameter λ, so that $E(X) = \lambda$ and $\sigma_X = \sqrt{\lambda}$. Then when λ is large, X has approximately a normal distribution with parameters $\mu = \lambda$ and $\sigma = \sqrt{\lambda}$.
2. Let X have a gamma distribution with parameters α and β, so that $E(X) = \alpha\beta$ and $\sigma_X^2 = \alpha\beta^2$. Then if α is large, X has approximately a normal distribution with parameters $\mu = \alpha\beta$ and $\sigma^2 = \alpha\beta^2$.

There is no simple answer to the question, "How large must n be before the approximation is accurate?" The accuracy depends on the set of values whose probability is desired.

Example 5.23   Suppose that the number of applications X for a credit card received by a department store during a given week has a Poisson distribution with $\lambda = 20$. What is the probability that between 15 and 25 applications are received during the week? To approximate, we need $\mu = \lambda = 20$ and $\sigma = \sqrt{\lambda} = 4.47$. Since X is integer valued, we again use the continuity correction:

$$P(15 \leq X \leq 25) \doteq P(14.5 \leq \text{normal variable} \leq 25.5)$$

$$= P\left(\frac{14.5 - 20}{4.47} \leq Z \leq \frac{25.5 - 20}{4.47}\right) = .7814$$

The exact probability is .783, so the approximation is very good. However, the exact probability $P(X \geq 30)$ is .022 while the approximation gives .017, so the relative error is much larger.

Exercises / Section 5.4

1. Let X_1, X_2, and X_3 be three independent normal random variables with expected values μ_1, μ_2, and μ_3 and variances σ_1^2, σ_2^2, and σ_3^2, respectively.
   a. If $\mu_1 = \mu_2 = \mu_3 = 100$ and $\sigma_1^2 = \sigma_2^2 = \sigma_3^2 = 12$, calculate $P(X_1 + X_2 + X_3 \leq 309)$ and $P(288 \leq X_1 + X_2 + X_3 \leq 312)$.
   b. Using the μ_i's and σ_i's of (a), calculate $P(105 \leq \bar{X})$ and $P(98 \leq \bar{X} \leq 102)$.
   c. Using the μ_i's and σ_i's of (a), calculate $P(-10 \leq X_1 - .5X_2 - .5X_3 \leq 5)$.

   d. If $\mu_1 = 90$, $\mu_2 = 100$, $\mu_3 = 110$, $\sigma_1^2 = 10$, $\sigma_2^2 = 12$, and $\sigma_3^2 = 14$, calculate $P(X_1 + X_2 + X_3 \leq 306)$ and $P(98 \leq \bar{X} \leq 102)$.

2. Five automobiles of the same type are to be driven on a 300-mile trip. The first two will use an economy brand of gasoline and the other three will use a name brand. Let X_1, X_2, X_3, X_4, and X_5 be the observed number of miles per gallon for the five cars, and suppose that these variables are indepen-

dent and normally distributed with $\mu_1 = \mu_2 = 20$, $\mu_3 = \mu_4 = \mu_5 = 21$, and $\sigma^2 = 4$ for all five variables. Define a r.v. Y by

$$Y = \frac{X_1 + X_2}{2} - \frac{X_3 + X_4 + X_5}{3}$$

so that Y is a measure of the difference in efficiency between economy gas and name brand gas. Compute $P(0 \leq Y)$ and $P(-1 \leq Y \leq 1)$.

3. Let $X_1, X_2, \ldots, X_{100}$ denote the actual net weights of 100 randomly selected 50 lb bags of fertilizer.

  a. If the expected weight of each bag is 50 and the variance is 1, calculate $P(49.75 \leq \bar{X} \leq 50.25)$ (approximately) using the C.L.T.

  b. If the expected weight is 49.8 lbs rather than 50 lbs so that on average bags are underfilled, calculate $P(49.75 \leq \bar{X} \leq 50.25)$.

4. There are 40 students in an elementary statistics class. On the basis of years of experience, the instructor knows that the time needed to grade a randomly chosen first examination paper is a random variable with an expected value of six minutes and a standard deviation of six minutes.

  a. If grading times are independent and the instructor begins grading at 6:50 P.M. and grades continuously, what is the (approximate) probability that he is through grading before the 11:00 P.M. TV news begins?

  b. If the sports report begins at 11:10, what is the probability that he misses part of the report if he waits until grading is done before turning on the TV?

5. Suppose that when the pH of a certain chemical compound is 5.00, the pH measured by a randomly selected beginning chemistry student is a random variable with mean 5.00 and standard deviation .3. A large batch of the compound is subdivided and a sample given to each student in a morning lab and each student in an afternoon lab. Let $\bar{X} =$ the average pH as determined by the morning students and $\bar{Y} =$ the average pH as determined by the afternoon students.

  a. If pH is a normal variable and there are 25 students in each lab, compute $P(-.1 \leq \bar{X} - \bar{Y} \leq .1)$. *Hint:* $\bar{X} - \bar{Y}$ is a linear combination of normal variables, so is normally distributed. Compute $\mu_{\bar{X}-\bar{Y}}$ and $\sigma_{\bar{X}-\bar{Y}}$.

  c. If there are 36 students in each lab, but pH determinations are not assumed normal, calculate (approximately) $P(-.1 \leq \bar{X} - \bar{Y} \leq .1)$.

6. Let X be a binomial random variable with $n = 100$ and $p = .8$. Use the normal approximation to compute the following probabilities both with and without the continuity correction.

  a. $P(75 \leq X \leq 85)$ **b.** $P(X \leq 90)$
  c. $P(.7 \leq X/100 \leq .9)$

7. The first assignment in an introductory computer programming class involves keypunching a short program. If past experience indicates that 40% of all beginning students will make no keypunching errors, compute the (approximate) probability that in a class of 50 students

  a. at least 25 will make no errors. *Hint:* normal approximation to the binomial.

  b. between 15 and 25 will make no errors

8. If two loads are applied to a cantilever beam as shown in the accompanying drawing, the bending moment at 0 due to the loads is $a_1 X_1 + a_2 X_2$.

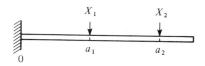

  a. Suppose that X_1 and X_2 are independent random variables with means 2 and 4 kips, respectively and standard deviations .5 and 1.0 kip, respectively. If $a_1 = 5$ ft and $a_2 = 10$ ft, what is the expected bending moment and what is the standard deviation of the bending moment?

  b. If X_1 and X_2 are normally distributed, what is the probability that the bending moment will exceed 75 kip-ft?

  c. Suppose that the positions of the two loads are random variables. Denoting them by A_1 and A_2, assume that these variables have means of 5 and 10 ft, respectively, that each has a standard deviation of .5, and that all A_i's and X_i's are independent of one another. What is the expected moment now?

  d. For the situation of (c), what is the variance of the bending moment?

  e. If the situation is as described in (a) except that Corr $(X_1, X_2) = .5$ (so that the two loads are not independent), what is the variance of the bending moment?

9. In the previous problem the weight of the beam it-self contributes to the bending moment. Assume that the beam is of uniform thickness and density, so that the resulting load is uniformly distributed on the beam. If the weight of the beam is random, the resulting load from the weight is also random; denote this load by W (kips/ft).
   a. If the beam is 12 ft long, W has mean 1.5 and standard deviation .25, and the fixed loads are as described in (a) of the previous problem, what are the expected value and variance of the bending moment? *Hint:* If the load due to the beam were w kips/ft, the contribution to the bending moment would be $w \int_0^{12} x \, dx$.
   b. If all three variables (X_1, X_2, and W) are nor-mally distributed, what is the probability that the bending moment will be at most 200 kip-ft?

10. I have three errands to take care of in the Admin-istration Building. Let X_i = the time that it takes for the ith errand ($i = 1, 2, 3$) and let X_4 = the total time that I spend walking to and from the building and between each errand. Suppose that the X_i's are independent, normally distributed, with the following means and standard deviations: $\mu_1 = 15$ min, $\sigma_1 = 4$, $\mu_2 = 5$, $\sigma_2 = 1$, $\mu_3 = 8$, $\sigma_3 = 2$, $\mu_4 = 12$, $\sigma_4 = 3$. I plan to leave my office at precisely 10:00 A.M. and wish to post a note on my door which reads, "I will return by t A.M." What time t should I write down if I want the probability of my arriving after t to be .01?

11. Suppose that the expected tensile strength of type-A steel is 106 k.s.i. and the standard deviation of tensile strength is 8 k.s.i. For type-B steel suppose that the expected tensile strength and standard de-viation of tensile strength are 104 k.s.i. and 6 k.s.i., respectively. Let $\bar{X}$ = the sample average tensile strength of a random sample of 40 type-A specimens, and let $\bar{Y}$ = the sample average tensile strength of a random sample of 35 type-B specimens.
   a. What is the approximate distribution of $\bar{X}$? Of $\bar{Y}$?
   b. What is the approximate distribution of $\bar{X} - \bar{Y}$? Justify your answer.
   c. Calculate (approximately) $P(-1 \le \bar{X} - \bar{Y} \le 1)$.
   d. Calculate $P(\bar{X} - \bar{Y} \ge 6)$. If you actually ob-served $\bar{X} - \bar{Y} \ge 6$, would you doubt that $\mu_1 - \mu_2 = 2$?

12. In an area having sandy soil, 50 small trees of a certain type were planted, and another 50 trees were planted in an area having clay soil. Let X = the number of trees planted in sandy soil which survive one year, and Y = the number of trees planted in clay soil which survive one year. If the probability that a tree planted in sandy soil will survive one year is .7 and the probability of one-year survival in clay soil is .6, compute (ap-proximately) $P(-5 \le X - Y \le 5)$ (do not bother with the continuity correction).

13. The number of parking tickets issued in a certain city on any given weekday has a Poisson distribu-tion with parameter $\lambda = 50$.
   a. What is the approximate probability that be-tween 35 and 70 tickets are given out on a par-ticular day?
   b. What is the approximate probability that the total number of tickets given out during a five-day week is between 225 and 275?

14. Suppose that the distribution of the time X (hours) spent by students at a certain university on their senior projects is gamma with parameters $\alpha = 50$ and $\beta = 2$. Use the normal approximation to com-pute the probability that a randomly selected stu-dent spends at most 125 hours on the senior project.

Supplementary Exercises / Chapter 5

1. A restaurant serves three fixed-price dinners cost-ing $7, $9, and $10. For a randomly selected cou-ple dining at this restaurant, let X = the cost of the man's dinner and Y = the cost of the woman's dinner. Suppose that the joint p.m.f. of X and Y is given at the right.

$p(x, y)$		7	$\overset{y}{9}$	10
	7	.05	.05	.10
x	9	.05	.10	.35
	10	0	.20	.10

a. Compute the marginal p.m.f. of X and of Y.

b. What is the probability that the man's and the woman's dinner cost at most \$9 each?

c. Are X and Y independent? Justify your answer.

d. What is the expected total cost of the dinner for the two people?

e. Suppose that when a couple opens fortune cookies at the conclusion of the meal, they find the message "You will receive as a refund the difference between the cost of the more expensive and the less expensive meal that you have chosen." How much does the restaurant expect to refund?

2. A health food store stocks two different brands of a certain type of grain. Let $X =$ the amount (lbs) of brand A on hand, and $Y =$ the amount of brand B on hand. Suppose that the joint p.d.f. of X and Y is

$$f(x, y) = \begin{cases} kxy & x \geq 0, y \geq 0, 20 \leq x + y \leq 30 \\ 0 & \text{otherwise} \end{cases}$$

a. Draw the region of positive density and determine the value of k.

b. Are X and Y independent? Answer by first deriving the marginal p.d.f. of each variable.

c. Compute $P(X + Y \leq 25)$.

d. What is the expected total amount of this grain on hand?

e. Compute $\text{Cov}(X, Y)$ and $\text{Corr}(X, Y)$.

f. What is the variance of the total amount of grain on hand?

3. Let $X_1, X_2, \ldots, X_n$ be random variables denoting n independent bids for an item which is for sale. Suppose that each X_i is uniformly distributed on the interval [100, 200]. If the seller sells to the highest bidder, how much can he expect to earn on the sale? *Hint:* Let $Y = \max(X_1, X_2, \ldots, X_n)$. First find $F_Y(y)$ by noting that $Y \leq y$ iff each X_i is $\leq y$. Then obtain the p.d.f. and $E(Y)$.

4. Suppose that my calorie intake at breakfast is a random variable with expected value 500 and standard deviation 50, my calorie intake at lunch is random with expected value 800 and standard deviation 100, and my calorie intake at dinner is a random variable with expected value 1700 and standard deviation 200. Assuming that intakes on different meals are independent of one another and that I eat three meals per day each day, what is the probability that my average calorie intake per day over the next (365-day) year is between 2950 and 3050? *Hint:* Let X_i, Y_i, and Z_i denote the three calorie intakes on day i. Then total intake $= \Sigma (X_i + Y_i + Z_i)$.

5. Suppose that the proportion of rural voters in a certain state who favor a particular gubernatorial candidate is .45, and the proportion of suburban and urban voters favoring the candidate is .60. If a sample of 200 rural voters and 300 urban and suburban voters is obtained, what is the approximate probability that at least 250 of these voters favor this candidate?

6. Let μ denote the true pH of a chemical compound. A sequence of n independent sample pH determinations will be made. Suppose that each sample pH is a random variable with expected value μ and standard deviation .1. How many determinations are required if we wish the probability that the sample average is within .02 of the true pH to be at least .95? What theorem justifies your probability calculation?

7. If the amount of soft drink that I consume on any given day is independent of consumption on any other day and is normally distributed with $\mu = 13$ oz and $\sigma = 2$, and if I currently have two six-packs of 16-oz bottles, what is the probability that I still have some soft drink left at the end of two weeks (14 days)?

8. A student has a class which is supposed to end at 9:00 A.M. and another which is supposed to begin at 9:10 A.M. Suppose that the actual ending time of the 9 A.M. class is a normally distributed r.v. X_1 with mean 9:02 and standard deviation 1.5 min, and that the starting time of the next class is also a normally distributed r.v. X_2 with mean 9:10 and standard deviation 1 min. Suppose also that the time necessary to get from one classroom to the other is a normally distributed r.v. X_3 with mean 6 min and standard deviation 1 min. What is the probability that the student makes it to the second class before the lecture starts? (Assume independence of X_1, X_2, and X_3, which is reasonable if the student pays no attention to the finishing time of the first class.)

9. a. Use the general formula for the variance of a linear combination to write an expression for $\text{Var}(aX + Y)$. Then let $a = \sigma_Y/\sigma_X$ and show that $\rho \geq -1$. *Hint:* Variance is always ≥ 0, and $\text{Cov}(X, Y) = \sigma_X \cdot \sigma_Y \cdot \rho$.

b. By considering $\text{Var}(aX - Y)$, conclude that $\rho \leq 1$.

c. Use the fact that $\text{Var}(W) = 0$ only if W is a constant to show that $\rho = 1$ only if $Y = aX + b$.

10. Suppose that a randomly chosen individual's verbal score X and quantitative score Y on a nationally administered aptitude examination have joint p.d.f.

$$f(x, y) = \begin{cases} \frac{2}{5}(2x + 3y) & 0 \leq x \leq 1, 0 \leq y \leq 1 \\ 0 & \text{otherwise} \end{cases}$$

You are asked to provide a prediction t of the individual's total score $X + Y$. The error of prediction is the mean square error $E[(X + Y - t)^2]$. What value of t minimizes the error of prediction?

11. a. Let X_1 have a chi-squared distribution with parameter ν_1 (see Section 4.4), and let X_2 be independent of X_1 and have a chi-squared distribution with parameter ν_2. Use the technique of Example 5.13 to show that $X_1 + X_2$ has a chi-squared distribution with parameter $\nu_1 + \nu_2$.

b. In Exercise 4.4.12 you were asked to show that if Z is a standard normal r.v., then Z^2 has a chi-squared distribution with $\nu = 1$. Let Z_1, $Z_2, \ldots, Z_n$ be n independent standard normal r.v.'s. What is the distribution of $Z_1^2 + \cdots + Z_n^2$? Justify your answer.

c. Let $X_1, \ldots, X_n$ be a random sample from a normal distribution with mean μ and variance σ^2. What is the distribution of $Y = \sum_{i=1}^{n} [(X_i - \mu)/\sigma]^2$? Justify your answer.

12. a. Show that $\text{Cov}(X, Y + Z) = \text{Cov}(X, Y) + \text{Cov}(X, Z)$.

b. Let X_1 and X_2 be quantitative and verbal scores on one aptitude exam, and Y_1 and Y_2 be corresponding scores on another exam. If $\text{Cov}(X_1, Y_1) = 5$, $\text{Cov}(X_1, Y_2) = 1$, $\text{Cov}(X_2, Y_1) = 2$, and $\text{Cov}(X_2, Y_2) = 8$, what is the covariance between the two total scores $X_1 + X_2$ and $Y_1 + Y_2$?

c. Let $X_1, \ldots, X_n$ and $Y_1, \ldots, Y_n$ be random variables and $a_1, \ldots, a_n$ and $b_1, \ldots, b_n$ be constants. Show how $\text{Cov}(\sum a_i X_i, \sum b_j Y_j)$ can be expressed in terms of the quantities $\text{Cov}(X_i, Y_j)$, $i = 1, \ldots, n; j = 1, \ldots, n$.

Bibliography   Derman, Cyrus, Glaser, Leon, and Olkin, Ingram, *Probability Models and Applications*, Macmillan, New York, 1980. Contains a careful and comprehensive exposition of joint distributions, rules of expectation, and limit theorems.

Larsen, Richard and Marx, Morris, *Introduction to Mathematical Statistics*, Prentice-Hall, Englewood Cliffs, N.J., 1980. More limited coverage than in the book by Derman et al., but well written and readable.

CHAPTER 6

Point Estimation

Introduction

Given a parameter of interest, such as a population mean μ or population proportion p, the objective of point estimation is to use a sample to compute a number which represents in some sense a good guess for the true value of the parameter. The resulting number is called a point estimate. In Section 6.1 we present some general concepts of point estimation. Section 6.2 describes and illustrates two important methods for obtaining point estimates, the method of moments and the method of maximum likelihood, and closes with a discussion of a method for obtaining "robust" estimates which are reliable for a wide variety of underlying population models.

6.1 Some General Concepts of Point Estimation

Inference is almost always directed toward drawing some type of conclusion about one or more parameters (population characteristics). *The objective of point estimation is to compute from the sample a single number which will be our best guess for the true value of the parameter under investigation.* Parameters of frequent interest in applied problems include

1. a single population mean μ,
2. a population median $\tilde{\mu}$,
3. a single population proportion p,
4. a population variance σ^2 or population standard deviation σ,
5. the difference $\mu_1 - \mu_2$ between the means of two different populations, and
6. the difference $p_1 - p_2$ between proportions within two different populations.

Often concepts and results will apply simultaneously to many different parameters, so we will sometimes use the Greek letter θ as generic notation for an arbitrary parameter. In a particular context, θ may represent μ or σ or $p_1 - p_2$, and so on.

Definition: A **point estimate** of a parameter θ is a single number, computed from sample information, which serves as a guess for the true value of θ. The general rule, or function of the observations, which is used to obtain a point estimate is called a **point estimator.**

For example, we may wish to estimate the mean μ of a population by using the sample mean $\overline{X}$ as the point estimator. Then if the observed values are $x_1 = 3.2$, $x_2 = 4.8$, and $x_3 = 4.3$, the point estimate is $\overline{x} = (3.2 + 4.8 + 4.3)/3 = 4.10$. Using the same estimator for the sample $x_1 = 3.0$, $x_2 = 4.1$, $x_3 = 3.8$ yields the estimate $\overline{x} = 3.63$.

The distinction between the estimator and the estimate is the same as the distinction between a random variable X and the observed or computed value x of the variable. We have previously distinguished between the variable and a particular value by using uppercase and lowercase letters. Here, however, when estimating a parameter θ we shall use the notation $\hat{\theta}$ (read "theta hat") both for the estimator of θ and for a particular estimate. In the example above, the estimator of μ was $\hat{\mu} = \overline{X}$ and the first estimate was $\hat{\mu} = 4.10$.

For each of the six parameters (1)–(6) above, a plausible point estimator is

1. parameter $= \mu$, estimator $= \hat{\mu} = \overline{X}$, the sample mean;
2. parameter $= \widetilde{\mu}$, estimator $= \widehat{\widetilde{\mu}} = \widetilde{X}$, the sample median;
3. parameter $= p$, estimator $= \hat{p} = X/n$, the sample proportion of successes;
4. parameter $= \sigma^2$, estimator $= \hat{\sigma}^2 = S^2$, the sample variance;
   parameter $= \sigma$, estimator $= \hat{\sigma} = S$, the sample standard deviation;
5. parameter $= \mu_1 - \mu_2$, estimator $= (\hat{\mu_1 - \mu_2}) = \overline{X} - \overline{Y}$, the difference between two sample means computed from two independent random samples $X_1, \ldots, X_m$ and $Y_1, \ldots, Y_n$, respectively;
6. parameter $= p_1 - p_2$, estimator $= (\hat{p_1 - p_2}) = X/m - Y/n$, the difference between the corresponding sample proportions, where X and Y are the numbers of "successes" in independent random samples of size m and n, respectively.

If for each parameter of interest there were only one reasonable point estimator, there would not be much to point estimation. In most problems, though, there will be more than one reasonable estimator.

Example 6.1   A consumer organization wishes to obtain a point estimate for the true average lifetime (hours of writing) for an inexpensive brand of ball-point pen. Suppose that the distribution of lifetime is normal with parameters μ and σ, so that μ is both the expected lifetime and median lifetime (since the normal p.d.f. is symmetric about μ). The organization purchases $n = 10$ such pens and inserts each in a specially constructed machine which will cause the pen to write continuously until the ink runs out. Let $X_1, X_2, \ldots, X_{10}$ denote the lifetimes, assumed to be a random sample from the normal distribution with parameters μ and σ. Suppose that the observed lifetimes are $x_1 = 26.3$, $x_2 = 35.1$, $x_3 = 23.0$, $x_4 = 28.4$, $x_5 = 31.6$, $x_6 = 30.9$, $x_7 = 25.2$, $x_8 = 28.0$, $x_9 = 27.3$, and $x_{10} = 29.2$. Consider the following estimators and resulting estimates for μ:

a. estimator $= \bar{X}$, estimate $= \bar{x} = \Sigma x_i/10 = 28.50$
b. estimator $= \tilde{X}$, estimate $= \tilde{x} = (28.0 + 28.4)/2 = 28.20$
c. estimator $= [\min (X_i) + \max (X_i)]/2 =$ the average of the two extreme lifetimes, estimate $= [\min (x_i) + \max (x_i)]/2 = (23.0 + 35.1)/2 = 29.05$
d. estimator $= \bar{X}_{tr(10)}$, the 10% trimmed mean (discard the smallest and largest 10% of the sample and then average), estimate $=$

$$\bar{X}_{tr(10)} = \frac{25.2 + 26.3 + 27.3 + 28.0 + 28.4 + 29.2 + 30.9 + 31.6}{8} = 28.36$$

Each one of the estimators (a)–(d) uses a different measure of the center of the sample to estimate μ. Which of the estimates is closest to the true value? We can't answer this without knowing the true value. A question which can be answered is, "Which estimator, when used on other samples of X_i's, will tend to produce estimates closest to the true value?" We will shortly consider this type of question.

Example 6.2

An automobile manufacturer has developed a new type of bumper which is supposed to absorb impacts with less damage than previous bumpers. The manufacturer has used this bumper in a sequence of 25 controlled crashes against a wall, each at 10 m.p.h., using one of its compact car models. Let $X =$ the number of crashes which result in no visible damage to the automobile. The parameter to be estimated is $p =$ the proportion of all such crashes which result in no damage [alternatively, $p = P(\text{no damage in a single crash})$]. If X is observed to be $x = 15$, the most reasonable estimator and estimate are

$$\text{estimator } \hat{p} = \frac{X}{n}, \quad \text{estimate} = \frac{x}{n} = \frac{15}{25} = .60$$

Example 6.3

Officials of a paint company have been concerned about the variability in drying time of the company's top-quality interior latex paint. Let X be the drying time of a paint sample on a test board, and let $\sigma^2 = \text{Var}(X)$ (the variance of the population of all such drying times). If n test boards are set up and the drying times are X_1, $X_2, \ldots, X_n$, then one estimator of σ^2 is the sample variance:

$$\hat{\sigma}^2 = S^2 = \frac{\Sigma (X_i - \bar{X})^2}{n - 1} = \frac{\Sigma X_i^2 - (\Sigma X_i)^2/n}{n - 1}$$

If $n = 10$ and the observed x_i's are as in Example 6.1 (with minutes replacing hours), then the corresponding estimate is

$$\hat{\sigma}^2 = s^2 = \frac{\sum x_i^2 - (\sum x_i)^2/10}{9} = \frac{(26.3)^2 + \cdots + (29.2)^2 - (285)^2/10}{9}$$

$$= 11.90$$

The estimate of σ would then be $\hat{\sigma} = s = \sqrt{11.90} = 3.45$.

An alternative estimator would result from using divisor n instead of $n - 1$ (that is, the average squared deviation):

$$\hat{\sigma}^2 = \frac{\sum (X_i - \bar{X})^2}{n}, \quad \text{estimate} = \frac{107.10}{10} = 10.71$$

We will shortly indicate why many statisticians prefer S^2 to the estimator with divisor n.

In the best of all possible worlds, we could find an estimator $\hat{\theta}$ for which $\hat{\theta} = \theta$ always. However, $\hat{\theta}$ is a function of the sample X_i's, so is itself a random variable. For some samples $\hat{\theta}$ will yield a value larger than θ, while for other samples $\hat{\theta}$ will underestimate θ. If we write

$$\hat{\theta} = \theta + \text{error of estimation}$$

then an accurate estimator would be one resulting in small estimation errors, so that estimated values will be near the true value. An estimator which has the properties of unbiasedness and minimum variance will often be accurate in this sense.

Unbiased Estimators

Suppose that we have two measuring instruments; one instrument has been accurately calibrated, but the other systematically gives readings under the true value being measured. When each instrument is used repeatedly on the same object, because of measurement error the observed measurements will not be identical. However, the measurements produced by the first instrument will be distributed about the true value in such a way that on average this instrument measures what it purports to measure, so is called an unbiased instrument. The second instrument yields observations which have a systematic error component or bias.

Definition: A point estimator $\hat{\theta}$ is said to be an **unbiased estimator** of θ if $E(\hat{\theta}) = \theta$ for every possible value of θ. If $\hat{\theta}$ is not unbiased, the difference $E(\hat{\theta}) - \theta$ is called the bias of $\hat{\theta}$.

Thus $\hat{\theta}$ is unbiased if its distribution is "centered" at the true value of the parameter. Figure 6.1 pictures the distributions of several biased and unbiased estimators. Note that "centered" here means that the expected value, not the median, of the distribution of $\hat{\theta}$ is equal to θ.

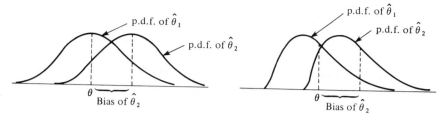

Figure 6.1 P.d.f.'s of an unbiased estimator $\hat{\theta}_1$ and a biased estimator $\hat{\theta}_2$ for a parameter θ

Principle of unbiased estimation: When choosing between several different estimators of θ, select one which is unbiased.

In Example 6.2 the sample proportion X/n was used as an estimator of p, where X, the number of sample successes, had a binomial distribution with parameters n and p. Thus

$$E(\hat{p}) = E\left(\frac{X}{n}\right) = \frac{1}{n}E(X) = \frac{1}{n}(np) = p$$

Proposition: When X is a binomial random variable with parameters n and p, the sample proportion $\hat{p} = X/n$ is an unbiased estimator of p.

No matter what the true value of p is, the distribution of the estimator $\hat{p}$ will be centered at the true value.

Consider now the problem of estimating σ^2.

Proposition: Let $X_1, X_2, \ldots, X_n$ be a random sample from a distribution with mean μ and variance σ^2. Then the estimator

$$\hat{\sigma}^2 = S^2 = \frac{\Sigma (X_i - \bar{X})^2}{n - 1}$$

is an unbiased estimator of σ^2.

Proof: For any r.v. Y, $\text{Var}(Y) = E(Y^2) - [E(Y)]^2$, so $E(Y^2) = \text{Var}(Y) + [E(Y)]^2$. Applying this to

$$S^2 = \frac{1}{n - 1}\left[\sum X_i^2 - \frac{(\Sigma X_i)^2}{n}\right]$$

gives

$$E(S)^2 = \frac{1}{n-1}\left\{\sum E(X_i^2) - \frac{1}{n}E[(\sum X_i)^2]\right\}$$

$$= \frac{1}{n-1}\left\{\sum (\sigma^2 + \mu^2) - \frac{1}{n}\{\text{Var}(\sum X_i) + [E(\sum X_i)]^2\}\right\}$$

$$= \frac{1}{n-1}\left\{n\sigma^2 + n\mu^2 - \frac{1}{n}n\sigma^2 - \frac{1}{n}(n\mu)^2\right\}$$

$$= \frac{1}{n-1}\{n\sigma^2 - \sigma^2\} = \sigma^2 \quad \text{as desired}$$

The estimator which uses divisor n can be expressed as $(n-1)S^2/n$, so

$$E\left[\frac{(n-1)S^2}{n}\right] = \frac{n-1}{n}E(S^2) = \frac{n-1}{n}\sigma^2$$

This estimator is therefore not unbiased. The bias is $(n-1)\sigma^2/n - \sigma^2 = -\sigma^2/n$. Because the bias is negative, the estimator with divisor n tends to underestimate σ^2, and this is why the $n-1$ divisor is preferred by many statisticians (though when n is large, the bias is small and there is little difference between the two).

Although S^2 is unbiased for σ^2, S is a biased estimator of σ. While it is possible to derive an unbiased estimator of σ, it is not as convenient to use as S, so we shall continue to use $\hat{\sigma} = S$.

In Example 6.1 we proposed several different estimators for the mean μ of a normal distribution. If there were a unique unbiased estimator for μ, the estimation problem would be resolved by using that estimator. Unfortunately this is not the case.

Proposition: If $X_1, X_2, \ldots, X_n$ is a random sample from a distribution with mean μ, then $\bar{X}$ is an unbiased estimator of μ. If in addition the distribution is continuous and symmetric, then $\widetilde{X}$ and any trimmed mean are also unbiased estimators of μ.

The fact that $\bar{X}$ is unbiased is just a restatement of one of our rules of expected value: $E(\bar{X}) = \mu$ for every possible value of μ (for discrete as well as continuous distributions). The unbiasedness of the other estimators is more difficult to verify.

According to this proposition, the principle of unbiasedness by itself does not always allow us to select a single estimator. When the underlying population is normal, even the third estimator in Example 6.1 is unbiased, and there are many other unbiased estimators. What we now need is a way of selecting among unbiased estimators.

Estimators with Minimum Variance

Suppose that $\hat{\theta}_1$ and $\hat{\theta}_2$ are two estimators of θ which are both unbiased. Then although the distribution of each estimator is centered at the true value of θ, the spread of the distributions about the true value may be different.

> **Principle of minimum variance unbiased estimation:** Among all estimators of θ which are unbiased, choose the one which has minimum variance. The resulting $\hat{\theta}$ is called the minimum variance unbiased estimator (MVUE) of θ.

Figure 6.2 pictures the p.d.f.'s of two unbiased estimators, with $\hat{\theta}_1$ having smaller variance than $\hat{\theta}_2$. Then $\hat{\theta}_1$ is more likely than $\hat{\theta}_2$ to produce an estimate close to the true θ. The MVUE is in a certain sense the most likely among all unbiased estimators to produce an estimate close to the true θ.

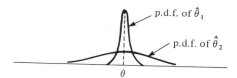

Figure 6.2 Graphs of the p.d.f.'s of two different unbiased estimators

One of the triumphs of mathematical statistics has been the development of methodology for identifying the MVUE in a wide variety of situations. The most important result of this type for our purposes concerns estimating the mean μ of a normal distribution.

> **Theorem:** Let $X_1, \ldots, X_n$ be a random sample from a normal distribution with parameters μ and σ. Then the estimator $\hat{\mu} = \bar{X}$ is the MVUE for μ.

The proof of this result can be found in the book *Introduction to Mathematical Statistics* by Hogg and Craig. Whenever we are convinced that the population being sampled is normal, the result says that $\bar{X}$ should be used to estimate μ. In Example 6.1, then, our estimate would be $\bar{x} = 28.50$.

In some situations it is possible to obtain an estimator with small bias which would be preferred to the best unbiased estimator. This is illustrated in Figure 6.3.

Figure 6.3 A biased estimator is preferable to the MVUE

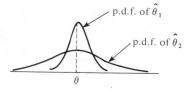

However, MVUE's are often easier to obtain than the type of biased estimator whose distribution is pictured.

Some Complications

The last theorem does not say that in estimating a population mean μ, the estimator $\bar{X}$ should be used irrespective of the distribution being sampled.

Example 6.4

Suppose that we wish to estimate the thermal conductivity μ of a certain material. Using standard measurement techniques, we will obtain a random sample $X_1, \ldots,$ X_n of n thermal conductivity measurements. Let us assume that the p.d.f. of each of these measurements belongs to one of the following three families:

$$f(x) = \frac{1}{\sqrt{2\pi\sigma^2}} \, e^{-(1/2\sigma^2)(x-\mu)^2} \qquad -\infty < x < \infty \tag{6.1}$$

$$f(x) = \frac{1}{\pi[1 + (x - \mu)^2]} \qquad -\infty < x < \infty \tag{6.2}$$

$$f(x) = \begin{cases} \dfrac{1}{2c} & -c \leq x - \mu \leq c \\ 0 & \text{otherwise} \end{cases} \tag{6.3}$$

The p.d.f. (6.1) is the normal distribution, (6.2) is called the Cauchy distribution, and (6.3) is a uniform distribution. All three of the distributions are symmetric about μ, and in fact the Cauchy distribution is bell-shaped but with much heavier tails (more probability further out) than the normal curve. The uniform distribution has no tails. The four estimators for μ considered earlier are $\bar{X}$, $\tilde{X}$, $\bar{X}_e$ (the average of the two extreme observations), and $\bar{X}_{tr(10)}$, a trimmed mean.

The very important moral here is that the best estimator for μ depends crucially on which distribution is being sampled. In particular

1. If the random sample comes from a normal distribution, then $\bar{X}$ is the best of the four estimators, since it has minimum variance among all unbiased estimators.

2. If the random sample comes from a Cauchy distribution, then $\bar{X}$ and $\bar{X}_e$ are terrible estimators for μ, while $\tilde{X}$ is quite good (the MVUE is not known); $\bar{X}$ is bad because it is very sensitive to outlying observations, and the heavy tails of the Cauchy distribution make a few such observations likely to appear in any sample.

3. If the underlying distribution is uniform, the best estimator is $\bar{X}_e$; this estimator is greatly influenced by outlying observations, but the lack of tails makes such observations impossible.

4. *The trimmed mean is best in none of these three situations, but works reasonably well in all three. That is, $\bar{X}_{tr(10)}$ does not suffer too much in comparison with the best procedure in any of the three situations.*

More generally, recent research in statistics has established that when estimating a point of symmetry μ of a continuous probability distribution, a trimmed mean with trimming proportion 10% or 20% (from each end of the sample) produces reasonably behaved estimates over a very wide range of possible models. For this reason a trimmed mean with small trimming percentage is said to be a **robust estimator.**

In some situations the choice is not between two different estimators constructed from the same sample, but instead between estimators based on two different experiments.

Example 6.5

Suppose that a certain type of component has a lifetime distribution which is exponential with parameter λ, so that expected lifetime is $\mu = 1/\lambda$. A sample of n such components is selected, and each is put into operation. If the experiment is continued until all n lifetimes $X_1, \ldots, X_n$ have been observed, then $\bar{X}$ is an unbiased estimator of μ.

In some experiments, though, the components are left in operation only until the time of the rth failure, where $r < n$. This procedure is referred to as **censoring.** Let Y_1 denote the time of the first failure (the minimum lifetime among the n components), Y_2 denote the time at which the second failure occurs (the second smallest lifetime), and so on. Since the experiment terminates at time Y_r, the total accumulated lifetime at termination is

$$T_r = \sum_{i=1}^{r} Y_i + (n - r) Y_r$$

We now demonstrate that $\hat{\mu} = T_r/r$ is an unbiased estimator for μ. To do so, we need two properties of exponential variables:

1. The memoryless property (see Section 4.4), which says that at any time point, remaining lifetime has the same exponential distribution as original lifetime.
2. If $X_1, \ldots, X_k$ are independent, each exponential with parameter λ, then min $(X_1, \ldots, X_k)$ is exponential with parameter $k\lambda$ and has expected value $1/k\lambda$.

Since all n components last until Y_1, $n - 1$ last an additional $Y_2 - Y_1$, $n - 2$ an additional $Y_3 - Y_2$ amount of time, and so on, another expression for T_r is

$$T_r = nY_1 + (n - 1)(Y_2 - Y_1) + (n - 2)(Y_3 - Y_2) + \cdots$$
$$+ (n - r + 1)(Y_r - Y_{r-1})$$

But Y_1 is the minimum of n exponential variables, so $E(Y_1) = 1/n\lambda$. Similarly, $Y_2 - Y_1$ is the smallest of the $n - 1$ remaining lifetimes, each exponential with parameter λ (by the memoryless property), so $E(Y_2 - Y_1) = 1/(n - 1)\lambda$. Continuing, $E(Y_{i+1} - Y_i) = 1/(n - i)\lambda$, so

$$E(T_r) = nE(Y_1) + (n - 1) E(Y_2 - Y_1) + \cdots + (n - r + 1) E(Y_r - Y_{r-1})$$

$$= n \cdot \frac{1}{n\lambda} + (n - 1) \cdot \frac{1}{(n - 1)\lambda} + \cdots + (n - r - 1)$$

$$\cdot \frac{1}{(n - r - 1)\lambda} = \frac{r}{\lambda}$$

Therefore, $E(T_r/r) = (1/r) E(T_r) = 1/r \cdot r/\lambda = 1/\lambda = \mu$ as claimed.

As an example, suppose that 20 components are put on test and $r = 10$. Then if the first 10 failure times are 11, 15, 29, 33, 35, 40, 47, 55, 58, and 72, the estimate of μ is

$$\hat{\mu} = \frac{11 + 15 + \cdots + 72 + (10)(72)}{10} = 111.5$$

The advantage of the experiment with censoring is that it terminates more quickly than the uncensored experiment. However, it can be shown that $\text{Var}(T_r/r) = \lambda^2/r$, which is larger than λ^2/n, the variance of $\bar{X}$ in the uncensored experiment.

Reporting a Point Estimate: The Standard Error

In addition to reporting the value of a point estimate, some indication of its precision should be given. The usual measure of precision is the standard error of the estimator used.

> **Definition:** The **standard error** of an estimator $\hat{\theta}$ is its standard deviation $\sigma_{\hat{\theta}} = \sqrt{\text{Var}(\hat{\theta})}$. If the standard error itself involves unknown parameters whose values can be estimated, substitution of these estimates into $\sigma_{\hat{\theta}}$ yields the **estimated standard error** $\hat{\sigma}_{\hat{\theta}}$.

Example 6.1
(continued)

Assuming that writing lifetime is normally distributed, $\hat{\mu} = \bar{X}$ is the best estimator of μ. If the value of σ is known to be 3.5, the standard error of $\bar{X}$ is $\sigma_{\bar{X}} = \sigma/\sqrt{n} = 3.5/\sqrt{10} = 1.11$. If, as is usually the case, the value of σ is unknown, the estimate $\hat{\sigma} = s = 3.45$ is substituted into $\sigma_{\bar{X}}$ to obtain the estimated standard error $s/\sqrt{n} = 1.09$.

Example 6.2
(continued)

The standard error of $\hat{p} = X/n$ is

$$\sigma_{\hat{p}} = \sqrt{\text{Var}(X/n)} = \sqrt{\frac{\text{Var}(X)}{n^2}} = \sqrt{\frac{npq}{n^2}} = \sqrt{\frac{pq}{n}}$$

Since p (and $q = 1 - p$) are unknown (else why estimate), we substitute $\hat{p} = x/n$ and $\hat{q} = 1 - x/n$ into $\sigma_{\hat{p}}$, yielding the estimated standard error $\hat{\sigma}_{\hat{p}} =$

$\sqrt{\hat{p}\hat{q}/n} = \sqrt{(.6)(.4)/25} = .098$. Alternatively, since the largest value of pq is attained when $p = q = .5$, an upper bound on the standard error is $\sqrt{1/4n} = .10$.

When the point estimator $\hat{\theta}$ has approximately a normal distribution, which will often be the case when n is large, then we can be reasonably confident that the true value of θ lies within approximately two standard errors (standard deviations) of $\hat{\theta}$. Thus in the pen lifetime example, if $n = 36$, $\hat{\mu} = \bar{x} = 28.50$, and $s = 3.60$, then $s/\sqrt{n} = .60$, so within two estimated standard errors of $\hat{\mu}$ translates to the interval $28.50 \pm (2)(.60) = (27.20, 29.70)$.

If $\hat{\theta}$ is not necessarily approximately normal but is unbiased, then it can be shown that the estimate will deviate from θ by as much as four standard errors at most 6% of the time. We would then expect the true value to lie within four standard errors of $\hat{\theta}$ (and this is a very conservative statement, since it applies to *any* unbiased $\hat{\theta}$).

Summarizing, the standard error tells us roughly within what distance of $\hat{\theta}$ we can expect the true value of θ to lie.

Exercises / Section 6.1

1. The interpupillary distance was measured for each of 10 randomly selected adult males, yielding the following measurements (mm): 64.9, 62.1, 70.8, 65.3, 68.7, 57.2, 62.5, 69.8, 64.3, 71.6.
   a. Assuming that interpupillary distance is normally distributed, use the sample mean $\bar{X}$ and the sample standard deviation S to obtain estimates of μ and σ.
   b. Supposing only that the distribution is continuous, use the sample median $\tilde{X}$ to obtain an estimate of the population median $\tilde{\mu}$. Why in this situation wouldn't you use $\bar{X}$ to estimate $\tilde{\mu}$?
   c. Assuming that interpupillary distance has a symmetric, but not necessarily normal, distribution, use a 10% trimmed mean to obtain an estimate of μ.
   d. Let p denote the proportion of all males having interpupillary distance between 60 and 70. Determine the number of sample x_i's which satisfy $60 \le x_i \le 70$, and use it to estimate the population proportion p.

2. A sample of 20 students who had recently taken elementary statistics yielded the following information on brand of calculator owned (T = Texas Instruments, H = Hewlett-Packard, C = Casio, S = Sharp): T, T, H, T, C, T, T, S, C, H, S, S,

T, H, C, T, T, T, H, T.
   a. Estimate the true proportion of all such students who own a Texas Instruments calculator.
   b. Among the brands owned, only Hewlett-Packard uses reverse Polish logic. Estimate the proportion of all such students who own a calculator which does not use reverse Polish logic.
   c. Three of the four Hewlett-Packard calculators in the sample were programmable. Estimate the proportion of all students who own a calculator which is both made by H-P and is programmable.

3. Twelve samples of a certain brand of white bread were analyzed and the carbohydrate content (percentage of nitrogen-free extract) of each sample was determined, yielding the following data: 76.93, 76.88, 77.07, 76.68, 76.39, 75.09, 76.88, 77.67, 78.15, 76.50, 77.16, 76.42.
   a. Assuming that the distribution of carbohydrate content is normal, estimate the true average carbohydrate content for this brand. What estimator did you use?
   b. Making no assumption whatsoever about the distribution of carbohydrate content, estimate the value below which (and above which) half of all carbohydrate contents will fall. What estimator did you use?

c. Again making no assumption about the underlying distribution, estimate the true proportion of samples whose carbohydrate content would not exceed 76%.

d. What is the estimated standard error of the estimator that you used in (a)?

4. In Exercise 3, let $X_1, \ldots, X_m$ denote the random variables observed, assumed to be a random sample from a distribution with mean μ_1 and variance σ_1^2. For n samples of a second brand, let $Y_1, \ldots, Y_n$ denote the carbohydrate contents, assumed to be a random sample from a distribution with mean μ_2 and variance σ_2^2.

a. Show using rules of expected value that $\overline{X} - \overline{Y}$ is an unbiased estimator of $\mu_1 - \mu_2$. If the observed $\overline{Y}$ is $\bar{y} = 74.28$ for $n = 14$ samples, what is the point estimate?

b. Use the rules of variance from Section 5.3 to obtain an expression for the variance and standard deviation (standard error) of $\overline{X} - \overline{Y}$.

c. If σ_1 and σ_2 are unknown, but $s_1 = .75$, $s_2 = .70$, compute the estimated standard error of $\overline{X} - \overline{Y}$.

5. Each of 150 newly manufactured items is examined and the number of scratches per item is recorded (the items are supposed to be free of scratches), yielding the following data.

Number of scratches per item	0	1	2	3	4	5	6	7
Observed frequency	18	37	42	30	13	7	2	1

Let X = the number of scratches on a randomly chosen item, and assume that X has a Poisson distribution with parameter λ.

a. Find an unbiased estimator of λ and compute the estimate for the above data. *Hint:* $E(X) = \lambda$ for X Poisson, so $E(\overline{X}) = ?$

b. What is the standard deviation (standard error) of your estimator? Compute the estimated standard error. *Hint:* $\sigma_X^2 = \lambda$ for X Poisson.

6. Using a long rod which has length μ, you are going to lay out a square plot in which the length of each side is μ. Thus the area of the plot will be μ^2. However, you don't know the value of μ, so you decide to make n independent measurements $X_1, X_2, \ldots, X_n$ of the length. Assume that each X_i has mean μ (unbiased measurements) and variance σ^2

a. Show that $\overline{X}^2$ is not an unbiased estimator for μ^2. *Hint:* For any r.v. Y, $E(Y^2) = \text{Var}(Y) + [E(Y)]^2$. Apply this with $Y = \overline{X}$.

b. For what value of k is the estimator $\overline{X}^2 - kS^2$ unbiased for μ^2? *Hint:* Compute $E(\overline{X}^2 - kS^2)$.

7. Of n_1 randomly selected male smokers, X_1 smoked filter cigarettes, while of n_2 randomly selected female smokers, X_2 smoked filter cigarettes. Let p_1 and p_2 respectively denote the probabilities that a randomly selected male and female smoke filter cigarettes,

a. Show that $(X_1/n_1) - (X_2/n_2)$ is an unbiased estimator for $p_1 - p_2$. *Hint:* $E(X_i) = n_i p_i$ for $i = 1, 2$.

b. What is the standard error of the estimator in (a)?

c. How would you use the observed values x_1 and x_2 to estimate the standard error of your estimator?

d. If $n_1 = n_2 = 200$, $x_1 = 127$, $x_2 = 176$, use the estimator of (a) to obtain an estimate of $p_1 - p_2$.

e. Use the result of (c) and the data of (d) to estimate the standard error of the estimator.

8. Suppose that a certain type of fertilizer has an expected yield per acre of μ_1 with variance σ^2, while the expected yield for a second type of fertilizer is μ_2 with the same variance σ^2. Let S_1^2 and S_2^2 denote the sample variances of yields based on sample sizes n_1 and n_2, respectively, of the two fertilizers. Show that the pooled (combined) estimator

$$\hat{\sigma}^2 = \frac{(n_1 - 1)S_1^2 + (n_2 - 1)S_2^2}{n_1 + n_2 - 2}$$

is unbiased for σ^2 (this estimator will be used in Chapters 8 and 9).

9. Suppose that a bus arrives at my bus stop every θ minutes, but that I don't know θ. If I arrive at the stop at a randomly selected moment, then my waiting time is uniformly distributed on $[0, \theta]$. Let $X_1, X_2, \ldots, X_5$ denote my waiting times on five separate occasions, with observed values 3.7, 5.1, 1.2, 4.6, and 6.3. Use the estimator $\hat{\theta} = \max (X_1, \ldots, X_5)$ to obtain an estimate of θ. Can $\hat{\theta}$ ever overestimate θ? Do you think that $\hat{\theta}$ is an unbiased estimator? Explain.

10. Suppose that the true average growth μ of one type of plant during a one-year period is identical to that of a second type, but the variance of growth for the first type is σ^2 while for the second type the variance is $4\sigma^2$. Let $X_1, \ldots, X_m$ be m independent growth observations on the first type [so $E(X_i) = \mu$, $\text{Var}(X_i) = \sigma^2$], and let $Y_1, \ldots, Y_n$ be n independent growth observations on the second type [$E(Y_i) = \mu$, $\text{Var}(Y_i) = 4\sigma^2$].

   a. Show that for any δ between 0 and $1, \hat{\mu} = \delta \bar{X} + (1 - \delta) \bar{Y}$ is unbiased for μ.

   b. For fixed m and n, compute $\text{Var}(\hat{\mu})$ and then find the value of δ which minimizes $\text{Var}(\hat{\mu})$. *Hint:* Differentiate $\text{Var}(\hat{\mu})$ with respect to δ.

11. In Chapter 3 we defined a negative binomial r.v. as the number of failures which occur before the rth success in a sequence of independent and identical success/failure trials. The p.m.f. of X is

$$nb(x; r, p)$$

$$= \begin{cases} \dbinom{x + r - 1}{x} p^r (1 - p)^x & x = 0, 1, 2, \ldots \\ 0 & \text{otherwise} \end{cases}$$

   a. Suppose that $r \geq 2$. Show that $\hat{p} = (r - 1)/(X + r - 1)$ is an unbiased estimator for p. *Hint:* Write out $E(\hat{p})$ and cancel $x + r - 1$ inside the sum.

   b. A reporter wishes to interview five individuals who support a certain candidate, so begins asking people whether (S) or not (F) they support the candidate. If the sequence of responses is *SFFSFFFSS*, estimate p = the true proportion who support the candidate.

12. Let $X_1, X_2, \ldots, X_n$ be a random sample from a p.d.f. $f(x)$ which is symmetric about μ, so that $\widetilde{X}$ is an unbiased estimator of μ. If n is large, it can be shown that $\text{Var}(\widetilde{X}) \doteq 1/4n[f(\mu)]^2$.

   a. Compare $\text{Var}(\widetilde{X})$ to $\text{Var}(\bar{X})$ when the underlying distribution is normal.

   b. When the underlying p.d.f. is Cauchy (see Example 6.4), $\text{Var}(\bar{X}) = \infty$, so $\bar{X}$ is a terrible estimator. What is $\text{Var}(\widetilde{X})$ in this case when n is large?

13. An investigator wishes to estimate the proportion of students at a certain university who have violated the honor code. Having obtained a random sample of n students, she realizes that asking each, "Have you violated the honor code?" will probably

result in some untruthful responses. Consider the following scheme, called a **randomized response** technique. The investigator makes up a deck of 100 cards, of which 50 are of type I and 50 are of type II.

   Type I: Have you violated the honor code (yes or no)?

   Type II: Is the last digit of your telephone number a 0, 1, or 2 (yes or no)?

Each student in the random sample is asked to mix the deck, draw a card, and answer the resulting question truthfully. Because of the irrelevant question on type-II cards, a yes response no longer stigmatizes the respondent, so we assume that responses are truthful (this depends on having enough type-II cards in the deck). Let p denote the proportion of honor code violators (that is, the probability of a randomly selected student being a violator), and let $\lambda = P$ (yes response). The accompanying tree diagram summarizes the response possibilities.

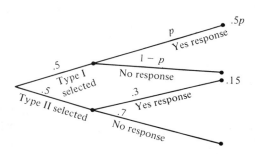

$$\Longrightarrow \lambda = P(\text{yes response}) = .5p + .15$$

   a. Let Y denote the number of yes responses, so $Y \sim \text{Bin}(n, \lambda)$. Thus Y/n is an unbiased estimator of λ. Derive an estimator for p based on Y. If $n = 80$ and $y = 20$, what is your estimate? *Hint:* Solve $\lambda = .5p + .15$ for p and then substitute Y/n in place of λ.

   b. Use the fact that $E(Y/n) = \lambda$ to show that your estimator $\hat{p}$ is unbiased.

   c. If there were 70 type-I and 30 type-II cards, what would your estimator for p be?

6.2 Methods of Point Estimation

There are many estimation problems in which an unbiased estimator cannot be obtained. We now discuss two "constructive" methods for obtaining point estimators, the method of moments and method of maximum likelihood. By constructive we mean that the general definition of each type of estimator suggests explicitly how to obtain the estimator in any specific problem. While maximum likelihood estimators are generally preferable to moment estimators because of certain efficiency properties, they often require significantly more computation than do moment estimators.

The Method of Moments

The basic idea of this method is to equate certain sample characteristics, such as the mean, to the corresponding population expected values. Then solving these equations for unknown parameter values yields the estimators.

Definition: Let $X_1, \ldots, X_n$ be a random sample from a p.m.f. or p.d.f. $f(x)$. For $k = 1, 2, 3, \ldots$ the **kth population moment**, or **kth moment of the distribution $f(x)$**, is $E(X^k)$. The **kth sample moment** is $(1/n) \sum_{i=1}^{n} X_i^k$.

Definition: Let $X_1, X_2, \ldots, X_n$ be a random sample from a distribution with p.m.f. or p.d.f. $f(x; \theta, \ldots, \theta_m)$, where $\theta_1, \ldots, \theta_m$ are parameters whose values are unknown. Then the **moment estimators** $\hat{\theta}_1, \ldots, \hat{\theta}_m$ are obtained by equating the first m sample moments to the corresponding first m population moments and solving for $\theta_1, \ldots, \theta_m$.

If, for example, $m = 2$, $E(X)$ and $E(X^2)$ will be functions of θ_1 and θ_2. Setting $E(X) = (1/n) \sum X_i (=\bar{X})$ and $E(X^2) = (1/n) \sum X_i^2$ gives two equations in θ_1 and θ_2. The solution then defines the estimators. For estimating a population mean μ, the method gives $\mu = \bar{X}$, so the estimator is the sample mean.

Example 6.6

Let $X_1, X_2, \ldots, X_n$ represent a random sample of service times of n customers at a certain facility, where the underlying distribution is assumed exponential with parameter λ. Since there is only one parameter to be estimated, the estimator is obtained by equating $E(X)$ to $\bar{X}$. Since $E(X) = 1/\lambda$ for an exponential distribution, this gives $1/\lambda = \bar{X}$ or $\lambda = 1/\bar{X}$. The moment estimator of λ is then $\hat{\lambda} = 1/\bar{X}$.

Example 6.7

Let $X_1, \ldots, X_n$ be a random sample from a gamma distribution with parameters α and β. From Chapter 4, $E(X) = \alpha\beta$ and $E(X^2) = \beta^2 \Gamma(\alpha + 2)/\Gamma(\alpha) = \beta^2(\alpha + 1)\alpha$. The moment estimators of α and β are obtained from solving

$$\bar{X} = \alpha\beta, \quad \frac{1}{n} \sum X_i^2 = \alpha(\alpha + 1)\beta^2$$

Since $\alpha(\alpha+1)\beta^2 = \alpha^2\beta^2 + \alpha\beta^2$ and the first equation implies $\alpha^2\beta^2 = \bar{X}^2$, the second equation becomes

$$\frac{1}{n}\sum X_i^2 - \bar{X}^2 = \alpha\beta^2$$

Now dividing each side of this second equation by the corresponding side of the first equation and substituting back gives the estimators

$$\hat{\alpha} = \frac{\bar{X}^2}{(1/n)\sum X_i^2 - \bar{X}^2}, \quad \hat{\beta} = \frac{(1/n)\sum X_i^2 - \bar{X}^2}{\bar{X}}$$

To illustrate, the survival time data mentioned in Example 4.19 is

152, 115, 109, 94, 88, 137, 152, 77, 160, 165, 125, 40, 128, 123, 136, 101, 62, 153, 83, 69

with $\bar{x} = 113.5$ and $(1/20)\sum x_i^2 = 14{,}087.8$. The estimates are

$$\hat{\alpha} = \frac{(113.5)^2}{14{,}078.8 - (113.5)^2} = 10.8, \quad \hat{\beta} = \frac{14{,}078.8 - (113.5)^2}{113.5} = 10.5$$

These estimates of α and β differ from the values suggested by Gross and Clark because they used a different estimation technique.

Example 6.8

Let $X_1, \ldots, X_n$ be a random sample from a generalized negative binomial distribution with parameters r and p (Section 3.7). Since $E(X) = r(1-p)/p$ and $\mathrm{Var}(X) = r(1-p)/p^2$, $E(X^2) = \mathrm{Var}(X) + [E(X)]^2 = r(1-p)(r-rp+1)/p^2$. Equating $E(X)$ to $\bar{X}$ and $E(X^2)$ to $(1/n)\sum X_i^2$ gives eventually

$$\hat{p} = \frac{\bar{X}}{(1/n)\sum X_i^2 - \bar{X}^2}, \quad \hat{r} = \frac{\bar{X}^2}{(1/n)\sum X_i^2 - \bar{X}^2 - \bar{X}}$$

As an illustration, Reep, Pollard, and Benjamin ("Skill and Chance in Ball Games," *J. Royal Stat. Soc.*, 1971, pp. 623–629) consider the negative binomial distribution as a model for the number of goals per game scored by National Hockey League teams. The data for 1966–1967 follows (420 games):

goals	0	1	2	3	4	5	6	7	8	9	10
freq.	29	71	82	89	65	45	24	7	4	1	3

Then

$$\bar{x} = \sum x_i/420 = [(0)(29) + (1)(71) + \cdots + (10)(3)]/420 = 2.98$$

and

$$\sum x_i^2/420 = [(0)^2(29) + (1)^2(71) + \cdots + (10)^2(3)]/420 = 12.40$$

Thus

$$\hat{p} = \frac{2.98}{12.40 - (2.98)^2} = .85, \quad \hat{r} = \frac{(2.98)^2}{12.40 - (2.98)^2 - 2.98} = 16.5$$

Although r by definition must be positive, the denominator of $\hat{r}$ could be negative, indicating that the negative binomial distribution is not appropriate.

Maximum Likelihood Estimation

The method of maximum likelihood was first introduced by R. A. Fisher, a geneticist and statistician, in the 1920s. Most statisticians recommend this method, at least when the sample size is large, since the resulting estimators have certain desirable efficiency properties.

Example 6.9   A sample of 10 new phonograph records pressed by a certain record company is obtained. Upon opening each record jacket, it is found that the first, third, and tenth records are warped while the others are not. Let $p = P(\text{warped record})$ and define $X_1, \ldots, X_{10}$ by $X_i = 1$ if the ith record is warped and 0 otherwise. Then the observed x_i's are 1, 0, 1, 0, 0, 0, 0, 0, 0, 1, so the joint p.m.f. of the sample is

$$f(x_1, x_2, \ldots, x_{10}; p) = p(1 - p)p \cdots p = p^3(1 - p)^7 \tag{6.4}$$

We now ask "for what value of p is the observed sample most likely to have occurred?" That is, we wish to find the value of p which maximizes (6.4), or equivalently maximizes the natural log of (6.4).* Since

$$\ln[f(x_1, \ldots, x_{10}; p)] = 3\ln(p) + 7\ln(1 - p) \tag{6.5}$$

which is a differentiable function of p, equating the derivative of (6.5) to zero gives the maximizing value:[†]

$$\frac{d}{dp}\ln[f(x_1, \ldots, x_{10}; p)] = \frac{3}{p} - \frac{7}{1 - p} = 0 \Rightarrow p = \frac{3}{10} = \frac{x}{n}$$

where x is the observed number of successes (warped records). The estimate of p is now $\hat{p} = \frac{3}{10}$. It is called the maximum likelihood estimate because for fixed $x_1, \ldots, x_{10}$, it is the parameter value which maximizes the likelihood (joint p.m.f.) of the observed sample.

Note that if we had been told only that among the 10 records there were 3 which were warped, (6.4) would be replaced by the binomial p.m.f. $\binom{10}{3}p^3(1 - p)^7$, which is also maximized for $\hat{p} = \frac{3}{10}$.

---

* Since $\ln[g(x)]$ is a monotonic function of $g(x)$, finding x to maximize $\ln[g(x)]$ is equivalent to maximizing $g(x)$ itself. In statistics, taking the logarithm frequently changes a product to a sum, which is easier to work with.

[†] This conclusion requires checking the second derivative, but the details are omitted.

> **Definition:** Let $X_1, X_2, \ldots, X_n$ have joint p.m.f. or p.d.f.
>
> $$f(x_1, x_2, \ldots, x_n; \theta_1, \ldots, \theta_m), \tag{6.6}$$
>
> where the parameters $\theta_1, \ldots, \theta_m$ have unknown values. When $x_1, \ldots, x_n$ are the observed samples values and (6.6) is regarded as a function of $\theta_1, \ldots, \theta_m$, it is called the **likelihood function.** The maximum likelihood estimates (m.l.e.'s) $\hat{\theta}_1, \ldots, \hat{\theta}_m$ are those values of the θ_i's which maximize the likelihood function, so that
>
> $$f(x_1, \ldots, x_n; \hat{\theta}_1, \ldots, \hat{\theta}_m) \geq f(x_1, \ldots, x_n; \theta_1, \ldots, \theta_m) \text{ for all } \theta_1, \ldots, \theta_m.$$
>
> When the X_i's are substituted in place of the x_i's, the **maximum likelihood estimators** result.

The likelihood function tells us how likely the observed sample is as a function of the possible parameter values. Maximizing the likelihood gives the parameter values for which the observed sample is most likely to have been generated—that is, the parameter values which "agree most closely" with the observed data.

Example 6.10 Suppose that $X_1, X_2, \ldots, X_n$ is a random sample from an exponential distribution with parameter λ. Because of independence, the likelihood function is a product of the individual p.d.f.'s:

$$f(x_1, \ldots, x_n; \lambda) = (\lambda e^{-\lambda x_1}) \cdots (\lambda e^{-\lambda x_n}) = \lambda^n e^{-\lambda \Sigma x_i}$$

The ln (likelihood) is

$$\ln[f(x_1, \ldots, x_n; \lambda)] = n \ln(\lambda) - \lambda \sum x_i$$

Equating $d/d\lambda$ ln (likelihood) to zero results in $n/\lambda - \Sigma x_i = 0$, or $\lambda = n/\Sigma x_i = 1/\bar{x}$. Thus the maximum likelihood estimator is $\hat{\lambda} = 1/\bar{X}$; it is identical to the method of moments estimator [but is not an unbiased estimator, since $E(1/\bar{X}) \neq 1/E(\bar{X})$].

Example 6.11 Let $X_1, \ldots, X_n$ be a random sample from a normal distribution. The likelihood function is

$$f(x_1, \ldots, x_n; \mu, \sigma^2) = \frac{1}{\sqrt{2\pi\sigma^2}} e^{-(1/2\sigma^2)(x_1-\mu)^2} \cdots \frac{1}{\sqrt{2\pi\sigma^2}} e^{-(1/2\sigma^2)(x_n-\mu)^2}$$

$$= \left(\frac{1}{2\pi\sigma^2}\right)^{n/2} e^{-(1/2\sigma^2)\Sigma(x_i-\mu)^2}$$

so

$$\ln[f(x_1, \ldots, x_n; \mu, \sigma^2)] = -\frac{n}{2}\ln(2\pi\sigma^2) - \frac{1}{2\sigma^2}\sum(x_i - \mu)^2$$

To find the maximizing values of μ and σ^2, we must take the partial derivatives of ln (f) with respect to μ and σ^2, equate them to zero, and solve the resulting two equations. Omitting the details, the resulting m.l.e.'s are

$$\hat{\mu} = \bar{X}, \quad \hat{\sigma}^2 = \frac{\sum (X_i - \bar{X})^2}{n}$$

The m.l.e. of σ^2 is not the unbiased estimator, so two different principles of estimation (unbiasedness and maximum likelihood) yield two different estimators.

Example 6.12   In Chapter 3 we discussed the use of the Poisson distribution for modeling the number of "events" which occur in a two-dimensional region. Assume that when the region R being sampled has area $a(R)$, the number of events X occurring in R has a Poisson distribution with parameter $\lambda a(R)$ (where λ is the expected number of events per unit area), and also that non-overlapping regions yield independent X's.

Suppose that an ecologist selects n nonoverlapping regions $R_1, \ldots, R_n$ and counts the number of plants of a certain species found in each region. The joint p.m.f. (likelihood) is then

$$p(x_1, \ldots, x_n; \lambda) = \frac{[\lambda \cdot a(R_1)]^{x_1} e^{-\lambda \cdot a(R_1)}}{x_1!} \cdots \frac{[\lambda \cdot a(R_n)]^{x_n} e^{-\lambda \cdot a(R_n)}}{x_n!}$$

$$= \frac{[a(R_1)]^{x_1} \cdots [a(R_n)]^{x_n} \cdot \lambda^{\sum x_i} \cdot e^{-\lambda \sum a(R_i)}}{x_1! \cdots x_n!}$$

The ln (likelihood) is

$$\ln [p(x_1, \ldots, x_n; \lambda)]$$

$$= \sum x_i \cdot \ln [a(R_i)] + \ln (\lambda) \cdot \sum x_i - \lambda \sum a(R_i) - \sum \ln (x_i!)$$

Taking $\dfrac{d}{d\lambda}$ ln (p) and equating it to zero yields

$$\frac{\sum x_i}{\lambda} - \sum a(R_i) = 0$$

so

$$\lambda = \frac{\sum x_i}{\sum a(R_i)}$$

The maximum likelihood estimator is then $\hat{\lambda} = \sum X_i / \sum a(R_i)$. This is intuitively reasonable because λ is the true density (plants per unit area), while $\hat{\lambda}$ is the sample density since $\sum a(R_i)$ is just the total area sampled. Because $E(X_i) = \lambda \cdot a(R_i)$, the estimator is unbiased.

Sometimes an alternative sampling procedure is used. Instead of fixing regions to be sampled, the ecologist will select n points in the entire region of interest and let y_i = the distance from the ith point to the nearest plant. The c.d.f. of Y = distance to the nearest plant is

$$F_Y(y) = P(Y \le y) = 1 - P(Y > y) = 1 - P \begin{pmatrix} \text{no plants in a} \\ \text{circle of radius } y \end{pmatrix}$$

$$= 1 - \frac{e^{-\lambda \pi y^2} (\lambda \pi y^2)^0}{0!} = 1 - e^{-\lambda \cdot \pi y^2}$$

Taking the derivative of $F_Y(y)$ with respect to y yields

$$f_Y(y; \lambda) = \begin{cases} 2\pi\lambda y e^{-\lambda \pi y^2} & y \ge 0 \\ 0 & \text{otherwise} \end{cases}$$

If we now form the likelihood $f_Y(y_1; \lambda) \cdots f_Y(y_n; \lambda)$, differentiate ln (likelihood), and so on, the resulting m.l.e. is

$$\hat{\lambda} = \frac{n}{\pi \Sigma Y_i^2} = \frac{\text{number of plants observed}}{\text{total area sampled}},$$

which is also a sample density. It can be shown that in a sparse environment (small λ), the distance method is in a certain sense better, while in a dense environment the first sampling method is better.

Example 6.13   Let $X_1, \ldots, X_n$ be a random sample from a Weibull p.d.f.

$$f(x; \alpha, \beta) = \begin{cases} \dfrac{\alpha}{\beta^\alpha} \cdot x^{\alpha-1} \cdot e^{-(x/\beta)^\alpha} & x \ge 0 \\ 0 & \text{otherwise} \end{cases}$$

Writing the likelihood and ln (likelihood), then setting $\partial/\partial\alpha [\ln(f)] = 0$, $\partial/\partial\beta [\ln(f)] = 0$ yields the equations

$$\alpha = \left[\frac{\Sigma x_i^\alpha \cdot \ln(x_i)}{\Sigma x_i^\alpha} - \frac{\Sigma \ln(x_i)}{n} \right]^{-1}$$

$$\beta = \left(\frac{\Sigma x_i^\alpha}{n} \right)^{1/\alpha}$$

These two equations cannot be solved explicitly to give general formulas for the m.l.e.'s $\hat{\alpha}$ and $\hat{\beta}$. Instead, for each sample $x_1, \ldots, x_n$, the equations must be solved using an iterative numerical procedure. Even moment estimators of α and β are somewhat complicated (see Exercise 2).

Estimating Functions of Parameters

In Example 6.11 we obtained the maximum likelihood estimator of σ^2 when the underlying distribution was normal. The m.l.e. of $\sigma = \sqrt{\sigma^2}$, as well as many other m.l.e.'s, can be easily derived using the following proposition.

Proposition: Let $\hat{\theta}_1, \hat{\theta}_2, \ldots, \hat{\theta}_m$ be the m.l.e.'s of the parameters $\theta_1, \theta_2, \ldots, \theta_m$. Then the m.l.e. of any function $h(\theta_1, \theta_2, \ldots, \theta_m)$ of these parameters is the function $h(\hat{\theta}_1, \hat{\theta}_2, \ldots, \hat{\theta}_m)$ of the m.l.e.'s.

Example 6.11
(continued)

The m.l.e.'s of μ and σ^2 were $\hat{\mu} = \bar{X}$ and $\hat{\sigma}^2 = \Sigma (X_i - \bar{X})^2/n$. To obtain the m.l.e. of the function $h(\mu, \sigma^2) = \sqrt{\sigma^2} = \sigma$, substitute the m.l.e.'s into the function:

$$\hat{\sigma} = \sqrt{\hat{\sigma}^2} = \left[\frac{1}{n} \sum (X_i - \bar{X})^2 \right]^{1/2}$$

The m.l.e. of σ is not the sample standard deviation S, though they are close unless n is quite small.

Example 6.13
(continued)

The mean value of a r.v. X which has a Weibull distribution is

$$\mu = \beta \cdot \Gamma(1 + 1/\alpha)$$

The m.l.e. of μ is therefore $\hat{\mu} = \hat{\beta} \Gamma(1 + 1/\hat{\alpha})$, where $\hat{\alpha}$ and $\hat{\beta}$ are the m.l.e.'s of α and β. In particular, $\bar{X}$ is not the m.l.e. of μ, though it is an unbiased estimator. At least for large n, $\hat{\mu}$ is a better estimator than $\bar{X}$.

The proof of the above proposition can be found in the mathematical statistics book by DeGroot. The proposition itself is usually called the **invariance principle** for m.l.e.'s.

A Desirable Property of the M.L.E.

While the principle of maximum likelihood estimation has considerable intuitive appeal, the following proposition provides additional rationale for the use of m.l.e.'s.

Proposition: Under very general conditions on the joint distribution of the sample, when the sample size n is large, the maximum likelihood estimator of any parameter θ is approximately unbiased $[E(\hat{\theta}) \doteq \theta]$ and has variance which is nearly as small as can be achieved by any estimator. Stated another way, the m.l.e. $\hat{\theta}$ is approximately the MVUE of θ.

Because of this result and the fact that calculus-based techniques can usually be used to derive the m.l.e.'s (though often numerical methods, such as Newton's method, are necessary), maximum likelihood estimation is the most widely used estimation technique among statisticians. Many of the estimators used in the remainder of the book are m.l.e.'s. Obtaining an m.l.e., however, does require that the underlying distribution be specified. If the investigator is not sure which family the underlying distribution belongs to, then an estimator such as a trimmed mean may be appropriate (though without a specific family, one must think carefully about which parameter should be estimated). Statisticians are now seeking "robust" estimators which perform well for a variety of underlying models. We will shortly describe one such method related to maximum likelihood estimation.

Some Complications

Sometimes calculus cannot be used to obtain m.l.e.'s.

Example 6.14 Suppose that my waiting time for a bus is uniformly distributed on $[0, \theta]$, and that the results $x_1, \ldots, x_n$ of a random sample from this distribution have been observed. Since $f(x; \theta) = 1/\theta$ for $0 \le x \le \theta$ and 0 otherwise,

$$f(x_1, \ldots, x_n; \theta) = \begin{cases} \dfrac{1}{\theta^n} & 0 \le x_1 \le \theta, \ldots, 0 \le x_n \le \theta \\ 0 & \text{otherwise} \end{cases}$$

As long as $\max(x_i) \le \theta$, the likelihood is $1/\theta^n$, which is positive, but as soon as $\theta < \max(x_i)$, the likelihood drops to 0. This is illustrated in Figure 6.4. Calculus won't work because the maximum of the likelihood occurs at a point of discontinuity, but the figure shows that $\hat{\theta} = \max(X_i)$. Thus if my waiting times are 2.3, 3.7, 1.5, .4, and 3.2, then the maximum likelihood estimate is $\hat{\theta} = 3.7$.

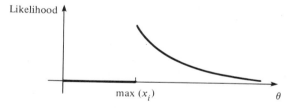

Figure 6.4 The likelihood function for Example 6.14

Example 6.15 A method which is often used to estimate the size of a wildlife population involves performing a capture/recapture experiment. In this experiment an initial sample of M animals is captured, each of these animals is tagged, and the animals are then returned to the population. After allowing enough time for the tagged individuals to mix into the population, another sample of size n is captured. With $X =$ the number of tagged animals in the second sample, the objective is to use the observed x to estimate the population size N.

The parameter of interest is $\theta = N$, which can assume only integer values, so even after determining the likelihood function (p.m.f. of X here), using calculus to obtain N would present difficulties. If we think of a success as a previously tagged animal being recaptured, then sampling is without replacement from a population containing M successes and $N - M$ failures, so that X is a hypergeometric r.v. and the likelihood function is

$$p(x; N) = h(x; n, M, N) = \frac{\binom{M}{x} \cdot \binom{N - M}{n - x}}{\binom{N}{n}}$$

The integer valued nature of N notwithstanding, it would be quite difficult to take the derivative of $p(x; N)$. However, if we consider the ratio of $p(x, N)$ to $p(x; N - 1)$, we have

$$\frac{p(x; N)}{p(x; N - 1)} = \frac{(N - M) \cdot (N - n)}{N(N - M - n + x)}$$

This ratio is larger than 1 if and only if $N < Mn/x$. The value of N for which $p(x; N)$ is maximized is therefore the largest integer less than Mn/x. If we use standard mathematical notation $[r]$ for the largest integer less than r, the maximum likelihood estimate of N is $\hat{N} = [Mn/x]$. As an illustration, if $M = 200$ fish are taken from a lake and tagged, subsequently $n = 100$ fish are recaptured, and among the 100 there are $x = 11$ tagged fish, then $\hat{N} = [(200)(100)/11] = [1818.18] = 1818$. The estimate is actually rather intuitive; x/n is the proportion of the recaptured sample which is tagged, while M/N is the proportion of the entire population which is tagged. The estimate is obtained by equating these two proportions (estimating a population proportion by a sample proportion).

M-Estimation

Suppose that the population distribution under investigation is symmetric about an unknown value θ (the median, and mean if it exists). The p.d.f. can be written as $f(x - \theta)$, where $f(x)$ is a p.d.f. which is symmetric about zero. Examples appear in the table below and are illustrated in Figure 6.5; $f(x - \theta)$ is shifted by the amount θ to the right of $f(x)$.

Distribution	$f(x)$	$f(x - \theta)$
(a) Normal $(\theta, 1)$	$\dfrac{1}{\sqrt{2\pi}} e^{-(1/2)x^2}$	$\dfrac{1}{\sqrt{2\pi}} e^{-(1/2)(x - \theta)^2}$
(b) Double Exponential	$\frac{1}{2} e^{-\lvert x \rvert}$	$\frac{1}{2} e^{-\lvert x - \theta \rvert}$
(c) Uniform $(-\frac{1}{2}, \frac{1}{2})$	$\begin{cases} 1 & -\frac{1}{2} \leq x \leq \frac{1}{2} \\ 0 & \text{otherwise} \end{cases}$	$\begin{cases} 1 & -\frac{1}{2} \leq x - \theta \leq \frac{1}{2} \\ 0 & \text{otherwise} \end{cases}$

If the $f(x)$ function is unknown, we previously suggested using a trimmed mean to estimate θ. We now describe a different method of robust estimation, M-estimation, which is important in part because it can be used in regression analysis.

Suppose then that $X_1, X_2, \ldots, X_n$ is a random sample from the p.d.f. $f(x - \theta)$. The joint p.d.f. (likelihood) is the product $f(x_1 - \theta) \cdots f(x_n - \theta)$, so the log

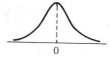 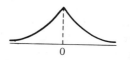 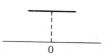

Figure 6.5 $f(x)$ function for three symmetric distributions

likelihood is $\Sigma \ln f(x_i - \theta)$. With known f, the maximum likelihood estimate $\hat{\theta}$ is that value of θ which maximizes the log likelihood, or equivalently which minimizes $\Sigma \rho(x_i - \theta)$ where $\rho(\cdot) = -\ln(\cdot)$. Assuming differentiability of $\rho(\cdot)$, $\hat{\theta}$ is the solution to

$$\frac{d}{d\theta} \sum_i \rho(x_i - \theta) = \sum_i \frac{d}{d\theta} \rho(x_i - \theta) = \sum_i \psi(x_i - \theta) = 0 \qquad (6.7)$$

where $\psi(x_i - \theta) = \dfrac{d}{d\theta} \rho(x_i - \theta)$. As examples,

	Normal $(\sigma = 1)$	*Double Exponential*
$f(x)$	$\dfrac{1}{\sqrt{2\pi}} e^{-(1/2)x^2}$	$\dfrac{1}{2} e^{-|x|}$
$\rho(x)$	$\dfrac{x^2}{2} + c$	$|x| + c$
$\psi(x)$	x	$\begin{cases} -1 & x < 0 \\ 1 & x > 0 \end{cases}$
$\hat{\theta}$	$\bar{x}$	$\tilde{x} = \text{median } (x_i\text{'s})$

Not knowing the f function, we could try to select a ψ function in (6.7) which yields good estimates for a wide variety of distributions. In particular, we might like $\psi(x)$ to yield a θ close to $\bar{x}$ when the distribution is normal but not give much weight to outlying values when the underlying distribution has heavy tails.

Definition: An **M-estimator** corresponding to a function $\psi(x)$ is that $\hat{\theta}$ which satisfies $\Sigma \psi(X_i - \hat{\theta}) = 0$.

Since the sample mean puts too much weight on outlying values and the sample median often puts too little weight on values away from the middle of the sample, a reasonable choice of ψ function is a compromise between those for the normal and double exponential distributions:

Figure 6.6

$$\psi(x) = \begin{cases} -a & x < -a \\ x & |x| \le a \\ a & x > a \end{cases} \qquad (6.8)$$

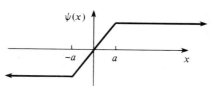

The resulting M-estimator is "robust" in that it compares favorably with $\overline{X}$ (for an appropriate choice of a) when the underlying distribution is normal and considerably improves on $\overline{X}$ for heavy-tailed distributions. Unfortunately there is no nice formula for the resulting $\hat{\theta}$, so that (6.7) must be solved by numerical methods for any particular sample. We shall comment on this shortly.

One difficulty with $\psi(x)$ in (6.8) is that if after obtaining $\hat{\theta}$ for a sample $x_1, \ldots, x_n$, each x_i is multiplied by c, the new estimate will in general not equal $c\hat{\theta}$—it is not "scale invariant." A related problem is that (6.8) does not allow the determination of whether or not x_i is an outlier to depend on the spread of the distribution. For example, if the distribution is normal with σ quite large relative to a, then $\hat{\theta}$ will be quite different from $\overline{X}$. To remedy this, let

$$s_1 = \frac{\text{median of } |x_1 - \tilde{x}|, |x_2 - \tilde{x}|, \ldots, |x_n - \tilde{x}|}{.6745}$$

so that s_1 is a constant times the median deviation from the median. When the distribution is normal, s_1 estimates σ; the sample standard deviation s is not used because it gives full weight to every squared deviation so is not robust. Then a **scale invariant M-estimate** is given by the $\hat{\theta}$ which satisfies

$$\sum_i \psi\left(\frac{x_i - \hat{\theta}}{cs_1}\right) = 0 \qquad (6.9)$$

For example, the choice $a = 1$ in (6.8) and $c = 1.5$ in (6.9) implies that if all x_i's are within $1.5s_1$ of $\overline{x}$, then $\hat{\theta} = \overline{x}$; thus $a = 1.5$ gives a $\hat{\theta}$ close to $\overline{X}$ when the distribution is actually normal.

Another choice of ψ which works well is the **biweight function**

$$\psi(x) = \begin{cases} x(1 - x^2)^2 & |x| \le 1 \\ 0 & |x| > 1 \end{cases} \qquad (6.10)$$

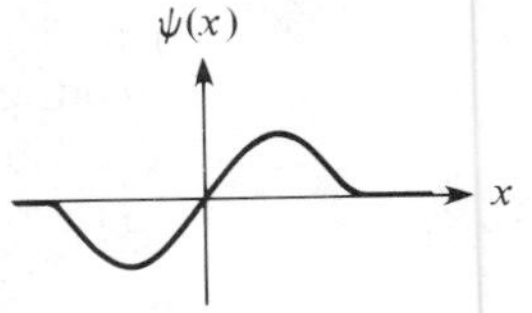

Figure 6.7

with $c = 4$ in (6.9). Even when a single observation x_i is extremely far to the left or right of the other x_i's, its value will still influence the sample median $\tilde{x}$ in that deletion of x_i from the sample will change $\tilde{x}$. The biweight estimate, though, will treat a sufficiently discrepant x_i as though it were not even in the sample, while still being sensitive to changes in x_i in the middle range of sample values. The book by Mosteller and Tukey contains an excellent exposition of properties of the biweight estimate.

Equation (6.9) must be solved numerically by using a technique, such as Newton's method, which produces a sequence of values $\theta_1, \theta_2, \theta_3, \ldots$ converging to the estimate of θ. Let the initial value be $\theta_1 = \tilde{x}$, the sample median. Then a modification of Newton's method yields the second value

$$\theta_2 = \theta_1 + \frac{cs_1 \sum \psi[x_i - \theta_1)/cs_1]}{\sum \dfrac{\psi[(x_i - \theta_1)/cs_1]}{(x_i - \theta_1)/cs_1}} = \theta_1 + \frac{cs_1 \sum \psi[(x_i - \theta_1)/cs_1]}{\sum w_i} \qquad (6.11)$$

where the w_i's are the "weights" in the denominator of the middle expression in (6.11). But then

$$\theta_2 = \theta_1 + \frac{\sum w_i(x_i - \theta_1)}{\sum w_i} = \frac{\sum w_i x_i}{\sum w_i} \qquad (6.12)$$

which shows that *the second value θ_2 is a weighted average of the x_i's where the weights depend on the first value. After computing θ_2, the weights are recomputed and (6.11) is again used to compute θ_3, and so on.* Thus (6.9) can be solved by using "iteratively reweighted least squares," since at each stage the next θ is found by minimizing $\sum w_i(x_i - \theta)^2$. The advantage of this over Newton's method is that the derivative of ψ at each θ_i is not required.

Example 6.16 Example 1.12 presented data on cerebral metabolic rate for glucose in rhesus monkeys. Assuming the underlying distribution to be symmetric about θ, we shall estimate θ using the M-estimate computed from the biweight function (6.10). For $|x| \le 1$, $\psi(x)/x = (1 - x^2)^2$, so the ith weight for any trial θ is

$$w_i = \begin{cases} \left[1 - \left(\dfrac{x_i - \theta}{4s_1} \right)^2 \right]^2 & |x_i - \theta| \le 4s_1 \\ \\ 0 & \text{otherwise} \end{cases}$$

To begin, we need $\theta_1 = \tilde{x} = 4.93$ and $s_1 = \text{median} |x_i - \tilde{x}|/.6745 = .34/.6745 = .5041$, so $4s_1 = 2.0164$.

x_i	4.51	4.59	4.90	4.93	6.80	5.08	5.67	$\theta_1 = 4.93$
$\|x_i - \theta_1\|$	.42	.34	.03	.00	1.87	.15	.74	
w_i	.957	.972	1.000	1.000	.140	.995	.865	$\theta_2 = \dfrac{\sum w_i x_i}{\sum w_i} = 4.98$
$\|x_i - \theta_2\|$	.47	.39	.08	.05	1.82	.10	.69	
w_i	.946	.963	.998	.999	.185	.998	.883	$\theta_3 = 5.00$
$\|x_i - \theta_3\|$	.49	.41	.10	.07	1.80	.08	.67	
w_i	.941	.959	.998	.999	.203	.998	.890	$\theta_4 = 5.00$

The M-estimate is $\hat{\theta} = 5.00$. Note that $\bar{x} = 5.211$, while the average of the five middle observations (a trimmed mean) is 5.03.

In Chapter 13 we shall briefly indicate how this methodology can be applied to regression problems. It is quite likely that in the near future packaged computer programs for calculating M-estimates will be widely available.

Exercises / Section 6.2

1. Suppose that I select n records manufactured by a particular company, and let X = the number among the n which are warped, $p = P$(warped record). Assume that only X is observed, rather than the sequence of S's and F's.
   a. Derive the maximum likelihood estimator of p. If $n = 20$ and $x = 3$, what is the estimate?
   b. Is the estimator of (a) unbiased?
   c. If $n = 20$ and $x = 3$, what is the m.l.e. of the probability $(1 - p)^5$ that none of the next five records examined is warped? *Hint:* the invariance principle.

2. Let X have a Weibull distribution with parameters α and β, so
$$E(X) = \beta \cdot \Gamma(1 + 1/\alpha),$$
$$\text{Var}(X) = \beta^2 \{\Gamma(1 + 2/\alpha) - [\Gamma(1 + 1/\alpha)]^2\}$$

   a. Based on a random sample $X_1, \ldots, X_n$, write equations for the method of moments estimators of α and β. Show that once the estimate of β has been obtained, the estimate of α can be found from a table of the gamma function, and that the estimate of β is the solution to a complicated equation involving the gamma function.
   b. If $n = 20$, $\bar{x} = 28.0$, and $\Sigma x_i^2 = 16,500$, compute the estimates. *Hint:* $[\Gamma(1.2)]^2/\Gamma(1.4) = .95$.

3. Let X denote the proportion of allotted time that a randomly selected student spends working on a certain aptitude test, and suppose that the p.d.f. of X is
$$f(x; \theta) = \begin{cases} (\theta + 1)x^\theta & 0 \le x \le 1 \\ 0 & \text{otherwise} \end{cases}$$
   where $-1 < \theta$. A random sample of 10 students yielded data $x_1 = .92$, $x_2 = .79$, $x_3 = .90$, $x_4 = .65$, $x_5 = .86$, $x_6 = .47$, $x_7 = .73$, $x_8 = .97$, $x_9 = .94$, $x_{10} = .77$.
   a. Use the method of moments to obtain an estimator of θ, and then compute the estimate for this data.
   b. Obtain the maximum likelihood estimator of θ, and then compute the estimate for the given data.

4. Two different computer systems are monitored for a total of n weeks. Let X_i denote the number of breakdowns of the first system during the ith week, and suppose that the X_i's are independent and drawn from a Poisson distribution with parameter λ_1. Similarly let Y_i denote the number of breakdowns of the second system during the ith week, and assume independence with each Y_i Poisson with parameter λ_2. Derive the m.l.e.'s of λ_1, λ_2, and $\lambda_1 - \lambda_2$. *Hint:* Using independence, write the joint p.m.f. (likelihood) of the X_i's and Y_i's together.

5. In the warped record problem (Exercise 1), instead of selecting $n = 20$ records to examine, suppose that I examine records in succession until I have found $r = 3$ warped ones. If the twentieth record is the third warped one (so that the number of records examined which were not warped is $x = 17$), what is the maximum likelihood estimate of p? Is this the same as the estimate in Exercise 1 (a)? Why or why not? Is it the same as the estimate computed from the unbiased estimator of Exercise 11 in Section 6.1?

6. The shear strength of each of 10 test spot welds is determined, yielding the following data (p.s.i.):
   392, 376, 401, 367, 389, 362, 409, 415, 358, 375
   a. Assuming that shear strength is normally distributed, estimate the true average shear strength and standard deviation of shear strength using the method of maximum likelihood.
   b. Again assuming a normal distribution, estimate the strength value below which 95% of all welds will have their strength. *Hint:* What is the ninety-fifth percentile in terms of μ and σ? Now use the invariance principle.

7. Referring back to Exercise 6, suppose that we decide to examine another test spot weld. Let X = the shear strength of the weld. Use the given data to obtain the m.l.e. of $P(X \le 400)$. *Hint:* $P(X \le 400) = \Phi((400 - \mu)/\sigma)$.

8. Let $X_1, \ldots, X_n$ be a random sample from a gamma distribution with parameters α and β.
   a. Derive the equations whose solution yields the maximum likelihood estimators of α and β. Do you think that they can be solved explicitly?
   b. Show that the m.l.e. of $\mu = \alpha\beta$ is $\hat{\mu} = \bar{X}$.

9. At time $t = 0$, 20 identical components are put on test. The lifetime distribution of each is exponential with parameter λ. The experimenter then leaves the test facility unmonitored. On his return 24 hours later, the experimenter immediately terminates the test after noticing that $y = 15$ of the 20 components are still in operation (so five have failed). Derive the maximum likelihood estimate of λ. *Hint:* Let $Y = $ the number which survive 24 hours. Then $Y \sim$ Bin (n, p). What is the m.l.e. of p? Now notice that $p = P(X_i \geq 24)$ where X_i is exponentially distributed. This relates λ to p, so the former can be estimated once the latter has been.

10. A miles per gallon determination was made for each of seven randomly selected cars of a given type, resulting in the data

26.8,   26.9,   27.3,   27.4,   30.2,   27.6,   28.5

Assuming that the mpg distribution is symmetric about θ, use the M-estimation technique with the biweight function (as illustrated in Example 6.16) to estimate θ.

Supplementary Exercises / Chapter 6

1. An estimator $\hat{\theta}$ is said to be **consistent** if for any $\epsilon > 0$, $P(|\hat{\theta} - \theta| > \epsilon) \to 0$ as $n \to \infty$. That is, $\hat{\theta}$ is consistent if, as the sample size gets larger, it is less and less likely that $\hat{\theta}$ will be further than ϵ from the true value of θ. Show that $\overline{X}$ is a consistent estimator of μ when the sample comes from a normal distribution. *Hint:* $\overline{X}$ is then normally distributed with mean μ and variance σ^2/n. Write $P(|\overline{X} - \mu| > \epsilon)$ as a sum of two integrals involving the standard normal p.d.f. (more generally, as long as μ is finite, $\overline{X}$ will be a consistent estimator for *any* underlying distribution).

2. **a.** Let $X_1, \ldots, X_n$ be a random sample from a uniform distribution on $[0, \theta]$. Then the m.l.e. of θ is $\hat{\theta} = Y = \max(X_i)$. Use the fact that $Y \leq y$ iff each $X_i \leq y$ to derive the c.d.f. of Y. Then show that the p.d.f. of $Y = \max(X_i)$ is

$$f_Y(y) = \begin{cases} \dfrac{ny^{n-1}}{\theta^n} & 0 \leq y \leq \theta \\ 0 & \text{otherwise} \end{cases}$$

b. Use the result of (a) to compute $E(\hat{\theta})$ [that is, $E(Y)$]. Is $\hat{\theta}$ an unbiased estimator? Is $K\hat{\theta}$ unbiased for an appropriate choice of K? What choice?

3. At time $t = 0$ there is one individual alive in a certain population. A **pure birth process** then unfolds as follows. The time until the first birth is exponentially distributed with parameter λ. After the first birth there are two individuals alive. The time until the first gives birth again is exponential with parameter λ, and similarly for the second individual. Therefore the time until the next birth is the minimum of two exponential (λ) variables, which is exponential with parameter 2λ. Similarly, once the second birth has occurred, there are three individuals alive, so the time until the next birth is an exponential r.v. with parameter 3λ, and so on (the memoryless property of the exponential distribution is being used here). Suppose that the process is observed until the sixth birth has occurred, and the successive birth times are 25.2, 41.7, 51.2, 55.5, 59.5, 61.8 (from which you should calculate the times between successive births). Derive the m.l.e. of λ. *Hint:* The likelihood is a product of exponential terms.

4. The **mean square error** of an estimator $\hat{\theta}$ is MSE $(\hat{\theta}) = E(\hat{\theta} - \theta)^2$. If $\hat{\theta}$ is unbiased, then MSE $(\hat{\theta}) = $ Var $(\hat{\theta})$, but in general MSE $(\hat{\theta}) = $ Var $(\hat{\theta}) + $ (bias)^2. Consider the estimator $\hat{\sigma}^2 = KS^2$, where $S^2 = $ sample variance. What value of K minimizes the mean square error of this estimator? *Hint:* It can be shown that $E[(S^2)^2] = (n + 1)\sigma^4/(n - 1)$ [in general it is quite difficult to find $\hat{\theta}$ to minimize MSE $(\hat{\theta})$, which is why we look only at unbiased estimators and minimize Var $(\hat{\theta})$].

5. Let $X_1, \ldots, X_n$ be a random sample from a p.d.f. which is symmetric about μ. An estimator for μ which has been found to perform well for a variety of underlying distributions is the *Hodges-Lehmann estimator*. To define it, first compute for each $i \leq j$ and each $j = 1, 2, \ldots, n$ the pairwise aver-

age $\bar{X}_{i,j} = (X_i + X_j)/2$. Then the estimator is $\hat{\mu}$ = the median of the $\bar{X}_{i,j}$'s. Compute the value of this estimate using the data of Exercise 1 in Section 1.4. *Hint:* Construct a square table with the x_i's listed on the left margin and on top. Then compute averages on and above the diagonal.

6. The runners in a marathon race have been randomly assigned (without regard to running ability) the numbers $1, 2, \ldots, N$. An apartment dweller at an intermediate point on the course looks out the window and observes the numbers on n of these runners. On the basis of this information, how can N be estimated?

   a. For $i = 1, 2, \ldots, n$, let X_i = the number on the ith runner observed (so with $n = 4$, we might have $x_1 = 231$, $x_2 = 146$, $x_3 = 27$, $x_4 = 171$). Then each X_i has p.m.f. $f(x) = 1/N$ for $x = 1, 2, \ldots, N$. Compute $E(X_i)$ and then show that $\hat{N} = 2\bar{X} - 1$ is an unbiased estimator for N.

   b. To obtain a different estimator for N, let $1 \le Y_1 \le Y_2 \le \cdots \le Y_n \le N$ denote the ordered values of the X_i's [so for the sample in (a), $y_1 = 27$, $y_2 = 146$, $y_3 = 171$, and $y_4 = 231$]. Since the n observed numbers are distributed uniformly throughout the N possible values, the expected length of each of the $n + 1$ gaps $Y_1 - 1$, $Y_2 - Y_1 - 1$, ..., $Y_n - Y_{n-1} - 1$, and $N - Y_n$ between successive observations (that is, expected number of numbers between successive ordered values) is the same. The sum of these gaps is $N - n$, so E (length of any particular gap) = $(N - n)/(n + 1)$. Since one such gap is $N - Y_n$ (the difference between N and the largest sample number), this gives $E(N - Y_n) = (N - n)/(n + 1)$. Use this result to show that Y_n (the largest sample number) is not an unbiased estimator for N (it never overestimates N), but that $\hat{N} = [(n + 1)Y_n/n] - 1$ is unbiased.

   c. Since $Y_1 - 1$ is another of the $n + 1$ gaps, $E(Y_1 - 1) = (N - n)/(n + 1)$. Use this and the result of (b) to show that $\hat{N} = Y_1 - 1 + Y_n$ is an unbiased estimator of N.

   d. For the four observed values of (a), what are the three estimates of N? It can be shown that the unbiased estimator of (b) is actually the MVUE of N, but the proof involves some advanced methods from mathematical statistics. Unfortunately, this $\hat{N}$ is not integer valued.

Bibliography

DeGroot, Morris, *Probability and Statistics*, Addison-Wesley, Reading, Mass., 1974. Includes an excellent discussion of both general properties and methods of point estimation; of particular interest are examples showing how general principles and methods can yield unsatisfactory estimators in particular situations.

Hogg, Robert and Craig, Allen, *Introduction to Mathematical Statistics* (3rd ed.), Macmillan, New York, 1970. A good discussion of unbiasedness.

Larsen, Richard and Marx, Morris, *Introduction to Mathematical Statistics*, Prentice-Hall, Englewood Cliffs, N.J., 1980. A very good discussion of point estimation from a slightly more mathematical perspective than the present text.

Mosteller, Frederick and Tukey, John, *Data Analysis and Regression*, Addison-Wesley, Reading, Mass., 1977. Includes some discussion of robust estimation; the authors have been among the pioneers in this area of statistics.

CHAPTER 7

Hypothesis-Testing Procedures Based on a Single Sample

Introduction

In Chapter 3 we introduced procedures based on the binomial distribution for testing hypotheses about a population proportion. Here we focus on the development of procedures whose structure depends in some way on the normal distribution. The discussion begins with hypothesis tests concerning μ when it is assumed both that the population is normal and that σ is known. In succeeding sections we identify situations in which one or both of these assumptions are unnecessary. Tests for hypotheses about a population proportion are then developed for the case of a large sample size, and procedures for testing hypotheses about σ are presented. The chapter closes with remarks concerning test selection and the issue of statistical versus practical significance.

7.1 Tests About the Mean of a Normal Population

The assumption of a normal population is often realistic, since many population distributions are well approximated by a normal curve. In addition we first assume that the value of σ is known. Although this is very infrequently the case in practice, it simplifies the presentation of test procedures and properties of tests. When the sample size n is large, neither assumption is necessary; this is discussed in Section 7.2.

Example 7.1 A manufacturer of a particular type and size of automobile tire uses a machine to study the wear characteristics of tires. The machine exerts uniform pressure on a particular surface, allowing comparisons between tires which are not affected by factors present under typical driving conditions. Based on extensive experiments with the current version of this tire, the true average tread life is believed to be 16,000 miles, with a standard deviation of 1500 miles. A change in the manufacturing process is being contemplated; it should not affect the standard deviation but will, if successful, increase average tread life. A sample of 25 tires manufactured experimentally using the new process has been tested, yielding a sample average tread life of 17,400 miles. Because implementation of the new process is costly, the manufacturer will consider a change only if the true average life for the new process is more than 17,000 miles. Does the sample average of 17,400 miles justify such a conclusion?

The population under investigation in Example 7.1 is a conceptual population consisting of tread lives for all possible tires which could be manufactured using the new process. Our basic assumption is that the population distribution is normal with mean μ and standard deviation σ, with σ for the new process having the same value as for the current process. Formally stated,

> **Assumptions for Example 7.1:** The 25 tread lives constitute a random sample $X_1, X_2, \ldots, X_{25}$ from a normal distribution with mean μ and standard deviation $\sigma = 1500$.

The two contrasting claims in our testing problem are then $\mu \leq 17,000$ and $\mu > 17,000$. For the moment we use the simpler hypothesis $\mu = 17,000$ in place of $\mu \leq 17,000$. One of the contrasting claims becomes the null hypothesis H_0 and the other becomes the alternative H_a. In any testing problem, H_0 is favored and the burden of proof is on H_a; H_0 will not be rejected in favor of H_a unless experimental evidence strongly favors H_a. With the choice $H_0: \mu = 17,000$ and $H_a: \mu > 17,000$, the current process will then be retained unless data strongly suggests that H_a is true. The value $\mu = 17,000$ which separates H_0 from H_a is called the **null value** of the parameter.*

Motivation for the Choice of a Test Statistic

The next step is to look for a test statistic—that is, a function of the sample observations which will be used as a decision maker. Since μ is both the mean and median of the population, we might think of using either the sample mean or the sample median to decide between H_0 and H_a. When the population is normal, however, the fact that $\bar{X}$ is the minimum variance unbiased estimator of μ suggests that we use a procedure based on it. Intuitively the data will strongly favor H_a only if the observed value of $\bar{X}$ is considerably larger than the null or boundary value of 17,000, so we wish to reject H_0 in favor of H_a only if $\bar{X} - 17,000$ is a large positive number. But how large is large? In our example the observed sample mean is $\bar{x} = 17,400$, so $\bar{x} - 17,000 = 400$. That is, the distance between the observed

*It is recommended that the reader consult Section 3.5 for more detailed comments on the basic terminology of hypothesis testing.

sample mean and the null value is 400. Is 400 a large enough difference to suggest rejection of H_0? The answer to this question depends crucially on the shape, and in particular the spread, of the distribution of $\bar{X}$ values. Because the X_i's are normal, $\bar{X}$ is normal with mean μ and standard deviation $\sigma/\sqrt{n}$. Now if $\sigma/\sqrt{n}$ is large, an observed $\bar{x}$ far above the null value would not be very surprising even when H_0 is true, as pictured in Figure 7.1(a). On the other hand, if $\sigma/\sqrt{n}$ is quite small, then an $\bar{x}$ value far above 17,000 would be inconsistent with H_0.

Figure 7.1 (a) Distribution of $\bar{X}$, H_0 true, $\sigma/\sqrt{n}$ large, (b) distribution of $\bar{X}$, H_0 true, $\sigma/\sqrt{n}$ small

To decide whether $\bar{X} - 17,000$ is large, let us express this distance *in units of standard deviations of* $\bar{X}$. Then a large number of standard deviations will be inconsistent with H_0, while a small number would not be. This line of reasoning yields

Test statistic: $Z = \dfrac{\bar{X} - 17,000}{\sigma/\sqrt{n}} = \dfrac{\bar{X} - 17,000}{300}$

Rejection region: $Z \geq c$

Specifying the Rejection Region

To completely specify the procedure, the critical value c must be determined. This is where controlling the probability of a type I error enters the picture. A type I error consists of rejecting H_0 when H_0 is true. Denoting this error probability by α,

$$\alpha = P(\text{rejecting } H_0 \text{ when } H_0 \text{ is true}) = P(Z \geq c \text{ when } \mu = 17,000) \quad (7.1)$$

But when H_0 is true, the test statistic Z has a standard normal distribution because $\bar{X}$ has been standardized using the null value 17,000. If an α of .05 is desired, (7.1) becomes

$$.05 = P\left(Z \geq c \quad \text{when} \quad Z \sim N(0, 1)\right)$$

From Table 4.1 it follows that $c = z_{.05} = 1.645$. The test procedure with **level of significance** (maximum type I error probability) .05 is now given by

---

Test statistic: $Z = \dfrac{\bar{X} - 17,000}{300}$ (7.2)

Rejection region for level .05 test: $Z \geq 1.645$

Because the observed value of $\overline{X}$ in the example is $\overline{x} = 17,400$, the computed value of the test statistic is $z = 400/300 = 1.33$. Because 1.33 is not in the rejection region, the conclusion is that the new process does not yield an improvement in tread life of the desired magnitude. The intuitive reason for this decision is that the observed mean 17,400 is only 1.33 standard deviations (of $\overline{X}$) to the right of the null value, which is not a large distance.

Calculation of Type II Error Probabilities

The value 1.645 in the rejection region was chosen to control the probability of a type I error. To decide whether or not the test adequately discriminates between H_0 and H_a, the probability of a type II error (not rejecting H_0 when H_0 is false) should also be calculated for various alternative values of μ. For many tests such a calculation is very difficult, but for the test (7.2) it is not. When H_0 is false, Z no longer has a standard normal distribution, because in standardizing $\overline{X}$ an alternative value of μ should be subtracted. The inequality in (7.2) is equivalent to

$$\overline{X} \geq (300)(1.645) + 17,000 = 17,493.5 \tag{7.3}$$

The probability of a type II error for the alternative value $\mu = 18,000$ is

$$\beta(18,000) = P(\text{not rejecting } H_0 \text{ when } \mu = 18,000)$$
$$= P(\overline{X} < 17,493.5 \text{ when } \mu = 18,000) \tag{7.4}$$
$$= \Phi\left(\frac{17,493.5 - 18,000}{300}\right) = \Phi(-1.69) = .0455$$

where $\Phi(\cdot)$ is the standard normal c.d.f. For the alternative $\mu = 17,600$

$$\beta(17,600) = P(\overline{X} < 17,493.5 \text{ when } \mu = 17,600)$$
$$= \Phi\left(\frac{17,493.5 - 17,600}{300}\right) = .3594$$

The Relationship Between α, β, and Sample Size

While $\beta(18,000)$ is reasonably small, the manufacturer may feel that a value of $\mu = 17,600$ for the new process represents a significant improvement over the null value of 17,000, so that the probability .3594 of not detecting this improvement is too large. If n remains fixed at 25, then a type II error probability can be decreased only by increasing α. The manufacturer may, though, wish to specify β for the alternative $\mu = 17,600$ and let this requirement determine n. Consider the restriction that $\beta(17,600) = .1$. For a level .05 test, not rejecting H_0 when $Z < 1.645$ is equivalent to

$$\text{don't reject } H_0 \text{ when } \overline{X} < \frac{(1500)(1.645)}{\sqrt{n}} + 17,000 \text{, or}$$

$$\overline{X} < \frac{2467.5}{\sqrt{n}} + 17,000 \tag{7.5}$$

Then we wish

$$.1 = \beta(17{,}600) = P\left(\bar{X} < \frac{2467.5}{\sqrt{n}} + 17{,}000 \quad \text{when} \quad \mu = 17{,}600\right)$$

$$= \Phi\left(1.645 - .4\sqrt{n}\right) \tag{7.6}$$

where the last equality in (7.6) results from standardizing by subtracting 17,600 and dividing by $1500/\sqrt{n}$. Because the value below which .1 of the area under the z curve lies is $-z_{.1} = -1.28$, (7.6) implies that $-1.28 = 1.645 - .4\sqrt{n}$, from which $n = (2.925/.4)^2 = 53.48$. Because n must be an integer, $n = 54$ is needed for our level .05 test to also satisfy $\beta(17{,}600) = .1$.

Implication of a More Realistic H_0

The test procedure "reject H_0 if $Z \geq 1.645$" is still appropriate if $H_0: \mu = 17{,}000$ is replaced by $H_0: \mu \leq 17{,}000$. For $\mu = 17{,}000$, $\alpha = .05$, while it can be shown that $\alpha(\mu) < .05$ when $\mu < 17{,}000$. That is, the test still has level of significance .05, since $\alpha(\mu) \leq .05$ for every $\mu \leq 17{,}000$. The boundary value 17,000 is the "worst case," so we pay most attention to it in selecting the critical value c. In general the level α test for a simple H_0 consisting of an equality statement will satisfy $P(\text{type I error}) \leq \alpha$ for the more realistic H_0 containing an appropriate inequality. In the sequel, therefore, H_0 will appear as an equality statement.

The Power Function

The type I and type II error calculations can be combined by studying the **power function** $\pi(\mu)$ of the test. $\pi(\mu)$ is defined for each possible value μ by

$$\pi(\mu) = P(\text{rejecting } H_0 \text{ when true value} = \mu) \tag{7.7}$$

Because H_0 is true when $\mu \leq 17{,}000$, we want low power for any such μ, while the power should be high when $\mu > 17{,}000$. Since

$$\pi(\mu) = \begin{cases} \alpha(\mu) & \text{if} \quad \mu \leq 17{,}000 \\ 1 - \beta(\mu) & \text{if} \quad \mu > 17{,}000 \end{cases}$$

small α and β yield a power function of the desired type. For the test [(7.2) or (7.3)] under consideration

$$\pi(\mu) = P(\bar{X} \geq 17{,}493.5 \quad \text{when true value} = \mu)$$

$$= 1 - \Phi\left(\frac{17{,}493.5 - \mu}{300}\right)$$

The graph of $\pi(\mu)$ is sketched in Figure 7.2.

Figure 7.2 Power function for the test (7.3)

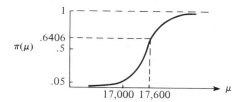

While Figure 7.2 shows that there are values greater than 17,000 for which the power is not very high (such as $\mu = 17{,}001$), we really want high power only for μ's which represent a significant practical departure from H_0. For the sample size fixed at $n = 25$, it can be shown that the test (7.3) is most powerful among all level .05 tests. That is, among all tests with power function $\pi(\mu) \leq .05$ for $\mu \leq 17{,}000$, the test (7.3) has maximum possible power for any $\mu > 17{,}000$. Thus this test is better than any test based on the sample median $\tilde{X}$ provided that the assumption of normality is valid.

General Description of the Test Procedures

Assumptions: $X_1, X_2, \ldots, X_n$ constitute a random sample from a normal distribution with unknown mean μ and known standard deviation σ.

The null hypothesis will always be stated in the form $H_0: \mu = \mu_0$, where μ_0 is a fixed number, the null value of the parameter. When H_0 is true, standardizing $\bar{X}$ by subtracting μ_0 and dividing by $\sigma/\sqrt{n}$ yields a standard normal r.v. This leads to the following test procedures for the three commonly encountered alternative hypotheses:

Null hypothesis: H_0: $\mu = \mu_0$

Test statistic: $Z = \dfrac{\bar{X} - \mu_0}{\sigma/\sqrt{n}}$

Alternative Hypothesis	*Rejection Region for Level α Test*
H_a: $\mu > \mu_0$	$Z \geq z_\alpha$
H_a: $\mu < \mu_0$	$Z \leq -z_\alpha$
H_a: $\mu \neq \mu_0$	either $Z \geq z_{\alpha/2}$ or $Z \leq -z_{\alpha/2}$

The first alternative hypothesis consists of μ values above μ_0, so the rejection region is **upper-tailed,** consisting only of large positive values of Z. For the second

Figure 7.3 (*a*) Rejection region for level (*a*) upper-tailed test, (*b*) lower-tailed test, (*c*) two-tailed test

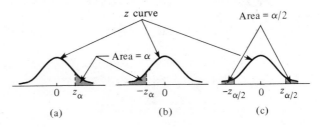

alternative a **lower-tailed** region is appropriate, consisting only of Z values which are sufficiently negative. In the third we wish to reject H_0 if $\bar{X}$ falls too far to *either* side of μ_0, yielding a **two-tailed** region consisting of both large positive and large negative value of Z.

In any particular problem involving the mean μ of a normal population when σ is known, the following steps should be followed in reaching a decision.

1. Identify the parameter of interest (a first step in any inference problem).
2. Identify the null value μ_0.
3. Identify the appropriate alternative hypothesis.
4. Write the test statistic Z (substituting μ_0, n, and σ).
5. Write the rejection region corresponding to the appropriate alternative.
6. For the chosen α, obtain z_α or $z_{\alpha/2}$ and substitute into the rejection region.
7. Compute Z.
8. Decide whether or not H_0 should be rejected, and describe the conclusion in the context of the problem.

A similar sequence of steps is appropriate in other testing problems.

Example 7.2

The drying time of a particular brand and type of paint is known to be normally distributed with $\mu = 75$ min and $\sigma = 9.4$ min. In an attempt to improve the drying time, a new additive has been developed. Use of the additive in 100 test samples of the paint yields an observed sample average drying time of $\bar{x} = 68.5$. Assuming that drying times with the additive are still normally distributed with $\sigma = 9.4$, does the experimental evidence indicate that the additive improves true average drying time? Because of the expense associated with using the additive, evidence should strongly suggest an improvement in drying time before such a conclusion is adopted.

1. The parameter of interest here is μ = true average drying time of the paint with additive.
2. The null value is $\mu_0 = 75$, and the null hypothesis is H_0: $\mu = 75$.
3. The alternative hypothesis is H_a: $\mu < 75$ (the improvement claim), so a lower-tailed test is appropriate.
4. Test statistic $= Z = \dfrac{\bar{X} - 75}{9.4/\sqrt{100}} = \dfrac{\bar{X} - 75}{.94}$
5. Reject H_0 if $Z \le -z_\alpha$
6. For $\alpha = .01$ (a level .01 test), $-z_{.01} = -2.33$, so reject H_0 if $Z \le -2.33$
7. The computed Z is

$$z = \frac{\bar{x} - 75}{.94} = \frac{68.5 - 75}{.94} = -6.91$$

8. Since -6.91 is ≤ -2.33, H_0 is rejected at level .01, and we conclude that the additive does produce a significant improvement in true average drying time.

Example 7.3

A manufacturer of sprinkler systems used in office buildings to protect against fires claims that its sprinklers are activated at a temperature of 130 °F. A sample of $n = 9$

systems, when tested, yields a sample average activation temperature of 131.08 °F. If activation times are normally distributed with true standard deviation $\sigma = 1.5$ °F, does the data contradict the company's claim at level $\alpha = .05$?

Here μ = true average activation temperature, and $\mu_0 = 130$. Because a departure from μ_0 in either direction is of concern (the true average activation temperature should be neither too high nor too low), the relevant hypotheses are

$$H_0: \mu = 130 \quad \text{versus} \quad H_a: \mu \neq 130$$

The alternative is two-sided, so the rejection region is two-tailed.

$$\text{Test statistic} = Z = \frac{\overline{X} - 130}{1.5/\sqrt{9}} = \frac{\overline{X} - 130}{.5}$$

Reject H_0 if either $Z \geq z_{\alpha/2}$ or $Z \leq -z_{\alpha/2}$.

Since $\alpha = .05$, $\alpha/2 = .025$, so $z_{\alpha/2} = z_{.025} = 1.96$. H_0 will be rejected if either $Z \geq 1.96$ or if $Z \leq -1.96$. The computed value of Z is

$$z = \frac{\overline{x} - 130}{.5} = \frac{131.08 - 130}{.5} = 2.16$$

Since $2.16 \geq 1.96$, H_0 is rejected in favor of H_a, because the observed $\overline{x}$ is more than 1.96 standard deviations (of $\overline{X}$) above the null value. True average activation temperature does appear to differ from the value claimed by the company.

The Power Function and $\beta(\mu)$

Recall that $\pi(\mu) = P(\text{rejecting } H_0 \text{ when } \mu = \text{true value})$, and for any alternative value of μ, $\beta(\mu) = 1 - \pi(\mu)$. By expressing each rejection region in terms of an inequality or inequalities involving $\overline{X}$ isolated on one side, and then restandardizing using value μ (as in Example 7.1), the power function for each of the three tests is as follows:

Alternative Hypothesis	Power Function $\pi(\mu)$ for Level α Test
$H_a: \mu > \mu_0$	$1 - \Phi\left(z_\alpha + \dfrac{\mu_0 - \mu}{\sigma/\sqrt{n}}\right)$
$H_a: \mu < \mu_0$	$\Phi\left(-z_\alpha + \dfrac{\mu_0 - \mu}{\sigma/\sqrt{n}}\right)$
$H_a: \mu \neq \mu_0$	$1 - \Phi\left(z_{\alpha/2} + \dfrac{\mu_0 - \mu}{\sigma/\sqrt{n}}\right)$
	$+ \Phi\left(-z_{\alpha/2} + \dfrac{\mu_0 - \mu}{\sigma/\sqrt{n}}\right)$

where $\Phi(z)$ = c.d.f. of the standard normal distribution

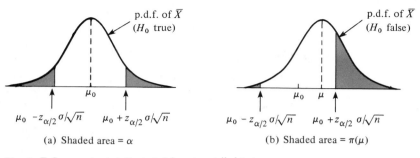

(a) Shaded area = α (b) Shaded area = $\pi(\mu)$

Figure 7.4 α and $\pi(\mu)$ illustrated for a two-tailed test

Figure 7.4(a) pictures the p.d.f. of $\overline{X}$ when H_0 is true and (b) pictures the p.d.f. of $\overline{X}$ where H_0 is false and the true value is μ. For a two-tailed test the shaded area in (a) is α and the shaded area in (b) is $\pi(\mu)$.

Example 7.2
(continued)

The alternative hypothesis in the paint-drying example was $H_a\colon \mu < 75$. For $\mu = 70$,

$$\pi(70) = \Phi\left(-2.33 + \frac{75 - 70}{9.4/\sqrt{100}}\right) = \Phi\,(2.99) = .9986$$

The probability of a type II error when $\mu = 70$ is $\beta(70) = 1 - .9986 = .0014$, which is quite small, as desired.

Example 7.3
(continued)

The alternative was $H_a\colon \mu \neq 130$ in the sprinkler example, so

$$\pi(130.5) = 1 - \Phi\left(1.96 + \frac{130 - 130.5}{1.5/\sqrt{9}}\right) + \Phi\left(-1.96 + \frac{130 - 130.5}{1.5/\sqrt{9}}\right)$$

$$= 1 - \Phi\,(.96) + \Phi\,(-2.96)$$

$$= 1 - .8315 + .0015 = .1700$$

The power is small, and $\beta(130.5) = 1 - .1700 = .8300$ is large, because $\mu = 130.5$ is not very far from $\mu_0 = 130$ (since $\sigma/\sqrt{n} = .5$, 130.5 is only one standard deviation of $\overline{X}$ from μ_0). Similarly $\pi(129.5) = .1700$.

Figure 7.5 illustrates the power functions of tests whose rejection regions are upper-tailed, lower-tailed, and two-tailed.

Choosing the Sample Size

Once n and α have been selected, the power function of the test, and therefore $\beta(\mu)$, is fixed for each μ. Instead of fixing n beforehand, we can fix α and $\beta(\mu)$ for some alternative value of μ, and then solve for the necessary sample size n. An argument

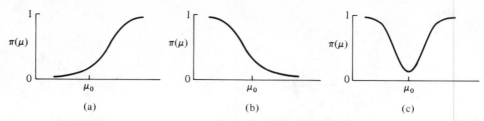

Figure 7.5 Power function for (a) upper-tailed test, (b) lower-tailed test, (c) two-tailed test

parallel to the one given for Example 7.1 yields the *sample size n for which the level α test also has P(type II error when true value = μ) = β*:

$$n = \begin{cases} \left[\dfrac{\sigma(z_\alpha + z_\beta)}{\mu_0 - \mu} \right]^2 & \text{for a one-tailed (upper or lower) test} \\[2em] \left[\dfrac{\sigma(z_{\alpha/2} + z_\beta)}{\mu_0 - \mu} \right]^2 & \text{for a two-tailed test} \end{cases}$$

The *n* for the two-tailed test is only approximate, since it is obtained by neglecting the smaller of the two resulting probabilities in the $\beta(\mu)$ calculation.

Example 7.2
(continued)

For $n = 100$ in the paint-drying example, $\beta(72) = 1 - \pi(72) = .1949$. If we wish $\beta(72) = .05$ for a level $\alpha = .01$ test, then

$$n = \left[\frac{9.4(2.33 + 1.645)}{75 - 72} \right]^2 \doteq 156$$

The *P*-Value of a Test

Suppose that in the tire problem of Example 7.1, the observed $\bar{X}$ had been $\bar{x} = 17,555$. Then computed value of Z would be $z = (17,555 - 17,000)/300 = 1.85$. The rejection region is $Z \geq z_\alpha$, so if we choose $\alpha = .05$, then $z_{.05} = 1.645$, $1.85 \geq 1.645$, and H_0 is rejected. However, if $\alpha = .01$, then 1.85 is not ≥ 2.33, so H_0 is not rejected. The decision depends critically on the chosen level α. As α decreases, z_α increases, so there will be a smallest value of α for which $1.85 \geq z_\alpha$ and H_0 is rejected. This smallest value is the α for which $1.85 = z_\alpha$. This equation says that α is the area underneath the standard normal curve to the right of 1.85, as illustrated in Figure 7.6. Since the area under the curve to the right of 1.85 is $1 - \Phi(1.85) = .0322$, the smallest level of significance at which H_0 is rejected when $z = 1.85$ is $\alpha = .0322$. If the chosen α exceeds .0322, H_0 is rejected, while if $\alpha < .0322$, then H_0 cannot be rejected.

> **Definition:** Given a test procedure and the computed value of the test statistic, the **prob value** or **P-value** of the test, denoted by P, is the smallest value of α for which the use of a level α test results in the rejection of H_0.

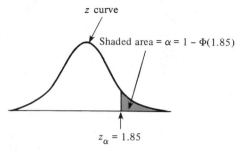

z curve

Shaded area = $\alpha = 1 - \Phi(1.85)$

$z_\alpha = 1.85$

Figure 7.6 Determining the smallest α for which H_0 is rejected when $z = 1.85$

Example 7.3 (continued)

In the sprinkler system problem the computed Z was $z = 2.16$. Since the rejection region is two-tailed, H_0 is rejected at level α if $2.16 \geq z_{\alpha/2}$. The smallest α for which H_0 is rejected is the value for which $z_{\alpha/2} = 2.16$, so that $\alpha/2$ is the area under the standard normal curve to the right of 2.16. This gives $\alpha/2 = 1 - \Phi(2.16) = 1 - .9846 = .0154$, so $\alpha = 2(.0154) = .0308$. The P-value here is thus $P = .0308$. Any α larger than .0308, such as $\alpha = .05$, results in rejection of H_0 (since then $2.16 \geq z_{\alpha/2}$), while any smaller α results in acceptance of H_0 (since $2.16 < z_{\alpha/2}$).

It is customary to call the data **significant** when H_0 is rejected and not significant otherwise, so the definition can be paraphrased by saying that *the P-value is the smallest level α at which the data is significant*. Rather than reporting a statement such as "H_0 was rejected at level .05" or "H_0 was not rejected at level .01," it is more informative to report the P-value itself. Once the P-value is known, any individual looking at the report can decide for himself or herself just how significant the data is without having the investigator impose a level of significance. In particular, once you have been told the P-value, you can reach a conclusion at your selected level α by comparing α to the P-value, as illustrated in Figure 7.7.

Figure 7.7 Making a decision using the P-value

If chosen α is in (a), do not reject H_0 ($\alpha < P$-value)
If chosen α is in (b), reject H_0 ($\alpha \geq P$-value)

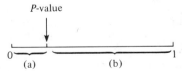

P-value

0   1
(a)   (b)

For many tests the exact P-value is difficult to compute, but for the Z tests discussed here, it is easy. Let z be the computed value of the test statistic Z. Then

$$P\text{-value} = P = \begin{cases} 1 - \Phi\,(z) & \text{for an upper-tailed test} \\ \Phi\,(z) & \text{for a lower-tailed test} \\ 2[1 - \Phi\,(|\,z\,|)] & \text{for a two-tailed test} \end{cases}$$

Example 7.2
(continued)

The rejection region in the paint-drying example was lower-tailed. If $\bar{x} = 72.42$, then $z = (72.42 - 75)/.94 = -2.75$, so $P = \Phi\,(-2.75) = 1 - \Phi\,(2.75) = .0030$. At level $\alpha = .01$, H_0 would be rejected, while at level .001 it would not be rejected.

The P-value is really the chance, computed assuming H_0 true, of obtaining a more extreme value of the test statistic than the value actually obtained. Thus if $P = .003$, only .3% of the time would a more extreme value be observed if H_0 is true. This is why a very small P-value suggests that H_0 should be rejected.

Exercises / Section 7.1

1. Answer the following questions for the tire problem in Example 7.1
   a. If $\bar{x} = 17{,}752$ and a level $\alpha = .01$ test is used, what is the decision?
   b. If a level .01 test is used, what is $\beta(18{,}000)$?
   c. If a level .01 test is used and it is also required that $\beta(17{,}500) = .05$, what sample size n is necessary?
   d. If $\bar{x} = 17{,}752$, what is the smallest α at which H_0 is rejected (using $n = 25$)?
   e. If $\bar{x} = 17{,}800$, what is the P-value (using $n = 25$)?

2. Answer the following questions for the paint-drying problem in Example 7.2.
   a. How many standard deviations (of $\bar{X}$) below the null value is $\bar{x} = 72.3$?
   b. If $\bar{x} = 72.3$, what is the conclusion using $\alpha = .01$?
   c. What is α for the test procedure which rejects H_0 when $Z \le -2.88$?
   d. For the test procedure of (c), what is $\pi(70)$? What is $\beta(70)$?
   e. If the test procedure of (c) is used, what n is necessary to ensure that $\beta(70) = .01$?
   f. What is the P-value if $\bar{x} = 72.3$?
   g. If a level .01 test is used with $n = 100$, what is the probability of a type I error when $\mu = 76$?

3. Answer the following questions for the sprinkler problem in Example 7.3.
   a. What is the rejection region for a level .01 test?
   b. If $\bar{x} = 128.75$, what is the conclusion at level of significance .01?
   c. What is α for the test which rejects H_0 if either $Z \ge 2.81$ or $Z \le -2.81$?
   d. What is $\beta(130.5)$ for the test of (c)?
   e. If $n = 25$ and the test of (c) is used, what is $\beta(130.5)$?
   f. Using $n = 9$, what is the P-value when $\bar{x} = 128.75$?
   g. If someone reported to you that the P-value was .0232 and you then decided that a level .01 test was appropriate, what would you conclude?
   h. If the P-value is reported as .0232, what was the computed $\bar{x}$ value?

4. The melting point of each of 16 samples of a certain brand of hydrogenated vegetable oil was determined, resulting in $\bar{x} = 94.32$. Assume that the distribution of melting point is normal with $\sigma = 1.20$.
   a. Test $H_0: \mu = 95$ versus $H_a: \mu \ne 95$ using a two-tailed level .01 test.
   b. If the true value of μ is 94 and a level .01 test is used, what is $\pi(94)$, the power of the test

when $\mu = 94$, and what is $\beta(94)$, the probability of a type II error?

c. What value of n is necessary to ensure that $\beta(94) = .1$ when $\alpha = .01$?

d. What is the P-value of the data?

e. If instead of reporting $\bar{x}$, the P-value had been reported as .0236, what would you conclude at level .01? Explain your reasoning.

5. A television manufacturer claims that at most 250 microamperes of current are needed to attain a certain brightness level with a particular type of set. A sample of 20 sets yields a sample average current of $\bar{x} = 257.3$. Let μ denote the true average current necessary to achieve the desired brightness with sets of this type, and assume that μ is the mean of a normal population with $\sigma = 15$.

a. Test at level .05 the null hypothesis that μ is at most 250 against the appropriate alternative.

b. What is the P-value for the above data?

c. If $\mu = 260$, what is the power of the test, and what is the probability of a type II error?

d. Suppose that it was decided that the manufacturer's claim should not be accepted unless the data strongly supports it. Would you choose as your null hypothesis the H_0 suggested in (a), or would you select a different H_0? Explain. *Hint:* From a practical viewpoint, the claim that at most 250 microamps are required is the same as the claim that fewer than 250 microamps are required.

6. It is desired that a mixture of pulverized fuel ash and Portland cement to be used for grouting should have a compressive strength of more than 1300 KN/m². Suppose that compressive strength is normally distributed with $\sigma = 60$. Since the mixture will not be used unless the evidence is conclusive, we wish to test H_0: $\mu = 1300$ (actually ≤ 1300) versus H_a: $\mu > 1300$, where $\mu =$ true average compressive strength.

a. If the average compressive strength for 20 specimens is computed to be $\bar{x} = 1329.2$, should the grouting mixture be used? Test the above hypotheses using $\alpha = .01$.

b. For $n = 20$ and $\alpha = .01$, what is the probability of deciding to use the mixture when $\mu = 1325$?

c. If you wish to use $\alpha = .01$ and also have $\beta(1325) = .01$, what is the smallest sample size you can use?

d. What is the smallest α at which H_0 is rejected (that is, the P-value) if $\bar{x} = 1329.2$ and $n = 20$?

7. The desired percentage of SiO_2 in a certain type of aluminous cement is 5.5. To test whether the true average percentage is 5.5 for a particular production facility, 16 independently obtained samples are analyzed. Suppose that the percentage of SiO_2 in a sample is normally distributed with $\sigma = .3$, and that $\bar{x} = 5.25$.

a. Does this indicate conclusively that the true average percentage differs from 5.5? Carry out the analysis using the sequence of steps suggested in the text.

b. If the true average percentage is $\mu = 5.6$ and a level $\alpha = .01$ test based on $n = 16$ is used, what is the probability of detecting this departure from H_0?

c. What is the P-value when $\bar{x} = 5.25$?

8. A structure is to be built in a certain area, and the design engineers are concerned about the resulting amount of settlement into the soil. A particular design is judged satisfactory if the true average amount of settlement for a beam supporting a certain weight is at most 3 cm. Assume that settlement amount is a normally distributed variable with $\sigma = .6$ cm. The average amount of settlement for 10 randomly placed beams is 3.15 cm. Because it would be expensive to change the design, it will be judged satisfactory unless the data suggests that the true average amount of settlement is too great. What would you conclude using a level .01 test?

9. A spectrophotometer used for measuring CO concentration [ppm (parts per million) by volume] is checked for accuracy by taking readings on a manufactured gas (called span gas) in which the CO concentration is very precisely controlled at 70 ppm. If the readings suggest that the spectrophotometer is not working properly, it will have to be recalibrated. Assume that if properly calibrated, measured concentration for span gas samples is normally distributed with $\sigma = 5$. On the basis of the six readings, 85, 77, 82, 68, 72, 69, is recalibration necessary? Test using $\alpha = .05$.

10. Show that for any $\Delta > 0$, the power function of the two-tailed test satisfies $\pi(\mu_0 - \Delta) = \pi(\mu_0 + \Delta)$, so that $\pi(\mu)$ is symmetric about μ_0.

11. For a fixed alternative value μ, show that $\beta(\mu) \to 0$ as $n \to \infty$ for either a one-tailed or a two-tailed test.

12. Consider a fixed α and alternative value μ, and suppose that $\sigma_1 > \sigma_2$. For which value of σ, σ_1, or σ_2, is $\beta(\mu)$ larger? Justify your assertion.

13. The fact that the power function of the two-tailed test is symmetric about μ_0 (Exercise 10) implies in Example 7.3 that $\beta(129.5) = \beta(130.5)$, $\beta(129) = \beta(131)$, and so on. Yet one might argue that the alternative $\mu = 131$ is of more concern than the alternative $\mu = 129$ (a low activation temperature might be preferred to a high one).
    a. Show that the two-tailed test which rejects H_0 if either $Z \geq 1.88$ or $Z \leq -2.05$ has $\alpha = .05$.
    b. Compute $\pi(129)$ and $\pi(131)$ for the procedure of (a). Which is larger? If we are more concerned with a departure above $\mu_0 = 130$ than below, which one do we want to be larger?

Hint: To compute the probability of rejecting H_0, first express the procedure in the form "reject H_0 if either $\overline{X} \geq c_1$ or $\overline{X} \leq c_2$," and then standardize.
    c. Suppose that you wanted to use a level .05 test but that you wanted $\pi(131)$ to be even larger than what you calculated in (b). Suggest an appropriate two-tailed test.

14. Let γ be such that $0 < \gamma < \alpha$. For testing H_0: $\mu = \mu_0$ versus H_a: $\mu \neq \mu_0$, consider the test which rejects H_0 if either $Z \geq z_\gamma$, or $Z \leq -z_{\alpha - \gamma}$, where $Z = (\overline{X} - \mu_0)/(\sigma/\sqrt{n})$.
    a. Show that $P(\text{type I error}) = \alpha$.
    b. Derive an expression for the power function $\pi(\mu)$ of this test. *Hint:* Express the test in the form "reject H_0 if either $\overline{X} \geq c_1$ or $\leq c_2$."
    c. Let $\Delta > 0$. For what values of γ (relative to α) will $\pi(\mu_0 + \Delta) > \pi(\mu_0 - \Delta)$?

7.2 Large-Sample Tests for a Population Mean

In the first section of the chapter, hypothesis-testing procedures for deciding between hypotheses involving a population mean μ were presented. The validity of these procedures depended on two rather restrictive assumptions:

1. The observations come from a normal population or distribution.
2. The standard deviation σ is known.

In this section we indicate that when the sample size n is large, the tests previously described remain valid even when assumption 1 and/or 2 are/is not satisfied. It should be emphasized that assumption 1 is often a very reasonable assumption to make, and in fact will be quite crucial to our discussion of a small-sample procedure in Section 7.4. However, assumption 2 is rarely satisfied, because when the true value of a mean is unknown, it is unrealistic to suppose that the value of σ is known.

Example 7.4

The members of a particular academic department make heavy use of copying machines in their work. Because much time was being wasted in traveling to and from the nearest machines and in waiting in line to use these machines, the department submitted a request for authorization to purchase a small copier to be located in the departmental office. The administration decided to monitor the department's use of copiers over a 50-day period; only if the data indicated that the average daily usage was more than 250 copies would the purchase request be approved. At the end of the monitoring period, the sample average number of copies per day was computed to be 265, with a sample standard deviation of 42. If the administration used a level .05 test to reach a decision, did the department get its copier?

Defining μ = true average number of copies per day made for the department, the appropriate null (simplified) and alternative hypotheses are H_0: $\mu = 250$ versus H_a: $\mu > 250$. The administration will reject H_0 and approve the purchase only if the experimental evidence is reasonably compelling. The parameter of interest, H_0, and H_a fit exactly the discussion of the last section, but in the problem description neither normality nor known σ is presumed. First note that the sample size $n = 50$ is large enough for the Central Limit Theorem to be invoked. Supposing for the moment that σ is known, the variable

$$Z = \frac{\bar{X} - 250}{\sigma/\sqrt{50}} \tag{7.8}$$

has (when H_0 is true) approximately a standard normal distribution. If we therefore reject H_0 when $Z \geq 1.645$, the procedure will determine an *approximate* level .05 test. In other words, *the only modification that we need make to the procedures of Section 7.1 when normality of the distribution is not assumed but n is large is to insert the word "approximate" in front of "level α."*

We are still faced with the problem of unknown σ. Here again a large sample size comes to our rescue. Because $n = 50$ is large, the sample standard deviation S will tend to be quite close to the true value of σ. In this case, it can be shown that

$$Z = \frac{\bar{X} - 250}{S/\sqrt{n}} \tag{7.9}$$

also has approximately a standard normal distribution. If we now reject H_0 in favor of H_a when Z of (7.9) is ≥ 1.645, the resulting test will have approximate level .05. Notice that whereas the original test statistic Z specified the *true* number of standard deviations separating $\bar{X}$ and μ_0, the new statistic gives the *estimated* number of standard deviations separating these two quantities.

For the given data, $\bar{x} = 265$ and $s = 42$, so that the computed value of Z is

$$z = \frac{265 - 250}{42/\sqrt{50}} = \frac{15}{5.94} = 2.53$$

Since $2.53 \geq 1.645$, H_0 is rejected at (approximate) level .05 in favor of the alternative that $\mu > 250$, so the department is awarded its copier.

General Description of the Test Procedures

Example 7.4 shows that when assumptions of normality and known σ are not appropriate, if n is large then a minor modification of the tests of Section 7.1 yields procedures for which the type I error probability is controlled.

Test procedures for H_0: $\mu = \mu_0$ when $n > 30$:

Test statistic: $Z = \dfrac{\overline{X} - \mu_0}{S/\sqrt{n}}$

Alternative Hypothesis	*Rejection Region for Approximate Level α Test*
H_a: $\mu > \mu_0$	$Z \geq z_\alpha$
H_a: $\mu < \mu_0$	$Z \leq -z_\alpha$
H_a: $\mu \neq \mu_0$	either $Z \geq z_{\alpha/2}$ or $Z \leq -z_{\alpha/2}$

Example 7.5   Returning to the paint-drying-time example in the previous section, in which the value of σ was assumed known and equal to 9.4, if we now assume that σ is unknown but that the *sample* standard deviation is computed to be $s = 9.4$, then because $n = 100$ the large sample procedure can be applied to yield the same value of Z and the same conclusion as in the original example.

After describing the test procedures in the last section, we went on to the computation of β and the determination of appropriate sample size for the tests. The value of σ was needed to obtain these quantities, so that analogous computations are considerably more difficult in the absence of such knowledge. We comment again on this point in Section 7.4.

Exercises / Section 7.2

1. Reconsider Exercise 7.1.4 involving the melting point of vegetable oil.
   a. If a sample of size 34 resulted in $\overline{x} = 94.45$ and $s = 1.37$, test H_0: $\mu = 95$ versus H_a: $\mu \neq 95$ at level .01. Is it necessary to assume here that the melting point distribution is normal?
   b. What is the P-value for the data and Z test?

2. In Exercise 7.1.5 regarding the current necessary for a certain TV brightness level, suppose that a sample of 36 sets resulted in $\overline{x} = 247.2$ and $s = 10.5$. Test H_0: $\mu = 250$ versus H_a: $\mu > 250$ at level .05.

3. To obtain information on the corrosion resistance properties of a certain type of steel conduit, 35 specimens are buried in soil for a two-year period. The maximum penetration (in mils) for each specimen is then measured, yielding a sample average penetration of $\overline{x} = 52.7$ and a sample standard deviation of $s = 4.8$. The conduits were manufactured with the specification that true average penetration be at most 50 mils. To see whether experimental data indicates that specifications have not been met, test H_0: $\mu = 50$ versus H_a: $\mu > 50$ using a large sample test with $\alpha = .05$. In reaching a conclusion, follow the sequence of steps given in Section 7.1.

4. A sample of 40 speedometers of a particular brand is obtained, and each is calibrated to check for accuracy at 55 mph. The resulting sample average and sample standard deviation are $\overline{x} = 53.8$ and $s = 1.3$, respectively. Let μ = the true average reading when actual speed is 55 mph. Does the sample evidence suggest strongly that $\mu \neq 55$? Use a level .01 test.

5. The Chappy V-notch impact test is the basis for studying many material toughness criteria. This test was applied to 32 samples of a particular alloy at 110 °F. The sample average amount of transverse lateral expansion was computed to be 73.1 mils, and the sample standard deviation was $s = 5.9$ mils. To be suitable for a particular application, the true average amount of expansion should be less than 75 mils. The alloy will not be used unless the sample provides strong evidence of this criteria having been met. Test the relevant hypotheses using $\alpha = .01$ to decide whether the alloy is suitable. What is the P-value of the data?

6. The relative conductivity of a semiconductor device is determined by the amount of impurity "doped" into the device during its manufacture. A silicon diode to be used for a specific purpose requires a cut-on voltage of .60 volt, and if this is not achieved then the mechanism governing the amount of impurity must be adjusted. A sample of 120 such diodes yielded a sample average voltage of .62 volt and sample standard deviation of .11.

  a. At level .001, does the data indicate that the true average cut-on voltage is something other than .60?

  b. What is the P-value of the data?

7. A particular type of automobile engine has been designed so that the minimum octane number of gasoline needed to prevent knocking is 88. Each of 36 engines was tested to determine the octane rating necessary to prevent knocking, resulting in a sample average octane number of 87.3 and sample standard deviation of 2.3. It will be claimed that the design specification has been met unless the data strongly indicates otherwise.

  a. Use a level .01 test to decide whether or not the company can make the claim that the specification has been met.

  b. What is the P-value of the data?

8. An aspirin manufacturer fills bottles by weight rather than by count. Since each bottle should contain 100 tablets, the average weight per tablet should be five grains. Each of 100 tablets taken from a very large lot is weighed, resulting in a sample average weight per tablet of 4.87 grains and a sample standard deviation of .35 grain. Does this information provide strong evidence for concluding that the company is not filling its bottles as advertised? Test the appropriate hypotheses using $\alpha = .01$, and also compute the P-value of the data.

9. The owner of a gas station which makes headlight inspections for the state is trying to decide whether or not to discontinue the service. He figures that to make the operation profitable, the station must average in excess of 15 inspections per week. Unless data indicates strongly that this is the case, inspections will be discontinued. Data for the past year (52 weeks) yields a sample average of 16.7 inspections per week with a sample standard deviation of 4.5. Is this strong enough evidence to cause the owner to retain the inspection service? Test the relevant hypotheses.

7.3 Large-Sample Tests for Population Proportions

The large-sample tests of the previous section can be fit into a general framework which allows for the straightforward description of such tests in a number of different situations. In the present section we first describe this general framework, and then apply it to tests about population proportions. More applications will appear in subsequent sections.

Construction of a Large-Sample Test Procedure

As in Section 6.1, let θ denote a parameter of interest. We wish to test the null hypothesis H_0: $\theta = \theta_0$ (θ_0, the null value, is a specified number) against one of the three standard alternatives. Let $\hat{\theta}$ be an estimator of θ based on a large sample, and suppose that $\hat{\theta}$ satisfies the following two properties:

1. $\hat{\theta}$ is an unbiased estimator of θ—that is, $E(\hat{\theta}) = \theta$ (at least approximately).
2. $\hat{\theta}$ has approximately a normal distribution.

Denote the standard deviation (standard error) of $\hat{\theta}$ by $\sigma_{\hat{\theta}}$. It follows from 1 and 2 that when H_0 is true,

$$Z = \frac{\text{estimator} - \text{null value}}{\text{standard deviation of estimator}} = \frac{\hat{\theta} - \theta_0}{\sigma_{\hat{\theta}}} \qquad (7.10)$$

has approximately a standard normal distribution, so an approximate level α test can be obtained by using an appropriate Z critical value to specify the rejection region. Unfortunately it is frequently the case that $\sigma_{\hat{\theta}}$ will involve unknown parameters, so that this Z must be modified. Let $\hat{\sigma}_{\hat{\theta}}$ denote an appropriate estimator of $\sigma_{\hat{\theta}}$, and define a test statistic by

$$Z = \frac{\text{estimator} - \text{null value}}{\text{estimated standard deviation of estimator}} = \frac{\hat{\theta} - \theta_0}{\hat{\sigma}_{\hat{\theta}}} \qquad (7.11)$$

Because of the assumption of a large sample size, this modified Z will also have approximately a standard normal distribution. Either Z of (7.10) or Z of (7.11) is the test statistic.

Alternative Hypothesis	*Rejection Region for Approximate Level α Test*
$H_a: \theta > \theta_0$	$Z \geq z_{\alpha}$
$H_a: \theta < \theta_0$	$Z \leq -z_{\alpha}$
$H_a: \theta \neq \theta_0$	either $Z \geq z_{\alpha/2}$ or $Z \leq -z_{\alpha/2}$

Example 7.6

Let the parameter of interest be μ. Then the sample mean $\bar{X}$ is an unbiased estimator of μ, and when n is large has (by the Central Limit Theorem) approximately a normal distribution. The standard error of $\bar{X}$ is $\sigma_{\bar{X}} = \sigma/\sqrt{n}$, so when σ is not known we use the estimator $\hat{\sigma}_{\bar{X}} = S/\sqrt{n}$. The test statistic Z for testing $H_0: \mu = \mu_0$ becomes

$$Z = \frac{\bar{X} - \mu_o}{S/\sqrt{n}}$$

which is exactly the statistic of the last section. Hence the large-sample tests of Section 7.2 are subsumed under the general framework described above.

Application to a Population Proportion

Example 7.7

A number of years ago, the College of Arts and Sciences of a major university eliminated general breadth requirements. Recently the faculty has become concerned over a possible lack of diversity in selection of courses by many students. It has therefore been decided to check the transcripts of a randomly selected group of 100 recent graduates to make an inference about the true proportion of recent graduates who would have satisfied the abandoned requirements. If experimental evidence indicates that fewer than 80% of all recent graduates would have satisfied the requirements, then the faculty will press for reinstitution of the requirements. Because of adverse student reaction, the data should conclusively indicate that reinstitution is called for before such action is contemplated. If only 74 of the 100 former students in the sample would have satisfied the previous requirements, what action should the faculty take?

The parameter of interest is

p = the true proportion of recent graduates who would have satisfied the abandoned requirements

The wording suggests that our null hypothesis should be identified with the "no reinstitution is necessary" claim, with H_0 being rejected only for strongly contradictory data. More precisely, we wish to test

$$H_0: p = .8 \quad \text{versus} \quad H_a: p < .8$$

where again H_0 has been simplified to an equality claim. Let X = the number of students among the 100 sampled who would have satisfied the requirements. Then X is a binomial random variable (at least approximately), so that the problem looks exactly like those encountered in the first treatment of hypothesis testing in Chapter 3. To proceed as we did there, though, a binomial table for $n = 100$ is necessary, which is not available to us (though such a table does exist). However, because $n = 100$ is a large sample size, a large sample test derived from the general framework will be essentially equivalent to a test based on a binomial table.

Description of the Procedure

A point estimator for p is $\hat{p} = X/n$, which is unbiased. Further, when n is large and p is not too near 0 or 1, X has approximately a normal distribution, so X/n does also. The standard error of $\hat{p}$ is $\sigma_{\hat{p}} = \sqrt{p(1-p)/n}$. When the null hypothesis $H_0: p = p_0$ is true, $\sigma_{\hat{p}} = \sqrt{p_0(1-p_0)/n}$, so in standardizing $\hat{p}$ to get an approximately standard normal statistic, we do not have to estimate $\sigma_{\hat{p}}$. The large sample test statistic is

$$Z = \frac{\hat{p} - p_0}{\sqrt{p_0(1-p_0)/n}}$$

For each alternative to H_0: $p = p_0$ we can now specify the rejection region for an approximate level α test when both $np_0 \geq 5$ and $n(1 - p_0) \geq 5$.

Alternative Hypothesis	Rejection Region
H_a: $p > p_0$	$Z \geq z_\alpha$
H_a: $p < p_0$	$Z \leq -z_\alpha$
H_a: $p \neq p_0$	either $Z \geq z_{\alpha/2}$ or $Z \leq -z_{\alpha/2}$

Example 7.7 (continued)

X has observed value $x = 74$, so $\hat{p} = 74/100$. The null value is $p_0 = .8$ and the appropriate rejection region is $Z \leq -z_\alpha$. The computed Z is

$$z = \frac{.74 - .80}{\sqrt{(.8)(.2)/100}} = \frac{-.06}{.04} = -1.5$$

The P-value is $\Phi(-1.5) = .0668$, so even at level .05 the data is not significant. In the absence of strong evidence, the null hypothesis cannot be rejected, so the students are saved!

Power, β, and Choice of Sample Size

To compute $\pi(p) = P(\text{rejecting } H_0 \text{ when true parameter value } = p)$, the inequality in the rejection region is first rewritten so that $\hat{p}$ appears by itself on one side [for example, $Z \geq z_\alpha$ is equivalent to $\hat{p} \geq p_0 + z_\alpha\sqrt{p_0(1 - p_0)/n}$]. Now $\hat{p}$ can be standardized by subtracting p and dividing by $\sqrt{p(1 - p)/n}$. This yields

Alternative Hypothesis	Power Function $\pi(p)$
H_a: $p > p_0$	$1 - \Phi\left[\dfrac{p_0 - p + z_\alpha\sqrt{p_0(1 - p_0)/n}}{\sqrt{p(1 - p)/n}}\right]$
H_a: $p < p_0$	$\Phi\left[\dfrac{p_0 - p - z_\alpha\sqrt{p_0(1 - p_0)/n}}{\sqrt{p(1 - p)/n}}\right]$
H_a: $p \neq p_0$	$1 - \Phi\left[\dfrac{p_0 - p + z_{\alpha/2}\sqrt{p_0(1 - p_0)/n}}{\sqrt{p(1 - p)/n}}\right]$
	$+ \Phi\left[\dfrac{p_0 - p - z_{\alpha/2}\sqrt{p_0(1 - p_0)/n}}{\sqrt{p(1 - p)/n}}\right]$

Type II error probabilities are computed from $\beta(p) = 1 - \pi(p)$.

Example 7.7
(continued)

Using $\alpha = .05$ with $p_0 = .8$ and $p = .7$,

$$\pi(.7) = \Phi\left[\frac{.8 - .7 - 1.645\sqrt{(.8)(.2)/(100)}}{\sqrt{(.7)(.3)/(100)}}\right] = \Phi(.74) = .7704$$

Thus $\beta(.7) = 1 - .7704 = .2296$. This error probability is not as small as we might like, but $n = 100$ is not very large and .7 is not too far from the null value .8.

To find the sample size n for which the level α test also satisfies $\beta(p) = \beta$, equate $1 - \pi(p)$ to β and solve for n as in Section 7.1. This gives

$$n = \begin{cases} \left[\dfrac{z_\alpha\sqrt{p_0(1 - p_0)} + z_\beta\sqrt{p(1 - p)}}{p - p_0}\right]^2 & \text{one-tailed test} \\[3em] \left[\dfrac{z_{\alpha/2}\sqrt{p_0(1 - p_0)} + z_\beta\sqrt{p(1 - p)}}{p - p_0}\right]^2 & \text{two-tailed test} \end{cases}$$

Example 7.7
(continued)

For $n = 100$, $\beta(.7) = .2296$. If we wish $\beta(.7) = .1$, then $z_\beta = z_{.1} = 1.28$, so

$$n = \left[\frac{1.645\sqrt{(.8)(.2)} + 1.28\sqrt{(.7)(.3)}}{.7 - .8}\right]^2 \doteq 155$$

Exercises / Section 7.3

1. Let p denote the true proportion of Budweiser drinkers who can distinguish their beer from Schlitz. If X denotes the number of correct identifications in a sample of 100 Bud drinkers and we observe $x = 57$, test $H_0: p = .5$ versus $H_a: p \neq .5$ using a level .05 test.

2. The city council in a large city has become concerned about the trend toward exclusion of renters with children in apartments within the city. The housing coordinator has decided to select a random sample of 125 apartments and determine for each whether or not children would be permitted. Let p = the true proportion of apartments which prohibit children. If $p > .75$ the council will consider appropriate legislation.
   a. If 102 of the 125 sampled exclude renters with children, test $H_0: p = .75$ versus $H_a: p > .75$

using a level .05 test.
   b. What is the probability of a type II error when $p = .8$ (not rejecting H_0 when $p = .8$)?
   c. How many apartments would have to be sampled to ensure that $\beta(.8) = .1$?

3. A telephone company is trying to decide whether some new lines in a large community should be installed underground. Because a small surcharge will be added to telephone bills to pay for the extra installation costs, the company has decided to survey customers and proceed only if the survey strongly indicates that more than 60% of all customers favor underground installation. If 118 of 160 customers surveyed favor underground installation in spite of the surcharge, what should the company do? Test the relevant hypotheses using $\alpha = .05$.

4. A university library ordinarily has a complete shelf

inventory done once every year. Because of new shelving rules instituted the previous year, the head librarian feels that it may be possible to save money by postponing the inventory. The librarian decides to select at random 800 books from the library's collection and have them searched in a preliminary manner. If evidence indicates strongly that the true proportion of misshelved or unlocatable books is less than .02, then the inventory will be postponed.

a. Among the 800 books searched, 12 were misshelved or unlocatable. Test the relevant hypothesis and advise the librarian what to do (use $\alpha = .05$).

b. If the true proportion of misshelved and lost books is actually .01, what is the probability that the inventory will be (unnecessarily) taken?

c. If the true proportion is .05, what is the probability that the inventory will be postponed?

5. A package delivery service advertises that at least 90% of all packages brought in by 9:00 A.M. for delivery in the same city will be delivered by noon.

a. If only 65 of 80 independently mailed packages are delivered within the promised time period, does this strongly contradict the advertised claim? Test the relevant hypothesis using $\alpha = .01$.

b. What is the P-value of the data in (a)?

c. If when the true percentage of deliveries within the promised period is only 75%, we wish to detect such a departure from the service's claim with probability .95 (using a level .01 test), how many packages should be mailed?

6. The incidence of a certain type of chromosome error in the U.S. adult male population is believed to be 1 in 80. A random sample of 600 individuals in U.S. penal institutions reveals 12 who have such errors. Can it be concluded that the incidence rate of such errors among prisoners differs from the presumed rate for the entire adult male population? Use a level .001 test. What is the P-value of the data?

7. Because of variability in the manufacturing process, the actual yielding point of a sample of mild steel subjected to increasing stress will usually differ from the theoretical yielding point. Let p denote the true proportion of samples which yield before their theoretical yielding point. If on the basis of a sample it can be concluded that more than 20% of all specimens yield before the theoretical point, the production process will have to be modified.

a. If 15 of 60 specimens yield before the theoretical point, what is the P-value when the appropriate test is used, and what would you advise the company to do?

b. If the true percentage of "early yields" is actually 50% (so that the theoretical point is the median of the yield distribution) and a level .01 test is used, what is the probability that the company concludes a modification of the process is necessary?

8. As discussed in Exercise 6.1.13, a randomized response technique is often used when truthful responses might stigmatize respondents. To see whether or not homeowners in a large city were adhering to regulations on water usage during a drought, a sample of 200 homeowners was obtained. Each was allowed to select a card from a well-mixed deck; half the cards asked, "Have you violated the regulations?" and the other half asked, "Is the last digit of your social security number one of the digits 0, 1, or 2?" With p = the true proportion of violators, $\lambda = P(\text{yes response})$, and Y = the number of yes responses (a binomial r.v. with parameters $n = 200$ and λ), $p = 2\lambda - .3$, so p can be estimated by $\hat{p} = 2\hat{\lambda} - .3$ where $\hat{\lambda} = Y/n$. Use this information to describe a large sample test for $H_0: p = .1$ versus $H_a: p > .1$, and carry out the test at level .05 when the number of yes responses is 50.

7.4 The t Test

When testing hypotheses about a population mean μ with the true value of σ unknown, the test procedures of Section 7.2 can be used provided that n is large. Those procedures are valid (approximate level α) whether or not the population being sampled is normal. In expensive or time-consuming experiments it is frequently not possible to obtain a large sample size. Then since the CLT can no longer be used, to obtain a valid test procedure we must make some specific assumptions about the nature of the underlying population distribution. The most reasonable assumption in a wide variety of applications is that of normality.

Assumption: The population of interest is normal, so that $X_1, \ldots, X_n$ constitutes a random sample from a normal distribution with both μ and σ unknown.

Since many population distributions are quite well approximated by a normal curve, a procedure derived under this assumption will have wide applicability. If this assumption is unreasonable, there are several possible courses of action. One can either specify another distributional family (Weibull, gamma, lognormal, Poisson, and the like) in place of the normal and use some general principles of test construction to obtain appropriate procedures, or else one can use a distribution-free test (see Chapter 15) that is valid for many underlying distributions.

The *t* Distribution

The key result underlying the procedures in Section 7.2 was that for large n, the r.v. $Z = (\bar{X} - \mu)/(S/\sqrt{n})$ has approximately a standard normal distribution. When n is small, S is no longer likely to be close to σ, so the variability in the distribution of Z arises from randomness both in the numerator and in the denominator. This implies that the probability distribution of $(\bar{X} - \mu)/(S/\sqrt{n})$ will be more spread out than the standard normal distribution. The result on which test procedures are based introduces a new family of probability distributions called the family of t distributions.

Theorem: When $\bar{X}$ is the mean of a random sample of size n from a normal distribution with mean μ, the random variable

$$T = \frac{\bar{X} - \mu}{S/\sqrt{n}} \tag{7.12}$$

has a probability distribution called a t distribution with $n - 1$ degrees of freedom.

Properties of *t* Distributions

Before applying this theorem, we must first discuss properties of t distributions. While the variable of interest is still $(\bar{X} - \mu)/(S/\sqrt{n})$, we now denote it by T to emphasize that it does not have a standard normal distribution when n is small. Recall that a normal distribution is governed by two parameters, the mean μ and the standard deviation σ. A t distribution is governed by only one parameter, called the **number of degrees of freedom** of the distribution, abbreviated by d.f. We denote this parameter by the Greek letter ν. Possible values of ν are the positive integers $1, 2, 3, \ldots$. Each different value of ν corresponds to a different t distribution. The term "degrees of freedom" for t refers to the fact that the estimator S in the denominator of t involves the n quantities $X_1 - \overline{X}, X_2 - \overline{X}, \ldots, X_n - \overline{X}$; only $n - 1$ of these are actually independently determined, since $\Sigma (X_i - \overline{X}) = 0$, so it is said that the estimator S is based on only $n - 1$ d.f.

For any fixed value of the parameter ν, the density function which specifies the associated t curve has an even more complicated appearance than the normal density function. Fortunately we need concern ourselves only with several of the more important features of these curves.

Properties of t Distributions: Let t_ν denote the density function curve for ν degrees of freedom.

   1. Each t_ν curve is bell-shaped and centered at 0.
   2. As ν increases, the spread of the corresponding t_ν curve decreases.
   3. Each t_ν curve is more spread out than the standard normal (z) curve.
   4. As $\nu \rightarrow \infty$, the sequence of t_ν curves approaches the standard normal curve (so the z curve is often called the t curve with d.f. $= \infty$).

Figure 7.8 illustrates several of these properties for selected values of ν.

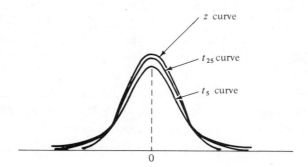

z curve

t_{25} curve

t_5 curve

0

Figure 7.8 t_ν and z curves

Critical Values of t Distributions

Since we want to use a T statistic to determine a level α test in the same way that Z was previously used, it is necessary to establish notation analogous to z_α for the t distribution.

Notation: Let $t_{\alpha,\nu}$ = the number on the measurement axis for which the area under the t curve with ν degrees of freedom to the right of $t_{\alpha,\nu}$ is α.

Thus $t_{\alpha,\nu}$ is the $100(1 - \alpha)$ percentile of the t distribution with ν d.f. A pictorial definition of $t_{\alpha,\nu}$ appears in Figure 7.9.

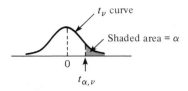

Figure 7.9 A pictorial definition of $t_{\alpha,\nu}$

Table A.5 in the Appendix gives $t_{\alpha,\nu}$ for selected values of α and ν. The columns of the table correspond to different values of α. To obtain $t_{.05,15}$, go to the $\alpha = .05$ column, look down to the $\nu = 15$ row, and read $t_{.05,15} = 1.753$. Similarly, $t_{.05,22} = 1.717$ (.05 column, $\nu = 22$ row), and $t_{.01,22} = 2.508$.

The values of $t_{\alpha,\nu}$ exhibit regular behavior as we move across a row or down a column. For fixed ν, $t_{\alpha,\nu}$ increases as α decreases, since we must move further to the right of zero to capture area α in the tail. For fixed α, as ν is increased (that is, as we look down any particular column of the t table) the value of $t_{\alpha,\nu}$ decreases. This is because a larger value of ν implies a t distribution with smaller spread, so it is not necessary to go so far from zero to capture tail area α. Furthermore, $t_{\alpha,\nu}$ decreases more slowly as ν increases. Consequently, the table skips from $\nu = 30$ to 40 to 60 to ∞; many tables go directly from $\nu = 30$ to ∞. Because t_{∞} is the standard normal curve, the familiar z_{α} values appear in the last row of the table. The rule of thumb suggested earlier for deciding whether the Central Limit Theorem could be used (if $n > 30$) comes from the approximate equality of the standard normal and t distributions for $\nu \geq 30$.

Description of the t Test

Under the basic assumption of this section, a level α test for $H_0: \mu = \mu_0$ consists of using

$$\text{Test statistic: } T = \frac{\bar{X} - \mu_0}{S/\sqrt{n}} \tag{7.13}$$

along with the appropriate rejection region for each alternative of interest:

Alternative Hypothesis	Rejection Region for a Level α Test
$H_a: \mu > \mu_0$	$T \geq t_{\alpha,n-1}$
$H_a: \mu < \mu_0$	$T \leq -t_{\alpha,n-1}$
$H_a: \mu \neq \mu_0$	either $T \geq t_{\alpha/2,n-1}$ or $T \leq -t_{\alpha/2,n-1}$

That these are level α tests follows directly from the theorem of this section. Also, when n is large the t tests become the Z tests of Section 7.2, since then $t_{\alpha,\,n-1} \doteq z_{\alpha}$.

Example 7.8

In order to test gasoline mileage performance for a new version of one of its compact cars, an automobile manufacturer selected six nonprofessional drivers who were willing to drive a car from Phoenix to Los Angeles. The miles per gallon figures for each of the six cars at the conclusion of the trip were 27.2, 29.3, 31.5, 28.7, 30.2, and 29.6. Based on this data, the manufacturer wishes to advertise that this type of car should get at least 30 mpg on long trips. Assuming that the mpg distribution is normal, would the above data contradict this claim?

The parameter of interest is

μ = true average mpg for this type of car on such a trip

The manufacturer will claim that $\mu \geq 30$ unless the data indicates otherwise, so the appropriate alternative is H_a: $\mu < 30$. The summary statistics are $\Sigma\, x_i = 176.5$ and $\Sigma\, x_i^2 = 5202.47$, so

$$\bar{x} = \frac{176.5}{6} = 29.42, \quad s^2 = \frac{5202.47 - (176.5)^2/6}{6-1} = 2.086, \quad s = 1.44$$

The test statistic is

$$T = \frac{\bar{X} - \mu_0}{S/\sqrt{n}} = \frac{\bar{X} - 30}{S/\sqrt{6}}$$

At level $\alpha = .05$, H_0 is rejected if $T \leq -t_{\alpha,n-1} = -t_{.05,5} = -2.015$. Since the computed value of T is $-.99$, which is not ≤ -2.015, H_0 would not be rejected at level .05 (nor even at level .1, since $-t_{.1,5} = -1.476$), so the company can make the desired claim.

Example 7.9

A chain of high fidelity stores has received an advanced shipment of a new model of stereo receiver. To determine how actively to promote the receiver, its specified performance characteristics are studied. One of the manufacturer's specifications is that the receiver generates a maximum power of 65 watts at 8 ohms. A sample of eight receivers yields a sample average maximum power of 63.1 watts with a sample standard deviation of 1.7 watts. Does this data suggest that the true average maximum power is something other than what the manufacturer has specified? Assume that the distribution being sampled is normal. The null value of μ = true average maximum power is $\mu_0 = 65$, and we are concerned with a departure from this value in either direction, so the alternative is H_a: $\mu \neq 65$. The appropriate rejection region is two-tailed; at level .05, H_0 is rejected if either $T \geq t_{.025,7}$ or $T \leq -t_{.025,7}$. Since $t_{.025,7} = 2.365$ and the computed value of T is

$$t = \frac{\bar{x} - \mu_0}{s/\sqrt{n}} = \frac{63.1 - 65}{1.7/\sqrt{8}} = -3.16$$

H_0 is rejected at level .05 (because $-3.16 \leq -2.365$), and we conclude that $\mu \neq 65$.

P-Values for the t Test

Because for each ν the t table contains only seven $t_{\alpha,\nu}$ values, exact P-values cannot be computed (though there are more extensive t tables). Instead, an upper and lower bound can be computed as follows. If $|t|$ falls between values in the α_1 and α_2 columns ($\alpha_1 < \alpha_2$) of the $\nu = n - 1$ row of the table, then $\alpha_1 < P < \alpha_2$ if the test is one-tailed or $2\alpha_1 < P < 2\alpha_2$ if the test is two-tailed. If $|t| > t_{.0005,n-1}$, we can state only that $P < .0005$ (one-tailed) or $.001$ (two-tailed), while if $|t| < t_{.1,n-1}$, then $P > .1$ (one-tailed) or $.2$ (two-tailed).

Example 7.9
(continued)

With $\nu = n - 1 = 7$, $|t| = 3.16$, which satisfies $2.998 < 3.16 < 3.499$. Since the test was one-tailed, we conclude that $.005 < P < .01$.

Power, β, and Choice of Sample Size

For the Z tests in Section 7.1, derived under the assumption that σ was known, formulas were given for $\pi(\mu)$ and the sample size n necessary to ensure that the level α test also had $\beta(\mu) = \beta$. Unfortunately there are no such simple formulas for the t test. This is because the probability distribution of t when $\mu \neq \mu_0$ is very complicated.

However, research workers have been able to construct graphs of the power functions of the one- and two-tailed t tests. These curves appear in Figure 7.10. To understand how these curves are used, first note that in Section 7.1 both power and sample size for an alternative value μ are functions not just of the absolute difference $|\mu_0 - \mu|$ but of $d = |\mu_0 - \mu|/\sigma$. The quantity d is the distance between μ_0 and μ measured in units of standard deviation. Intuitively, when $|\mu_0 - \mu| = 10$, the power will be higher for $\sigma = 2$ than $\sigma = 5$, since in the first case the alternative of interest lies 5 standard deviations from μ_0 while in the second case the distance is only 2.

The fact that power depends upon d rather than just $|\mu - \mu_0|$ is unfortunate, since to use the power curves one must have some preliminary idea of the true value of σ. A conservative (large) guess for σ will yield a conservative value of the power and a conservative estimate of the sample size necessary to control $\beta(\mu)$. Such a guess is often the result of prior information about or experience with similar situations. Because when n is large the t tests become the large sample Z tests, the power curves of Figure 7.10 can be used in connection with these latter tests.

Example 7.10

An engineering company located in a large city has run out of space at its present location, so is interested in obtaining space for expansion at a nearby location. The company is looking at several existing buildings. Because there will be a good deal of interaction between engineers at the existing facility and at the new facility, the company plans to operate a shuttle bus between the two facilities during working hours. It is therefore decided that any prospective location for which the average shuttle time would exceed 20 min should be eliminated. For a prospective location, letting $\mu = $ true average shuttle time, the company would like to test $H_0: \mu = 20$ versus $H_a: \mu > 20$ by sending a shuttle bus on some trial runs. The

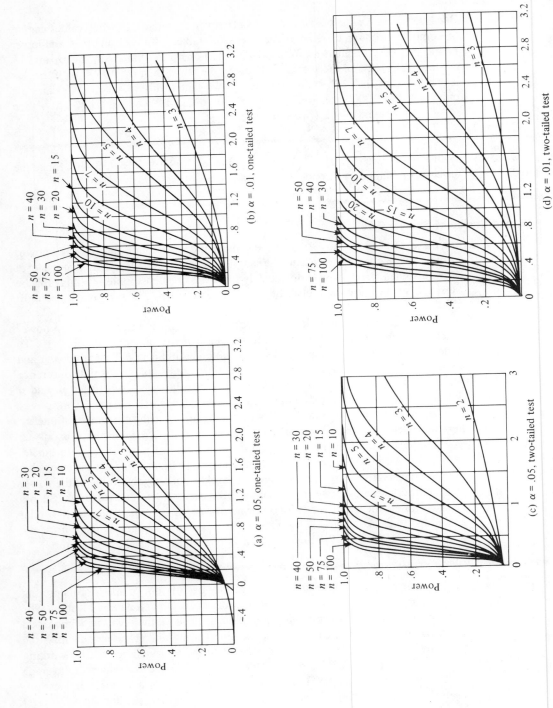

Figure 7.10 Power curves for the *t* test (Adapted from "Operating Characteristics for the Common Statistical Tests of Significance," Annals Math. Stat., 1946).

company would like to use $\alpha = .01$, and would also like $\beta(25) = .05$ in order to ensure that if the true average commuting time is as much as 25 min, it is quite likely that H_0 will be rejected and the prospective location eliminated from consideration. How many trial runs of the shuttle bus are necessary?

Suppose that from past experience a value of $\sigma = 5$ minutes is reasonable. Then $d = |20 - 25|/5 = 1$. Each curve in the figures is the graph of the power function of the t test for a different sample size. Since power $= 1 - \beta(\mu)$, we therefore wish power $= .95$ when $d = 1$. The horizontal axis specifies d and the vertical axis specifies power, so after locating the point $(1, .95)$ on Figure 7.10(a) the curve which comes closest to this point determines the sample size (if the point lies between two curves, an intermediate sample size should be selected by interpolation). For the present example, the $n = 20$ curve passes through this point, so 20 trial runs of the shuttle bus are necessary to achieve the desired α and β.

Exercises / Section 7.4

1. Determine the values of the following quantities:
   a. $t_{.1,14}$ b. $t_{.05,14}$ c. $t_{.05,20}$ d. $t_{.05,40}$
   e. the ninety-ninth percentile of the t distribution with 20 d.f.
   f. the fifth percentile of the t distribution with 24 d.f.
   g. $P(T \leq 3.408)$ when T has a t distribution with 28 d.f.
   h. $P(-2.681 \leq T \leq 2.681)$ when T has a t distribution with 12 d.f.

2. The amount of shaft wear (.0001 in.) after a fixed mileage was determined for each of $n = 8$ internal combustion engines having copper lead as a bearing material, resulting in $\bar{x} = 3.72$ and $s = 1.25$.
   a. Assuming that the distribution of shaft wear is normal with mean μ, use the t test at level .05 to test $H_0: \mu = 3.50$ versus $H_a: \mu > 3.50$.
   b. Using $\sigma = 1.25$, what is the power $\pi(\mu)$ and the type II error probability $\beta(\mu)$ of the test for the alternative $\mu = 4.00$?

3. The following radiation readings (milliroentgens per hour) were obtained from television display areas in a sample of 10 department stores ("Many Set Color TV Lounges Show Highest Radiation," *J. Environmental Health*, 1969, pp. 359–360):

   .40, .48, .60, .15, .50, .80, .50, .36, .16, .89

   The recommended limit for this type of radiation exposure is .5 mr/hr. Assuming that the observations come from a normal distribution with mean

μ (the true average amount of radiation in television display areas in all department stores), test $H_0: \mu = .5$ versus $H_a: \mu > .5$ using a level .1 test.

4. In an experiment designed to measure the time necessary for an inspector's eyes to become used to the reduced amount of light necessary for penetrant inspection, the sample average time for $n = 9$ inspectors was 6.32 sec with a sample standard deviation of 1.65 sec. It has previously been assumed that the average adaptation time was at least 7 sec. Assuming adaptation time to be normally distributed, does the data contradict prior belief? Use the t test with $\alpha = .1$.

5. The times of first sprinkler activation for a series of tests with fire prevention sprinkler systems using an aqueous film-forming foam were (in sec)

   27, 41, 22, 27, 23, 35, 30, 33, 24, 27, 28, 22, 24

   (see "Use of AFFF in Sprinkler Systems," *Fire Technology*, 1976, p. 5). The system has been designed so that true average activation time is at most 25 sec under such conditions. Does the data strongly contradict the validity of this design specification?
   a. Test the relevant hypotheses using $\alpha = .05$. What assumption are you making about the distribution of activation time?
   b. Obtain an upper and lower bound for the P value of the data.
   c. Suppose that before the experiment had been

performed, the experimenter has guessed σ to be approximately 6 sec. Approximately what sample size would then be appropriate for achieving power .95 for the alternative $\mu = 30$ (using a level .05 test)?

6. A sequence of tests designed to study dielectric failure due to partial discharges (corona) occurring in electrical equipment yielded the following observations on discharge maintenance voltage (mJ per cycle):

3.1, 3.3, 2.8, 2.6, 3.4, 3.2, 3.0, 2.5, 3.5

a. Does the data suggest that the true average discharge maintenance voltage is something other than 3.00? Use the t test at level .05.

b. What can be said about the P-value of the data?

7. In an experiment to study the effects of longitudinal acceleration on passengers in ground transportation vehicles, the acceleration before which loss of balance occurred was measured for a sample of 12 individuals standing facing forward without support. The sample average was 5.3 ft/sec with a sample standard deviation of 1.2. Does this data strongly contradict the prior belief that on average loss of balance would occur with an acceleration of at most 5.0? Use a level .01 test.

8. The National Bureau of Standards had previously reported the value of Se content in NBS orchard leaves to be .080 ppm. The paper "A Neutron Acti-

vation Method for Determining Submicrogram Selenium in Forage Grasses" (*Soil Science Soc. Amer. J.*, 1978, pp. 57–60) reported the following Se content for six different determinations:

.072, .073, .080, .078, .088, .080

Does the data contradict the previously reported value?

a. Test the relevant hypotheses using $\alpha = .01$.

b. Suppose that prior to the experiment, a value of $\sigma = .005$ had been assumed. How many determinations would then have been appropriate to obtain power .90 for the alternative $\mu = .075$?

9. A particular type of car is supposed to be brought back to the dealer for free servicing after 3000 miles. The dealer has become concerned that new car owners may be waiting too long to bring them in, so decides to select a random sample of recent purchases. Let μ = the true average mileage before the first service. If data strongly indicates that $\mu > 3000$, then the dealer will begin refusing to service cars which have too many more miles than 3000.

a. If $n = 41$, $\bar{x} = 3119$, $s = 295$, use a level .01 test to decide what the dealer should do. What is the approximate P-value?

b. Is it really necessary to assume that mileage before first service has approximately a normal distribution? Justify your assertion.

7.5 Test Procedures for a Population Variance

Although hypotheses concerning a population variance or standard deviation are usually of less interest than hypotheses about a mean, there are occasions when procedures for testing such hypotheses are needed. We will present both a procedure based on the assumption of normality and a large sample procedure. An important pedogogical reason for studying the normal case is that it involves use of the

Figure 7.11 Graphs of chi-squared density functions

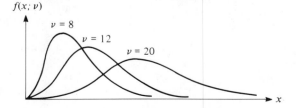

$f(x; \nu)$

$\nu = 8$

$\nu = 12$

$\nu = 20$

x

chi-squared distribution, which is needed in connection with other inferential procedures.

The Assumption of Normality and the Chi-Squared Distribution

When the population is normal, test procedures for σ^2 are based on the following result concerning the sample variance S^2.

Theorem: Let $X_1, X_2, \ldots, X_n$ be a random sample from a normal distribution with parameters μ and σ^2. Then the random variable

$$\frac{(n-1)S^2}{\sigma^2} = \frac{\Sigma\,(X_i - \bar{X})^2}{\sigma^2} \tag{7.14}$$

has a chi-squared (χ^2) probability distribution with $n - 1$ degrees of freedom (d.f.).

As discussed in Section 4.4, the chi-squared distribution is a continuous probability distribution with a single parameter ν, called the number of degrees of freedom, with possible values 1, 2, 3, ... The graphs of several χ^2 p.d.f.'s are illustrated in Figure 7.11. Each p.d.f. $f(x; \nu)$ is positive only for $x > 0$, and each has a positive skew (long upper tail), though the distribution moves rightward and becomes more symmetric as ν increases.

To specify inferential procedures which use the chi-squared distribution, we need notation analogous to $t_{\alpha,\nu}$ for the t distribution.

Notation: Let $\chi^2_{\alpha,\nu}$ denote the point on the measurement axis such that α of the area under the chi-squared curve with ν d.f. lies to the right of $\chi^2_{\alpha,\nu}$.

Thus $\chi^2_{\alpha,\nu}$ is the $100(1 - \alpha)$ percentile of the chi-squared distribution with ν d.f., as illustrated in Figure 7.12.

Figure 7.12 $\chi^2_{\alpha,\nu}$ notation illustrated

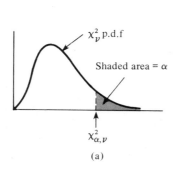

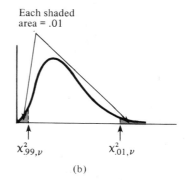

(a) (b)

Because the t distribution is symmetric, it was necessary to tabulate only upper-tail critical values ($t_{\alpha,\nu}$ for small values of α). The chi-squared distribution is not symmetric, so Appendix Table A.6 contains values of $\chi^2_{\alpha,\nu}$ both for α near 0 and near 1, as illustrated in Figure 7.12(b). For example, $\chi^2_{.025,14} = 26.119$ and $\chi^2_{.95,20}$ (the fifth percentile) $= 10.851$.

Test Procedures When the Population Is Normal

According to the above theorem, when the sample is drawn from a normal population and the null hypothesis $H_0: \sigma^2 = \sigma_0^2$ is true (equivalently, $H_0: \sigma = \sigma_0$), the **test statistic**

$$\chi^2 = \frac{(n-1)S^2}{\sigma_0^2} \tag{7.15}$$

has a χ^2 distribution with $n - 1$ d.f. The rejection regions which specify level α tests for the three commonly encountered alternatives are as follows:

Alternative Hypothesis	Rejection Region
$H_a: \sigma^2 > \sigma_0^2$	$\chi^2 \geq \chi^2_{\alpha,n-1}$
$H_a: \sigma^2 < \sigma_0^2$	$\chi^2 \leq \chi^2_{1-\alpha,n-1}$
$H_a: \sigma^2 \neq \sigma_0^2$	either $\chi^2 \geq \chi^2_{\alpha/2,n-1}$ or $\chi \leq \chi^2_{1-\alpha/2,n-1}$

For hypotheses about μ, we looked at the *difference* between $\bar{X}$ and μ_0, whereas for σ^2 we examine the *ratio* of S^2 to σ_0^2, rejecting H_0 if the value of this ratio differs too much from what we expect when H_0 is true.

Example 7.11   To ensure reasonably uniform characteristics for a particular application, it is desired that the true standard deviation of the softening point of a certain type of petroleum pitch be at most .50 °C. The softening points of 10 different samples were determined, yielding a sample standard deviation of .58 °C. Does this strongly contradict the uniformity specification?

To answer this question, we use a level .01 test for $H_0: \sigma^2 = .25$ versus $H_a: \sigma^2 > .25$, assuming that softening point is normally distributed. H_0 will be rejected if $\chi^2 \geq \chi^2_{.01,9} = 21.67$. Since $s^2 = (.58)^2 = .336$,

$$\chi^2 = \frac{(n-1)s^2}{\sigma_0^2} = \frac{(9)(.336)}{.25} = 12.11$$

This computed value does not exceed the critical value 21.67, so at level .01 H_0 is not rejected. The true standard deviation of the softening point appears to be at most .50 °C.

A set of curves giving P(type II error) for these chi-squared tests appears in *Engineering Statistics* by Bowker and Lieberman.

Large-Sample Procedures

When the underlying population is not necessarily normal but n is large, test procedures are based on the following theoretical result.

> **Theorem:** Let $X_1, \ldots, X_n$ be a random sample from a distribution with standard deviation σ. Then if n is large, the sample standard deviation S has approximately a normal distribution with $E(S) \doteq \sigma$ and $\text{Var}(S) \doteq \sigma^2/2n$, so that
>
> $$Z = \frac{S - \sigma}{\sigma/\sqrt{2n}} \tag{7.16}$$
>
> has approximately a standard normal distribution.

For testing $H_0: \sigma^2 = \sigma_0^2$, σ_0 is now used in place of σ in (7.16).

> Test statistic: $Z = \dfrac{S - \sigma_0}{\sigma_0/\sqrt{2n}}$
>
> | *Alternative Hypothesis* | *Rejection Region* |
> |---|---|
> | $H_a: \sigma^2 > \sigma_0^2$ | $Z \geq z_\alpha$ |
> | $H_a: \sigma^2 < \sigma_0^2$ | $Z \leq -z_\alpha$ |
> | $H_a: \sigma^2 \neq \sigma_0^2$ | either $Z \geq z_{\alpha/2}$ or $Z \leq -z_{\alpha/2}$ |

Example 7.12 Before agreeing to purchase a large order of polyethylene sheaths for a particular type of high-pressure oil-filled submarine power cable, a company wants to see conclusive evidence that the true standard deviation of thickness is at most .050 mm. Hence the burden of proof is on the manufacturer, so the relevant hypotheses are $H_0: \sigma^2 = .0025$ versus $H_a: \sigma^2 < .0025$. A sample of 40 sheaths is obtained, and the sample standard deviation is found to be $s = .043$ mm. For $\alpha = .01$, $-z_{.01} = -2.33$, and

$$z = \frac{s - .05}{.05/\sqrt{80}} = -1.25$$

Because -1.25 is not ≤ -2.33, H_0 is not rejected, so the evidence is not conclusive enough to justify the purchase.

Exercises / Section 7.5

1. Determine the values of the following quantities:
   a. $\chi^2_{.1,15}$
   b. $\chi^2_{.1,25}$
   c. $\chi^2_{.01,25}$
   d. $\chi^2_{.005,25}$
   e. $\chi^2_{.99,25}$
   f. $\chi^2_{.995,25}$
   g. The ninety-fifth percentile of the chi-squared distribution with $\nu = 10$.
   h. The fifth percentile of the chi-squared distribution with $\nu = 10$.
   i. $P(10.98 \leq \chi^2 \leq 36.78)$ where χ^2 is a chi-squared r.v. with $\nu = 22$.

2. Twenty refrigerator/freezers of a certain brand were randomly selected and a temperature measurement made in each at the center of the freezer compartment. The sample standard deviation of the 20 measurements was computed to be 1.35 °F. Does this data provide strong evidence for concluding that the true standard deviation of temperature is more than 1.00 °F? Test using $\alpha = .05$, making explicit any assumptions necessary to carry out the analysis.

3. Does the sprinkler activation time data of Exercise 7.4.5 provide strong evidence for concluding that the true standard deviation of activation time is less than 7.5 sec? Use the χ^2 test with $\alpha = .05$.

4. A new version of a qualifying exam for certain government jobs has been designed so that the scores are normally distributed with one standard deviation equaling 25 points. A random sample of 29 individuals taking the exam yielded a sample standard deviation of $s = 21.3$. At level $\alpha = .1$, does this data allow one to conclude that the true standard deviation is something other than 25?

5. Determination of the lead concentration in air samples from a downtown parking garage on each of $n = 32$ different weekdays yielded a sample standard deviation of 3.3 $\mu g/m^3$. Can it be concluded that the true standard deviation of lead concentration exceeds 2.5? Use the large sample test with $\alpha = .001$.

6. a. What is the P-value for the data in Exercise 5 analyzed using the large sample test?
   b. Suppose that the χ^2 test is used on this data. Use the χ^2 table to make an educated guess of the P-value.
   c. Does it look as though the two tests provide roughly the same amount of evidence against H_0? Why or why not?

7. The following observations were made on fracture toughness of base plate of 18% nickel maraging steel ["Fracture Testing of Weldments," *ASTM Special Publ. No. 381*, 1965, pp. 328–356 (in ksi$\sqrt{\text{in.}}$, given in increasing order)]:

   69.5, 71.9, 72.6, 73.1, 73.3, 73.5, 75.5, 75.7, 75.8, 76.1, 76.2, 76.2, 77.0, 77.9, 78.1, 79.6, 79.7, 79.9, 80.1, 82.2, 83.7, 93.7

   Suppose that for a particular application, the true standard deviation should be reasonably small, and it is decided that 5 ksi$\sqrt{\text{in.}}$ is the dividing point between satisfactory and unsatisfactory values of σ. It would be a serious error to use the steel if the true value of σ were unsatisfactory, so the data should be quite convincing before it is used. Assuming fracture toughness to be normally distributed, test the appropriate hypotheses.

7.6 Some Comments on Selecting a Test Procedure

Once the experimenter has decided on the question of interest and the method for gathering data (the design of the experiment), construction of an appropriate test procedure consists of three distinct steps:

1. Specify a test statistic (the function of the observed values which will serve as the decision maker).
2. Decide on the general form of the rejection region (typically reject H_0 for suitably large values of the test statistic, reject for suitably small values, or reject for both small and large values).
3. Select the specific numerical critical value or values which will separate the rejection region from the acceptance region (by obtaining the distribution of the test statistic when H_0 is true, and then selecting a level of significance).

In the examples thus far, both steps 1 and 2 above were carried out in an ad hoc manner through intuition. For example, when the underlying population was assumed normal with mean μ and known σ, we were led from $\bar{X}$ to the standardized test statistic

$$Z = \frac{\bar{X} - \mu_0}{\sigma/\sqrt{n}}$$

For testing $H_0: \mu = \mu_0$ versus $H_a: \mu > \mu_0$, intuition then suggested rejecting H_0 when Z was large. Finally the critical value was determined by specifying the level of significance α and using the fact that Z has a standard normal distribution when H_0 is true. The reliability of the test in reaching a correct decision can be assessed by studying the power function.

Issues which ought to be considered in carrying out steps 1–3 encompass the following questions:

1. What are the practical implications and consequences of choosing a particular level of significance once the other aspects of a test procedure have been determined?
2. Does there exist a general principle, not dependent just on intuition, which can be used to obtain best or good test procedures?
3. When there exist two or more tests which are appropriate in a given situation, how can the tests be compared to decide which should be used?
4. If a test is derived under specific assumptions about the distribution or population being sampled, how well will the test procedure work when the assumptions are violated?

Statistical versus Practical Significance

While the process of reaching a decision by using the methodology of classical hypothesis testing involves selecting a level of significance and then accepting or rejecting H_0 at that level α, simply reporting the α used and the decision reached conveys little of the information contained in the sample data. Especially when the results of an experiment are to be communicated to a large audience, rejection of H_0 at level .05 will be much more convincing if the observed value of the test statistic greatly exceeds the 5% critical value than if it barely exceeds that value. This is precisely what led to the notion of P-value as a way of reporting significance without imposing a particular α on others who might wish to draw their own conclusions.

Even if a P-value is included in a summary of results, however, there may be difficulty in interpreting this value and in making a decision. This is because a small P-value, which would ordinarily indicate **statistical significance** in that it would strongly suggest rejection of H_0 in favor of H_a, may be the result of a large sample size in combination with a departure from H_0 which has little **practical significance.** In many experimental situations, only departures from H_0 of large magnitude would be worthy of detection, whereas a small departure from H_0 would have little practical significance.

Consider as an example testing $H_0: \mu = 100$ versus $H_a: \mu > 100$ where μ is the mean of a normal population with $\sigma = 10$. Suppose that a true value of $\mu = 101$ would not represent a serious departure from H_0 in the sense that not

Table 7.1 *An Illustration of the Effect of Large Sample Size on P-Values and Power*

n	P-value when $\bar{x} = 101$	$\pi(101)$ for level .01 test
25	.3085	.0418
100	.1587	.0918
400	.0228	.3707
900	.0013	.7486
1600	.0000335	.9525
2500	.000000297	.9962
10000	7.69×10^{-24}	.9989

rejecting H_0 when $\mu = 101$ would be a relatively inexpensive error. For a reasonably large sample size n, this μ would lead to an $\bar{x}$ value near 101, so we would not want this sample evidence to argue strongly for rejection of H_0 when $\bar{x} = 101$ is observed. For various sample sizes, Table 7.1 records both the P-value when $\bar{x} = 101$ and also the probability of rejecting H_0 at level .01 when $\mu = 101$ [the power function $\pi(\mu)$ evaluated at $\mu = 101$].

The second column in Table 7.1 shows that even for moderately large sample sizes, the P-value of $\bar{x} = 101$ argues very strongly for rejection of H_0, whereas the observed $\bar{x}$ itself suggests that in practical terms the true value of μ differs little from the null value $\mu_0 = 100$. The third column points out that even when there is little practical difference between the true μ and the null value, for a fixed level of significance a large sample size will almost always lead to rejection of the null hypothesis at that level. To summarize, *one must be especially careful in interpreting evidence when the sample size is large, since any small departure from* H_0 *will almost surely be detected by a test, yet such a departure may have little practical significance.*

The Likelihood Ratio Principle

We saw in Chapter 6 that once an appropriate family of distributions is chosen as a set of population models, the method of maximum likelihood can be used to construct estimators of parameters. In a similar fashion, the likelihood ratio principle provides a general method for constructing test procedures.

Example 7.13 Let X denote the number of TV sets received by a repair shop during a one-week period which need a particular part. Suppose that X has a Poisson distribution with parameter λ, so that the p.m.f. of X is

$$p(x; \lambda) = \begin{cases} \dfrac{e^{-\lambda}\lambda^x}{x!} & x = 0, 1, 2, \ldots \\ 0 & \text{otherwise} \end{cases}$$

Let $X_1, X_2, \ldots, X_{10}$ represent the numbers of such sets received during 10 different weeks, assumed to be a random sample from the distribution $p(x; \lambda)$. On the basis of observed values $X_1, \ldots, X_{10}$ we wish to test $H_0: \lambda = 1$ versus $H_a: \lambda = 2$

(we will shortly consider the more realistic hypotheses $H_0: \lambda \leq 1$ versus $H_a: \lambda > 1$).

The joint p.m.f. of the sample is, by independence,

$$f(x_1, \ldots, x_{10}; 1) = \frac{(e^{-1})^{10} (1)^{x_1+\cdots+x_{10}}}{x_1! \cdots x_{10}!} \quad \text{when} \quad \lambda = 1, \text{ and}$$

$$f(x_1, \ldots, x_{10}; 2) = \frac{(e^{-2})^{10} (2)^{x_1+\cdots+x_{10}}}{x_1! \cdots x_{10}!} \quad \text{when} \quad \lambda = 2$$

The **likelihood ratio principle** says to form the ratio of $f(x_1, \ldots, x_{10}; 2)$ to $f(x_1, \ldots, x_{10}; 1)$ (the ratio of likelihoods or likelihood ratio) and reject H_0 when the ratio is "sufficiently large." This is reasonable because a ratio considerably larger than 1 indicates that the observed sample is much more likely to have been obtained under H_a than under H_0 (though just how large is sufficient must still be determined). Here the likelihood ratio is

$$\frac{f(x_1, \ldots, x_{10}; 2)}{f(x_1, \ldots, x_{10}; 1)} = e^{-10} \cdot 2^{x_1+\cdots+x_{10}}$$

which is large when $x_1 + \cdots + x_{10}$ is large. The likelihood ratio principle thus leads to the procedure

reject H_0 if $X_1 + \cdots + X_{10} \geq c$

The critical value c should now be determined to control $\alpha = P(\text{type I error})$. It can be shown that the sum of n independent Poisson variables with parameters $\lambda_1, \ldots, \lambda_n$ has a Poisson distribution with parameter $\lambda_1 + \cdots + \lambda_n$. This implies that when H_0 is true in our problem, the test statistic $X_1 + \cdots + X_{10}$ has a Poisson distribution with parameter $10\lambda = 10$. Now from the appendix table of Poisson probabilities, $F(15; 10) = .951$, so for X a Poisson variable with $\lambda = 10$, $P(X \geq 16) = .049 \doteq .05$. Thus for testing $H_0: \lambda = 1$ versus $H_a: \lambda = 2$, the likelihood ratio test with $\alpha = .05$ is

reject H_0 if $X_1 + \cdots + X_{10} \geq 16$

If the total number of sets needing the specified part was $x_1 + \cdots + x_{10} = 21$, then at level .05 H_0 would be rejected in favor of H_a.

Often H_0 will specify many different parameter values (for example, $H_0: \lambda \leq 1$ rather than $H_0: \lambda = 1$ in Example 7.13), and this will virtually always be the case for H_a ($H_a: \lambda > 1$ rather than $H_a: \lambda = 2$). In such situations we compute

LRS = likelihood ratio statistic

$$= \frac{\text{maximum likelihood, computed assuming } H_a \text{ true}}{\text{maximum likelihood, computed assuming } H_0 \text{ true}} \tag{7.17}$$

That is, as a measure of the evidence favoring $H_0(H_a)$, we use the largest possible likelihood assuming that H_0 (H_a) is true. The null hypothesis is then rejected if the LRS is large, since this indicates that H_a is more plausible than H_0.

**Example 7.13
(continued)**

Consider testing H_0: $\lambda \le 1$ versus H_a: $\lambda > 1$. Since $E(X) = \lambda$, H_a is certainly not more plausible than H_0 if $\bar{x} \le 1$. In fact, it can be shown that LRS ≤ 1 if $\bar{x} \le 1$, so the LRS is certainly not large in this case. Because the maximum likelihood estimate of λ is $\hat{\lambda} = \bar{x} = \Sigma\, x_i/10$, when $\bar{x} > 1$ the numerator of (7.17) is

$$f(x_1, \ldots, x_{10}; \hat{\lambda}) = f(x_1, \ldots, x_{10}; \bar{x})$$

while the denominator is $f(x_1, \ldots, x_{10}; 1)$. The ratio is then

$$\text{LRS} = \frac{f(x_1, \ldots, x_{10}; \bar{x})}{f(x_1, \ldots, x_{10}; 1)} = \frac{(e^{-\bar{x}})^{10}\,(\bar{x})^{\Sigma\, x_i}}{(e^{-1})^{10}\,(1)^{\Sigma\, x_i}} = e^{10} \cdot \left(\frac{\Sigma\, x_i}{10e}\right)^{\Sigma\, x_i}$$

Because LRS is large when $\Sigma\, x_i$ itself is large, the likelihood ratio test is again

reject H_0 if $X_1 + \cdots + X_{10} \ge c$

For the choice $c = 16$ (as in Example 7.13), P(type I error) $\le .05$ for every $\lambda \le 1$. If as before $\Sigma\, x_i = 21$, then H_0: $\lambda \le 1$ is rejected in favor of H_a: $\lambda > 1$.

In Example 7.13 both the test statistic and form of the rejection region could have been derived using intuition. There are many more complicated statistical situations, though, in which the likelihood ratio principle will yield a procedure when intuition fails. In addition to their providing a general method for the construction of test procedures in both simple and complex problems, another very important justification for the use of likelihood ratio tests is that many of the likelihood ratio tests have maximum power (minimum type II error probability) among *all* level α tests that might reasonably be considered as possible procedures. That is, many likelihood ratio tests are best tests in a very strong sense. This is true of the *t* test, of the tests about population proportions and variances (these tests are likelihood ratio tests), and of the two-sample *t* tests, regression tests, and analysis of variance tests to be presented in subsequent chapters.

A practical limitation on the use of the likelihood ratio principle is that in order to construct the likelihood ratio test statistic, the form of the probability distribution from which the sample comes must be specified. To derive the *t* test from the likelihood ratio principle, the investigator must assume a normal p.d.f. If an investigator is willing to assume that the distribution is symmetric but does not want to be specific about its exact form (such as normal, uniform, or Cauchy), then the principle fails because there is no way to write a joint p.d.f. simultaneously valid for all symmetric distributions. In Chapter 15 we will present several **distribution-free** test procedures, so called because the probability of a type I error is controlled simultaneously for many different underlying distributions. These procedures are useful when the investigator has limited knowledge of the underlying distribution. We shall also say more about issues 3 and 4 listed at the outset of this section.

Exercises / Section 7.6

1. Reconsider the paint drying problem discussed in Example 7.2. The hypotheses were $H_0: \mu = 75$ versus $H_a: \mu < 75$, with σ assumed to have value 9.4. Consider the alternative value $\mu = 74$, which in the context of the problem would presumably not be a practically significant departure from H_0.

   a. For a level .01 test, compute the power at this alternative for sample sizes $n = 100$, 900, and 2500.

   b. If the observed value of $\bar{X}$ is $\bar{x} = 74$, what can you say about the resulting P-value when $n = 2500$? Is the data statistically significant at any of the standard values of α?

   c. Would you really want to use a sample size of 2500 along with a level .01 test (leaving aside the cost of such an experiment)? Explain.

2. Consider the large sample level .01 test in Section 7.3 for test $H_0: p = .2$ against $H_a: p > .2$.

   a. For the alternative value $p = .21$, compute the power $\pi(.21)$ for sample sizes $n = 100$, 2500, 10,000, 40,000, and 90,000.

   b. For $\hat{p} = x/n = .21$, compute the P-value when $n = 100$, 2500, 10,000 and 40,000.

   c. In most situations, would it be reasonable to use a level .01 test in conjunction with a sample size of 40,000? Why or why not?

3. After playing with each of two different brands of tennis ball, each of n randomly selected individuals is asked which he or she prefers. Letting $X =$ number who prefer brand A, X has a binomial distribution with parameters n and p. Show that the likelihood ratio test for $H_0: p = .5$ versus $H_a: p \neq .5$ is to reject H_0 if X is either $\geq c$ or $\leq n - c$. *Hint:* The m.l.e. of p is $\hat{p} = x/n$. Note also that the function $u^u(1 - u)^{1-u}$ is symmetric about $u = .5$; a function of this form should appear in the likelihood ratio.

4. Suppose that a certain type of component is known to have an exponentially distributed lifetime with parameter λ. Let $X_1, X_2, \ldots, X_n$ be a random sample of such lifetimes. Show that the likelihood ratio test for $H_0: \lambda \geq \lambda_0$ versus $H_a: \lambda < \lambda_0$ based on this random sample rejects H_0 if $\Sigma X_i \geq c$. *Hint:* Because $E(X) = \mu = 1/\lambda$, $H_0: \lambda \geq \lambda_0$ is equivalent to $H_0: \mu \leq \mu_0 (\mu_0 = 1/\lambda_0)$, and the rejection region is $\bar{X} \geq c/n$. You do not have to obtain the value of c for which the test has level α.

Supplementary Exercises for Chapter 7

1. A small dog food company being considered for acquisition by a large conglomerate asserts that at least half the grocery stores in its market area stock its brand. To see whether this is the case, a random sample of 200 stores is selected, of which 78 are found to carry the brand.

   a. Does this data contradict the dog food company's claim? Test using $\alpha = .05$.

   b. What is the P-value of the data?

2. A sample of 50 lenses used in eyeglasses yields a sample mean thickness of 3.05 mm and a sample standard deviation of .34 mm. The desired true average thickness of such lenses is 3.20 mm. Does the data strongly suggest that the true average thickness of such lenses is something other than what is desired? Test using $\alpha = .05$.

3. In Exercise 2 above, suppose that the experimenter had believed before collecting the data that the value of σ was approximately .30. If the experimenter wished the probability of a type II error to be .05 when $\mu = 3.00$, was a sample size 50 unnecessarily large?

4. In an experiment to test the effects of hormones on growth of beef cattle, 200 mg of progesterone and 20 mg of estradiol benzoate are implanted in the outer ear of 16 randomly selected steers, each weighing approximately 500 lb. The sample average weight gain per day for a certain number of days is found to be 2.79 lb with a sample standard deviation of .41 lb per day. Does this data strongly suggest that the true average daily weight gain for steers treated with the hormone implant

exceeds 2.50? Making any necessary assumptions, test the relevant hypotheses using $\alpha = .001$.

5. In an investigation of the toxin produced by a certain poisonous snake, a researcher prepared 26 different vials, each containing 1 gm of the toxin, and then determined the amount of antitoxin needed to neutralize the toxin. The sample average amount of antitoxin necessary was found to be 1.89 mg with a sample standard deviation of .42. Previous research had indicated that the true average neutralizing amount was 1.75 mg per gram of toxin. Does the new data contradict the value suggested by prior research?

   a. Test the relevant hypotheses at level .05.
   b. What can you say about the P-value of the data?

6. Does the data in Exercise 5 above strongly indicate that the true standard deviation of neutralizing amount of antitoxin per gram of toxin exceeds .25 mg? Use a level .05 test.

7. The sample unrestrained compressive strength for 45 specimens of a particular type of brick was computed to be 3107 psi with a sample standard deviation of 188. The distribution of unrestrained compressive strength may be somewhat skewed. Does the data strongly indicate that the true average unrestrained compressive strength is less than the design value of 3200? Test using $\alpha = .001$.

8. To test the ability of auto mechanics to identify simple engine problems, an automobile with a single such problem was taken in turn to 72 different car repair facilities. Only 42 of the 72 mechanics who worked on the car correctly identified the problem. Does this strongly indicate that the true proportion of mechanics who could identify this problem is less than .75? Compute the P-value and reach a conclusion accordingly.

9. When $X_1, X_2, \ldots, X_n$ are independent Poisson variables, each with parameter λ, and n is large, the sample mean $\bar{X}$ has approximately a normal distribution with $\mu = E(\bar{X}) = \lambda$ and $\sigma^2 = \text{Var}(\bar{X}) = \lambda/n$. This implies that

$$Z = \frac{\bar{X} - \lambda}{\sqrt{\lambda/n}}$$

has approximately a standard normal distribution. For testing $H_0: \lambda = \lambda_0$, we can replace λ by λ_0 in the equation for Z to obtain a test statistic. This

statistic is actually preferred to the large sample statistic with denominator $S/\sqrt{n}$ (when the X_i's are Poisson) because it is tailored explicitly to the Poisson assumption. If the number of requests for consulting received by a certain statistician during a five-day work week has a Poisson distribution and the total number of consulting requests during a 36-week period is 160, does this suggest that the true average number of weekly requests exceeds 4.0? Test using $\alpha = .02$.

10. The July 3, 1978, *Los Angeles Times* reported that under conditions of the desegregation plan to be implemented, any students who travel more than 45 min one way on a bus would be required to have only three years of integrated schooling, while those traveling at most 45 min would face a five-year requirement. Suppose that travel time between two particular locations is normal, and that a random sample of 20 travel times yields observations

    47.3, 44.1, 44.5, 43.7, 46.2, 45.3, 47.0, 42.5,
    41.9, 47.6, 47.1, 45.8, 45.7, 47.8, 46.6, 44.2,
    47.2, 43.6, 46.5, 45.9

    Does this data indicate that only three years of integrated schooling is required (five years will be required unless the data indicates otherwise)? Use $\alpha = .10$.

11. In Exercise 10 above, does the data indicate that the true standard deviation of travel time between the two locations is different from 2.5?

    a. Test the appropriate hypotheses using $\alpha = .01$.
    b. What can you say about the P-value of the data?

12. Aphid infestation of fruit trees can be controlled either by spraying with pesticide or by inundation with ladybugs. In a particular area, four different groves of fruit trees are selected for experimentation. The first three groves are sprayed with pesticides 1, 2, and 3, respectively and the fourth is treated with ladybugs, with the following results on yield:

Pesticide	n_i = number of trees	$\bar{x}_i$ (bushels/tree)	s_i
1	100	10.5	1.5
2	90	10.0	1.3
3	100	10.1	1.8
4	120	10.7	1.6

Let μ_i and σ_i^2 denote the true average yield and variance of yield for treatment i ($i = 1, 2, 3, 4$).

a. The "parameter" $\theta = \frac{1}{3}(\mu_1 + \mu_2 + \mu_3) - \mu_4$ measures the difference in true average yields between pesticides and ladybugs. Write an unbiased estimator $\hat{\theta}$ for θ, and use the rules of expected value to verify that it is unbiased. What is the estimate for the given data?

b. Use the rules of variance to compute the standard deviation of $\hat{\theta}(\sigma_{\hat{\theta}})$, and then compute the estimated standard deviation of $\hat{\theta}(\sigma_{\hat{\theta}})$.

c. Use (a) and (b) along with the general method for constructing large sample tests (Section 7.3) to test $H_0: \theta = 0$ versus $H_a: \theta \neq 0$ using $\alpha = .05$.

13. From Section 7.5 when n is large, the sample standard deviation S has approximately a normal distribution with $E(S) \doteq \sigma$, $\text{Var}(S) \doteq \sigma^2/2n$. From the CLT, we already know that when n is large, $\overline{X}$ is approximately normal with $E(\overline{X}) = \mu$ and $\text{Var}(\overline{X}) = \sigma^2/n$.

a. Assuming that the underlying distribution is normal, what is an approximately unbiased estimator of the ninety-ninth percentile $\theta = \mu + 2.33\sigma$?

b. When the X_i's are normal, it can be shown that $\overline{X}$ and S are independent r.v.'s (one measures location while the other measures spread). Use this to compute $\text{Var}(\hat{\theta})$ and $\sigma_{\hat{\theta}}$ for the estimator $\hat{\theta}$ of (a). What is the estimated standard error $\hat{\sigma}_{\hat{\theta}}$?

c. Write a test statistic for testing $H_0: \theta = \theta_0$ which has approximately a standard normal distribution when H_0 is true. If soil pH is normally distributed in a certain region and 64 soil samples yield $\overline{x} = 6.33$, $s = .16$, does this provide strong evidence for concluding that at least 95% of all possible samples would have a pH of less than 6.75? Test using $\alpha = .01$.

14. Let $X_1, X_2, \ldots, X_n$ be a random sample from an exponential distribution with parameter λ. Then it can be shown that $2\lambda \Sigma X_i$ has a chi-squared distribution with $\nu = 2n$ (by first showing that $2\lambda X_i$ has a chi-squared distribution with $\nu = 2$).

a. Use this fact to obtain a test statistic and rejection region which together specify a level α test for $H_0: \mu = \mu_0$ versus each of the three commonly encountered alternatives. *Hint:* $E(X_i) = \mu = 1/\lambda$, so $\mu = \mu_0$ is equivalent to $\lambda = 1/\mu_0$.

b. Suppose that 10 identical components, each having exponentially distributed time until failure, are tested. The resulting failure times are

95, 16, 11, 3, 42, 71, 225, 64, 87, 123

Use the test procedure of (a) to decide if the data strongly suggests that the true average lifetime is less than the previously claimed value of 75.

Bibliography   DeGroot, Morris, *Probability and Statistics,* Addison-Wesley, Reading, Mass., 1974. A very good exposition of the general principles of hypothesis testing, including applications of the likelihood ratio method.

Larsen, Richard and Marx, Morris, *Introduction to Mathematical Statistics,* Prentice-Hall, Englewood Cliffs, N.J., 1980. Similar to DeGroot's presentation, but slightly less mathematical.

McClave, James and Dietrich, Frank, *Statistics,* Dellen, San Francisco, 1978. A very good introduction to methods of hypothesis testing in a book written at a precalculus level.

CHAPTER 8

Hypothesis Testing Procedures Based on Two Samples

Introduction

In Chapter 7 we presented methods for testing various hypotheses about parameters of a single population (μ, p, σ^2). Here we focus on methods for testing hypotheses about the means, proportions, and variances of two different populations or distributions.

8.1 *Z* Tests for Differences Between Two Population Means

The hypotheses to be tested in this section are all phrased in terms of a difference $\mu_1 - \mu_2$ between the means of two different population distributions. We might, for example, wish to test hypotheses about the difference between true average breaking strengths of two different types of corrugated fiberboards; one such hypothesis would state that $\mu_1 - \mu_2 = 0$—that is, that $\mu_1 = \mu_2$. To do so, a sample of breaking strengths for each type is needed.

Basic assumptions:

  1. $X_1, X_2, \ldots, X_m$ is a random sample from a population with mean μ_1 and variance σ_1^2.
  2. $Y_1, Y_2, \ldots, Y_n$ is a random sample from a population with mean μ_2 and variance σ_2^2.
  3. The X and Y samples are independent of one another.

The natural estimator of $\mu_1 - \mu_2$ is $\bar{X} - \bar{Y}$, the difference between the corresponding sample means. The test statistic results from standardizing this estimator, so we need expressions for the expected value and standard deviation of $\bar{X} - \bar{Y}$.

> **Proposition:** The expected value of $\bar{X} - \bar{Y}$ is $\mu_1 - \mu_2$, so $\bar{X} - \bar{Y}$ is an unbiased estimator of $\mu_1 - \mu_2$. The standard deviation of $\bar{X} - \bar{Y}$ is
>
> $$\sigma_{\bar{X}-\bar{Y}} = \sqrt{\frac{\sigma_1^2}{m} + \frac{\sigma_2^2}{n}}$$

Proof: Both these results depend on the rules of expected value and variance presented in Chapter 5. Since the expected value of a difference is the difference of expected values,

$$E(\bar{X} - \bar{Y}) = E(\bar{X}) - E(\bar{Y}) = \mu_1 - \mu_2$$

Because the X and Y samples are independent, $\bar{X}$ and $\bar{Y}$ are independent quantities; a rule from Section 5.3 then states that the variance of the difference is the *sum* of $\text{Var}(\bar{X})$ and $\text{Var}(\bar{Y})$:

$$\text{Var}(\bar{X} - \bar{Y}) = \text{Var}(\bar{X}) + \text{Var}(\bar{Y}) = \frac{\sigma_1^2}{m} + \frac{\sigma_2^2}{n}$$

The standard deviation of $\bar{X} - \bar{Y}$ is the square root of this expression.

If we think of $\mu_1 - \mu_2$ as a parameter θ, then its estimator is $\hat{\theta} = \bar{X} - \bar{Y}$ with standard deviation $\sigma_{\hat{\theta}}$ given by the above proposition. When σ_1^2 and σ_2^2 both have known values, the test statistic will have the form $(\hat{\theta} - \text{null value})/\sigma_{\hat{\theta}}$; this form of a test statistic was used in several one-sample problems in the previous chapter. When σ_1^2 and σ_2^2 are unknown, the sample variances must be used to estimate $\sigma_{\hat{\theta}}$.

Test Procedures for Normal Populations with Known Variances

The null hypothesis to be tested is $H_0: \mu_1 - \mu_2 = \Delta_0$, where Δ_0 is a given number (the null value of the difference in population means). In most problems $\Delta_0 = 0$, so that the null hypothesis becomes $H_0: \mu_1 = \mu_2$. The test procedures to be studied first are based on the following two assumptions:

> **Assumption 1:** Both populations are normal, so $X_1, \ldots, X_m$ is a normal random sample, and so is $Y_1, \ldots, Y_n$.
>
> **Assumption 2:** The values of both population variances σ_1^2 and σ_2^2 are known.

Assumption 2 is the much less palatable of the two in most situations, but when sample sizes are large the variances can be replaced by their estimates.

Since the X_i's constitute a normal sample, $\bar{X}$ has a normal distribution and so does $\bar{Y}$. Thus $\bar{X} - \bar{Y}$ is a linear combination of normal variables, so by a result from Section 5.4, $\bar{X} - \bar{Y}$ has a normal distribution. Standardizing $\bar{X} - \bar{Y}$ gives the standard normal variable

$$Z = \frac{\bar{X} - \bar{Y} - (\mu_1 - \mu_2)}{\sqrt{\dfrac{\sigma_1^2}{m} + \dfrac{\sigma_2^2}{n}}} \tag{8.1}$$

When the null hypothesis is true, $\mu_1 - \mu_2 = \Delta_0$, so substitution of Δ_0 into (8.1) in place of $\mu_1 - \mu_2$ gives a statistic which has a standard normal distribution when H_0 is true.

To complete the description of the test procedure, the set of Z values for which H_0 is to be rejected must be specified. For the alternative $H_a: \mu_1 - \mu_2 > \Delta_0$, only a value of $\bar{X} - \bar{Y}$ much larger than Δ_0 would cast sufficient doubt on H_0 to suggest its rejection. But such a value of $\bar{X} - \bar{Y}$ is equivalent to a large positive value of Z. We can now obtain a level α test by rejecting H_0 if $Z \geq z_\alpha$, since under H_0 the test statistic has a standard normal distribution. Our reasoning here is the same as in the one sample case.

Null hypothesis: $H_0: \mu_1 - \mu_2 = \Delta_0$

Test statistic: $Z = \dfrac{\bar{X} - \bar{Y} - \Delta_0}{\sqrt{\dfrac{\sigma_1^2}{m} + \dfrac{\sigma_2^2}{n}}}$

Alternative Hypothesis	*Rejection Region for Level α Test*
$H_a: \mu_1 - \mu_2 > \Delta_0$	$Z \geq z_\alpha$
$H_a: \mu_1 - \mu_2 < \Delta_0$	$Z \leq -z_\alpha$
$H_a: \mu_1 - \mu_2 \neq \Delta_0$	either $Z \geq z_{\alpha/2}$ or $Z \leq -z_{\alpha/2}$

Example 8.1   A letter in the May 19, 1978, *Journal of the American Medical Association* reported that of 215 male physicians who were Harvard graduates and died between November 1974 and October 1977, the 125 in full-time practice lived an average of 48.9 years beyond graduation, while the 90 with academic affiliations lived an average of 43.2 years beyond graduation. Does the data suggest that the mean lifetime after graduation for doctors in full-time practice differs from the mean lifetime for those who obtain an academic affiliation (if so, those medical students who say that they are "dying to obtain an academic affiliation" may be closer to the truth than they realize; put another way, is "publish or perish" really "publish and perish")?

Let μ_1 denote the true average number of years lived beyond graduation for physicians in full-time practice, and let μ_2 denote the same quantity for physicians with academic affiliations. Assuming the 125 and 90 physicians to be random samples from populations 1 and 2, respectively (which may not be reasonable if there is reason to believe that Harvard graduates have special characteristics which differentiate them from all other physicians—in this case inferences would be restricted just to the "Harvard populations"), we have $\bar{x} = 48.9$ and $\bar{y} = 43.2$. In order to apply the above test procedure, we must know σ_1^2 and σ_2^2. The letter from which the data was taken gave no information about variances, so for illustration assume that $\sigma_1 = 14.6$ and $\sigma_2 = 14.4$. The hypotheses are $H_0: \mu_1 - \mu_2 = 0$ versus $H_a: \mu_1 - \mu_2 \neq 0$, so Δ_0 is zero. H_a is two sided, so a two-tailed test is appropriate. If the desired level (type I error probability) is $\alpha = .01$, then H_0 is rejected if either $Z \geq 2.58$ or $Z \leq -2.58$. The computed value of Z is

$$z = \frac{48.9 - 43.2}{\sqrt{\frac{(14.6)^2}{125} + \frac{(14.4)^2}{90}}} = \frac{5.70}{\sqrt{1.70 + 2.30}} = 2.85$$

Since $2.85 \geq 2.58$, H_0 is rejected in favor of the conclusion that $\mu_1 \neq \mu_2$. In fact, for a two-tailed Z test, the P-value is $P = 2[1 - \Phi(2.85)] = .0032$. The letter reported the P value to be less than .005, so the assumed values of σ_1 and σ_2 are consistent with the conclusion stated in the letter.

Using a Comparison To Identify Causality

Investigators are often interested in comparing either the effects of two different treatments on a response or the response after treatment with the response after no treatment (treatment versus control). If the individuals or objects to be used in the comparison are not assigned by the investigators to the two different conditions, the study is said to be **observational.** The data in Example 8.1 is the result of a **retrospective** observational study; the investigator did not start out by selecting a sample of doctors and assigning some to the "academic affiliation" treatment and the others to the "full-time practice" treatment, but instead identified members of the two groups by looking backward in time (through obituaries!) to past records.

The difficulty with drawing conclusions based on an observational study is that while statistical analysis may indicate a significant difference in response between the two groups, the difference may be due to some underlying factors which had not been controlled rather than to any difference in treatments. Was the observed difference in Example 8.1 really due to a difference in the type of medical practice after graduation, or is there some other underlying factor which might also furnish a plausible explanation for the difference? Observational studies have been used to argue for a causal link between smoking and lung cancer. There are many studies which show that the incidence of lung cancer is significantly higher among smokers than among nonsmokers. However, individuals had decided whether or not to become smokers long before investigators arrived on the scene, and factors in making this decision may have played a causal role in the contraction of lung cancer.

A **randomized controlled experiment** results when investigators assign subjects to the two treatments in a random fashion. When statistical significance is observed in such an experiment, the investigator and other interested parties will have more confidence in the conclusion that the difference in response has been caused by a difference in treatments. A very famous example of this type of experiment and conclusion is the Salk polio vaccine experiment described in Section 8.4. These issues are discussed at greater length in the (nonmathematical) books by Moore and by Freedman, Purves, and Pisani listed in the Chapter 1 references.

Power, β, and Choice of Sample Sizes

Using Assumptions 1 and 2, the power $\pi(\mu_1 - \mu_2) = P$(rejecting H_0 when true difference $= \mu_1 - \mu_2$) is easily calculated for each Z test. Consider, for example, the alternative $\mu_1 - \mu_2 > \Delta_0$. Then the rejection region $Z \geq z_\alpha$ can be expressed as

$$\text{reject } H_0 \text{ if } \quad \overline{X} - \overline{Y} \geq \Delta_0 + z_\alpha \sqrt{\frac{\sigma_1^2}{m} + \frac{\sigma_2^2}{n}} \tag{8.2}$$

To compute $\pi(\mu_1 - \mu_2)$, restandardize each side of the inequality in (8.2) by first subtracting $\mu_1 - \mu_2$ and then dividing by the square root. The other two alternatives are handled in a similar fashion.

Alternative Hypothesis	*Power* $= \pi(\mu_1 - \mu_2)$
$H_a: \mu_1 - \mu_2 > \Delta_0$	$1 - \Phi\left(z_\alpha - \dfrac{\mu_1 - \mu_2 - \Delta_0}{\sigma} \right)$
$H_a: \mu_1 - \mu_2 < \Delta_0$	$\Phi\left(-z_\alpha - \dfrac{\mu_1 - \mu_2 - \Delta_0}{\sigma} \right)$
$H_a: \mu_1 - \mu_2 \neq \Delta_0$	$1 - \Phi\left(z_{\alpha/2} - \dfrac{\mu_1 - \mu_2 - \Delta_0}{\sigma} \right)$ $+ \Phi\left(-z_{\alpha/2} - \dfrac{\mu_1 - \mu_2 - \Delta_0}{\sigma} \right)$

where $\sigma = \sqrt{\dfrac{\sigma_1^2}{m} + \dfrac{\sigma_2^2}{n}}$ and $\Phi(\cdot)$ is the standard normal c.d.f.

The probability of a type II error is $\beta(\mu_1 - \mu_2) = 1 - \pi(\mu_1 - \mu_2)$.

Example 8.1
(continued)

In the physician's age problem, $\Delta_0 = 0$ and $\sigma = 2$. The probability of rejecting H_0 when there is actually a five-year difference in true average number of years lived beyond graduation is

$$\pi(5) = 1 - \Phi\left(2.58 - \frac{5 - 0}{2} \right) + \Phi\left(-2.58 - \frac{5 - 0}{2} \right)$$

$$= 1 - \Phi(.08) + \Phi(-5.08) \doteq 1 - .5319 = .4681$$

so $\beta(5) = 1 - \pi(5) = .5319$. Because σ_1 and σ_2 are rather large while m and n are not, the test is not all that likely to reject H_0 even when there is quite a significant departure from H_0 (evidence of a five-year difference might discourage many doctors from seeking academic affiliations).

As in Chapter 7, sample sizes m and n can be obtained for which a level α test also satisfies $\beta(\Delta_0 + d) = \beta$ for the alternative $\mu_1 - \mu_2 = \Delta_0 + d$ (specified type II error probability or power at a fixed alternative value of $\mu_1 - \mu_2$). This is done by setting $1 - \pi(\Delta_0 + d) = \beta$ and using the given expressions for power. For an upper-tailed test, this yields

$$\frac{\sigma_1^2}{m} + \frac{\sigma_2^2}{n} = \frac{d^2}{(z_\alpha + z_\beta)^2}$$

When $m = n$ (equal sample sizes), this equation gives

$$m = n = \frac{(\sigma_1^2 + \sigma_2^2)(z_\alpha + z_\beta)^2}{d^2}$$

These expressions are correct also for a lower-tailed test, whereas for a two-tailed test α is replaced by $\alpha/2$.

Example 8.1
(continued)

If when $\mu_1 - \mu_2 = 5$ years, we wish to detect such a difference with probability .95 (power = .95), then $\beta = .05$ for $d = 5$. Using $\alpha = .01$, $z_{\alpha/2} = 2.58$ and $z_\beta = 1.645$, so for equal sample sizes

$$m = n = \frac{[(14.6)^2 + (14.4)^2](2.58 + 1.645)^2}{5^2} = 300$$

In the next section we will use the power curves from Chapter 7 to accomplish these tasks for the t test, where hand calculations would be prohibitive.

While the two-sample Z test was developed using heuristic arguments (that is, first basing the test on $\overline{X} - \overline{Y}$ and then rejecting if Z was too large for the one-sided alternative), it can also be derived using the likelihood ratio principle. In addition, among all level α tests, it is the one which has maximum power. In a very strong sense it is a best test when the underlying assumptions are satisfied.

Large-Sample Tests When Variances Are Unknown

While Assumption 1 is often a reasonable approximation in practice, it is almost never true that σ_1^2 and σ_2^2 are known. However, when both sample sizes are large ($m > 30$ and $n > 30$), a modification of the above Z test can be used even when neither assumption is reasonable.

If m is large, the C.L.T. tells us that even if the X_i's are not themselves normal variables, the sample mean $\overline{X}$ will have an approximately normal distribution. Similarly, if n is large, then $\overline{Y}$ will be approximately normal even if the underlying distribution is not. Therefore $\overline{X} - \overline{Y}$ will be approximately normal, so the test statistic Z we previously used will have approximately a standard normal distribution when H_0 is true. The use of Z along with the appropriate one of the three rejection regions then yields an approximate level α test whether or not the underlying populations are approximately normal. Put another way, if $m > 30$, and $n > 30$, then the Z test above is valid whether or not Assumption 1 is justified.

When the sample sizes are large and σ_1^2 and σ_2^2 are unknown, it can be shown that substitution of the sample variances S_1^2 and S_2^2 in place of σ_1^2 and σ_2^2, respectively, results in the **large-sample test statistic**

$$Z = \frac{\overline{X} - \overline{Y} - \Delta_0}{\sqrt{\dfrac{S_1^2}{m} + \dfrac{S_2^2}{n}}} \tag{8.3}$$

which has an approximately standard normal distribution when H_0 is true. The denominator of (8.3) is an estimator of the standard deviation $\sigma_{\bar{X}-\bar{Y}}$ rather than $\sigma_{\bar{X}-\bar{Y}}$ itself. When Z is used in conjunction with any one of the usual alternatives and rejection regions, the resulting test has approximate level α even when the populations are not normal.

Example 8.2

Does proximity of place of residence to heavily traveled roads result in higher blood lead levels? The paper "Blood Lead of Persons Living Near Freeways" (*Arch. Environmental Health*, 1967, pp. 695–702) reported that in a sample of 30 women who did not live near a freeway, the sample average blood lead level was 9.9 with a sample standard deviation of 4.9, while a second sample of 35 females who did live near a freeway resulted in a sample average and sample standard deviation of 16.7 and 7.0, respectively (a histogram of the first sample suggests that the underlying distribution is reasonably close to normal, so we regard the sample size of 30 as sufficiently large to allow use of the large sample test).

With μ_1 (μ_2) denoting the true average blood lead level for females who do not (do) live close to a freeway, the hypotheses to be tested are H_0: $\mu_1 - \mu_2 = 0$ versus H_a: $\mu_1 - \mu_2 < 0$. At level .01, the lower-tailed test rejects H_0 if $Z \leq -2.33$. The computed value of Z is

$$z = \frac{9.9 - 16.7}{\sqrt{\dfrac{(4.9)^2}{30} + \dfrac{(7.0)^2}{35}}} = \frac{-6.80}{1.483} = -4.58$$

Since -4.58 is ≤ -2.33, H_0 is rejected at level .01. The data suggests that for females there is an *association* between proximity to freeways and blood lead level—but this is far from saying that proximity *causes* increased blood lead. A convincing argument for causality would ultimately rest on much more data from carefully designed experiments.

Example 8.3

In selecting a sulphur concrete for roadway construction in regions which experience heavy frost, it is important that the chosen concrete have a low value of thermal conductivity in order to minimize subsequent damage due to changing temperatures. Suppose that two types of concrete, a graded aggregate and a no-fines aggregate, are being considered for a certain road. Because using the no-fines aggregate involves a higher initial cost, it is decided that the graded aggregate will be used unless experimental data shows conclusively that its thermal conductivity is more than .1 w/mK higher than that of the no-fines concrete. Based on the accompanying sum-

Table 8.1

Type	Sample size	Sample average conductivity	Sample SD
Graded	35	.497	.187
No-fines	35	.359	.158

mary of experimental data, is use of the no-fines concrete justified?

Let μ_1 and μ_2 denote the true average thermal conductivity for the graded aggregate and no-fines aggregate concrete, respectively. The two hypotheses are $H_0: \mu_1 - \mu_2 = .1$ versus $H_a: \mu_1 - \mu_2 > .1$; only if H_0 can be conclusively rejected in favor of H_a will the no-fines concrete be used. The test is upper-tailed, and at level .05 H_0 is rejected if $Z \geq 1.645$. With $\Delta_0 = .1$, the computed value of Z is

$$z = \frac{(.497 - .359) - .1}{\sqrt{\dfrac{(.187)^2}{35} + \dfrac{(.158)^2}{35}}} = \frac{.038}{.041} = .93$$

Since .93 is less than the critical value 1.645, H_0 cannot be rejected. We conclude that the use of the more expensive no-fines concrete is not justified.

Exercises / Section 8.1

1. Persons having Reynaud's syndrome are apt to suffer a sudden impairment of blood circulation in fingers and toes. In an experiment to study the extent of this impairment, each subject immersed a forefinger in water and the resulting heat output (cal/cm^2/min) was measured. For $m = 10$ subjects with the syndrome, the average heat output was $\bar{x} = .64$, and for $n = 10$ nonsufferers, the average output was 2.05. Let μ_1 and μ_2 denote the true average heat outputs for the two types of subjects. Assume that the two distributions of heat output are normal with $\sigma_1 = .2$ and $\sigma_2 = .4$.

a. Test $H_0: \mu_1 - \mu_2 = -1.0$ versus $H_a: \mu_1 - \mu_2 < -1.0$ at level .01 (H_a says that the calorie output for sufferers is more than 1 cal/cm^2/min below that for nonsufferers).

b. Compute the P-value for the value of Z obtained in (a).

c. What is the probability of a type II error when the actual difference between μ_1 and μ_2 is $\mu_1 - \mu_2 = -1.2$?

d. Assuming that $m = n$, what sample sizes are necessary to ensure that $\beta = .1$ when $\mu_1 - \mu_2 = -1.2$ (so that $d = -.2$)?

2. An experiment to compare the tension bond strength of polymer latex modified mortar (portland cement mortar to which polymer latex emulsions have been added during mixing) to that of unmodified mortar resulted in $\bar{x} = 18.12$ kgf/cm^2 for the modified mortar ($m = 40$) and $\bar{y} = 16.87$ kgf/cm^2 for the unmodified mortar ($n = 32$). Let

μ_1 and μ_2 be the true average tension bond strengths for the modified and unmodified mortars, respectively.

a. Assuming that $\sigma_1 = 1.6$ and $\sigma_2 = 1.4$, test $H_0: \mu_1 - \mu_2 = 0$ versus $H_a: \mu_1 - \mu_2 > 0$ at level .01.

b. Compute the probability of a type II error for the test of (a) when $\mu_1 - \mu_2 = 1$.

c. Suppose that the investigator decided to use a level .05 test and wished $\beta = .10$ when $\mu_1 - \mu_2 = 1$. If $m = 40$, what value of n is necessary?

d. How would the analysis and conclusion of (a) change if σ_1 and σ_2 were unknown but $s_1 = 1.6$ and $s_2 = 1.4$?

3. The question, "Do you think that advertising is a good source of information?" was asked of each individual in a sample of 50 members of civic groups such as Rotary and Kiwanis, and also of 40 members of at least one consumer activist organization. The responses were in the form of a six-point Likert scale with 1 = strongly agree to 6 = strongly disagree. The results follow:

Civic: $m = 50$, $\bar{x} = 2.43$, $s_1 = 1.06$

Activist: $n = 40$, $\bar{y} = 3.51$, $s_2 = 1.60$

Does the data indicate that on average, activists regard advertising as less informative than do members of civic organizations? Test the relevant hypotheses using $\alpha = .10$. (Adapted from "The Development of a Scale to Measure Consumer

Discontent," *J. Marketing Research*, 1976, pp. 373–381).

4. An article in the *Journal of Educational Psychology* (1976) reported results of an experiment to investigate possible differences due to sex in classroom attentiveness among first graders. A behavior observation schedule was used to obtain the attentiveness scores.

Boys: $m = 53, \bar{x} = .76, s_1 = .13$

Girls: $n = 35, \bar{y} = .84, s_2 = .10$

Let μ_1 and μ_2 denote the true average attentiveness scores for boys and girls, respectively. Use the large-sample Z test to test $H_0: \mu_1 - \mu_2 = 0$ versus $H_a: \mu_1 - \mu_2 \neq 0$ at level .05.

5. An experiment was performed to compare the fracture toughness of high-purity 18 Ni maraging steel with commercial purity steel of the same type (*Corrosion Science*, 1971, pp. 723–736). For $m = 32$ specimens, the sample average toughness was $\bar{x} = 65.6$ for the high-purity steel, while $\bar{y} = 59.8$ for $n = 38$ specimens of commercial steel. Because the high-purity steel is more expensive, its use for a certain application can be justified only if its fracture toughness exceeds that of commercial purity steel by more than 5.
   a. Assuming that $\sigma_1 = 1.2$ and $\sigma_2 = 1.1$, test the relevant hypotheses using $\alpha = .001$.
   b. Compute the power and β for the test of (a) when $\mu_1 - \mu_2 = 6$.
   c. If $s_1 = 1.2$ and $s_2 = 1.1$ (rather than σ_1 and σ_2), how does the analysis and conclusion of (a) change?

6. In a study carried out to compare the effectiveness of a contract grading method with the traditional method of grading, ninth-grade students were tested for retention of material five weeks after completing a venereal disease unit in a health course.

Contract: $m = 32, \bar{x} = 30.41, s_1 = 8.00$

Traditional: $n = 31, \bar{y} = 31.50, s_2 = 8.12$

(A pretest revealed virtually no difference in knowledge level prior to the unit.) Does the data suggest that there is a difference in true average retention level for the two methods?
   a. Test using $\alpha = .01$.
   b. Compute the P-value.
      ("Retention of Knowledge: Grade Contract

Method Compared to the Traditional Grading Method," *J. Experimental Education*, 1974, pp. 92–96.)

7. The amount of serum cholesterol (mg/100cc) was determined both for a sample of 64 full-blooded American Indians and for a sample of 45 Indians having mixed blood ("Serum Cholesterol Level in American Indians," *U.S. Public Health Reports*, May 1959) in order to discover whether the low incidence of heart disease among Indians was related to genetic or environmental factors.

Full-blooded: $m = 64, \bar{x} = 216.5, s_1 = 45.8$

Mixed-blood: $n = 45, \bar{y} = 209.8, s_2 = 37.7$

   a. Does this data suggest a difference in true average serum cholesterol level between full-blooded Indians and Indians of mixed blood? Use a level .1 test.
   b. Suppose that on the basis of extensive experience with serum cholesterol distributions, the values of σ_1 and σ_2 had both been assumed to be 40 prior to the experiment. If the difference in true average serum cholesterol level was actually 10, what would the probability of a type II error be for a level .1 test based on the above sample sizes?
   c. Assuming $\sigma_1 = \sigma_2 = 40$ and equal sample sizes, what value of n achieves $\beta(10) = .1$?

8. An article in the 1974 *Journal of Personality and Social Psychology* reported the following data on self-esteem scores for a sample of Blacks and another sample of whites.

Blacks: $m = 157, \bar{x} = 38.1, s_1 = 5.16$

Whites: $n = 51, \bar{y} = 38.4, s_2 = 5.88$

Does this data indicate that on average, Blacks have a lower degree of self-esteem than do whites? Use a level .05 test. Then compute the P-value for the data.

9. A mechanical engineer wishes to compare strength properties of steel beams with similar beams made with a particular alloy. The same number of beams, n, of each type will be tested. Each beam will be set in a horizontal position with a support on each end, a force of 2500 lbs will be applied at the center, and the deflection will be measured. From past experience with such beams, the engineer is willing to assume that the true standard deviation of deflection for both types of beam is

.05 in. Because the alloy is more expensive, the engineer wishes to test at level .01 whether or not it has smaller average deflection than the steel beam. What value of n is appropriate if the desired type II error probability is .05 when the difference in true average deflection favors the alloy by .04 in.?

10. The level of monoamine oxidase activity in blood platelets (nm/mg protein/h) was determined for each individual in a sample of 33 chronic schizophrenics, resulting in $\bar{x} = 2.69$ and $s_1 = 2.30$, as well as for 35 normal subjects, resulting in $\bar{y} = 6.35$, $s_2 = 4.03$. Does this data strongly suggest that true average MAO activity for normal subjects is more than twice the activity level for schizophrenics? Derive a test procedure and carry out the test using $\alpha = .01$. *Hint:* H_0 and H_a here have a different form from the three standard cases. Let μ_1 and μ_2 refer to true average MAO activity for schizophrenics and normals, respectively, and consider the parameter $\theta = 2\mu_1 - \mu_2$. Write H_0 and H_a in terms of θ, estimate θ, and derive $\hat{\sigma}_{\hat{\theta}}$. ("Reduced Monoamine Oxidase Activity in Blood Platelets from Schizophrenic Patients," *Nature,*

July 28, 1972, pp. 225–226).

11. Show for the upper-tailed test with σ_1, σ_2 known that as either m or n is increased, the power increases when $\mu_1 - \mu_2 > \Delta_0$.

12. For the case of equal sample sizes ($m = n$) and fixed α, what happens to the necessary sample size n as β is decreased, where β is the desired type II error probability at a fixed alternative?

13. To decide whether or not two different types of steel have the same true average fracture toughness values, n samples of each type are tested, yielding the following results:

Type	Sample Average	Sample SD
1	60.1	1.0
2	59.9	1.0

Calculate the P-value for the appropriate two-sample Z test assuming that the above data was based on $n = 100$. Then repeat the calculation for $n = 400$. Is the small P-value for $n = 400$ indicative of a difference which has practical significance? Would you have been satisfied with just a report of the P-value? Comment briefly.

8.2 The Two-Sample t Test

It is virtually always the case in real problems that the values of the population variances are unknown. In the previous section we illustrated for large sample sizes the use of a test in which the sample variances were used in place of the population variances. In fact, for large samples, the Central Limit Theorem allows us to use the test even when the two populations of interest are not normal.

There are many problems, though, in which at least one sample size is small and the population variances have unknown values. We now develop a test which is appropriate in this situation. However, because sample sizes are small, we will not have the C.L.T. working for us, so to produce a valid test we must make stronger assumptions about the underlying populations. The use of the test is then restricted to situations in which these assumptions are at least approximately satisfied, so that the two-sample t test is not as broadly applicable as the large-sample Z test. There are two assumptions necessary for the development.

Assumption 1: Both populations are normal, so that $X_1, X_2, \ldots, X_m$ is a normal random sample and so is $Y_1, \ldots, Y_n$ (with the X's and Y's independent of one another).

Assumption 2: The values of the two population variances σ_1^2 and σ_2^2 are equal, so that their common value can be denoted by σ^2 (which is unknown).

The first assumption is of course analogous to what we assumed in connection with the one-sample t test, but the second assumption is new. Remember that the mean μ of a normal population specifies the location or center of the bell-shaped curve, while the variance σ^2 controls its spread. Assumption 2 then says that we are not comparing the locations of just any two normal populations, but only populations which are known (or believed) a priori to have the same spread. When Assumption 2 is not reasonable, there is no test procedure which is known to have good properties (controlled α and good power) even when both populations are normal. One possible approach is briefly described at the end of this section. Some authors recommend the use of a formal preliminary test for $\sigma_1^2 = \sigma_2^2$, but there are technical difficulties associated with this. Our approach is simply to "eyeball" the two sample variances; if they are of roughly the same order of magnitude, then one can be comfortable in using the test.

The Pooled Estimator of σ^2

The natural estimator of $\mu_1 - \mu_2$ is still $\overline{X} - \overline{Y}$, but now the variance of this estimator can be expressed as

$$\text{Var}(\overline{X} - \overline{Y}) = \frac{\sigma^2}{m} + \frac{\sigma^2}{n} = \sigma^2\left(\frac{1}{m} + \frac{1}{n}\right) \tag{8.4}$$

Because the populations are both normal, $\overline{X} - \overline{Y}$ is too, so if we standardize using the square root of (8.4), the result is a standard normal variable. The major result in this section will be that if σ in the denominator of the standardized variable is replaced by an appropriate estimator, the resulting variable will have a t distribution.

Because σ^2 is the variance of both the X distribution and the Y distribution, the best estimator should depend on both the X_i's and the Y_j's. Furthermore, more weight should be given to the sample corresponding to the larger of the two sample sizes. Both S_1^2 and S_2^2, the two sample variances, are estimators of σ^2; a better estimator than either one individually is the following weighted average of the two.

Definition: The **pooled estimator** of the common variance σ^2, denoted by S_p^2, is defined by

$$S_p^2 = \frac{(m-1)}{m+n-2}S_1^2 + \frac{(n-1)}{m+n-2}S_2^2$$

$$= \frac{(m-1)S_1^2 + (n-1)S_2^2}{m+n-2} \tag{8.5}$$

The pooled estimator of σ is S_p.

The middle expression in (8.5) shows that S_p^2 is of the form $\lambda S_1^2 + (1 - \lambda)S_2^2$ where $0 < \lambda < 1$, so that S_p^2 is a weighted average. If, for example, m is much larger than n, then much more weight will be placed on S_1^2 than on S_2^2 (if $m = 15$ and $n = 7$, then $\lambda = .7$ and $1 - \lambda = .3$). Only if $m = n$ will $\lambda = 1 - \lambda = .5$, in which case S_p^2 is the ordinary unweighted average of S_1^2 and S_2^2.

Since $(m - 1)S_1^2 = \Sigma (X_i - \overline{X})^2$ and $(n - 1)S_2^2 = \Sigma (Y_j - \overline{Y})^2$, the numerator of S_p^2 is $\Sigma (X_i - \overline{X})^2 + \Sigma (Y_j - \overline{Y})^2$ (this would not be the case if the weighting factors were m and n, respectively). The X sample contributes $m - 1$ degrees of freedom to the estimator and the Y sample contributes $n - 1$ degrees of freedom, for a total of $m + n - 2$ d.f.

The t Statistic and Test

Because of (8.4) the variable

$$Z = \frac{\overline{X} - \overline{Y} - (\mu_1 - \mu_2)}{\sigma \sqrt{\dfrac{1}{m} + \dfrac{1}{n}}}$$

has a standard normal distribution. Substitution of S_p in place of σ yields

---

Theorem: Under Assumptions 1 and 2 of this section

$$T = \frac{\overline{X} - \overline{Y} - (\mu_1 - \mu_2)}{S_p \sqrt{\dfrac{1}{m} + \dfrac{1}{n}}} \tag{8.6}$$

has a t distribution with $m + n - 2$ degrees of freedom.

The two-sample t statistic for testing H_0: $\mu_1 - \mu_2 = \Delta_0$ results from replacing $\mu_1 - \mu_2$ in (8.6) by Δ_0. The rejection regions for the various alternatives are similar to those in the one-sample case; each region uses a t critical value based on $m + n - 2$ d.f.

Null hypothesis: H_0: $\mu_1 - \mu_2 = \Delta_0$

Test statistic: $T = \dfrac{\overline{X} - \overline{Y} - \Delta_0}{S_p \sqrt{\dfrac{1}{m} + \dfrac{1}{n}}}$ $\tag{8.7}$

Alternative Hypothesis	Rejection Region for a Level α Test
H_a: $\mu_1 - \mu_2 > \Delta_0$	$T \geq t_{\alpha, m+n-2}$
H_a: $\mu_1 - \mu_2 < \Delta_0$	$T \leq -t_{\alpha, m+n-2}$ $\qquad$ (8.8)
H_a: $\mu_1 - \mu_2 \neq \Delta_0$	either $T \geq t_{\alpha/2, m+n-2}$ or $T \leq -t_{\alpha/2, m+n-2}$

The test procedures summarized as (8.7) and (8.8) can be derived by using the likelihood ratio principle discussed in Chapter 7. It can also be shown that among all reasonable tests for H_0 which have level of significance α, the two-sample t test has maximum power. Because of these results, statisticians refer to the two-sample t test as a best test.

Example 8.4

The paper "Production of Soluble Organic Nitrogen During Activated Sludge Treatment" (*J. Water Pollution Control Fed.*, 1981, pp. 99–112) reported that for a sample consisting of 14 observations from one type of activated sludge culture, the sample average soluble chemical oxygen demand (SCOD, in mg/l) was 18.1 with a sample standard deviation of 6.0, while for a sample of 16 observations from another type of culture, the sample average SCOD was 15.9 with a sample standard deviation of 5.0. Let μ_1 and μ_2 denote the true average SCOD values for these two types of cultures. Assuming that the sampled distributions are both normal with $\sigma_1 = \sigma_2$, let us use a level .05 test to decide whether sample data strongly suggests that μ_1 and μ_2 differ from one another.

The hypotheses of interest are $H_0: \mu_1 - \mu_2 = 0$ versus $H_a: \mu_1 - \mu_2 \neq 0$. The test statistic is

$$T = \frac{\overline{X} - \overline{Y}}{S_p\sqrt{\dfrac{1}{m} + \dfrac{1}{n}}}$$

and H_0 is rejected if either $T \geq t_{\alpha/2,\, m+n-2}$ or $T \leq -t_{\alpha/2,\, m+n-2}$. With $\alpha = .05$, $m = 14$, $n = 16$, $t_{\alpha/2,\, m+n-2} = t_{.025,\, 28} = 2.048$; so H_0 will be rejected if $T \geq 2.048$ or $T \leq -2.048$. Since $s_1 = 6.0$ and $s_2 = 5.0$

$$s_p^2 = \frac{(14 - 1)(6.0)^2 + (16 - 1)(5.0)^2}{14 + 16 - 2} = 30.11$$

so $s_p = 5.49$ (almost the average of s_1 and s_2, but this will not always be the case). The computed value of T is then

$$t = \frac{18.1 - 15.9}{5.49\sqrt{\frac{1}{14} + \frac{1}{16}}} = \frac{2.2}{2.01} = 1.10$$

Since 1.10 is neither ≥ 2.048 nor ≤ -2.048, H_0 cannot be rejected at level .05.

Example 8.5

In an experiment to study the effects of exposure to ozone, 20 rats were exposed to ozone in the amount of 2 parts per million for a period of 30 days. The average lung volume for these rats was determined to be 9.28 ml with a standard deviation of .37, while the average lung volume for a control group of 17 rats with similar initial characteristics was 7.97 ml with a standard deviation of .41. Does this data indicate that there is an increase in true average lung volume due to ozone? ("Effect of Chronic Ozone Exposure on Lung Elasticity in Young Rats," *J. Applied Physiology*, 1974, pp. 92–97).

Letting μ_1 and μ_2 be the true average lung volumes for the exposed and unexposed condition, respectively, we wish to test $H_0: \mu_1 - \mu_2 = 0$ versus $H_a: \mu_1 - \mu_2 > 0$. The alternative is one sided, and we use an upper-tailed rejection region. The sample summary statistics are $\bar{x} = 9.28$, $s_1 = .37$, $\bar{y} = 7.97$, and $s_2 = .41$. The computed value of the pooled estimate is

$$s_p^2 = \frac{(m-1)s_1^2 + (n-1)s_2^2}{m+n-2} = \frac{19(.37)^2 + 16(.41)^2}{35} = .151$$

so $s_p = .39$. The computed value of T is

$$t = \frac{\bar{x} - \bar{y} - 0}{s_p\sqrt{\frac{1}{m} + \frac{1}{n}}} = \frac{9.28 - 7.97}{.39\sqrt{\frac{1}{20} + \frac{1}{17}}} = \frac{1.31}{(.39)(.33)} = 10.18$$

The number of degrees of freedom for the test is $m + n - 2 = 35$ (which exceeds 30 even though both sample sizes are small), so for a level .01 test $t_{.01, 35} \doteq 2.440$ (interpolating between 30 and 40 d.f.). Since $10.18 \geq 2.440$, H_0 is rejected at level .01, and we conclude that exposure to ozone increases true average lung volume.

Power and Type II Error

Just as was the case in the one-sample problem, the power for a particular alternative value of $\mu_1 - \mu_2 - \Delta_0$ (or, equivalently, type II error probability for this value) is much more difficult to calculate for the t test than for the Z test. However, for purposes of calculating the power at a particular alternative value or choosing sample sizes to yield a specified power at that value, the power curves of Section 7.4 can be used. There is, though, one important restriction: The sample sizes m and n must be equal. When this is the case, the power at a particular alternative value is obtained by calculating the value of $d = |\mu_1 - \mu_2 - \Delta_0|/2\sigma$ (so that one must make a guess as to the value of σ), going up from d on the horizontal axis to the curve corresponding to sample size $2n - 1$, and reading the power from the vertical axis. Alternatively, if the power at alternative $\mu_1 - \mu_2 - \Delta_0$ is to be $1 - \beta$, then the curve closest to the desired point $(d, 1 - \beta)$ is located. If the sample size on this curve is denoted by n', then the common value of m and n is $n = (n' + 1)/2$. Because when σ is increased, the power is decreased, the effect of guessing a large value of σ is to reduce the power or increase the sample size—that is, the resulting quantity will be on the pessimistic or conservative side.

Example 8.4
(continued)

The test for equality of true average SCOD values was two-tailed at level .05. Suppose that the intention of the experimenter had been to use sample sizes $m = n = 15$ and that prior knowledge had suggested $\sigma = 5$ as a reasonable value; what is the power of the test when $\mu_1 - \mu_2 = 2.5$? Here $d = |\mu_1 - \mu_2 - \Delta_0|/10 = .25$, and the curve from which the power is read corresponds to a sample size of $2(15) - 1 = 29$. The resulting power is approximately .25. A power value of .7 when $d = .25$ requires that $n' = 100$, so that sample sizes of $m = n = 51$ are necessary. For $\mu_1 - \mu_2 = 4$, $\sigma = 5$, and $m = n = 15$, $d = .4$ and the power of the test is approximately .55, which is still not very high.

A Test Procedure When $\sigma_1^2 \neq \sigma_2^2$

When Assumptions 1 and 2 are satisfied, there is a general consensus among statisticians that the two-sample *t* test described above should be used. In fact, it has been shown that if the distributions being sampled are not too nonnormal and/or the two variances are not too different from one another, then the *t* test works reasonably well in the sense that the actual level of significance is approximately the specified α, and the test continues to have good power against alternatives. Statisticians customarily summarize these properties by saying that the *t* test is robust in the presence of mild departures from assumptions.

If, however, σ_1^2 and σ_2^2 are very different from one another, then using the *t* test (which involves a pooled estimate of a single parameter σ^2) will tend to yield erroneous conclusions. Unfortunately, even when the populations are still assumed normal, there is no test procedure which is known to have very good properties. The following procedure (called the Smith-Satterthwaite test) is known to be approximately a level α test, but its power properties have proved difficult to study, so it is not known whether the test is in any sense a best test. The test statistic has approximately a *t* distribution when H_0 is true, but the number of degrees of freedom ν is estimated from the data (ν will not usually be an integer, so must be rounded to obtain a critical value from the *t* table).

Test statistic: $T' = \dfrac{\overline{X} - \overline{Y} - \Delta_0}{\sqrt{\dfrac{S_1^2}{m} + \dfrac{S_2^2}{n}}}$

Degrees of freedom: $\nu = \dfrac{\left(\dfrac{S_1^2}{m} + \dfrac{S_2^2}{n}\right)^2}{\dfrac{(S_1^2/m)^2}{m-1} + \dfrac{(S_2^2/n)^2}{n-1}}$

Alternative Hypothesis	*Rejection Region for Approximate Level α Test*
$H_a: \mu_1 - \mu_2 > \Delta_0$	$T' \geq t_{\alpha,\nu}$
$H_a: \mu_1 - \mu_2 < \Delta_0$	$T' \leq -t_{\alpha,\nu}$
$H_a: \mu_1 - \mu_2 \neq \Delta_0$	either $T' \geq t_{\alpha/2,\nu}$ or $T' \leq -t_{\alpha/2,\nu}$

Example 8.6   A paper in the *Journal of Nervous and Mental Disorders* (1968, vol. 146, pp. 136–146) reported the following data on the amount of dextroamphetamine excreted by a sample of children having organically related disorders and a sample of children with nonorganic disorders (dextroamphetamine is a drug commonly used to treat hyperkinetic children).

Organic: 17.53, 20.60, 17.62, 28.93, 27.10

Nonorganic: 15.59, 14.76, 13.32, 12.45, 12.79

(observations refer to percentage of recovery of the drug seven hours after its administration)

The summary values are $\bar{x} = 22.36$, $\bar{y} = 13.78$, $s_1^2 = 28.63$, and $s_2^2 = 1.80$. The data suggests that there is much less variability in percentage of recovery for "nonorganic" children than for "organic" children. To use T' to test H_0: $\mu_1 - \mu_2 = 0$ versus H_a: $\mu_1 - \mu_2 \neq 0$, where μ_1 and μ_2 refer to the true average percentages of recovery for the two conditions, we need

$$\nu = \frac{\left(\dfrac{28.63}{5} + \dfrac{1.80}{5}\right)^2}{\dfrac{(28.63/5)^2}{4} + \dfrac{(1.80/5)^2}{4}} = \frac{37.04}{8.20 + .03} = 4.50 \doteq 5$$

$$t' = \frac{22.36 - 13.78}{\sqrt{\dfrac{28.63}{5} + \dfrac{1.80}{5}}} = \frac{8.58}{2.47} = 3.47$$

Since $t_{.005,5} = 4.032$, the data is not significant at level .01. If the pooled two-sample t test had been used, the computed t value would be $t = 4.9$, which would exceed $t_{.005, 8}$.

When it is reasonable to assume that the two populations do have the same shape and spread (so $\sigma_1^2 = \sigma_2^2$) but may differ with respect to location, yet the investigator does not wish to impose the assumption of normality, there is a distribution-free test which performs quite well. This test, the Wilcoxon rank-sum test, will be discussed in Chapter 15.

Exercises / Section 8.2

1. The insulin-binding capacity (pmol/mg protein) was measured for a sample of diabetic rats treated with a low dose of insulin and another sample of diabetic rats treated with a high dose, yielding the following data:

   Low dose: $m = 8$, $\bar{x} = 1.98$, $s_1 = .51$

   High dose: $n = 12$, $\bar{y} = 1.30$, $s_2 = .35$

   (*J. Clinical Investigation*, 1978, pp. 552–560).
   a. Compute s_p^2, the pooled estimate of σ^2, and s_p.
   b. Does the data indicate that there is any difference in true average insulin-binding capacity due to the dosage level? Use the two-sample t test with $\alpha = .001$.
   c. Between what two values does the P-value lie?

2. Borderline and mildly retarded children attending a hospital developmental evaluation clinic were divided into two groups on the basis of the presence or absence of a probable aetiological factor causing the retardation. Blood-lead concentration was measured for each child, yielding the following data:

   Aetiology unknown: 25.5, 23.2, 27.6, 24.3, 26.1, 25.0

   Probable aetiology: 21.2, 19.8, 20.3, 21.0, 19.6

   Does the data indicate any difference in true average blood-lead concentration for the two types of children? Test the appropriate hypotheses using the two-sample t test with $\alpha = .05$.

3. To study the immune response during aging, the level of anti-A isoglutinins was measured for a sample of 22 adults over 60 years old and 15 adults whose age was between 18 and 40. The data appears below. Does the data suggest that the true average anti-A isoglutinins level is smaller for those over 60 than for younger adults? Test using $\alpha = .05$, and then place bounds on the P-value.

> 60: $m = 22, \bar{x} = 19.19, s_1 = 5.99$

18–40: $n = 15, \bar{y} = 25.33, s_2 = 4.51$

4. In order to investigate whether or not learning could be transferred by nucleic acid, 10 rats were trained on a light/dark discrimination apparatus until they had achieved 20 straight days of criterion performance. They were then killed, and nucleic acid extracted from their brains was injected into 10 test animals. Simultaneously, nucleic acid from the brains of 10 untrained rats was injected into 10 other test animals. For each of these 20 test animals, the number of errors in 20 trials with the original training apparatus was recorded, yielding the following data:

Trained injection: 7, 9, 6, 11, 13, 8, 7, 13, 12, 9

Untrained injection: 12, 8, 9, 13, 14, 9, 8, 10, 7, 15

Use the two-sample t test with $\alpha = .01$ to decide whether transfer of learning can occur via nucleic acid. (*Science*, vol. 151, p. 834).

5. Serum-prolactin was measured using radio-immunoassay during the menstrual cycle of each of 28 women who had the premenstrual syndrome (cyclical recurrence or complaints of tension and headaches only during the premenstrual period, and dramatic and complete relief of symptoms when full menstrual flow begins) and also for 21 women in a control group. Measurements refer to average concentration during the cycle, in ng/ml.

Premenstrual syndrome: $m = 28, \sum x_i = 986.44,$

$$\sum x_i^2 = 36,030.31$$

Control: $n = 21, \sum y_i = 557.55,$

$$\sum y_i^2 = 15,877.53$$

Does this data suggest any difference in the true mean level of serum-prolactin between women with the premenstrual syndrome and those without it?

6. The flatwise bending strength (lb/in.) was determined for dowel joints of two different types used in wood furniture frame construction, with the second type having greater rail thickness (1.5 in.) than the first type (1.25 in.)

1.25 in.: $m = 10, \bar{x} = 1215.6, s_1 = 137.4$

1.5 in.: $n = 10, \bar{y} = 1376.4, s_2 = 155.0$

a. Does the larger rail thickness of the second type of dowel result in a greater true average bonding strength than for dowels of the first type? Test the relevant hypotheses at level .05.

b. Place upper and lower bounds on the P-value of the data.

c. Suppose that prior to the experiment, the investigator had believed $\sigma = 150$. What would the power of the test be for $\mu_1 - \mu_2 = 100$?

d. What value of n would be required to yield power .50 when $\mu_1 - \mu_2 = 100$?

7. Most educators believe that the most desirable testing environment is one in which there are no disruptions (quiet conditions). To ascertain the effect of systematic planned disruptions on test performance in a situation in which students were not aware that they were actually being tested, 14 students were given the Torrance Test of Creative Thinking (not easily recognizable as a test) under quiet conditions and another 13 took the test during a period of planned disruptions.

Quiet: $m = 14, \bar{x} = 54.0, s_1 = 6.43$

Disruptive: $n = 13, \bar{y} = 46.8, s_2 = 5.89$

Does the data strongly indicate that planned systematic disruption lowers the true average TTCT score by more than five points? Test using $\alpha = .01$.

8. High nitrate intake in food consumption has been shown to have a number of deleterious effects, including lower thyroxin production, increased incidence of cyanosis in newborns, and lower milk production in dairy cows. The following data is the result of an experiment to measure the percentage of weight gain for young laboratory mice given a standard diet and mice given 2000 ppm nitrate in their drinking water.

Nitrate: 12.7, 19.3, 20.5, 10.5, 14.0, 10.8, 16.6, 14.0, 17.2

Control: 18.2, 32.9, 10.0, 14.3, 16.2, 27.6, 15.7

Use the test procedure with statistic T' to decide whether the data indicates at level .01 that a heavy dose of nitrate retards true average percentage weight gain in mice.

9. An experiment is to be performed to assess the effect of companionship at mealtime on food consumption. The amount of food consumed by each of n mice eating alone will be determined. Then each of n similar mice will be allowed to eat with a companion mouse trained to eat continuously. The research hypothesis is that the presence of such a companion will increase true average food consumption, so the investigators plan to use a level .01 upper-tailed test. What value of n is appropriate if σ is taken to be 1.0 and it is desired that the type II error probability should be $\beta = .10$ when the increase in true average food consumption is as much as 1?

8.3 Analysis of Paired Data

In Sections 8.1 and 8.2 we considered testing for a difference between two means μ_1 and μ_2. This was done by utilizing the results of a random sample $X_1, X_2, \ldots, X_m$ from the distribution with mean μ_1 and a completely independent (of the X's) sample $Y_1, \ldots, Y_n$ from the distribution with mean μ_2. That is, either m individuals were selected from population one and n different individuals from population two, or m individuals (or experimental objects) were given one treatment and another set of n individuals were given the other treatment. In contrast, there are a number of experimental situations in which there is only one set of n individuals or experimental objects, and two observations are made on each individual or object.

Example 8.7

An article in the 1949 *Journal of Applied Physiology* ("Interval of Useful Consciousness at Various Altitudes") reported on the effect of a transfusion of a large amount of blood on the ability to perform a simple task at high altitude. The high-altitude condition (35,000 ft) was achieved by using a gas mask. The investigator believed that more blood would increase an individual's interval of useful consciousness following interruption of normal oxygen supply. Does the accompanying (slightly modified) data support this belief?

Table 8.2

Subject	1	2	3	4	5
Seconds of useful consciousness before transfusion	38	65	68	59	67
Seconds of useful consciousness after transfusion	45	77	81	70	73

Basic assumption: The data consisis of n pairs (X_1, Y_1), (X_2, Y_2), ..., (X_n, Y_n). Both the X_i's and Y_i's are assumed to be normally distributed with means μ_1 and μ_2, respectively. The variables within *different* pairs are assumed to be independent of one another.

We are again interested in testing hypotheses about the difference $\mu_1 - \mu_2$. The denominator of the two-sample t test was obtained by first applying the rule $\text{Var}(\overline{X} - \overline{Y}) = \text{Var}(\overline{X}) + \text{Var}(\overline{Y})$. However, with paired data the X and Y observations within each pair are often not independent, so that $\overline{X}$ and $\overline{Y}$ are not independent of one another, and the rule is not valid. We must therefore abandon the two-sample t test and look for an alternative method of analysis.

The Paired t Test

Consider the n differences $D_1 = X_1 - Y_1, D_2 = X_2 - Y_2, \ldots, D_n = X_n - Y_n$. The D_i's are normally distributed (since the X_i's and Y_i's are), and because different pairs are independent, the D_i's are independent of one another. If we let $D = X - Y$, where X and Y are the first and second observation, respectively, within an arbitrary pair, then the expected difference is

$$\mu_D = E(X - Y) = E(X) - E(Y) = \mu_1 - \mu_2$$

(the rule of expected values used here is valid even when X and Y are dependent). Thus any hypothesis about $\mu_1 - \mu_2$ can be phrased as a hypothesis about the mean difference μ_D. But since the D_i's constitute a normal random sample (of differences) with mean μ_D, hypotheses about μ_D can be tested using a one-sample t test. That is, *to test hypotheses about $\mu_1 - \mu_2$ when data is paired, form the differences $D_1, D_2, \ldots, D_n$ and carry out a one-sample t test (based on $n - 1$ d.f.) on the differences.*

Null hypothesis: $H_0: \mu_D = \Delta_0$,   where $D = X - Y$ is the difference between the first and second observations within a pair ($\mu_D = \mu_1 - \mu_2$).

Test statistic: $T_{\text{paired}} = \dfrac{\overline{D} - \Delta_0}{S_D / \sqrt{n}}$ (where $\overline{D}$ and S_D are the sample mean and standard deviation, respectively, of the D_i's)

Alternative Hypothesis	*Rejection Region for Level α Test*
$H_a: \mu_D > \Delta_0$	$T_{\text{paired}} \geq t_{\alpha, n-1}$
$H_a: \mu_D < \Delta_0$	$T_{\text{paired}} \leq -t_{\alpha, n-1}$
$H_a: \mu_D \neq \Delta_0$	either $T_{\text{paired}} \geq t_{\alpha/2, n-1}$ or $T_{\text{paired}} \leq -t_{\alpha/2, n-1}$

This paired test is valid even if $\sigma_1^2 \neq \sigma_2^2$, since the differences are still normally distributed and S_D^2 estimates $\sigma_D^2 = \text{Var}(X - Y)$. While a two-sample t test would be based on $2n - 2$ d.f., the paired t test uses only $n - 1$ d.f. To ensure a correct analysis, we give up $n - 1$ degrees of freedom. There are situations in which an experiment can be run in either a paired or an unpaired (independent samples) manner; we will shortly discuss the issues involved in choosing between the two.

Example 8.6
(continued)

If the investigator's theory is correct, we expect the interval of useful consciousness Y after a transfusion to be longer than the interval X before the transfusion. That is, we expect the difference $X - Y$ to be negative. With $\mu_D = E(X - Y)$ and the investigator's theory identified with the alternative hypothesis, we wish to test $H_0: \mu_D = 0$ versus $H_a: \mu_D < 0$ $(\mu_1 - \mu_2 = 0$ versus $\mu_1 - \mu_2 < 0)$. At level .01, because $n = 5$, H_0 will be rejected if $T_{\text{paired}} \leq -t_{.01,4} = -3.747$.

The observed differences $d_i = x_i - y_i$ are $d_1 = -7$, $d_2 = -12$, $d_3 = -13$, $d_4 = -11$, and $d_5 = -6$, so $\Sigma\, d_i = -49$ and $\Sigma\, d_i^2 = 519$. Then $\bar{d} = -\frac{49}{5} = -9.8$ and $s_D^2 = [\Sigma\, d_i^2 - (\Sigma\, d_i)^2/n]/(n - 1) = [519 - (49)^2/5]/4 = 9.70$ so $s_D = \sqrt{9.70} = 3.11$. This gives

$$t_{\text{paired}} = \frac{\bar{d} - 0}{s_D/\sqrt{n}} = \frac{-9.8}{3.11/\sqrt{5}} = -7.05$$

Since $-7.05 \leq -3.747$, H_0 is rejected at level .01, and the data does give strong support to the investigator's theory.

Example 8.8

An article in *Applied Spectroscopy* ("The Determination of Total Iron in Venezualan Laterites," 1978, pp. 57–62) discussed determination of the percentage of iron in laterites using both an atomic absorption (A.A.) method and a classical volumetric technique (V). The data appears below; does there appear to be any difference on average between the two measurement techniques?

Table 8.3

Sample number	1	2	3	4	5	6	7	8
x_i (A.A.)	30.99	31.47	30.00	30.64	35.25	30.62	31.91	31.37
y_i (V)	30.05	31.75	28.50	31.18	35.12	30.55	31.88	31.05
d_i	.94	−.28	1.50	−.54	.13	.07	.03	.32

Sample number	9	10	11	12	13	14	15	16
x_i	13.22	21.14	27.21	28.27	29.75	24.90	27.86	31.33
y_i	12.97	21.92	27.26	28.14	30.05	25.10	27.72	31.30
d_i	.25	−.78	−.05	.13	−.30	−.20	.14	.03

The summary quantities are $\bar{d} = .087$ and $s_D = .538$, so the computed value of the statistic is

$$t_{\text{paired}} = \frac{.087 - 0}{.538/\sqrt{16}} = .647$$

For a two-sided alternative $H_a: \mu_D \neq 0$, a level .1 test requires $t_{.05,15} = 1.753$; because .647 is neither ≤ -1.753 nor ≥ 1.753, H_0 is not rejected at level .1, and we conclude that there is no significant difference between the two measurement techniques.

When the number of pairs is more than 30, the assumption of normality is not necessary, since the C.L.T. can be used to validate the resulting Z test.

Paired Data and the Two-Sample t Test

To see what happens if the two-sample t test is used (incorrectly) to analyze paired data in which there is dependence within pairs, consider again Example 8.7. The summary quantities are $\bar{x} = 59.40$, $\bar{y} = 69.20$, $s_1^2 = 155.30$, and $s_2^2 = 200.20$, so $s_p^2 = 177.75$ and $s_p = 13.33$. The computed value of the two-sample t statistic is

$$ t = \frac{\bar{x} - \bar{y} - 0}{s_p \sqrt{\dfrac{1}{n} + \dfrac{1}{n}}} = \frac{-9.8}{13.33\sqrt{2/5}} = -1.16 $$

At level .01, t is compared to $t_{.01, 2n-2} = -t_{.01,8} = -2.896$. Since -1.16 is greater than -2.896, H_0 would not be rejected, so the two-sample t test yields a conclusion opposite to that suggested by the correct paired analysis.

Notice that the numerators of t and t_{paired} are identical. This is true in general, since $\bar{d} = \Sigma\, d_i/n = [\Sigma\,(x_i - y_i)]/n = (\Sigma\,x_i)/n - (\Sigma\,y_i)/n = \bar{x} - \bar{y}$. The difference between the two test statistics is due entirely to the denominators. Each test statistic is obtained by standardizing $\bar{X} - \bar{Y}$ ($=\bar{D}$), but in the presence of dependence the standardization in t is incorrect. To see this, recall from Section 5.3 that

$$ \text{Var}(X \pm Y) = \text{Var}(X) + \text{Var}(Y) \pm 2\,\text{Cov}(X, Y) $$

Since the correlation between X and Y is $\rho = \text{Corr}(X, Y) = \text{Cov}(X, Y)/\sqrt{\text{Var}(X)} \cdot \sqrt{\text{Var}(Y)}$,

$$ \text{Var}(X - Y) = \sigma^2 + \sigma^2 - 2\rho\sigma^2 = 2\sigma^2(1 - \rho) $$

Applying this to $\bar{X} - \bar{Y}$,

$$ \boxed{\;\text{Var}(\bar{X} - \bar{Y}) = \text{Var}(\bar{D}) = \text{Var}\!\left(\frac{1}{n}\sum D_i\right) = \frac{\text{Var}(D_i)}{n} = \frac{2\sigma^2(1 - \rho)}{n}\;} $$

The two-sample t test is based on the assumption of independence, in which case $\rho = 0$. But in many paired experiments, there will be a strong *positive* dependence between X and Y (large X associated with large Y), so that ρ will be positive and the variance of $\bar{X} - \bar{Y}$ will be smaller than $2\sigma^2/n$. Thus, *whenever there is positive dependence within pairs, the denominator for t_{paired} should be smaller than for t of the independent samples test.* Often t will be much closer to zero than t_{paired}, considerably understating the significance of the data.

Paired Versus Unpaired Experiments

In our examples paired data resulted from two observations on the same individual (Example 8.7) or experimental object (ore sample in Example 8.8). Even when this cannot be done, paired data with dependence within pairs can be obtained by matching individuals or objects on one or more characteristics which are thought to influence responses. For example, in a medical experiment to compare the efficacy of two drugs for lowering blood pressure, the experimenter's budget might allow for the treatment of 20 patients. If 10 patients are randomly selected for treatment with the first drug and another 10 independently selected for treatment with the second drug, an independent-samples experiment results.

However, the experimenter, knowing that blood pressure is influenced by age and weight, might decide to pair off patients so that within each of the resulting 10 pairs, age and weight were approximately equal (though there might be sizable differences between pairs). Then each drug would be given to a different patient within each pair for a total of 10 observations on each drug.

Without this matching (or "blocking"), one drug might appear to outperform the other just because patients in one sample were lighter and younger and thus more susceptible to a decrease in blood pressure than the heavier and older patients in the second sample. However, there is a price to be paid for pairing—a smaller number of degrees of freedom for the paired analysis—so we must ask when one type of experiment should be preferred to the other.

There is no straightforward and precise answer to this question, but there are some useful guidelines. The first point is that if we have a choice between two t tests which are both valid (and carried out at the same level of significance α), we should prefer the test which has the larger number of degrees of freedom. The reason for this is that a larger number of d.f. means greater power against any fixed alternative value of the parameter or parameters. That is, for a fixed type I error probability, the probability of a type II error is decreased by increasing d.f. This is readily verified by examining the power curves in Chapter 7.

However, if the experimental units are quite heterogeneous in their responses (large σ^2), this will make it difficult to detect small but significant differences between two treatments. This is essentially what happened in the data set in Example 8.7—for both "treatments" (before transfusion and after transfusion), there is great between-subject variability which tends to mask differences between treatments within subjects. If there is a large correlation within subjects, then the variance of $\overline{D} = \overline{X} - \overline{Y}$ will be $2\sigma^2 (1 - \rho)/n$, which will be much smaller than the unpaired variance $2\sigma^2/n$. Because of this reduced variance due to pairing, it will be easier to detect a difference than if independent samples are used. The pros and cons of pairing can now be summarized as follows:

1. If there is great heterogeneity between subjects (large σ^2) and a large correlation within subjects (large positive ρ), then the loss in degrees of freedom will be compensated for by the increased precision associated with pairing, so a paired experiment is preferable to an independent-samples experiment.
2. If the experimental units are relatively homogeneous (small σ^2) and the correlation within pairs is not large, the gain in precision due to pairing will be outweighed by the decrease in degrees of freedom, so an independent-samples experiment should be used.

Of course, the magnitude of σ^2 and ρ will not usually be known very precisely, so an investigator will be required to make a seat-of-the-pants judgment as to whether 1 or 2 obtains. In general, if the number of observations which can be obtained is large, then a loss in degrees of freedom (for example, from 40 to 20) will not be serious, but if the number is small, then the loss (say, from 16 to eight) because of pairing may be serious if not compensated for by increased precision. In making such a decision, the power curves for the *t* tests should be consulted.

Exercises / Section 8.3

1. The paper "A Supplementary Behavioral Program To Improve Deficient Reading Performance" (*J. Abnormal Child Psychology*, 1973, pp. 390–399) reported the results of an experiment in which seven pairs of children reading below grade level were obtained by matching so that within each pair the two children were equally deficient in reading ability. Then one child from each pair received experimental training, while the other received standard training. Based on the accompanying improvement scores, does the experimental training appear to be superior to the standard training? Use the paired *t* test to test $H_0: \mu_D = 0$ versus $H_a: \mu_D > 0$ at level .1.

Pair:	1	2	3	4	5	6	7
Experimental (*x*):	.5	1.0	.6	.1	1.3	.1	1.0
Control (*y*):	.8	1.1	−.1	.2	.2	1.5	.8

2. Two types of fish attractors, one made from vitrified clay pipes and the other from cement blocks and brush, were used during 16 different time periods spanning four years at Lake Tohopekaliga, Florida ("Two Types of Fish Attractors Compared in Lake Tohopekaliga, Florida," *Trans. Amer. Fisheries Soc.*, 1978, pp. 689–695). The following observations are of fish caught per fishing day.

Period:	1	2	3	4	5	6	7	8
Pipe:	6.64	7.89	1.83	.42	.85	.29	.57	.63
Brush:	9.73	8.21	2.17	.75	1.61	.75	.83	.56

Period:	9	10	11	12	13	14	15	16
Pipe:	.32	.37	.00	.11	4.86	1.80	.23	.58
Brush:	.76	.32	.48	.52	5.38	2.33	.91	.79

Does one attractor appear to be more effective on average than the other?

a. Use the paired *t* test with $\alpha = .01$ to test $H_0: \mu_D = 0$ versus $H_a: \mu_D \neq 0$.
b. What happens if the two-sample *t* test is used ($s_1 = 2.48$ and $s_2 = 2.91$)?

3. The paper "Relative Controllability of Dissimilar Cars" (*Human Factors*, 1962, pp. 375–380) reported results of an experiment to compare handling ability for two cars having quite different lengths, wheelbases, and turning radii. The observations are time in seconds required for subjects to parallel park each car.

Subject:	1	2	3	4	5	6	7
Car A:	37.0	25.8	16.2	24.2	22.0	33.4	23.8
Car B:	17.8	20.2	16.8	41.4	21.4	38.4	16.8

Subject:	8	9	10	11	12	13	14
Car A:	58.2	33.6	24.4	23.4	21.2	36.2	29.8
Car B:	32.2	27.8	23.2	29.6	20.6	32.2	53.8

Does the data suggest that the average person will more easily handle one car than the other? Test the relevant hypotheses using $\alpha = .10$.

4. In an experiment designed to study the effects of illumination level on task performance ("Performance of Complex Tasks Under Different Levels of Illumination," *J. Illuminating Eng.*, 1976, pp. 235–242), subjects were required to insert a fine-tipped probe into the eyeholes of 10 needles in rapid succession both for a low light level with a black background and a higher level with a white background.

Subject:	1	2	3	4	5
White:	25.85	28.84	32.05	25.74	20.89
Black:	18.23	20.84	22.96	19.68	19.50

Subect:	6	7	8	9
White:	41.05	25.01	24.96	27.47
Black:	24.98	16.61	16.07	24.59

Does the data indicate that the higher level of illumination yields a decrease of more than 5 in true average task completion time?

5. Tardive dyskinesia denotes a syndrome comprising a variety of abnormal involuntary movements assumed to follow long-term use of antipsychotic drugs. In an experiment to see whether the drug

deanol produced a greater improvement from base-line scores than a placebo treatment ("Double Blind Evaluation of Deanol in Tardive Dyskinesia," *J. Amer. Med. Assn.*, 1978, pp. 1997–1998), the two treatments were administered for four weeks each in random order to 14 patients, resulting in the following total severity index (TSI) scores:

Patient:	1	2	3	4	5	6	7
Deanol:	12.4	6.8	12.6	13.2	12.4	7.6	12.1
Placebo:	9.2	10.2	12.2	12.7	12.1	9.0	12.4

Patient:	8	9	10	11	12	13	14
Deanol:	5.9	12.0	1.1	11.5	13.0	5.1	9.6
Placebo:	5.9	8.5	4.8	7.8	9.1	3.5	6.4

(The investigators attributed placebo improvement to frequent attention and the nature of the experiment.) Does the data indicate that on average deanol yields a higher TSI than the placebo treatment? Test using $\alpha = .10$.

6. The urinary fluoride concentration (ppm) was determined for 11 randomly chosen livestock both at the beginning of and in the middle of their grazing period in a region which had previously been exposed to fluoride pollution ("Fluoride Pollution Caused by a Brickworks in the Flemish Countryside of Belgium," *Int. J. Environmental Studies*, 1978, pp. 245–252).

Subject:	1	2	3	4	5	6
Beginning:	24.7	46.1	18.5	29.5	26.3	33.9
Middle:	12.4	14.1	7.6	9.5	19.7	10.6

Subject:	7	8	9	10	11
Beginning:	23.1	20.7	18.0	19.3	23.0
Middle:	9.1	11.5	13.3	8.3	15.0

Does the data suggest that there has been a decrease in the true average urinary fluoride concentration during the period under consideration? *Note:* $s_1 = 8.31$ and $s_2 = 3.53$, suggesting that $\sigma_1 \neq \sigma_2$, but the paired t test is still valid.

7. Construct a paired data set for which $t_{\text{paired}} = \infty$, so that the data is highly significant when the correct analysis is used, yet t for the two-sample t test is quite near zero, so the incorrect analysis yields an insignificant result.

8. To assess the effects of two different coatings, A and B, on the corrosion of pipe, n pipe samples are coated with A and another n with B. The samples are then paired, with an A and a B in each pair, and each of the n pairs is buried in a different location (different soil, depth, and the like). Let $X_1, \ldots, X_n$ denote the amount of corrosion for the A-coated pipes after a specified time period, and let $Y_1, \ldots, Y_n$ denote the amount of corrosion for the B-coated pipes. A standard model for the X's and Y's (which will appear later in connection with the analysis of variance) is

$$X_1, \ldots, X_n \quad \text{independent, normal, } \text{Var}(X_i)$$
$$= \sigma^2, E(X_i) = \tau_i + \delta_1$$
$$Y_1, \ldots, Y_n \quad \text{independent, normal, } \text{Var}(Y_i)$$
$$= \sigma^2, E(Y_i) = \tau_i + \delta_2$$

The expected value is then the sum of two contributions, one due to the location (τ_i) and the other due to the coating (δ_j). In terms of these parameters, how would you state and test the hypothesis that on average there is no difference between the two coatings versus the alternative that there is a difference? *Hint:* To eliminate the "irrelevant" τ_i's, take differences. The expressions provide an alternative model (to that of dependence within pairs) for which the paired t test is appropriate.

8.4 Testing for Differences Between Population Proportions

Having presented in previous sections methods for comparing the means of two different populations, we now turn to the comparison of two population proportions. The notation for this problem is an extension of the notation used in the corresponding one-population problem. We let p_1 and p_2 denote the proportion of individuals in population one and two, respectively, who possess a particular characteristic. Alternatively, if we use the label S for an individual who possesses the characteristic of interest (does favor a particular proposition, is a member of a particular political party, has read at least one book within the last month, and so on), then p_1 and p_2 represent the probabilities of seeing the label S on a randomly chosen individual from populations one and two, respectively.

Analogously to the hypotheses for $\mu_1 - \mu_2$, the most general null hypothesis that we might wish to consider would be of the form $H_0 : p_1 - p_2 = \Delta_0$, where Δ_0 is again a prespecified number. Although for population means the case $\Delta_0 \neq 0$ presented no difficulties, it turns out that for population proportions $\Delta_0 = 0$ and $\Delta_0 \neq 0$ must be considered separately. Since the vast majority of actual problems of this sort involve $\Delta_0 = 0$, we focus primarily on this case. An additional complication is that although the test procedure when sample sizes are large is easily derived, understood, applied, and used by virtually all statisticians, the case of small sample sizes is not as easily handled. Because large sample sizes arise much more frequently than not, we develop the large-sample procedure first, and then present one possible procedure for small samples.

Description of the Large-Sample Test Procedure for $H_0 : p_1=p_2$

We shall assume the availability of a sample of m individuals from the first population and n from the second. The variables X and Y will represent the number of individuals in each sample possessing the characteristic which defines p_1 and p_2. Provided the population sizes are much larger than the sample sizes, the distribution of X can be taken to be binomial with parameters m and p_1, and similarly Y is taken to be a binomial variable with parameters n and p_2. Furthermore, the samples are assumed to be independent of one another, so that X and Y are independent random variables.

The obvious estimator for $p_1 - p_2$, the difference in population proportions, is the corresponding difference in sample proportions $X/m - Y/n$. With $\hat{p}_1 = X/m$ and $\hat{p}_2 = Y/n$, the estimator of $p_1 - p_2$ can be expressed as $\hat{p}_1 - \hat{p}_2$. The large-sample test statistic results from properly standardizing this estimator.

Proposition: Let $X \sim \text{Bin}\,(m, p_1)$ and $Y \sim \text{Bin}\,(n, p_2)$ with X and Y independent variables. Then

$$E\left(\frac{X}{m} - \frac{Y}{n}\right) = p_1 - p_2 \quad , \quad \text{so} \quad \frac{X}{m} - \frac{Y}{n}$$

is an unbiased estimator of $p_1 - p_2$, and

$$\text{Var}\left(\frac{X}{m} - \frac{Y}{n}\right) = \frac{p_1 q_1}{m} + \frac{p_2 q_2}{n} \qquad (\text{where } q_i = 1 - p_i) \qquad (8.9)$$

Proof: Since $E(X) = mp_1$ and $E(Y) = np_2$,

$$E\left(\frac{X}{m} - \frac{Y}{n}\right) = \frac{1}{m}E(X) - \frac{1}{n}E(Y) = \frac{1}{m}mp_1 - \frac{1}{n}np_2 = p_1 - p_2$$

Since $\text{Var}(X) = mp_1 q_1$, $\text{Var}(Y) = np_2 q_2$, and X and Y are independent,

$$\text{Var}\left(\frac{X}{m} - \frac{Y}{n}\right) = \text{Var}\left(\frac{X}{m}\right) + \text{Var}\left(\frac{Y}{n}\right) = \frac{1}{m^2}\,\text{Var}(X) + \frac{1}{n^2}\,\text{Var}(Y)$$

$$= \frac{p_1 q_1}{m} + \frac{p_2 q_2}{n}$$

The null hypothesis is $H_0 : p_1 = p_2$, so when H_0 is true there is a common value for the two population proportions, which we denote by p. In the one sample problem, we standardized $\hat{p}$ assuming that H_0 was true, which meant dividing by $\sqrt{p_0 q_0 / n}$. To standardize $\hat{p}_1 - \hat{p}_2$ assuming H_0 to be true, we substitute the common value p into (8.9) in place of p_1 and p_2, yielding

$$\begin{array}{c} \text{variance of } \hat{p}_1 - \hat{p}_2 \\ \text{when } H_0 \text{ is true} \end{array} = \frac{pq}{m} + \frac{pq}{n} = pq\left(\frac{1}{m} + \frac{1}{n}\right) \tag{8.10}$$

If both m and n are large, then both $\hat{p}_1$ and $\hat{p}_2$ have approximately normal distributions, so $\hat{p}_1 - \hat{p}_2$ (being the difference of two approximately normal variables) does, too. When $H_0 : p_1 = p_2$ is true, the standard deviation of $\hat{p}_1 - \hat{p}_2$ is the square root of (8.10), so the variable

$$\frac{\hat{p}_1 - \hat{p}_2 - 0}{\sqrt{pq(1/m + 1/n)}} \tag{8.11}$$

has approximately a standard normal distribution. Unfortunately, (8.11) cannot serve as a test statistic, since the denominator involves p. To obtain a test statistic having approximately a standard normal distribution when H_0 is true, the common proportion p must be estimated.

Assuming then that $p_1 = p_2 = p$, instead of separate samples of size m and n from two different populations (two different binomial distributions), we really have a single sample of size $m + n$ from one population with proportion p. Since the total number of individuals in this combined sample having the characteristic of interest is $X + Y$, the estimator of p is

$$\hat{p} = \frac{X + Y}{m + n} = \frac{m}{m + n}\hat{p}_1 + \frac{m}{m + n}\hat{p}_2 \tag{8.12}$$

The far right hand side of (8.12) shows that p is actually a weighted average of estimators $\hat{p}_1$ and $\hat{p}_2$ obtained from the two samples. If we take (8.12) (with $\hat{q} = 1 - \hat{p}$) and substitute back into (8.11), the resulting statistic has approximately a standard normal distribution when H_0 is true.

Null hypothesis: $H_0 : p_1 - p_2 = 0$

Test statistic (large samples): $Z = \dfrac{\hat{p}_1 - \hat{p}_2}{\sqrt{\hat{p}\hat{q}(1/m + 1/n)}}$

Alternative Hypothesis	*Rejection Region for Approximate Level α Test*
$H_a : p_1 - p_2 > 0$	$Z \geq z_\alpha$
$H_a : p_1 - p_2 < 0$	$Z \leq -z_\alpha$
$H_a : p_1 - p_2 \neq 0$	either $Z \geq z_{\alpha/2}$ or $Z \leq -z_{\alpha/2}$

Example 8.9   The paper "Evaluation of Telephone Energy Information Centers in Minnesota" (J. *Environmental Systems*, 1980, pp. 229–248) reported that in a random sample of 248 callers to a center operated by the Northern States Power Co., 84 had purchased at least one new appliance rated as energy efficient, while in a control sample of 270 individuals who had not called the center, 59 had purchased such an appliance. Test at level .05 to see if the data indicates that the true proportion of callers who purchased such an appliance exceeds the true proportion of noncallers who made this type of purchase.

With p_1 and p_2 denoting the true proportion of all callers and noncallers, respectively, who purchased an energy-efficient appliance, the hypotheses of interest are $H_0 : p_1 - p_2 = 0$ versus $H_a : p_1 - p_2 > 0$. At level .05, H_0 will be rejected if $Z \geq z_{.05} = 1.645$. The necessary sample quantities are $\hat{p}_1 = x/m = 84/248 = .339$, $\hat{p}_2 = y/n = 59/270 = .219$, and $\hat{p} = (x + y)/(m + n) = (84 + 59)/(248 + 270) = 133/518 = .257$, so the computed value of Z is

$$z = \frac{.339 - .219}{\sqrt{(.257)(.743)\left(\frac{1}{248} + \frac{1}{270}\right)}} = \frac{.120}{.0384} = 3.12$$

Since 3.12 is ≥ 1.645, H_0 is rejected in favor of the conclusion that $p_1 > p_2$.

Example 8.10   In an experiment to investigate whether the response rate on a questionnaire could be improved by using a semipersonal as opposed to a form covering letter ("Type of Transmittal Letter and Questionnaire Color as Two Variables Influencing Response Rate in a Mail Survey," *J. Applied Psychology*, 1974, pp. 535–536), of the 1018 subjects receiving the semipersonal letter, 325 responded, while only 225 of the 1022 receiving the form letter responded. Is the extra expense associated with a semipersonal letter justified?

Let p_1 and p_2 be the true response proportions (probabilities) for semipersonal and form letters, respectively. Then because of the extra expense, we want to conclude that $p_1 - p_2 > 0$ only if the data strongly supports this claim. Putting the burden of proof on the alternative, we wish to test $H_0 : p_1 - p_2 = 0$ versus $H_a : p_1 - p_2 > 0$. Because H_a is one sided, an upper-tailed test (reject H_a if $Z \geq z_\alpha$) is appropriate. Computations give

$$\hat{p}_1 = \frac{325}{1018} = .319, \hat{p}_2 = \frac{225}{1022} = .220, \hat{p} = \frac{550}{2040} = .270$$

so

$$z = \frac{.319 - .220}{\sqrt{(.270)(.730)\left(\frac{1}{1018} + \frac{1}{1022}\right)}} = \frac{.099}{.0197} = 5.04$$

Since $5.04 \geq 3.08$, H_0 is rejected at level .001, and we conclude that a semipersonal letter brings a significantly higher response rate (at least for questionnaires resembling the one used in this experiment).

Power, Type II Error Probabilities, and Sample Sizes

The power (or β) of the test is somewhat more difficult to compute than was the case for other large-sample tests. The reason is that the denominator of Z is an estimate of the standard deviation of $\hat{p}_1 - \hat{p}_2$ assuming that $p_1 = p_2 = p$. When H_0 is false, $\hat{p}_1 - \hat{p}_2$ must be restandardized using

$$\sigma_{\hat{p}_1 - \hat{p}_2} = \sqrt{\frac{p_1 q_1}{m} + \frac{p_2 q_2}{n}} \tag{8.13}$$

Because the power is not a function of just the difference $p_1 - p_2$, we denote it by $\pi(p_1, p_2)$.

Alternative Hypothesis	Power $= \pi(p_1, p_2)$
$H_a : p_1 - p_2 > 0$	$1 - \Phi\left[\dfrac{z_\alpha \sqrt{\bar{p}\bar{q}\,(1/m + 1/n)} - (p_1 - p_2)}{\sigma}\right]$
$H_a : p_1 - p_2 < 0$	$\Phi\left[\dfrac{-z_\alpha \sqrt{\bar{p}\bar{q}\,(1/m + 1/n)} - (p_1 - p_2)}{\sigma}\right]$
$H_a : p_1 - p_2 \neq 0$	$1 - \Phi\left[\dfrac{z_{\alpha/2} \sqrt{\bar{p}\bar{q}\,(1/m + 1/n)} - (p_1 - p_2)}{\sigma}\right]$
	$\quad + \Phi\left[\dfrac{-z_{\alpha/2} \sqrt{\bar{p}\bar{q}\,(1/m + 1/n)} - (p_1 - p_2)}{\sigma}\right]$

where $\bar{p} = (mp_1 + np_2)/(m + n)$, $\bar{q} = (mq_1 + nq_2)/(m + n)$, and σ is given by (8.13)

Proof: For the upper-tailed test ($H_a : p_1 - p_2 > 0$)

$$\pi(p_1, p_2) = P\left[\hat{p}_1 - \hat{p}_2 \geq z_\alpha \sqrt{\hat{p}\hat{q}\,(1/m + 1/n)}\right]$$

$$= P\left[\frac{\hat{p}_1 - \hat{p}_2 - (p - p_2)}{\sigma} \geq \frac{z_\alpha \sqrt{\hat{p}\hat{q}\,(1/m + 1/n)} - (p_1 - p_2)}{\sigma}\right]$$

When m and n are both large, $\hat{p} = (m\hat{p}_1 + n\hat{p}_2)/(m + n) \doteq (mp_1 + np_2)/(m + n) = \bar{p}$, and $\hat{q} \doteq \bar{q}$, which yields the above (approximate) expression for $\pi(p_1, p_2)$.

As before, $\beta(p_1, p_2) = 1 - \pi(p_1, p_2)$. Alternatively, for specified p_1, p_2 with $p_1 - p_2 = d$, the sample sizes necessary to achieve $\beta(p_1, p_2) = \beta$ can be determined. For example, for the upper-tailed test, we equate $-z_\beta$ to the argument of $\Phi(\cdot)$ (that is, what's inside the parentheses) in the above table. If $m = n$, there is a simple expression for the common value.

For the case $m = n$, the level α test has type II error probability β at the alternative values p_1, p_2 with $p_1 - p_2 = d$ when

$$n = \frac{\left[z_\alpha \sqrt{(p_1 + p_2)(q_1 + q_2)/2} + z_\beta \sqrt{p_1 q_1 + p_2 q_2}\right]^2}{d^2} \qquad (8.14)$$

for an upper- or lower-tailed test, with $\alpha/2$ replacing α for a two-tailed test.

Example 8.11   One of the truly impressive applications of statistics occurred in connection with the design of the 1954 Salk Polio Vaccine experiment and analysis of the resulting data. Part of the experiment focused on the efficacy of the vaccine in combating paralytic polio. Because it was felt that without a control group of children, there would be no sound basis for assessment of the vaccine, it was decided to administer the vaccine to one group and a placebo injection (visually indistinguishable from the vaccine but known to have no effect) to a control group. For ethical reasons and also because it was thought that knowledge of vaccine administration might have an effect on treatment and diagnosis, the experiment was conducted in a **double blind** manner. That is, neither the individuals receiving injections nor those administering them actually knew who was receiving vaccine and who was receiving the placebo (samples were numerically coded)—remember, at that point it was not at all clear whether the vaccine was beneficial.

Letting p_1 and p_2 be the probabilities of a child getting paralytic polio for the control and treatment condition, respectively, the objective was to test $H_0 : p_1 - p_2 = 0$ versus $H_a : p_1 - p_2 > 0$ (the alternative states that a vaccinated child is less likely to contract polio than an unvaccinated child). Supposing that the true value of p_1 is .0003 (an incidence rate of 30 per 100,000), the vaccine would be a significant improvement if the incidence rate was halved—that is, $p_2 = .00015$. Using a level $\alpha = .05$ test, it would then be reasonable to ask for sample sizes for which $\beta = .1$ when $p_1 = .0003$ and $p_2 = .00015$. Assuming equal sample sizes, the required n is obtained from (8.14) as

$$n = \frac{\left[1.645 \sqrt{(.5)(.00045)(1.99955)} + 1.28 \sqrt{(.00015)(.99985) + (.003)(.9997)}\right]^2}{(.0003 - .00015)^2}$$

$$= [(.0349 + .0272)/.00015]^2 \doteq 171,400$$

The actual data for this experiment appears below. Sample sizes of approximately 200,000 were used. The reader can easily verify that $z = 6.47$, a highly significant value. The vaccine was judged a resounding success! An excellent expository article describing the experiment and results appears in *Statistics: A Guide to the Unknown* edited by Judith Tanur.

placebo: $m = 201,229$ $x =$ number of cases of paralytic polio $= 110$

vaccine: $n = 200,745$ $y = 33$

Test Procedure for $H_0 : p_1 - p_2 = \Delta_0$

When the null hypothesis is $H_0 : p_1 - p_2 = \Delta_0$ with $\Delta_0 \neq 0$, the above Z statistic is not appropriate, because $\hat{p}_1 - \hat{p}_2$ was standardized assuming that $p_1 = p_2$. In the case under consideration, it is customary to standardize $\hat{p}_1 - \hat{p}_2$ using (8.9) with the estimates substituted in place of p_1 and p_2, yielding the test statistic

$$Z = \frac{\hat{p}_1 - \hat{p}_2 - \Delta_0}{\sqrt{\dfrac{\hat{p}_1 \hat{q}_1}{m} + \dfrac{\hat{p}_2 \hat{q}_2}{n}}}$$

The test statistic is now used in conjunction with the usual rejection regions to yield approximate level α tests. Power calculations and sample size determinations can be made, but since $\Delta_0 \neq 0$ arises much less frequently than $\Delta_0 = 0$, the details are omitted.

A Small-Sample Test for $H_0 : p_1 - p_2 = 0$

Although the overwhelming majority of problems involving a comparison of p_1 and p_2 yield large sample sizes, on occasion a small-sample-size problem must be confronted. In this case either $\hat{p}_1$ or $\hat{p}_2$ (or both) cannot be assumed to have approximately a normal distribution, so the Z tests just discussed are not valid. In Chapter 3, for a small-sample single population problem the test statistic was X itself, which had a binomial distribution. The test statistic proposed here for the two-sample problem will have a hypergeometric distribution when H_0 is true. The test procedure is called the Fisher-Irwin test, named after the two individuals who first proposed it. To fix ideas, consider the following example.

Example 8.12 In a study on the effect of monosodium glutamate (MSG) on reproductive ability ("Monosodium Glutamate Administration to the Newborn Reduces Reproductive Ability in Female and Male Mice," *Science*, vol. 196, pp. 452–454), MSG-treated and control female mice were mated with control males. Of the 10 control females, nine subsequently became pregnant, while only three of the MSG-treated females became pregnant. Does this data indicate that MSG treatment reduces the likelihood of pregnancy among mice?

If we let p_1 be the probability that a randomly chosen control mouse will subsequently become pregnant and p_2 be the probability of a pregnancy for an MSG treated mouse (so p_1 and p_2 are proportions in hypothetical populations), then the question of interest can be answered by testing $H_0 : p_1 - p_2 = 0$ versus $H_a : p_1 - p_2 > 0$. Following previously defined notation, $m = n = 10$ and $x = 9$, $y = 3$.

The development of the test procedure requires that we regard the total number of successes $X + Y$ as being fixed at the value t. Then given that $X + Y = t$, large values of X (along with small values of Y) support H_a, while small or moderate values of X suggest that H_0 should be retained. For fixed t, we therefore want to reject H_0 when X is large (its upper limit will be m or t, whichever is smaller), so we seek the probability distribution of X when H_0 is true given that $X + Y = t$. In Example 8.11, given that the total number of pregnancies was 12, we wish the distribution of the number of pregnant control mice X.

The Conditional Distribution of X and the Test Procedure

Given that the combined sample contains t successes, when H_0 is true any of these successes is no more or no less likely to be from the first population than it is to be from the second. This implies that all ways of breaking up the $m + n$ responses in the combined sample into a first sample of size m and a second of size n are equally likely. That is, any of the $\binom{m + n}{m}$ ways of selecting the m responses for the first sample has the same probability under H_0. The number of ways of selecting the m responses for the first sample to include exactly x successes (leaving $t - x$ for the second sample) is $\binom{t}{x} \cdot \binom{m + n - t}{m - x}$. This is because for each way of choosing exactly x of the t successes, there are $m + n - t$ failures from which the remaining sample one responses are chosen. Because outcomes are equally likely, the probability of exactly x sample one successes is the ratio of the number of outcomes having x sample one successes to the total number of outcomes:

$$P(X = x \mid t \text{ successes in } m + n \text{ responses})$$

$$= \frac{\binom{t}{x}\binom{m + n - t}{m - x}}{\binom{m + n}{m}} \quad \text{when } H_0 \text{ is true} \qquad (8.15)$$

By comparison with (1) of Section 3.6, (8.15) can be recognized as a probability having the form of a hypergeometric distribution. This can be paraphrased by saying *"conditional on t successes among the m + n sample responses, the number of successes X in the first sample has a hypergeometric distribution when H_0 is true."*

In the MSG example, we wish to reject H_0 if X is too large. The observed value of X is $x = 9$. To decide whether this is too large, we select a level of significance α and use (8.15) to determine the value of c for which $P(X \geq c \mid t \text{ successes}) \doteq \alpha$. If the observed x is $\geq c$, then H_0 is rejected at level α. Put another way, we can compute the probability of a value at least as extreme as the observed x—this is the P-value. If this P-value is too small, then H_0 is rejected. For our example, $m = n = 10$, $t = 9 + 3 = 12$, and the observed value of X is $x = 9$; the only value more extreme than 9 is 10, so

$$P(X = 9 \mid 12 \quad \text{successes}) = \frac{\binom{12}{9}\binom{8}{1}}{\binom{20}{10}} = .0095$$

$$P(X = 10 \mid 12 \quad \text{successes}) = \frac{\binom{12}{10}\binom{8}{0}}{\binom{20}{10}} = .0003$$

The P-value for the data is $.0095 + .0003 = .0098$, so at level $\alpha = .01$, H_0 is rejected in favor of H_a.

If the alternative is $H_a : p_1 - p_2 < 0$, then H_0 is rejected when X is too small. Expression (8.15) is then used to compute the probability of each value $\leq$ the observed x; if this P-value is less than the desired α, H_0 is rejected. When H_a is two sided, a two-tailed rejection region is appropriate. Then (8.15) is used to compute probabilities for both small and large x values; the rejection region is determined by putting in successively those x values whose probabilities as computed by (8.15) are larger and larger, until no more x values can be entered without the cumulative probability exceeding α. If the observed x is in the resulting region, H_0 is rejected. Since a description using symbols is a bit cumbersome, we content ourselves with this word description.

Because the test depends on using (8.15) to determine α, and (8.15) hinges on supposing $X + Y$ to be fixed at t, some statisticians have quarreled with the use of the Fisher-Irwin test when $X + Y$ is not actually fixed by the sampling scheme (which it was not in our example). However, in the absence of any more desirable competitors, the Fisher-Irwin test is often used whether or not $X + Y$ is fixed in advance. Power, type II error probabilities, and sample sizes considerations are beyond the scope of the text. For a more comprehensive discussion of the test see either the book by Daniel or the book by Marasciulo and McSweeny, both listed in the bibliography for Chapter 15.

Exercises / Section 8.4

1. A sample of 300 urban adult residents of a particular state revealed 63 who favored increasing the highway speed limit from 55 to 65 mph, while a sample of 180 rural residents yielded 75 who favored the increase. Does this data indicate that the sentiment for increasing the speed limit is stronger among rural than among urban residents?
   a. Test $H_0 : p_1 = p_2$ versus $H_a : p_1 < p_2$ using $\alpha = .05$, where p_1 refers to the urban population.
   b. If the true proportions favoring the increase are actually $p_1 = .20$ (urban) and $p_2 = .40$ (rural),

what is the probability that H_0 will be rejected using a level .05 test with $m = 300$, $n = 180$?

2. A study of the relationship between level of occupational prestige and self-evaluation of competence yielded the following data, where success here refers to an individual who had a high self-competence rating ("The Influence of Educational Attainment on Self-Evaluation of Competence," *Sociology Education*, 1972, pp. 303–311).

Low occupational prestige: $m = 175$, $x = 77$

High occupational prestige: $n = 194$, $y = 107$

Does the data indicate that a person with high occupational prestige is more likely to exhibit high self-esteem than an individual with low occupational prestige? Test the relevant hypotheses using $\alpha = .05$.

3. A random sample of 5726 telephone numbers taken in March 1970 yielded 1105 which were unlisted, and one year later a sample of 5384 yielded 980 unlisted numbers.
   a. Test at level .10 to see if there is a difference in true proportions of unlisted numbers between the two years.
   b. If $p_1 = .20$ and $p_2 = .18$, what sample sizes ($m = n$) would be necessary to detect such a difference with probability .90?

4. A recent study (*Lancet*, October 1976) reported the following data on survival rate among patients suffering cardiac arrest both when resuscitation was started by (trained) lay people and when it was delayed until the arrival of an ambulance crew.

Lay: $m = 75$, number survived $= x = 27$

Ambulance: $n = 556$, number survived $= y = 43$

Does the data suggest that a program to train lay people to perform resuscitation would be beneficial?

5. To study the extent of fundamentalism among participating Catholic and Protestant adolescents, samples of 150 Catholics and 400 Protestants were obtained. Each individual was asked, "Do you believe that God sends bad luck and sickness to people or punishes them when they do wrong?" A yes answer to this question was interpreted as evidence of fundamental beliefs. Among the Catholic youths there were 69 yes responses, while among the Protestants there were 238 yes responses. Use a level .01 test to see if there is any difference between Catholic

and Protestant adolescents in their degree of fundamentalism.

6. A study sponsored by the National Institute of Allergy and Infectious Diseases (*Newsweek*, August 22, 1977) involved 28 patients stricken by herpes encephalites, a viral disease involving severe inflammation of the brain. Of the $m = 10$ patients who received a placebo, $x = 7$ died, while of 18 who received an experimental drug called ara-A, $y = 5$ died. Let p_1 and p_2 be the true death proportions for the placebo and drug, respectively. Regarding $X + Y = 12$ as fixed, use the Fisher-Irwin procedure to test $H_0 : p_1 = p_2$ versus $H_a : p_1 > p_2$ by computing the P-value and rejecting H_0 if $P < .01$.

7. The paper "Discrimination Behavior and Hybridization of the Blue-Winged and Golden-Winged Warblers" (*Evolution*, 1972, pp. 282–293) reported the following data on ability of two different species to discriminate between songs of their own and other species.

	Discrimination	Nondiscrimination
Blue-winged	4	6
Golden-winged	3	9

Use the Fisher-Irwin test to decide whether the data suggests any difference between the true proportions of discriminators for the two species.

8. Sometimes experiments involving success/failure responses are run in a paired or before/after manner. Suppose that before a major policy speech by a political candidate, n individuals are selected and asked whether (S) or not (F) they favor the candidate. Then after the speech the same n people are asked the same question. The responses can be entered in a square table as follows:

		After	
		S	*F*
Before	*S*	X_1	X_2
	F	X_3	X_4

where $X_1 + X_2 + X_3 + X_4 = n$

Let p_1, p_2, p_3, and p_4 denote the four cell probabilities, so that $p_1 = P(S$ before and S after), and so on. We wish to test the hypothesis that the true proportion of supporters (S) after the speech has not increased against the alternative that it has increased.

a. State the two hypotheses of interest in terms of p_1, p_2, p_3, and p_4.

b. Construct an estimator for the after/before difference in success probabilities.

c. When n is large, it can be shown that the r.v. $(X_i - X_j)/n$ has approximately a normal distribution with variance $[p_i + p_j - (p_i - p_j)^2]/n$.

Use this to construct a test statistic which has approximately a standard normal distribution when H_0 is true (the result is called McNemar's test).

d. If $x_1 = 350$, $x_2 = 150$, $x_3 = 200$, and $x_4 = 300$, what do you conclude?

8.5 Tests for the Equality of Population Variances

Tests for the equality of two population variances (or standard deviations) are occasionally needed, though such problems arise much less frequently than those involving means or proportions. We consider both the case in which the populations under investigation are normal and the case in which the sample sizes are both large. To obtain test procedures for the first case, a new family of probability distributions must be introduced and used.

The F Distribution

The F probability distribution has two parameters, denoted by ν_1 and ν_2. The parameter ν_1 is called the number of numerator degrees of freedom, and ν_2 is the number of denominator degrees of freedom; here ν_1 and ν_2 are positive integers. A random variable which has an F distribution cannot assume a negative value. Since the density function is complicated and will not be used explicitly, we omit the formula. There is an important connection between an F variable and chi-squared variables. If X_1 and X_2 are independent chi-squared r.v.'s with ν_1 and ν_2 degrees of freedom, respectively, then the r.v.

$$F = \frac{X_1/\nu_1}{X_2/\nu_2} \tag{8.16}$$

the ratio of the two chi-squared variables divided by their d.f.'s, can be shown to have an F distribution.

Figure 8.1 illustrates the graph of a typical F density function. Analogous to the notation $t_{\alpha, \nu}$ and $\chi^2_{\alpha, \nu}$, we use F_{α, ν_1, ν_2} for the point on the axis which captures α of the area under the F density function with ν_1 and ν_2 d.f. in the upper tail. The density function is not symmetric, so it would seem that both upper- and lower-tail critical values must be tabulated. This is not necessary, though, because

Figure 8.1 An F density function

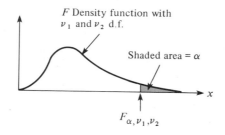

$$F_{1-\alpha, \nu_1, \nu_2} = 1/F_{\alpha, \nu_2, \nu_1} \qquad (8.17)$$

Appendix Table A.7 gives F_{α, ν_1, ν_2} for $\alpha = .05$ and $.01$, and various values of ν_1 (in different columns of the table) and ν_2 (in different rows of the table). For example, $F_{.05, 6, 10} = 3.22$ and $F_{.05, 10, 6} = 4.06$. To obtain $F_{.95, 6, 10}$, the number which captures $.95$ of the area to its right (and thus $.05$ to the left) under the F curve with $\nu_1 = 6$ and $\nu_2 = 10$, we use (8.17): $F_{.95, 6, 10} = 1/F_{.05, 10, 6} = 1/4.06 = .246$.

Tests When the Populations Are Normal

> **Theorem:** Let $X_1, \ldots, X_m$ be a random sample from a normal distribution with variance σ_1^2, let $Y_1, \ldots, Y_n$ be another random sample (independent of the X_i's) from a normal distribution with variance σ_2^2, and let S_1^2 and S_2^2 denote the two sample variances. Then the random variable
>
> $$F = \frac{S_1^2/\sigma_1^2}{S_2^2/\sigma_2^2}$$
>
> has an F distribution with $\nu_1 = m - 1$ and $\nu_2 = n - 1$.

This theorem results from combining (8.16) with the fact that $(m - 1)S_1^2/\sigma_1^2$ and $(n - 1)S_2^2/\sigma_2^2$ each have a chi-squared distribution with $m - 1$ and $n - 1$ d.f., respectively (see Section 7.5). Because F involves a ratio rather than a difference, the test statistic is the ratio of sample variances; the claim that $\sigma_1^2 = \sigma_2^2$ is then rejected if the ratio differs by too much from 1.

Null hypothesis: $H_0 : \sigma_1^2 = \sigma_2^2$
Test statistic: $F = S_1^2/S_2^2$

Alternative Hypothesis	*Rejection Region for a Level α Test*
$H_a : \sigma_1^2 > \sigma_2^2$	$F \geq F_{\alpha, m-1, n-1}$
$H_a : \sigma_1^2 < \sigma_2^2$	$F \leq F_{1-\alpha, m-1, n-1}$
$H_a : \sigma_1^2 \neq \sigma_2^2$	either $F \geq F_{\alpha/2, m-1, n-1}$ or $F \leq F_{1-\alpha/2, m-1, n-1}$

Since critical values are tabled only for $\alpha = .05$ and $.01$, the two-tailed test can be performed only at levels $.10$ or $.02$. More extensive F tables for testing at other levels are available elsewhere.

Example 8.13 On the basis of data reported in a 1979 article in the *Journal of Gerontology* ("Serum Ferritin in an Elderly Population," pp. 521–24), the authors concluded that the ferritin distribution in the elderly had a smaller variance than in the younger adults. (Serum ferritin is used in diagnosing iron deficiency.) For a sample of 28 elderly males the sample standard deviation of serum ferritin (mg/ℓ) was $s_1 = 52.6$, and for

26 young males the sample standard deviation was $s_2 = 84.2$. Does this data support the above conclusion as applied to males?

Let σ_1^2 and σ_2^2 denote the variance of the serum ferritin distributions for elderly males and young males, respectively. We wish to test $H_0 : \sigma_1^2 = \sigma_2^2$ versus $H_a : \sigma_1^2 < \sigma_2^2$. At level .05, H_0 will be rejected if $F \leq F_{.95, 27, 25}$. To obtain the critical value, we need $F_{.05, 25, 27}$. From the F table, $F_{.05, 25, 27} \doteq 2.54$ (since $F_{.05, 24, 27} = 2.55$), so $F_{.95, 27, 25} \doteq 1/2.54 = .394$. The computed value of F is $(52.6)^2/(84.2)^2 = .390$. Since $.390 \leq .394$, H_0 is rejected at level .05 in favor of H_a, so variability does appear to be greater in young males than in elder males.

The book by Bowker and Lieberman contains graphs of curves giving P (type II error) for the F test.

The validity of the two-sample t test rested on the assumption that $\sigma_1^2 = \sigma_2^2$. It is tempting to use the F test as a preliminary test to decide whether or not the two-sample t test should be used (use it if $H_0 : \sigma_1^2 = \sigma_2^2$ cannot be rejected in favor of $H_a : \sigma_1^2 \neq \sigma_2^2$), but we do not recommend this. The reason is that the F test may yield a significant value of F not only when H_0 is false but also when the underlying populations are not normal—it is sensitive to departures from normality as well as from H_0. Since the t test is a good test as long as the underlying populations are not too nonnormal, using a preliminary F test might result in not using the t test for the wrong reason. An additional complication is that the compound procedure (F at level α followed by t at level α) will not necessarily have level of significance α.

Large-Sample Tests for $H_0 : \sigma_1^2 = \sigma_2^2$

When both m and n are large, a test procedure which does not presume normal populations can be based on the fact that S_1 and S_2 have approximately normal distributions with means σ_1 and σ_2, respectively, and variances $\sigma_1^2/2m$ and $\sigma_2^2/2n$, respectively. When $H_0 : \sigma_1^2 = \sigma_2^2$ is true, the r.v.

$$Z = \frac{S_1 - S_2}{\sigma\sqrt{1/2m + 1/2n}} \tag{8.18}$$

has approximately a standard normal distribution, where σ is the common value of σ_1 and σ_2. Since σ is unknown, we obtain a test statistic by using S_p, the pooled estimator of σ, in place of σ in (8.18). Rejection regions for the standard alternatives then have exactly the same form as for other Z tests. We omit the details.

Exercises / Section 8.5

1. Obtain or compute the following quantities:
   a. $F_{.05, 5, 8}$ **b.** $F_{.05, 8, 5}$ **c.** $F_{.95, 5, 8}$ **d.** $F_{.95, 8, 5}$
   e. the ninety-ninth percentile of the F distribution with $\nu_1 = 10$, $\nu_2 = 12$
   f. the first percentile of the F distribution with $\nu_1 = 10$, $\nu_2 = 12$

   g. $P(F \leq 6.16)$ for $\nu_1 = 6$, $\nu_2 = 4$
   h. $P(.177 \leq F \leq 4.74)$ for $\nu_1 = 10$, $\nu_2 = 5$

2. The sample standard deviation of sodium concentration in whole blood (m-equiv./ℓ) for $m = 20$

marine eels was found to be $s_1 = 40.5$, while the sample standard deviation of concentration for $n = 20$ fresh water eels was $s_2 = 32.1$ ("Ionic Composition of the Plasma and Whole Blood of Marine and Fresh Water Eels," *Comp. Biochemistry and Physiology*, 1974, pp. 541–544). Assuming normality of the two concentration distributions, test at level .10 to see if the data suggests any difference between concentration variances for the two types of eels.

3. A general contractor is considering purchasing lumber from one of two different suppliers. A sample of 12 2 in. $\times$ 4 in.'s of a certain length is obtained from each supplier and the length of each board is measured. The sample standard deviation of length for the first supplier's boards is found to be $s_1 = .13$ in., while $s_2 = .17$ in. for the second supplier. Does this data indicate that the lengths of one supplier's 2 in. $\times$ 4 in.'s are subject to more variability than those of the other supplier? Test using $\alpha = .02$, assuming normality.

4. The fat content (%) was determined for $m = 10$ samples of ground chuck and $n = 16$ samples of ground beef purchased at a certain market, yielding $s_1 = 1.4$ and $s_2 = 2.2$. Does this data indicate that there is more variability in the fat content of ground beef than of ground chuck? Assuming normality, test the relevant hypotheses using $\alpha = .05$.

5. In a study of copper deficiency in cattle, the copper values (μg Cu/100 ml blood) were determined both for cattle grazing in an area known to have a well-defined molybdenum anomalies (metal values in excess of the normal range of regional variation) and for cattle grazing in a nonanomalous area ("An Investigation into Copper Deficiency in Cattle in the Southern Pennines," *J. Agricultural Soc. Cambridge*, 1972, pp.157–163), resulting in $s_1 = 21.5$ ($m = 48$) for the anomalous condition and $s_2 = 19.45$ ($n = 45$) for the nonanomalous condition. Use the large-sample procedure to test for the equality versus inequality of population variances ($\alpha = .05$).

Supplementary Exercises / Chapter 8

1. The molar percentage of bile salts was determined both for a sample of 12 children suffering from cystic fibrosis who were receiving a pancreatic enzyme treatment and for a second sample of 14 who had ceased taking the treatment for one week. Does the data suggest that cessation of the treatment leads to an increase in the true average molar percentage of bile salts? Test the relevant hypotheses at level .05; be sure to state any assumptions which underlie your analysis.

Treated: 54.8, 70.8, 54.3, 64.5, 50.5, 73.3, 86.4, 67.4, 82.2, 55.1, 60.6, 50.8

Untreated: 61.5, 54.8, 62.4, 74.3, 80.3, 77.7, 78.2, 74.0, 48.3; 61.0, 36.5, 57.0, 40.9, 43.7

("Abnormal Biliary Lipid Composition in Cystic Fibrosis," *New England J. Medicine*, December 15, 1977, pp. 1301–1305.)

2. In Exercise 1 above, describe how the experiment could be done in a paired fashion using 13 children. What are the advantages and disadvantages of such an approach? Would pairing in this particular situation appear to be a good strategy? Why or why not?

3. A famous paper on the effects of marijuana smoking ("Clinical and Psychological Effects of Marijuana in Man," *Science*, December 1968, pp. 1234–1241) described the results of an experiment in which the change in heartbeat rate was measured for $n = 9$ naive subjects (who had never before used marijuana) both 15 minutes after smoking at a low dose level and 15 minutes after smoking a placebo (untreated) cigarette.

Subject:	1	2	3	4	5	6	7	8	9
Placebo:	16	12	8	20	8	10	4	−8	8
Low dose:	20	24	8	8	4	20	28	20	20

Does the data suggest that marijuana smoking leads to a greater increase in heartbeat rate than does smoking a placebo cigarette? Test using $\alpha = .01$.

4. In the article referenced in Exercise 3, data was also given for $m = 8$ chronic users of marijuana whose change in heartbeat rate was measured 15 minutes after smoking a low dose:

32, 36, 20, 8, 32, 54, 24, 60

Does the data indicate that chronic users experience on average a greater change in heartbeat rate than do naive subjects for the same experimental conditions? *Hint:* For naive subjects, $s_1 = 8.19$, and for chronic users, $s_2 = 17.14$.

5. Does the data of Exercises 3 and 4 suggest that there is more variability in change in heartbeat rate for chronic users than for naive subjects? Test using $\alpha = .01$. Does your conclusion here have any bearing on the test procedure appropriate in Exercise 4? Explain.

6. In the article referenced in Exercise 1 of Section 8.2, there were actually four experimental treatments: control nondiabetic, untreated diabetic, low-dose diabetic, and high-dose diabetic. Denote the sample size for the ith treatment by n_i and the sample variance by S_i^2 ($i = 1, 2, 3, 4$). Assuming that the true variance for each treatment is σ^2, construct a pooled estimator of σ^2 which is unbiased, and verify using rules of expected value that it is indeed unbiased. What is your estimate for the following actual data?

Treatment:	1	2	3	4
Sample size:	16	18	8	12
Sample s.d.:	.64	.81	.51	.35

7. Suppose that a level .05 test of $H_0 : \mu_1 - \mu_2 = 0$ versus $H_a : \mu_1 - \mu_2 > 0$ is to be performed assuming that $\sigma_1 = \sigma_2 = 10$ and normality of both distributions using equal sample sizes ($m = n$). Evaluate the probability of a type II error when $\mu_1 - \mu_2 = 1$ and $n = 25, 100, 2500$, and $10,000$. Can you think of real problems in which the difference $\mu_1 - \mu_2 = 1$ has little practical significance? Would sample sizes of $n = 10,000$ be desirable in such problems?

8. The following data refers to airborne bacteria count (number of colonies/cu. ft.) both for $m = 8$ carpeted hospital rooms and for $n = 8$ uncarpeted rooms ("Microbial Air Sampling in a Carpeted Hospital," *J. Environmental Health*, 1968, p. 405). Does there appear to be a difference in true average bacteria count between carpeted and uncarpeted rooms?

Carpeted: 11.8, 8.2, 7.1, 13,0, 10.8, 10.1, 14.6, 14.0
Uncarpeted: 12.1, 8.3, 3.8, 7.2, 12.0, 11.1, 10.1, 13.7

Suppose that you later learned that all carpeted rooms were in a veterans hospital while all uncarpeted rooms were in a children's hospital.

Would you be able to assess the effect of carpeting? Comment.

9. In a study of the relationship between nightmare frequency and sex, 160 men and 196 women were each asked whether nightmares occurred often (at least one per month) or seldom, with the following results:

Men: $m = 160$, $x = 55$ (often = success)
Women: $n = 192$, $y = 60$

Does the data indicate at level .05 that there is any difference in the true proportions of men and women who often have nightmares? ("Personality Characteristics of Nightmare Sufferers," *J. Nervous and Mental Diseases*, 1971, pp. 29–31).

10. In a sample of 500 male and 500 female high school students enrolled in a college preparatory curriculum, it was found that 230 of the males and only 197 of the females had been encouraged by their male counselors to go on to a four-year college. Does the data suggest that the difference in proportions between males and females encouraged to go on to a four-year college is more than .1? Test using $\alpha = .05$.

11. McNemar's test, developed in Exercise 8.4.8, can also be used when individuals are paired off (matched) to yield n pairs and then one member of each pair is given treatment 1 and the other is given treatment 2. Then X_1 is the number of pairs in which both treatments were successful, and similarly for X_2, X_3, and X_4. The test statistic for testing equal efficacy of the two treatments is $(X_2 - X_3)/\sqrt{(X_2 + X_3)}$, which has approximately a standard normal distribution when H_0 is true. Use this to test whether or not the drug ergotamine is effective in the treatment of migraine headaches.

		Ergotamine	
		S	F
Placebo	S	44	34
	F	46	30

The data is fictitious, but the conclusion agrees with that in the paper "Controlled Clinical Trial of Ergotamine Tartrate" (*British Med. J.*, 1970, pp. 325–327).

12. Let $X_1, \ldots, X_m$ be a random sample from a Poisson distribution with parameter λ_1, and let $Y_1, \ldots, Y_n$ be a random sample from another Poisson distribution with parameter λ_2. We wish to

test $H_0 : \lambda_1 - \lambda_2 = 0$ against one of the three standard alternatives. Since $\mu = \lambda$ for a Poisson distribution, when m and n are large the large sample Z test of Section 8.1 can be used. However, the fact that $\text{Var}(\overline{X}) = \lambda/n$ suggests that a different denominator should be used in standardizing $\overline{X} - \overline{Y}$. Develop a large-sample test procedure appropriate to this problem, and then apply it to the following data to test whether or not the plant densities for a particular species are equal in two different regions (where each observation is the number of plants found in a randomly located square sampling quadrat having area 1m², so for region 1, there were 40 quadrats in which one plant was observed, and so on).

Frequency:	0	1	2	3	4	5	6	7	
Region 1:	28	40	28	17	8	2	1	1	$m = 125$
Region 2:	14	25	30	18	49	2	1	1	$n = 100$

Bibliography   See the bibliography at the end of Chapter 7.

CHAPTER 9

Interval Estimation

Introduction

Almost any parameter that we might wish to estimate has as its set of possible values an entire interval of numbers. If, for example, we wish to estimate the true average net weight μ of fertilizer bags having nominal weight 50 lb, then μ might be any number between, say, 45 and 55. Even if the variable of interest is discrete, as with X = the number of phonograph records purchased by a randomly chosen customer leaving a certain record store, the true average number of records purchased per customer (μ) could be any number between, say, .5 and 4.0, and the true proportion p of customers purchasing at least one record might be any number between 0 and 1.

The fact that the set of possible parameter values is a continuum implies that a point estimate, though it will represent our best guess for the true value of the parameter, may be close to that true value but will virtually never actually equal it. Because of this, reporting only the estimated value is generally unsatisfactory; some measure of how close the point estimate is likely to be to the true value is required. One way to do this, as suggested in Section 6.1, is to report both the estimate and its standard deviation (standard error). Then, if the estimator has at least approximately a normal distribution, we can be quite confident that the true value lies within two or three standard deviations of the estimated value. This amounts to replacing the point estimate, a single number, by an entire interval of plausible values, and that is exactly what an interval estimate or confidence interval is—an interval of plausible values for the parameter being estimated. The degree of plausibility will be specified by a confidence level, so that we will speak of a 95% confidence interval (confidence level 95%) or a 99% interval.

In this chapter we obtain confidence intervals for parameters μ, p, and σ of a single population, and also for $\mu_1 - \mu_2$, $p_1 - p_2$, and σ_1/σ_2 when two populations are being studied and compared.

317

9.1 A Confidence Interval for a Mean of a Normal Population

As with hypothesis testing in Chapter 7, the first confidence interval considered here will be for the mean μ of a normal population when the value of σ is known. We then consider situations in which the assumptions of known σ and/or normality are not necessary. In particular, given the results $x_1, x_2, \ldots, x_n$ of a random sample $X_1, X_2, \ldots, X_n$ taken from a normal distribution with mean μ and known standard deviation σ, we wish to obtain an interval estimate for μ.

Example 9.1

The Institute of Nutrition of Central America and Panama (INCAP) has carried out extensive dietary studies and research projects in Central America. In one study reported in the November 1964 issue of the *American Journal of Clinical Nutrition* ("The Blood Viscosity of Various Socioeconomic Groups in Guatamala"), serum total cholesterol measurements for a sample of 49 low-income rural Indians were reported as follows (in mg/L):

204, 108, 140, 152, 158, 129, 175, 146, 157, 174, 192, 194, 144, 152, 135, 223, 145, 231, 115, 131, 129, 142, 114, 173, 226, 155, 166, 220, 180, 172, 143, 148, 171, 143, 124, 158, 144, 108, 189, 136, 136, 197, 131, 95, 139, 181, 165, 142, 162

Derive a confidence interval for the true average serum total cholesterol level μ for all low-income rural Indians residing in Guatamala. Assume that serum cholesterol level is normally distributed with $\sigma = 30$ (when n is large, the normality assumption is unnecessary, and subsequently s will be used in place of the assumed value of σ).

A 95% Confidence Interval

Because we have assumed that the X_i's constitute a random sample from a normal distribution, the sample mean $\overline{X}$ has a normal distribution with expected value μ and standard deviation $\sigma/\sqrt{n}$. This in turn implies that

$$Z = \frac{\overline{X} - \mu}{\sigma/\sqrt{n}} \tag{9.1}$$

has a standard normal distribution, where (9.1) results from standardizing $\overline{X}$. Recalling that the area under the standard normal curve between -1.96 and 1.96 equals .95,

$$P\left(-1.96 \le \frac{\overline{X} - \mu}{\sigma/\sqrt{n}} \le 1.96\right) = .95 \tag{9.2}$$

The expression within parentheses in (9.2) is composed of two inequalities, with μ appearing in the middle along with $\overline{X}$ and $\sigma/\sqrt{n}$. The objective now is to manipulate these inequalities to produce a set of equivalent inequalities $A \le \mu \le B$ where the endpoints A and B involve $\overline{X}$ and $\sigma/\sqrt{n}$. In (9.3) we give a sequence of sets of inequalities, each equivalent to that in (9.2):

$$-1.96 \cdot \frac{\sigma}{\sqrt{n}} \le \overline{X} - \mu \le 1.96 \cdot \frac{\sigma}{\sqrt{n}} \quad \text{(after multiplying through by } \sigma/\sqrt{n})$$

$$-\overline{X} - 1.96 \cdot \frac{\sigma}{\sqrt{n}} \le -\mu \le -\overline{X} + 1.96 \cdot \frac{\sigma}{\sqrt{n}} \quad \text{(after subtracting } \overline{X} \text{ from each term)} \tag{9.3}$$

$$\overline{X} - 1.96 \cdot \frac{\sigma}{\sqrt{n}} \le \mu \le \overline{X} + 1.96 \cdot \frac{\sigma}{\sqrt{n}} \quad \text{(after multiplying through by } -1)$$

The transition from the next to last to the last line of (9.3) may appear puzzling; remember that when an inequality is multipled through by a negative number, the direction of the inequality is reversed.

Because each set of inequalities in (9.3) is equivalent to the original one, the probability associated with each is .95. That is, for the last set in (9.3),

$$P\left(\overline{X} - 1.96 \frac{\sigma}{\sqrt{n}} \le \mu \le \overline{X} + 1.96 \frac{\sigma}{\sqrt{n}}\right) = .95 \tag{9.4}$$

The event inside the parentheses in (9.4) has a somewhat unfamiliar appearance; always before the random quantity has appeared in the middle with constants on both ends, as in $a \le Y \le b$. In (9.4) the random quantity appears on the two ends while the unknown constant μ appears in the middle. To interpret (9.4), think of a **random interval** having left endpoint $\overline{X} - 1.96 \cdot \sigma/\sqrt{n}$ and right endpoint $\overline{X} + 1.96 \cdot \sigma/\sqrt{n}$, which in interval notation is

$$\left(\overline{X} - 1.96 \cdot \frac{\sigma}{\sqrt{n}}, \quad \overline{X} + 1.96 \cdot \frac{\sigma}{\sqrt{n}}\right) \tag{9.5}$$

The interval (9.5) is random because the two endpoints of the interval involve a random variable. Note that the interval is centered at the sample mean $\overline{X}$, and extends $1.96 \, \sigma/\sqrt{n}$ to either side of $\overline{X}$. Thus the length of the interval is $2 \cdot (1.96) \cdot \sigma/\sqrt{n}$, which is not random; only the location of the interval (its midpoint $\overline{X}$) is random (Figure 9.1). Now (9.4) can be paraphrased by "*the probability is .95 that the random interval (9.5) includes or covers the true value of μ.*" Before any

Figure 9.1 The random interval (9.5) centered at $\overline{X}$

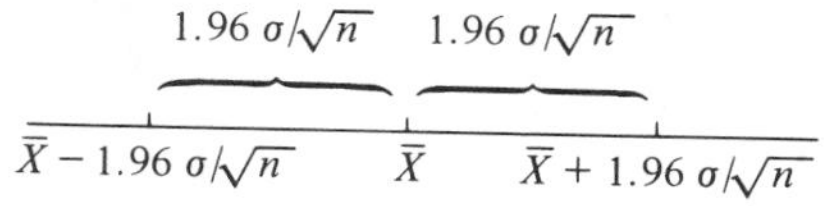

experiment is performed and any data is gathered, it is quite likely (probability .95) that μ will lie inside (9.5).

Definition: If after observing $X_1 = x_1$, $X_2 = x_2$, . . ., $X_n = x_n$, we compute the observed sample mean $\bar{x}$ and then substitute $\bar{x}$ into (9.5) in place of $\bar{X}$, the resulting fixed interval is called a **95% confidence interval for μ.** This confidence interval can be expressed either as

$$\left(\bar{x} - 1.96 \cdot \frac{\sigma}{\sqrt{n}}, \bar{x} + 1.96 \cdot \frac{\sigma}{\sqrt{n}}\right) \quad \begin{array}{l} \text{is a 95\% confidence} \\ \text{interval for } \mu \end{array}$$

or as

$$\bar{x} - 1.96 \cdot \frac{\sigma}{\sqrt{n}} < \mu < \bar{x} + 1.96 \cdot \frac{\sigma}{\sqrt{n}} \quad \text{with 95\% confidence}$$

Example 9.1 (continued)

The sample size is $n = 49$ and σ is assumed equal to 30. The computed sample mean is $\bar{x} = 157.02$, so the endpoints of the 95% interval are

$$\text{left endpoint} = \bar{x} - 1.96 \cdot \frac{\sigma}{\sqrt{n}} = 157.02 - (1.96)\frac{30}{\sqrt{49}}$$

$$= 157.02 - 8.40 = 148.62$$

$$\text{right endpoint} = \bar{x} + 1.96 \cdot \frac{\sigma}{\sqrt{n}} = 157.02 + (1.96)\frac{30}{\sqrt{49}}$$

$$= 157.02 + 8.40 = 165.42$$

The 95% interval is thus (148.62, 165.42) or $148.62 < \mu < 165.42$. This is our interval of plausible values (interval estimate) for μ at confidence level 95%.

Interpreting a Confidence Interval

The confidence level 95% for the interval defined above was inherited from the probability .95 for the random interval (9.5). Intervals having other levels of confidence will be introduced shortly. For now, though, consider how 95% confidence can be interpreted.

Because we started with an event whose probability was .95—that the random interval (9.5) would capture the true value of μ—and then used the data in Example 9.1 to compute the fixed interval (148.62, 165.42), it is tempting to conclude that μ is within this fixed interval with probability .95. But by substituting $\bar{x} = 157.02$ in for $\bar{X}$, all randomness disappears; the interval (148.62, 165.42) is not a random interval, and μ is a constant (unfortunately unknown to us), so it is *incorrect* to write $P(\mu \text{ lies in } (148.62, 165.42)) = .95$.

If a probability statement involving the fixed interval is not appropriate, how can we give meaning to "95% confidence"? The answer lies in recalling the long-run frequency interpretation of probability: To say that an event has probability .95 is to say that if the experiment on which an event A is defined is performed over and over again, in the long run A will occur 95% of the time. Suppose that we obtain another 49 observations of serum cholesterol level, and compute another 95% interval. Then suppose we consider repeating this for a third sample, a fourth sample, and so on. Letting A be the event that $\overline{X} - 1.96 \cdot \sigma/\sqrt{n} < \mu < \overline{X} + 1.96 \cdot \sigma/\sqrt{n}$, since $P(A) = .95$, in the long run 95% of our computed confidence intervals will contain μ. This is illustrated in Figure 9.2, where the vertical line cuts the measurement axis at the true (but unknown) value of μ. Notice that of the 11 intervals pictured, only intervals 3 and 11 fail to contain μ. In the long run, only 5% of the intervals so constructed would fail to contain μ.

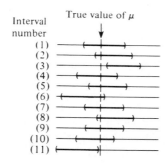

Figure 9.2 Repeated construction of 95% confidence intervals

According to this interpretation, the confidence level 95% is not so much a statement about any particular interval such as (148.62, 165.42), but pertains to what would happen if a very large number of like intervals were to be constructed. While this may seem unsatisfactory, the root of the difficulty lies with our interpretation of probability—it applies to a long sequence of replications of an experiment rather than just a single replication. There is another approach to the construction of and interpretation of confidence intervals which uses the notion of subjective probability and Bayes' theorem, but the technical details are beyond the scope of this text; the book by Winkler (see the Chapter 2 bibliography) is a good source. The interval presented here (as well as each interval presented subsequently) is called a "classical" confidence interval because its interpretation rests on the classical notion of probability (though the main ideas were developed as recently as the 1930s).

Other Levels of Confidence

Suppose that we wish a 99% confidence interval rather than a 95% interval. Then rather than starting with a probability of .95, we must begin with a probability of .99. Since the area under the standard normal curve between -2.58 and 2.58 equals .99, replacing 1.96 with 2.58 in the definition yields a 99% interval.

This suggests that any desired level of confidence can be achieved by replacing

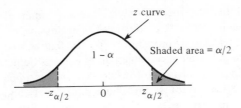

Figure 9.3 $P(-z_{\alpha/2} \le Z \le z_{\alpha/2}) = 1 - \alpha$

1.96 or 2.58 with the appropriate standard normal critical value. As Figure 9.3 shows, a probability of $1 - \alpha$ is achieved by using $z_{\alpha/2}$ in place of 1.96.

Definition: A **100 $(1-\alpha)$% confidence interval** for the mean μ of a normal population when the value of σ is known is given by

$$\left(\overline{x} - z_{\alpha/2} \cdot \frac{\sigma}{\sqrt{n}}, \overline{x} + z_{\alpha/2} \cdot \frac{\sigma}{\sqrt{n}} \right) \tag{9.6}$$

Example 9.2   A savings and loan association which finances a large number of home purchases in a particular region wanted information on the extent to which mortgage payments reduced the amount of disposable income during the initial year of occupancy for first-time home buyers. A sample of 35 recently granted mortgage applications was obtained, and for each the amount of the monthly mortgage payment as a percentage of take-home income (100 · payment/income) was computed. The sample average percentage was found to be 24.7. Assuming that σ is known to be 3.0 and that percentage is a normally distributed variable (among all successful applicants), compute a 90% confidence interval for the true average percentage μ.

From the definition above, a 90% interval requires that $100 (1 - \alpha) = 90$, so that $\alpha = .10$; then $z_{\alpha/2} = z_{.05} = 1.645$. Substituting this into (9.6) along with $\overline{x} = 24.7$, $\sigma = 3.0$, and $n = 35$ yields the interval

$$\left(24.7 - (1.645)\frac{(3.0)}{\sqrt{35}}, 24.7 + (1.645)\frac{(3.0)}{\sqrt{35}} \right) = (23.9, 25.5)$$

Although (9.6) can be used to obtain an interval with any desired degree of confidence, there are only three levels of confidence which appear at all frequently in the literature. These are 99%, 95% and 90%, and the intervals are obtained by using 2.58, 1.96, and 1.645, respectively, in place of $z_{\alpha/2}$.

Confidence Level, Precision, and Choice of Sample Size

In deciding whether to compute a 95% confidence interval or a 99% interval, it may seem ridiculous to settle for a lower level of confidence when a higher level can be obtained. However, the old adage "you must give up something to get something" remains true here. What do we give up by using the 99% interval? Recall that because the 95% interval extends $1.96 \cdot \sigma/\sqrt{n}$ to either side of $\bar{x}$, the length of the interval is $2(1.96) \cdot \sigma/\sqrt{n} = 3.92 \cdot \sigma/\sqrt{n}$. Similarly, the length of the 99% interval is $2(2.58) \cdot \sigma/\sqrt{n} = 5.16 \cdot \sigma/\sqrt{n}$. That is, we have more confidence in the 99% interval precisely because it is longer. The higher the desired degree of confidence, the longer the resulting interval. In fact the only 100% confidence interval for μ is $(-\infty, \infty)$, which is not terribly informative since even before sampling, we knew that this interval covered μ.

If we think of the length of the interval as specifying its precision or accuracy, then the confidence level (or reliability) of the interval is inversely related to its precision. A highly reliable interval estimate may be quite imprecise in that the endpoints of the interval may be far apart, while a precise interval may entail relatively low reliability. Thus it cannot be said unequivocally that a 99% interval is to be preferred to a 95% interval; the gain in reliability entails a loss in precision.

The astute reader may by this time have observed that there is the possibility of "having one's cake and eating it too." Until now we have proceeded as though the sample size n was fixed before consideration of interval length. But if we wish our 95% interval to have a prescribed length, the sample size can be chosen to achieve this.

Example 9.1 (continued)

With a sample size of $n = 49$ and $\sigma = 30$, the length of the 95% interval is $165.42 - 148.62 = 16.80$. Suppose we had wanted the sample size for which interval length was at most 5. Then with length $= L$, we wish

$$L = 5 = 2 \cdot (1.96) \cdot \frac{30}{\sqrt{n}}, \quad \text{that is}$$

$$\sqrt{n} = 2(1.96)\frac{(30)}{5} = 23.52 \quad \text{so} \quad n = 553.19$$

Because n must be an integer, a sample size of 554 is required.

The general formula for the sample size n necessary to ensure an interval length L is obtained from $L = 2 \cdot z_{\alpha/2} \cdot \sigma/\sqrt{n}$ as

$$n = \left(2z_{\alpha/2} \cdot \frac{\sigma}{L}\right)^2 \tag{9.7}$$

where if the right-hand side of (9.7) is not an integer, it must be rounded up. Notice how n behaves as a function of the length L, standard deviation σ, and confidence level $100(1 - \alpha)$:

1. As the desired length L is decreased, n increases.
2. As σ (the amount of underlying variability in the population) increases, n increases.
3. As the confidence level is increased, α is decreased, so $z_{\alpha/2}$ must increase (to increase the "capture probability" $1 - \alpha$), so n increases.

Exercises / Section 9.1

1. A confidence interval is desired for the true average stray load loss μ (watts) for a certain type of induction motor when the line current is held at 10 amps for a speed of 1500 rpm. Assume that stray load loss is normally distributed with $\sigma = 3.0$.
   a. Compute a 95% confidence interval for μ when $n = 25$ and $\bar{x} = 58.3$.
   b. Compute a 95% confidence interval for μ when $n = 100$ and $\bar{x} = 58.3$.
   c. Compute a 99% confidence interval for μ when $n = 100$ and $\bar{x} = 58.3$.
   d. Compute an 82% confidence interval for μ when $n = 100$ and $\bar{x} = 58.3$.
   e. How large must n be if the length of the 99% interval for μ is to be 1.0?

2. Assume that the helium porosity (in percentage) of coal samples taken from any particular seam is normally distributed with true standard deviation .75.
   a. Compute a 95% confidence interval for the true average porosity of a certain seam if the average porosity for 20 specimens from the seam was 4.85.
   b. Compute a 98% confidence interval for true average porosity of another seam based on 16 specimens with a sample average porosity of 4.56.
   c. How large a sample size is necessary if the length of the 95% interval is to be .40?

3. On the basis of extensive tests, the yield point of a particular type of mild steel reinforcing bar is known to be normally distributed with $\sigma = 100$. The composition of the bar has been slightly modified, but the modification is not believed to have affected either the normality or the value of σ.
   a. Assuming this to be the case, if a sample of 25 modified bars resulted in a sample average yield point of 8439 lb, compute a 90% confidence interval for the true average yield point of the modified bar.

   b. How would you modify the interval in (a) to obtain a confidence level of 92%?

4. By how much must the sample size n be increased if the length of the confidence interval (9.6) is to be halved? If the sample size is increased by a factor of 25, what effect will this have on the length of the interval? Justify your assertions.

5. Let $\alpha_1 > 0$, $\alpha_2 > 0$, with $\alpha_1 + \alpha_2 = \alpha$. Then
$$P\left(-z_{\alpha_1} \le \frac{\bar{X} - \mu}{\sigma/\sqrt{n}} \le z_{\alpha_2}\right) = 1 - \alpha$$
   a. Use this equation to derive a more general expression for a $100(1 - \alpha)\%$ confidence interval for μ of which the interval (9.6) is a special case.
   b. Let $\alpha = .05$ and $\alpha_1 = \alpha/4$, $\alpha_2 = 3\alpha/4$. Does this result in a shorter or longer interval than the interval (9.6)?

6. a. Under the same conditions as those leading to the interval (9.6), $[P(\bar{X} - \mu)/(\sigma/\sqrt{n}) \le z_\alpha] = 1 - \alpha$. Use this to derive a one-sided interval for μ which has infinite length and provides a lower bound on μ. What is this 95% interval for the data in Exercise 2(a) above?
   b. What is an analogous interval to that of (a) which provides an upper bound on μ? Compute this 99% interval for the data of Exercise 1(a) above.

7. When $X_1, \ldots, X_n$ is a random sample from a normal distribution having variance σ^2, the variable $(n - 1)S^2/\sigma^2$ has a chi-squared distribution with $n - 1$ degrees of freedom, so that
$$P\left(\chi^2_{1-\alpha/2,n-1} \le \frac{(n - 1)S^2}{\sigma^2} \le \chi^2_{\alpha/2,n-1}\right) = 1 - \alpha$$
(See Section 7.5 for an explanation of the $\chi^2_{\alpha,\nu}$ notation.)
   a. Use this equation to derive a random interval for

σ^2 having coverage probability $1 - \alpha$, and then obtain a $100(1 - \alpha)\%$ confidence interval for σ^2.

b. What is a $100(1 - \alpha)\%$ confidence interval for σ?

c. The amount of lateral expansion (mils) was de-

termined for a sample of $n = 9$ pulsed-power gas metal arc welds used in LNG ship containment tanks. The resulting sample standard deviation was $s = 2.81$ mils. Assuming normality, derive a 95% confidence interval for σ^2 and for σ.

9.2 Large-Sample Intervals for Population Means

We began our discussion of confidence intervals in the last section by imposing two assumptions, the first that the population being sampled is normal and the second that σ is known. In hypothesis testing we were able to relax these assumptions when the sample size (or sizes) was (were) large, and we now do the same for confidence intervals. While the assumption of normality is frequently reasonable, the true value of σ is rarely known; when n is large, the computed sample standard deviation s can be used in place of σ, and the normality of the population is unnecessary.

A Large-Sample Interval for μ

Let $X_1, X_2, \ldots, X_n$ be a random sample from a population having mean μ and standard deviation σ. Then provided that n is large (> 30), even if the underlying population is not normal, $\bar{X}$ has approximately a normal distribution (by the C.L.T.), so the standardized variable

$$Z = \frac{\bar{X} - \mu}{\sigma/\sqrt{n}}$$

has approximately a standard normal distribution. Now the only difference between this situation and the one faced at the outset of Section 9.1 is the presence of the word "approximately." If we then follow through the argument leading to the 95% interval or, more generally, the $100(1 - \alpha)\%$ interval (9.6), every step is valid provided that "approximately" is inserted in the appropriate place.

Summarizing, even if the population is not normal, provided that $n > 30$ the interval $(\bar{x} - z_{\alpha/2} \cdot \sigma/\sqrt{n}, \bar{x} + z_{\alpha/2} \cdot \sigma/\sqrt{n})$ is a confidence interval for μ with approximate confidence level $100(1 - \alpha)\%$. But computation of the interval still requires knowledge of σ. The result on which the "unknown σ" interval is based states that if n is large and S is the sample standard deviation, then the variable

$$Z = \frac{\bar{X} - \mu}{S/\sqrt{n}}$$

has approximately a standard normal distribution. If we use this Z as was done in Section 9.1 along with the qualifier "approximately" in the subsequent derivation, the validity of the following proposition becomes apparent.

Proposition: If the sample size n exceeds 30, then an approximate $100(1 - \alpha)\%$ confidence interval for a population mean μ is given by

$$(\bar{x} - z_{\alpha/2} \cdot s/\sqrt{n}, \bar{x} + z_{\alpha/2} \cdot s/\sqrt{n}) \tag{9.8}$$

Example 9.3   A sample of 56 research cotton samples resulted in a sample average percentage elongation of 8.17 and a sample standard deviation of 1.42 ("An Apparent Relation Between the Spiral Angle ϕ, the Percent Elongation E_1, and the Dimensions of the Cotton Fiber," *Textile Research J.*, 1978, pp. 407–410). A 95% large sample confidence interval for the true average percent elongation μ is then

$$\left(\bar{x} - \frac{1.96s}{\sqrt{n}}, \bar{x} + \frac{1.96s}{\sqrt{n}} \right)$$

$$= \left(8.17 - \frac{(1.96)(1.42)}{\sqrt{56}}, 8.17 + \frac{(1.96)(1.42)}{\sqrt{56}} \right)$$

$$= (8.17 - .37, 8.17 + .37) = (7.80, 8.54)$$

Example 9.4   In the article "Petitioners Under Chapter XIII of the Bankruptcy Act" (J. *Consumer Affairs,* vol. 3, Summer 1969), it was reported that in a sample of 250 petitioners who filed for nonbusiness bankruptcy between January 1964 and December 1966 (out of a total of 31,956 such petitioners), the sample average number of months taken to acquire the debts listed on the petitions was 35.41 with a sample standard deviation of 21.34. Compute a 99% confidence interval for the true average number of months taken by bankruptcy petitioners to acquire debts listed on their petitions.

A 99% interval requires $100(1 - \alpha) = 99$, so $\alpha = .01$ and $z_{\alpha/2} = z_{.005} = 2.58$. Substituting into (9.8) gives

$$\left(35.41 - \frac{(2.58)(21.34)}{\sqrt{250}}, 35.41 + \frac{(2.58)(21.34)}{\sqrt{250}} \right) = (31.93, 38.89)$$

Note that if we go two estimated standard deviations $(2s)$ to the left of $\bar{x}$, we are below zero. Since zero is a lower bound for number of months while there is effectively no upper bound, the actual distribution of number of months would appear to be at least somewhat positively skewed rather than normal. But of course the large sample interval does not require that the underlying distribution be normal.

Unfortunately the choice of sample size to yield a desired interval length is not as straightforward here as it was for the case of known σ. This is because the length of (9.8) is $2 \cdot z_{\alpha/2} \cdot s/\sqrt{n}$; since s is not known before the data has been gathered, the length of the interval cannot be determined solely by choice of n. The only option for an investigator who wishes to specify a desired length is to make a guess as to what the value of s (or σ) might be—by being conservative and guessing at a large value of s, an n larger than necessary will be chosen.

While the interval (9.8) has approximate confidence level $100(1 - \alpha)\%$ even when the underlying distribution is not normal, this does not say that it is the shortest $100(1 - \alpha)\%$ interval. It can be shown that when the underlying distribution is normal, (9.8) is the shortest (and therefore most precise) $100(1 - \alpha)\%$ interval.

However, if the experimenter is willing to specify another distribution in place of the normal, it may be possible to obtain an interval shorter than (9.8). Alternatively, a distribution-free interval can be used, and if the population being sampled is quite nonnormal (still symmetric but with "heavy tails" compared to the normal), then this interval will tend to be shorter than (9.8); we will expand on this in Chapter 15.

A Large-Sample Interval for $\mu_1 - \mu_2$

While interest in many two-sample problems centers on testing hypotheses about $\mu_1 - \mu_2$, sometimes an interval estimate is desired. Here we assume that we have a random sample $X_1, \ldots, X_m$ from a population with a mean μ_1 and variance σ_1^2, and another random sample $Y_1, \ldots, Y_n$, independent of the X_i's, from a population with mean μ_2 and variance σ_2^2. Because $E(\bar{X} - \bar{Y}) = \mu_1 - \mu_2$, $\mathrm{Var}(\bar{X} - \bar{Y}) = \sigma_1^2/m + \sigma_2^2/n$, and $\bar{X}$ and $\bar{Y}$ both have approximately normal distributions when m and n are both large, the standardized variable

$$Z = \frac{\bar{X} - \bar{Y} - (\mu_1 - \mu_2)}{\sqrt{\sigma_1^2/m + \sigma_2^2/n}} \tag{9.9}$$

has approximately a standard normal distribution. Manipulating the inequalities $-z_{\alpha/2} \leq Z \leq z_{\alpha/2}$ as before yields, when σ_1^2 and σ_2^2 are both known, the approximate $100(1 - \alpha)\%$ interval

$$\left(\bar{x} - \bar{y} - z_{\alpha/2} \sqrt{\frac{\sigma_1^2}{m} + \frac{\sigma_2^2}{n}}, \ \bar{x} - \bar{y} + z_{\alpha/2} \sqrt{\frac{\sigma_1^2}{m} + \frac{\sigma_2^2}{n}} \right) \tag{9.10}$$

In most problems σ_1^2 and σ_2^2 will not be known, but large sample sizes permit the substitution of sample variances s_1^2 and s_2^2 for σ_1^2 and σ_2^2.

> **Proposition:** Let $\bar{x}$ and $\bar{y}$ be the computed sample means from two independent random samples of size m and n, respectively, with computed sample variances s_1^2 and s_2^2. Then provided that m and n both exceed 30, an approximate $100(1 - \alpha)\%$ confidence interval for $\mu_1 - \mu_2$ is
>
> $$\left(\bar{x} - \bar{y} - z_{\alpha/2} \sqrt{\frac{s_1^2}{m} + \frac{s_2^2}{n}}, \ \bar{x} - \bar{y} + z_{\alpha/2} \sqrt{\frac{s_1^2}{m} + \frac{s_2^2}{n}} \right) \tag{9.11}$$

Example 9.5

Tensile strength tests were carried out on two different grades of wire rod ("Fluidized Bed Patenting of Wire Rods," *Wire J.*, June 1977, pp. 56–61), resulting in the following data.

Table 9.1

Grade	Sample size	Sample mean (kg/mm^2)	Sample st. dev.
AISI 1064	$m = 129$	$\bar{x} = 107.6$	$s_1 = 1.3$
AISI 1078	$n = 129$	$\bar{y} = 123.6$	$s_2 = 2.0$

Let μ_1 and μ_2 denote the true average tensile strengths for grades 1064 and 1078 respectively. Then a 95% confidence interval for $\mu_1 - \mu_2$ is

$$\left(107.6 - 123.6 - 1.96 \sqrt{\frac{(1.3)^2}{129} + \frac{(2.0)^2}{129}}, \right.$$

$$\left. 107.6 - 123.6 + 1.96 \sqrt{\frac{(1.3)^2}{129} + \frac{(2.0)^2}{129}} \right)$$

$$= (-16.0 - .41, -16.0 + .41) = (-16.41, -15.59)$$

If the variances σ_1^2 and σ_2^2 are at least approximately known and the investigator uses equal sample sizes, then from (9.10) the sample size n for each sample which yields a $100(1 - \alpha)$% interval of length L is

$$n = \frac{4z_{\alpha/2}^2(\sigma_1^2 + \sigma_2^2)}{L^2} \tag{9.12}$$

where (9.12) will generally have to be rounded up to an integer.

One other experimental situation in which a confidence interval for a difference in means might be desired is the case of paired data, which we originally introduced in Chapter 8. Because pairing is most frequently associated with a small number of observations (say, before/after experiments carried out over time, which are expensive), we postpone discussion on this problem to Section 9.4.

A General Formula for a Large-Sample Confidence Interval

Other large-sample confidence intervals have the same form as those of this section. Let θ be the parameter of interest (we have so far considered $\theta = \mu$ and $\theta = \mu_1 - \mu_2$) with an estimator $\hat{\theta}$ satisfying.

1. $\hat{\theta}$ has (at least approximately) a normal distribution.
2. $E(\hat{\theta}) = \theta$ (at least approximately).
3. $\hat{\sigma}_{\hat{\theta}}$ is an estimate of the standard deviation of $\hat{\theta}$.

Then *a large-sample $100(1 - \alpha)$% confidence interval for θ is*

$$\text{estimate} \pm z_{\alpha/2} \cdot \left(\begin{array}{c} \text{estimated standard} \\ \text{deviation of } \hat{\theta} \end{array} \right)$$

$$= (\hat{\theta} - z_{\alpha/2} \cdot \hat{\sigma}_{\hat{\theta}}, \hat{\theta} + z_{\alpha/2} \cdot \hat{\sigma}_{\hat{\theta}}) \tag{9.13}$$

For example, if $\theta = \mu$, then the estimate is $\hat{\theta} = \bar{x}$ with estimated standard deviation $s/\sqrt{n}$, yielding the interval (9.8).

Exercises / Section 9.2

1. A random sample of 200 First National Bank customers having free checking accounts (which require a $300 minimum balance) revealed that during a particular month, the sample average number of checks written was 38.7 with a sample standard deviation of 8.3. Construct a 95% confidence interval for μ, the true average number of checks written during the month by all customers having such accounts. What assumptions are you making about the distribution of the number of checks written?

2. The article "Song Dialects and Colonization in the House Finch" (*Condor*, 1975, pp. 407–422) reported that the sample average duration for $n = 175$ songs of black house finches was 1.47 sec with a sample standard deviation of .95 sec. Compute a 90% confidence interval for the true average duration for black house finch songs. Does the data suggest that an assumption of normality for the distribution of song duration is plausible?

3. The sample average ultimate tensile strength for a sample of 35 high-strength magnetic alloy steel rings used in turbine generators was 152.3 ksi, while the sample standard deviation was 4.8 ksi. Obtain a 99% confidence interval for the true average ultimate tensile strength of such rings.

4. It was stated in Section 7.6 that when n is large, the sample standard deviation S has approximately a normal distribution with expected value σ and standard deviation $\sigma/\sqrt{2n}$. Use this to derive a large-sample $100(1 - \alpha)\%$ confidence interval for σ. What is the 95% confidence interval for σ in Exercise 2? *Hint:* Let $\theta = \sigma$, $\hat{\theta} = S$, and use the large-sample interval (9.13).

5. Most educators feel that at the fourth-grade level, children working simple arithmetic problems should show decreasing dependence on techniques such as counting aloud, making tally marks, and the like. A sample of 63 counters (identified by observation) yielded a sample average IQs of 95.7 and a sample standard deviation of 13.0, while a sample of 55 noncounters yielded a sample average IQ of 100.5 with a sample standard deviation of 13.6. Construct a 95% confidence interval for the difference $\mu_1 - \mu_2$ in true average IQs between counters and noncounters at this grade level.

6. A study of reading habits of both college graduates and nongraduates is to be undertaken. An equal number of individuals in the two categories will be selected, and each will be asked to report on the number of books that he or she reads during the next year. On the basis of past studies of reading habits, the investigators feel that values $\sigma_1 = 20$ (grads) and $\sigma_2 = 15$ are conservative (though ultimately s_1 and s_2 will be used). What should the sample sizes be if the 95% confidence interval for the difference in true average number of books read between those with a college degree and those without one is to have length at most five?

7. Example 9.1 presented data from an article in the *American Journal of Clinical Nutrition;* the sample mean and sample standard deviation of serum cholesterol for $m = 49$ low-income Guatemalan rural Indians were $\bar{x} = 157.02$ and $s_1 = 31.75$, respectively. The same article listed serum total cholesterol values for $n = 49$ low-income rural Black Caribs, with $\bar{y} = 195.06$ and $s_2 = 59.97$ (this second group subsisted on a diet much higher in protein and fat than that of the first group). Construct a 99% confidence interval for the difference in true average serum cholesterol values between the Indian and Black Carib populations. Does the computed interval include zero? Does this suggest a conclusion if $H_0: \mu_1 = \mu_2$ is tested against $H_0: \mu_1 \neq \mu_2$ at level .01?

8. Supplementary Exercise 12 of Chapter 7 presented data on average yield of fruit trees after treatment with three different pesticides and a fourth treatment involving ladybugs. Let μ_i = the true average yield (bushels/tree) after receiving the ith treatment. Then

$$\theta = \tfrac{1}{3}(\mu_1 + \mu_2 + \mu_3) - \mu_4$$

measures the difference in true average yields between treatment with pesticides and treatment with ladybugs. When n_1, n_2, n_3, and n_4 are all large, the estimator $\hat{\theta}$ obtained by replacing each μ_i by $\bar{X}_i$ is approximately normal. Use this to derive a large sample $100(1 - \alpha)\%$ confidence interval for θ, and compute the 95% interval for the given data.

9.3 Confidence Intervals for Population Proportions

It is not difficult to derive a confidence interval for a single population proportion p when the sample size is small, but because most proportion problems involve large samples, we pass immediately to this case; for the small sample procedure, consult the book by Steel and Torrie (Chapter 11 bibliography).

A Large-Sample Confidence Interval for *p*

The information available for making an inference about p is a collection of n independent success/failure responses. Then X, the number of successes among the n responses, has a binomial distribution with parameters n and p. The point estimator of p is $\hat{p} = X/n$, with $E(\hat{p}) = p$ and $\text{Var}(\hat{p}) = pq/n$. Further, when n is large, $\hat{p}$ has approximately a normal distribution. A confidence interval for p can now be obtained by one of the following two methods.*

1. Identify the parameter of interest p with θ, so that $\hat{\theta} = \hat{p} = X/n$ and $\hat{\sigma}_{\hat{\theta}}$, the estimated standard deviation of $\hat{\theta}$, is $\sqrt{\hat{p}\hat{q}/n}$. Then form the large-sample interval $\hat{\theta} \pm z_{\alpha/2} \cdot \hat{\sigma}_{\hat{\theta}}$ discussed at the end of the previous section.

2. The large sample $100(1 - \alpha)\%$ confidence interval for μ was based on the probability statement

$$P\left(-z_{\alpha/2} \leq \frac{\overline{X} - \mu}{S/\sqrt{n}} \leq z_{\alpha/2}\right) \doteq 1 - \alpha$$

Replacing $\overline{X}$ and S by $\overline{x}$ and s, respectively, the interval consists of all values of μ for which the inequalities inside the parentheses are both satisfied. The upper and lower endpoints result from replacing each inequality by $=$ and solving for μ (for example, $-z_{\alpha/2} = \frac{(\overline{x} - \mu)}{(s/\sqrt{n})}$ yields $\mu = $ upper endpoint $= \overline{x} + z_{\alpha/2} \cdot s/\sqrt{n}$).

The inequalities in the probability statement

$$P\left(-z_{\alpha/2} \leq \frac{\hat{p} - p}{\sqrt{p(1 - p)/n}} \leq z_{\alpha/2}\right) \doteq 1 - \alpha$$

can be similarly manipulated. Replacing each $\leq$ by $=$ yields a quadratic equation in p whose solutions are

$$p = \frac{\hat{p} + \frac{z_{\alpha/2}^2}{2n} \pm z_{\alpha/2}\sqrt{\frac{\hat{p}\hat{q}}{n} + \frac{z_{\alpha/2}^2}{4n^2}}}{1 + (z_{\alpha/2}^2)/n}$$

The $+$ sign goes with the upper endpoint of the interval and the $-$ sign with the lower endpoint. Since n is large, the three terms involving $z_{\alpha/2}^2$ are negligible

---

*Method 2 is actually more general than method 1, because it can be used for small- as well as large-sample problems.

compared to other terms. Ignoring them gives exactly the same interval as is obtained from method 1:

Proposition: A large-sample $100(1 - \alpha)\%$ confidence interval for a population proportion p based on n independent dichotomous (success/failure) responses is

$$\left(\hat{p} - z_{\alpha/2} \sqrt{\frac{\hat{p}\hat{q}}{n}}, \, \hat{p} + z_{\alpha/2} \sqrt{\frac{\hat{p}\hat{q}}{n}} \right) \qquad (9.13)$$

where $\hat{p} = x/n$ and x is the observed number of success responses (that response defining p). This interval can be used whenever $n\hat{p} \geq 5$ and $n\hat{q} \geq 5$.

Example 9.6

An article in the November 24, 1978, *Los Angeles Times* reported that in a sample of 244 doctors, 184 said that they would object to the sale of human organs for transplants (giving such reasons as the possibility of a sudden epidemic in body-snatching, murder for salable parts, and the poor being sold for spare parts). Obtain a 90% confidence interval for the true proportion p of all doctors who would object to such sales.

The sample size is $n = 244$ and the number of "success" responses is $x = 184$, so $\hat{p} = 184/244 = .754$ and $\hat{q} = .246$. The 90% interval is therefore

$$\left(.754 - 1.645 \sqrt{\frac{(.754)(.246)}{244}}, \, .754 + 1.645 \sqrt{\frac{(.754)(.246)}{244}} \right)$$

$$= (.709, .799)$$

Precision and Sample Size

The interval (9.13) extends a distance $z_{\alpha/2} \sqrt{\hat{p}\hat{q}/n}$ both to the left and to the right of the point estimate $\hat{p}$, so the length of the interval is

$$L = 2z_{\alpha/2} \sqrt{\frac{\hat{p}\hat{q}}{n}} \qquad (9.14)$$

Because the right-hand side of (9.14) involves the unknown $\hat{p}$, the length is not determined solely by specification of n. Solving for n yields

---

$$n = \frac{4z_{\alpha/2}^2 \cdot \hat{p}\hat{q}}{L^2} = \frac{4z_{\alpha/2}^2 \hat{p}(1 - \hat{p})}{L^2} \qquad (9.15)$$

There are now several possible approaches to a choice of n. One approach is to specify approximately what you expect the value of $\hat{p}$ to be, and use that in (9.15).

Alternatively, a choice of n can be made by taking advantage of the fact that $\hat{p}(1 - \hat{p})$ is maximized for $\hat{p} = \frac{1}{2}$ and decreases as $\hat{p}$ moves away from $\frac{1}{2}$ in either direction. Suppose, for example, that p_0 represents an upper bound on what we believe the value of $\hat{p}$ will be, and that $p_0 \leq \frac{1}{2}$. Then use of p_0 in (9.15) in place of $\hat{p}$ will give the maximum value of n necessary to arrive at length L. *The most conservative approach is to use $\hat{p} = \frac{1}{2}$, for then the length will be $\leq L$ no matter what $\hat{p}$ is actually observed.*

Example 9.6 (continued)

For $n = 244$ the 90% interval had length $.799 - .709 = .09$. Suppose that we had wanted an interval length of $L = .05$ and did not want to make a prior guess at $\hat{p}$. Then using $\hat{p} = \frac{1}{2}$ in (9.15) gives

$$n = \frac{4(1.645)^2 \cdot (.25)}{(.05)^2} = 1082.4$$

so a sample size of $n = 1083$ would be necessary.

Then if $\hat{p}$ actually equaled $.754$ for this sample size, the interval would be $(.732, .776)$. This interval has length $.44$; only if $\hat{p}$ happened to be $.5$ would the length actually equal the upper bound of $L = .05$.

A Large-Sample Confidence Interval for $p_1 - p_2$

As with means, most two-sample problems involve the objective of comparison through hypothesis testing, but sometimes an interval estimate for $p_1 - p_2$ is appropriate. Both $\hat{p}_1 = X/m$ and $\hat{p}_2 = Y/n$ have approximate normal distributions when m and n are both large. If we identify θ with $p_1 - p_2$, then $\hat{\theta} = \hat{p}_1 - \hat{p}_2$ satisfies the conditions necessary for obtaining a large-sample confidence interval. In particular the estimated standard deviation of $\hat{\theta}$ is $\sqrt{(\hat{p}_1\hat{q}_1/m) + (\hat{p}_2\hat{q}_2/n)}$. The $100(1 - \alpha)\%$ interval $\hat{\theta} \pm z_{\alpha/2} \cdot \hat{\sigma}_{\hat{\theta}}$ then becomes

$$\left(\hat{p}_1 - \hat{p}_2 - z_{\alpha/2} \sqrt{\frac{\hat{p}_1\hat{q}_1}{m} + \frac{\hat{p}_2\hat{q}_2}{n}}, \; \hat{p}_1 - \hat{p}_2 + z_{\alpha/2} \sqrt{\frac{\hat{p}_1\hat{q}_1}{m} + \frac{\hat{p}_2\hat{q}_2}{n}} \right) \quad (9.17)$$

Example 9.7

An article which appeared in the June 13, 1976, *Cleveland Plain Dealer* reported the results of a study of the Lamaze method of prepared childbirth. The study used a group of 129 women trained in the Lamaze technique and a group of equal size consisting of women who had not had the training. While 104 of the Lamaze women had spontaneous, natural deliveries, only 61 of the untrained women did. The sample sizes here are $m = n = 129$, and the summary statistics are $x = 104$ and $y = 61$. With $p_1 =$ the true proportion of all Lamaze-trained women who have spontaneous, natural deliveries, and p_2 defined analogously for untrained women, $\hat{p}_1 = .81$ and $\hat{p}_2 = .47$, so a 95% confidence interval for $p_1 - p_2$ is

$$\left(.81 - .47 - 1.96 \sqrt{\frac{(.81)(.19)}{129} + \frac{(.47)(.53)}{129}},\right.$$

$$\left..81 - .47 + 1.96 \sqrt{\frac{(.81)(.19)}{129} + \frac{(.47)(.53)}{129}}\right)$$

$$= (.34 - .11, .34 + .11) = (.23, .47)$$

Because the sample sizes in Example 9.7 are relatively small, the interval is rather wide. The length of the interval (9.17) involves the unknowns $\hat{p}_1$ and $\hat{p}_2$, but because $\hat{p}_1\hat{q}_1$ and $\hat{p}_2\hat{q}_2$ are both $\leq \frac{1}{4}$, an argument analogous to that in the one-sample case can be used to determine equal sample sizes so that the length is $\leq L$.

Exercises / Section 9.3

1. In a survey of 277 randomly selected adult female shoppers, 69 stated that whenever an advertised item is unavailable at their local supermarket, they request a raincheck ("Consumer Attitudes Toward Unavailability and Mispricing of Advertised Items by Grocery Stores," *J. Consumer Affairs*, 1977, no. 1, pp. 158–166). Obtain a 99% confidence interval for the true proportion p of adult female shoppers who request a raincheck in such situations.

2. An article in the December 12, 1977, *Los Angeles Times* reported that a new technique, graphic stress telethermometry (GST), accurately detected 23 out of 29 known breast cancer cases. Construct a 90% confidence interval for the true proportion of breast cancers which would be detected by the GST technique (because n is small, the interval will be quite wide).

3. A state legislator wishes to survey residents of her district to see what proportion of the electorate is aware of her position on using state funds to pay for abortions.
   a. What sample size is necessary if the 95% confidence interval for p is to have length at most .10 irrespective of p (that is, what n is necessary for the legislator to be 95% confident that $\hat{p}$ is within .05 of the true proportion)?
   b. If the legislator has strong reason to believe that at least $\frac{2}{3}$ of the electorate knows of her position, how large a sample size would you recommend?

4. The superintendent of large school district, having once had a course in probability and statistics, believes that the number of teachers absent on any given day has a Poisson distribution with parameter λ. Use the accompanying data on absences for 50 days to derive a large sample confidence interval for λ. *Hint:* The mean and variance of a Poisson variable both equal λ, so

$$Z = \frac{\bar{X} - \lambda}{\sqrt{\lambda/n}}$$

has approximately a standard normal distribution. Now proceed as in the derivation of the interval for p by making a probability statement (with probability $1 - \alpha$) and using either method 1 or method 2.

Number of absences	0	1	2	3	4	5	6	7	8	9	10
Frequency	1	4	8	10	8	7	5	3	2	1	1

5. To study the effect of wording on sentiment for gun registration, two different samples of voters were obtained, and each member of the first (second) sample was asked to respond to question A (B) below.

   A: Would you favor or oppose a law which would require a person to obtain a police permit before he could buy a gun?

   B: Would you . . . a gun, or do you think that such a law would interfere too much with the right of citizens to own guns? (where . . . is worded identically to A).

Question A: $m = 615$, number who favor $= 463$
Question B: $m = 585$, number who favor $= 403$

Compute a 90% confidence interval for the difference $p_1 - p_2$ in true proportions of individuals who would respond, "I favor such a law." ("Attitude Measurement and the Gun Control Paradox," *Public Opinion Q.*, Winter 1977–1978, pp. 427–438.)

6. Two different types of alloy, A and B, have been used to manufacture experimental specimens of a small tension link to be used in a certain engineering application. The ultimate strength (ksi) of each specimen was determined, and the results are summarized in the accompanying frequence distribution.

	A	B
26–under 30	6	4
30–under 34	12	9
34–under 38	15	19
38–under 42	7	10
	$m = 40$	$n = 42$

Compute a 95% confidence interval for the difference between the true proportion of all specimens of alloys A and B which have an ultimate strength of at least 34 ksi.

7. A random sample of 120 individuals who had each taken a drivers' education course while in high school revealed that 96 had not been cited for a moving violation in the five-year period subsequent to obtaining a license. Another sample of 100 individuals who had not taken such a course yielded 72 who had not had such a violation in the five-year period. Compute a 98% confidence interval for the difference in true proportions cited for moving violations between all those having such a course and all those not having the course.

8. Suppose that the data in Exercise 4 above came from an urban school district, and that a similar survey of teacher attendance patterns in a suburban district having roughly the same number of teachers yielded the following data.

Number of absences	0	1	2	3	4	5	6	7	8	9	10	11	12
Frequency	1	2	5	10	12	7	6	2	2	1	1	0	1

Assuming that the number of absences on any given day is a Poisson variable for each district with parameters λ_1 and λ_2, respectively, derive a large-sample confidence interval for the difference in true average number of absences per day between the two districts.

9. Suppose that a 95% confidence interval for $p_1 - p_2$ is to be constructed based on equal sample sizes from the two populations. For what value of $n (= m)$ will the resulting interval have length at most .1 irrespective of the results of the sampling?

9.4 Small-Sample Intervals for Means of Normal Populations

When sample sizes are small, the Central Limit Theorem cannot be invoked to obtain approximate probabilities and confidence intervals. We faced this problem in hypothesis testing, and to derive valid procedures there we imposed some extra structure on the problem by assuming that underlying populations were normal. The same approach will be used here to obtain intervals for μ and $\mu_1 - \mu_2$ when population variances are unknown. Chapter 15 treats the same problems without the assumption of normality, and the resulting intervals are distribution free. When the populations are normal, though, the t intervals of this section are known to be shortest $100(1 - \alpha)\%$ confidence intervals, so are better than the distribution-free intervals.

The *t* Interval for μ

Let $X_1, X_2, \ldots, X_n$ be a random sample from a normal distribution with mean μ, and let $\overline{X}$ and S denote the sample mean and sample standard deviation. Then (see Section 7.4) the variable

$$T = \frac{\overline{X} - \mu}{S/\sqrt{n}}$$

has a t distribution with $n - 1$ degrees of freedom, so that

$$P(-t_{\alpha/2,n-1} \leq T \leq t_{\alpha/2,n-1}) = 1 - \alpha \tag{9.18}$$

Expression (9.18) differs from expressions in earlier sections in that T and $t_{\alpha/2,n-1}$ are used in place of Z and $z_{\alpha/2}$, but it can be manipulated in the same manner to obtain a confidence interval for μ.

Proposition: Let $\bar{x}$ and s be the sample mean and sample standard deviation computed from the results of a random sample from a normal population with mean μ. Then a $100(1 - \alpha)\%$ confidence interval for μ is

$$\left(\bar{x} - t_{\alpha/2,n-1} \cdot \frac{s}{\sqrt{n}} \,, \ \bar{x} + t_{\alpha/2,n-1} \cdot \frac{s}{\sqrt{n}} \right) \tag{9.19}$$

Example 9.8

The results of a Wagner turbidity test performed on 15 samples of standard Ottawa testing sand were (in microamperes)

26.7, 25.8, 24.0, 24.9, 26.4, 25.9, 24.4, 21.7, 24.1, 25.9, 27.3 26,9, 27.3, 24.8, 23.6

("Influence of Washing on Properties of Standard Ottawa Testing Sand," *J. Materials,* 1971, pp. 218–233). To obtain a 95% confidence interval for μ = the true average Wagner turbidity of all such sand samples, we first compute $\bar{x} = 25.31$ and $s = 1.58$. With $n = 15$ and $\alpha = .05$, $t_{\alpha/2,n-1} = t_{.025,14} = 2.145$. The desired interval is then

$$\bar{x} \pm t_{\alpha/2,n-1} \frac{s}{\sqrt{n}} = 25.31 \pm \frac{(2.145)(1.58)}{\sqrt{15}} = 25.31 \pm .88 = (24.43, 26.19)$$

Unfortunately it is not easy to select n to control the length of the t interval. This is because the length involves the unknown (before the data is collected) s, and because n enters not only through $1/\sqrt{n}$ but also through $t_{\alpha/2,n-1}$. As a result, an appropriate n can be obtained only by trial and error.

The t Interval for $\mu_1 - \mu_2$

To obtain a confidence interval for the difference between means of two normal populations when at least one sample size is small, we assume that the two population variances are equal. Then the variable

$$T = \frac{\bar{X} - \bar{Y} - (\mu_1 - \mu_2)}{S_p \sqrt{1/m + 1/n}}$$

has a t distribution with $m + n - 2$ degrees of freedom (where S_p^2 is the pooled estimator of the common variance σ^2), so

$$P(-t_{\alpha/2, m+n-2} \le T \le t_{\alpha/2, m+n-2}) = 1 - \alpha \tag{9.20}$$

Manipulation of the inequalities inside the parentheses in (9.20) in the same manner as before yields the desired interval.

Proposition: Let $\bar{x}$ and s_1^2 be the mean and variance computed for a random sample of size m from a normal population with mean μ_1 and variance σ^2, and let $\bar{y}$ and s_2^2 be computed from a random sample of size n, also from a normal population with mean μ_2 and the same variance σ^2. With $s_p^2 = [(m - 1)s_1^2 + (n - 1)s_2^2]/(m + n - 2)$ as the pooled estimate of σ^2, a $100(1 - \alpha)\%$ confidence interval for $\mu_1 - \mu_2$ is

$$\left(\bar{x} - \bar{y} - t_{\alpha/2, m+n-2} \cdot s_p \sqrt{\frac{1}{m} + \frac{1}{n}}, \right. \tag{9.21}$$
$$\left. \bar{x} - \bar{y} + t_{\alpha/2, m+n-2} \cdot s_p \sqrt{\frac{1}{m} + \frac{1}{n}} \right)$$

Example 9.9   Pesticides, particularly the organochlorines, have contributed to the decline of several species of birds. The objective of the research reported in "Aerial Pesticide Applications and Ring Necked Pheasants" (*J. Wildlife Mgmt.*, vol. 38, pp. 679–685) was to study the effects of several such pesticides on pheasants. In 1970 several applications of these pesticides were made over a small area in Idaho. Subsequently 20 wild pheasants were captured in the area. The average amount of plasma cholinesterase (an enzyme which affects muscle control) for the 20 (in μ moles of acetycholine hydrolized per ml plasma/min) was found to be 1.55 with a sample standard deviation of .39. Nine wild pheasants captured outside the area (the control group) had an average cholinesterase amount of 1.60, with a sample standard deviation of .40. With $m = 20$, $\bar{x} = 1.55$, $s_1 = .39$, $n = 9$, $\bar{y} = 1.60$, and $s_1 = .40$, a 95% interval for the true average difference in plasma cholinesterase is

$$1.55 - 1.60 \pm (2.052)(.39) \sqrt{\tfrac{1}{20} + \tfrac{1}{9}}$$

$$= (-.05 \pm .32)$$

$$= (-.37, .27)$$

A Confidence Interval When Data Is Paired

When a pair of observations is taken from each individual or experimental object, or when individuals or objects are matched on related characteristics, the sample has the form $(X_1, Y_1), \ldots, (X_n, Y_n)$. While different pairs are independent, within a pair the X and Y measurements are not assumed independent, so the independent-samples interval is not appropriate. Assume that both the X distribution and the Y distribution are normal, and let $\mu_D = \mu_1 - \mu_2$. Then, with $D_i = X_i - Y_i$ and $\overline{D}$ and S_D denoting the sample mean and standard deviation of the n differences,

$$T = \frac{\overline{D} - \mu_D}{S_D/\sqrt{n}}$$

has a t distribution with $n - 1$ degrees of freedom. It follows that a $100(1 - \alpha)\%$ confidence interval for μ_D, the true mean difference within a pair is

$$\left(\overline{d} - t_{\alpha/2,n-1} \cdot s_D/\sqrt{n}, \quad \overline{d} + t_{\alpha/2,n-1} \cdot s_D/\sqrt{n}\right) \tag{9.22}$$

Example 9.10 In an experiment reported in the article "Emotional Attributes of Color: A Comparison of Violet and Green" (*Perceptual and Motor Skills*, 1971, pp. 403–406), 14 subjects were exposed to 60-sec intervals of alternating violet and green light for a total of six minutes. Galvanic skin response during the first 12 sec of exposure (calibrations) was measured and averaged over exposure intervals. The results appear below.

Table 9.2

Subject	1	2	3	4	5	6	7	8	9	10	11	12	13	14
Violet	3.0	3.7	4.0	3.2	3.6	3.5	4.2	3.8	3.7	3.4	3.6	3.8	3.4	3.4
Green	2.2	2.7	3.1	2.9	3.3	2.6	2.9	2.8	3.2	2.5	3.5	3.0	2.3	3.5
d_i	.8	1.0	.9	.3	.3	.9	1.3	1.0	.5	.9	.1	.8	1.1	−.1

The computed summary quantities are $\overline{d} = .70$ and $s_D = .41$. With $t_{.025,13} = 2.160$, a 95% confidence interval for μ_D is

$$\left(.70 - (2.160)(.41)/\sqrt{13}, \ .70 + (2.160)(.41)/\sqrt{13}\right) = (.45, \ .95)$$

If n is large, $t_{\alpha/2,n-1}$ can be approximated by $z_{\alpha/2}$. In this case the normality of X's and Y's need not be assumed, since the C.L.T. can be invoked.

Exercises / Section 9.4

1. In a study of the flammability of material used in children's sleepwear, the char length (inches) for five samples of washed acetate/nylon brushed tricot fabric was measured. The resulting observations were

8.1, 10.4, 9.5, 8.9, 10.7

Assuming that char length is a normally distributed

variable, compute a 95% confidence interval for the true average char length for this fabric.

2. Trisodium carboxymethylaxysuccinate (NaCMOS) is a chemical compound being considered for inclusion in various detergent products. Because it would eventually become a component of wastewater, an experiment was performed to assess its effects on aquatic life. Following exposure to 100 mg NaCMOS/liter for a 16-hour period, the NaCMOS content was determined for each of six different goldfish. The resulting sample mean and sample standard deviation were 9.1 and 3.2, respectively. Compute a 90% confidence interval for the true average NaCMOS content for goldfish undergoing such exposure. ("Acute Fish Toxicity and Absorption Tests of an Experimental Detergent Builder, Trisodium Carboxymethyloxysuccinate," *J. Testing and Evaluation,* 1979, pp. 16–17.)

3. For each of 18 preserved cores from oil-wet carbonate reservoirs, the amount of residual gas saturation after a solvent injection was measured at water flood-out. Observations, in percent of pore volume, were

23.5, 31.5, 34.0, 46.7, 45.6, 32.5, 41.4, 37.2, 42.5, 46.9, 51.5, 36.4, 44.5, 35.7, 33.5, 39.3, 22.0, 51.2

Compute a 98% confidence interval for the true average amount of residual gas saturation. ("Relative Permeability Studies of Gas-Water Flow Following Solvent Injection in Carbonate Rocks," *Soc. Petroleum Engineers J.,* 1976, pp. 23–30.)

4. For some time researchers have been unsure as to what actually causes the pharmacological and behavioral effects resulting from smoking marijuana. The two major contenders are Δ^9tetrahydrocannabinol (Δ^9THC) and one of its metabolites, 11-hydroxy-Δ^9tetrahydrocannabinol (11-OH-Δ^9THC). The article "Intravenous Injection in Man of Δ^9THC and 11-OH-Δ^9THC" (*Science,* 1972, p. 633) reported the results of an experiment in which subjects were given infusions of either Δ^9THC or 11-OH-Δ^9THC and asked to report the moment at which the effect of the drug was first perceived. The (unpaired) data below refers to the necessary dose in micrograms per kilogram of body weight.

Δ^9THC: 19.54, 14.47, 16.00, 24.83, 26.39, 11.49

11-OH-Δ^9THC: 15.95, 25.89, 20.53, 15.52, 14.18, 16.00

Compute a 90% confidence interval for the difference between true average dose-to-perception for Δ^9THC and for 11-OH-Δ^9THC.

5. The article "Mercury in Aquatic Birds at Clay Lake, Western Ontario" (*J. Wildlife Mgmt.,* 1973, pp. 58–61) reported the following data on mercury residues in breast muscles.

Mallard ducks: $m = 16,\ \bar{x} = 6.13,$
$s_1 = 2.40$

Blue-winged teals: $n = 17,\ \bar{y} = 6.46,$
$s_2 = 1.73$

Compute a 95% confidence interval for the difference in true average mercury residues between these two types of bird in the region of interest.

6. In an experiment to study the effects of liming and urea fertilizer application on dimethoate (a pesticide) retention by loamy soil, the following percentages of dimethoate recovery were observed:

Soil treated with lime: 28.5, 24.7, 26.2, 23.9, 29.6

Soil treated with urea: 38.7, 41.6, 35.9, 41.8, 43.2

Compute a 95% confidence interval for the difference in true average percentage of dimethoate recovery between the two treatments.

7. Testosterone is a drug which has been taken by athletes with the intention of increasing muscle size and strength. An article in *Lancet* (October 1976) reported the results of an experiment in which 11 athletic men were given doses of methandienone (a derivative of testosterone) during a six-week period and a placebo during another six-week period. Body weight and composition, muscle strength, and the like were studied. After each period muscle width change (in mm, determined by X ray) was measured for each subject.

Subject:	1	2	3	4	5	6
Control period change in muscle width:	3.7	5.2	4.0	4.7	4.3	3.9
Drug period change in muscle width:	13.1	16.5	15.3	15.7	14.1	15.0

Subject: 7 8 9 10 11

Control period change in muscle width: 4.2 4.9 5.1 4.1 4.0

Drug period change in muscle width: 15.5 16.1 15.8 14.3 15.2

Construct a 95% confidence interval for the difference in change in muscle width between the control and drug period (these numbers are hypothetical).

8. The paper "Selection of a Method to Determine Residual Chlorine in Sewage Effluents" (*Water and Sewage Works*, 1971, pp. 360–364) reported the results of an experiment in which two different methods for determining chlorine content were used on samples of Cl_2-demand-free water for various doses and contact times. Observations are in mg/1.

Sample:	1	2	3	4
MSI method:	.39	.84	1.76	3.35
SIB method:	.36	1.35	2.56	3.92

Sample:	5	6	7	8
MSI method:	4.69	7.70	10.52	10.92
SIB method:	5.35	8.33	10.70	10.91

Construct a 99% confidence interval for the difference in true average residual chlorine readings between the two methods.

9. When $X_1, \ldots, X_m$ and $Y_1, \ldots, Y_n$ are both random samples from normal distributions with variances σ_1^2 and σ_2^2, respectively, the random variable

$$F = \frac{S_1^2/\sigma_1^2}{S_2^2/\sigma_2^2}$$

has an F distribution with $\nu_1 = m - 1$ d.f. and $\nu_2 = n - 1$ d.f. (see Section 8.5 for a discussion of the F distribution). Use this result to construct a $100(1 - \alpha)\%$ confidence interval for σ_2^2/σ_1^2 and for σ_2/σ_1. Hint: Manipulate a probability statement to isolate σ_2^2/σ_1^2.

10. The article "Enhancement of Compressive Properties of Failed Concrete Cylinders with Polymer Impregnation" (*J. Testing and Evaluation*, 1977, pp. 333–337) reported the following data on impregnated compressive modulus (psi $\times 10^6$) when two different polymers were used to repair cracks in failed concrete.

Epoxy: 1.75, 2.12, 2.05, 1.97
MMA prepolymer: 1.77, 1.59, 1.70, 1.69

Use the result of Exercise 9 to compute a 90% confidence interval for the ratio of variances.

Supplementary Exercises / Chapter 9

1. Analysis of the venom of seven eight-day-old worker bees yielded the following observations on histamine content (nanograms):

649, 832, 418, 530, 384, 899, 755

Compute a 90% confidence interval for the true average histamine content for all worker bees of this age. Were any assumptions necessary to justify your computation?

2. A sample of 40 transistors of a certain type yielded a sample average saturation current of 9.46 milliamps with a sample standard deviation of .58. Compute a 95% confidence interval for the true average saturation current for transistors of this type. What assumptions, if any, are you making?

3. The financial manager of a large department store chain selected a random sample of 200 of its credit card customers and found that 136 had incurred an interest charge during the previous year because of an unpaid balance.

a. Compute a 90% confidence interval for the true proportion of credit card customers who incurred an interest charge during the previous year.

b. If the desired length of the 90% interval is .05, what sample size is necessary to ensure this?

c. Compute an 82% confidence interval for the true proportion.

4. A study of ventilation during sleep for both normal infants and infants who had experienced an aborted form of the sudden-infant-death-syndrome (requiring at least two resuscitations during sleep) yielded the following data on partial pressure of carbon dioxide (mm of Hg):

Aborted SIDS: $m = 11$, $\bar{x} = 38.9$, $s_1 = 3.5$
Normal: $n = 12$, $\bar{y} = 35.1$, $s_2 = 1.9$

Compute a 99% confidence interval for ~~the~~ difference between true average partial pressure
SIDS sufferers and true average partial pressure
for normal infants. ("Abnormal Regulation of Ventilation in Infants at Risk for Sudden-Infant-Death-
Syndrome," *New England J. Medicine*, 1977, vol.
297, pp. 747–750.)

5. It is important that face masks used by firefighters
be able to withstand high temperatures, since
firefighters commonly work in temperatures of 200
to 500 °F. In a test of one type of mask, 11 of 35
had their lenses pop off at 250°. Construct a 90%
confidence interval for the true proportion of
masks of this type whose lenses would pop out at
250°.

6. A random sample of 200 drivers from a particular
county who drive a foreign car yielded 115 who
use their seatbelts regularly, while another sample
of 300 drivers who drive a domestic model yielded
154 who use their seatbelts regularly. Derive a
99% confidence interval for the difference in proportions of those using seatbelts regularly between
drivers of foreign cars and drivers of domestic
cars.

7. The reaction time (RT) to a stimulus is the interval
of time commencing with stimulus presentation
and ending with the first discernible movement of
a certain type. The paper "Relationship of Reaction Time and Movement Time in a Gross Motor
Skill" (*Perceptual and Motor Skills*, 1973, pp.
453–454) reported that the sample average RT for
16 experienced swimmers to a pistol start was
.214 seconds with a sample standard deviation of
.036 seconds. Making any necessary assumptions,
derive a 90% confidence interval for true average
RT for all experienced swimmers.

8. Courtship in fiddler crabs is initiated by the male
and consists of certain movements of the chelae
known as "waving." The article "Coastal Distribution, Display, and Sound Production by Florida
Fiddler Crabs" (*Animal Behavior*, 1967, pp.
449–459) reported on duration time for a sequence
of five waves in a series both when females were
present and when they were absent. Use the accompanying data to compute a 99% confidence interval for the difference in true average duration
between the "female-absent" condition and
"female-present" condition.

Absent: $m = 12$, $\bar{x} = 25.2$ sec, $s_1 = 6.9$
Present: $n = 12$, $\bar{y} = 15.8$ sec, $s_2 = 5.7$

9. The article "Sex and Race Discrimination in the
New-Car Showroom: A Fact or Myth: (*J. Consumer Affairs*, 1977, pp. 107–113) reported the results of an experiment in which individuals of different races and sexes visited car dealerships to request the best possible deal on a certain car. Consider the following representative (hypothetical)
data:

Dealership:	1	2	3	4	5
Black female:	4459	4320	4268	4585	4736
White male:	4348	4385	4231	4516	4550

Dealership:	6	7	8	9
Black female:	4262	4440	4398	4823
White male:	4203	4285	4408	4570

Compute a 95% confidence interval for the true
average difference between best price offered
black females and white males.

10. A sample of both surface soil and subsoil was
taken from eight randomly selected agricultural locations in a particular county. The soil samples
were analyzed to determine both surface pH and
subsoil pH, with the following results:

Location:	1	2	3	4
Surface pH:	6.55	5.98	5.59	6.17
Subsoil pH:	6.78	6.14	5.80	5.91

Location:	5	6	7	8
Surface pH:	5.92	6.18	6.43	5.68
Subsoil pH:	6.10	6.01	6.18	5.88

Compute a 90% confidence interval for the true
average difference between surface and subsoil pH
for agricultural land in this county. What assumptions have you made about the underlying pH distributions?

11. An experimenter wishes to obtain a confidence interval for the difference between true average
breaking strength for cables manufactured by company I and by company II. Suppose that breaking
strength is normally distributed for both types of
cable with $\sigma_1 = 30$ psi and $\sigma_2 = 20$ psi.
   a. If costs dictate that the sample size for the type
   I cable should be three times the sample size
   for the type II cable, how many observations
   are required if the 99% confidence interval is to
   be no longer than 20 psi?

b. Suppose that a total of 200 observations is to be made. How many of the observations should be made on cable-I samples if the length of the resulting interval is to be a minimum?

12. The small samples confidence interval for $\mu_1 - \mu_2$ which was presented in Section 9.4 is valid when $\sigma_1 = \sigma_2$. This interval was derived from the same t variable that was used in Section 8.2 to obtain hypothesis-testing procedures. In that section we also described a procedure for testing H_0: $\mu_1 - \mu_2 = \Delta_0$ based on a variable T' having approximately a t distribution with d.f. ν estimated from the data.

   a. Use T' to derive an approximate $100(1 - \alpha)\%$ confidence interval for $\mu_1 - \mu_2$ when $\sigma_1 \neq \sigma_2$ (but the populations are still normal).

   b. Compute the 95% interval of (a) for the data in Exercise 10 of Section 9.4, in which μ_1 and μ_2 refer to the true average compressive modulus for epoxy polymer and MMA polymer, respectively, used to repair failed concrete.

13. Let $X_1, X_2, \ldots, X_n$ be a random sample from an exponential distribution with parameter λ (see Section 4.4). Then it can be shown that $Y = 2\lambda \sum X_i$ has a chi-squared distribution with $2n$ degrees of freedom, so that

$$P(\chi^2_{1-\alpha/2,2n} \leq Y \leq \chi^2_{\alpha/2,2n}) = 1 - \alpha$$

   a. Recalling that $E(X) = \mu = 1/\lambda$, use the probability statement to obtain a confidence interval for λ and also a confidence interval for μ.

   b. A sample of 10 batteries used in hand calculators is put on test and each time to failure is recorded. The sample average failure time is computed to be 38.4 hours. Assuming failure time to be exponentially distributed, compute a 95% confidence interval for the true average failure time of such batteries.

   c. Derive a confidence interval for the median $\tilde{\mu}$ of an exponential distribution, and then compute the 95% interval for $\tilde{\mu}$ based on the results of (b). *Hint:* Recall that $\tilde{\mu}$ satisfies $P(X \leq \tilde{\mu}) = .5$. Solve for $\tilde{\mu}$ in terms of λ and use the result of (a).

14. Let $X_1, X_2, \ldots, X_n$ be a random sample from a continuous probability distribution having median $\tilde{\mu}$ (so that $P(X_i \leq \tilde{\mu}) = P(X_i \geq \tilde{\mu}) = .5$).

   a. Show that

$$P(\min(X_i) \leq \tilde{\mu} \leq \max(X_i)) = 1 - (\tfrac{1}{2})^{n-1}$$

so that $(\min(x_i), \max(x_i))$ is a $100(1 - \alpha)\%$ confidence interval for $\tilde{\mu}$ with $\alpha = (\tfrac{1}{2})^{n-1}$. *Hint:* The complement of the event $\{\min(X_i) \leq \tilde{\mu} \leq \max(X_i)\}$ is $\{\max(X_i) \leq \tilde{\mu}\} \cup \{\min(X_i) \geq \tilde{\mu}\}$. But $\max(X_i) \leq \tilde{\mu}$ iff $X_i \leq \tilde{\mu}$ for all i.

   b. For each of six normal male infants, the amount of the amino acid alanine (mg/100 ml) was determined while on an isoleucine-free diet, resulting in the following data:

   2.84, 3.54, 2.80, 1.44, 2.94, 2.70

   Compute a 97% confidence interval for the true median amount of alanine for infants on such a diet. ("The Essential Amino Acid Requirements of Infants," *Amer. J. Nutrition*, 1964, pp. 322–330).

   c. Let $x_{(2)}$ denote the second smallest of the x_i's and $x_{(n-1)}$ denote the second largest of the x_i's. What is the confidence coefficient of the interval $(x_{(2)}, x_{(n-1)})$ for $\tilde{\mu}$?

15. Let $X_1, X_2, \ldots, X_n$ be a random sample from a uniform distribution on the interval $[0, \theta]$, so that

$$f(x) = \begin{cases} \dfrac{1}{\theta} & 0 \leq x \leq \theta \\ 0 & \text{otherwise} \end{cases}$$

Then if $Y = \max(X_i)$, it can be shown that the random variable $U = Y/\theta$ has density function

$$f_U(u) = \begin{cases} nu^{n-1} & 0 \leq u \leq 1 \\ 0 & \text{otherwise} \end{cases}$$

   a. Use $f_U(u)$ to verify that

$$P\left((\alpha/2)^{1/n} \leq \frac{Y}{\theta} \leq (1 - \alpha/2)^{1/n}\right) = 1 - \alpha$$

and use this to derive a $100(1 - \alpha)\%$ confidence interval for θ.

   b. Verify that $P(\alpha^{1/n} \leq Y/\theta \leq 1) = 1 - \alpha$, and derive a $100(1 - \alpha)\%$ confidence interval for θ based on this probability statement.

   c. Which of the two intervals derived above is shorter? If my waiting time for a morning bus is uniformly distributed and observed waiting times are $x_1 = 4.2$, $x_2 = 3.5$, $x_3 = 1.7$, $x_4 = 1.2$, and $x_5 = 2.4$, derive a 95% confidence interval for θ by using the shorter of the two intervals.

16. Let $0 \le \gamma \le \alpha$. Then a $100(1 - \alpha)\%$ confidence interval for μ when n is large is

$$\left(\bar{x} - z_\gamma \cdot \frac{s}{\sqrt{n}}, \quad \bar{x} + z_{\alpha-\gamma} \cdot \frac{s}{\sqrt{n}} \right)$$

The choice $\gamma = \alpha/2$ yields the usual interval derived in Section 9.1; if $\gamma \ne \alpha/2$, the above interval is not symmetric about $\bar{x}$. The length of this interval is $L = s(z_\gamma + z_{\alpha-\gamma})/\sqrt{n}$. Show that L is minimized for the choice $\gamma = \alpha/2$, so that the symmetric interval is the shortest. *Hints:* (a) By definition of z_α, $\Phi(z_\alpha) = 1 - \alpha$, so that $z_\alpha = \Phi^{-1}(1 - \alpha)$; (b) The relationship between the derivative of a function $y = f(x)$ and its inverse $x = f^{-1}(y)$ is $(d/dy)f^{-1}(y) = 1/f'(x)$.

Bibliography   See the bibliography at the end of Chapter 7.

CHAPTER 10

The Analysis of Variance

Introduction

In studying methods for the analysis of quantitative data, we first focused on problems involving a single sample of numbers and then turned to a comparative analysis of two different such samples. In one-sample problems the data consisted of observations on or responses from individuals or experimental objects randomly selected from a single population. In two-sample problems either the two samples were drawn from two different populations and the parameters of interest were the population means, or else two different treatments were applied to experimental units (individuals or objects) selected from a single population; in this latter case the parameters of interest were referred to as true treatment means.

The analysis of variance, or more briefly **ANOVA,** refers broadly to a collection of experimental situations and statistical procedures for the analysis of quantitative responses from experimental units. The simplest analysis of variance problem is referred to variously as a **single-factor, single-classification,** or **one-way** ANOVA and involves the analysis either of data sampled from more than two numerical populations (distributions) or data from experiments in which more than two treatments have been used. The characteristic which differentiates the treatments or populations from one another is called the **factor** under study, and the different treatments or populations are referred to as the **levels** of the factor. Examples of such situations include

1. an experiment to study the effects of five different brands of gasoline on automobile engine operating efficiency (mpg)
2. an experiment to study the effects of the presence of four different sugar solutions (glucose, sucrose, fructose, and a mixture of the three) on bacterial growth

3. an experiment to assess the effects of different amounts of a particular psyche-delic drug on manual dexterity
4. an experiment to decide whether varying the tracking weight of a stereo cartridge has any effect on record lifetime

In 1 the factor of interest is gasoline brand, and there are five different levels of the factor. In 2 the factor is sugar, with four levels (or five, if a control solution containing no sugar is used). In both 1 and 2 the factor is qualitative in nature, and the levels correspond to possible categories of the factor. In 3 and 4 the factors are concentration of drug and tracking weight, respectively, and both these factors are quantitative in nature, so that the levels identify different settings of the factor. When the factor of interest is quantitative, statistical techniques from regression analysis (discussed in Chapters 12 and 13) can also be used to analyze the data.

The present chapter focuses on single-factor ANOVA. The question of central interest here is whether or not there are differences in true averages associated with the different treatments or levels of the factor. The null hypothesis states that there are no differences between any of the μ_i's (population means or expected treatment responses), and the alternative hypothesis says that at least two μ_i's differ from one another. The F test for testing H_0 versus H_a is presented in Section 10.1. If H_0 is rejected, further analysis to identify differences is usually appropriate, so multiple comparison methods for detecting such differences are discussed in Section 10.2. The last section covers some other aspects of single-factor ANOVA. The next chapter introduces ANOVA experiments involving more than a single factor.

10.1   Single-Factor Analysis of Variance

Example 10.1, which describes an experiment similar to one reported on in the 1971 volume of *The Research Quarterly,* will be used to illustrate the notation and calculations of single-factor ANOVA.

Example 10.1   A physical education researcher wished to study the effect of four different teaching techniques on swimming proficiency for individuals completing a course in beginning swimming. The techniques used were (a) verbal cues, (b) verbal cues and videotaped feedback, (c) videotaped feedback, and (d) the control treatment (neither verbal cues nor videotaped feedback). A group of 24 nonswimming college women was randomly divided into four groups of six subjects each. Each group of subjects was taught using one of the four techniques, and at the end of the course was timed while swimming 10 yards using the butterfly stroke. The results are shown in Table 10.1.

Table 10.1

Treatment	Time (in seconds)						Sample mean
1. Verbal	18.7	21.1	17.9	19.5	22.1	18.3	19.60
2. Verbal + videotaped	19.9	17.6	18.2	20.0	16.9	17.5	18.35
3. Videotaped	18.6	20.3	21.7	19.7	20.9	20.8	20.33
4. Control	19.1	18.9	18.4	18.8	17.7	20.5	18.90

Does the data indicate that the different teaching techniques affect the true average swimming time?

The Hypotheses To Be Tested

To set up the appropriate null and alternative hypotheses, let μ_1, μ_2, μ_3, and μ_4 denote the true average swimming times for beginning swimmers for teaching techniques 1, 2, 3, and 4. The null hypothesis in single-factor ANOVA is that the different settings or levels of the factor have no effect on true average response. In the swimming example this claim becomes

$$H_0: \mu_1 = \mu_2 = \mu_3 = \mu_4$$

The alternative claim is that there are differences between the true average responses for the different levels; this can be stated as

$$H_a: \text{at least two of the } \mu_i\text{'s are unequal}$$

Note that in a two-sample, two-sided problem, there are only two population means, and H_a states that they are not equal. The alternative hypothesis for ANOVA is more complicated than this, since H_a could be true if $\mu_1 = 20$, $\mu_2 = 19.5$, $\mu_3 = 19$, and $\mu_4 = 18.5$, or if $\mu_1 = 20$ and $\mu_2 = \mu_3 = \mu_4 = 19$, or for a variety of other possible configurations of the μ_i's. H_a specifies that at least two are unequal, but there are many possible groupings of μ_i's for which H_a would be true.

ANOVA Notation

Two different letters x and y were used to label variables and data in a two-sample problem, but it is cumbersome to associate a different letter with each sample when three or more populations or treatments are under study. To represent data in ANOVA experiments, we shall use variables with two subscripts. The first subscript i will identify the population or treatment, and the second subscript j will identify the position of the observation in the sample from population i. The data will appear in a rectangular table (as in Example 10.1) with values in the ith sample (from population or treatment i) placed in row i of the table. Let

$X_{i,j} = $ the random variable which denotes the jth measurement taken from the ith population, or the measurement taken on the jth experimental unit which receives the ith treatment, and

$x_{i,j} = $ the observed value of $X_{i,j}$ when the experiment is performed.

In the swimming example, $X_{2,3}$ would represent the swimming time of the third student to be taught using the second technique, and the observed value of $X_{2,3}$ was $x_{2,3} = 18.2$. The first subscript i refers to the population sampled or treatment administered, and the second subscript identifies the position of the observation in the ith sample, sometimes referred to as the jth replication of the ith treatment. In cases in which there is no ambiguity, we will use x_{ij} rather than $x_{i,j}$; this would not work for, say, x_{112} ($x_{11,2}$ or $x_{1,12}$?), but in the swimming example the omission of the comma causes no difficulty.

In some experiments, the number of observations in a particular row of the table may not be the same for all rows. For example, some populations may be more difficult to sample than others, or some treatments may be more difficult to apply, or some experimental units may subsequently prove unsuitable or unresponsive. While unequal sample sizes cause no complication in methodology for single-factor ANOVA, the notation is more straightforward for the case of equal sample sizes (as in the swimming problem). The case of unequal sample sizes is dealt with in Section 10.3. With equal sample sizes, let

I = the number of populations or treatments under study

J = the number of observations in each sample

For the swimming problem $I = 4$ and $J = 6$.

The mean of the observations for the ith sample is given by

$$\overline{X}_{i\cdot} = \frac{\sum_{j=1}^{J} X_{ij}}{J}$$

for $i = 1, 2, \ldots, I$. The dot in place of the second subscript in $\overline{X}_{i\cdot}$ signifies that we have added over all values of that subscript while holding the first subscript fixed, and the horizontal bar signifies division by J to obtain an average. For the swimming experiment the observed values of the sample means are $\overline{x}_{1\cdot} = 19.60$, $\overline{x}_{2\cdot} = 18.35$, $\overline{x}_{3\cdot} = 20.33$, and $\overline{x}_{4\cdot} = 18.90$. The average of all observed values in the experiment is

$$\overline{X}_{\cdot\cdot} = \frac{\sum_{i=1}^{I} \sum_{j=1}^{J} X_{ij}}{IJ}$$

where again a dot indicates that the subscript was summed over all values and a bar that an average was then taken. $\overline{X}_{\cdot\cdot}$ is often referred to as the **grand mean.** For the swimming data, $\overline{x}_{\cdot\cdot} = 19.30$.

Motivation for the Test Statistic

To decide whether population means differed in the two-sample problem, we examined the difference $\overline{X} - \overline{Y}$ between the sample means. To investigate possible differences among the μ_i's, we look for a way of measuring differences among the I sample means. If for the moment we forget that each $\overline{X}_{i\cdot}$ is actually an average of J variables, a natural way of measuring variability among $\overline{X}_{1\cdot}, \overline{X}_{2\cdot}, \ldots, \overline{X}_{I\cdot}$ is to examine the sample variance of these sample means:

$$S_{\overline{X}}^{2} = \frac{\sum_{i=1}^{I} (\overline{X}_{i\cdot} - \overline{X}_{\cdot\cdot})^{2}}{I - 1}$$

If all the observed $\overline{x}_{i\cdot}$'s are close to one another, and thus close to the grand mean $\overline{x}_{\cdot\cdot}$, then the computed sample variance $s_{\overline{X}}^{2}$ will be relatively small, lending credence to H_0. On the other hand, discrepancies in the values of the $\bar{x}_i.$'s, which would suggest that H_0 is false, lead to a relatively large value of $s_{\bar{X}}^2$. But what is a large value of $s_{\bar{X}}^2$ and what is a small value? To proceed further, we make our first assumption about the distribution of the X_{ij}'s:

Assumption 1: The variables X_{ij} ($i = 1, \ldots, I$ and $j = 1, \ldots, J$) are independent of one another with

$$E(X_{ij}) = \mu_i, \qquad \text{Var}(X_{ij}) = \sigma^2$$

This assumption is analogous to the one which prefaced the two-sample t test. There we assumed that both populations had the same variance σ^2, while here we are assuming that all I of the populations have the same variance σ^2. Another way to express this assumption is to write

$$X_{ij} = \mu_i + \epsilon_{ij}$$

where each ϵ_{ij} is a random error term having mean 0 and variance σ^2, so that ϵ_{ij} is the amount by which jth observed value deviates from its mean μ_i.

Now because each $\overline{X}_i.$ is the sample mean of a random sample of size J,

$$\text{Var}(\overline{X}_i.) = \frac{\text{Var}(X_{ij})}{J} = \frac{\sigma^2}{J}$$

When H_0 is true, all of the μ_i's are equal, so we can denote their common value by μ. Thus when H_0 is true,

$$E(\overline{X}_i.) = \mu \quad \text{and} \quad \text{Var}(\overline{X}_i.) = \frac{\sigma^2}{J} \quad (i = 1, \ldots, I)$$

so the $\overline{X}_i.$'s can be regarded as a sample of size I from a population (of sample means) having mean μ and variance σ^2/J. Since $S_{\bar{X}}^2$ is the sample variance of the $\overline{X}_i.$'s, $S_{\bar{X}}^2$ is an unbiased estimator of the "population variance" σ^2/J. Thus $JS_{\bar{X}}^2$ (which, like $S_{\bar{X}}^2$ itself, measures the spread among the $\overline{X}_i.$'s) is an unbiased estimator of σ^2 when H_0 is true. Because this estimator is based on sample means for the I samples, it is often called the **between-samples estimator** of σ^2.

When H_0 is not true, the $\overline{X}_i.$'s do not all have the same mean value, so tend to be more spread out than when H_0 is true. In particular, it can be shown that $JS_{\bar{X}}^2$ will tend to *overestimate* σ^2 in this case. If the true value of σ^2 were known, we could compare $JS_{\bar{X}}^2$ to σ^2 and reject H_0 if the former greatly exceeds the latter. While we don't know σ^2, there is a way of estimating it which yields an unbiased estimator whether or not H_0 is true. The sample variance S_i^2 of the ith sample (whose value is computed from the observations in the ith row of the data table) is, for each i, an unbiased estimator for σ^2. Because there are J observations in each sample, we combine the S_i^2's by taking their unweighted average to obtain an unbiased estimator of σ^2. This estimator is called the **within-samples estimator** of σ^2, since it is obtained by combining (pooling) estimators from the individual samples.

Proposition: Let $\overline{X}_i$. and S_i^2 $(i = 1, \ldots, I)$ denote the sample mean and variance of the ith sample. Define the **between-samples estimator** $\hat{\sigma}_B^2$ by

$$\hat{\sigma}_B^2 = JS_{\overline{X}}^2 = \frac{J\sum\limits_{i=1}^{I} (\overline{X}_i. - \overline{X}..)^2}{I - 1} = \frac{\sum\limits_{i=1}^{I} \sum\limits_{j=1}^{J} (\overline{X}_i. - \overline{X}..)^2}{I - 1} \qquad (10.1)$$

and the **within-sample estimator** $\hat{\sigma}_W^2$ by

$$\hat{\sigma}_W^2 = \frac{\sum\limits_{i=1}^{I} S_i^2}{I} = \frac{1}{I}\left[\sum_{i=1}^{I} \frac{1}{J-1} \sum_{j=1}^{J} (X_{ij} - \overline{X}_i.)^2\right] = \frac{\sum\limits_{i=1}^{I} \sum\limits_{j=1}^{J} (X_{ij} - \overline{X}_i.)^2}{I(J - 1)} \qquad (10.2)$$

Then $\hat{\sigma}_B^2$ is an unbiased estimator of σ^2 when H_0 is true, but $E(\hat{\sigma}_B^2) > \sigma^2$ when H_0 is false, while $\hat{\sigma}_W^2$ is unbiased for σ^2 whether or not H_0 is true.

The decision as to whether or not H_0 should be rejected will be made by comparing the two estimators of σ^2. To this end, define

$$F = \frac{\hat{\sigma}_B^2}{\hat{\sigma}_W^2} = \frac{J \sum (\overline{X}_i. - \overline{X}..)^2/(I - 1)}{\sum\sum (X_{ij} - \overline{X}_i.)^2/I(J - 1)} \qquad (10.3)$$

F will be our test statistic, and H_0 will be rejected in favor of H_a when the computed value of F is sufficiently large. To give a more precise description, one more assumption is necessary,

Assumption 2: All I populations are normal populations, so each X_{ij} is a normally distributed variable.

The F Distribution

A level α test can be specified if the probability distribution of F when H_0 is true can be determined. It can be shown that when H_0 is true, the statistic F has a probability distribution called an F distribution. The F distribution has two parameters, one called the number of degrees of freedom of the numerator of F (numerator d.f.) and the other called the number of degrees of freedom of the denominator. Let ν_1 = numerator d.f. and ν_2 = denominator d.f. Then F_{α, ν_1, ν_2} will denote the upper-tail critical value for the F curve with ν_1 and ν_2 d.f., as pictured in Figure 10.1. Notice that F cannot be negative, because it is a ratio of two nonnegative quantities. Because the distribution has two parameters, it is difficult to tabulate F_{α, ν_1, ν_2} for many values of α. Since the ANOVA F test is upper-tailed and the traditional levels of significance are $\alpha = .05$ and $\alpha = .01$, Appendix Table A.7 gives F_{α, ν_1, ν_2} only for these two values of α. Part (a) of the table gives $\alpha = .05$ values and part (b)

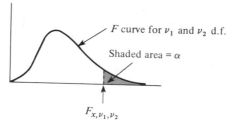

Figure 10.1 The F curve and critical value F_{α, ν_1, ν_2}

gives $\alpha = .01$ values. The numerator d.f. ν_1 is found along the top margin of the table (column) and ν_2 is found along the left-hand margin (row). For example, $F_{.05, 3, 20} = 3.10$.

The Test Procedure

The description of the test procedure will be complete when the rejection region is specified. First numerator d.f. and denominator d.f. must be identified. The numerator of F is computed from the I $(\overline{X}_i. - \overline{X}..)$'s; since $\Sigma (\overline{X}_i. - \overline{X}..) = 0$, the numerator has $I - 1$ d.f. Each sample contributes $J - 1$ d.f. to the denominator, for a total of $I(J - 1)$ denominator d.f.

> **Theorem:** In single-factor ANOVA with I populations or treatments and J observations from each population, when H_0 is true and Assumptions 1 and 2 are satisfied, F of (10.3) has an F distribution with $\nu_1 = I - 1$ and $\nu_2 = I(J - 1)$. The level α test for H_0 versus H_a is given by
>
> reject H_o in favor of H_a if $F \geq F_{\alpha, I-1, I(J-1)}$ (10.4)

The F test is an upper-tailed test, since large values of F suggest that $\hat{\sigma}_B^2$ has overestimated σ^2 and thus that the μ_i's are not all equal.*

Example 10.1 (continued)

For the swimming data, $I - 1 = 3$ and $I(J - 1) = (4)(5) = 20$, so $F_{.05, 3, 20} = 3.10$ specifies the critical value for a level .05 test. We will shortly present computational formulas for the numerator and denominator of F, which yield $\hat{\sigma}_B^2 = 4.44$ and $\hat{\sigma}_W^2 = 1.63$. The computed value of F is $f = 4.44/1.63 = 2.72$, which does not exceed 3.10, so H_0 is not rejected and the data is not significant at level .05.

Computational Formulas for ANOVA

The calculations leading to F can be done efficiently by using formulas similar to the computing formula for a single sample variance s^2. Let X_i. represent the *sum* (not

---

*While our development of the F test has been based on heuristics, the test procedure can be derived from the likelihood ratio principle.

average, since there is no bar) of the X_{ij}'s for i fixed (sum of the numbers in the ith row of the table) and $X..$ denote the sum of *all* the X_{ij}'s (the **grand total**).

Definition: The **total sum of squares (SST)**, **treatment sum of squares (SSTr)**, and **error sum of squares (SSE)** are given by

$$SST = \sum_{i=1}^{I} \sum_{j=1}^{J} (X_{ij} - \overline{X}..)^2 = \sum_{i=1}^{I} \sum_{j=1}^{J} X_{ij}^2 - \frac{1}{IJ} X^2..$$

$$SSTr = \sum_{i=1}^{I} \sum_{j=1}^{J} (\overline{X}_i. - \overline{X}..)^2 = \frac{1}{J} \sum_{i=1}^{I} X_i^2. - \frac{1}{IJ} X^2.. \qquad (10.5)$$

$$SSE = \sum_{i=1}^{I} \sum_{j=1}^{J} (X_{ij} - \overline{X}_i.)^2, \text{ where } X_i. = \sum_{j=1}^{J} X_{ij}, X.. = \sum_{i=1}^{I} \sum_{j=1}^{J} X_{ij}$$

Referring back to expression (10.3) for the test statistic, *SSTr* appears in the numerator of F, while *SSE* appears in the denominator of F; the reason for defining *SST* will be apparent shortly.

The expressions on the far right-hand side of *SST* and *SSTr* are the computing formulas for these sums of squares. Both *SST* and *SSTr* involve $X^2../IJ$ (the square of the grand total divided by IJ), which is usually called the **correction factor for the mean.** After computing the correction factor, *SST* is obtained by squaring each number in the data table, adding these squares together, and subtracting the correction factor. *SSTr* results from squaring each row total, summing them, dividing by J, and subtracting the correction factor.

Example 10.1 (continued) $x.. = 463.1$, so the correction factor is $(463.1)^2/24 = 8935.90$. Then

$$SST = \sum\sum x_{ij}^2 - \frac{1}{IJ} x^2..$$

$$= (18.7)^2 + (21.1)^2 + \cdots + (20.5)^2 - 8935.90$$

$$= 8981.77 - 8935.90 = 45.87$$

and

$$SSTr = \frac{1}{4}[(117.6)^2 + (110.1)^2 + (122.0)^2 + (113.4)^2] - 8935.90$$

$$= 13.32$$

The reason that no computing formula appears for *SSE* is that there is a simple relationship between the three sums of squares.

> ## The fundamental identity of single-factor ANOVA:
>
> $$SST = SSTr + SSE$$
>
> (10.6)

Thus if any two of the sums of squares are computed, the third can be obtained through (10.6). SST and $SSTr$ are easiest to compute, and then $SSE = SST - SSTr$, which gives $SSE = 45.87 - 13.32 = 32.55$ for the swimming data.

Proof: The fundamental identity follows from the relation

$$x_{ij} - \bar{x}.. = (x_{ij} - \bar{x}_i.) + (\bar{x}_i. - \bar{x}..)$$

(10.7)

Squaring both sides of (10.7) yields

$$(x_{ij} - \bar{x}..)^2 = (x_{ij} - \bar{x}_i.)^2 + 2(x_{ij} - \bar{x}_i.)(\bar{x}_i. - \bar{x}..) + (\bar{x}_i. - \bar{x}..)^2$$

(10.8)

Upon summing both sides of (10.8) over all i and j, SST appears on the left, while the squared terms on the right give SSE and $SSTr$. The middle term is

$$2\sum_i \sum_j (x_{ij} - \bar{x}_i.)(\bar{x}_i. - \bar{x}..) = 2\sum_i (\bar{x}_i. - \bar{x}..)\left[\sum_j (x_{ij} - \bar{x}_i.)\right]$$

$$= 2\sum_i (\bar{x}_i. - \bar{x}..) \cdot 0 = 0$$

The interpretation of the fundamental identity is an important aid to an understanding of ANOVA. SST is a measure of the total variation in the data—the sum of all squared deviations about the grand mean. The identity says that this total variation can be partitioned into two pieces. SSE measures variation which would be present (within rows) even if H_0 is true, and is thus the part of total variation which is *unexplained* by the truth or falsity of H_0. $SSTr$ is the amount of variation (between rows) which can be *explained* by possible differences in the μ_i's. If explained variation is large relative to unexplained variation, then H_0 is rejected in favor of H_a.

Once $SSTr$ and SSE are computed, F is the ratio of two quantities called **mean squares,** which are simply the two sums of squares divided by their appropriate number of degrees of freedom:

$$MSTr = \frac{SSTr}{I - 1}, \quad MSE = \frac{SSE}{I(J - 1)}, \quad F = \frac{MSTr}{MSE}$$

(10.9)

For the swimming data, $MSTr = 13.32/3 = 4.44$, $MSE = 32.55/20 = 1.63$, so $f = 4.44/1.63 = 2.72$. The computations are summarized in a format called an **ANOVA table,** as illustrated in Table 10.2.

Table 10.2

Source of variation	d.f.	Sum of squares	Mean square	F
Treatments	$I - 1 = 3$	$SSTr = 13.32$	$MSTr = SSTr/(I - 1) = 4.44$	$\dfrac{MSTr}{MSE} = 2.72$
Error	$I(J - I) = 20$	$SSE = 32.55$	$MSE = SSE/I(J - 1) = 1.63$	
Total	$IJ - 1 = 23$	$SST = 45.87$		

Example 10.2   A biologist wished to study the effects of ethanol on sleep time. A sample of 20 rats, matched for age and other characteristics, was selected, and each rat was given an oral injection having a particular concentration of ethanol per body weight. The rapid eye movement (REM) sleep time for each rat was then recorded for a 24-hour period, with the following results:

Treatment (concentration of ethanol)

						$x_i.$	$\bar{x}_i.$
0 (control)	88.6	73.2	91.4	68.0	75.2	396.4	79.28
1 gm/kg	63.0	53.9	69.2	50.1	71.5	307.7	61.54
2 gm/kg	44.9	59.5	40.2	56.3	38.7	239.6	47.92
4 gm/kg	31.0	39.6	45.3	25.2	22.7	163.8	32.76

$$x.. = 1107.5, \ \bar{x}.. = 55.375$$

Does the data indicate that the true average REM sleep time depends on the concentration of ethanol? (This example is based on an experiment reported in "Relationship of Ethanol Blood Level to REM and Non-REM Sleep Time and Distribution in the Rat," *Life Sciences,* 1978, pp. 839–846).

The $\bar{x}_i.$'s differ rather substantially from one another, but there is also a great deal of variability within each sample, so to answer the question precisely we must carry out the analysis of variance. With $\Sigma \Sigma x_{ij}^2 = 68,697.61$ and $x^2../IJ = (1107.5)^2/20 = 61,327.81$, the computing formulas (10.5) yield

$$SST = 68,697.61 - 61,327.81 = 7369.8$$

$$SSTr = \tfrac{1}{5}[(396.40)^2 + (307.70)^2 + (239.60)^2 + (163.80)^2] - 61,327.81$$

$$= 67,210.21 - 61,327.81 = 5882.4$$

and

$$SSE = 7369.8 - 5882.4 = 1487.4$$

The ANOVA table is

Table 10.3

Source	d.f.	Sum of squares	Mean square	F
Treatments	3	5882.4	1960.8	21.09
Error	16	1487.4	93.0	
Total	19	7369.8		

Since $F_{.05,3,16} = 3.24$, the computed value 21.09 exceeds this critical value, so that the data is significant at level .05. We conclude that true average REM sleep time does indeed depend on the concentration of ethanol in the blood.

When the F test causes H_0 to be rejected, the experimenter will often be interested in further analysis to decide which μ_i's differ from which others. Procedures for doing this are called multiple comparison procedures, and several are described in the next two sections.

Exercises / Section 10.1

1. In an experiment to compare the tensile strengths of $I = 5$ different types of copper wire, $J = 4$ samples of each type were used. The between-samples and within-samples estimates of σ^2 were computed as $\hat{\sigma}_B^2 = 2573.3$ and $\hat{\sigma}_W^2 = 1394.2$, respectively. Use the F test at level .05 to test $H_0 : \mu_1 = \mu_2 = \mu_3 = \mu_4 = \mu_5$ versus H_a : at least two μ_i's are unequal.

2. The lumen output was determined for each of $I = 3$ different brands of 60-watt soft-white light-bulbs, with $J = 8$ bulbs of each brand tested. The sums of squares were computed as $SSE = 4773.3$ and $SSTr = 591.2$. State the hypotheses of interest (including word definitions of parameters), and use the F test of ANOVA ($\alpha = .05$) to decide whether or not there are any differences in true average lumen outputs between the three brands for this type of bulb.

3. In a study to assess the effects of malaria infection on mosquito hosts ("Plasmodium Cynomolgi: Effects of Malaria Infection on Laboratory Flight Performance of Anopheles Stephensi Mosquitos," *Experimental Parasitology*, 1977, pp. 397–404), mosquitos were fed on either infective or non-infective rhesus monkeys. Subsequently the distance they flew during a 24-hour period was measured using a flight mill. The mosquitos were divided into four groups of eight mosquitos each: infective rhesus and sporozites present (IRS), infective rhesus and oocysts present (IRD), infective rhesus and no infection developed (IRN), and non-infective (C). The summary data values are $\bar{x}_1. = 4.39$ (IRS), $\bar{x}_2. = 4.52$ (IRD), $\bar{x}_3. = 5.49$ (IRN), $\bar{x}_4. = 6.36$ (C), $\bar{x}.. = 5.19$, and $\Sigma\Sigma x_{ij}^2 = 911.91$. Use the ANOVA F test at level

.05 to decide whether or not there are any differences between true average flight times for the four treatments.

4. The article "Origin of Precambrian Iron Formations" (*Econ. Geology*, 1964, pp. 1025–1057) reported the following data on total Fe for four types of iron formation (1 = carbonate, 2 = silicate, 3 = magnetite, 4 = hematite).

(1)	20.5,	28.1,	27.8,	27.0,	28.0,
	25.2,	25.3,	27.1,	20.5,	31.3
(2)	26.3,	24.0,	26.2,	20.2,	23.7,
	34.0,	17.1,	26.8,	23.7,	24.9
(3)	29.5,	34.0,	27.5,	29.4,	27.9,
	26.2,	29.9,	29.5,	30.0,	35.6
(4)	36.5,	44.2,	34.1,	30.3,	31.4,
	33.1,	34.1,	32.9,	36.3,	25.5

a. Compute the four sample means $\bar{x}_1.$, $\bar{x}_2.$, $\bar{x}_3.$, $\bar{x}_4.$, and also the grand mean $\bar{x}...$

b. Compute the sample variance of the sample means, (that is, $s_{\bar{x}}^2$) and then multiply it by $J = 10$ to obtain $MSTr$, the estimate of σ^2 in the numerator of the F statistic.

c. Compute the sample variance separately for each of the $I = 4$ samples, and then average them to obtain MSE, the estimator of σ^2 in the denominator of F.

d. Use the results of (b) and (c) to compute F, and then use a level $\alpha = .01$ test to decide whether or not the four population means differ from one another.

e. Construct an ANOVA table to display the results of the analysis.

5. In an experiment to investigate the performance of four different brands of sparkplugs intended for

use on a 125-cc two-stroke motorcycle, five plugs of each brand were tested and the number of miles (at a constant speed) until failure was observed. The partial ANOVA table for the data appears below. Fill in the missing entries, state the relevant hypotheses, and carry out a test.

Source	d.f.	Sum of squares	Mean square	F
Brand				
Error		235,419.04		
Total		310,500.76		

6. Six samples of each of four types of cereal grain grown in a certain region were analyzed to determine thiamin content, resulting in the following data (micrograms/gram):

Wheat: 5.2, 4.5, 6.0, 6.1, 6.7, 5.8
Barley: 6.5, 8.0, 6.1, 7.5, 5.9, 5.6
Maize: 5.8, 4.7, 6.4, 4.9, 6.0, 5.2
Oats: 8.3, 6.1, 7.8, 7.0, 5.5, 7.2

Does this data suggest that at least two of the grains differ with respect to true average thiamin content? Use a level $\alpha = .05$ test.

7. A study of the properties of metal plate-connected trusses used for roof support ("Modeling Joints Made with Light-Gauge Metal Connector Plates," *Forest Products J.*, 1979, pp. 39–44) yielded the following observations on axial stiffness index (kips/in) for plate lengths of 4, 6, 8, 10, and 12 in.

4: 309.2, 409.5, 311.0, 326.5, 316.8, 349.8, 309.7
6: 402.1, 347.2, 361.0, 404.5, 331.0, 348.9, 381.7
8: 392.4, 366.2, 351.0, 357.1, 409.9, 367.3, 382.0
10: 346.7, 452.9, 461.4, 433.1, 410.6, 384.2, 362.6
12: 407.4, 441.8, 419.9, 410.7, 473.4, 441.2, 465.8

Does variation in plate length have any effect on true average axial stiffness? State and test the relevant hypotheses using analysis of variance with $\alpha = .01$. Display your results in an ANOVA table. *Hint:* $\Sigma\Sigma x_{ij}^2 = 5,241,420.79$.

8. An article in the British scientific journal *Nature* ("Sucrose Induction of Hepatic Hyperplasis in the Rat," August 25, 1972, p. 461) reported on an experiment in which each of five groups consisting of six rats was put on a diet with a different carbohydrate. At the conclusion of the experiment the DNA content of the liver of each rat was deter-

mined (mg/g liver), with the following results:

Carbo-hydrate	Starch	Sucrose	Fructose	Glucose	Maltose
$\bar{x}_i.$	2.58	2.63	2.13	2.41	2.49

Assuming also that $\Sigma\Sigma x_{ij}^2 = 183.4$, does the data indicate that true average DNA content is affected by the type of carbohydrate in the diet? Construct an ANOVA table and use a .05 level of significance.

9. The article "The Effect of Enzyme Inducing Agents on the Survival Times of Rats Exposed to Lethal Levels of Nitrogen Dioxide" (*Toxicology and Applied Pharmacology*, 1978, pp. 169–174) reported the following data on survival times for rats exposed to nitrogen dioxide (70 ppm) via different injection regimens. There were $J = 14$ rats in each group.

Regimen	$\bar{x}_i.$(min)	s_i
(1) Control	166	32
(2) 3 Methylcholanthrene	303	53
(3) Allylisopropylacetamide	266	54
(4) Phenobarbital	212	35
(5) Chlorpromazine	202	34
(6) p Aminobenzoic Acid	184	31

a. Compute s_i^2 ($i = 1, \ldots, 6$) and then *MSE*.
b. Compute *SSTr* and *MSTr* using the $\bar{x}_i.$'s.
c. Test the null hypothesis that true average survival time does not depend on injection regimen against the alternative that there is some dependence on injection regimen using $\alpha = .01$.

10. In single-factor ANOVA with I treatments and J observations per treatment, let $\mu = \frac{1}{I}\Sigma \mu_i$.

a. Express $E(\bar{X}..)$ in terms of μ. *Hint:*
$$\bar{X}.. = \frac{1}{I}\Sigma\bar{X}_i..$$

b. Compute $E(\bar{X}_i.^2)$. *Hint:* For any r.v. Y, $E(Y^2) = \text{Var}(Y) + [E(Y)]^2$.
c. Compute $E(\bar{X}..^2)$.
d. Compute $E(SSTr)$ and then show that
$$E(MSTr) = \sigma^2 + \frac{J}{I-1}\Sigma(\mu_i - \mu)^2.$$

e. Using the result of (d), what is $E(MSTr)$ when H_0 is true? When H_0 is false, how does $E(MSTr)$ compare to σ^2?

10.2 Multiple Comparisons in ANOVA

When the computed value of the F statistic in single-factor ANOVA is not significant, the analysis is terminated because no differences between the μ_i's have been identified. But when H_0 is rejected, the investigator will usually want to know which of the μ_i's are different from one another. A method for carrying out this further analysis is called a multiple comparisons procedure. There are a number of such procedures in the statistics literature. Here we present two which many statisticians recommend for deciding for each i and j whether $\mu_i = \mu_j$.

Tukey's Procedure (The *T* Method)

Tukey's procedure involves the use of a new statistic Q called a **studentized range statistic**. The distribution of Q is the **studentized range distribution** and depends on two parameters: a numerator d.f. m and a denominator d.f. ν. We will shortly discuss the definition of Q, but first we show how this distribution can be applied to perform multiple comparisons. Let $Q_{\alpha, m, \nu}$ denote the upper-tail α critical value of the studentized range distribution with m numerator d.f. and ν denominator d.f. (analogous to F_{α, ν_1, ν_2}). Values of $Q_{\alpha, m, \nu}$ are given in Appendix Table A.8. Then $Q_{\alpha, I, I(J-1)}$ can be used to obtain simultaneous confidence intervals for all pairwise differences $\mu_i - \mu_j$. Notice that numerator d.f. for the appropriate Q critical value is I, the number of treatments or populations, and not $I - 1$, as it was for F.

Proposition: With probability $1 - \alpha$,

$$\overline{X}_{i\cdot} - \overline{X}_{j\cdot} - Q_{\alpha, I, I(J-1)} \sqrt{MSE/J} \leq \mu_i - \mu_j \leq \overline{X}_{i\cdot} - \overline{X}_{j\cdot}$$
$$+ Q_{\alpha, I, I(J-1)} \sqrt{MSE/J} \quad (10.10)$$

for every i and j $(i = 1, \ldots, I$ and $j = 1, \ldots, I)$.

When the computed $\overline{x}_{i\cdot}$, $\overline{x}_{j\cdot}$, and MSE are substituted into (10.10), the result is a collection of **simultaneous confidence statements** about the true values of all differences $\mu_i - \mu_j$ between true treatment means. *Each interval from (10.10) which does not include zero yields the conclusion that μ_i and μ_j differ significantly at level α.*

The computation of all intervals in (10.10) looks complicated, but a straightforward sequence of steps culminates in a pictorial summary of the conclusions:

The *T* method for identifying significantly different μ_i's

1. Select α and find $Q_{\alpha, I, I(J-1)}$ from Appendix Table A.8.
2. Determine $w = Q_{\alpha, I, I(J-1)} \cdot \sqrt{MSE/J}$.
3. List the sample means in increasing order and underline those pairs which differ by less than w. Any pair of sample means not underscored by the same line corresponds to a pair of true treatment means which are judged significantly different.

Example 10.3 It is widely believed that although "house" brands (those sold only through a particular chain of stores) and regionally distributed brands of canned goods tend to be less expensive than their nationally distributed counterparts, they also tend to be less desirable in terms of quantity and quality of contents. To see whether such beliefs have any basis in fact, five different brands of fruit cocktail were chosen: two national brands (1 and 2), a regional brand (3), and a house brand from each of two large supermarket chains (4 and 5). Nine 16-oz cans of each brand were purchased, opened and drained of liquid, and their net contents weighed. The sample average weights for the five brands were $\bar{x}_1. = 14.5$, $\bar{x}_2. = 13.8$, $\bar{x}_3. = 13.3$, $\bar{x}_4. = 14.3$, and $\bar{x}_5. = 13.1$. The ANOVA table summarizing the first part of the analysis appears as Table 10.4.

Table 10.4

Source of variation	d.f.	Sum of squares	Mean square	F
Treatments (brands)	4	13.32	3.33	37.84
Error	40	3.53	.088	
Total	44	16.85		

Since $F_{.05, 4, 40} = 2.61$, H_0 is rejected (decisively) at level .05. We now use Tukey's procedure to look for significant differences among the μ_i's. From Appendix Table A.8, $Q_{.05, 5, 40} = 4.04$ (the second subscript on Q is I and not $I - 1$ as in F), so $w = 4.04 \sqrt{.088/9} = .4$. Arranging the five sample means in increasing order, every pair differing by less than .4 is underscored:

$$\begin{array}{ccccc} \bar{x}_5. & \bar{x}_3. & \bar{x}_2. & \bar{x}_4. & \bar{x}_1. \\ \underline{13.1 \quad\quad 13.3} & & 13.8 & \underline{14.3 \quad\quad 14.5} \end{array}$$

Thus brands 1 and 4 are not significantly different from one another, but are significantly higher than the other three brands in their true average contents. Brand 2 is significantly better than 3 and 5 but worse than 1 and 4, and brands 3 and 5 do not differ significantly. Note that one house brand (4) is in the top group, while a national brand is in the lowest group.

In Example 10.3, if $\bar{x}_2. = 14.15$ rather than 13.8 with the same computed w, then the configuration of underscored means would be

$$\begin{array}{ccccc} \bar{x}_5. & \bar{x}_3. & \bar{x}_2. & \bar{x}_4. & \bar{x}_1. \\ \underline{1.31 \quad\quad 13.3} & & \underline{14.15 \quad\quad 14.3 \quad\quad 14.5} \end{array}$$

When there are no significant differences among a group of more than two means, it is customary to draw a continuous line underneath the entire group.

Example 10.4 Consider again the ethanol/sleep-time data introduced in the previous section, in which H_0 was rejected at level .05. There were $I = 4$ treatments and 16 d.f. for error, so $Q_{.05, 4, 16} = 4.05$ and $w = 4.05\sqrt{93.0/5} = 17.47$. Ordering the means and underscoring yields

$$
\begin{array}{cccc}
\bar{x}_4 . & \bar{x}_3 . & \bar{x}_2 . & \bar{x}_1 . \\
32.76 & 47.92 & 61.54 & 79.28
\end{array}
$$

The interpretation of this underscoring must be done with care, since we seem to have concluded that treatments 2 and 3 do not differ, 3 and 4 do not differ, yet 2 and 4 do differ. The suggested way of expressing this is to say that while evidence allows us to conclude that treatments 2 and 4 differ from one another, neither has been shown to be significantly different from 3. Treatment 1 has a significantly higher true average REM sleep time than any of the other treatments.

The Interpretation of α in Multiple Comparison

Tukey's method involves the simultaneous construction of confidence intervals for all differences $\mu_i - \mu_j$ of pairs of treatment means. In the construction of confidence intervals in the previous chapter, $\alpha = .05$ or 95% confidence referred to the error rate or confidence level for the individual statement about the parameter of interest. If an ANOVA experiment involves comparison of four treatments, then Tukey's procedure obtains simultaneously six different intervals (since there are six different pairs of treatment means). The confidence level $\alpha = .05$ no longer refers to a particular interval, but instead to the experiment as a whole, so is called an **experimentwise error rate.** If Tukey's method were used on a great many different ANOVA data sets, then in approximately 95% of these experiments no erroneous claim would be made about any of the $\mu_i - \mu_j$'s, while in only 5% would at least one incorrect claim be made. That is, in 95% of all experiments, every confidence interval constructed from (10.10) would include the true value of $\mu_i - \mu_j$, and only 5% of the time would at least one interval fail to cover the true value of a $\mu_i - \mu_j$. Because the confidence level for the entire set of comparisons of means is 95%, the confidence level for any particular comparison is larger than 95% and increases as the number of comparisons (that is, treatment means) increases. This distinction between the experimentwise error rate or confidence level and a "per comparison" error rate or confidence level is important, and should be kept in mind when interpreting the results of any multiple comparison analysis.

Multiple Comparisons and t Intervals

A t interval for a difference $\mu_1 - \mu_2$ was presented in Chapter 9. The interval depended on using the pooled estimator S_p for σ and was based on the t distribution with $n_1 + n_2 - 2$ d.f. In one-way ANOVA, we can replace S_p by $\sqrt{MSE}$ and $n_1 + n_2 - 2$ by $I(J - 1)$ to obtain a $100(1 - \alpha)\%$ confidence interval for a particular difference $\mu_i - \mu_j$:

$$
\bar{x}_i . - \bar{x}_j . \pm t_{\alpha/2, I(J-1)} \sqrt{MSE\left(\frac{1}{J} + \frac{1}{J}\right)} \tag{10.11}
$$

If several 95% intervals ($\alpha = .05$) of the form (10.11) are computed for different

$\mu_i - \mu_j$'s, the simultaneous confidence level will be less than 95% (for two completely independent intervals, the simultaneous confidence level would be $100(.95)^2 \doteq 90\%$, but because MSE appears in each interval computed from (10.11), different intervals are not independent). However, when several intervals are computed, there is a simple lower bound on the simultaneous confidence level.

Proposition: If an interval of the form (10.11) is computed for each i and j with $i < j$, resulting in $k = \binom{I}{2} = I(I-1)/2$ intervals, the simultaneous confidence that all k intervals cover the true $\mu_i - \mu_j$'s is at least $100(1 - k\alpha)$.*

The multiple comparison procedure resulting from these intervals, called the **Bonferroni** procedure, is to identify μ_i and μ_j as significantly different from one another if the interval for $\mu_i - \mu_j$ does not include zero. While the intervals tend to be somewhat longer than the Tukey intervals (resulting in fewer significantly different μ_i's), the intervals are valid for unequal sample sizes (Section 10.3), whereas the Tukey intervals cannot be used in this case. Each interval (10.11) has the same length (because of the equal sample sizes), so with $w = t_{\alpha/2, I(J-1)} \sqrt{2MSE/J}$, the same underscoring method used in the Tukey procedure works here: After the $\bar{x}_i.$'s are ordered and those which differ by less than w are underscored, pairs not underscored correspond to significantly different $\mu_i - \mu_j$'s.

Example 10.4 (continued)

There were four different treatments in the ethanol/sleep-time experiment, so $k = 4(3)/2 = 6$. If a 99% interval is computed for each of the six $\mu_i - \mu_j$'s ($\alpha = .01$), the simultaneous confidence level will be at least $100(1 - 6\alpha)\% = 94\%$. With $t_{.005, 16} = 2.921$, $MSE = 93.0$, and $J = 5$, $w = 2.921 \sqrt{2(93.0)/5} = 17.82$.

$\bar{x}_4.$	$\bar{x}_3.$	$\bar{x}_2.$	$\bar{x}_1.$
32.76	47.92	61.54	79.28

Because w is slightly larger than for the T method, the pair corresponding to $i = 1$, $j = 2$ is now underscored, so only the pairs (μ_1, μ_3), (μ_1, μ_4), and (μ_2, μ_4) are judged significantly different.

Confidence Intervals for Other Parametric Functions

In some situations a confidence interval is desired for a function of the μ_i's more complicated than a difference $\mu_i - \mu_j$. Let $\theta = \Sigma c_i \mu_i$, where the c_i's are constants.

---

* To obtain a simultaneous confidence level of at least $100(1 - \alpha)\%$ for k comparisons, values of $t_{\alpha/2k, \nu}$ are required. These appear in "Tables of the Bonferroni t Statistic" (*J. Amer. Stat. Assn.*, 1977, pp. 469–478).

One such function is $\frac{1}{2}(\mu_1 + \mu_2) - \frac{1}{3}(\mu_3 + \mu_4 + \mu_5)$, which in the context of Example 10.3 measures the difference between fruit cocktail can contents for national and nonnational brands. Because the X_{ij}'s are normally distributed with $E(X_{ij}) = \mu_i$, $\text{Var}(X_{ij}) = \sigma^2$, $\hat{\theta} = \sum_i c_i \overline{X}_i.$ is normally distributed, unbiased for θ, and

$$\text{Var}(\hat{\theta}) = \text{Var}(\sum_i c_i \overline{X}_i.) = \sum_i c_i^2 \, \text{Var}(\overline{X}_i.) = \frac{\sigma^2}{J} \sum_i c_i^2$$

Estimating σ^2 by *MSE* and forming $\hat{\sigma}_{\hat{\theta}}$ results in a t variable $(\hat{\theta} - \theta)/\hat{\sigma}_{\hat{\theta}}$, which can be manipulated to obtain the following $100(1 - \alpha)\%$ confidence interval for $\sum c_i \mu_i$:

$$\sum c_i \overline{x}_i. \pm t_{\alpha/2, I(J-1)} \sqrt{\frac{MSE \sum c_i^2}{J}} \tag{10.12}$$

**Example 10.3
(continued)** For

$$\theta = \frac{1}{2}(\mu_1 + \mu_2) - \frac{1}{3}(\mu_3 + \mu_4 + \mu_5),$$

$$\sum c_i^2 = (\tfrac{1}{2})^2 + (\tfrac{1}{2})^2 + (-\tfrac{1}{3})^2 + (-\tfrac{1}{3})^2 + (-\tfrac{1}{3})^2 = \tfrac{5}{6}$$

With $\hat{\theta} = \frac{1}{2}(\overline{x}_1. + \overline{x}_2.) - \frac{1}{3}(\overline{x}_3. + \overline{x}_4. + \overline{x}_5.) = .583$ and $MSE = .088$, a 95% interval is

$$.583 \pm 2.201 \sqrt{5(.088)/(6)(9)} = .583 \pm .182 = (.401, .765).$$

If confidence intervals for $k \, (\geq 2)$ parametric functions are desired, the Bonferroni method can be applied using the k intervals computed from (10.12), and the resulting simultaneous confidence level is at least $100(1 - k\alpha)\%$.

The Studentized Range Statistic and Simultaneous Intervals

Definition: Let $Y_1, \ldots, Y_m$ be independent normal random variables, each with mean 0 and standard deviation γ, and let S be an estimator of γ based on ν degrees of freedom which is independent of the Y_i's. The **studentized range statistic** is defined by

$$Q = \frac{\max_{i,j} |Y_i - Y_j|}{S}$$

The word "range" comes from the fact that the numerator of Q is max $(Y_1, \ldots, Y_m) - \min(Y_1, \ldots, Y_m)$, the range of the Y_i's, while "studentizing" refers to division by an estimator of γ.

The identification of the Y_i's, m, γ, S, and ν with the ANOVA quantities is as follows:

$$Y_i = \overline{X}_i. - \mu_i \quad [\text{so } E(Y_i) = 0], \quad m = I \text{ (the number of levels of the factor)},$$

$$\gamma = \sqrt{\text{Var}(\overline{X}_i. - \mu_i)} = \sqrt{\sigma^2/J}, \; S = \sqrt{MSE/J}, \; \nu = I(J - 1) \qquad (10.13)$$

By definition of Q and Q_α

with probability $\quad 1 - \alpha, \quad \dfrac{\displaystyle\max_{i,j} |Y_i - Y_j|}{S} \leq Q_{\alpha, m, \nu}$

This is equivalent to

with probability $\quad 1 - \alpha, \quad \dfrac{|Y_i - Y_j|}{S} \leq Q_{\alpha, m, \nu} \quad$ for all i and j $\qquad (10.14)$

Substitution of the quantities (10.13) into (10.14) and some further manipulation yields the intervals (10.10).

Exercises / Section 10.2

1. An experiment to compare the spreading rates of five different brands of yellow interior latex paint available in a particular area used four gallons ($J = 4$) of each paint. The sample average spreading rate (sq ft/gal) for the five brands were $\overline{x}_1. = 462.0$, $\overline{x}_2. = 512.8$, $\overline{x}_3. = 437.5$, $\overline{x}_4. = 469.3$, and $\overline{x}_5. = 532.1$. The computed value of F was found to be significant at level $\alpha = .05$. With $MSE = 272.8$, use Tukey's procedure to investigate significant differences in the true average spreading rates between brands.

2. In Exercise 1 suppose that $\overline{x}_3. = 427.5$. Now which true average spreading rates differ significantly from one another? Be sure to use the method of underscoring to illustrate your conclusions, and also write a paragraph summarizing your results.

3. Repeat Exercise 2 supposing that $\overline{x}_2. = 502.8$ in addition to $\overline{x}_3. = 427.5$.

4. Use Tukey's procedure on the data in Exercise 10.1.3 to identify differences in true average flight times among the four types of mosquitos.

5. Use Tukey's procedure on the data of Exercise 10.1.4 to identify differences in true average total Fe between the four types of formations (use $MSE = 15.64$).

6. Use Tukey's procedure to investigate differences in true average axial stiffness for the data in Exercise 10.1.7.

7. Use Tukey's procedure to investigate any differences in true average survival times due to the 6 different injection regimens for the data in Exercise 10.1.9.

8. Repeat Exercise 1 using the Bonferroni procedure in place of Tukey's procedure.

9. Repeat Exercise 5 using the Bonferroni procedure (using $\alpha = .001$ in (10.11)).

10. Repeat Exercise 7 using the Bonferroni procedure.

11. Referring to Exercise 10.1.8, construct a t confidence for

$$\theta = \mu_1 - (\mu_2 + \mu_3 + \mu_4 + \mu_5)/4$$

which measures the difference between the average DNA content for the starch diet and the combined average for the four other diets. Does the resulting interval include zero?

12. Consider a single-factor ANOVA experiment in which $I = 3$, $J = 5$, $\overline{x}_1. = 10$, $\overline{x}_2. = 12$, and $\overline{x}_3. = 20$. Find a value of SSE for which $F > F_{.05, 2, 12}$, so that $H_0 : \mu_1 = \mu_2 = \mu_3$ is rejected, yet when Tukey's procedure is applied none of the μ_i's can be said to differ significantly from one another. This shows that occasionally the F test and Tukey's procedure yield contrary results, which is the result of the two procedures being based on different probability distributions.

13. Referring to Exercise 12, suppose that $\overline{x}_1. = 10$,

$\bar{x}_2. = 15$, and $\bar{x}_3. = 20$. Can you now find a value of SSE which produces such a contradiction between the F test and Tukey's procedure?

14. Referring to Exercise 10.1.9, compute confidence intervals with simultaneous confidence level at least 98% for $\mu_1 - \frac{1}{5}(\mu_2 + \mu_3 + \mu_4 + \mu_5 + \mu_6)$ and $\frac{1}{4}(\mu_2 + \mu_3 + \mu_4 + \mu_5) - \mu_6$.

10.3 More on Single-Factor ANOVA

In this section we briefly consider some issues relating to single-factor ANOVA that were not dealt with in the first two sections. These include an alternative description of the model parameters, the power of the F test, the relationship of the test to procedures previously considered, sensitivity (robustness) of the test to assumptions, and formulas for the case of unequal sample sizes.

An Alternative Description of the ANOVA Model

Rather than use μ_i to denote the true mean for the ith treatment, there is another set of parameters which is sometimes used to describe the model and which suggests appropriate generalizations to models involving more than one factor. Define a parameter μ by

$$\mu = \frac{1}{I} \sum_{i=1}^{I} \mu_i \tag{10.15}$$

and the parameters $\alpha_1, \ldots, \alpha_I$ by

$$\alpha_i = \mu_i - \mu \quad i = 1, \ldots, I \tag{10.16}$$

Then the treatment mean μ_i can be written as $\mu + \alpha_i$, where μ represents the true average overall response in the experiment, and α_i is the effect, measured as a departure from μ, due to the ith treatment. Whereas we initially had I parameters, we now have $I + 1$ ($\mu, \alpha_1, \ldots, \alpha_I$). However, because $\Sigma \alpha_i = 0$ (the average departure from the overall mean response is zero), only I of these new parameters are independently determined, so there are as many independent parameters as there were before. In terms of μ and the α_i's, the model becomes

$$X_{ij} = \mu + \alpha_i + \epsilon_{ij} \quad i = 1, \ldots, I \quad j = 1, \ldots, J \tag{10.17}$$

In Chapter 11 we will develop analogous models for multifactor ANOVA. The claim that the μ_i's are identical is equivalent to the equality of the α_i's, and because $\Sigma \alpha_i = 0$, the null hypothesis becomes

$$H_0: \alpha_1 = \alpha_2 = \cdots = \alpha_I = 0$$

In Section 10.1 it was stated that $MSTr$ is a good estimator of σ^2 when H_0 is true, but otherwise tends to overestimate σ^2. More precisely,

$$E(MSTr) = \sigma^2 + \frac{J}{I-1} \sum \alpha_i^2 \tag{10.18}$$

When H_0 is true, $\Sigma \alpha_i^2 = 0$ so $MSTr$ is an unbiased estimator for σ^2 (MSE is unbiased whether or not H_0 is true). If $\Sigma \alpha_i^2$ is used as a measure of the extent to which H_0 is false, then a larger value of $\Sigma \alpha_i^2$ will result in a greater tendency for $MSTr$ to overestimate σ^2. In the next chapter, formulas for expected mean squares for multi-factor models will be used to suggest how to form F ratios to test various hypotheses.

Proof of the formula for $E(MSTr)$:

For any random variable Y, $E(Y^2) = \text{Var}(Y) + [E(Y)]^2$, so

$$E(SSTr) = E\left(\frac{1}{J}\sum_i X_i^2. - \frac{1}{IJ}X^2..\right) = \frac{1}{J}\sum_i E(X_i^2.) - \frac{1}{IJ}E(X^2..)$$

$$= \frac{1}{J}\sum_i\left\{\text{Var}(X_i.) + [E(X_i.)]^2\right\} - \frac{1}{IJ}\left\{\text{Var}(X..) + [E(X..)]^2\right\}$$

$$= \frac{1}{J}\sum_i\left\{J\sigma^2 + [J(\mu + \alpha_i)]^2\right\} - \frac{1}{IJ}[IJ\sigma^2 + (IJ\mu)^2]$$

$$= I\sigma^2 + IJ\mu^2 + 2\mu J\sum_i \alpha_i + J\sum_i \alpha_i^2 - \sigma^2 - IJ\mu^2$$

$$= (I - 1)\sigma^2 + J\sum_i \alpha_i^2 \quad (\text{since } \sum \alpha_i = 0)$$

The result then follows from the relationship $MSTr = SSTr/(I - 1)$.

Power of the F Test

Recall that the power of a test is, for a fixed set of parameter values, the probability of rejecting H_0 when that set is the set of true values. Fortunately, the power of the F test depends on the α_i's and σ^2 only through $\Sigma \alpha_i^2/\sigma^2$, so the power for many different alternatives can be simultaneously evaluated. For example, $\Sigma \alpha_i^2 = 4$ for each of the following sets of α_i's for which H_0 is false, so the power is identical for all three alternatives.

$$\alpha_1 = -1, \quad \alpha_2 = -1, \quad \alpha_3 = 1, \quad \alpha_4 = 1$$
$$\alpha_1 = -\sqrt{2}, \quad \alpha_2 = \sqrt{2}, \quad \alpha_3 = 0 \quad \alpha_4 = 0$$
$$\alpha_1 = -\sqrt{3}, \quad \alpha_2 = \sqrt{1/3}, \quad \alpha_3 = \sqrt{1/3}, \quad \alpha_4 = \sqrt{1/3}$$

The quantity $J\Sigma \alpha_i^2/\sigma^2$ is called the **noncentrality parameter** for one-way ANOVA (because when H_0 is false the test statistic has a *noncentral F* distribution with this as one of its parameters), and the power of the test is an increasing function of the value of this parameter. Thus for fixed values of σ^2 and J, the null hypothesis is more likely to be rejected for alternatives far from H_0 (large $\Sigma \alpha_i^2$) than for alternatives close to H_0. For a fixed value of $\Sigma \alpha_i^2$, the power increases as the sample size J on each treatment increases and decreases as the variance σ^2 increases (since more underlying variability makes it more difficult to detect any given departure from H_0).

Hand computations of power and sample size determinations for the F test are (as with t tests) quite difficult, so statisticians have constructed sets of power curves from which values of the power can be obtained. With ν_1 denoting numerator d.f., Figures 10.2* and 10.3* give power curves for $\nu_1 = 3$ ($I = 4$ in one-way ANOVA)

Figure 10.2   Power function for the ANOVA F test ($\nu_1 = 3$)

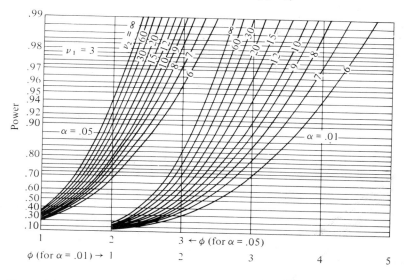

Figure 10.3   Power function for the ANOVA F test ($\nu_1 = 4$)

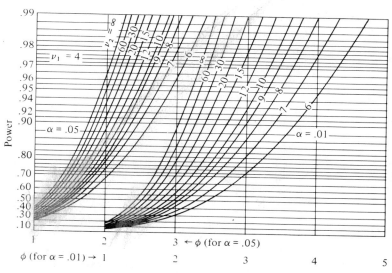

* Reproduced with permission from E. S. Pearson and H. O. Hartley, "Charts of the Power Function for Analysis of Variance Tests, Derived from the Non-central F Distribution," *Biometrika*, vol. 38, 1951, p. 112.

and $\nu_1 = 4$ ($I = 5$), respectively. Once ν_1 and α have been determined, the value of $\phi^2 = (J/I) \sum \alpha_i^2 / \sigma^2$ is computed and the appropriate figure is entered at the value of ϕ on the horizontal axis. The power is then read from the vertical axis by finding the curve corresponding to $\nu_2 = $ denominator d.f. $= I(J - 1)$.

Example 10.5

The effects of four different heat treatments on yield point (tons/in.^{2}) of steel ingots are to be investigated. A total of eight ingots will be cast using each treatment. Supposing that the true standard deviation of yield point for any of the four treatments is $\sigma = 1$, how likely is it that H_0 will be rejected at level .05 if three of the treatments have the same expected yield point and the other treatment has an expected yield point which is 1 ton/in.^{2} greater than the common value of the other three (that is, the fourth yield is on average one s.d. above those for the first three treatments)?

Supposing that $\mu_1 = \mu_2 = \mu_3$ and $\mu_4 = \mu_1 + 1$, $\mu = (\sum \mu_i)/4 = \mu_1 + \frac{1}{4}$. Then $\alpha_1 = \mu_1 - \mu = -\frac{1}{4}$, $\alpha_2 = -\frac{1}{4}$, $\alpha_3 = -\frac{1}{4}$, $\alpha_4 = \frac{3}{4}$ so $\phi^2 = \frac{8}{4} [(-\frac{1}{4})^2 + (-\frac{1}{4})^2 + (-\frac{1}{4})^2 + (\frac{3}{4})^2] = \frac{3}{2}$ and $\phi = 1.22$. The degrees of freedom are $\nu_1 = I - 1 = 3$ and $\nu_2 = I(J - 1) = 28$, so interpolating visually between $\nu_2 = 20$ and $\nu_2 = 30$ gives a power of approximately .47. This power is rather small, so we might decide to increase the value of J. How many ingots of each type would be required to yield a power of .95 for the alternative under consideration? By trying different values of J we can verify that $J = 24$ will meet the requirement, but any smaller J will not.

Relationship of the *F* Test to the *t* Test

When the number of treatments or populations is $I = 2$, all formulas and results connected with the F test still make sense, so ANOVA can be used to test $H_0 : \mu_1 = \mu_2$ versus $H_a : \mu_1 \neq \mu_2$. In this case a two-tailed two-sample t test can also be used. Which test should be used? Somewhat surprisingly, the two are equivalent. It can be verified from the t and F tables that $t_{\alpha/2, \nu}^2 = F_{\alpha, 1, \nu}$, and a bit of algebra shows that the relationship between the two test statistics is also $T^2 = F$ in this case. Thus the same conclusion (reject or accept H_0) will be reached irrespective of which test is used. Of course, if the alternative is one sided, then the one-tailed t test must be used, since the ANOVA alternative is always $H_a : \mu_1 \neq \mu_2$.

Single-Factor ANOVA When Sample Sizes Are Unequal

When the sample sizes from each population or treatment are not equal, let J_1, $J_2, \ldots, J_I$ denote the I sample sizes and let $n = \sum_i J_i$ denote the total number of observations. Define sums of squares (along with their computing formulas) by

$$SST = \sum_{i=1}^{I} \sum_{j=1}^{J_i} (X_{ij} - \overline{X}..)^2 = \sum_{i=1}^{I} \sum_{j=1}^{J_i} X_{ij}^2 - \frac{1}{n} X_{..}^2.$$

$$SSTr = \sum_{i=1}^{I} \sum_{j=1}^{J_i} (\overline{X}_i. - \overline{X}..)^2 = \sum_{i=1}^{I} \frac{1}{J_i} X_{i.}^2 - \frac{1}{n} X_{..}^2.$$

$$SSE = \sum_{i=1}^{I} \sum_{j=1}^{J_i} (X_{ij} - \overline{X}_{i}.)^2 = SST - SSTr$$

$SSTr$ and SSE have $I - 1$ and $n - I$ $[=\Sigma (J_i - 1)]$ d.f., respectively. With $MSTr = SSTr/(I - 1)$ and $MSE = SSE/(n - I)$, the F statistic is again the ratio of these two mean squares.

Theorem: In one-way ANOVA with sample sizes $J_1, J_2, \ldots, J_I$ and $n = \Sigma J_i$, when all population or treatment distributions are normal with variance σ^2, a level α test for $H_0 : \mu_1 = \cdots = \mu_I$ versus $H_a :$ at least two μ_i's differ is given by

Test statistic: $\quad F = \dfrac{MSTr}{MSE} \quad$ where $MSTr = \dfrac{SSTr}{I - 1}, MSE = \dfrac{SSE}{n - I}$

Rejection region: $\quad F \geq F_{\alpha, I-1, n-I}$

Example 10.6   When a lightbulb is switched on and off at low frequency, the "flicker" is easily detected, while to most observers a high-frequency flicker will be perceived as emanating light continuously without flickering. The critical flicker frequency (cff) is the highest frequency (in cycles per second) at which a flicker is detected. The cff depends on illumination conditions and intensity of the source, and varies from individual to individual. In an experiment to determine whether the amount of pigmentation in the eye, as indicated by iris color, would affect the cff, 22 subjects were used, and a cff measurement was obtained for each subject (each of which was an average of four independent determinations, making the data more nearly normal). Nine subjects had brown irises, six had green, and seven had blue. The observed cff's were

brown:	24.0	28.2	25.3	24.8	26.6	25.1	26.0	27.1	24.5
green:	26.4	24.2	29.0	26.9	28.5	24.1			
blue:	27.1	25.3	29.8	28.4	26.6	29.3	28.2		

Letting μ_1, μ_2, and μ_3 be the true average cff for the three iris colors, we wish to test $H_0 : \mu_1 = \mu_2 = \mu_3$ against the alternative that at least two μ_i's are unequal.

The summary statistics for this data are $x_1. = 231.93$, $x_2. = 159.12$, $x_3. = 194.70$, $x_.. = 585.75$, and $\Sigma\Sigma x_{ij}^2 = 15,645.26$. The ANOVA table is

Table 10.4

Source	d.f.	Sum of squares	Mean square	F
Treatments (iris color)	2	16.55	8.28	4.75
Error	19	33.12	1.74	
Total	21	49.67		

Because $F_{.05, 2, 19} = 3.13$, H_0 is rejected at level .05 in favor of the conclusion that iris color does affect true average cff (the numbers are fictitious here, but the conclusion agrees with that reported in "The Effect of Iris Color on Critical Flicker Frequency," *J. General Psychology*, 1973, pp. 91–95).

Multiple Comparisons When Sample Sizes Are Unequal

There is as yet no suitable modification of the T method for multiple comparisons when the sample sizes are unequal. While the Bonferroni method could be used (with $n - I$ d.f. for error), there is another method, due to Scheffé, which tends to produce shorter intervals when many confidence intervals are to be constructed. If $c_1, \ldots, c_I$ are constants with $\Sigma c_i = 0$, then $\Sigma c_i \mu_i$ is called a **constrast** in the μ_i's. Any pairwise difference, such as $\mu_1 - \mu_2$, is a contrast ($c_1 = 1$, $c_2 = -1$, $c_3 = \cdots = c_I = 0$), and so are $\mu_1 - \frac{1}{3}(\mu_2 + \mu_3 + \mu_4)$ and $\frac{1}{2}(\mu_1 + \mu_2) - \frac{1}{2}(\mu_3 + \mu_4)$. One might examine a particular contrast in order to compare averages of two different sets of μ_i's. Scheffé's method gives simultaneous confidence intervals for all possible contrasts in the μ_i's. Each interval is centered at the corresponding estimate $\Sigma c_i \bar{x}_i$.

Proposition: With simultaneous confidence $100(1 - \alpha)\%$, for every set of constants $c_1, \ldots, c_I$ with $\Sigma c_i = 0$,

$$\sum_{i=1}^{I} c_i \bar{x}_i. - \sqrt{\sum_{i=1}^{I} \frac{c_i^2}{J_i}} \cdot \sqrt{(I-1)(MSE) F_{\alpha, I-1, n-I}} \leq \sum_{i=1}^{I} c_i \mu_i$$

$$\leq \sum_{i=1}^{I} c_i \bar{x}_i. + \sqrt{\sum_{i=1}^{I} \frac{c_i^2}{J_i}} \cdot \sqrt{(I-1)(MSE) F_{\alpha, I-1, n-I}}$$

(10.19)

Every interval of the form (10.19) that does not include zero is said to correspond to a constrast which differs significantly from zero. Consideration of the intervals for all $\mu_i - \mu_j$'s enables multiple comparisons to be made.

Example 10.6 (continued)

For the cff data, the sample means are $\bar{x}_1. = 25.77$ ($J_1 = 9$), $\bar{x}_2. = 26.52$ ($J_2 = 6$), and $\bar{x}_3. = 27.81$ ($J_3 = 7$). At 95% confidence, the interval for $\mu_1 - \mu_2$ is $(-2.49, .99)$, the interval for $\mu_2 - \mu_3$ is $(-3.12, .54)$, and the interval for $\mu_1 - \mu_3$ is $(-3.70, -.38)$. Because the first two intervals include zero, μ_1 and μ_2 are judged not significantly different and μ_2 and μ_3 are judged not significantly different. Because the third interval does not include zero, μ_1 and μ_3 are said to differ significantly from one another. The confidence interval for $\mu_1 - \frac{1}{2}(\mu_2 + \mu_3)$ is (with $c_1 = 1$, $c_2 = c_3 = -\frac{1}{2}$)

$$25.77 - \tfrac{1}{2}(26.52 + 27.81) \pm \sqrt{\left(\frac{1}{9}\right)^2 + \frac{(-\frac{1}{2})^2}{6} + \frac{(-\frac{1}{2})^2}{7}} \cdot \sqrt{2(1.74)(3.13)}$$

$$= -1.40 \pm 1.43 = (-2.83, .03)$$

Scheffé's method can be used when sample sizes are equal, but the resulting intervals for the $\mu_i - \mu_j$'s tend to be longer than those obtained from the T method; of course Scheffé's method also yields intervals for other contrasts.

Data Transformation

The use of ANOVA methods can be invalidated by substantial differences in the variances $\sigma_1^2, \ldots, \sigma_I^2$ (which until now have been assumed equal with common value σ^2). It sometimes happens that $\text{Var}(X_{ij}) = \sigma_i^2 = g(\mu_i)$, a known function of μ_i (so that when H_0 is false, the variances are not equal). For example, if X_{ij} has a Poisson distribution with parameter λ_i (approximately normal if $\lambda_i \geq 10$), then $\mu_i = \lambda_i$ and $\sigma_i^2 = \lambda_i$ so $g(\mu_i) = \mu_i$ is the known function. In such cases one can often transform the X_{ij}'s to $h(X_{ij})$'s so that they will have approximately equal variances (while leaving the transformed variables approximately normal), and then the F test can be used on the transformed observations. The key idea in choosing a transformation $h(\cdot)$ is that often $\text{Var}[h(X_{ij})] \doteq \text{Var}(X_{ij}) \cdot [h'(\mu_i)]^2 = g(\mu_i) \cdot [h'(\mu_i)]^2$. We wish the function $h(\cdot)$ for which $g(\mu_i) \cdot [h(\mu_i)]^2 = c$ (a constant) for every i.

> **Proposition:** If $\text{Var}(X_{ij}) = g(\mu_i)$, a known function of μ_i, then a transformation $h(X_{ij})$ which "stabilizes the variance" so that $\text{Var}[h(X_{ij})]$ is approximately the same for each i is given by $h(x) \propto \int [g(x)]^{-1/2} dx$.

In the Poisson case, $g(x) = x$, so $h(x)$ should be proportional to $\int x^{-1/2} dx = 2x^{1/2}$. Thus Poisson data should be transformed to $h(x_{ij}) = \sqrt{x_{ij}}$ before the analysis.

A Random Effects Model

The single-factor problems considered so far have all been assumed to be examples of a **fixed effects** ANOVA model. By this we mean that the chosen levels of the factor under study are the only ones considered relevant by the experimenter. The single-factor fixed effects model is

$$X_{ij} = \mu + \alpha_i + \epsilon_{ij}, \quad \sum \alpha_i = 0 \tag{10.20}$$

where both μ and the α_i's are fixed parameters whose values are unknown and the ϵ_{ij}'s are random.

In some single-factor problems the particular levels studied by the experimenter are chosen, either by design or through sampling, from a large population of levels. For example, to study the effects on task performance time of using different operators on a particular machine, a sample of five operators might be chosen from a large pool of operators. Similarly, the effect of soil pH on the yield of maize plants might be studied by using soils with four specific pH values chosen from among the many possible pH levels. When the levels used are selected at random from a larger population of possible levels, the factor is said to be random rather than fixed, and the fixed effects model (10.20) is no longer appropriate. An analogous **random effects** model is obtained by replacing the fixed α_i's in (10.20) by random variables.

The resulting model description is

$$X_{ij} = \mu + A_i + \epsilon_{ij} \quad \text{with} \quad E(A_i) = E(\epsilon_{ij}) = 0$$

$$\text{Var}(\epsilon_{ij}) = \sigma^2, \text{Var}(A_i) = \sigma_A^2 \tag{10.21}$$

all A_i's and ϵ_{ij}'s normally distributed and independent of one another.

The condition $E(A_i) = 0$ in (10.21) is similar to the condition $\Sigma \alpha_i = 0$ in (10.20); it states that the expected or average effect of the ith level measured as a departure from μ is zero.

For the random effects model (10.21), the hypothesis of no effects due to different levels is $H_0 : \sigma_A^2 = 0$, which says that different levels of the factor contribute nothing to variability of the response. *Although the hypotheses in the single-factor fixed and random effects models are different, they are tested in exactly the same way,* by forming $F = MSTr/MSE$ and rejecting H_0 if $F \geq F_{\alpha, I-1, n-I}$. This can be justified intuitively by noting that $E(MSE) = \sigma^2$ (as for fixed effects), while

$$E(MSTr) = \sigma^2 + \frac{1}{I-1}\left(n - \frac{\Sigma J_i^2}{n}\right)\sigma_A^2 \tag{10.22}$$

(where $J_1, J_2, \ldots, J_I$ are the samples sizes and $n = \sum J_i$).

The factor in parentheses on the right side of (10.22) is nonnegative, so again $E(MSTr) = \sigma^2$ if H_0 is true and $E(MSTr) > \sigma^2$ if H_0 is false.

Because the computations and ANOVA table are identical for the two models, we omit an example. Random effects will be considered again in Chapter 11 in connection with multifactor experiments.

Exercises / Section 10.3

1. The following data refers to yield of tomatoes (kg/plot) for four different levels of salinity; salinity level here refers to electrical conductivity (EC), where the chosen levels were EC = 1.6, 3.8, 6.0, and 10.2 nmhos/cm.

    1.6:   59.5, 53.3, 56.8, 63.1, 58.7
    3.8:   55.2, 59.1, 52.8, 54.5
    6.0:   51.7, 48.8, 53.9, 49.0
    10.2:   44.6, 48.5, 41.0, 47.3, 46.1

    Using the F test at level $\alpha = .05$ to test for any differences in true average yield due to the different salinity levels.

2. Apply Scheffé's method to the data in Exercise 1 to identify significant differences among the μ_i's.

3. The following partial ANOVA table is taken from the article "Perception of Spatial Incongruity" (*J. Nervous and Mental Disease*, 1961, p. 222) in which the abilities of three different groups to identify a perceptual incongruity were assessed and compared. All individuals in the experiment had been hospitalized to undergo psychiatric treatment. There were 21 individuals in the depressive group, 32 individuals in the functional "other" group, and 21 individuals in the brain-damaged group. Complete the ANOVA table and carry out the F test at level $\alpha = .01$.

Source	d.f.	Sum of squares	Mean square	F
Groups		152.18		
Error				
Total		1123.14		

4. An article in the *Canadian Entomologist* ("Influence of Natural Diets and Larval Density on Gypsy Moth, Lymantria Dispor, Egg Mass Characteristics," 1977, pp. 1313–1318) reported the following data on egg mass diameters for moths reared on five different diets.

Diet	n_i	$\bar{x}_i.$ (mm)	s_i
Red maple '74	13	1.134	.0252
Red oak/red maple	10	1.148	.0253
Red maple '75	20	1.159	.0179
Red oak	16	1.191	.0200
Red oak/white pine	16	1.217	.0160

a. Compute the grand mean $\bar{x}..$, $SSTr$, and $MSTr$. *Hint:* $x.. = \Sigma n_i \bar{x}_i.$.

b. Compute SSE and MSE. *Hint:* $SSE = \Sigma (n_i - 1) s_i^2$.

c. Does the data suggest that there are any differences between true average egg diameters for the different diets? Test using $\alpha = .05$.

d. Apply Scheffé's method to look for means which differ significantly from one another.

5. The article "Major Appliance Prices in the Chicago Area," (*J. Business*, 1977, pp. 231–235) reported the following data on prices for a certain size of refrigerator at a sample of stores in the Chicago area.

Brand	Sample size	$\bar{x}_i.$	s_i
Frigidaire	18	423.21	18.71
GE	28	418.13	21.15
Whirlpool	19	421.27	17.87

Does this data suggest any differences in the true average prices charged for the three brands?

6. Samples of six different brands of diet/imitation margarine were analyzed to determine the level of physiologically active polyunsaturated fatty acids (PAPFUA, in percentages), resulting in the following data:

Imperial:	14.1, 13.6, 14.4, 14.3
Parkay:	12.8, 12.5, 13.4, 13.0, 12.3
Blue Bonnet:	13.5, 13.4, 14.1, 14.3
Chiffon:	13.2, 12.7, 12.6, 13.9
Mazola:	16.8, 17.2, 16.4, 17.3, 18.0

Fleischmann's: 18.1, 17.2, 18.7, 18.4

(The above numbers are fictitious, but the sample means agree with data reported in the January 1975 issue of *Consumer Reports*.)

a. Use ANOVA to test for differences among the true average PAPFUA percentages for the different brands.

b. Use Scheffé's method to compute confidence intervals for all $(\mu_i - \mu_j)$'s.

c. Mazola and Fleischmann's are corn based while the others are soybean based. Compute a Scheffé interval for the contrast

$$\frac{(\mu_1 + \mu_2 + \mu_3 + \mu_4)}{4} - \frac{(\mu_5 + \mu_6)}{2}$$

7. For a single-factor ANOVA with sample sizes J_i ($i = 1, 2, \ldots, I$), show that $SSTr = \Sigma J_i(\bar{X}_i. - \bar{X}..)^2 = \Sigma_i J_i \bar{X}_i.^2 - n\bar{X}..^2$, where $n = \Sigma J_i$.

8. When sample sizes are equal ($J_i = J$), the parameters $\alpha_1, \alpha_2, \ldots, \alpha_I$ of the alternative parameterization are restricted by $\Sigma \alpha_i = 0$. For unequal sample size, the most natural restriction is $\Sigma J_i \alpha_i = 0$. Use this to show that

$$E(MSTr) = \sigma^2 + \frac{1}{I - 1} \Sigma J_i \alpha_i^2$$

What is $E(MSTr)$ when H_0 is true? [This expectation is correct if $\Sigma J_i \alpha_i = 0$ is replaced by the restriction $\Sigma \alpha_i = 0$ (or any other single linear restriction on the α_i's used to reduce the model to I independent parameters), but $\Sigma J_i \alpha_i = 0$ simplifies the algebra and yields natural estimates for the model parameters (in particular, $\hat{\alpha}_i = \bar{x}_i. - \bar{x}..$)].

9. Suppose that the x_{ij}'s are "coded" by $y_{ij} = cx_{ij} + d$. How does the value of the F statistic computed from the y_{ij}'s compare to the value computed from the x_{ij}'s? Justify your assertion.

10. Reconsider Example 10.5 involving an investigation of the effects of different heat treatments on the yield point of steel ingots.

a. If $J = 8$ and $\sigma = 1$, what is the power of the level .05 F test when $\mu_1 = \mu_2$, $\mu_3 = \mu_1 - 1$, $\mu_4 = \mu_1 + 1$?

b. For the alternative of (a), what value of J is necessary to obtain a power of .95?

c. If there are $I = 5$ heat treatments, $J = 10$, and $\sigma = 1$, what is the power of the level .05

F test when four of the μ_i's are equal and the fifth differs by 1 from the other four?

11. Referring back to Exercise 10.1.8, what is the power of the test when true average DNA content is identical for three of the starches and falls below this common value by one standard deviation (σ) for the other two starches?

12. When sample sizes are not equal, the noncentrality parameter is $\Sigma J_i \alpha_i^2 / 2\sigma^2$ and $\phi^2 = (1/I) \Sigma J_i \alpha_i^2 / \sigma^2$. Referring to Exercise 1 above, what is the power of the test when $\mu_2 = \mu_3$, $\mu_1 = \mu_2 - \sigma$, $\mu_4 = \mu_2 + \sigma$?

13. In an experiment to compare the quality of four different brands of reel-to-reel recording tape, five 2400-ft reels of each brand were selected and the number of flaws in each reel was determined.

Brand		
A	10, 5, 12, 14, 8
B	14, 12, 17, 9, 8
C	13, 18, 10, 15, 18
D	17, 16, 12, 22, 14

It is believed that the number of flaws has approximately a Poisson distribution for each brand. Analyze the data at level .01 to see whether the expected number of flaws per reel is the same for each brand.

14. Suppose that X_{ij} is a binomial variable with parameters n and p_i (so approximately normal when $np_i \geq 5$ and $nq_i \geq 5$). Then since $\mu_i = np_i$, $\text{Var}(X_{ij}) = \sigma_i^2 = np_i(1 - p_i) = \mu_i(1 - \mu_i/n)$. How should the X_{ij}'s be transformed so as to stabilize the variance? *Hint:* $g(\mu_i) = \mu_i(1 - \mu_i/n)$.

15. Simplify $E(MSTr)$ for the random effects model when $J_1 = J_2 = \cdots = J_I = J$.

Bibliography   Dunn, Olive Jean and Clark, Virginia, *Applied Statistics: Analysis of Variance and Regression,* John Wiley, New York, 1974. Contains a good introductory survey of ANOVA models and methods of analysis.

Neter, John and Wasserman, William, *Applied Linear Statistical Models,* Richard D. Irwin, Homewood, Ill., 1974. The second half of this book contains a very well-presented survey of ANOVA; the level is comparable to that of the present text, but the discussion is more comprehensive, making the book an excellent reference.

Ott, Lyman, *An Introduction to Statistical Methods and Data Analysis,* Duxbury Press, Boston, 1977. Includes several chapters on ANOVA methodology which can profitably be read by students desiring a very nonmathematical exposition; there is a good chapter on various multiple comparison methods.

Walpole, Ronald and Myers, Raymond, *Probability and Statistics for Engineers and Scientists* (2nd ed.), Macmillan, New York, 1978. Covers roughly the same ground as does the present text, with slightly more material on the power of ANOVA tests.

CHAPTER 11

Multifactor Analysis of Variance

Introduction

In the previous chapter we used the analysis of variance to test for equality of either I different population means or the true average responses associated with I different levels of a single factor (alternatively referred to as I different treatments). In many experimental situations there are two or more factors which are of simultaneous interest. This chapter extends the methods of Chapter 10 to investigate such multifactor situations.

In the first two sections we concentrate on the case of two factors of interest. We shall use I to denote the number of levels of the first factor (A) and J to denote the number of levels of the second factor (B). Then there are IJ possible combinations consisting of one level of factor A and one of factor B; each such combination is called a treatment, so that there are IJ different treatments. The number of observations made on treatment (i, j) will be denoted by K_{ij}. In Section 11.1 we present the model and analysis when $K_{ij} = 1$. An important special case of this type is a randomized block design, in which a single factor A is of primary interest but another factor, "blocks," is created in order to control for extraneous variability in experimental units or subjects. In Section 11.2 we focus on the case $K_{ij} = K > 1$, and mention briefly the difficulties associated with unequal K_{ij}'s.

Section 11.3 discusses experiments involving more than two factors, including a Latin square design which controls for the effects of two extraneous factors thought to influence the response variable. When the number of factors is large, an experiment consisting of at least one observation for each treatment would be quite expensive and time consuming. An important special case, which we discuss in the last section, is that in which there are p factors, each of which has two levels. There are then 2^p different treatments, and we consider both the case in which observations are made on all these treatments (a complete design) and the case in which observations are made for only a selected subset of treatments (an incomplete design).

11.1 Two-Factor ANOVA with $K_{ij} = 1$

When factor A consists of I levels and factor B consists of J levels, there are IJ different combinations (pairs) of levels of the two factors, each called a treatment. With K_{ij} = the number of observations on the treatment consisting of factor A at level i and factor B at level j, we focus in this section on the case $K_{ij} = 1$, so that the data consists of IJ observations. We shall first discuss the fixed-effects model, in which the only levels of interest for the two factors are those actually represented in the experiment. The case in which one or both factors are random is discussed briefly at the end of the section.

Example 11.1 In a study on automobile traffic and air pollution reported in the *International Journal of Environmental Studies* ("Automobile Traffic and Air Pollution in a Developing Country," 1977, pp. 197–203), air samples taken at four different times and at five different locations were analyzed to obtain the amount of particulate matter present in the air (mg/m³).

			Factor B: Location				
			1	2	3	4	5
	Oct.	1975	76	67	81	56	51
Factor A:	Jan.	1976	82	69	96	59	70
Time	May	1976	68	59	67	54	42
	Sept.	1976	63	56	64	58	37

Is there any difference in true average amount of particulate matter present in the air due to either different sampling times or different locations?

The Notation

As in single-factor ANOVA, double subscripts are used to identify random variables and observed values. Let

X_{ij} = the random variable which denotes the measurement when factor A is held at level i and factor B is held at level j

x_{ij} = the observed value of X_{ij}

The x_{ij}'s are usually presented in a two-way table in which the ith row contains the observed values on factor A held at level i and the jth column contains the observed values on factor B held at level j. In the air pollution experiment of Example 11.1, the number of levels of factor A is $I = 4$ and the number of levels of factor B is $J = 5$.

Whereas in single-factor ANOVA we were interested only in row means and the grand mean, here we are interested also in column means. Let

$$\overline{X}_{i\cdot} = \frac{\text{the average of measurements obtained when}}{\text{factor } A \text{ is held at level } i} = \frac{\sum_{j=1}^{J} X_{ij}}{J}$$

$$\overline{X}_{\cdot j} = \begin{array}{c} \text{the average of measurements obtained when} \\ \text{factor } B \text{ is held at level } j \end{array} = \frac{\displaystyle\sum_{i=1}^{I} X_{ij}}{I}$$

$$\overline{X}_{\cdot\cdot} = \qquad\qquad \text{the grand mean} \qquad\qquad = \frac{\displaystyle\sum_{i=1}^{I}\sum_{j=1}^{J} X_{ij}}{IJ}$$

with observed values $\overline{x}_{i\cdot}$, $\overline{x}_{\cdot j}$, and $\overline{x}_{\cdot\cdot}$. Totals rather than averages are denoted by omitting the horizontal bar (so $x_{\cdot j} = \sum_i x_{ij}$, and so on). Intuitively, to see whether there is any effect due to the levels of factor A, we should compare the observed $\overline{x}_{i\cdot}$'s with one another, while information about the different levels of factor B should come from the $\overline{x}_{\cdot j}$'s.

The Model

Proceeding by analogy to single-factor ANOVA, one's first inclination in specifying a model is to let $\mu_{ij} =$ the true average response when factor A is at level i and factor B at level j, giving IJ mean parameters. Then let

$$X_{ij} = \mu_{ij} + \epsilon_{ij}$$

where ϵ_{ij} is the random amount by which the observed value differs from its expectation and the ϵ_{ij}'s are assumed normal and independent with common variance σ^2. Unfortunately there is no valid test procedure for this choice of parameters. The reason is that under the alternative hypothesis of interest, the μ_{ij}'s are free to take on any values whatsoever, while σ^2 can be any value greater than zero, so that there are $IJ + 1$ freely varying parameters. But there are only IJ observations, so after using each x_{ij} as an estimate of μ_{ij}, there is no way to estimate σ^2.

To rectify this problem of a model having more parameters than observed values, we must specify a model which is realistic yet which involves relatively few parameters. Suppose that we assume the existence of I parameters $\alpha_1, \alpha_2, \ldots, \alpha_I$ and J parameters $\beta_1, \beta_2, \ldots, \beta_J$ such that

$$X_{ij} = \alpha_i + \beta_j + \epsilon_{ij} \quad i = 1, \ldots, I; \quad j = 1, \ldots, J \qquad (11.1)$$

That is, we now assume that each μ_{ij} can be written in the form

$$\mu_{ij} = \alpha_i + \beta_j \qquad (11.2)$$

Including σ^2, there are now $I + J + 1$ model parameters, so if $I \geq 3$ and $J \geq 3$, then there will be fewer parameters than observations (in fact, we will shortly modify (11.2) so that even $I = 2$ and/or $J = 2$ will be accommodated).

The model specified in (11.1) and (11.2) is called an **additive model,** because each mean response μ_{ij} is the sum of an effect due to factor A at level i (α_i) and an effect due to factor B at level $j(\beta_j)$. The difference between mean responses for

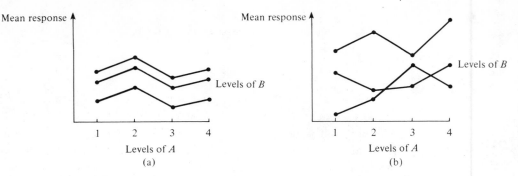

Figure 11.1 Mean responses for (*a*) an additive, and (*b*) a nonadditive model

factor A at level i and level i' when B is held at level j is $\mu_{ij} - \mu_{i'j}$. When the model is additive,

$$\mu_{ij} - \mu_{i'j} = (\alpha_i + \beta_j) - (\alpha_{i'} + \beta_j) = \alpha_i - \alpha_{i'}$$

which is independent of the level j of the second factor. A similar result holds for $\mu_{ij} - \mu_{ij'}$. Thus additivity means that the difference in mean responses for two levels of one of the factors is the same for all levels of the other factor. Figure 11.1(a) shows a set of mean responses which satisfy the condition of additivity, while Figure 11.1(b) shows a nonadditive configuration of mean responses.

Example 11.1 (continued)

If we plot the observed x_{ij}'s in a manner analogous to that of Figure 11.1, the result is shown in Figure 11.2. While there is some "crossing over" in the observed x_{ij}'s, the configuration is reasonably representative of what would be expected under additivity with just one observation per treatment.

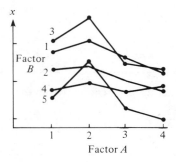

Figure 11.2 Plot of data from Example 1.1

Expression (11.2) is not quite the final model description, because the α_i's and β_j's are not uniquely determined. Pictured below are two different configurations of the α_i's and β_j's which yield the same additive μ_{ij}'s.

	$\beta_1 = 1$	$\beta_2 = 4$
$\alpha_1 = 1$	$\mu_{11} = 2$	$\mu_{12} = 5$
$\alpha_2 = 2$	$\mu_{21} = 3$	$\mu_{22} = 6$

	$\beta_1 = 2$	$\beta_2 = 5$
$\alpha_1 = 0$	$\mu_{11} = 2$	$\mu_{12} = 5$
$\alpha_2 = 1$	$\mu_{21} = 3$	$\mu_{22} = 6$

By subtracting any constant c from all α_i's and adding c to all β_j's, other configurations corresponding to the same additive model are obtained. This non-uniqueness is eliminated by use of the model

$$X_{ij} = \mu + \alpha_i + \beta_j + \epsilon_{ij}, \quad \text{where} \quad \sum_{i=1}^{I} \alpha_i = 0 \quad \text{and} \quad \sum_{j=1}^{J} \beta_j = 0 \quad (11.3)$$

which is analogous to the alternative choice of parameters for single-factor ANOVA discussed in Section 10.3. It is not difficult to verify that (11.3) is an additive model in which the parameters are uniquely determined (for example, for the μ_{ij}'s mentioned above, $\mu = 4$, $\alpha_1 = -.5$, $\alpha_2 = .5$, $\beta_1 = -1.5$, and $\beta_2 = 1.5$). Notice that there are only $I - 1$ independently determined α_i's and $J - 1$ independently determined β_j's, so (including μ) (11.3) specifies $I + J - 1$ mean parameters.

The interpretation of the parameters of (11.3) is straightforward: μ is the true grand mean (mean response averaged over all levels of both factors), α_i is the effect of factor A at level i (measured as a deviation from μ), and β_j is the effect of factor B at level j. Unbiased estimators for these parameters are

$$\hat{\mu} = \overline{X}.., \quad \hat{\alpha}_i = \overline{X}_{i\cdot} - \overline{X}.., \quad \text{and} \quad \hat{\beta}_j = \overline{X}_{\cdot j} - \overline{X}..$$

These are also maximum likelihood estimators when the ϵ_{ij}'s are normal.

The Hypotheses

There are two different hypotheses of interest in a two-factor experiment with $K_{ij} = 1$. The first, denoted by H_{0A}, states that the different levels of factor A have no effect on true average response, while the second, denoted by H_{0B}, states that there is no factor-B effect:

$$H_{0A}: \alpha_1 = \alpha_2 = \cdots = \alpha_I = 0 \quad \text{versus} \quad H_{aA}: \text{at least one } \alpha_i \neq 0$$
$$H_{0B}: \beta_1 = \beta_2 = \cdots = \beta_J = 0 \quad \text{versus} \quad H_{aB}: \text{at least one } \beta_j \neq 0 \quad (11.4)$$

(No factor A effect implies all α_i's are equal, so they must all be 0 since they sum to 0, and similarly for the β_j's.)

The Test Procedures

The description and analysis now follow closely that for single-factor ANOVA. The relevant sums of squares and their computing forms are given by

$$SST = \sum_{i=1}^{I} \sum_{j=1}^{J} (X_{ij} - \overline{X}..)^2 = \sum_{i=1}^{I} \sum_{j=1}^{J} X_{ij}^2 - \frac{1}{IJ} X_{..}^2$$

$$SSA = \sum_{i=1}^{I} \sum_{j=1}^{J} (\overline{X}_{i\cdot} - \overline{X}..)^2 = \frac{1}{J} \sum_{i=1}^{I} X_{i\cdot}^2 - \frac{1}{IJ} X_{..}^2$$

(11.5)

$$SSB = \sum_{i=1}^{I} \sum_{j=1}^{J} (\overline{X}_{\cdot j} - \overline{X}..)^2 = \frac{1}{I} \sum_{j=1}^{J} X_{\cdot j}^2 - \frac{1}{IJ} X_{..}^2$$

and

$$SSE = \sum_{i=1}^{I} \sum_{j=1}^{J} (X_{ij} - \overline{X}_{i\cdot} - \overline{X}_{\cdot j} + \overline{X}..)^2$$

(SSE comes from substituting parameter estimates into $\sum \sum [X_{ij} - (\mu + \alpha_i + \beta_j)]^2$.) The fundamental identity for two-factor ANOVA with one observation per treatment is

$$SST = SSA + SSB + SSE$$

(11.6)

Thus *SSE* can be obtained by subtraction once the other three sums of squares have been computed. As in single-factor ANOVA, total variation is split into a part (*SSE*) which is not explained by either the truth or the falsity of H_{0A} or H_{0B} and two parts which can be explained by possible falsity of the two null hypotheses.

The numbers of degrees of freedom for *SSA* and *SSB* are $I - 1$ and $J - 1$ respectively. There are IJ observations, but $(I - 1) + (J - 1) + 1$ mean parameters are independently estimated, leaving $IJ - [1 + (I - 1) + (J - 1)] = (I - 1)(J - 1)$ d.f. for error. There are then three mean squares, with $MS = SS/\text{d.f.}$ The test procedures for the two pairs of hypotheses are

Hypotheses: H_{0A} versus H_{aA}

Test statistic: $F_A = \dfrac{MSA}{MSE}$

Rejection region: $F_A \geq F_{\alpha, I-1, (I-1)(J-1)}$

Hypotheses: H_{0B} versus H_{aB}

Test statistic: $F_B = \dfrac{MSB}{MSE}$

Rejection region: $F_B \geq F_{\alpha, J-1, (I-1)(J-1)}$

The results of the analysis are usually displayed in an ANOVA table.

Example 11.1
(continued)

The computed summary statistics are $\Sigma \Sigma x_{ij}^2 = 84{,}853$, $x_1. = 331$, $x_2. = 376$, $x_3. = 290$, $x_4. = 278$, $x_{.1} = 289$, $x_{.2} = 253$, $x_{.3} = 308$, $x_{.4} = 227$, $x_{.5} = 200$, and $x.. = 1275$. Thus

$$SST = 84{,}853 - \tfrac{1}{20}(1275)^2 = 3571.75$$

$$SSA = \tfrac{1}{5}[(331)^2 + (376)^2 + (290)^2 + (278)^2] - \tfrac{1}{20}(1275)^2 = 1182.95$$

$$SSB = \tfrac{1}{4}[(289)^2 + (253)^2 + (308)^2 + (227)^2 + (200)^2] - \tfrac{1}{20}(1275)^2$$
$$= 2199.50$$

and

$$SSE = 3571.75 - 1182.95 - 2199.50 = 189.30$$

Table 11.1

Source of variation	d.f.	Sum of squares	Mean square	F
Factor A (time)	$I - 1 = 3$	$SSA = 1182.95$	$MSA = 394.32$	$f_A = 24.99$
Factor B (location)	$J - 1 = 4$	$SSB = 2199.50$	$MSB = 549.88$	$f_B = 34.85$
Error	$(I - 1)(J - 1) = 12$	$SSE = 189.30$	$MSE = 15.78$	
Total	$IJ - 1 = 19$	$SST = 3571.75$		

With $F_{.05,3,12} = 3.49$, $f_A = 24.99 \geq 3.49$, so H_{0A} is rejected at level .05 in favor of the claim that the different sampling times do have different effects on pollution. Similarly, $F_{.05,4,12} = 3.26$ and $34.85 \geq 3.26$, so H_{0B} is rejected in favor of the claim that pollution varies with sampling location.

Expected Mean Squares

The plausibility of using the F tests just described is demonstrated by computing the expected mean squares. After some tedious algebra,

$$E(MSE) = \sigma^2 \quad \text{(when the model is additive)}$$

$$E(MSA) = \sigma^2 + \frac{J}{I - 1} \sum_{i=1}^{I} \alpha_i^2$$

$$E(MSB) = \sigma^2 + \frac{I}{J - 1} \sum_{j=1}^{J} \beta_j^2$$

When H_{0A} is true, MSA is an unbiased estimator of σ^2, so F is a ratio of two unbiased estimators of σ^2. When H_{0A} is false, MSA tends to overestimate σ^2, so H_{0A} should be rejected when the ratio F_A is too large. Similar comments apply to MSB and H_{0B}.

Multiple Comparisons in Two-Factor ANOVA

When either H_{0A} or H_{0B} has been rejected, Tukey's procedure can be used to identify significant differences between the levels of the factor under investigation. The steps in the analysis are identical to those for a single-factor ANOVA:

1. For comparing levels of factor A, obtain $Q_{\alpha, I, (I-1)(J-1)}$.
   For comparing levels of factor B, obtain $Q_{\alpha, J, (I-1)(J-1)}$.

2. Compute $w = Q \cdot$ (estimated standard deviation of the sample means being compared)

$$= Q_{\alpha, I, (I-1)(J-1)} \cdot \sqrt{MSE/J} \quad \text{for factor } A \text{ comparisons}$$

$$\text{or } Q_{\alpha, J, (I-1)(J-1)} \cdot \sqrt{MSE/I} \quad \text{for factor } B \text{ comparisons}$$

(because, for example, the standard deviation of $\overline{X}_{i \cdot}$ is $\sigma/\sqrt{J}$).

3. Arrange the sample means in increasing order, underscore those pairs differing by less than w, and identify pairs not underscored by the same line as corresponding to significantly different levels of the given factor.

Example 11.1
(continued)

For factor A (time) in the air pollution problem, $Q_{.05, 4, 12} = 4.20$ and $w = (4.20) \cdot \sqrt{15.78/5} = 7.48$. We now underscore all $\overline{x}_{i \cdot}$'s differing by less than 7.48.

$\overline{x}_{4 \cdot}$	$\overline{x}_{3 \cdot}$	$\overline{x}_{1 \cdot}$	$\overline{x}_{2 \cdot}$
55.60	58.00	66.20	75.20

Thus each pair of sampling times except for times 3 and 4 yields significantly different amounts of particulate matter. For factor B (location), $w = 8.97$, and underscoring yields

$\overline{x}_{\cdot 5}$	$\overline{x}_{\cdot 4}$	$\overline{x}_{\cdot 2}$	$\overline{x}_{\cdot 1}$	$\overline{x}_{\cdot 3}$
50.00	56.75	63.25	72.75	77.00

from which significant differences in location can be identified.

Randomized Block Experiments

In using single-factor ANOVA to test for the presence of effects due to the I different treatments under study, once the IJ subjects or experimental units have been chosen, treatments should be allocated in a completely random fashion. That is, J subjects should be chosen at random for the first treatment, then another sample of J chosen at random from the remaining $IJ - J$ subjects for the second treatment, and so on.

It frequently happens, though, that subjects or experimental units exhibit heterogeneity with respect to other variables which may affect the observed responses. When this is the case, the presence or absence of a significant F value may be due to this extraneous variation rather than to the presence or absence of factor effects. This was the reason for introducing a paired experiment in Chapter 8. The analogy to a paired experiment when $I > 2$ is called a **randomized block** experiment. An extraneous factor, "blocks," is constructed by dividing the IJ units into J groups with

I units in each group. This grouping or blocking is done in such a way that within each block, the *I* units are homogeneous with respect to other factors thought to affect the responses. Then within each homogeneous block, the *I* treatments are randomly assigned to the *I* units or subjects in the block.

Example 11.2   A consumer product-testing organization wished to compare the annual power consumption for five different brands of dehumidifier. Because power consumption depends on the prevailing humidity level, it was decided to monitor each brand at four different levels ranging from moderate to heavy humidity (thus blocking on humidity level). Within each level, brands were randomly assigned to the five selected locations. The resulting amount of power consumption (annual kwh) appears in Table 11.2.

Table 11.2

Treatments (brands)	Blocks (humidity level)				$x_{i\cdot}$	$\bar{x}_{i\cdot}$
	1	2	3	4		
1	685	792	838	875	3190	797.50
2	722	806	893	953	3374	843.50
3	733	802	880	941	3356	839.00
4	811	888	952	1005	3656	914.00
5	828	920	978	1023	3749	937.25
$x_{\cdot j}$	3779	4208	4541	4797	17,325	

Since $\Sigma \Sigma x_{ij}^2 = 15{,}178{,}901.00$ and $x_{\cdot\cdot}^2/IJ = 15{,}007{,}781.25$

$$SST = 15{,}178{,}901.00 - 15{,}007{,}781.25 = 171{,}119.75$$
$$SSA = \tfrac{1}{4}[60{,}244{,}049] - 15{,}007{,}781.25 = 53{,}231.00$$
$$SSB = \tfrac{1}{5}[75{,}619{,}995] - 15{,}007{,}781.25 = 116{,}217.75$$

and

$$SSE = 171{,}119.75 - 53{,}231.00 - 116{,}217.75 = 1671.00$$

Table 11.3

Source of variation	d.f.	Sum of squares	Mean square	F
Treatments (brands)	4	53,231.00	13,307.75	$f_A = 95.57$
Blocks	3	116,217.75	38,739.25	$f_B = 278.20$
Error	12	1671.00	139.25	
Total	19	171,119.75		

Since $F_{.05,4,12} = 3.26$ and $f_A = 95.57 > 3.26$, H_0 is rejected in favor of H_a, and we conclude that power consumption does depend on the brand of humidifier. To identify significantly different brands, we use Tukey's procedure. $Q_{.05,5,12} = 4.52$ and $w = 4.52\sqrt{139.25/4} = 26.67$.

$\bar{x}_{1.}$	$\bar{x}_{3.}$	$\bar{x}_{2.}$	$\bar{x}_{4.}$	$\bar{x}_{5.}$
797.50	839.00	843.50	914.00	937.25

The underscoring indicates that the brands can be divided into three groups with respect to power consumption.

Because the block factor is of secondary interest, $F_{.05,3,12}$ is not needed, though the computed value of F_B is clearly highly significant.

In many experimental situations in which treatments are to be applied to subjects, a single subject can receive all I of the treatments. Blocking is then often done on the subjects themselves to control for variability between subjects; each subject is then said to act as its own control. Social scientists sometimes refer to such experiments as repeated-measures designs. The "units" within a block are then the different "instances" of treatment application. Similarly, blocks are often taken as different time periods, locations, or observers.

Example 11.3

The accompanying data appeared in the paper "Compounding of Discriminative Stimuli from the Same and Different Sensory Modalities" (*J. Experimental Analysis Behavior*, 1971, pp. 337–342). Rat response was maintained by fixed interval schedules of reinforcement in the presence of a tone or two separate lights. The lights were either of moderate ($L1$) or of low intensity ($L2$). Observations are given as the mean number of responses emitted by each subject during single and compound stimuli presentations over a four-day period.

Table 11.4

Stimulus	Subject 1	2	3	4	$x_{i.}$	$\bar{x}_{i.}$
$L1$	8.0	17.3	52.0	22.0	99.3	24.8
$L2$	6.9	19.3	63.7	21.6	111.5	27.9
Tone (T)	9.3	18.8	60.0	28.3	116.4	29.1
$L1 + L2$	9.2	24.9	82.4	44.9	161.4	40.3
$L1 + T$	12.0	31.7	83.8	37.4	164.9	41.2
$L2 + T$	9.4	33.6	96.6	40.6	180.2	45.1
$x_{.j}$	54.8	145.6	438.5	194.8	833.7	

With $\Sigma \Sigma x_{ij}^2 = 44{,}614.21$, $SST = 15{,}653.56$, $SSA = 1428.28$, $SSB = 13{,}444.63$, and $SSE = 780.65$

Table 11.5

Source of variation	d.f.	Sum of squares	Mean square	F
Stimuli (A)	5	1428.28	285.60	$f_A = 5.49$
Subjects (B)	3	13,444.63	4481.54	$f_B = 86.12$
Error	15	780.65	52.04	
Total	23	15,653.56		

Since $F_{.05,5,15} = 2.90$ and $5.49 \geq 2.90$, we conclude that there are differences in the true average responses associated with the different stimuli. For Tukey's procedure, $w = 4.59\sqrt{52.04/4} = 16.56$.

$\bar{x}_{1.}$	$\bar{x}_{2.}$	$\bar{x}_{3.}$	$\bar{x}_{4.}$	$\bar{x}_{5.}$	$\bar{x}_{6.}$
24.8	27.9	29.1	40.3	41.2	45.1

Thus both $L1$ and $L2$ are significantly different from $L2 + T$, and there are no other significant differences between the stimuli.

In most randomized block experiments in which subjects serve as blocks, the subjects actually participating in the experiment are selected from a large population. The subjects then contribute random rather than fixed effects. This does not affect the procedure for comparing treatments when $K_{ij} = 1$ (one observation per "cell," as in this section), but the procedure is altered if $K_{ij} = K > 1$. We shall shortly consider two-factor models in which effects are random.

More on Blocking

When $I = 2$, either the F test or the paired differences t test can be used to analyze the data. The resulting conclusion will not depend on which procedure is used, since $T^2 = F$ and $t^2_{\alpha/2,\nu} = F_{\alpha,1,\nu}$.

Just as with pairing, blocking entails both a potential gain and a potential loss in precision. If there is a great deal of heterogeneity in experimental units, the value of the variance parameter σ^2 in the one-way model will be large. The effect of blocking is to filter out the variation due to differences in experimental units, resulting in a smaller value of the variance σ^2 in the two-way model appropriate for a randomized block experiment. Other things being equal, a smaller value of σ^2 results in a test which is more likely to detect departures from H_0 (that is, a test with greater power).

However, other things are not equal here, since the single-factor F test is based on $I(J - 1)$ degrees of freedom for error, while the two-factor F test is based on $(I - 1)(J - 1)$ degrees of freedom for error. Fewer degrees of freedom for error results in a decrease in power, essentially because the denominator estimator of σ^2 is not as precise. This loss in degrees of freedom can be especially serious if the experimenter can afford only a small number of observations. Nevertheless, if it appears that blocking will significantly reduce variability, it is probably worth the loss in degrees of freedom. The book by Bowker and Lieberman presents a thorough discussion of power calculations for the two different F tests, from which a more quantitative assessment of gain versus loss can be made.

Models for Random Effects

In many experiments the actual levels of a factor used in the experiment, rather than being the only ones of interest to the experimenter, have been selected from a much larger population of possible levels of the factor. In a two-factor situation, when this is the case for both factors, a **random effects** model is appropriate. The case in which the levels of one factor are the only ones of interest and the levels of the other factor are selected from a population of levels leads to a **mixed effects** model. The two-factor random effects model when $K_{ij} = 1$ is

$$X_{ij} = \mu + A_i + B_j + \epsilon_{ij} \quad i = 1, \ldots, I; j = 1, \ldots, J$$

where the A_i's, B_j's, and ϵ_{ij}'s are all independent, normally distributed random variables with mean zero and variances σ_A^2, σ_B^2, and σ^2, respectively. The hypotheses of interest are then $H_{0A} : \sigma_A^2 = 0$ (expected response does not depend on which level of factor A is selected) versus $H_{aA} : \sigma_A^2 > 0$ and $H_{0B} : \sigma_B^2 = 0$ versus $H_{aB} : \sigma_B^2 > 0$. While $E(MSE) = \sigma^2$ as before, the expected mean squares for factor A and B are now

$$E(MSA) = \sigma^2 + J\sigma_A^2, \qquad E(MSB) = \sigma^2 + I\sigma_B^2$$

Thus when $H_{0A}(H_{0B})$ is true, $F_A(F_B)$ is still a ratio of two unbiased estimators of σ^2. It can be shown that a level α test for H_{0A} versus H_{aA} still rejects H_{0A} if $F_A \geq F_{\alpha, I-1, (I-1)(J-1)}$, and similarly the same procedure as before is used to decide between H_{0B} and H_{aB}.

For the case in which factor A is fixed and factor B is random, the mixed model is

$$X_{ij} = \mu + \alpha_i + B_j + \epsilon_{ij} \quad i = 1, \ldots, I; j = 1, \ldots, J$$

where $\Sigma \alpha_i = 0$ and the B_j's and ϵ_{ij}'s are normally distributed with mean zero and variances σ_B^2 and σ^2, respectively. Now the two null hypotheses are

$$H_{0A} : \alpha_1 = \cdots = \alpha_I = 0 \quad \text{and} \quad H_{0B} : \sigma_B^2 = 0,$$

with expected mean squares

$$E(MSE) = \sigma^2, E(MSA) = \sigma^2 + \frac{J}{I-1} \sum \alpha_i^2, E(MSB) = \sigma^2 + I\sigma_B^2$$

The test procedures for H_{0A} versus H_{aA} and H_{0B} versus H_{aB} are exactly as before. For example, in the analysis of the air pollution data in Example 11.1, if the five locations were randomly selected, then because $F_B = 34.85$ and $F_{.05,4,12} = 3.26$, $H_{0B} : \sigma_B^2 = 0$ is rejected in favor of $H_{aB} : \sigma_B^2 > 0$. An estimate of the "variance component" σ_B^2 is then given by $(MSB - MSE)/J = 106.82$.

Summarizing, when $K_{ij} = 1$, although the hypotheses and expected mean squares differ from the case of both effects fixed, the test procedures are identical.

Exercises / Section 11.1

1. The number of miles of useful tread wear (in thousands) was determined for tires of each of five different makes of subcompact car (factor A, with $I = 5$) in combination with each of four different brands of radial tire (factor B, with $J = 4$), resulting in $IJ = 20$ observations. The values

$SSA = 30.6$, $SSB = 44.1$, and $SSE = 59.2$ were then computed. Assume that an additive model is appropriate.

a. Test $H_0 : \alpha_1 = \alpha_2 = \alpha_3 = \alpha_4 = \alpha_5 = 0$ (no differences in true average tire lifetime due to makes of cars)   versus H_a : at least one $\alpha_i \neq 0$ using a level .05 test.

b. Test $H_0 : \beta_1 = \beta_2 = \beta_3 = \beta_4 = 0$ (no differences in true average tire lifetime due to brands of tires)   versus H_a : at least one $\beta_j \neq 0$ using a level .05 test.

2. Four different coatings are being considered for corrosion protection of metal pipe. The pipe will be buried in three different types of soil. To investigate whether the amount of corrosion depends either on the coating or on the type of soil, 12 pieces of pipe are selected. Each piece is coated with one of the four coatings and buried in one of the three types of soil for a fixed time, after which the amount of corrosion (depth of maximum pits, in .0001 in.) is determined. The data appears in the accompanying table.

Soil type (B)

		1	2	3
	1	64	49	50
Coating (A)	2	53	51	48
	3	47	45	50
	4	51	43	52

a. Assuming the validity of the additive model, carry out the ANOVA analysis using an ANOVA table to see whether the amount of corrosion depends on either the type of coating used or the type of soil. Use $\alpha = .05$.

b. Compute $\hat{\mu}$, $\hat{\alpha}_1$, $\hat{\alpha}_2$, $\hat{\alpha}_3$, $\hat{\alpha}_4$, $\hat{\beta}_1$, $\hat{\beta}_2$, and $\hat{\beta}_3$.

3. The paper "Adiabatic Humidification of Air with Water in a Packed Tower" (*Chem. Eng. Prog.*, 1952, pp. 362–370) reported data on gas film heat transfer coefficient (Btu/hr ft^2 in °F) as a function of gas rate (factor A) and liquid rate (factor B).

B

	1(190)	2(250)	3(300)	4(400)
1(200)	200	226	240	261
2(400)	278	312	330	381
3(700)	369	416	462	517
4(1100)	500	575	645	733

A (applied to left column)

a. After constructing an ANOVA table, test at level .01 both the hypothesis of no gas rate effect against the appropriate alternative and the hypothesis of no liquid rate effect against the appropriate alternative.

b. Use Tukey's procedure to investigate differences in expected heat transfer coefficient due to different gas rates.

c. Repeat (b) for liquid rates.

4. In an experiment to see whether the amount of coverage of light blue interior latex paint depended either on the brand of paint or on the brand of roller used, one gallon of each of four brands of paint was applied using each of three brands of roller, resulting in the following data (number of square feet covered).

Roller brand

		1	2	3
	1	454	446	451
Paint	2	446	444	447
brand	3	439	442	444
	4	444	437	443

a. Construct the ANOVA table. *Hint:* The computations can be expedited by subtracting 400 (or any other convenient number) from each observation. This does not affect the final results.

b. State and test hypotheses appropriate for deciding whether paint brand has any effect on coverage. Use $\alpha = .05$.

c. Repeat (b) for brand of roller.

d. Use Tukey's method to identify significant differences among brands. Is there one brand which seems clearly preferable to the others?

5. In an experiment to assess the effect of the angle of pull on the force required to cause separation in electrical connectors, four different angles (factor A) were used and each of a sample of five connectors (factor B) was pulled once at each angle ("A Mixed Model Factorial Experiment in Testing Electrical Connectors," *Industrial Quality Control*, 1960, pp. 12–16). The data appears in the accompanying table.

B

		1	2	3	4	5
	0°	45.3	42.2	39.6	36.8	45.8
	2°	44.1	44.1	38.4	38.0	47.2
A	4°	42.7	42.7	42.6	42.2	48.9
	6°	43.5	45.8	47.9	37.9	56.4

Does the data suggest that true average separation force is affected by the angle of pull? State and test the appropriate hypotheses at level .01 by first constructing an ANOVA table.

6. A particular county employs three assessors who are responsible for determining the value of residential property in the county. To see whether or not these assessors differ systematically in their assessments, five houses are selected and each assessor is asked to determine the market value of each house. With factor A denoting assessors ($I = 3$) and factor B denoting houses ($J = 5$), suppose that $SSA = 11.7$, $SSB = 113.5$, and $SSE = 25.6$.
   a. Test $H_0 : \alpha_1 = \alpha_2 = \alpha_3 = 0$ at level .05 (H_0 states that there are no systematic differences between assessors).
   b. Explain why a randomized block experiment with only five houses was used rather than a one-way ANOVA experiment involving a total of 15 different houses with each assessor asked to assess five different houses (a different group of five for each assessor).

7. The article "Rate of Stuttering Adaptation Under Two Electro-Shock Conditions" (*Behavior Research Therapy*, 1967, pp. 49–54) gave adaptation scores for three different treatments: no shock (1), shock following each stuttered word (2), and shock during each moment of stuttering (3). These treatments were used on each of 18 stutterers.
   a. Summary statistics include $x_{1.} = 905$, $x_{2.} = 913$, $x_{3.} = 936$, $x_{..} = 2754$, $\sum_j x_{.j}^2 = 430{,}295$, and $\sum \sum x_{ij} = 143{,}930$. Construct the ANOVA table and test at level .05 to see whether true average adaptation score depends on the treatment given.
   b. Judging from the F ratio for subjects (factor B), do you think that blocking on subjects was effective in this experiment? Explain.

8. The accompanying table gives plasma epinephrine concentration for 10 experimental subjects during (1) isoflurane, (2) halothane, and (3) cyclopropane anesthesia ("Sympathoadrenal and Hemodynamic Effects of Isoflurane, Halothane, and Cyclopropane in Dogs," *Anesthesiology*, 1974, pp. 465–470).
   a. Does the choice of anesthetic affect true average concentration? Test $H_0 : \alpha_1 = \alpha_2 = \alpha_3 = 0$ at level .05 after constructing the ANOVA table.

		1	2	3	4	5
Anesthetic (A)	1	.28	.51	1.00	.39	.29
	2	.30	.39	.63	.38	.21
	3	1.07	1.35	.69	.28	1.24

		6	7	8	9	10
Anesthetic (A)	1	.36	.32	.69	.17	.33
	2	.88	.39	.51	.32	.42
	3	1.53	.49	.56	1.02	.30

Subject (B)

$$\sum \sum x_{ij}^2 = 13.7980$$

   b. Use Tukey's procedure to investigate significant differences among the anesthetics.

9. Suppose that in the experiment described in Exercise 6 the five houses had actually been selected at random from among those of a certain age and size, so that factor B is random rather than fixed. Test $H_0 : \sigma_B^2 = 0$ versus $H_a : \sigma_B^2 > 0$ using a level .01 test.

10. a. Show that a constant d can be added to (or subtracted from) each x_{ij} without affecting any of the ANOVA sums of squares.
    b. Suppose that each x_{ij} is multiplied by a nonzero constant c. How does this affect the ANOVA sums of squares? How does this affect the values of the F statistics F_A and F_B? What affect does "coding" the observations by $y_{ij} = cx_{ij} + d$ have on the conclusions resulting from the ANOVA procedures?

11. Use the fact that $E(X_{ij}) = \mu + \alpha_i + \beta_j$ with $\sum \alpha_i = \sum \beta_j = 0$ to show that $E(\bar{X}_{i.} - \bar{X}_{..}) = \alpha_i$, so that $\hat{\alpha}_i = \bar{X}_{i.} - \bar{X}_{..}$ is an unbiased estimator for α_i.

12. The power curves of Figure 10.2 can be used to obtain the power $[P(\text{rejecting } H_0)]$ of the F test for two-factor ANOVA. For fixed values of α_1, $\alpha_2, \ldots, \alpha_I$, the quantity $\phi^2 = (J/I) \sum \alpha_i^2 / \sigma^2$ is computed, then the figure corresponding to $\nu_1 = I - 1$ is entered on the horizontal axis at the value ϕ, and the power is read on the vertical axis from the curve labeled $\nu_2 = (I - 1)(J - 1)$.
    a. For the corrosion experiment described in Exercise 2, find the power when $\alpha_1 = 4$, $\alpha_2 = 0$, $\alpha_3 = \alpha_4 = -2$, and $\sigma = 4$. Repeat if $\alpha_1 = 6$, $\alpha_2 = 0$, $\alpha_3 = \alpha_4 = -3$, and $\sigma = 4$.
    b. By symmetry, what is the power of the test for H_{0B} versus H_{aB} in Example 1.1 when $\beta_1 = 4$, $\beta_2 = \beta_3 = \beta_4 = \beta_5 = -1$, and $\sigma = 4$?

11.2 Two-Factor ANOVA with $K_{ij} > 1$

In Section 11.1 we analyzed data from a two-factor experiment in which there was one observation for each of the IJ combinations of levels of the two factors. To obtain valid test procedures, the μ_{ij}'s were assumed to have an additive structure with $\mu_{ij} = \mu + \alpha_i + \beta_j$, $\Sigma \alpha_i = \Sigma \beta_j = 0$. Additivity means that the difference in true average responses for any two levels of the factors is the same for each level of the other factor. For example, $\mu_{ij} - \mu_{i'j} = (\mu + \alpha_i + \beta_j) - (\mu + \alpha_{i'} + \beta_j) = \alpha_i - \alpha_{i'}$ independent of the level j of the second factor. This is shown in Figure 11.1a, in which the lines connecting true average responses are parallel.

Figure 11.1(b) depicts a set of true average responses which does not have additive structure. The lines connecting these μ_{ij}'s are not parallel, which means that the difference in true average responses for different levels of one factor does depend on the level of the other factor. When additivity does not hold, we say that there is **interaction** between the different levels of the factors. The assumption of additivity allowed us in Section 11.1 to obtain an estimator of the random error variance σ^2 (*MSE*) which was unbiased whether or not either null hypothesis of interest was true. When $K_{ij} > 1$ for at least one (i, j) pair, a valid estimator of σ^2 can be obtained without assuming additivity. In specifying the appropriate model and deriving test procedures, we will focus primarily on the case $K_{ij} = K > 1$, so the number of observations per "cell" (for each combination of levels) is constant.

Parameters for the Fixed Effects Model with Interaction

Rather than use the μ_{ij}'s themselves as model parameters, it is usual to use an equivalent set which reveals more clearly the role of interaction. Let

$$\mu = \frac{1}{IJ}\sum_i \sum_j \mu_{ij}, \quad \mu_{i\cdot} = \frac{1}{J}\sum_j \mu_{ij}, \quad \text{and} \quad \mu_{\cdot j} = \frac{1}{I}\sum_i \mu_{ij} \qquad (11.7)$$

Thus μ is the expected response averaged over all levels of both factors (the true grand mean), $\mu_{i\cdot}$ is the expected response averaged over levels of the second factor when the first factor A is held at level i, and similarly for $\mu_{\cdot j}$. Now define

$$\alpha_i = \mu_{i\cdot} - \mu = \text{the effect of factor } A \text{ at level } i$$
$$\beta_j = \mu_{\cdot j} - \mu = \text{the effect of factor } B \text{ at level } j \qquad (11.8)$$
and
$$\gamma_{ij} = \mu_{ij} - (\mu + \alpha_i + \beta_j)$$

Then

$$\mu_{ij} = \mu + \alpha_i + \beta_j + \gamma_{ij} \qquad (11.9)$$

and the model is additive if and only if all γ_{ij}'s $= 0$. The γ_{ij}'s are referred to as the **interaction parameters.** The α_i's are called the **main effects for factor A**, while the β_j's are the **main effects for factor B**. Although there are I α_i's, J β_j's, and IJ γ_{ij}'s in addition to μ, the conditions $\Sigma \, \alpha_i = 0$, $\Sigma \, \beta_j = 0$, $\Sigma \, \gamma_{ij} = 0$ for any i, and $\underset{i}{\Sigma} \, \gamma_{ij} = 0$ for any j [all by virtue of (11.7) and (11.8)] imply that only IJ of these new parameters are independently determined: μ, $I - 1$ of the α_i's, $J - 1$ of the β_j's, and $(I - 1)(J - 1)$ of the γ_{ij}'s.

There are now three sets of hypotheses which will be considered:

$$H_{0AB} : \gamma_{ij} = 0 \quad \text{for all } i, j \quad \text{versus} \quad H_{aAB} : \text{at least one } \gamma_{ij} \neq 0$$

$$H_{0A} : \alpha_1 = \cdots = \alpha_I = 0 \quad \text{versus} \quad H_{aA} : \text{at least one } \alpha_i \neq 0$$

$$H_{0B} : \beta_1 = \cdots = \beta_J = 0 \quad \text{versus} \quad H_{aB} : \text{at least one } \beta_j \neq 0$$

The no-interaction hypothesis H_{0AB} is usually tested first. If H_{0AB} is not rejected, then the other two hypotheses can be tested to see whether or not the main effects are significant. If H_{0AB} is rejected and H_{0A} is then tested and accepted, the resulting model $\mu_{ij} = \mu + \beta_j + \gamma_{ij}$ does not lend itself to straightforward interpretation. In such a case it is best to construct a picture similar to that of Figure 11.1(*b*) to try to visualize the way in which the factors interact.

Notation, Model, and Analysis

We now use triple subscripts for both random variables and observed values, with X_{ijk} and x_{ijk} referring to the kth observation (replication) when factor A is at level i and factor B is at level j. The model is then

$$X_{ijk} = \mu + \alpha_i + \beta_j + \gamma_{ij} + \epsilon_{ijk}$$
$$i = 1, \ldots, I; j = 1, \ldots, J; k = 1, \ldots, K \qquad (11.10)$$

where the ϵ_{ij}'s are independent and normally distributed, each with mean zero and variance σ^2.

Again a dot in place of a subscript means that we have summed over all values of that subscript, while a horizontal bar denotes averaging. Thus $X_{ij.}$ is the total of all K observations made for factor A at level i and factor B at level j [all observations in the (i, j)th cell] and $\overline{X}_{ij.}$ is the average of these K observations.

Example 11.4 Three different varieties of tomato (Harvester, Pusa Early Dwarf, and Ife No. 1) and four different plant densities (10, 20, 30, and 40 thousand plants per hectare) are being considered for planting in a particular region. To see whether either variety or plant density affects yield, each combination of variety and plant density is used in three different plots, resulting in the following data on yields (based on the article "Effects of Plant Density on Tomato Yields in Western Nigeria," *Experimental Agriculture*, 1976, pp. 43–47):

Table 11.6

Variety	Planting density (B) 10,000	20,000	30,000	40,000	$x_{i..}$	$\bar{x}_{i..}$
H	10.5, 9.2, 7.9	12.8, 11.2, 13.3	12.1, 12.6, 14.0	10.8, 9.1, 12.5	136.0	11.33
(A) Ife	8.1, 8.6, 10.1	12.7, 13.7, 11.5	14.4, 15.4, 13.7	11.3, 12.5, 14.5	146.5	12.21
P	16.1, 15.3, 17.5	16.6, 19.2, 18.5	20.8, 18.0, 21.0	18.4, 18.9, 17.2	217.5	18.13
$x_{.j.}$	103.3	129.5	142.0	125.2	500.00	
$\bar{x}_{.j.}$	11.48	14.39	15.78	13.91		13.89

Here $I = 3$, $J = 4$, and $K = 3$, for a total of $IJK = 36$ observations.

To test the hypotheses of interest, we again define sums of squares and present computing formulas:

$$SST = \sum_i \sum_j \sum_k (X_{ijk} - \bar{X}...)^2 = \sum_i \sum_j \sum_k X_{ijk}^2 - \frac{1}{IJK} X^2...$$

$$SSE = \sum_i \sum_j \sum_k (X_{ijk} - \bar{X}_{ij.})^2 = \sum_i \sum_j \sum_k X_{ijk}^2 - \frac{1}{K} \sum_i \sum_j X_{ij.}^2$$

$$SSA = \sum_i \sum_j \sum_k (\bar{X}_{i..} - \bar{X}...)^2 = \frac{1}{JK} \sum_i X_{i..}^2 - \frac{1}{IJK} X^2...$$

$$SSB = \sum_i \sum_j \sum_k (\bar{X}_{.j.} - \bar{X}...)^2 = \frac{1}{IK} \sum_j X_{.j.}^2 - \frac{1}{IJK} X^2...$$

$$SSAB = \sum_i \sum_j \sum_k (\bar{X}_{ij.} - \bar{X}_{i..} - \bar{X}_{.j.} + \bar{X}...)^2$$

$SSAB$ is called the **interaction sum of squares;** there is no computing formula for it because the fundamental ANOVA identity enables it to be obtained by subtraction once the other SS's have been computed:

$$SST = SSA + SSB + SSAB + SSE$$

Total variation is thus partitioned into four pieces: unexplained (SSE—which would be present whether or not any of the three null hypotheses was true) and three pieces which may be explained by the truth or falsity of the three H_0's.

Each SS has associated with it a number of degrees of freedom: $IJK - 1$ for SST, $IJ(K - 1)$ for SSE, $I - 1$ for SSA, $J - 1$ for SSB, and $(I - 1)(J - 1)$ for $SSAB$. Each of four mean squares is defined by $MS = SS/\text{d.f.}$ The expected mean squares suggest that each set of hypotheses should be tested using the appropriate ratio of mean squares with *MSE* in the denominator:

$$E(MSE) = \sigma^2$$

$$E(MSA) = \sigma^2 + \frac{JK}{I-1} \sum_{i=1}^{I} \alpha_i^2$$

$$E(MSB) = \sigma^2 + \frac{IK}{J-1} \sum_{j=1}^{J} \beta_j^2$$

$$E(MSAB) = \sigma^2 + \frac{K}{(I-1)(J-1)} \sum_{i=1}^{I} \sum_{j=1}^{J} \gamma_{ij}^2$$

Each of the three mean square ratios can be shown to have an *F* distribution when the associated H_0 is true, which yields the following level α test procedures:

Hypotheses	*Test Statistic*	*Rejection Region*
H_{0A} versus H_{aA}	$F_A = \dfrac{MSA}{MSE}$	$F_A \geq F_{\alpha, I-1, IJ(K-1)}$
H_{0B} versus H_{aB}	$F_B = \dfrac{MSB}{MSE}$	$F_B \geq F_{\alpha, J-1, IJ(K-1)}$
H_{0AB} versus H_{aAB}	$F_{AB} = \dfrac{MSAB}{MSE}$	$F_{AB} \geq F_{\alpha, (I-1)(J-1), IJ(K-1)}$

As before, the results of the analysis are summarized in an ANOVA table.

Example 11.4
(continued)

From the given data, $x^2_{\cdots} = (500)^2 = 250{,}000$,

$$\sum_i \sum_j \sum_k x_{ijk}^2 = (10.5)^2 + (9.2)^2 + \cdots + (18.9)^2 + (17.2)^2 = 7404.80$$

$$\sum_i x_{i\cdots}^2 = (136.0)^2 + (146.5)^2 + (217.5)^2 = 87{,}264.50$$

and

$$\sum_j x_{\cdot j\cdot}^2 = 63{,}280.18$$

The cell totals ($x_{ij\cdot}$'s) are

	10,000	20,000	30,000	40,000
H	27.6	37.3	38.7	32.4
Ife	26.8	37.9	43.5	38.3
P	48.9	54.3	59.8	54.5

from which $\sum_i \sum_j x_{ij}^2 = (27.6)^2 + \cdots + (54.5)^2 = 22{,}100.28$. Then

$$SST = 7404.80 - \tfrac{1}{36}(250{,}000) = 7404.80 - 6944.44 = 460.36$$

$$SSA = \tfrac{1}{12}(87{,}264.50) - 6944.44 = 327.60$$

$$SSB = \tfrac{1}{9}(63{,}280.18) - 6944.44 = 86.69$$

$$SSE = 7404.80 - \tfrac{1}{3}(22{,}100.28) = 38.04$$

and

$$SSAB = 460.36 - 327.60 - 86.69 - 38.04 = 8.03$$

Table 11.7 now summarizes the computations:

Table 11.7

Source of variation	d.f.	Sum of squares	Mean square	F
Varieties	2	327.60	163.8	$f_A = 103.02$
Density	3	86.69	28.9	$f_B = 18.18$
Interaction	6	8.03	1.34	$f_{AB} = .84$
Error	24	38.04	1.59	
Total	35	460.36		

Since $F_{.01,6,24} = 3.63$ and $f_{AB} = .84$ is not ≥ 3.63, H_{0AB} cannot be rejected at level .01, so we conclude that the interaction effects are not significant. Now the presence or absence of main effects can be investigated. Since $F_{.01,2,24} = 5.61$ and $f_A = 103.2 \geq 5.61$, H_{0A} is rejected at level .01 in favor of the conclusion that different varieties do affect the true average yields. Similarly, $f_B = 18.18 \geq 4.24 = F_{.01,3,24}$, so we conclude that true average yield also depends on plant density.

Multiple Comparisons

When the no-interaction hypothesis H_{0AB} is not rejected and at least one of the two main effect null hypotheses is rejected, Tukey's method can be used to identify significant differences in levels. For identifying differences among the α_i's when H_{0A} is rejected,

1. Obtain $Q_{\alpha,I,IJ(K-1)}$, where the second subscript I identifies the number of levels being compared and the third subscript refers to the number of degrees of freedom for error.
2. Compute $w = Q\sqrt{MSE/JK}$, where JK is the number of observations averaged to obtain each of the $\bar{x}_{i\cdot\cdot}$'s compared in step 3.
3. Order the $\bar{x}_{i\cdot\cdot}$'s from smallest to largest and, as before, underscore all pairs which differ by less than w. Pairs not underscored correspond to significantly different levels of factor A.

To identify different levels of factor B when H_{0B} is rejected, replace the second subscript in Q by J, replace JK by IK in w, and replace $\bar{x}_{i\cdot\cdot}$ by $\bar{x}_{\cdot j\cdot}$.

Example 11.4
(continued)

For factor A (varieties), $I = 3$, so with $\alpha = .01$ and $IJ(K - 1) = 24$, $Q_{.01,3,24} = 4.55$. Then $w = 4.55\sqrt{1.59/12} = 1.66$, so ordering and underscoring gives

$\bar{x}_{1..}$	$\bar{x}_{2..}$	$\bar{x}_{3..}$
11.33	12.21	18.13

The Harvester and Ife varieties do not appear to differ significantly from one another in effect on true average yield, but both differ from the Pusa variety.

For factor B (density), $J = 4$ so $Q_{.01,4,24} = 4.91$ and $w = 4.91\sqrt{1.59/9} = 2.06$.

$\bar{x}_{.1.}$	$\bar{x}_{.4.}$	$\bar{x}_{.2.}$	$\bar{x}_{.3.}$
11.48	13.91	14.39	15.78

Thus with experimentwise error rate .01, which is quite conservative, only the lowest density appears to differ significantly from all others. Even with $\alpha = .05$ (so that $w = 1.64$), densities 2 and 3 cannot be judged significantly different from one another in their effect on yield.

Models with Mixed and Random Effects

In some problems the levels of either factor may have been chosen from a large population of possible levels, so that the effects contributed by the factor are random rather than fixed. As in Section 11.1, if both factors contribute random effects, the model is referred to as a random effects model, while if one factor is fixed and the other is random, a mixed effects model results. We shall consider here the analysis for a mixed effects model in which factor A (rows) is the fixed factor and factor B (columns) is the random factor. The case in which both factors are random is dealt with in the exercises.

The mixed effects model in this situation is

$$X_{ijk} = \mu + \alpha_i + B_j + G_{ij} + \epsilon_{ijk}$$
$$i = 1, \ldots, I; j = 1, \ldots, J; k = 1, \ldots, K$$

where μ and α_i's are constants with $\Sigma \, \alpha_i = 0$, and the B_j's, G_{ij}'s, and ϵ_{ijk}'s are normally distributed random variables with expected value 0 and variances σ_B^2, σ_G^2, and σ^2, respectively. Because the I levels of the fixed factor A are the only ones under consideration, we also assume that $\Sigma_i \, G_{ij} = 0$ (this implies that the G_{ij}'s for fixed j are not independent of one another but are negatively correlated). The three hypotheses of interest are then

$$H_{0A} : \alpha_1 = \alpha_2 = \cdots = \alpha_I = 0 \quad \text{versus} \quad H_{aA} : \text{at least one } \alpha_i \neq 0$$
$$H_{0B} : \sigma_B^2 = 0 \quad\quad\quad\quad\quad\quad\quad \text{versus} \quad H_{aB} : \sigma_B^2 > 0$$
$$H_{0G} : \sigma_G^2 = 0 \quad\quad\quad\quad\quad\quad\quad \text{versus} \quad H_{aG} : \sigma_G^2 > 0$$

It is customary to test H_{0A} and H_{0B} only if the no-interaction hypothesis H_{0G} cannot be rejected.

The relevant sums of squares and mean squares needed for the test procedures are defined and computed exactly as in the fixed effects case. The expected mean squares for the mixed model are

$$E(MSE) = \sigma^2$$

$$E(MSA) = \sigma^2 + \frac{IK}{I-1}\sigma_G^2 + \frac{JK}{I-1}\sum \alpha_i^2$$

$$E(MSB) = \sigma^2 + IK\sigma_B^2$$

and

$$E(MSAB) = \sigma^2 + \frac{IK}{I-1}\sigma_G^2$$

Thus to test the no-interaction hypothesis, the ratio $F_{AB} = MSAB/MSE$ is again appropriate, with H_{0G} rejected if $F_{AB} \geq F_{\alpha,(I-1)(J-1),IJ(K-1)}$. Similarly, the F ratio and test procedure for testing H_{0B} are exactly as they were in the fixed effects case. However, for testing H_{0A} versus H_{aA}, the expected mean squares suggest that while the numerator of the F ratio should still be MSA, the denominator should be $MSAB$ rather than MSE, and this is indeed the case:

> For testing H_{0A} versus H_{aA} (factors A fixed, B random), the test statistic is $F_A = MSA/MSAB$, and the rejection region is $F_A \geq F_{\alpha,I-1,(I-1)(J-1)}$.

Example 11.5 A study of lifetimes of stereo cartridges carried out by staff members of a magazine devoted to high fidelity equipment focused on four different brands of cartridges, each designed to track at two grams. To see whether lifetime was affected by choice of record label, three different labels were randomly selected from the population of all labels, and two lifetime observations (hours of playing time) were obtained for each combination of brand of cartridge and record label.

Table 11.8

Brand of cartridge	Record label			
	1	2	3	$x_{i\cdot\cdot}$
1	697, 658	718, 688	640, 679	4080
2	635, 684	700, 736	696, 665	4116
3	670, 696	693, 659	675, 703	4096
4	651, 678	668, 709	715, 687	4108
$x_{\cdot j\cdot}$	5369	5571	5460	16,400

Additional summary quantities are $\sum_i \sum_j \sum_k x_{ijk}^2 = 11,220,964$ and $\sum_i \sum_j x_{ij\cdot}^2 = 22,427,518$.

Factor A, cartridges, is fixed here, since there are only four brands of interest to the experimenters and they are all represented in the experiment. Factor B, however, is random, since it is not the three specific labels actually chosen that are of interest; instead the question is whether or not there is variability in lifetime associated with the population of all labels. The three null hypotheses are then $H_{0A}: \alpha_1 = \alpha_2 = \alpha_3 = \alpha_4 = 0$, $H_{0B}: \sigma_B^2 = 0$, and $H_{0G}: \sigma_G^2 = 0$. The ANOVA table is Table 11.9; remember that $F_A = MSA/MSAB$, not MSA/MSE as with fixed effects.

Table 11.9

Source of variation	d.f.	Sum of squares	Mean square	F
Cartridges (A)	3	122.66	40.89	$f_A = .055$
Labels (B)	2	2558.58	1279.29	$f_B = 2.13$
Interaction	6	4411.09	735.18	$f_{AB} = 1.22$
Error	12	7205	600.42	
Total	23	14,297.33		

None of the three computed F's exceeds $F_{.05}$ for appropriate numerator and denominator degrees of freedom, so none of the three null hypotheses can be rejected. In particular the data suggests no differences in lifetime between the four brands of cartridge and no variability in lifetime due to different record labels.

Unequal K_{ij}'s

When at least two of the K_{ij}'s are unequal, the ANOVA computations are much more complex than for the case $K_{ij} = K$, and there are no nice formulas for the appropriate test statistics. Assuming that no cells are empty ($K_{ij} \neq 0$ for any i, j), let

$$SSE = \sum_{i=1}^{I} \sum_{j=1}^{J} \sum_{k=1}^{K_{ij}} (X_{ijk} - \overline{X}_{ij\cdot})^2 \quad \text{with d.f.} = \sum_i \sum_j K_{ij} - IJ$$

Then $E(MSE) = E(SSE/\text{d.f.}) = \sigma^2$ regardless of whether or not any of the null hypotheses of interest is true or false, and MSE appears in the denominator of each test statistic (for the fixed effects model). However, the previously defined sums of squares for main effects (SSA and SSB) and interaction are no longer the correct quantities for the numerators of the F statistics.

To obtain a test for $H_{0AB}: \gamma_{ij} = 0$ for all i, j versus H_{aAB}, we proceed as follows. Under H_{0AB}, $\mu_{ij} = \mu + \alpha_i + \beta_j$ with restrictions $\Sigma \alpha_i = \Sigma \beta_j = 0$. Consider

$$SS(\mu, \alpha, \beta) = \min_{\mu, \alpha_i\text{'s}, \beta_j\text{'s}} \left\{ \sum_i \sum_j \sum_k [X_{ijk} - (\mu + \alpha_i + \beta_j)]^2 \right\}$$

(where only α_i's and β_j's with $\Sigma \alpha_i = \Sigma \beta_j = 0$ are considered). $SS(\mu, \alpha, \beta)$ is the smallest sum of squared deviations about the μ_{ij}'s which is possible considering only μ_{ij}'s for which H_{0AB} is true. Similarly, with restrictions $\Sigma \alpha_i = \Sigma \beta_j = 0$, $\sum_i \gamma_{ij} = \sum_j \gamma_{ij} = 0$ for all i, j, let

$$SS(\mu, \alpha, \beta, \gamma) = \min_{\mu, \alpha_i's, \beta_j's, \gamma_{ij}'s} \left\{ \sum_i \sum_j \sum_k [X_{ijk} - (\mu + \alpha_i + \beta_j + \gamma_{ij})]^2 \right\}$$

the smallest sum of squared deviations possible under either H_{0AB} or H_{aAB}. The difference $SS(\mu, \alpha, \beta) - SS(\mu, \alpha, \beta, \gamma)$ is always nonnegative, since we can get at least as good a fit to the observations by including the γ_{ij}'s. If the difference is quite large, then the fit is much better under H_{aAB} than under H_{0AB}, suggesting that H_{0AB} should be rejected.

Whether or not $K_{ij} = K$, $SS(\mu, \alpha, \beta, \gamma) = SSE$. When $K_{ij} = K$, it can be shown that $SS(\mu, \alpha, \beta) - SS(\mu, \alpha, \beta, \gamma) = SSAB$, the interaction sum of squares (amount of variation in the X_{ijk}'s explained by the possible presence of interaction effects), and the previous test statistic is

$$F_{AB} = \frac{MSAB}{MSE} = \frac{[SS(\mu, \alpha, \beta) - SSE]/(I - 1)(J - 1)}{MSE}$$

However, if the K_{ij}'s are unequal, while F_{AB} still has this form, the value of $SS(\mu, \alpha, \beta)$ is quite difficult to determine. The minimizing values of μ, the α_i's and the β_j's must be obtained by solving a complicated system of equations.

The numerators of the F ratios for testing H_{0A} and H_{0B} can be obtained in a similar manner. In fact, many ANOVA tests involve in the numerator of the F ratio a difference $SS_1 - SS_2$, where SS_1 is the minimizing sum of squares under the null hypothesis being considered and SS_2 is the minimizing sum of squares under either H_0 or H_a. These tests can also be shown to be likelihood ratio tests.

If the K_{ij}'s do not differ too much from one another, an approximate method of analysis called the method of unweighted means can be used. A good elementary reference for this method and other issues relating to unequal K_{ij}'s is Chapter 20 of the book *Applied Regression Analysis and Other Multivariate Methods* by Kleinbaum and Kupper.

Exercises / Section 11.2

1. In an experiment to assess the effects of curing time (factor A) and type of mix (factor B) on the compressive strength of hardened cement cubes, three different curing times were used in combination with four different mixes, with three observations obtained for each of the 12 curing time/mix combinations. The resulting sums of squares were computed to be $SSA = 30,763.0$, $SSB = 34,185.6$, $SSE = 97,436.8$ and $SST = 205,966.6$.

   a. Construct an ANOVA table.

   b. Test at level .05 the null hypothesis H_{0AB}: all γ_{ij}'s $= 0$ (no interaction of factors) against H_{aAB}: at least one $\gamma_{ij} \neq 0$.

   c. Test at level .05 the null hypothesis H_{0A}: $\alpha_1 = \alpha_2 = \alpha_3$ (factor A main effects are

   absent) against H_{aA}: at least one $\alpha_i \neq 0$.

   d. Test H_{0B}: $\beta_1 = \beta_2 = \beta_3 = \beta_4 = 0$ versus H_{aB}: at least one $\beta_j \neq 0$ using a level .05 test.

   e. The values of the $\bar{x}_{i..}$'s were $\bar{x}_{1..} = 4010.88$, $\bar{x}_{2..} = 4029.10$, and $\bar{x}_{3..} = 3960.02$. Use Tukey's procedure to investigate significant differences among the three curing times.

2. The accompanying data table gives observations on total acidity of coal samples of three different types, with determinations made using three different concentrations of ethanolic NaOH ("Chemistry of Brown Coals," *Australian J. Applied Science*, 1958, pp. 375–379).

Type of coal

		Morwell	Yallourn	Maddingley
NaOH conc.	.404N	8.27, 8.17	8.66, 8.61	8.14, 7.96
	.626N	8.03, 8.21	8.42, 8.58	8.02, 7.89
	.786N	8.60, 8.20	8.61, 8.76	8.13, 8.07

Additionally, $\sum_i \sum_j \sum_k x_{ijk}^2 = 1240.1525$ and $\sum_i \sum_j x_{ij.}^2$.
$= 2479.9991$.

a. Assuming both effects to be fixed, construct an ANOVA table, test for the presence of interaction, and then test for the presence of main effects for each factor (all using level .01).

b. Use Tukey's procedure to identify significant differences among the types of coal.

3. The current (in microamperes) necessary to produce a certain level of brightness of a television tube was measured for two different types of glass and three different types of phosphor, resulting in the accompanying data ("Fundamentals of Analysis of Variance," *Industrial Quality Control*, 1956, pp. 5–8).

Phosphor type

	1	2	3
Glass type 1	280, 290, 285	300, 310, 295	270, 285, 290
2	230, 235, 240	260, 240, 235	220, 225, 230

Assuming that both factors are fixed, test H_{0AB} versus H_{aAB} at level .01. Then if H_{0AB} cannot be rejected, test the two sets of main effect hypotheses.

4. In an experiment to investigate the effect of "cement factor" (number of sacks of cement per cubic yard) on flexural strength of the resulting concrete ("Studies of Flexural Strength of Concrete. Part 3: Effects of Variation in Testing Procedure," *Proceedings ASTM*, 1957, pp. 1127–1139), $I = 3$ different factor values were used, $J = 5$ different batches of cement were selected, and $K = 2$ beams were cast from each cement factor/batch combination. Summary values include $\sum \sum \sum x_{ijk}^2 = 12,280,103$, $\sum \sum x_{ij.}^2 = 24,529,699$, $\sum x_{i..}^2 = 122,380,901$, $\sum x_{.j.}^2 = 73,427,483$, and $x... = 10,143$.

a. Construct the ANOVA table.

b. Assuming a mixed model with cement factor (A) fixed and batches (B) random, test the three pairs of hypotheses of interest at level .05.

5. The accompanying data was obtained in an experiment to investigate whether compressive strength of concrete cylinders depended on the type of capping material used or variability in different batches ("The Effect of Type of Capping Material on the Compressive Strength of Concrete Cylinders," *Proceedings ASTM*, 1958, pp. 1166–1186). Each number is a cell total ($x_{ij.}$) based on $K = 3$ observations.

Batch

		1	2	3	4	5
Capping material	1	1847	1942	1935	1891	1795
	2	1779	1850	1795	1785	1626
	3	1806	1892	1889	1891	1756

In addition, $\sum \sum \sum x_{ijk}^2 = 16,815,853$ and $\sum \sum x_{ij.}^2 = 50,433,409$. Obtain the ANOVA table, and then test at level .01 the hypotheses H_{0G} versus H_{aG}, H_{0A} versus H_{aA}, and H_{0B} versus H_{aB} assuming that capping is a fixed effect and batches is a random effect.

6. Show that $E(\overline{X}_{i..} - \overline{X}...) = \alpha_i$, so that $\overline{X}_{i..} - \overline{X}...$ is an unbiased estimator for α_i (in the fixed effects model).

7. With $\hat{\gamma}_{ij} = \overline{X}_{ij.} - \overline{X}_{i..} - \overline{X}_{.j.} + \overline{X}...$, show that $\hat{\gamma}_{ij}$ is an unbiased estimator for γ_{ij} (in the fixed effects model).

8. Show how a $100(1 - \alpha)\%$ t confidence interval for $\alpha_i - \alpha_{i'}$ can be obtained. Then compute a 95% interval for $\alpha_2 - \alpha_3$ using the data from Exercise 2. *Hint:* With $\theta = \alpha_2 - \alpha_3$, the result of Exercise 6 indicates how to obtain $\hat{\theta}$. Then compute $\text{Var}(\hat{\theta})$ and $\sigma_{\hat{\theta}}$, and obtain an estimate of $\sigma_{\hat{\theta}}$ by using $\sqrt{MSE}$ to estimate σ (which identifies the appropriate number of d.f.).

9. When both factors are random in a two-way ANOVA experiment with K replications per combination of factor levels, the expected mean squares are $E(MSE) = \sigma^2$, $E(MSA) = \sigma^2 + K\sigma_G^2 + JK\sigma_A^2$, $E(MSB) = \sigma^2 + K\sigma_G^2 + IK\sigma_B^2$, and $E(MSAB) = \sigma^2 + K\sigma_G^2$.

a. What F ratio is appropriate for testing $H_{0G} : \sigma_G^2 = 0$ versus $H_{aG} : \sigma_G^2 > 0$?

b. Answer (a) for testing $H_{0A} : \sigma_A^2 = 0$ versus $H_{aA} : \sigma_A^2 > 0$ and $H_{0B} : \sigma_B^2 = 0$ versus $H_{aB} : \sigma_B^2 > 0$.

11.3 Three-Factor ANOVA

To indicate the nature of models and analyses when ANOVA experiments involve more than two factors, we shall focus here on the case of three fixed factors—A, B, and C. The numbers of levels of the three factors will be denoted by I, J, and K, respectively, and L_{ijk} = the number of observations made with factor A at level i, factor B at level j, and factor C at level k. As with two-factor ANOVA, the analysis is quite complicated when the L_{ijk}'s are not all equal, so we further specialize to $L_{ijk} = L$. X_{ijkl} and x_{ijkl} then denote the observed value, before and after the experiment is performed, of the lth replication ($l = 1, 2, \ldots, L$) when the three factors are fixed at levels i, j, and k.

To understand the parameters which will appear in the three-factor ANOVA model, first recall that in two-factor ANOVA with replications, $E(X_{ijk}) = \mu_{ij} = \mu + \alpha_i + \beta_j + \gamma_{ij}$, where the restrictions $\sum_i \alpha_i = \sum_j \beta_j = 0$, $\sum_i \gamma_{ij} = 0$ for every j, and $\sum_j \gamma_{ij} = 0$ for every i were necessary to obtain a unique set of parameters. If we use dot subscripts on the μ_{ij}'s to denote averaging (rather than summation), then

$$\mu_{i\cdot} - \mu_{\cdot\cdot} = \frac{1}{J} \sum_j \mu_{ij} - \frac{1}{IJ} \sum_i \sum_j \mu_{ij} = \alpha_i$$

is the effect of factor A at level i averaged over levels of factor B, while

$$\mu_{ij} - \mu_{\cdot j} = \mu_{ij} - \frac{1}{I} \sum_i \mu_{ij} = \alpha_i + \gamma_{ij}$$

is the effect of factor A at level i specific to factor B at level j. If the effect of A at level i depends on the level of B, then there is interaction between the factors, and the γ_{ij}'s are not all zero. In particular

$$\mu_{ij} - \mu_{\cdot j} - \mu_{i\cdot} + \mu_{\cdot\cdot} = \gamma_{ij} \tag{11.11}$$

The Three-Factor Model

The model for three-factor ANOVA with $L_{ijk} = L$ is

$$X_{ijkl} = \mu_{ijk} + \epsilon_{ijkl} \qquad \begin{aligned} & i = 1, \ldots, I; j = 1, \ldots, J; \\ & k = 1, \ldots, K; l = 1, \ldots, L \end{aligned} \tag{11.12}$$

where the ϵ_{ijkl}'s are independent and normally distributed with mean zero and variance σ^2, and

$$\mu_{ijk} = \mu + \alpha_i + \beta_j + \delta_k + \gamma_{ij}^{AB} + \gamma_{ik}^{AC} + \gamma_{jk}^{BC} + \gamma_{ijk} \tag{11.13}$$

The restrictions necessary to obtain uniquely defined parameters are that the sum over any subscript of any parameter on the right-hand side of (11.13) equal zero.

The parameters γ_{ij}^{AB}, γ_{ik}^{AC}, and γ_{jk}^{BC} are called two-factor interactions, while γ_{ijk} is called a three-factor interaction; the α_i's, β_j's, and δ_k's are the main effects parameters. For any fixed level k of the third factor, analogous to (11.11),

$$\mu_{ijk} - \mu_{i \cdot k} - \mu_{\cdot jk} + \mu_{\cdot \cdot k} = \gamma_{ij}^{AB} + \gamma_{ijk}$$

is the interaction of the ith level of A with the jth level of B specific to the kth level of C, while

$$\mu_{ij \cdot} - \mu_{i \cdot \cdot} - \mu_{\cdot j \cdot} + \mu_{\cdot \cdot \cdot} = \gamma_{ij}^{AB}$$

is the interaction between A at level i and B at level j averaged over levels of C. If the interaction of A at level i and B at level j does not depend on k, then all γ_{ijk}'s equal 0. Thus nonzero γ_{ijk}'s represent nonadditivity of the two-factor γ_{ij}^{AB}'s over the various levels of the third factor C. If the experiment included more than three factors, there would be corresponding higher-order interaction terms with analogous interpretations. Note that in the above argument, if we had considered fixing the level of either A or B (rather than C, as was done) and examining the γ_{ijk}'s, their interpretation would be the same—if any of the interactions of two factors depend on the level of the third factor, then there are nonzero γ_{ijk}'s.

The Analysis of a Three-Factor Experiment

When $L > 1$, there is a sum of squares for each main effect, each two-factor interaction, and the three-factor interaction. To write these in a way which indicates how sums of squares are defined when there are more than three factors, note that any of the model parameters in (11.13) can be estimated unbiasedly by averaging X_{ijkl} over appropriate subscripts and taking differences. Thus

$$\hat{\mu} = \overline{X}_{\cdots\cdot}, \quad \hat{\alpha}_i = \overline{X}_{i\cdots} - \overline{X}_{\cdots\cdot}, \quad \hat{\gamma}_{ij}^{AB} = \overline{X}_{ij\cdot\cdot} - \overline{X}_{i\cdots} - \overline{X}_{\cdot j\cdot\cdot} + \overline{X}_{\cdots\cdot},$$

$$\hat{\gamma}_{ijk} = \overline{X}_{ijk\cdot} - \overline{X}_{ij\cdot\cdot} - \overline{X}_{i\cdot k\cdot} - \overline{X}_{\cdot jk\cdot} + \overline{X}_{i\cdots} + \overline{X}_{\cdot j\cdot\cdot} + \overline{X}_{\cdot\cdot k\cdot} - \overline{X}_{\cdots\cdot}.$$

with other main effects and interaction estimators obtained by symmetry. Then sums of squares are

$$SST = \sum_i \sum_j \sum_k \sum_l (X_{ijkl} - \overline{X}_{\cdots\cdot})^2 = \sum_i \sum_j \sum_k \sum_l X_{ijkl}^2 - \frac{X_{\cdots\cdot}^2}{IJKL}$$

$$SSA = \sum_i \sum_j \sum_k \sum_l \hat{\alpha}_i^2 = JKL \sum_i (\overline{X}_{i\cdots} - \overline{X}_{\cdots\cdot})^2$$

$$= \frac{1}{JKL} \sum_i X_{i\cdots}^2 - \frac{X_{\cdots\cdot}^2}{IJKL}$$

$$SSAB = \sum_i \sum_j \sum_k \sum_l (\hat{\gamma}_{ij}^{AB})^2$$

$$= \frac{1}{KL} \sum_i \sum_j X_{ij\cdot\cdot}^2 - \frac{1}{JKL} \sum_i X_{i\cdots}^2 - \frac{1}{IKL} \sum_j X_{\cdot j\cdot\cdot}^2 + \frac{X_{\cdots\cdot}^2}{IJKL}$$

$$SSABC = \sum_i \sum_j \sum_k \sum_l \hat{\gamma}_{ijk}^2, \quad SSE = \sum_i \sum_j \sum_k \sum_l (X_{ijkl} - \overline{X}_{ijk\cdot})^2$$

$$= \sum_i \sum_j \sum_k \sum_l X_{ijkl}^2 - \frac{1}{L} \sum_i \sum_j \sum_k X_{ijk\cdot}^2$$

with the other main effect and two factor interaction SS's obtained by symmetry.

The fundamental ANOVA identity here is

$$SST = SSA + SSB + SSC + SSAB$$
$$+ SSAC + SSBC + SSABC + SSE \tag{16.14}$$

so that $SSABC$ can be obtained by subtraction. Each main effect and interaction SS has associated with it a number of degrees of freedom equal to the number of independently determined parameters which define the SS. Thus SSA has $I - 1$ d.f., $SSAB$ has $(I - 1)(J - 1)$ d.f., $SSABC$ has $(I - 1)(J - 1)(K - 1)$ d.f., and so on, while SST has $IJKL - 1$ d.f. and SSE has $IJK(L - 1)$ d.f.

Each SS (excepting SST) when divided by its d.f. gives a mean square, with

$$E(MSE) = \sigma^2$$

$$E(MSA) = \sigma^2 + \frac{JKL}{I - 1} \sum_i \alpha_i^2$$

$$E(MSAB) = \sigma^2 + \frac{KL}{(I - 1)(J - 1)} \sum_i \sum_j (\gamma_{ij}^{AB})^2$$

$$E(MSABC) = \sigma^2 + \frac{L}{(I - 1)(J - 1)(K - 1)} \sum_i \sum_j \sum_k (\gamma_{ijk})^2$$

and similar expressions for the other expected mean squares. Main effect and interaction hypotheses are tested by forming F ratios with MSE in each denominator:

Null Hypothesis	Test Statistic	Rejection Region
H_{0A}: all α_i's $= 0$	$F_A = \dfrac{MSA}{MSE}$	$F_A \geq F_{\alpha, I-1, IJK(L-1)}$
H_{0AB}: all γ_{ij}^{AB}'s $= 0$	$F_{AB} = \dfrac{MSAB}{MSE}$	$F_{AB} \geq F_{\alpha, (I-1)(J-1), IJK(L-1)}$
H_{0ABC}: all γ_{ijk}'s $= 0$	$F_{ABC} = \dfrac{MSABC}{MSE}$	$F_{ABC} \geq F_{\alpha, (I-1)(J-1)(K-1), IJK(L-1)}$

Usually the main effect hypotheses are tested only if all interactions are judged not significant.

The above analysis assumes that $L_{ijk} = L > 1$. If $L = 1$, then as in the two-factor case, the highest-order interactions must be assumed to equal zero in order to obtain an MSE which estimates σ^2. Setting $L = 1$ and disregarding the fourth subscript summation over l, the above formulas for SS's are still valid, and $SSE = \sum_i \sum_j \sum_k \hat{\gamma}_{ijk}^2$ with $\bar{X}_{ijk.} = X_{ijk}$ in the expression for $\hat{\gamma}_{ijk}$.

Example 11.6 The following observations (body temperature $-$ 100 °F) were reported in an experiment to study heat tolerance of cattle ("The Significance of the Coat in Heat Tolerance of Cattle," *Australian J. Agriculture Research*, 1959, pp. 744–748). Measurements were made at four different periods (factor A, with $I = 4$) on two different strains of cattle (factor B, with $J = 2$) having four different types of coat (factor C, with $K = 4$); $L = 3$ observations were made for each of the $4 \times 2 \times 4 = 32$ combinations of levels of the three factors.

		B 1				B 2		
	C 1	C 2	C 3	C 4	C 1	C 2	C 3	C 4
A 1	3.6	3.4	2.9	2.5	4.2	4.4	3.6	3.0
	3.8	3.7	2.8	2.4	4.0	3.9	3.7	2.8
	3.9	3.9	2.7	2.2	3.9	4.2	3.4	2.9
A 2	3.8	3.8	2.9	2.4	4.4	4.2	3.8	2.0
	3.6	3.9	2.9	2.2	4.4	4.3	3.7	2.9
	4.0	3.9	2.8	2.2	4.6	4.7	3.4	2.8
A 3	3.7	3.8	2.9	2.1	4.2	4.0	4.0	2.0
	3.9	4.0	2.7	2.0	4.4	4.6	3.8	2.4
	4.2	3.9	2.8	1.8	4.5	4.5	3.3	2.0
A 4	3.6	3.6	2.6	2.0	4.0	4.0	3.8	2.0
	3.5	3.7	2.9	2.0	4.1	4.4	3.7	2.2
	3.8	3.9	2.9	1.9	4.2	4.2	3.5	2.3

The table of cell totals ($X_{ijk.}$'s) for all combinations of the three factors is

$X_{ijk.}$	B 1				B 2			
	C 1	C 2	C 3	C 4	C 1	C 2	C 3	C 4
A 1	11.3	11.0	8.4	7.1	12.1	12.5	10.7	8.7
A 2	11.4	11.6	8.6	6.8	13.4	13.2	10.9	7.7
A 3	11.8	11.7	8.4	5.9	13.1	13.1	11.1	6.4
A 4	10.9	11.2	8.4	5.9	12.3	12.6	11.0	6.5

There are three tables for cell totals of different pairs of factors:

$X_{ij..}$	$B1$	$B2$	$X_{i...}$
$A1$	37.8	44.0	81.8
$A2$	38.4	45.2	83.6
$A3$	37.8	43.7	81.5
$A4$	36.4	42.4	78.8
$X_{.j..}$	150.4	175.3	325.7

$X_{i.k.}$	$C1$	$C2$	$C3$	$C4$	$X_{i...}$
$A1$	23.4	23.5	19.1	15.8	81.8
$A2$	24.8	24.8	19.5	14.5	83.6
$A3$	24.9	24.8	19.5	12.3	81.5
$A4$	23.2	23.8	19.4	12.4	78.8
$X_{..k.}$	96.3	96.9	77.5	55.0	325.7

$X_{.jk.}$	$C1$	$C2$	$C3$	$C4$	$X_{.j..}$
$B1$	45.4	45.5	33.8	25.7	150.4
$B2$	50.9	51.4	43.7	29.3	175.3
$X_{..k.}$	96.3	96.9	77.5	55.0	325.7

The correction factor $x^2.../IJKL$ is $(325.7)^2/96 = 1105.01$ and $\sum_i \sum_j \sum_k \sum_l x^2_{ijkl} = 1166.17$, so $SST = 1166.17 - 1105.01 = 61.16$. The other sums of squares are

$$SSA = \frac{(81.8)^2 + (83.6)^2 + (81.5)^2 + (78.8)^2}{24} - 1105.01$$
$$= 1105.50 - 1105.01 = .49$$

$$SSB = 1111.46 = 1105.01 = 6.45, \quad SSC = 1153.94 - 1105.01 = 48.93$$

$$SSAB = 1111.97 - 1105.50 - 1111.46 + 1105.01 = .02$$

$$SSAC = 1.61, \quad SSBC = .88, \quad SSE = 1166.17 - 1163.64 = 2.53$$

and

$$SSABC = 61.16 - [.49 + 6.45 + 48.93 + .02 + 1.61 + .88 + 2.53]$$
$$= .25$$

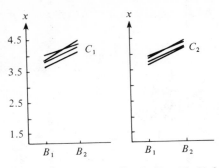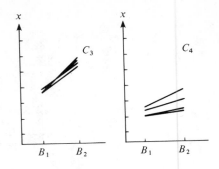

Figure 11.3 Plots of $\bar{x}_{ijk}$ for Example 11.6

Table 11.10

Source	d.f.	Sum of squares	Mean square	F
A	$I - 1 = 3$	.49	.163	4.13
B	$J - 1 = 1$	6.45	6.45	163.29
C	$K - 1 = 3$	48.93	16.31	412.91
AB	$(I - 1)(J - 1) = 3$	.02	.0067	.170
AC	$(I - 1)(K - 1) = 9$	1.61	.179	4.53
BC	$(J - 1)(K - 1) = 3$	.88	.293	7.42
ABC	$(I - 1)(J - 1)(K - 1) = 9$	.25	.0278	.704
Error	$IJK(L - 1) = 64$	2.53	.0395	
Total	$IJKL - 1 = 95$	61.16		

Since $F_{.01,9,64} \doteq 2.70$ and $f_{ABC} = MSABC/MSE = .704$ does not exceed 2.70, we conclude that three-factor interactions are not significant. However, although the AB interactions are also not significant, both AC and BC interactions as well as all main effects seem to be necessary in the model. When there are no ABC or AB interactions, a plot of the $\bar{x}_{ijk}$'s ($=\hat{\mu}_{ijk}$) separately for each level of C should reveal no substantial interactions (if only the ABC interactions are zero, plots are more difficult to interpret; see the article "Two-Dimensional Plots for Interpreting Interactions in the Three-Factor Analysis of Variance Model," *Amer. Statistician*, May 1979, pp. 63–69).

Tukey's procedure can be used in three-factor (or more) ANOVA. The second subscript on Q is the number of sample means being compared, while the third is degrees of freedom for error.

Models with random and mixed effects can also be analyzed. Sums of squares and degrees of freedom are identical to the fixed effects case, but expected mean squares are of course different for the random main effects or interactions. A good elementary reference is the book by Wine listed in the set of chapter references.

Latin Square Designs

When several factors are to be studied simultaneously, an experiment in which there is at least one observation for every possible combination of levels is referred to as a **complete layout**. If the factors are A, B, and C with I, J, and K levels, respectively, a complete layout requires at least IJK observations. Frequently an experiment of this size is either impracticable, because of cost, time, or space constraints, or literally impossible. For example, if the response variable is sales of a certain product and the factors are different display configurations, different stores, and different time periods, then only one display configuration can realistically be used in a given store during a given time period.

A three-factor experiment in which fewer than IJK observations are made is called an incomplete layout. There are some incomplete layouts in which the pattern of combinations of factors is such that the analysis is straightforward. One such three-factor design is called a **Latin square**. It is appropriate when $I = J = K$ (for example, four display configurations, four stores, and four time periods) and all two- and three-factor interaction effects are assumed absent. If the levels of factor A are identified with the rows of a two-way table and the levels of B with the columns of the table, then the defining characteristic of a Latin square design is that *every level of factor C appears exactly once in each row and exactly once in each column.* Pictured in Figure 11.4 are examples of a 3×3, 4×4, and 5×5 Latin square. There are 12 different 3×3 Latin squares, and the number of different $N \times N$ Latin squares increases rapidly with N (for example, every permutation of rows of a given Latin square yields a Latin square, and similarly for column permutations). It is recommended that the square actually used in a particular experiment be chosen at random from the set of all possible squares of the desired dimension; for further details, the book by Neter and Wasserman can be consulted.

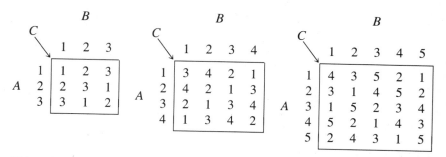

Figure 11.4 Examples of Latin squares

The Model and Analysis for Latin Squares

The letter N will be used to denote the common value of I, J, and K. Then a complete layout with one observation per combination would require N^3 observations, whereas a Latin square requires only N^2 observations. Once a particular square has been chosen, the value of k (the level of factor C) is completely determined by the values of i and j. To emphasize this, we use $x_{ij(k)}$ to denote the observed value when the three factors are at levels i, j, and k, respectively, with k taking on only one value for each i, j pair. The model is then

$$X_{ij(k)} = \mu + \alpha_i + \beta_j + \delta_k + \epsilon_{ij(k)} \qquad \begin{array}{l} i = 1, \ldots, N; j = 1, \ldots, N; \\ k = 1, \ldots, N \end{array}$$

where the $\epsilon_{ij(k)}$'s are independent and normally distributed with mean zero and variance σ^2, and $\Sigma \alpha_i = \Sigma \beta_j = \Sigma \delta_k = 0$.

We employ the following notation for totals and averages:

$$X_{i..} = \sum_j X_{ij(k)}, \quad X_{\cdot j\cdot} = \sum_i X_{ij(k)}, \quad X_{\cdot\cdot k} = \sum_{i,j} X_{ij(k)}, \quad X_{...} = \sum_i \sum_j X_{ij(k)},$$

$$\overline{X}_{i..} = \frac{X_{i..}}{N}, \qquad \overline{X}_{\cdot j\cdot} = \frac{X_{\cdot j\cdot}}{N}, \qquad \overline{X}_{\cdot\cdot k} = \frac{X_{\cdot\cdot k}}{N}, \qquad \overline{X}_{...} = \frac{X_{...}}{N^2}$$

Note that although $X_{i..}$ previously suggested a double summation, now it corresponds to a single sum over all j (and the associated values of k). The sums of squares used in the analysis are then

$$SST = \sum_i \sum_j (X_{ij(k)} - \overline{X}_{...})^2 = \sum_i \sum_j X_{ij(k)}^2 - \frac{X_{...}^2}{N}$$

$$SSA = \sum_i \sum_j (\overline{X}_{i..} - \overline{X}_{...})^2 = \frac{1}{N} \sum_i X_{i..}^2 - \frac{X_{...}^2}{N}$$

$$SSB = \sum_i \sum_j (\overline{X}_{\cdot j\cdot} - \overline{X}_{...})^2 = \frac{1}{N} \sum_j X_{\cdot j\cdot}^2 - \frac{X_{...}^2}{N}$$

$$SSC = \sum_i \sum_j (\overline{X}_{\cdot\cdot k} - \overline{X}_{...})^2 = \frac{1}{N} \sum_k X_{\cdot\cdot k}^2 - \frac{X_{...}^2}{N}$$

and

$$SSE = \sum_i \sum_j (X_{ij(k)} - \hat{\mu} - \hat{\alpha}_i - \hat{\beta}_j - \hat{\delta}_k)^2$$

$$= \sum_i \sum_j (X_{ij(k)} - \overline{X}_{i..} - \overline{X}_{\cdot j\cdot} - \overline{X}_{\cdot\cdot k} + 2\overline{X}_{...})^2$$

where $\hat{\mu}$, $\hat{\alpha}_i$, $\hat{\beta}_j$, and $\hat{\delta}_k$ are the usual unbiased estimators for the respective parameters.

An ANOVA identity is still valid here:

$$SST = SSA + SSB + SSC + SSE$$

so that SSE is computed by subtraction after the other four SS's have been computed. There are $N^2 - 1$ degrees of freedom for SST and $N - 1$ d.f. for each of the main effect SS's, so error d.f. $= N^2 - 1 - (N - 1) - (N - 1) - (N - 1) = (N - 1) \cdot (N - 2)$. Mean squares are then

$$MSA = \frac{SSA}{N - 1}, \quad MSB = \frac{SSB}{N - 1}, \quad MSC = \frac{SSC}{N - 1}$$

and

$$MSE = \frac{SSE}{(N - 1)(N - 2)}$$

For testing $H_{0C}: \delta_1 = \delta_2 = \cdots = \delta_N = 0$, the test statistic is $F_C = MSC/MSE$, with H_{0C} rejected if $F_C \geq F_{\alpha,N-1,(N-1)(N-2)}$. The other two main-effect null hypotheses are also rejected if the corresponding F ratio exceeds $F_{\alpha,N-1,(N-1)(N-2)}$.

If any of the null hypotheses is rejected, significant differences can be identified by using Tukey's procedure. After computing $w = Q_{\alpha,N,(N-1)(N-2)} \cdot \sqrt{MSE/N}$, pairs of sample means (the $\bar{x}_{i..}$'s, $\bar{x}_{.j.}$'s, or $\bar{x}_{..k}$'s) differing by more than w correspond to significant difference between associated factor effects (the α_i's, β_j's, or δ_k's).

The hypothesis H_{0C} is frequently the one of central interest, and a Latin square design is used to control for extraneous variation in the A and B factors, as was done by a randomized block design for the case of a single extraneous factor. Thus in the product sales example mentioned earlier, variation due to both stores and time periods is controlled by a Latin square design, enabling an investigator to test for the presence of effects due to different product display configurations.

Example 11.7 In an experiment to investigate the effect of relative humidity on abrasion resistance of leather cut from a rectangular pattern ("The Abrasion of Leather," *J. Inter. Soc. Leather Trades' Chemists*, 1946, p. 287), a 6 × 6 Latin square was used to control for possible variability due to row and column position in the pattern. The six levels of relative humidity studied were 1 = 25%, 2 = 37%, 3 = 50%, 4 = 62%, 5 = 75%, and 6 = 87%, with the following results.

		B (columns)						$x_{i..}$
		1	2	3	4	5	6	
	1	[3] 7.38	[4] 5.39	[6] 5.03	[2] 5.50	[5] 5.01	[1] 6.79	35.10
	2	[2] 7.15	[1] 8.16	[5] 4.96	[4] 5.78	[3] 6.24	[6] 5.06	37.35
A (rows)	3	[4] 6.75	[6] 5.64	[3] 6.34	[5] 5.31	[1] 7.81	[2] 8.05	39.90
	4	[1] 8.05	[3] 6.45	[2] 6.31	[6] 5.46	[4] 6.05	[5] 5.51	37.83
	5	[6] 5.65	[5] 5.44	[1] 7.27	[3] 6.54	[2] 7.03	[4] 5.96	37.89
	6	[5] 6.00	[2] 6.55	[4] 5.93	[1] 8.02	[6] 5.80	[3] 6.61	38.91
$x_{.j.}$		40.98	37.63	35.84	36.61	37.94	37.98	

Also, $x_{..1} = 46.10$, $x_{..2} = 40.59$, $x_{..3} = 39.56$, $x_{..4} = 35.86$, $x_{..5} = 32.23$, $x_{..6} = 32.64$, $x_{...} = 226.98$, and $\sum_i \sum_j x_{ij(k)}^2 = 1462.89$. Further computations are summarized in Table 11.11.

Table 11.11

Source of variation	d.f.	Sum of squares	Mean square	F
A (rows)	5	2.19	.438	2.50
B (columns)	5	2.57	.514	2.94
C (treatments)	5	23.53	4.706	26.89
Error	20	3.49	.175	
Total	35	31.78		

Since $F_{.05,5,20} = 2.71$ and $26.89 \geq 2.71$, H_{0C} is rejected in favor of the hypothesis that relative humidity does on average affect abrasion resistance.

To apply Tukey's procedure, $w = Q_{.05,6,20} \cdot \sqrt{MSE/6} = 4.45\sqrt{.175/6} = .76$. Ordering the $\bar{x}_{..k}$'s and underscoring yields

75%	87%	62%	50%	37%	25%
5.37	5.44	5.98	6.59	6.77	7.68

In particular, the lowest relative humidity appears to result in a true average abrasion resistance significantly higher than for any other relative humidity studied.

Exercises / Section 11.3

1. The output of a continuous extruding machine which coats steel pipe with plastic was studied as a function of the thermostat temperature profile (A, at three levels), type of plastic (B, at three levels), and the speed of the rotating screw which forces the plastic through a tube-forming die (C, at three levels). There were two replications (L = 2) at each combination of levels of the factor, yielding a total of 54 observations on output. The summary statistics include $x_{....} = 29,184$, $\sum x_{i...}^2 = 284,156,552$, $\sum x_{.j..}^2 = 284,001,155$, $\sum x_{..k.}^2 = 288,306,487$, $\sum \sum x_{ij..}^2 = 94,758,336$, $\sum \sum x_{i.k.}^2 = 96,187,405$, $\sum \sum x_{.jk.}^2 = 96,137,220$, $\sum \sum \sum x_{ijk.}^2 = 32,078,455$, and $\sum \sum \sum \sum x_{ijkl}^2 = 16,042,355$.

   a. Construct the ANOVA table.
   b. Use appropriate F tests to show that none of the F ratios for two- or three-factor interactions is significant at level .05.
   c. Which main effects appear significant?
   d. With $x_{..1.} = 8242$, $x_{..2.} = 9732$, and $x_{..3.} = 11,210$, use Tukey's procedure to identify significant differences among the levels of factor C.

2. To see whether thrust force in drilling is affected by drilling speed (A), feed rate (B), or material used (C), an experiment using four speeds, three rates, and two materials was performed, with two samples (L = 2) drilled at each combination of levels of the three factors. Summary statistics include $x_{....} = 17,709$, $\sum x_{i...}^2 = 78,631,967$, $\sum x_{.j..}^2 = 145,960,989$, $\sum x_{..k.}^2 = 160,582,841$, $\sum \sum x_{ij..}^2 = 36,779,799$, $\sum \sum x_{i.k.}^2 = 40,314,811$, $\sum \sum x_{.jk.}^2 = 74,975,035$, $\sum \sum \sum x_{ijk.}^2 = 18,919,719$, and $\sum \sum \sum \sum x_{ijkl}^2 = 9,516,679$. Construct the ANOVA table and identify significant interactions using $\alpha = .01$. Is there any single factor which appears to have no effect on thrust force (does any factor appear nonsignificant in every effect in which it appears)?

3. The paper "An Analysis of Variance Applied to Screw Machines" (*Industrial Quality Control*, 1956, pp. 8–9) described an experiment to investi-

gate how the length of steel bars was affected by time of day (A), heat treatment applied (B), and screw machine used (C). The three times were 8:00 A.M., 11:00 A.M., and 3:00 P.M., and there were two treatments and four machines (a $3 \times 2 \times 4$ factorial experiment), resulting in the accompanying data [coded as 1000(length $- 4.380$), which doesn't affect the analysis].

B_1

	C_1	C_2	C_3	C_4
A_1	6, 9, 1, 3	7, 9, 5, 5	1, 2, 0, 4	6, 6, 7, 3
A_2	6, 3, 1, -1	8, 7, 4, 8	3, 2, 1, 0	7, 9, 11, 6
A_3	5, 4, 9, 6	10, 11, 6, 4	-1, 2, 6, 1	10, 5, 4, 8

B_2

	C_1	C_2	C_3	C_4
A_1	4, 6, 0, 1	6, 5, 3, 4	-1, 0, 0, 1	4, 5, 5, 4
A_2	3, 1, 1, -2	6, 4, 1, 3	2, 0, -1, 1	9, 4, 6, 3
A_3	6, 0, 3, 7	8, 7, 10, 0	0, -2, 4, -4	4, 3, 7, 0

a. Construct the ANOVA table for this data.
b. Test to see whether any of the interaction effects are significant at level .05.
c. Test to see whether any of the main effects are significant at level .05 (that is, H_{0A} versus H_{aA}, and so on).
d. Use Tukey's procedure to investigate significant differences between the four machines.

4. The following summary quantities were computed from an experiment involving four levels of nitrogen (A), two times of planting (B), and two levels of potassium (C) ("Use and Misuse of Multiple Comparison Procedures," *Agronomy J.*, 1977, pp. 205–208). Only one observation (N content, in

percentages, of corn grain) was made for each of the 16 combinations of levels. Computed summary quantities are $x_{...} = 20.96$ (since $L = 1$, a fourth subscript is unnecessary), $\Sigma x_{i..}^2 = 110.7354$, $\Sigma x_{.j.}^2 = 219.6610$, $\Sigma x_{..k}^2 = 219.6896$, $\Sigma\Sigma x_{ij.}^2 = 55.3764$, $\Sigma\Sigma x_{i.k}^2 = 55.8451$, $\Sigma\Sigma x_{.jk}^2 = 109.8474$, and $\Sigma\Sigma\Sigma x_{ijk}^2 = 27.6960$.
a. Construct the ANOVA table.
b. Assume that there are no three-way interaction effects, so that $MSABC$ is a valid estimate of σ^2, and test at level .05 for interaction and main effects.
c. The nitrogen averages are $\bar{x}_{1..} = 1.1200$, $\bar{x}_{2..} = 1.3025$, $\bar{x}_{3..} = 1.3875$, and $\bar{x}_{4..} = 1.4300$. Use Tukey's method to examine differences in percentage N among the nitrogen levels.

5. The paper "Kolbe-Schmitt Carbonation of 2-Napthol" (*Industrial and Eng. Chemistry: Process and Design Development*, 1969, pp. 165–173) presented the accompanying data on percentage yield of BON acid as a function of reaction time (1, 2, and 3 hours), temperature (30, 70, and 100 °C), and pressure (30, 70, and 100 psi). Assume that there is no three-factor interaction, so that $SSE = SSABC$ provides an estimate of σ^2. Construct the ANOVA table, and carry out all appropriate tests.

	B_1			B_2		
	C_1	C_2	C_3	C_1	C_2	C_3
A_1	68.5	73.0	68.7	72.8	80.1	72.0
A_2	74.5	75.0	74.6	72.0	81.5	76.0
A_3	70.5	72.5	74.7	69.5	84.5	76.0

	B_3		
	C_1	C_2	C_3
A_1	72.5	72.5	73.1
A_2	75.5	70.0	76.0
A_3	65.0	66.5	70.5

6. When factors A and B are fixed but factor C is random, the expected mean squares are

$E(MSE) = \sigma^2$, $E(MSA) = \sigma^2 + JL\sigma^2_{AC} +$

$\dfrac{JKL}{I-1} \sum \alpha_i^2$, $E(MSB) = \sigma^2 + IL\sigma^2_{BC} + \dfrac{IKL}{J-1} \times$

$\sum \beta_j^2$, $E(MSC) = \sigma^2 + IJL\sigma^2_C$, $E(MSAB) =$

$\sigma^2 + L\sigma^2_{ABC} + \dfrac{KL}{(I-1)(J-1)} \sum_i \sum_j (\gamma^{AB}_{ij})^2$,

$E(MSAC) = \sigma^2 + JL\sigma^2_{AC}$, $E(MSBC) = \sigma^2 + IL\sigma^2_{BC}$, $E(MSABC) = \sigma^2 + L\sigma^2_{ABC}$.

a. Based on these expected mean squares, what F ratios would you use to test $H_0: \sigma^2_{ABC} = 0$; $H_0: \sigma^2_C = 0$; $H_0: \gamma^{AB}_{ij} = 0$ for all i, j; and $H_0: \alpha_1 = \cdots = \alpha_I = 0$?

b. In an experiment to assess the effects of age, type of soil, and day of production on compressive strength of cement/soil mixtures, two ages (A), four types of soil (B), and three days (C, assumed random) were used, with $L = 2$ observations made for each combination of factor levels. The resulting sums of squares were $SSA = 14,318.24$, $SSB = 9656.40$, $SSC = 2270.22$, $SSAB = 3408.93$, $SSAC = 1442.58$, $SSBC = 3096.21$, $SSABC = 2832.72$, and $SSE = 8655.60$. Obtain the ANOVA table and carry out all tests using level .01.

7. Because of potential variability in aging due to different castings and segments on the castings, a Latin square design with $N = 7$ was used to investigate the effect of heat treatment on aging. With $A = $ castings, $B = $ segments, $C = $ heat treatments, summary statistics include $x... = 3815.8$, $\sum x^2_{i..} = 297,216.90$, $\sum x^2_{.j.} = 297,200.64$, $\sum x^2_{..k} = 297,155.01$, and $\sum \sum x^2_{ij(k)} = 297,317.65$. Compute the ANOVA table and test at level .05 the hypothesis that heat treatment has no effect on aging.

8. The article "The Responsiveness of Food Sales to Shelf Space Requirements" (*J. Marketing Research*, 1964, pp. 63–67) reported the use of a Latin square design to investigate the effect of shelf space on food sales. The experiment was carried out over a six-week period using six different stores, resulting in the following data on sales of powdered coffee cream (with shelf space index in parentheses).

			Weeks			
	1	2	3	4	5	6
1	27 (5)	14 (4)	18 (3)	35 (1)	28 (6)	22 (2)
2	34 (6)	31 (5)	34 (4)	46 (3)	37 (2)	23 (1)
3	39 (2)	67 (6)	31 (5)	49 (4)	38 (1)	48 (3)
Stores 4	40 (3)	57 (1)	39 (2)	70 (6)	37 (4)	50 (5)
5	15 (4)	15 (3)	11 (1)	9 (2)	18 (5)	17 (6)
6	16 (1)	15 (2)	14 (6)	12 (5)	19 (3)	22 (4)

Construct the ANOVA table, and state and test at level .01 the hypothesis that shelf space does not affect sales against the appropriate alternative.

9. The article "Variation in Moisture and Ascorbic Acid Content from Leaf to Leaf and Plant to Plant in Turnip Greens" (*Southern Cooperative Services Bull.*, 1951, pp. 13–17) used a Latin square design in which factor A is plant, factor B is leaf size (smallest to largest), factor C is time of weighing, and the response variable is moisture content.

			Leaf size (B)			
Time (C, in parentheses)		1	2	3	4	5
	1	6.67 (5)	7.15 (4)	8.29 (1)	8.95 (3)	9.62 (2)
	2	5.40 (2)	4.77 (5)	5.40 (4)	7.54 (1)	6.93 (3)
Plant (A)	3	7.32 (3)	8.53 (2)	8.50 (5)	9.99 (4)	9.68 (1)
	4	4.92 (1)	5.00 (3)	7.29 (2)	7.85 (5)	7.08 (4)
	5	4.88 (4)	6.16 (1)	7.83 (3)	5.83 (2)	8.51 (5)

When all three factors are random, the expected mean squares are $E(MSA) = \sigma^2 + N\sigma_A^2$, $E(MSB) = \sigma^2 + N\sigma_B^2$, $E(MSC) = \sigma^2 + N\sigma_C^2$, and $E(MSE) = \sigma^2$. This implies that the F ratios for testing $H_{0A}: \sigma_A^2 = 0$, $H_{0B}: \sigma_B^2 = 0$, and $H_{0C}: \sigma_C^2 = 0$ are identical to those for fixed effects. Obtain the ANOVA table and test at level .05 to see if there is any variation in moisture content due to the factors.

10. Analogous to a Latin square, a Graeco-Latin square design can be used when it is suspected that three extraneous factors may effect the response variable and all four factors (the three extraneous ones and the one of interest) have the same number of levels. In a Latin square, each level of the factor of interest (C) appears once in each row (with each level of A) and once in each column (with each level of B). In a Graeco-Latin square, each level of factor D appears once in each row, in each column, and also with each level of the third extraneous factor C. Alternatively, the design can be used when the four factors are all of equal interest, the number of levels of each is N, and resources are available for only N^2 observations. A 5×5 square is pictured in (a) below, with (k, l) in each cell denoting the kth level of C and lth level of D. In (b) we present data on weight loss in silicon bars used for semiconductor material as a function of volume of etch (A), color of nitric acid in the etch solution (B), size of bars (C), and time in the etch solution (D) (from "Applications of Analytic Techniques to the Semiconductor Industry," Fourteenth Midwest Quality Control Conference, 1959).

			B		
(C, D)	1	2	3	4	5
1	(1, 1)	(2, 3)	(3, 5)	(4, 2)	(5, 4)
2	(2, 2)	(3, 4)	(4, 1)	(5, 3)	(1, 5)
A 3	(3, 3)	(4, 5)	(5, 2)	(1, 4)	(2, 1)
4	(4, 4)	(5, 1)	(1, 3)	(2, 5)	(3, 2)
5	(5, 5)	(1, 2)	(2, 4)	(3, 1)	(4, 3)

(a)

65	82	108	101	126
84	109	73	97	83
105	129	89	89	52
119	72	76	117	84
97	59	94	78	106

(b)

Let $X_{ij(kl)}$ denote the observed weight loss when factor A is at level i, B is at level j, C is at level k, and D is at level l. Assuming no interaction between factors, total sum of squares SST (with $N^2 - 1$ d.f.) can be partitioned into SSA, SSB, SSC, SSD, and SSE. Give expressions for these sums of squares, including computing formulas, obtain the ANOVA table for the above data, and test each of the four main effect hypotheses using $\alpha = .05$.

11.4 2^p Factorial Experiments

If an experimenter wishes to study simultaneously the effect of p different factors on a response variable and the factors have $I_1, I_2, \ldots, I_p$ levels, respectively, then a complete experiment requires at least $I_1 \cdot I_2 \cdots I_p$ observations. In such situations the experimenter can often perform a "screening experiment" with each factor at only two levels to obtain preliminary information about factor effects. An experiment in which there are p factors, each at two levels, is referred to as a **2^p factorial experiment**. The analysis of data from such an experiment is computationally simpler than for more general factorial experiments. In addition, a 2^p experiment provides a simple setting for introducing the important concepts of confounding and fractional replications.

2^3 Experiments

As in Section 11.3 we let X_{ijkl} and x_{ijkl} refer to the observation from the lth replication with factors A, B, and C at levels i, j, and k, respectively. The model for this situation is

$$X_{ijkl} = \mu + \alpha_i + \beta_j + \delta_k + \gamma_{ij}^{AB} + \gamma_{ik}^{AC} + \gamma_{jk}^{BC} + \gamma_{ijk} + \epsilon_{ijkl} \qquad (11.14)$$

for $i = 1, 2; j = 1, 2; k = 1, 2; l = 1, \ldots, n$. The ϵ_{ijkl}'s are assumed independent, normally distributed, with mean zero and variance σ^2. Because there are only two levels of each factor, the side conditions on the parameters of (11.14) which uniquely specify the model are simply stated: $\alpha_1 + \alpha_2 = 0, \ldots, \gamma_{11}^{AB} + \gamma_{21}^{AB} = 0$, $\gamma_{12}^{AB} + \gamma_{22}^{AB} = 0$, $\gamma_{11}^{AB} + \gamma_{12}^{AB} = 0$, $\gamma_{21}^{AB} + \gamma_{22}^{AB} = 0$, and the like. These conditions imply that there is only one functionally independent parameter of each type (for each main effect and interaction). For example, $\alpha_2 = -\alpha_1$, while $\gamma_{21}^{AB} = -\gamma_{11}^{AB}$, $\gamma_{12}^{AB} = -\gamma_{11}^{AB}$, and $\gamma_{22}^{AB} = \gamma_{11}^{AB}$. Because of this, each sum of squares in the analysis will have a single degree of freedom.

The parameters of the model can be estimated by taking averages over various subscripts of the X_{ijkl}'s and then forming appropriate linear combinations of the averages. For example,

$$\hat{\alpha}_1 = \overline{X}_{1\cdots} - \overline{X}_{\cdots\cdots}$$
$$= \frac{(X_{111\cdot} + X_{121\cdot} + X_{112\cdot} + X_{122\cdot} - X_{211\cdot} - X_{212\cdot} - X_{221\cdot} - X_{222\cdot})}{8n}$$

and

$$\hat{\gamma}_{11}^{AB} = \overline{X}_{11\cdot\cdot} - \overline{X}_{1\cdots} - \overline{X}_{\cdot1\cdot\cdot} + \overline{X}_{\cdots\cdots}$$
$$= \frac{(X_{111\cdot} - X_{121\cdot} - X_{211\cdot} + X_{221\cdot} + X_{112\cdot} - X_{122\cdot} - X_{212\cdot} + X_{222\cdot})}{8n}$$

Each estimator is, except for the factor $1/8n$, a linear function of the cell totals ($X_{ijk\cdot}$'s) in which each coefficient is $+1$ or -1, with an equal number of each; such functions are called **contrasts** in the $X_{ijk\cdot}$'s. Furthermore, the estimators satisfy the same side conditions satisfied by the parameters themselves. For example

$$\hat{\alpha}_1 + \hat{\alpha}_2 = \overline{X}_{1\cdots} - \overline{X}_{\cdots\cdots} + \overline{X}_{2\cdots} - \overline{X}_{\cdots\cdots} = \overline{X}_{1\cdots} + \overline{X}_{2\cdots} - 2\overline{X}_{\cdots\cdots}$$
$$= \frac{1}{4n} X_{1\cdots} + \frac{1}{4n} X_{2\cdots} - \frac{2}{8n} X_{\cdots\cdots} = \frac{1}{4n} X_{\cdots\cdots} - \frac{1}{4n} X_{\cdots\cdots} = 0$$

Example 11.8 In an experiment to investigate the compressive strength properties of cement/soil mixtures, two different aging periods were used in combination with two different aging temperatures and two different soils. Two replications were made for each combination of levels of the three factors, resulting in the following data.

Age	Temperature	Soil 1	Soil 2
1	1	471, 413	385, 434
	2	485, 552	530, 593
2	1	712, 637	770, 705
	2	712, 789	741, 806

The computed cell totals are $x_{111\cdot} = 884$, $x_{211\cdot} = 1349$, $x_{121\cdot} = 1037$, $x_{221\cdot} = 1501$, $x_{112\cdot} = 819$, $x_{212\cdot} = 1475$, $x_{122\cdot} = 1123$, and $x_{222\cdot} = 1547$, so $x_{\cdots\cdot} = 9735$. Then

$$\hat{\alpha}_1 = (884 - 1349 + 1037 - 1501 + 819 - 1475 + 1123 - 1547)/16$$
$$= -125.5625 = -\hat{\alpha}_2$$
$$\hat{\gamma}_{11}^{AB} = (884 - 1349 - 1037 + 1501 + 819 - 1475 - 1123 + 1547)/16$$
$$= -14.5625 = -\hat{\gamma}_{12}^{AB} = -\hat{\gamma}_{21}^{AB} = \hat{\gamma}_{22}^{AB}$$

The other parameter estimates can be computed in the same manner.

Sums of Squares and Analysis for a 2^3 Experiment

The reason for computing parameter estimates is that sums of squares for the various effects are easily obtained from the estimates. For example

$$SSA = \sum_i \sum_j \sum_k \sum_l \hat{\alpha}_i^2 = 4n \sum_{i=1}^2 \hat{\alpha}_i^2 = 4n[\hat{\alpha}_1^2 + (-\hat{\alpha}_1)^2] = 8n\hat{\alpha}_1^2$$

and

$$SSAB = \sum_i \sum_j \sum_k \sum_l (\hat{\gamma}_{ij}^{AB})^2$$
$$= 2n \sum_{i=1}^2 \sum_{j=1}^2 (\hat{\gamma}_{ij}^{AB})^2 = 2n[(\hat{\gamma}_{11}^{AB})^2 + (-\hat{\gamma}_{11}^{AB})^2 + (-\hat{\gamma}_{11}^{AB})^2 + (\hat{\gamma}_{11}^{AB})^2]$$
$$= 8n(\hat{\gamma}_{11}^{AB})^2$$

Since each estimate is a contrast in the cell totals multiplied by $1/8n$, each sum of squares has the form $(\text{contrast})^2/8n$. Thus to compute the various sums of squares, we need to know the coefficients ($+1$ or -1) of the appropriate contrasts. The signs ($+$ or $-$) on each $x_{ijk\cdot}$ in each effect contrast are most conveniently displayed in a table. We shall use the notation (1) for the experimental condition $i = 1$, $j = 1$, $k = 1$, a for $i = 2$, $j = 1$, $k = 1$, ab for $i = 2$, $j = 2$, $k = 1$, and so on. If level

Factorial Effect

Experimental Condition	Cell Total	A	B	C	AB	AC	BC	ABC
(1)	$x_{111\cdot}$	−	−	−	+	+	+	−
a	$x_{211\cdot}$	+	−	−	−	−	+	+
b	$x_{121\cdot}$	−	+	−	−	+	−	+
ab	$x_{221\cdot}$	+	+	−	+	−	−	−
c	$x_{112\cdot}$	−	−	+	+	−	−	+
ac	$x_{212\cdot}$	+	−	+	−	+	−	−
bc	$x_{122\cdot}$	−	+	+	−	−	+	−
abc	$x_{222\cdot}$	+	+	+	+	+	+	+

Figure 11.5 Table of signs for computing effect contrasts

1 is thought of as "low" and level 2 as "high," any letter which appears denotes a high level of the associated factor. In Figure 11.5 each column gives the signs for a particular effect contrast in the $x_{ijk\cdot}$'s associated with the different experimental conditions.

In each of the first three columns, the sign is + if the corresponding factor is at the high level and − if it is at the low level. Every sign in the AB column is then the "product" of the signs in the A and B columns, with $(+)(+) = (−)(−) = +$ and $(+)(−) = (−)(+) = −$, and similarly for the AC and BC columns. Finally, the signs in the ABC column are the products of AB with C (or B with AC or A with BC). Thus, for example,

$$AC \text{ contrast} = +x_{111\cdot} -x_{211\cdot} +x_{121\cdot} -x_{221\cdot} -x_{112\cdot} +x_{212\cdot} -x_{122\cdot} +x_{222\cdot}$$

Once the seven effect contrasts are computed,

$$SS(\text{effect}) = \frac{(\text{effect contrast})^2}{8n}$$

Even with a table of signs, calculation of the contrasts is tedious. An efficient computational technique, due to Yates, is as follows. Write in a column the eight cell totals in the **standard order** as given in the table of signs and establish three further columns. In each of these three columns, the first four entries are the sums of entries 1 and 2, 3 and 4, 5 and 6, 7 and 8 of the previous columns. The last four entries are the differences between entries 2 and 1, 4 and 3, 6 and 5, and 8 and 7 of the previous column. The last column then contains $x_{....}$ and the seven effect contrasts in standard order. Squaring each contrast and dividing by $8n$ then gives the seven sums of squares.

Example 11.8 (continued)   Since $n = 2$, $8n = 16$.

Table 11.12

Treatment condition	$x_{ijk.}$	1	2	Effect contrast	$SS = (\text{contrast})^2/16$
$(1) = x_{111.}$	884	2233	4771	9735	
$a = x_{211.}$	1349	2538	4964	2009	252,255.06
$b = x_{121.}$	1037	2294	929	681	28,985.06
$ab = x_{221.}$	1501	2670	1080	-233	3393.06
$c = x_{112.}$	819	465	305	193	2328.06
$ac = x_{212.}$	1475	464	376	151	1425.06
$bc = x_{122.}$	1123	656	-1	71	315.06
$abc = x_{222.}$	1547	424	-232	-231	3335.06
					292,036.42

From the original data, $\sum_i \sum_j \sum_k \sum_l x_{ijkl}^2 = 6{,}232{,}289$, and

$$\frac{x_{\dots}^2}{16} = 5{,}923{,}139.06$$

so

$$SST = 6{,}232{,}289 - 5{,}923{,}139.06 = 309{,}149.94$$
$$SSE = SST - [SSA + \cdots + SSABC] = 309{,}149.94 - 292{,}036.42$$
$$= 17{,}113.52$$

Table 11.13

Source of variation	d.f.	Sum of squares	Mean square	F
A	1	252,255.06	252,255.06	117.92
B	1	28,985.06	28,985.06	13.55
C	1	3393.06	3393.06	1.59
AB	1	2328.06	2328.06	1.09
AC	1	1425.06	1425.06	.67
BC	1	315.06	315.06	.15
ABC	1	3335.06	3335.06	1.56
Error	8	17,113.52	2139.19	
Total	15	309,149.94		

2^p Experiments for $p > 3$

Although the computations when $p > 3$ are quite tedious, the analysis parallels that of the three-factor case. For example, if there are four factors A, B, C, and D, there are 16 different experimental conditions. The first eight in standard order are exactly those already listed for a three-factor experiment, while the second eight are obtained by placing the letter d beside each condition in the first group. Yates' method is then initiated by computing totals across replications, listing these totals in standard order, and proceeding as before; with p factors, the pth column to the right of the treatment totals will give the effect contrasts.

For $p > 3$ there will often be no replications of the experiment (so only one complete replicate is available). To obtain an error sum of squares for testing various hypotheses, the higher-order interactions are usually assumed absent and their SS's added to obtain an SSE.

Confounding

It is often not possible to carry out all 2^p experimental conditions of a 2^p factorial experiment in a homogeneous experimental environment. In such situations it may be possible to separate the experimental conditions into 2^r homogeneous blocks $(r < p)$, so that there are 2^{p-r} experimental conditions in each block. The blocks may, for example, correspond to different laboratories, different time periods, or different operators or work crews. In the simplest case, $p = 3$ and $r = 1$, so that there are two blocks with each block consisting of four of the eight experimental conditions.

As always, blocking is effective in reducing variation associated with extraneous sources. However, when the 2^p experimental conditions are placed in 2^r blocks, the price paid for this blocking is that $2^r - 1$ of the factor effects cannot be estimated. This is because $2^r - 1$ factor effects (main effects and/or interactions) are mixed up or **confounded** with the block effects. The allocation of experimental conditions to blocks is then usually done so that only higher-level interactions are confounded, while main effects and low-order interactions remain estimable and can be tested for.

To see how allocation to blocks is accomplished, consider first a 2^3 experiment with two blocks ($r = 1$) and four treatments per block. Suppose that we select ABC as the effect to be confounded with blocks. Then any experimental condition having an odd number of letters in common with ABC, such as b (one letter) or abc (three letters), is placed in one block, while any condition having an even number of letters in common with ABC (where zero is even) goes in the other block. This gives

Block 1   Block 2

(1), *ab, ac, bc*   *a, b, c, abc*

Figure 11.6   Confounding ABC in a 2^3 experiment

as the allocation of treatments to the two blocks.

In the absence of replications, the data from such an experiment would usually be analyzed by assuming that there were no two-factor interactions (additivity) and using $SSE = SSAB + SSAC + SSBC$ with three degrees of freedom to test for the presence of main effects. Most frequently, though, there are replications when just three factors are being studied. Suppose that there are u replicates, resulting in a total of $2^r \cdot u$ blocks in the experiment. Then after subtracting from SST all sums of squares associated with effects not confounded with blocks (computed using Yates' method), the block sum of squares is computed using the $2^r \cdot u$ block totals and then subtracted to yield SSE (so there are $2^r \cdot u - 1$ d.f. for blocks).

Example 11.9 The article "Factorial Experiments in Pilot Plant Studies" (*Industrial and Eng. Chemistry*, 1951, pp. 1300–1306) reports the results of an experiment to assess the effects of reactor temperature (A), gas throughput (B), and concentration of active constituent (C) on strength of the product solution (measured in arbitrary units) in a recirculation unit. Two blocks were used, with the ABC effect confounded with blocks, and there were two replications, resulting in the following data.

Replication 1

Block 1		Block 2	
(1)	99	a	18
ab	52	b	51
ac	42	c	108
bc	95	abc	35

Replication 2

Block 1		Block 2	
(1)	46	a	18
ab	−47	b	62
ac	22	c	104
bc	67	abc	36

Figure 11.7

The four block × replication totals are 288, 212, 88, and 220, with a grand total of 808, so

$$SSBl = \frac{(288)^2 + (212)^2 + (88)^2 + (220)^2}{4} - \frac{(808)^2}{16} = 5204.00$$

The other sums of squares are computed by Yates' method using the eight experimental condition totals, resulting in the following ANOVA table:

Table 11.14

Source of variation	d.f.	Sum of squares	Mean square	F
A	1	12,996	12,996	39.82
B	1	702.25	702.25	2.15
C	1	2756.25	2756.25	8.45
AB	1	210.25	210.25	.64
AC	1	30.25	30.25	.093
BC	1	25	25	.077
Blocks	3	5204	1734.67	5.32
Error	6	1958	326.33	
Total	15	23,857		

By comparison with $F_{.05,1,6} = 5.99$, we conclude that only the main effects for A and C differ significantly from zero.

Confounding Using More Than Two Blocks

In the case $r = 2$ (four blocks), three effects are confounded with blocks. The experimenter first chooses two defining effects to be confounded. For example, in a five-factor experiment ($A, B, C, D,$ and E), the two three-factor interactions BCD

and *CDE* might be chosen for confounding. The third effect confounded is then the **generalized interaction** of the two, obtained by writing the two chosen effects side by side and then cancelling any letters common to both: *(BCD) (CDE)* = BE. Notice that if *ABC* and *CDE* are chosen for confounding, their generalized interaction is *(ABC) (CDE)* = *ABDE*, so that no main-effects or two-factor interactions are confounded.

Once the two defining effects have been selected for confounding, one block consists of all treatment conditions having an even number of letters in common with both defining effects, the second block consists of all conditions having an even number of letters in common with the first defining contrast and an odd number of letters in common with the second contrast, while the third and fourth blocks consist of the "odd/even" and "odd/odd" contrasts. In a five-factor experiment with defining effects *ABC* and *CDE*, this results in the following allocation to blocks (with the number of letters in common with each defining contrast appearing beside each experimental condition):

Block 1		*Block 2*		*Block 3*		*Block 4*	
(1)	(0, 0)	d	(0, 1)	a	(1, 0)	c	(1, 1)
ab	(2, 0)	e	(0, 1)	b	(1, 0)	ad	(1, 1)
de	(0, 2)	ac	(2, 1)	cd	(1, 2)	ae	(1, 1)
acd	(2, 2)	bc	(2, 1)	ce	(1, 2)	bd	(1, 1)
ace	(2, 2)	abd	(2, 1)	ade	(1, 2)	be	(1, 1)
bcd	(2, 2)	abe	(2, 1)	bde	(1, 2)	abc	(3, 1)
bce	(2, 2)	acde	(2, 3)	abcd	(3, 2)	cde	(1, 3)
abde	(2, 2)	bcde	(2, 3)	abce	(3, 2)	abcde	(3, 3)

Figure 11.8 Four blocks in a 2^5 factorial experiment with defining effects *ABC* and *CDE*

The block containing (1) is called the **principal block.** Once it has been constructed, a second block can be obtained by selecting any experimental condition not in the principal block and obtaining its generalized interaction with every condition in the principal block. The other blocks are then constructed in the same way by first selecting a condition not in a block already constructed and finding generalized interactions with the principal block.

For experimental situations with $p > 3$, there is often no replication, so sums of squares associated with nonconfounded higher-order interactions are usually pooled to obtain an error sum of squares which can be used in the denominators of the various F statistics. All computations can again be carried out using Yates' technique, with *SSBl* being the sum of *SS*'s associated with confounded effects.

When $r > 2$, one first selects r defining effects to be confounded with blocks, making sure that no one of the effects chosen is the generalized interaction of any other two selected. The additional $2^r - r - 1$ effects confounded with the blocks are then the generalized interactions of all effects in the defining set (including not only generalized interactions of pairs of effects, but also of sets of three, four, and so on). For more details the book by Johnson and Leone can be consulted.

Fractional Replication

When the number of factors p is large, even a single replicate of a 2^p experiment can be expensive and time consuming. For example, one replicate of a 2^6 factorial experiment involves an observation for each of the 64 different experimental conditions. An appealing strategy in such situations is to make observations for only a fraction of the 2^p conditions. Provided that care is exercised in the choice of conditions to be observed, much information about factor effects can still be obtained.

Suppose that we decide to include only 2^{p-1} (half) of the 2^p possible conditions in our experiment; this is usually called a **half-replicate.** The price paid for this economy is twofold. First, information about a single effect (determined by the 2^{p-1} conditions selected for observation) is completely lost to the experimenter in the sense that no reasonable estimate of the effect is possible. Second, the remaining 2^{p-2} main effects and interactions are paired up so that any one effect in a particular pair is confounded with the other effect in the same pair. For example, one such pair may be $\{A, BCD\}$, so that separate estimates of the A main effect and BCD interaction are not possible. It is desirable, then, to select a half-replicate for which main effects and low-order interactions are paired off (confounded) only with higher-order interactions rather than with one another.

The first step in selecting a half-replicate is to select a defining effect as the nonestimable effect. Suppose that in a five-factor experiment, $ABCDE$ is chosen as the defining effect. Now the $2^5 = 32$ possible treatment conditions are divided into two groups with 16 conditions each, one group consisting of all conditions having an odd number of letters in common with $ABCDE$ and the other containing an even number of letters in common with the defining contrast. Then either group of 16 conditions is used as the half-replicate. The "odd" group is

$$a,\ b,\ c,\ d,\ e,\ abc,\ abd,\ abe,\ acd,\ ace,\ ade,\ bcd,\ bce,\ bde,\ cde,\ abcde$$

Each main effect and interaction other than $ABCDE$ is then confounded with (**aliased** with) its generalized interaction with $ABCDE$. Thus $(AB)(ABCDE) = CDE$, so the AB interaction and CDE interaction are confounded with each other. The resulting **alias pairs** are

$$\{A, BCDE\},\ \{B, ACDE\},\ \{C, ABDE\},\ \{D, ABCE\},\ \{E, ABCD\},\ \{AB, CDE\},$$
$$\{AC, BDE\},\ \{AD, BCE\},\ \{AE, BCD\},\ \{BC, ADE\},\ \{BD, ACE\},\ \{BE, ACD\},$$
$$\{CD, ABE\},\ \{CE, ABD\},\ \{DE, ABC\}$$

Note in particular that every main effect is aliased with a four-factor interaction. Assuming these interactions to be negligible allows for testing for the presence of main effects.

To select a quarter-replicate of a 2^p factorial experiment (2^{p-2} of the 2^p possible treatment conditions), two defining effects must be selected. These two and their generalized interaction become the nonestimable effects. Instead of alias pairs as in the half-replicate, each remaining effect is now confounded with three other effects, each being its generalized interaction with one of the three nonestimable effects.

Example 11.10 The article "More on Planning Experiments to Increase Research Efficiency" (*Industrial and Eng. Chemistry*, 1970, pp. 60–65) reported on the results of a quarter-replicate of a 2^5 experiment in which the five factors were A = condensation temperature, B = amount of material B, C = solvent volume, D = condensation time, and E = amount of material E. The response variable was the yield of the chemical process. The chosen defining contrasts were ACE and BDE, with generalized interaction $(ACE)(BDE) = ABCD$. The remaining 28 main effects and interactions can now be partitioned into seven groups of four effects each such that the effects within a group cannot be assessed separately. For example, the generalized interactions of A with the nonestimable effects are $(A)(ACE) = CE$, $(A)(BDE) = ABDE$, and $(A)(ABCD) = BCD$, so one alias group is $\{A, CE, ABDE, BCD\}$. The complete set of alias groups is

$$\{A, CE, ABDE, BCD\}, \{B, ABCE, DE, ACD\}, \{C, AE, BCDE, ABD\},$$
$$\{D, ACDE, BE, ABC\}, \{E, AC, BD, ABCDE\}, \{AB, BCE, ADE, CD\},$$
$$\{AD, CDE, ABE, BC\}$$

Analysis of a Fractional Replicate

Once the defining contrasts have been chosen for a quarter-replicate, they are used as in the discussion of confounding to divide the 2^p treatment conditions into four groups of 2^{p-2} conditions each. Then any one of the four groups is selected as the set of conditions for which data will be collected. Similar comments apply to a $1/2^r$ replicate of a 2^p factorial experiment.

Having made observations for the selected treatment combinations, a table of signs similar to Figure 11.5 is constructed. The table contains a row only for each of the treatment combinations actually observed rather than the full 2^p rows, and there is a single column for each alias group (since each effect in the group would have the same set of signs for the treatment conditions selected for observation). The signs in each column indicate as usual how contrasts for the various sums of squares are computed. Yates' method can also be used, but the rule for arranging observed conditions in standard order must be modified.

The difficult part of a fractional replication analysis typically involves deciding what to use for error sum of squares. Since there will usually be no replication (though one could observe, for example, two replicates of a quarter-replicate), some effect SS's must be pooled to obtain an error sum of squares. In a half-replicate of a 2^8 experiment, for example, an alias structure can be chosen so that the eight main effects and 28 two-factor interactions are each confounded only with higher-order interactions and that there are an additional 27 alias groups involving only higher-order interactions. Assuming the absence of higher-order interaction effects, the resulting 27 SS's can then be added to yield an error sum of squares, allowing one-degree-of-freedom tests for all main-effects and two-factor interactions. However, in many cases tests for main effects can be obtained only by pooling some or all of the SS's associated with alias groups involving two factor interactions, while the corresponding two-factor interactions cannot be investigated.

**Example 11.10
(continued)** The set of treatment conditions chosen and resulting yields for the quarter-replicate of the 2^5 experiment were

e	ab	ad	bc	cd	ace	bde	$abcde$
23.2	15.5	16.9	16.2	23.8	23.4	16.8	18.1

The abbreviated table of signs is

	A	B	C	D	E	AB	AD
e	−	−	−	−	+	+	+
ab	+	+	−	−	−	+	−
ad	+	−	−	+	−	−	+
bc	−	+	+	−	−	−	+
cd	−	−	+	+	−	+	−
ace	+	−	+	−	+	−	−
bde	−	+	−	+	+	−	−
$abcde$	+	+	+	+	+	+	+

Figure 11.9

With *SSA* denoting the sum of squares for effects in the alias group {A, CE, $ABDE$, BCD},

$$SSA = \frac{(-23.2 + 15.5 + 16.9 - 16.2 - 23.8 + 23.4 - 16.8 + 18.1)^2}{8}$$

$$= 4.65$$

Similarly, $SSB = 53.56$, SSC $= 10.35$, SSD $= .91$, $SSE' = 10.35$ (the ′ differentiates this quantity from error sum of squares SSE), $SSAB = 6.66$, and $SSAD = 3.25$, giving $SST = 4.65 + 53.56 + \ldots + 3.25 = 89.73$. To test for main effects, we use $SSE = SSAB + SSAD = 9.91$ with two degrees of freedom.

Table 11.15

Source	d.f.	Sum of squares	Mean square	F
A	1	4.65	4.65	.94
B	1	53.56	53.56	10.80
C	1	10.35	10.35	2.09
D	1	.91	.91	.18
E	1	10.35	10.35	2.09
Error	2	9.91	4.96	
Total	7	89.73		

Since $F_{.05,1,2} = 18.51$, none of the five main effects can be judged significant. Of course, with only two degrees of freedom for error, the test is not very powerful (that is, it is quite likely to fail to detect the presence of effects). The article from *Industrial and Engineering Chemistry* from which the data came actually had an independent estimate of the standard error of the treatment effects based on prior experience, so used a somewhat different analysis. Our analysis was done here only for illustrative purposes, since one would ordinarily want many more than two degrees of freedom for error.

The subjects of factorial experimentation, confounding, and fractional replication encompass many models and techniques we have not discussed. For more information, the chapter references should be consulted.

Exercises / Section 11.4

1. The accompanying data resulted from a 2^3 experiment with three replications per combination of treatments which was designed to study the effects of concentration of detergent (A), concentration of sodium carbonate (B), and concentration of sodium carboxy-methyl cellulose (C) on cleaning ability of a solution in washing tests (a large number indicates better cleaning ability than a small number).

Factor levels			Condition	Observations
A	B	C		
1	1	1	(1)	106, 93, 116
2	1	1	a	198, 200, 214
1	2	1	b	197, 202, 185
2	2	1	ab	329, 331, 307
1	1	2	c	149, 169, 135
2	1	2	ac	243, 247, 220
1	2	2	bc	255, 230, 252
2	2	2	abc	383, 360, 364

a. After obtaining cell totals $x_{ijk\cdot}$, compute estimates of β_1, γ_{11}^{AC}, and γ_{21}^{AC}.

b. Use the cell totals along with Yates' method to compute the effect contrasts and sums of squares. Then construct an ANOVA table and test all appropriate hypotheses using $\alpha = .05$.

2. In a study of processes used to remove impurities from cellulose goods (Optimization of Rope-Range Bleaching of Cellulosic Fabrics," *Textile Research J.*, 1976, pp. 493–496), the following data resulted from a 2^4 experiment involving the desizing process. The four factors were enzyme concentration (A), pH (B), temperature (C), and time (D).

a. Use Yates' algorithm to obtain sums of squares and the ANOVA table.

b. Do there appear to be any second, third, or fourth-order interaction effects present? Explain your reasoning. Which main effects appear to be significant?

Treat-ment	Enzyme, g/l	pH	Temp, °C	Time, h	Starch % by weight 1st repl.	2nd repl.
(1)	.50	6.0	60.0	6	9.72	13.50
a	.75	6.0	60.0	6	9.80	14.04
b	.50	7.0	60.0	6	10.13	11.27
ab	.75	7.0	60.0	6	11.80	11.30
c	.50	6.0	70.0	6	12.70	11.37
ac	.75	6.0	70.0	6	11.96	12.05
bc	.50	7.0	70.0	6	11.38	9.92
abc	.75	7.0	70.0	6	11.80	11.10
d	.50	6.0	60.0	8	13.15	13.00
ad	.75	6.0	60.0	8	10.60	12.37
bd	.50	7.0	60.0	8	10.37	12.00
abd	.75	7.0	60.0	8	11.30	11.64
cd	.50	6.0	70.0	8	13.05	14.55
acd	.75	6.0	70.0	8	11.15	15.00
bcd	.50	7.0	70.0	8	12.70	14.10
$abcd$	.75	7.0	70.0	8	13.20	16.12

3. In Exercise 1, suppose that a low water temperature has been used to obtain the data. The entire experiment is then repeated with a higher water temperature to obtain the following data. Use Yates' algorithm on the entire set of 48 observations to obtain the sums of squares and ANOVA table, and then test appropriate hypotheses at level .05.

Condition	Observations
d	144, 154, 158
ad	239, 227, 244
bd	232, 242, 246
abd	364, 362, 346
cd	194, 162, 203
acd	284, 295, 291
bcd	291, 287, 297
abcd	411, 406, 395

4. The following data on power consumption in electric furnace heats (kilowatts consumed per ton of melted product) resulted from a 2^4 factorial experiment with three replicates ("Studies on a 10-cwt Arc Furnace," *J. Iron and Steel Institute*, 1956, p. 22). The factors were nature of roof A (low, high), power setting B (low, high), scrap used C (tube, plate), and charge D (700 lb, 1000 lb).

Treatment	x_{ijklm}	Treatment	x_{ijklm}
(1)	866, 862, 800	d	988, 808, 650
a	946, 800, 840	ad	966, 976, 876
b	774, 834, 746	bd	702, 658, 650
ab	709, 789, 646	abd	784, 700, 596
c	1017, 990, 954	cd	922, 808, 868
ac	1028, 906, 977	acd	1056, 870, 908
bc	817, 783, 771	bcd	798, 726, 700
abc	829, 806, 691	abcd	752, 714, 714

Construct the ANOVA table and test all hypotheses of interest using $\alpha = .01$.

5. The article "Statistical Design and Analysis of Qualification Test Program for a Small Rocket Engine" *(Industrial Quality Control, 1964, pp. 14–18)* presented data from an experiment to assess the effects of vibration (A), temperature cycling (B), altitude cycling (C), and temperature for altitude cycling and firing (D) on thrust duration. A subset of the data appears below (in the paper there were four levels of D rather than just two). Use the Yates method to compute sums of squares and the ANOVA table. Then assume that three- and four-factor interactions are absent, pool the corresponding sums of squares to obtain an estimate of σ^2, and test all appropriate hypotheses at level .05.

		D_1		D_2	
		C_1	C_2	C_1	C_2
A_1	B_1	21.60	21.60	11.54	11.50
	B_2	21.09	22.17	11.14	11.32
A_2	B_1	21.60	21.86	11.75	9.82
	B_2	19.57	21.85	11.69	11.18

6. **a.** In a 2^4 experiment, suppose that two blocks are to be used, and it is decided to confound the $ABCD$ interaction with the block effect. Which treatments should be carried out in the first block [the one containing the treatment (1)], and which treatments are allocated to the second block?

b. In an experiment to investigate niacin retention in vegetables as a function of cooking temperature (A), sieve size (B), type of processing (C), and cooking time (D), each factor was held at two levels. Two blocks were used, with the allocation of blocks as given in (a) in order to confound only the $ABCD$ interaction with blocks. Use Yates' procedure to obtain the ANOVA table for the accompanying data.

Treatment	x_{ijkl}	Treatment	x_{ijkl}
(1)	91	d	72
a	85	ad	78
b	92	bd	68
ab	94	abd	79
c	86	cd	69
ac	83	acd	75
bc	85	bcd	72
abc	90	abcd	71

c. Assume that all three-way interaction effects are absent, so that the associated sums of squares can be combined to yield an estimate of σ^2, and carry out all appropriate tests at level .05.

7. **a.** An experiment was carried out to investigate the effects on audio sensitivity of varying resistance (A), two capacitances (B, C), and inductance of a coil (D) in part of a television circuit. If four blocks were used with four treatments per block, and the defining effects for confounding were AB and CD, which treatments appeared in each block?

b. Suppose that two replications of the experiment described in (a) were performed, resulting in the accompanying data. Obtain the ANOVA table and test all relevant hypotheses at level .01.

Treatment	x_{ijkl1}	x_{ijkl2}	Treatment	x_{ijkl1}	x_{ijkl2}
(1)	618	598	d	598	585
a	583	560	ad	587	541
b	477	525	bd	480	508
ab	421	462	abd	462	449
c	601	595	cd	603	577
ac	550	589	acd	571	552
bc	505	484	bcd	502	508
abc	452	451	abcd	449	455

8. In an experiment involving four factors A, B, C, and D and four blocks, show that at least one main-effect or two-factor interaction effect must be confounded with the block effect.

9. **a.** In a seven-factor experiment $(A, \ldots, G)$, suppose that a quarter-replicate is actually carried out. If the defining effects are $ABCDE$ and $CDEFG$, what is the third nonestimable effect and what treatments are in the group containing (1)? What are the alias groups of the seven main effects?
   b. If the quarter-replicate is to be carried out using four blocks (with eight treatments per block), what are the blocks if the chosen confounding effects are ACF and BDG?

10. Suppose that in the rocket thrust problem of Exercise 5, enough resources had been available for only a half-replicate of the 2^4 experiment.
    a. If the effect $ABCD$ is chosen as the defining effect for the replicate and the group of eight treatments for which data is obtained includes treatment (1), what other treatments are in the observed group and what are the alias pairs?
    b. Suppose that the results of carrying out the experiment as described in (a) are as recorded below (given in standard order after deleting the half not observed). Assuming that two and three factor interactions are negligible, test at level .05 for the presence of main effects.

    19.09, 20.11, 21.66, 20.44, 13.72, 11.26, 11.72, 12.29

11. A half-replicate of a 2^5 experiment to investigate the effects of heating time (A), quenching time (B), drawing time (C), position of heating coils (D), and measurement position (E) on hardness of steel castings resulted in the accompanying data. Construct the ANOVA table and (assuming second- and higher-order interactions to be negligible) test at level .01 for the presence of main effects.

Treatment:	a	b	c	d	e	abc	abd	abe
Observation:	70.4	72.1	70.4	67.4	68.0	73.8	67.0	67.8
Treatment:	acd	ace	ade	bcd	bce	bde	cde	abcde
Observation:	66.6	67.5	64.0	66.8	70.3	67.9	65.9	68.0

Bibliography

Bowker, Albert and Lieberman, Gerald, *Engineering Statistics* (2nd ed.), Prentice-Hall, Englewood Cliffs, N.J., 1972. A good discussion of power, mixed effects, and random effects.

Box, George, Hunter, William, and Hunter, Stuart, *Statistics for Experimenters*, John Wiley, New York, 1978. Contains a wealth of suggestions and insights on data analysis based on the authors' extensive consulting experience.

Dunn, Olive Jean and Clark, Virginia, *Applied Statistics: Analysis of Variance and Regression*, John Wiley, New York, 1974. See the Chapter 10 bibliography.

Johnson, Norman and Leone, Frederick, *Statistics and Experimental Design in Engineering and the Physical Sciences*, vol. II (2nd ed.), John Wiley, New York, 1977. Somewhat difficult to read because of awkward notation, but contains much information on various experimental designs.

Kleinbaum, David and Kupper, Lawrence, *Applied Regression Analysis and Other Multivariable Methods*, Duxbury Press, Boston, 1978. Contains an especially good discussion of problems associated with analysis of "unbalanced data"—that is, unequal K_{ij}'s.

Mendenhall, William, *Introduction to Linear Models and the Design and Analysis of Experiments,* Wadsworth, Belmont, Ca., 1968. A broad and informative survey of experimental designs which focuses more on an intuitive understanding than on mathematical development.

Neter, John and Wasserman, William, *Applied Linear Statistical Models,* Richard D. Irwin, Homewood, Ill., 1974. See the Chapter 10 bibliography.

Ott, Lyman, *An Introduction to Statistical Methods and Data Analysis,* Duxbury Press, Boston, 1978. See the Chapter 10 bibliography.

Steel, Robert and Torrie, James, *Principles and Procedures of Statistics* (2nd ed.), McGraw-Hill, New York, 1979. A good reference book for those interested only in methods.

Walpole, Ronald and Myers, Raymond, *Probability and Statistics for Engineers and Scientists* (2nd ed.), Macmillan, New York, 1978. See the Chapter 10 bibliography.

Wine, Lowell, *Statistics for Scientists and Engineers,* Prentice-Hall, Englewood Cliffs, N.J., 1964. A good reference on mixed and random effects models.

CHAPTER 12

Simple Linear Regression and Correlation

Introduction

In the two-sample problems discussed in earlier chapters, we were interested in comparing values of parameters for the x distribution and the y distribution. Even when observations were paired, we did not try to use information about one of the variables in studying the other variable. This is precisely the objective of regression analysis: to exploit the relationship between two (or more) variables so that we can gain information about one of them through knowing values of the other(s).

Much of mathematics is devoted to studying variables which are *deterministically* related. Saying that x and y are related in this manner means that once we are told the value of x, the value of y is completely specified. For example, suppose we decide to rent a car for a weekend and that the rental cost is $15.00 plus $.10 per mile driven. If we let x = the number of miles driven and y = the amount we will pay the rental agency, then $y = 15 + .1x$. If we drive the car 100 miles ($x = 100$), then $y = 15 + .1(100) = 25$. As another example, if the initial velocity of a particle is v_o and it undergoes constant acceleration a, then distance traveled = $y = v_o x + \frac{1}{2}ax^2$ where x = time.

There are many variables x and y which would appear to be related to one another, but not in a deterministic fashion. A familiar example to many students is given by variables x = high school grade point average and y = college grade point average. The value of y cannot be determined just from knowledge of x, and two different students could have the same x value but have very different y values. Yet there is a tendency for those students who have high (low) high school g.p.a.'s to also have high (low) college g.p.a.'s. Knowledge of a student's high school g.p.a. should be quite helpful in enabling us to predict how that person will do in college.

Other examples of variables related in a nondeterministic fashion include x = age of a child and y = size of that child's vocabulary, x = size of an engine in cubic centimeters and y = miles per gallon for an automobile equipped with that engine, and x = applied tensile force and y = amount of elongation in a metal strip. Many other examples will undoubtedly occur to the reader in connection with his or her own discipline.

Regression analysis is the part of statistics which deals with investigation of the relationship between two or more variables related in a nondeterministic fashion. In this chapter we generalize the linear relation $y = \beta_0 + \beta_1 x$ to a linear probabilistic relationship, develop procedures for making inferences about the parameters of the model, and obtain a quantitative measure (the correlation coefficient) of the extent to which the two variables are related. In the next chapter we consider techniques for validating a particular model and investigate nonlinear relationships and relationships involving more than two variables.

12.1 A Simple Linear Probabilistic Model and the Principle of Least Squares

The simplest deterministic mathematical relationship between two variables x and y is a linear relationship $y = \beta_0 + \beta_1 x$. The set of pairs (x, y) for which $y = \beta_0 + \beta_1 x$ determines a straight line with slope β_1 and y intercept β_0.* The objective of this section is to develop a linear probabilistic model and show how the parameters of the model can be estimated.

If the two variables are not deterministically related, then for a fixed value of x the value of the second variable is random. For example, if we are investigating the relationship between age of child and size of vocabulary and decide to select a child of age $x = 5.0$ years, then before the selection is made, vocabulary size is a random variable Y. After a particular five-year-old child has been selected and tested, a vocabulary of 2000 words may result. We would then say that the observed value of Y associated with fixing $x = 5.0$ was $y = 2000$.

More generally the variable whose value is fixed by the experimenter will be denoted by x and will be called the **independent variable**. For fixed x the second variable will be random; we denote this random variable and its observed value by Y and y, respectively, and refer to it as the **dependent variable**.

Usually observations will be made for a number of settings of the independent variable. Let $x_1, x_2, \ldots, x_n$ denote values of the independent variable for which observations are made, and let Y_i and y_i respectively denote the random variable and observed value associated with x_i. The available data then consists of the n pairs $(x_1, y_1), (x_2, y_2), \ldots, (x_n, y_n)$.

Example 12.1 The paper "A Study of Stainless Steel Stress-Corrosion Cracking by Potential Measurements" (*Corrosion*, 1962, pp. 425–432) reported on the relationship between applied stress (the independent variable x, in kg/mm^2) and time to fracture (the dependent variable y, in hours) for 18-8 stainless steel under uniaxial tensile stress in

---

*The slope of a line is the change in y for a one-unit increase in x. For example, if $y = -3x + 10$, then y decreases by 3 when x increases by 1, so the slope is -3. The y intercept is the height at which the line crosses the vertical axis, and is obtained by setting $x = 0$ in the equation.

a 40% $CaCl_2$ solution at 100°C. Ten different settings of applied stress were used, and the resulting data values (as read from a graph which appeared in the paper) were

i :	1	2	3	4	5	6	7	8	9	10
x_i:	2.5	5	10	15	17.5	20	25	30	35	40
y_i:	63	58	55	61	62	37	38	45	46	19

In Example 12.1, the value of x_i increased as i increased, but this need not be the case. Furthermore, the values of the x_i's need not all be distinct. The investigators might have made several independent determinations of fracture time for a stress of 15 kg/mm^2.

A first step in regression analysis involving two variables is to construct a **scatter plot** of the observed pairs (x_i, y_i). Figure 12.1 pictures a scatter plot of the data from Example 12.1, and suggests that no simple curve will pass through all points. There is, however, a strong tendency for a large value of stress to yield a small value of time to fracture, which suggests a relationship of some sort.

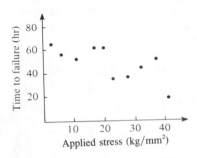

Figure 12.1 Scatter plot for data of Example 12.1

The Linear Probabilistic Model

For the deterministic model $y = \beta_0 + \beta_1 x$, the actual observed value of y is a linear function of x. The appropriate generalization of this to a probabilistic model assumes that *the expected value of Y is a linear function of x*, but that for fixed x, the variable Y differs from its expected value by a random amount.

Assumption 1: There exist parameters β_0 and β_1 such that for any fixed value of the independent variable x

$$Y = \beta_0 + \beta_1 x + \epsilon \tag{12.1}$$

where ϵ is a random variable with $E(\epsilon) = 0$ and $\text{Var}(\epsilon) = \sigma^2$.

For fixed x, let $\mu_{Y \cdot x}$ [or alternatively $E(Y|x)$] denote the expected value of Y.

If we think of a population of (x, y) pairs, then $\mu_{Y \cdot x}$ is the population mean of Y taken over all pairs having x as the value of the independent variable. If x = age and y = vocabulary size, then $\mu_{Y \cdot 5}$ is the average vocabulary size for all five-year-old children in the population. Similarly, let $\sigma^2_{Y \cdot x}$ denote the variance of Y in (12.1) for fixed x. Then

$$\mu_{Y \cdot x} = \beta_0 + \beta_1 x \quad \text{and} \quad \sigma^2_{Y \cdot x} = \sigma^2 \qquad (12.2)$$

The first part of expression (12.2) says precisely that the expected value of Y is a linear function of x, while the second part says that the spread of the distribution of Y about its expected value is the same for each x. Figure 12.2 illustrates model Assumption 1; the distribution of the random deviation ϵ is pictured in (a), and in (b) this distribution is turned on its side and centered at height $\beta_0 + \beta_1 x$ for three different x values.

The line whose equation is $y = \beta_0 + \beta_1 x$ will be called the **true regression line**, and β_0 and β_1 will be called the parameters of the true regression line. The interpretation of β_1 is that it equals the *expected change* (rather than the actual change) in Y for a one-unit increase in the value of x.

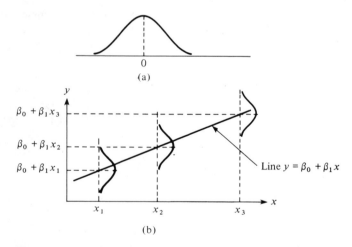

Figure 12.2   (a) Distribution of ϵ; (b) distribution of y for different values of x

---

Example 12.1
(continued)

Suppose that there is a linear probabilistic relationship between applied stress x and time to failure y, and that the equation of the true regression line is $y = 75 - 1.4x$. Then for an applied stress of $x = 25$ kg/mm², we expect a time to failure of $75 - (1.4)(25) = 40$ hr. The expected change in time to failure for a one-unit increase in applied stress is $\beta_1 = -1.4$, so we expect a 1.4-hr decrease for each unit increase in stress. This is illustrated in Figure 12.3.

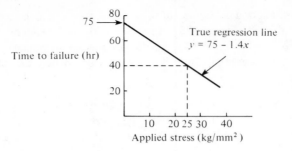

Figure 12.3 Use of the true regression line

The value $\beta_1 = 0$ postulates that $\mu_{Y \cdot x} = \beta_0$ independent of x, so that there is no linear relationship between the two variables.

Estimating the Regression Line

Before making inferences about the regression parameters β_0 and β_1, we must specify how the observed pairs $(x_1, y_1), \ldots, (x_n, y_n)$ are obtained from the model of Assumption 1.

Assumption 2: For fixed x values $x_1, x_2, \ldots, x_n$, the observations $y_1, \ldots, y_n$ are observed values of random variables $Y_1, Y_2, \ldots, Y_n$ generated independently by the model of Assumption 1. That is,

$$Y_1 = \beta_0 + \beta_1 x_1 + \epsilon_1$$
$$Y_2 = \beta_0 + \beta_1 x_2 + \epsilon_2$$
$$\vdots \qquad \vdots$$
$$Y_n = \beta_0 + \beta_1 x_n + \epsilon_n$$

where the ϵ_i's are n independent random deviations drawn from the same probability distribution having mean zero and variance σ^2.

Assumption 2 says that the pairs $(x_1, Y_1), \ldots, (x_n, Y_n)$ are distributed about the true regression line in a random manner. If the variance σ^2 is relatively large, then many observed pairs will lie far from the true regression line, but if σ^2 is relatively small, the pairs will cluster quite closely about the line $y = \beta_0 + \beta_1 x$. The limiting case is $\sigma^2 = 0$, so that all pairs lie exactly on the line—a deterministic relationship. Figure 12.4 illustrates a typical configuration of observed pairs for (a) small σ^2, and

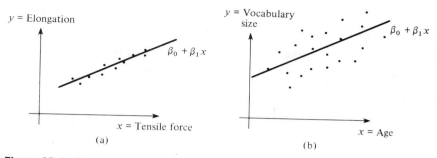

Figure 12.4 Typical sample for (a) small σ^2, (b) large σ^2

(b) large σ^2. In case (a), we gain a great deal of information about Y from knowing x, while in case (b) rather little is gained.

In practice the values of the parameters β_0 and β_1 will not be known, so the equation of the true regression line is also unknown. The data available for estimating β_0 and β_1 consists of the n observed pairs $(x_1, y_1), \ldots, (x_n, y_n)$. According to our linear probabilistic model, the observed points will be distributed about the true regression line in a random manner. Figure 12.5 shows a typical plot of observed pairs along with two candidates for the estimated regression line, $y = a_0 + a_1 x$ and $y = b_0 + b_1 x$. Intuitively the line $y = a_0 + a_1 x$ is not a reasonable estimate of the true line $y = \beta_0 + \beta_1 x$, since if $y = a_0 + a_1 x$ were the true line, the observed points would almost surely have been closer to this line. The line $y = b_0 + b_1 x$ is a more plausible estimate, since the observed points are scattered rather closely about this line.

Figure 12.5 and the foregoing discussion suggest that our estimate of $y = \beta_0 + \beta_1 x$ should be a line which provides in some sense a best fit to the observed data points. This is what motivates the principle of least squares, which can be traced back to the German mathematician Gauss (1777–1855). According to this principle, a line provides a good fit to the data if the vertical distances (deviations) from the observed points to the line are small (see Figure 12.6). The measure of the goodness of fit is the sum of the squares of these deviations. The best-fit line is then the one having the smallest possible sum of squared deviations.

Figure 12.5 Two different estimates of the true regression line

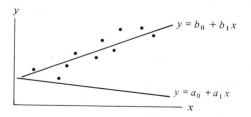

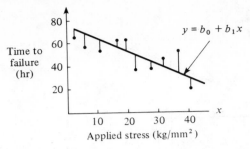

Figure 12.6 Deviations of Example 12.1 data from line $y = b_0 + b_1 x$

Principle of least squares: Among all straight lines $y = b_0 + b_1 x$, the **least squares line** or **estimated regression line** $y = \hat{\beta}_0 + \hat{\beta}_1 x$ is that line which minimizes the sum of squared deviations

$$f(b_0, b_1) = \sum_{i=1}^{n} [y_i - (b_0 + b_1 x_i)]^2 \tag{12.3}$$

That is, $f(\hat{\beta}_0, \hat{\beta}_1) \le f(b_0, b_1)$ for every b_0 and b_1.

The values of b_0 and b_1 for which (12.3) is minimized are $\hat{\beta}_0$ and $\hat{\beta}_1$, which are the point estimates of β_0 and β_1. Using the calculus to obtain the values of b_0 and b_1 which minimize $f(b_0, b_1)$ results in the system of **normal equations**

$$nb_0 + \left(\sum x_i\right)b_1 = \sum y_i$$

$$\left(\sum x_i\right)b_0 + \left(\sum x_i^2\right)b_1 = \sum x_i y_i \tag{12.4}$$

Derivation of the normal equations: To find the minimizing values of b_0 and b_1, take the partial derivative of $f(b_0, b_1)$ with respect to both b_0 and b_1 and set them both equal to zero (analogously to $f'(x) = 0$ in one-variable calculus):

$$\frac{\partial f(b_0, b_1)}{\partial b_0} = \sum 2(y_i - b_0 - b_1 x_i)(-1) = 0$$

$$\frac{\partial f(b_0, b_1)}{\partial b_1} = \sum 2(y_i - b_0 - b_1 x_i)(-x_i) = 0$$

Cancellation of the -2 factor and rearrangement gives the system (12.4).

The normal equations are linear in the unknowns b_0 and b_1; provided that not all x_i's are equal, the system has the unique solution

$$b_1 = \hat{\beta}_1 = \frac{\Sigma(x_i - \bar{x})(y_i - \bar{y})}{\Sigma(x_i - \bar{x})^2} = \frac{n\Sigma x_i y_i - (\Sigma x_i)(\Sigma y_i)}{n\Sigma x_i^2 - (\Sigma x_i)^2}$$

(12.5)

$$b_0 = \hat{\beta}_0 = \frac{\Sigma y_i - \hat{\beta}_1 \Sigma x_i}{n}$$

The expresssion for $\hat{\beta}_1$ on the right in (12.5) should be used for computation; it requires only the summary statistics Σx_i, Σy_i, Σx_i^2, $\Sigma x_i y_i$ (Σy_i^2 will be needed shortly), and minimizes the effects of roundoff. In computing $\hat{\beta}_0$, use extra digits in $\hat{\beta}_1$, since if Σx_i is large in magnitude rounding will affect the final answer. We emphasize that *before using (12.5), the scatter plot should be examined to see if a linear probabilistic model is plausible*. If the points do not tend to cluster about a straight line with roughly the same degree of spread for all x, other models should be investigated.

Example 12.1
(continued)

$n = 10$, $\Sigma x_i = 200$, $\Sigma x_i^2 = 5412.5$, $\Sigma y_i = 484$, and $\Sigma x_i y_i = 8407.5$, so

$$\hat{\beta}_1 = \frac{(10)(8407.5) - (200)(484)}{(10)(5412.5) - (200)^2} = \frac{-12725}{14125} = -.900885$$

and

$$\hat{\beta}_0 = \frac{484 - (-.900885)(200)}{10} = 66.417699$$

The equation of the estimated regression line is $y = 66.42 - .901x$. To obtain an estimate for expected time to failure when applied stress is 22.5, substitute $x = 22.5$ into the estimated regression equation; the resulting estimate is $66.42 - .901(22.5) = 46.1$. Note that the estimated expected failure time for $x = 30$ is 39.4, while the observed y for $x = 30$ was 45; these two values are not equal because the least squares line does not pass through the observed point (30, 45).

Whenever possible, it is preferable to do regression analysis using a standard statistical computer program package; in addition to $\hat{\beta}_0$ and $\hat{\beta}_1$, the resulting output will yield much more useful information. Typical output appears in the next chapter in the context of multiple regression.

Example 12.2

The paper "Root Regeneration and Early Growth of Red Oak Seedlings: Influence of Soil Temperature" (*Forest Science*, 1970, pp. 442–446) reported the results of a regression analysis in which the independent variable x was daily degree hours of soil heat and the dependent variable y was shoot elongation per seedling (cm). The accompanying data values were read from a graph which appeared in the paper.

x:	300	350	400	400	450	450	480	480
y:	5.8	4.5	5.9	6.2	6.0	7.5	6.1	8.6
x:	530	530	580	580	620	620	670	700
y:	8.9	8.2	14.2	11.9	11.1	11.5	14.5	14.8

Estimate the true average amount of shoot elongation when degree hours of soil heat equals 500 and when it equals 600.

To obtain the desired estimates, we first need $\hat{\beta}_0$ and $\hat{\beta}_1$. The summary statistics are $\Sigma x_i = 8140$, $\Sigma x_i^2 = 4,340,600$, $\Sigma y_i = 145.7$, and $\Sigma x_i y_i = 79,574$. Then

$$\hat{\beta}_1 = \frac{(16)(79,574) - (8140)(145.7)}{(16)(4,340,600) - (8140)^2} = \frac{87,186}{3,190,000} = .02733$$

$$\hat{\beta}_0 = \frac{145.7 - (.02733)(8140)}{16} = -4.79789$$

The estimated regression equation is $y = -4.798 + .0273x$, so the estimates of expected shoot elongation when $x = 500$ and $x = 600$ are $\hat{\mu}_{Y \cdot 500} = \hat{\beta}_0 + 500\,\hat{\beta}_1 = 8.87$ and $\hat{\mu}_{Y \cdot 600} = \hat{\beta}_0 + 600\,\hat{\beta}_1 = 11.60$, respectively.

Notice what happens in Example 12.2 if we try to estimate expected shoot elongation when $x = 150$: $\hat{\mu}_{Y \cdot 150} = \hat{\beta}_0 + 150\,\hat{\beta}_1 = -.69$, which is a ridiculous estimate. This is an extreme illustration of the **danger of extrapolation**—trying to make an inference for an x value well outside the range of x's for which observations have been made. The data provides no information about the nature of $\mu_{Y \cdot x}$ for such x's, and in the above example $\mu_{Y \cdot x}$ cannot reasonably be assumed to be linear for all $x > 0$.

Further Comments: Terminology and Scope of Regression Analysis

The phrase "regression analysis" was first used by Francis Galton in the late nineteenth century in connection with his work on the relationship between father's height x and son's height y. After collecting a number of pairs (x_i, y_i), Galton used the principle of least squares to obtain the equation of the estimated regression line with the objective of using it to predict son's height from father's height. In using the derived line, Galton found that if a father was above average in height, the son would also be expected to be above average in height, *but not by as much as the father was*. Similarly, the son of a shorter than average father would also be expected to be shorter than average, but not by as much as the father. Thus the predicted height of a son was "pulled back in" toward the mean; since "regression" means a coming or going back, Galton adopted the terminology "regression line." This phenomenon of being pulled back in toward the mean has been observed in many other situations (for example, batting averages from year to year in baseball), and is called **the regression effect**.

In the examples presented at the beginning of the section, the value of the independent variable was fixed, and only the dependent variable Y was random. This was not, however, the case with Galton's experiment; fathers' heights were not preselected, but instead both X and Y were random. Methods and conclusions of regression analysis can be applied both when the values of the independent variable are fixed in advance and when they are random variables, but because the derivations and interpretations are more straightforward for the fixed x case, we shall continue to work explicitly with it. For more commentary, see the excellent book by Neter and Wasserman listed in the bibliography.

Exercises / Section 12.1

1. The article "Some Field Experience in the Use of an Accelerated Method in Estimating 28-Day Strength of Concrete" (*J. Amer. Concrete Institute*, 1969, p. 895) considered regressing y = 28-day standard-cured strength (psi) against x = accelerated strength (psi). Suppose that the equation of the true regression line is $y = 1800 + 1.3x$.

   a. What is the expected value of 28-day strength when accelerated strength = 2500?

   b. By how much can we expect 28-day strength to change when accelerated strength increases by 1 psi?

   c. Answer (b) for an increase of 100 psi.

   d. Answer (b) for a decrease of 100 psi.

   e. Suppose that the error term ϵ has a normal distribution with $\sigma = 350$ psi. When accelerated strength is 2000, what is the probability that 28-day strength will exceed 5000 psi?

2. In an experiment to study the relationship between age x and vocabulary size y for young children, 10 children of particular ages were selected and vocabulary size (measured by testing) was determined for each child, resulting in the following data.

   x: 1.5 2.0 2.5 3.0 3.5 4.0 4.5 5.0 5.5 6.0
   y: 100 250 460 890 1210 1530 1840 2060 2300 2500

   (Data based on an article which appeared in *Scientific Amer.*, November 1978).

   a. Construct a scatter plot of the data (y versus x) to see that a linear probabilistic relationship is plausible.

   b. Verify that the summary statistics are
   $\Sigma x_i = 37.5$, $\Sigma y_i = 13,140$, $\Sigma x_i^2 = 161.25$,
   $\Sigma x_i y_i = 61,055$, and $\Sigma y_i^2 = 24,050,400$.

   c. Use the results of (b) to compute the equation of the least squares line.

   d. From (c), what would your estimate be for true average vocabulary size of all four-year-old children in the population sampled?

   e. If you were to select at random a three-year-old child from the population, what would be your prediction for vocabulary size of the selected child?

   f. By how much would you estimate that vocabulary size would change with a one-year age increase (within the age limits for the given data)?

   g. What happens if you use the estimated relationship to predict vocabulary size for a one-year-old child? Is this reasonable?

   h. Is it reasonable to suppose that y is linearly related to x for x values between 1.5 and 21? Discuss.

3. The March 29, 1975, issue of *Lancet* reported on the relationship between ages of a number of children who had high levels of lead absorption and a measure of wrist flexor and extensor muscle function for these children ("Neuropsychological Dysfunction in Children with Chronic Low-Level Lead Absorption"). The measure involved the number of taps with a stylus on a single metal plate during a 10-sec period. Representative data follows, with x = age in months and y = taps/10 sec.

   x: 73 84 98 112 116 132 150 160 164 180
   y: 35 40 50 42 46 41 52 52 51 66

   a. Write the system of normal equations (12.4) for this data.

   b. Compute $\hat{\beta}_0$ and $\hat{\beta}_1$ using (12.5), and verify that they solve the normal equations of (a).

   c. What would be your estimate of expected change in the measure of muscle function for a one-month increase in age?

d. Answer (c) for a one-year increase in age.

e. Estimate the true average measure of muscle function for 10-year-old children in the population sampled.

f. Compute the minimizing value of the sum of squared deviations about a straight line—that is, $f(\hat{\beta}_0, \hat{\beta}_1)$ for f given by expression (12.3).

4. The following summary statistics were obtained from a study which used regression analysis to investigate the relationship between pavement deflection and surface temperature of the pavement at various locations on a state highway. Here x = temperature (°F) and y = deflection adjustment factor.

$$n = 15, \Sigma x_i = 1425, \Sigma y_i = 10.68,$$
$$\Sigma x_i^2 = 139{,}037.25, \Sigma x_i y_i = 987.645,$$
$$\Sigma y_i^2 = 7.8518$$

(Many more than 15 observations were made in the study; the reference is "Flexible Pavement Evaluation and Rehabilitation," *Transportation Eng. J.*, 1977, pp. 75–85).

a. Compute $\hat{\beta}_1$, $\hat{\beta}_0$, and the equation of the estimated regression line. Graph the estimated line.

b. What is the estimate of expected change in the deflection adjustment factor when temperature is increased by 1 °F?

c. Suppose that temperature was measured in °C rather than °F. What would be the estimated regression line? Answer (b) for an increase of 1 °C. *Hint:* °F = $\frac{9}{5}$°C + 32; now substitute for the "old x" in terms of the "new x."

d. Supposing that a 200 °F surface temperature was within the realm of possibility, would you use the estimated line of (a) to predict deflection factor for this temperature? Why or why not?

5. The following data is representative of that reported in the article "An Experimental Correlation of Oxides of Nitrogen Emissions from Power Boilers Based on Field Data" (*J. Eng. for Power*, July 1973, pp. 165–170), with x = burner area liberation rate (MBtu/hr-ft²) and y = NO$_x$ emission rate (ppm).

x:	100	125	125	150	150	200	200
y:	150	140	180	210	190	320	280
x:	250	250	300	300	350	400	400
y:	400	430	440	390	600	610	670

a. Assuming that the simple linear regression model is valid, obtain the least squares estimate of the true regression line.

b. What is the estimate of expected NO$_x$ emission rate when burner area liberation rate equals 225?

c. By how much would you expect NO$_x$ emission rate to change when burner area liberation rate is decreased by 50?

d. For each x_i used in the experiment, compute the predicted value $\hat{y}_i = \hat{\beta}_0 + \hat{\beta}_1 x_i$ ($i = 1, \ldots, 14$). Then graph $\hat{y}$ versus y (predicted versus observed). If the linear relationship were actually deterministic (so predictions were perfect), what would this plot look like? Does the plot indicate that the prediction relationship is effective for the observed data?

6. The paper "Increased Oxygen Consumption During the Uptake of Water by the Eversible Vesicles of Petrobius Brevistylis" (J. *Insect Physiology*, 1977, pp. 1285–1294) presented the results of a regression of y = increased oxygen uptake (in μl) above the mean resting rate on x = weight increase (mg) when dehydrated insects were allowed access to distilled water. A sample size of n = 20 was used, and the computed summary statistics were (approximately, based on numbers read from a graph) $\Sigma x_i = 63.5$, $\Sigma y_i = 17.26$, $\Sigma x_i^2 = 311.74$, $\Sigma x_i y_i = 71.51$, and $\Sigma y_i^2 = 19.9625$.

a. Compute the equation of the estimated regression line.

b. There was only one observation made for an x value larger than 7: for $x_{20} = 9.8$, $y_{20} = 1.9$. The investigator would like to know whether the exclusion of this point greatly alters the estimated regression relationship. Compute the estimated regression line based just on the 19 pairs with (9.8, 1.9) deleted from the sample. What y would you predict using this new line when $x = 9.8$? *Hint:* First recompute the summary statistics; for example, new $\Sigma x_i = $ old $\Sigma x_i - 9.8$.

7. The paper "Effects of Bike Lanes on Driver and Bicyclist Behavior" (*ASCE Transportation Eng. J.*, 1977, pp. 243–256) reported the results of a regression analysis with x = available travel space in feet (a convenient measure of roadway width, defined as the distance between a cyclist and the roadway center line) and separation distance y between a bike and a passing car (determined by photography). The data, for 10 streets with bike lanes, appears below.

x: 12.8 12.9 12.9 13.6 14.5 14.6 15.1 17.5 19.5 20.8
y: 5.5 6.2 6.3 7.0 7.8 8.3 7.1 10.0 10.8 11.0

a. Verify that $\Sigma x_i = 154.20$, $\Sigma y_i = 80$, $\Sigma x_i^2 = 2452.18$, $\Sigma x_i y_i = 1282.74$, and $\Sigma y_i^2 = 675.16$.

b. Derive the equation of the estimated regression line.

c. What separation distance would you predict for another street which has 15.0 as its available travel space value?

d. What would be the estimate of expected separation distance for all streets having available travel space value 15?

e. Letting Y_{new} denote the actual separation distance for a new road with travel space value 15, what is the error of prediction in part (c)?

f. Write an expression for the error of estimation in (d), and compare with the error of prediction from (e). Which entails more uncertainty, estimation or prediction? Why?

8. Show that b_1 and b_0 of Expression (12.5) satisfy the normal equations (12.4).

9. Show that the "point of averages" $(\bar{x}, \bar{y})$ lies on the estimated regression line.

10. Suppose that an investigator has data on the amount of shelf space x devoted to display of a particular product and sales revenue y for that product. Then the investigator may wish to fit a model for which the true regression line passes through $(0, 0)$. The appropriate model is $Y = \beta_1 x + \epsilon$. Assume that $(x_1, y_1), \ldots, (x_n, y_n)$ are observed pairs generated from this model, and derive the least squares estimator of β_1. *Hint*: Write the sum of squared deviations as a function of b_1, a trial value, and use calculus to find the minimizing value of b_1.

11. a. Consider the data in Exercise 2. Suppose that instead of the least squares line passing through the points $(x_1, y_1), \ldots, (x_n, y_n)$ we wish the least squares line passing through $(x_1 - \bar{x}, y_1)$, $\ldots, (x_n - \bar{x}, y_n)$. Construct a scatter plot of the (x_i, y_i) points and then of the $(\bar{x}_i - \bar{x}, y_i)$ points. Use the plots to explain intuitively how the two least squares lines are related to one another.

b. Suppose that instead of the model $Y_i = \beta_0 + \beta_1 x_i + \epsilon_i$ $(i = 1, \ldots, n)$, we wish to fit a model of the form $Y_i = \beta_0^* + \beta_1^*(x_i - \bar{x}) + \epsilon_i$ $(i = 1, \ldots, n)$. What are the least squares estimators of β_0^* and β_1^*, and how do they relate to $\hat{\beta}_0$ and $\hat{\beta}_1$?

12.2 Inferences About the Slope Parameter β_1

In Section 12.1 we used the least squares criteria to obtain point estimates for the two parameters β_0 and β_1 of the true regression line. The values of the x_i's are assumed to be chosen before the experiment is performed, so only the Y_i's are random. The estimators for β_0 and β_1 are obtained by replacing y_i by Y_i in (12.5), yielding

$$\hat{\beta}_1 = \frac{n \Sigma x_i Y_i - (\Sigma x_i)(\Sigma Y_i)}{n \Sigma x_i^2 - (\Sigma x_i)^2}$$

$$\hat{\beta}_0 = \frac{\Sigma Y_i - \hat{\beta}_1 \Sigma x_i}{n}$$

(12.6)

The Mean and Variance of $\hat{\beta}_1$

In earlier chapters when we wanted to test a hypothesis about a single parameter, the test statistic was obtained by standardizing a point estimator. To accomplish this for β_1, it is necessary to know the expected value and variance of $\hat{\beta}_1$ (because β_1 is usually of more interest than β_0, procedures for making inferences about the latter parameter will not be discussed). In (12.6) the random variables $Y_1, \ldots, Y_n$ appear only in the numerator of $\hat{\beta}_1$, while the denominator is a constant. A bit of algebraic manipulation shows that $\hat{\beta}_1$ can be written as

$$\hat{\beta}_1 = \sum_{i=1}^{n} \frac{n}{c}(x_i - \bar{x})Y_i = \sum_{i=1}^{n} c_i Y_i \quad \left[\text{where} \quad c = n\sum x_i^2 - \left(\sum x_i^2\right)\right] \quad (12.7)$$

According to (12.7), $\hat{\beta}_1$ is a linear function of the Y_i's, so that the rules of expected value and variance from Chapter 5 can be applied to $\hat{\beta}_1$. This yields

---

The expected value of $\hat{\beta}_1$ is $\mu_{\hat{\beta}_1} = E(\hat{\beta}_1) = \beta_1$, so that $\hat{\beta}_1$ is an unbiased estimator of β_1, and the variance of $\hat{\beta}_1$ is

$$\sigma_{\hat{\beta}_1}^2 = \text{Var}(\hat{\beta}_1) = \frac{\sigma^2}{\sum (x_i - \bar{x})^2} = \frac{\sigma^2}{\sum x_i^2 - (\sum x_i)^2/n} \quad (12.8)$$

According to (12.8) the variance of $\hat{\beta}_1$ equals the variance σ^2 of the random error term—or equivalently of any Y_i—divided by $\sum (x_i - \bar{x})^2$. Because $\sum (x_i - \bar{x})^2$ is a measure of how spread out the x_i's are about $\bar{x}$, we conclude that making observations at x_i values which are quite spread out results in a more precise estimator of the slope parameter (smaller variance of $\hat{\beta}_1$) whereas values of x_i all close to one another imply a highly variable estimator.

To standardize $\hat{\beta}_1$, we can now subtract its expected value β_1, which for any particular null hypothesis would be the value specified by H_0, and then divide by the square root of (12.8). Unfortunately the resulting variable cannot usually be used as a test statistic, since it involves the unknown standard deviation σ. To obtain a test statistic or confidence interval, σ^2 must be estimated.

Estimating σ^2

The parameter σ^2 is a measure of the amount of variability inherent in the regression model. A large value of σ^2 will lead to observed (x_i, y_i)'s which are quite spread out about the true regression line, whereas when σ^2 is small the observed points will tend to fall very close to the true line (see Figure 12.4).

To estimate σ^2, we consider the extent to which the points in the sample deviate from the estimated line. Many large deviations (residuals) suggest a large σ^2, while all small deviations suggest a small σ^2.

Definition: The ith **fitted** (or **predicted**) **value**, denoted by $\hat{y}_i$, is given by $\hat{y}_i = \hat{\beta}_0 + \hat{\beta}x_i$ ($i = 1, \ldots, n$), and the ith **residual** is $y_i - \hat{y}_i$.

In words, the predicted value $\hat{y}_i$ is the value of y that we would predict or expect when using the estimated regression line with $x = x_i$; $\hat{y}_i$ is the height of the estimated regression line above the value x_i for which the ith observation was made. The residual $y_i - \hat{y}_i$ is the difference between the observed y_i and the predicted $\hat{y}_i$. If the residuals are all small in magnitude, then much of the variability in observed y values appears to be due to the linear relationship between x and y, while many large residuals suggest quite a bit of inherent variability in y relative to the amount due to the linear relation. Assuming that the line in Figure 12.6 is the least squares line, the residuals are identified by the vertical line segments from the observed points to the line.

Example 12.3   An investigation of the relationship between traffic flow x (1000's of cars per 24 hours) and lead content y of bark on trees near the highway (μg/g dry wt) yielded the data in the x_i and y_i columns below:

i	x_i	y_i	$\hat{y}_i$	$y_i - \hat{y}_i$
1	8.3	227	287.48	-60.48
2	8.3	312	287.48	24.52
3	12.1	362	424.98	-62.98
4	12.1	521	424.98	96.02
5	17.0	640	602.28	37.72
6	17.0	539	602.28	-63.28
7	17.0	728	602.28	125.72
8	24.3	945	866.42	78.58
9	24.3	738	866.42	-128.42
10	24.3	759	866.42	-107.42
11	33.6	1263	1202.93	60.07

The summary statistics are $\Sigma x_i = 198.3$, $\Sigma x_i^2 = 4198.03$, $\Sigma y_i = 7034$, $\Sigma y_i^2 = 5{,}390{,}382$, and $\Sigma x_i y_i = 149{,}354.4$, so

$$\hat{\beta}_1 = \frac{11(149{,}354.4) - (198.3)(7034)}{11(4198.03) - (198.3)^2} = 36.1838, \quad \hat{\beta}_0 = -12.8416$$

The estimated regression line is $y = -12.84 + 36.18x$. For numerical accuracy, the fitted values $\hat{y}_i$ are calculated from $\hat{y}_i = -12.8416 + 36.1838x_i$. The residuals should sum to zero; because of rounding, the sum here is .05.

The estimate of σ^2 is now calculated from the residuals as follows.

Definition: The **error sum of squares**, denoted by SSE, is

$$SSE = \Sigma (y_i - \hat{y}_i)^2 = \Sigma [y_i - (\hat{\beta}_0 + \hat{\beta}_1 x_i)]^2 \tag{12.9}$$

and the estimate of σ^2 is

$$\hat{\sigma}^2 = s^2 = \frac{SSE}{n - 2} = \frac{\Sigma (y_i - \hat{y}_i)^2}{n - 2} \tag{12.10}$$

The estimate s^2 used in a regression problem is different from $s^2 = \Sigma (x_i - \bar{x})^2 / (n - 1)$ used previously to estimate a variance. In a regression context, s^2 will now refer to (12.10).

The divisor $n - 2$ is used because the two parameters β_0 and β_1 have been estimated, resulting in a loss of two degrees of freedom; the number of degrees of freedom for SSE (or $\hat{\sigma}^2$) is then $n - 2$. In fact, it can be shown that the estimator S^2 is unbiased for σ^2 (so that a divisor of n or $n - 1$ would produce a biased estimator). Since s^2 is a sum of squares divided by an appropriate number of degrees of freedom, it is often called the **mean square for error** (*MSE*).

Example 12.3
(continued)

Squaring each residual computed earlier and adding gives $SSE = \Sigma \, (y_i - \hat{y}_i)^2 = 76{,}492.03$, so $s^2 = SSE / (n - 2) = 76{,}492.03/9 = 8499.11$ and $s = \hat{\sigma} = 92.19$. From (12.8) the variance of $\hat{\beta}_1$ is estimated by $\dfrac{s^2}{\Sigma \, x_i^2 - (\Sigma \, x_i)^2/n} = \dfrac{8499.11}{623.22} = 13.64$ and $\sigma_{\hat{\beta}_1}$ by $\sqrt{13.64} = 3.69$.

A Computational Formula for s^2

It is not necessary to compute the residuals in order to obtain s^2. By expanding $SSE = \Sigma \, [(y_i - (\hat{\beta}_0 + \hat{\beta}_1 x_i)]^2$, it can be shown that

$$s^2 = \frac{\Sigma \, y_i^2 - \hat{\beta}_0 \, \Sigma \, y_i - \hat{\beta}_1 \, \Sigma \, x_i y_i}{n - 2} \qquad (12.11)$$

In using (12.11) carrying extra digits in both $\hat{\beta}_0$ and $\hat{\beta}_1$ will avoid substantial roundoff error.

Example 12.3
(continued)

$n = 11$, $\Sigma \, y_i = 7034$, $\Sigma \, y_i^2 = 5{,}390{,}382$, $\Sigma \, x_i y_i = 149{,}354.4$, $\hat{\beta}_1 = 36.1838$, and $\hat{\beta}_0 = -12.8416$, so

$$SSE = 5{,}390{,}382 - (-12.8416)(7034) - (36.1838)(149{,}354.4)$$
$$= 76{,}500.075$$
$$s^2 = \frac{76{,}500.075}{9} = 8500.01, \quad s = 92.20$$

There is a slight discrepancy between *SSE* computed here and computed earlier using the residuals, which is due to roundoff in obtaining and squaring each residual, but the two values of s differ by only .01.

Testing Hypotheses About the Slope Parameter β_1

To this point the only assumptions we have made about the probability distribution of the Y_i's concerned independence, expected values, and variances. To test hypotheses and construct confidence intervals for regression parameters, a stronger assumption about the probability distribution of each Y_i is necessary.

Assumption 3: In addition to the independence of the Y_i's, $E(Y_i) = \beta_0 + \beta_1 x_i$, and $\text{Var}(Y_i) = \sigma^2$ for all i, we now assume that each Y_i has a normal distribution.

As in previous chapters, if the sample size n is large, then Assumption 3 is unnecessary, but in any problem in which n is small, the assumption must hold at least approximately for the following inferential procedures to be valid. Using Assumption 3, result (12.8), and the estimator S of σ, the estimator $\hat{\beta}_1$ can now be appropriately standardized.

Theorem: Under Assumptions 1–3 of the previous and present section, the variable

$$T = \frac{\hat{\beta}_1 - \beta_1}{S/\sqrt{\Sigma x_i^2 - (\Sigma x_i)^2/n}} \qquad (12.12)$$

has a t distribution with $n - 2$ degrees of freedom.

This theorem leads immediately to the description of a test procedure for testing hypotheses about β_1.

Null hypothesis: $H_0: \beta_1 = \beta_{10}$

Test statistic: $T = \dfrac{\hat{\beta}_1 - \beta_{10}}{S/\sqrt{\Sigma x_i^2 - (\Sigma x_i)^2/n}}$

Alternative Hypothesis	Rejection Region for Level α Test
$H_a: \beta_1 > \beta_{10}$	$T \geq t_{\alpha,n-2}$
$H_a: \beta_1 < \beta_{10}$	$T \leq -t_{\alpha,n-2}$
$H_a: \beta_1 \neq \beta_{10}$	either $T \geq t_{\alpha/2,n-2}$ or $T \leq -t_{\alpha/2,n-2}$

The most commonly encountered pair of hypotheses about β_1 is $H_0: \beta_1 = 0$ versus $H_a: \beta_1 \neq 0$. When this H_0 is true, $\mu_{Y \cdot x} = \beta_0$ independent of x, so knowledge of x gives no information about Y.

Example 12.4 A paper in the 1943 issue of the *Journal of Experimental Psychology* reported on an experiment to study the relationship between hypnotic susceptibility and intelligence. For each of 32 subjects both an intelligence score x and a hypnotic susceptibility score y were obtained. The computed values of the summary statistics were $\Sigma x_i = 3893$, $\Sigma x_i^2 = 478{,}537$, $\Sigma y_i = 290$, $\Sigma y_i^2 = 4160$, and $\Sigma x_i y_i = 36{,}473$. Assuming the linear model is appropriate, does the data argue strongly for the existence of a positive relationship between intelligence and hypnotic susceptibility? The question suggests testing $H_0: \beta_1 = 0$ versus $H_a: \beta_1 > 0$. We need

$$\hat{\beta}_1 = \frac{n\Sigma\, x_i y_i - (\Sigma\, x_i)(\Sigma\, y_i)}{n\Sigma\, x_i^2 - (\Sigma\, x_i)^2} = \frac{32(36{,}473) - (3893)(290)}{32(478{,}537) - (3893)^2} = .2420$$

$$\hat{\beta}_0 = \frac{290 - (.2420)(3893)}{32} = -20.3783$$

$$SSE = 4160 - (-20.3783)(29) - (.242)(36{,}473) = 1243.24$$

$$s^2 = \frac{1243.24}{30} = 41.44, \quad s = 6.44$$

The computed value of T is

$$t = \frac{.242 - 0}{6.44/\sqrt{4928.83}} = 2.64$$

For $\alpha = .01$, $t_{.01,n-2} = t_{.01,30} = 2.457$; since $2.64 \geq 2.457$, the data is significant at level .01 so H_0 is rejected, and we conclude that there does appear to be a positive relationship between intelligence and hypnotic susceptibility.

In Example 12.4 the fact that $\hat{\beta}_1$ was rather small in magnitude might have suggested that the t test would judge $\hat{\beta}_1$ not significantly different from zero, while exactly the reverse happened. A value of $\hat{\beta}_1$ near zero does not necessarily imply a weak relationship between x and y. In fact, $\hat{\beta}_1$ can be made very near zero by multiplying each y by a small number c, which amounts to changing the units of measurement on y (say, from inches to miles). The new $\hat{\beta}_1$ will be c times the old one, but the new S will also be c times the old one, so the t statistic will have the same value.

Example 12.5   In anthropological studies a characteristic of fossils which is of central importance is cranial capacity. Frequently skulls are at least partially decomposed, so it is necessary to use other characteristics to obtain information about capacity. One such measurement that has been used is the length of the lambda-opisthion chord (lambda is the point at which the occipital and the left and right parietal bones meet, while the opisthion is the most posterior point on the edge of the hole in the base of the skull through which the spinal cord passes). A paper which appeared in the 1971 *American Journal of Physical Anthropology* reported the following data for $n = 7$ *Homo erectus* fossils.

x (chord length in mm):	78	75	78	81	84	86	87
y (cranial capacity in cm^3):	850	775	750	975	915	1015	1030

The summary statistics are $\Sigma\, x_i = 569$, $\Sigma\, x_i^2 = 46{,}375$, $\Sigma\, y_i = 6310$, $\Sigma\, y_i^2 = 5{,}764{,}600$, and $\Sigma\, x_i y_i = 515{,}660$.

Suppose that from previous evidence, anthropologists had believed that for each 1-mm increase in chord length, cranial capacity would be expected to increase by 20 cm^3. Does this new experimental data strongly contradict prior belief? That is, should $H_0 : \beta_1 = 20$ be rejected in favor of $H_a : \beta_1 \neq 20$?

Because of the magnitude of the numbers, preliminary rounding should be avoided. We calculate

$$\hat{\beta}_1 = \frac{7(515,660) - (569)(6310)}{7(46,375) - (569)^2} = 22.25694 \quad \hat{\beta}_0 = -907.74306$$

$$SSE = 5,764,600 - (-907.74306)(6310) - (22.25694)(515,660)$$
$$= 15,445.03$$

$$s^2 = 3089.01 \quad \text{and} \quad s = 55.58$$

This gives

$$t = \frac{22.26 - 20}{55.58/\sqrt{123.43}} = .45$$

Since for $\alpha = .05$, $t_{\alpha/2,n-2} = t_{.025,5} = 2.571$ and $.45$ is neither ≥ 2.571 nor ≤ -2.571, H_0 cannot be rejected at level $.05$.

Finally, while the test procedures for hypotheses about β_1 were derived intuitively by standardizing $\hat{\beta}_1$, it can be shown that these procedures are all likelihood ratio tests. Exercise 12.2.14 discusses computation of the power of these t tests.

A Confidence Interval for β_1

The fact that T of expression (12.12) has a t distribution with $n - 2$ degrees of freedom is the key to constructing a confidence interval for β_1. As in Chapter 9, we start with the probability statement

$$P\left(-t_{\alpha/2,n-2} \leq \frac{\hat{\beta}_1 - \beta_1}{S/\sqrt{\Sigma x_i^2 - (\Sigma x_i)^2/n}} \leq t_{\alpha/2,n-2}\right) = 1 - \alpha$$

and manipulate the inequalities to isolate β_1 in the middle. The resulting $100(1 - \alpha)\%$ *confidence interval for* β_1 is

$$\left(\hat{\beta}_1 - t_{\alpha/2,n-2} \cdot \frac{s}{\sqrt{\Sigma x_i^2 - (\Sigma x_i)^2/n}},\right.$$
$$\left. \hat{\beta}_1 + t_{\alpha/2,n-2} \cdot \frac{s}{\sqrt{\Sigma x_i^2 - (\Sigma x_i)^2/n}}\right) \quad (12.13)$$

This interval has the general form *point estimate* $\pm$ *($t_{\alpha/2}$ value)(estimated SD of estimator)* as did the t intervals for μ and $\mu_1 - \mu_2$.

Example 12.5
(continued)

In the anthropology example, β_1 was the true average increase in cranial capacity for *Homo erectus* for a 1-mm increase in lambda-opisthion chord length. With $\hat{\beta}_1 = 22.26$, $s = 55.58$, $\Sigma x_i^2 - (\Sigma x_i)^2/n = 123.43$, and $t_{.025,5} = 2.571$, the 95% confidence interval for β_1 is

$$\left(22.26 - 2.571 \cdot \frac{55.58}{\sqrt{123.43}}, \ 22.26 + 2.571 \cdot \frac{55.58}{\sqrt{123.43}} \right)$$

$$= (22.26 - 12.86, 22.26 + 12.86) = (9.40, 35.12)$$

Exercises / Section 12.2

1. a. The estimated regression line for the age/vocabulary size data of Exercise 12.1.2 is $y = -827.81 + 571.15x$. Use this to compute the fitted values and residuals.

   b. Compute SSE, s^2, and s directly from the residuals of (a).

   c. Use the shortcut formula (12.11) along with $\hat{\beta}_1 = 571.15$ to compute s^2, and then obtain s. Then repeat for $\hat{\beta}_1 = 571.1515$. Do your answers for s differ significantly for practical purposes?

2. In a study of a reactive sputtering technique for the deposit of silicon nitride films on substrate material, the following measurements were obtained on P-etch rate (y) as a function of sputtering voltage (x) ("Preparation and Properties of Reactively Sputtered Silicon Nitride," *J. Vacuum Science and Technology*, 1967, pp. 37–40).

x:	400	600	800	800	1000
y:	44.0	39.9	35.0	33.8	29.1

Using first $\hat{\beta}_1 = -.0251$ and then $\hat{\beta}_1 = -.025$, compute SSE, s^2, and s using the shortcut computational formula (12.11). Do you think that use of one rather than the other might result in a different conclusion in making an inference?

3. a. Compute s^2 and s for the temperature-deflection factor data in Exercise 12.1.4.

   b. Use the results of (a) to estimate the variance and standard deviation of $\hat{\beta}_1$.

4. An article in a recent volume of *J. Public Health Eng.* reported the results of a regression analysis based on $n = 15$ observations in which $x =$ filter application temperature (°C) and $y = \%$ efficiency of BOD removal. Calculated quantities include

$\Sigma x_i = 402$, $\Sigma x_i^2 = 11,098$, $s = 3.725$, and $\hat{\beta}_1 = 1.7035$.

   a. Test at level .01 $H_0 : \beta_1 = 1$, which states that the expected increase in % BOD removal is 1 when filter application temperature increases by 1 °C, against the alternative $H_a : \beta_1 > 1$.

   b. Compute a 99% confidence interval for β_1, the expected increase in % BOD removal for a 1 °C increase in filter application temperature.

5. The article "Hydrogen, Oxygen, and Nitrogen in Cobalt Metal" (*Metallurgia*, 1969, pp. 121–127) contained a plot of the following data pairs, where $x =$ pressure of extracted gas (microns) and $y =$ extraction time (min).

x:	40	130	155	160	260	275	325	370	420	480
y:	2.5	3.0	3.1	3.3	3.7	4.1	4.3	4.8	5.0	5.4

   a. Estimate σ and the standard deviation of $\hat{\beta}_1$.

   b. Suppose that the investigators had believed prior to the experiment that $\beta_1 = .0060$. Does the data contradict this prior belief? Test using $\alpha = .10$.

6. Use the results in Exercise 5 to compute a 95% confidence interval for the slope of the true regression line.

7. Using the data in Exercise 2, compute a 99% confidence interval for the expected change in P-etch rate when sputtering voltage is increased by 1.

8. Refer back to Exercise 1. Does the data contradict prior belief that a one-year increase in age results in an expected vocabulary increase of at most 500 words (for children whose ages are within the range studied)? Test using $\alpha = .01$.

9. Does the data in Exercise 12.1.6 strongly indicate that the linear relationship between x and y can be used for prediction of y from a knowledge of x? Test the relevant hypotheses using $\alpha = .01$.

10. Use the data in Exercise 12.1.5 to compute a 95% confidence interval for the expected change in NO_x emission rate which results from an increase of 10 MBtu/hr-ft^2 in the burner area liberation rate.

11. Use the rules of expected value to show that $\hat{\beta}_0$ is an unbiased estimator for β_0.

12. **a.** Verify that $E(\hat{\beta}_1) = \beta_1$ by using the rules of expected value from Chapter 5.
    b. Use the rules of variance from Chapter 5 to verify the expression for $\text{Var}(\hat{\beta}_1)$ given in this section.

13. Verify that if each x_i is multiplied by a positive constant c, and each y_i is multiplied by another

positive constant d, the t statistic for testing $H_0 : \beta_1 = 0$ versus $H_a : \beta_1 \neq 0$ is unchanged in value.

14. The power of the t test for $H_0 : \beta_1 = \beta_{10}$ can be computed in the same manner as power was computed for the t tests of Chapters 7 and 8. If the alternative value of β_1 is denoted by β_{11}, the value of

$$ d = \frac{|\beta_{10} - \beta_{11}|}{\sigma \sqrt{\dfrac{n-1}{\sum x_i^2 - (\sum x_i)^2 / n}}} $$

is first calculated, then the appropriate power curve of Figure 7.10 is entered on the horizontal axis at the value of d, and the power is read from the curve for $n - 1$. Use this to compute the power of the test of Exercise 4 when $\beta_{11} = 2$ and $\sigma = 4$.

12.3 Inferences Concerning $\mu_{Y \cdot x}$ and Prediction of Future Values

The linear probabilistic model specifies that for any fixed value x of the independent variable, the mean or expected value of Y is given by $\mu_{Y \cdot x} = \beta_0 + \beta_1 x$. The quantity $\mu_{Y \cdot x}$ can be thought of as the average value of Y over all population units which have x as the value of the independent variable. By substituting the least squares estimates $\hat{\beta}_0$ and $\hat{\beta}_1$ in for β_0 and β_1, we obtain either

1. the estimate of $\mu_{Y \cdot x}$, $\hat{\mu}_{Y \cdot x} = \hat{\beta}_0 + \hat{\beta}_1 x$, or
2. the predicted y, $\hat{y} = \hat{\beta}_0 + \hat{\beta}_1 x$, associated with a particular population unit having independent variable value x.

Example 12.6 For the anthropology data in Example 12.5, in which $x =$ lambda-opisthion chord length and $y =$ cranial capacity for a *Homo erectus* skull, the estimates of the regression parameters were $\hat{\beta}_0 = -907.74$ and $\hat{\beta}_1 = 22.26$. Our point estimate for $\mu_{Y \cdot 80}$, the true average cranial capacity for all *Homo erectus* skulls having chord length 80, is $\hat{\mu}_{Y \cdot 80} = \hat{\beta}_0 + 80\hat{\beta}_1 = 873.06$. Similarly, if someone presented us with a single *Homo erectus* skull having chord length 80, we would predict a cranial capacity of 873.06 for that skull.

To make other inferences about $\mu_{Y \cdot x}$ and also construct a prediction interval for a future value of Y, we need the mean, variance, and probability distribution of $\hat{\beta}_0 + \hat{\beta}_1 x$.

Properties of $\hat{\beta}_0 + \hat{\beta}_1 x$

We have already noted that $E(\hat{\beta}_0) = \beta_0$ and $E(\hat{\beta}_1) = \beta_1$ (that is, $\hat{\beta}_0$ and $\hat{\beta}_1$ are unbiased estimators). Thus

$$\mu_{\hat{\beta}_0 + \hat{\beta}_1 x} = E(\hat{\beta}_0 + \hat{\beta}_1 x) = E(\hat{\beta}_0) + E(\hat{\beta}_1)x = \beta_0 + \beta_1 x.$$

Furthermore, some algebra yields

$$\hat{\beta}_0 + \hat{\beta}_1 x = \sum_{i=1}^{n} \left[\frac{1}{n} + \frac{n(x - \bar{x})(x_i - \bar{x})}{n \sum x_i^2 - (\sum x_i)^2} \right] Y_i = \sum_{i=1}^{n} d_i Y_i$$

so that $\hat{\beta}_0 + \hat{\beta}_1 x$ is a linear function of the Y_i's. The rules of Chapter 5 can then be used to calculate $\mathrm{Var}(\hat{\beta}_0 + \hat{\beta}_1 x)$. In addition, a linear function of normal random variables also has a normal distribution.

For the linear probabilistic model with normally distributed error terms, for fixed x

1. $\hat{\beta}_0 + \hat{\beta}_1 x$ has a normal distribution
2. $E(\hat{\beta}_0 + \hat{\beta}_1 x) = \beta_0 + \beta_1 x$, so $\hat{\beta}_0 + \hat{\beta}_1 x$ is an unbiased estimator of $\beta_0 + \beta_1 x$
3. $\mathrm{Var}(\hat{\beta}_0 + \hat{\beta}_1 x) = \sigma^2 \left[\dfrac{1}{n} + \dfrac{n(x - \bar{x})^2}{n \sum x_i^2 - (\sum x_i)^2} \right]$

The variance of $\hat{\beta}_0 + \hat{\beta}_1 x$ is smallest when $x = \bar{x}$ and increases as x moves away from $\bar{x}$ in either direction. Thus our estimator of the regression function is more precise near the center of the x_i's than far from the data. Standardizing $\beta_0 + \beta_1 x$ now leads to the result which underlies further inferential and prediction procedures.

Theorem: Under Assumptions 1–3 of the previous sections, the variable

$$T = \frac{\hat{\beta}_0 + \hat{\beta}_1 x - (\beta_0 + \beta_1 x)}{S \sqrt{\dfrac{1}{n} + \dfrac{n(x - \bar{x})^2}{n \sum x_i^2 - (\sum x_i)^2}}} \qquad (12.14)$$

has a t distribution with $n - 2$ degrees of freedom.

Testing Hypotheses Concerning $\mu_{Y \cdot x}$

Replacing $\beta_0 + \beta_1 x$ in (12.14) by its hypothesized value under H_0 produces a statistic for testing various hypotheses about $\mu_{Y \cdot x}$ for a fixed value of x.

Null hypothesis: $H_0 : \beta_0 + \beta_1 x = \mu_0$

Test statistic: $\quad T = \dfrac{\hat{\beta}_0 + \hat{\beta}_1 x - \mu_0}{S \sqrt{\dfrac{1}{n} + \dfrac{n(x - \bar{x})^2}{n \sum x_i^2 - (\sum x_i)^2}}}$

Alternative Hypothesis	Rejection Region for Level α Test
$H_a : \beta_0 + \beta_1 x > \mu_0$	$T \geq t_{\alpha, n-2}$
$H_a : \beta_0 + \beta_1 x < \mu_0$	$T \leq -t_{\alpha, n-2}$
$H_a : \beta_0 + \beta_1 x \neq \mu_0$	either $T \geq t_{\alpha/2, n-2}$ or $T \leq -t_{\alpha/2, n-2}$

Example 12.7 An article "Performance Test Conducted for a Gas Air-Conditioning System" (*Amer. Soc. Heating, Refrigerating, and Air Conditioning Eng.*, October 1969, p. 54) reported the following data on maximum outdoor temperature (x) and hours of chiller operation per day (y) for a 3-ton residential gas air-conditioning system.

x:	72	78	80	86	88	92
y:	4.8	7.2	9.5	14.5	15.7	17.9

Suppose that this system is actually a prototype model, which the manufacturer does not wish to produce unless the data strongly indicates that when maximum outdoor temperature is 82, the true average number of hours of chiller operation is less than 12. The appropriate hypotheses are then $H_0 : \beta_0 + 82\beta_1 = 12$ versus $H_a : \beta_0 + 82\beta_1 < 12$. The summary statistics are $\Sigma x_i = 496$, $\Sigma y_i = 69.60$, $\Sigma x_i^2 = 41,272$, $\Sigma y_i^2 = 942.28$, and $\Sigma x_i y_i = 5942.60$, from which we calculate $\hat{\beta}_1 = .702$, $\hat{\beta}_0 = -46.42$, and $s = .755$. The computed value of T is

$$t = \frac{-46.42 + (.702)(82) - 12}{.755 \sqrt{\dfrac{1}{6} + \dfrac{6(82 - 82.67)^2}{16,158}}} = \frac{-.856}{.755 \sqrt{.169}} = -2.76$$

For $\alpha = .01$, H_0 is rejected if $T \leq -t_{.01,4} = -3.474$. Since -2.76 is not ≤ -3.474, H_0 is not rejected at level .01; experimental evidence does not strongly indicate that when the maximum outdoor temperature is 82, true average daily chiller operation time is less than 12.

A Confidence Interval for $\mu_{Y \cdot x}$

As was done in Section 12.2 for β_1, a probability statement involving the T variable (12.14) can be manipulated to obtain a 100 $(1 - \alpha)$% confidence interval for $\mu_{Y \cdot x}$ (that is, for $\beta_0 + \beta_1 x$). The resulting interval is

$$\left(\hat{\beta}_0 + \hat{\beta}_1 x - t_{\alpha/2, n-2} \cdot s \sqrt{\frac{1}{n} + \frac{n(x - \bar{x})^2}{n \Sigma x_i^2 - (\Sigma x_i)^2}}, \right.$$

$$\left. \hat{\beta}_0 + \hat{\beta}_1 x + t_{\alpha/2, n-2} \cdot s \sqrt{\frac{1}{n} + \frac{n(x - \bar{x})^2}{n \Sigma x_i^2 - (\Sigma x_i)^2}} \right) \qquad (12.15)$$

Example 12.8   For a high-performance tissue machine used in processing of paper by paper mills, the following data was collected on machine speed x (m/min) and temperature in the drying hood y (°C) ("Gas Turbines for Process Improvement of Industrial Thermal Power Plants," *Combustion*, April 1976, pp. 35–41).

x:	1000	1100	1200	1250	1300	1400	1450
y:	220	280	350	375	450	470	500

To obtain a 99% confidence interval for true average hood temperature when machine speed equals 1200, we first calculate $\Sigma x_i = 8700$, $\Sigma x_i^2 = 10,965,000$, $\Sigma y_i = 2645$, $\Sigma y_i^2 = 1,063,325$, $\Sigma x_i y_i = 3,384,750$, $\hat{\beta}_1 = .640$, $\hat{\beta}_0 = -417.70$, and $s = 17.62$. Substituting $\hat{\beta}_0$, $\hat{\beta}_1$, and s into (12.15) along with $t_{.005,5} = 4.032$ and $x = 1200$ gives

$$\left(-417.70 + (.640)(1200) - (4.032)(17.62)\sqrt{\frac{1}{7} + \frac{7(1200 - 1242.86)^2}{1,065,000.02}}, \right.$$

$$\left. -417.70 + (.640)(1200) + (4.032)(17.62)\sqrt{\frac{1}{7} + \frac{7(1200 - 1242.86)^2}{1,065,000.02}} \right)$$

$$= (350.3 - 71.04\sqrt{.1549}, \; 350.3 + 71.04\sqrt{.1549}) = (322.34, 378.26)$$

as the 99% confidence interval for $\mu_{Y \cdot 1200}$. Note that for $x = 1200$, the predicted value $\hat{y} = 350.3$ and the observed value $y = 350$ are almost identical.

Because $(x - \bar{x})^2$ is smallest for $x = \bar{x}$ and increases as x moves away from $\bar{x}$, the length of the interval (12.15) for $\mu_{Y \cdot x}$ will increase as x gets further away from $\bar{x}$. In particular an x value quite far from the center of the x_i's for which observations were made will result in a very wide interval, which is reasonable in that little information about the regression relationship is available for such x's.

Simultaneous Confidence Intervals for Several $\mu_{Y \cdot x}$'s

In some situations a confidence interval is desired not just for a single x value but for two or more x values. Suppose that an investigator wishes a confidence interval for both $\mu_{Y \cdot v}$ and for $\mu_{Y \cdot w}$, where v and w are two different values of the independent variable. It is tempting to compute the interval (12.15) first for $x = v$ and then for $x = w$. Suppose that we use $\alpha = .05$ in each computation to get two 95% intervals. Then if the variables involved in computing the two intervals were independent of one another, the joint confidence coefficient would be $(.95) \cdot (.95) = .90$.

However, the intervals are not independent, because the same $\hat{\beta}_0$, $\hat{\beta}_1$, and S are used in each. We therefore cannot assert that the joint confidence coefficient for the two intervals is exactly 90%. It can be shown, though, that if the 100 $(1 - \alpha)\%$ confidence interval (12.15) is computed both for $x = v$ and for $x = w$ to obtain joint confidence intervals for $\mu_{Y \cdot v}$ and $\mu_{Y \cdot w}$, then *the joint confidence coefficient on the resulting pair of intervals is at least 100 $(1 - 2\alpha)\%$*. In particular using $\alpha = .05$ results in a joint confidence coefficient of *at least* 90%, while using $\alpha = .01$ results in at least 98% confidence.

Example 12.8
(continued)

The 99% interval for true average machine speed when hood temperature is 1200 was computed as (322.34, 378.26). The 99% interval when hood temperature equals 1350 is (413.11, 479.99). Thus we can be at least 98% confident that both $\mu_{Y \cdot 1200}$ lies in (322.34, 378.26) and $\mu_{Y \cdot 1350}$ lies in (413.11, 479.49).

The validity of these joint confidence intervals rests on a mathematical result called the Bonferroni inequality, so the joint confidence intervals are called Bonferroni intervals (the same type of intervals used in the second multiple comparisons method presented in Chapter 10). *The method can be generalized to yield joint intervals for k different $\mu_{Y \cdot x}$'s. Using the interval (12.15) for each different x results in a joint confidence coefficient of at least 100 (1 − kα)%.*

A Prediction Interval for a Future Value of Y

Analogous to the confidence interval (12.15) for $\mu_{Y \cdot x}$, one frequently wishes to obtain an interval of plausible values for the value of Y associated with some future experiment when the independent variable has value x. For instance, in the example in which vocabulary size y is related to the age x of a child, for $x = 6$ years (12.15) would be a confidence interval for the true average vocabulary size of all six-year-old children. Alternatively, we might wish an interval of plausible values for the vocabulary size of a particular six-year-old child.

A confidence interval refers to a parameter, or population characteristic, whose value is fixed but unknown to us. In contrast, a future value of Y is not a parameter but instead a random variable; for this reason we refer to an interval of plausible values for a future Y as a **prediction interval** rather than a confidence interval. The error of estimation is $\beta_0 + \beta_1 x - (\hat{\beta}_0 + \hat{\beta}_1 x)$, a difference between a fixed (but unknown) quantity and a random variable. The error of prediction is $Y - (\hat{\beta}_0 + \hat{\beta}_1 x)$, a difference between two random variables. There is thus more uncertainty in prediction than in estimation, so a prediction interval will be wider than a confidence interval. Because the future value Y is independent of the observed Y_i's,

$$\text{Var}[Y - (\hat{\beta}_0 + \hat{\beta}_1 x)] = \quad \text{variance of prediction error}$$

$$= \text{Var}(Y) + \text{Var}(\hat{\beta}_0 + \hat{\beta}_1 x) = \sigma^2 + \sigma^2 \left[\frac{1}{n} + \frac{n(x - \bar{x})^2}{n \Sigma x_i^2 - (\Sigma x_i)^2} \right] \quad (12.16)$$

$$= \sigma^2 \left[1 + \frac{1}{n} + \frac{n(x - \bar{x})^2}{n \Sigma x_i^2 - (\Sigma x_i)^2} \right]$$

Using (12.16), it can be shown that the interval

$$\left(\hat{\beta}_0 + \hat{\beta}_1 x - t_{\alpha/2, n-2} \cdot s \sqrt{1 + \frac{1}{n} + \frac{n(x - \bar{x})^2}{n \Sigma x_i^2 - (\Sigma x_i)^2}} \right.$$

$$\left. \hat{\beta}_0 + \hat{\beta}_1 x + t_{\alpha/2, n-2} \cdot s \sqrt{1 + \frac{1}{n} + \frac{n(x - \bar{x})^2}{n \Sigma x_i^2 - (\Sigma x_i)^2}} \right) \quad (12.17)$$

is a $100(1 - \alpha)\%$ prediction interval for a future value Y. That is, if (12.17) is used repeatedly, in the long run the resulting intervals will actually contain the observed y values $100(1 - \alpha)\%$ of the time. Notice that the one underneath the square root symbol makes the prediction interval (12.17) longer than the confidence interval (12.15), though the intervals are both centered at $\hat{\beta}_0 + \hat{\beta}_1 x$. Also, as $n \to \infty$ the length of the confidence interval approaches zero, while the length of the prediction interval does not (because even with perfect knowledge of β_0 and β_1, there will still be uncertainty in prediction).

Example 12.9 The article "The Incorporation of Uranium and Silver by Hydrothermally Synthesized Galena" (*Econ. Geology*, 1964, pp. 1003–1024) reported on the determination of silver content of galena crystals grown in a closed hydrothermal system over a range of temperature. With $x =$ crystallization temperature in °C and $y = Ag_2S$ in mol %, the data follows:

x:	398	292	352	575	568	450	550	408	484	350	503	600	600
y:	.15	.05	.23	.43	.23	.40	.44	.44	.45	.09	.59	.63	.60

From the summary values $\Sigma x_i = 6130$, $\Sigma x_i^2 = 3{,}022{,}050$, $\Sigma y_i = 4.73$, $\Sigma y_i^2 = 2.1785$, and $\Sigma x_i y_i = 2418.74$, the regression estimates are computed to be $\hat{\beta}_1 = .00143$, $\hat{\beta}_0 = -.311$, and $s = .131$. If the next crystallization temperature chosen is 500 °C, a 95% prediction interval for the resulting silver content is

$$-.311 + (.00143)(500) \pm (2.201)(.131)\sqrt{1 + \frac{1}{13} + \frac{13(500 - 471.54)^2}{1{,}709{,}750}}$$

$$= .40 \pm .30 = (.10, .70), \quad \text{a very wide interval.}$$

Exercises / Section 12.3

1. Reconsider the filter application temperature – % BOD removal experiment described in Exercise 12.2.4. In addition to information given there, $\hat{\beta}_0 = 8.2141$.
   a. Compute a 90% confidence interval for $\beta_0 + 25\beta_1$, the expected % BOD removal when filter application temperature is 25 °C.
   b. Test at level .10 the hypotheses $H_0 : \beta_0 + 25\beta_1 = 50$ versus $H_a : \beta_0 + 25\beta_1 > 50$ (the alternative hypothesis states that expected % BOD removal exceeds 50 when filter application temperature is 25).

2. The estimated regression line for the age/vocabulary size data of Exercise 12.1.2, is $y = -827.81 + 571.15x$, with $s = 83.86$.
   a. Interpret in words the quantity $\mu_{Y \cdot 3} = \beta_0 + 3\beta_1$. What is a point estimate of $\mu_{Y \cdot 3}$?
   b. What is the estimated standard deviation of the estimator $\hat{\mu}_{Y \cdot 3} = \hat{\beta}_0 + 3\hat{\beta}_1$ that you used in (a)?
   c. Obtain a 95% confidence interval for $\mu_{Y \cdot 3}$.
   d. Which confidence interval would be longer, the one for $\mu_{Y \cdot 3}$ or the one for $\mu_{Y \cdot 5}$? Answer without actually computing the intervals.
   e. Using level of significance .05, test $H_0 : \beta_0 + 3\beta_1 = 1000$ versus $H_a : \beta_0 + 3\beta_1 < 1000$. In words, what do H_0 and H_a say?

3. Based on the data in Example 12.9, does the expected silver content when temperature equals 400 °C appear to differ significantly from .250? Test the appropriate hypotheses at level .01. What can you say about the P-value for the resulting t?

4. Use the data in Example 12.7 to compute a 99% confidence interval for expected daily hours of chiller operation when maximum outdoor tem-

perature equals 85°. Repeat for 95°. Why is this second interval much wider?

5. An experiment to measure the macroscopic magnetic relaxation time in crystals (μ sec) as a function of the strength of the external biasing magnetic field (KG) yielded the following data ("An Optical Faraday Rotation Technique for the Determination of Magnetic Relaxation Times," *IEEE Trans. Magnetics*, June 1968, pp. 175–178, with data read from a graph which appeared in the paper).

x:	11.0	12.5	15.2	17.2	19.0	20.8
y:	187	225	305	318	367	365
x:	22.0	24.2	25.3	27.0	29.0	
y:	400	435	450	506	558	

The summary statistics are $\Sigma x_i = 223.2$, $\Sigma y_i = 4116$, $\Sigma x_i^2 = 4877\,50$, $\Sigma x_i y_i = 90{,}096.1$, and $\Sigma y_i^2 = 1{,}666{,}782$.

a. Compute a 95% confidence interval for expected relaxation time when field strength equals 18.
b. Compute a 95% prediction interval for future relaxation time when field strength equals 18.
c. Compute simultaneous confidence intervals for expected relaxation time when field strength equals both 15 and 20; your joint confidence coefficient should be at least 90%.

6. Exercise 12.2.5 presented data on x = pressure of extracted gas and y = extraction time.
   a. Does the data strongly contradict prior belief that when extraction pressure is 300, expected extraction time is at most 4 min? Use a level .01 test. What can you say about the P-value (upper and/or lower bound)?
   b. Extraction time is to be observed once for a pressure of 200 and once for a pressure of 300. Obtain a pair of prediction intervals for the future observed values such that the joint prediction coefficient is at least 90%.

7. Using the data in Example 12.8, obtain simultaneous confidence intervals for true average hood temperature when machine speed equals 1200, 1250, and 1300; your joint confidence coefficient should be at least 90%.

8. Based on the data in Example 12.2, can we conclude that expected amount of shoot elongation differs significantly from 10 when daily degree hours of soil heat equals 500? Use a level .05 test.

9. Verify that $\text{Var}(\hat{\beta}_0 + \hat{\beta}_1 x)$ is indeed given by the expression in the text. *Hint:* $\text{Var}(\Sigma d_i Y_i) = \Sigma d_i^2 \text{Var}(Y_i)$.

12.4 Correlation and the Coefficient of Determination

In Section 12.2 we saw that if in testing $H_0 : \beta_1 = 0$ versus $H_a : \beta_1 \neq 0$, H_0 is rejected in favor of H_a, then we can conclude that knowledge of x is useful in obtaining information about Y from the estimated regression model. However, we have not yet developed a precise quantitative measure of the strength of the relationship between the two variables. In fact there are many situations in which the objective in studying the joint behavior of two variables is to see whether or not they are related, rather than to use one to predict the value of the other. In this section we first develop the sample correlation coefficient r as a measure of how strongly related two variables x and y are in a sample, then relate r to the correlation coefficient ρ defined in Chapter 5, and finally show that r^2 provides information about the effectiveness of regression analysis.

The Sample Correlation Coefficient r

Given n pairs of observations $(x_1, y_1), (x_2, y_2), \ldots, (x_n, y_n)$, it is natural to speak of x and y having a positive relationship if large x's are paired with large y's and small x's with small y's. Similarly, if large x's are paired with small y's and small x's with large y's, then a negative relationship between the variables is implied. Consider the quantity

$$s_{xy} = \sum_{i=1}^{n} (x_i - \bar{x})(y_i - \bar{y})$$

Then if the relationship is strongly positive, an x_i above the mean $\bar{x}$ will tend to be paired with a y_i above the mean $\bar{y}$, so that $(x_i - \bar{x})(y_i - \bar{y}) > 0$, and this product will also be positive whenever both x_i and y_i are below their respective means. Thus a positive relationship implies that s_{xy} will be positive. An analogous argument shows that when the relationship is negative, s_{xy} will be negative, since most of the products $(x_i - \bar{x})(y_i - \bar{y})$ will be negative. This is illustrated in Figure 12.7.

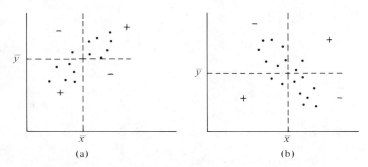

Figure 12.7 (a) Scatterplot with s_{xy} positive; (b) scatterplot with s_{xy} negative [+ means $(x_i - \bar{x})(y_i - \bar{y}) > 0$, and − means $(x_i - \bar{x})(y_i - \bar{y}) < 0$]

While s_{xy} seems a plausible measure of the strength of a relationship, we do not yet have any idea of how positive or negative it can be. Unfortunately, s_{xy} has a serious defect: By changing the units of measurement of either x or y, s_{xy} can be made either arbitrarily large in magnitude or arbitrarily close to zero. For example, if $s_{xy} = 25$ when x is measured in meters, then $s_{xy} = 25,000$ when x is measured in millimeters and $= .025$ when x is expressed in kilometers. A reasonable condition to impose on any measure of how strongly x and y are related is that the calculated measure should not depend on the particular units used to measure them. This condition is achieved by modifying s_{xy} to obtain the sample correlation coefficient.

Definition: The **sample correlation coefficient** for the n pairs $(x_1, y_1), \ldots, (x_n, y_n)$ is

$$
\begin{aligned}
r &= \frac{s_{xy}}{\sqrt{\sum (x_i - \bar{x})^2} \, \sqrt{\sum (y_i - \bar{y})^2}} \\
&= \frac{n \sum x_i y_i - (\sum x_i)(\sum y_i)}{\sqrt{n \sum x_i^2 - (\sum x_i)^2} \, \sqrt{n \sum y_i^2 - (\sum y_i)^2}}
\end{aligned}
\tag{12.18}
$$

The second expression in (12.18) is an efficient computing formula for r.

Example 12.10 The April 2, 1977, issue of the medical journal *Lancet* reported results on water/lead concentration in the maternal home during pregnancy and blood/lead concentration (both in μmol/liter) for a group of mentally retarded children. Representative data appears below. Calculate the sample correlation coefficient.

x (blood):	1.98	1.44	2.02	1.20	1.57	1.82	1.45	1.80
y (water):	5.6	7.7	8.8	5.1	6.8	3.9	4.5	5.8

According to (12.18) r can be calculated from the five quantities Σx_i, Σx_i^2, Σy_i, Σy_i^2, and $\Sigma x_i y_i$; these are 13.28, 22.63, 48.2, 309.44, and 80.81. Thus

$$r = \frac{8(80.81) - (13.28)(48.2)}{\sqrt{8(22.63) - (13.28)^2}\ \sqrt{8(309.44) - (48.2)^2}} = \frac{6.38}{(2.16)(12.34)} = .239$$

Properties of r

The most important properties of the sample correlation coefficient are

> **1.** r is independent of the units in which both x and y are measured
> **2.** $-1 \le r \le 1$
> **3.** $r = 1$ if and only if all (x_i, y_i) pairs lie on a straight line with positive slope, and $r = -1$ if and only if all (x_i, y_i) pairs lie on a straight line with negative slope.

Property 1 is equivalent to saying that r is unchanged if each x_i is replaced by cx_i and if each y_i is replaced by dy_i (a change in the scale of measurement), as well as if each x_i is replaced by $x_i - a$ and y_i by $y_i - b$ (which changes the location of zero on the measurement axis). This implies, for example, that r is the same whether temperature is measured in °F or °C.

The second property tells us that the maximum value of r, corresponding to the largest possible degree of positive relationship, is $r = 1$, while the most negative relationship is identified with $r = -1$. According to property 3 the largest positive and largest negative correlations are achieved only when all points lie along a straight line. Any other configuration of points, even if the configuration suggests a deterministic relationship between variables, will yield an r value less than one in absolute magnitude. Thus, r *measures the degree of linear relationship* among variables. A value of r near zero is not evidence of a lack of strong relationship, but only the absence of a linear relation, so that such a value of r must be interpreted with caution. Figure 12.8 illustrates several configurations of points associated with different values of r.

A frequently asked question is, "When can we say that there is a strong correlation, and when is the correlation weak?" A reasonable rule of thumb is to say that the correlation is weak if $0 \le |r| \le .5$, strong if $.8 \le |r| \le 1$, and moderate

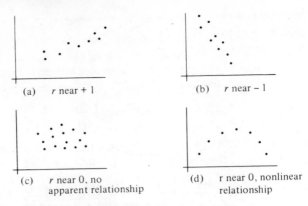

(a) r near $+ 1$ (b) r near $- 1$

(c) r near 0, no   (d)   r near 0, nonlinear
        apparent relationship   relationship

Figure 12.8   Data plots for different values of r

otherwise. In Example 12.10 the correlation between blood-lead concentration and water/lead concentration was weak.

The Population Correlation Coefficient ρ and Inferences About Correlation

The correlation coefficient r is a measure of how related x and y are in the observed sample. We can think of the pairs (x_i, y_i) as having been drawn from a bivariate population of pairs, with (X_i, Y_i) having joint probability distribution $f(x, y)$. In Chapter 5 we defined the correlation coefficient $\rho(X, Y)$ by

$$\rho = \rho(X, Y) = \frac{\text{Cov } (X, Y)}{\sigma_X \cdot \sigma_Y}, \quad \text{where}$$

$$\text{Cov } (X, Y) = \begin{cases} \displaystyle\sum_x \sum_y (x - \mu_X)(y - \mu_Y)f(x, y) & (X, Y) \text{ discrete} \\ \displaystyle\int_{-\infty}^{\infty} \int_{-\infty}^{\infty} (x - \mu_X)(y - \mu_Y) f(x, y) & (X, Y) \text{ continuous} \end{cases}$$

If we think of $f(x, y)$ as describing the distribution of pairs of values within the entire population, ρ becomes a measure of how strongly related x and y are in that population. In Chapter 5, we listed properties of ρ analogous to 1–3 above.

The population correlation coefficient ρ is a parameter or population characteristic, just as μ_X, μ_Y, σ_X, and σ_Y are, so we can use the sample correlation coefficient to make various inferences about ρ. In particular, r is a point estimate for ρ, and the corresponding estimator is

$$\hat{\rho} = R = \frac{\sum (X_i - \overline{X})(Y_i - \overline{Y})}{\sqrt{\sum (X_i - \overline{X})^2} \sqrt{\sum (Y_i - \overline{Y})^2}}$$

Example 12.11 In Example 12.4 we presented data on intelligence x and hypnotic susceptibility y. Although a regression equation was calculated there, rather than desiring to predict susceptibility from intelligence, the key issue is to assess how strongly the two variables are related. With $\Sigma x_i = 3893$, $\Sigma x_i^2 = 478,537$, $\Sigma y_i = 290$, $\Sigma y_i^2 = 4160$, $\Sigma x_i y_i = 36,473$, $n = 32$,

$$r = \frac{32(36,473) - (3893)(290)}{\sqrt{32(478,537) - (3893)^2} \sqrt{32(4160) - (290)^2}} = .433$$

Our point estimate of the population correlation coefficient ρ between intelligence and hypnotic susceptibility is $\hat{\rho} = r = .433$.

In our previous work concerning inference, we typically did not have to make strong assumptions about the population distribution in order to obtain point estimators and estimates of parameters. However, in order to construct confidence intervals and test procedures, we have usually assumed that the sample comes from a normal population. To test hypotheses about ρ, we must make an analogous assumption about the distribution of pairs of (x, y) values in the population. We are now assuming that *both* X and Y are random, whereas much of our regression work focused on x fixed by the experimenter.

Assumption: The joint probability distribution of (X, Y) is specified by

$$f(x, y)$$
$$= \frac{1}{2\pi \cdot \sigma_1 \sigma_2 \sqrt{1 - \rho^2}} \, e^{-\left[\left(\frac{x - \mu_1}{\sigma_1}\right)^2 - 2\rho\left(\frac{x - \mu_1}{\sigma_1}\right)\left(\frac{y - \mu_2}{\sigma_2}\right) + \left(\frac{y - \mu_2}{\sigma_2}\right)^2\right] \Big/ 2(1 - \rho^2)}$$

$$-\infty < x < \infty$$
$$-\infty < y < \infty \qquad\qquad (12.19)$$

where μ_1 and σ_1 are the mean and standard deviation of X and μ_2 and σ_2 are the mean and standard deviation of Y; $f(x, y)$ is called the **bivariate normal probability distribution.**

The bivariate normal distribution is obviously rather complicated, but for our purposes we only need a passing acquaintance with several of its properties. The surface determined by $f(x, y)$ lies entirely above the x-y plane $[f(x, y) \geq 0]$ and has a three-dimensional bell or mound-shaped appearance, as illustrated in Figure 12.9. If we slice through the surface with any plane perpendicular to the (x, y) plane and look at the shape of the curve sketched out on the "slicing plane," the result is a normal curve. More precisely, if X is fixed at value x, it can be shown that the (conditional) distribution of Y is normal with mean $\mu_{Y \cdot x} = \mu_2 - \rho\mu_1\sigma_2/\sigma_1 + \rho\sigma_2 x/\sigma_1$ and variance $(1 - \rho^2)\sigma_2^2$. This is exactly the model used in simple linear regression, with $\beta_0 = \mu_2 - \rho\mu_1\sigma_2/\sigma_1$, $\beta_1 = \rho\sigma_2/\sigma_1$, and $\sigma^2 = (1 - \rho^2)\sigma_2^2$ independent of x. The implication is that *if the observed pairs (x_i, y_i) are actually drawn from a bivariate normal distribution, then the simple linear regression model is an appropriate way of studying the behavior of Y for fixed x.* If $\rho = 0$, then $\mu_{Y \cdot x} = \mu_2$

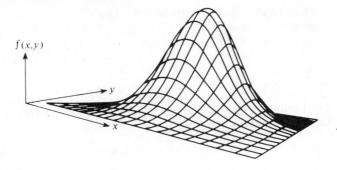

Figure 12.9   A plot of the bivariate normal p.d.f.

independent of x; in fact, when $\rho = 0$ the joint p.d.f. $f(x, y)$ of (12.19) can be factored into a part involving x only and a part involving y only, which implies that X and Y are independent variables.

Assuming that the pairs are drawn from a bivariate normal distribution allows us to test hypotheses about ρ and construct a confidence interval.

Procedure for Testing $H_0 : \rho = 0$

$$\text{Test statistic: } T = \frac{R\sqrt{n-2}}{\sqrt{1-R^2}}$$

When H_0 is true, T has a t distribution with $n-2$ degrees of freedom.

Alternative Hypothesis	Rejection Region for Level α Test
$H_a : \rho > 0$	$T \geq t_{\alpha, n-2}$
$H_a : \rho < 0$	$T \leq -t_{\alpha, n-2}$
$H_a : \rho \neq 0$	either $T \geq t_{\alpha/2, n-2}$ or $T \leq -t_{\alpha/2, n-2}$

Example 12.11 (continued) The computed correlation for the 32 observations on intelligence/hypnotic susceptibility was $r = .433$. To test H_0 versus $H_a : \rho > 0$ at level .01, we need $t_{.01, 30} = 2.457$. The computed value of the test statistic is

$$t = \frac{.433\sqrt{30}}{\sqrt{1 - (.433)^2}} = 2.63$$

Because $2.63 \geq 2.457$, the data is (barely) significant at level .01, so H_0 is rejected in favor of the conclusion that there is a somewhat positive relationship between intelligence and hypnotic susceptibility.

Looking back at Example 12.4, you will discover that we not only tested the same hypothesis about a positive relationship as we just did above, but that the computed value of t and the critical value $t_{.01, 30}$ were the same. This is no accident—the test procedure for $H_0 : \rho = 0$ is equivalent to the test procedure for $H_0 : \beta = 0$

because the two test statistics (and critical values for rejection regions) are identical. The implication is that you can test a relationship either by working with β_1 or by working with ρ. The former approach is consistent with regression analysis, while the latter is appropriate if interest lies only in the strength of relationship and not in prediction.

Other Inferences Concerning ρ

The procedure for testing $H_0 : \rho = \rho_0$ when $\rho_0 \neq 0$ is not equivalent to any procedure from regression analysis. The test statistic is based on a transformation of R called the Fisher transformation.

Proposition: When $(X_1, Y_1), \ldots, (X_n, Y_n)$ is a sample from a bivariate normal distribution, the random variable

$$V = \frac{1}{2} \ln \left(\frac{1 + R}{1 - R} \right) \tag{12.20}$$

has approximately a normal distribution with mean and variance

$$\mu_V = \frac{1}{2} \ln \left(\frac{1 + \rho}{1 - \rho} \right), \quad \sigma_V^2 = \frac{1}{n - 3} \tag{12.21}$$

The rationale for the transformation is to obtain a function of R which has a variance independent of ρ; this would not be the case with R itself. Also, the transformation should not be used if n is quite small, since the approximation will not be valid.

Test statistic: $Z = \dfrac{V - \frac{1}{2} \ln \left[(1 + \rho_0)/(1 - \rho_0) \right]}{1/\sqrt{n - 3}}$

where V is given by (12.20). When H_0 is true, Z has approximately a standard normal distribution.

Alternative Hypothesis	Rejection Region for Level α Test
$H_a : \rho > \rho_0$	$Z \geq z_\alpha$
$H_a : \rho < \rho_0$	$Z \leq -z_\alpha$
$H_a : \rho \neq \rho_0$	either $Z \geq z_{\alpha/2}$ or $Z \leq -z_{\alpha/2}$

Example 12.12 The paper "A Study of a Partial Nutrient Removal System for Wastewater Treatment Plants" (*Water Research*, 1972, pp. 1389–1397) reported on a method of nitrogen removal which involved the treatment of the supernatant from an aerobic digester. Both the influent total nitrogen x (mg/l) and the percentage of nitrogen removed were recorded for 20 days, with resulting summary statistics $\Sigma x_i = 285.9$, $\Sigma x_i^2 = 4409.55$, $\Sigma y_i = 690.30$, $\Sigma y_i^2 = 29040.29$, and $\Sigma x_i y_i = 10,869.71$. Does

the data indicate that influent total nitrogen and percentage of nitrogen removed are at least moderately positively correlated?

Our earlier interpretation of moderate positive correlation was $.5 < \rho < .8$, so we wish to test $H_0 : \rho = .5$ versus $H_a : \rho > .5$. The computed value of r is .773, so

$$\frac{1}{2} \ln \left(\frac{1 + .733}{1 - .733} \right) = .935, \quad \frac{1}{2} \ln \left(\frac{1 + .5}{1 - .5} \right) = .549$$

This gives $z = (.935 - .549) \sqrt{17} = 1.59$. Since $1.59 < 1.645$, at level .05, we cannot conclude that $\rho > .5$, so the relationship has not been proved to be even moderately strong (a somewhat surprising conclusion since $r = .73$, but when n is small a large r may result even when ρ is small).

To obtain a confidence interval for ρ, we first derive an interval for $\mu_V = \frac{1}{2} \ln [(1 + \rho)/(1 - \rho)]$. Standardizing V, writing a probability statement, and manipulating the resulting inequalities yields

$$\left(v - \frac{z_{\alpha/2}}{\sqrt{n - 3}}, \; v + \frac{z_{\alpha/2}}{\sqrt{n - 3}} \right) \tag{12.22}$$

as a $100 (1 - \alpha)\%$ interval for μ_V, where $v = \frac{1}{2} \ln [(1 + r)/(1 - r)]$. This interval can then be manipulated to yield a confidence interval for ρ:

The interval

$$\left(\frac{e^{2c_1} - 1}{e^{2c_1} + 1}, \; \frac{e^{2c_2} - 1}{e^{2c_2} + 1} \right) \tag{12.23}$$

is a $100 (1 - \alpha)\%$ confidence interval for ρ, where c_1 and c_2 are the left and right endpoints, respectively, of the interval (12.22).

Example 12.12 (continued) The sample correlation coefficient between influent nitrogen and percentage nitrogen removed was $r = .733$, giving $v = .935$. With $n = 20$, a 95% interval for μ_V is $(.935 - 1.96/\sqrt{17}, .935 + 1.96/\sqrt{17}) = (.460, 1.410) = (c_1, c_2)$. The 95% interval for ρ is

$$\left[\frac{e^{2(.46)} - 1}{e^{2(.46)} + 1}, \; \frac{e^{2(1.41)} - 1}{e^{2(1.41)} + 1} \right] \doteq (.43, .89)$$

The Coefficient of Determination

We have seen that a value of r near one in absolute magnitude is indicative of a strong linear relationship between the two variables under study, but have not yet related this to regression calculations. The square of the sample correlation coefficient, called the **coefficient of determination,** describes precisely how much of the variability in the observed y_i's is due to variation in the independent variable. Recalling that $\hat{y}_i = \hat{\beta}_0 + \hat{\beta}_1 x_i$ = the predicted y value when $x = x_i$, the error sum of squares $SSE = \Sigma (y_i - \hat{y}_i)^2$ measures how much variability in the y_i's is not explained by the regression relationship. If SSE is quite small, then almost all observed pairs lie near the least squares line, while if it is large, then there is much "residual variability" even after taking into account the possibility of a linear relationship. The total amount of variability in the y_i's can be measured by computing $SST = \Sigma (y_i - \bar{y})^2$, the total sum of squares of the y_i's about their mean. The ratio SSE/SST then represents the proportion of variability in the y_i's which is unexplained by the linear regression of y on x. If this proportion is nearly zero, then almost all the variation in y can be explained by a linear relationship between x and y and the fact that x is varying. This is where the correlation coefficient comes in.

Proposition: With $SST = \Sigma (y_i - \bar{y})^2 = \Sigma y_i^2 - (\Sigma y_i)^2/n$ and $SSE = \Sigma (y_i - \hat{y}_i)^2$

$$r^2 = \frac{SST - SSE}{SST} = 1 - \frac{SSE}{SST} \tag{12.24}$$

In words, the **coefficient of determination** r^2 equals the proportion of variation in the y_i's which is explained by the linear regression relationship between x and y, where variation is measured by a sum of squared deviations.

The proof of this proposition follows almost immediately from substituting the formula for $\hat{\beta}_1$ into an alternative expression for SSE:

$$SSE = \sum y_i^2 - \hat{\beta}_0 \sum y_i - \hat{\beta}_1 \sum x_i y_i$$

$$= SST - \hat{\beta}_1 \left[\frac{\sum x_i y_i - \left(\sum x_i\right)\left(\sum y_i\right)}{n} \right]$$

The property $-1 \le r \le 1$ also follows from (12.24).

Example 12.13 In Example 12.8 we examined the relationship between machine speed x and hood temperature y for a high-performance tissue processing machine. The correlation coefficient for the given data is $r = .988$, so $r^2 = .976$. Thus 97.6% of the sample variation in hood temperature can be explained by variation in machine speed through the linear regression relation between x and y.

The coefficient of determination is an important descriptive statistic in regression analysis because of its appealing interpretation as the proportion of explained variation. Most regression analysis computer printouts include the value of

r^2 in their output. We shall see shortly that there is a quantity analogous to r^2 for multiple linear regression (linear regression with two or more independent variables).

Regression and ANOVA

The splitting of the total sum of squares $\Sigma (y_i - \bar{y})^2$ into a part *SSE*, which measures unexplained variation, and a part $SST - SSE$, which measures variation explained by the linear relationship, is strongly reminiscent of one-way ANOVA. In fact, the null hypothesis $H_0 : \beta_1 = 0$ can be tested against $H_a : \beta_1 \neq 0$ by constructing an ANOVA table and rejecting H_0 if $F \geq F_{\alpha, 1, n-2}$.

Table 12.1

Source of variation	d.f.	Sum of squares	Mean square	F
Regression	1	$SSR = SST - SSE$	SSR	$\dfrac{SSR}{SSE/(n-2)}$
Error	$n - 2$	SSE	$\dfrac{SSE}{(n-2)}$	
Total	$n - 1$	SST		

The F test gives exactly the same result as the t test discussed previously because $T^2 = F$ and $t^2_{\alpha/2, n-2} = F_{\alpha, 1, n-2}$. Virtually all computer packages which have regression options include such an ANOVA table in the printout.

Example 12.13 (continued) The ANOVA table for the regression of hood temperature y on machine speed x is

Table 12.2

Source of variation	d.f.	Sum of squares	Mean square	F
Regression	1	62,345	62,345	201.11
Error	5	1548	310	
Total	6	63,893		

Since $F_{.01, 1, 5} = 16.3$, the computed F is highly significant (actually $F_{.001, 1, 5} = 47.2$).

Exercises / Section 12.4

1. Observations were made on $x =$ June–August precipitation (cm) and $y =$ perennial grass production (kg/ha) for certain pasture areas during each year of a 10-year period, with the following results:

x:	22.05	35.74	30.48	11.89	27.28
y:	291	629	823	307	660
x:	9.63	17.63	22.20	17.27	19.63
y:	263	375	366	563	558

("Influence of Precipitation on Perennial Grass Production in the Semidesert Southwest," *Ecology*, 1975, pp. 981–986).

a. Compute the sample correlation coefficient for this data.

b. If you were to do a regression analysis, what would the value of the coefficient of determination be?

2. Sixteen different air samples were obtained at Herald Square in New York City, and both the carbon monoxide concentration x (ppm) and benzo(a) pyrene concentration y ($\mu g/10^3 m^3$) measured for each sample ("Carcinogenic Air Pollutants in Relation to Automobile Traffic in New York City," *Environmental Science and Technology*, 1971, pp. 145–150).

x:	2.8	15.5	19.0	6.8	5.5	5.6	9.6	13.3
y:	.5	.1	.8	.9	1.0	1.1	3.9	4.0
x:	5.5	12.0	5.6	19.5	11.0	12.8	5.5	10.5
y:	1.3	5.7	1.5	6.0	7.3	8.1	2.2	9.5

a. Compute the sample correlation coefficient for this data.

b. Test the hypothesis $H_0 : \rho = 0$ against $H_a : \rho \neq 0$ at level .01.

3. Data on per capita disposable personal income x (in the United States) and per capita food expenditure y was obtained for a 24-year period. The summary statistics are $\Sigma x_i = 55,661$, $\Sigma y_i = 10,276$, $\Sigma x_i^2 = 144,051,189$, $\Sigma x_i y_i = 25,479,315$, and $\Sigma y_i^2 = 4,582,506$.

a. Compute r for this data.

b. If you decided to obtain an estimated regression line for this data, what proportion of variation in food expenditure would be explained by variation in disposable income?

4. The article "Pedunculate Oak Woodland in a Severe Environment" (*J. Ecology*, 1978, pp. 707–740) reports the following data on x = age (yrs) and y = annual trunk diameter growth increment (mm) for a sample of trees in a certain region.

x:	17	23	30	37.5	40
y:	2.20	1.25	.85	1.30	1.70
x:	46.5	50	54	55	93
y:	.75	.75	.50	1.00	.70

a. Does the data strongly indicate that the true

correlation coefficient ρ (for the population of all such trees in this area) differs from zero? Test using $\alpha = .05$. Place an upper and/or lower bound on the P-value.

b. Compute a 95% confidence interval for ρ.

5. An investigation of the relationship between water temperature x and calling rate y for a particular type of hybrid toad ("The Mating Call of Hybrids of the Fire-Bellied Toad and Yellow-Bellied Toad," *Oecologia*, 1974, pp. 61–71) yielded the following summary statistics: $n = 17$, $\Sigma x_i = 376.20$, $\Sigma y_i = 752$, $\Sigma x_i^2 = 8563.70$, $\Sigma x_i y_i = 17,140.40$, and $\Sigma y_i^2 = 34,496.00$. Does this data indicate that there is a strong positive correlation between water temperature and calling rate? Test the appropriate hypotheses using $\alpha = .05$.

6. The article "Increases in Steroid Binding Globulins Induced by Tamofixen in Patients with Carcinoma of the Breast" (*J. Endocrinology*, 1978, pp. 219–226) reported data on the effects of the drug tamofixen on change in the level of cortisol binding globulin of patients during treatment. With age $= x$ and $\triangle CBG = y$, summary values are $n = 26$, $\Sigma x_i = 1613$, $\Sigma (x_i - \bar{x})^2 = 3756.96$, $\Sigma y_i = 281.9$, $\Sigma (y_i - \bar{y})^2 = 465.34$, and $\Sigma x_i y_i = 16,731$.

a. Compute a 90% confidence interval for the true correlation coefficient ρ.

b. Test $H_0 : \rho = -.5$ versus $H_a : \rho < -.5$ at level .05.

c. In a regression analysis of y or x, what proportion of variation in change of cortisol binding globulin level could be explained by variation in patient age within the sample?

d. If you decided to perform a regression analysis with age as the dependent variable, what proportion of variation in age is explainable by variation in $\triangle CBG$?

7. For the soil heat/shoot elongation data in Example 12.2, what proportion of variation in shoot elongation is explained by variation in soil heat?

8. Verify that the t statistic for testing $H_0 : \beta_1 = 0$ in Section 12.2 is identical to the t statistic of this section for testing $H_0 : \rho = 0$.

9. Use the shortcut formula for computing *SSE* to verify that $r^2 = 1 - SSE/SST$.

10. Let s_x and s_y denote the sample standard deviations of the observed x's and y's, respectively [so $s_x^2 = \Sigma (x_i - \bar{x})^2/(n - 1)$ and similarly for s_y].

a. Show that an alternative expression for the estimated regression line $y = \hat{\beta}_0 + \hat{\beta}_1 x$ is

$$y = \bar{y} + r \cdot \frac{s_y}{s_x} (x - \bar{x})$$

b. This expression for the regression line can be interpreted as follows. Suppose that $r = .5$. What then is the predicted y for an x which lies one standard deviation (s_x units) above the mean of the x_i's? If r were 1, the prediction would be for y to lie one standard deviation above its mean $\bar{y}$, but since $r = .5$, we predict a y which is only .5 standard deviations (s_y units) above $\bar{y}$. Using the data in Exercise 6 for a patient whose age is one standard deviation below the average age in the sample, by how many standard deviations is the patient's predicted $\triangle$CBG above or below the average $\triangle$CBG for the sample?

Bibliography

Draper, Norman and Smith, Harry, *Applied Regression Analysis* (2nd ed.), John Wiley, New York, 1981. The most comprehensive and authoritative book on regression analysis currently in print.

McClave, James and Dietrich, Frank, *Statistics,* Dellen, San Francisco, 1978. Contains several very well-written chapters on regression analysis aimed at an audience having relatively little mathematical background.

Neter, John and Wasserman, William, *Applied Linear Statistical Models,* Richard D. Irwin, Homewood, Ill., 1974. The first 12 chapters constitute an extremely readable and informative survey of regression analysis.

Younger, Mary Sue, *A Handbook for Linear Regression,* Duxbury Press, Boston, 1979. A readable presentation of regression models and methodology; its discussion of the capabilities of the BMD, SAS, and SPSS "canned" statistical computer packages is especially enlightening.

CHAPTER 13

Nonlinear and Multiple Regression

Introduction

The probabilistic model studied in Chapter 12 specified that the observed value of the dependent variable Y deviated from the linear regression function $\mu_{Y \cdot x} = \beta_0 + \beta_1 x$ by a random amount. Here we consider two ways of generalizing the simple linear regression model. The first way is to replace $\beta_0 + \beta_1 x$ by a nonlinear function of x, while the second is to use a regression function involving more than a single independent variable. After fitting a regression function of the chosen form to the given data, it is of course important to have methods available for making inferences about the parameters of the chosen model. Before these methods are used, though, the data analyst should first assess the validity of the chosen model. In Section 13.1 we discuss methods, based primarily on a graphical analysis of the residuals (observed minus predicted y's), for checking the aptness of a fitted model.

In Section 13.2 we consider nonlinear regression functions of a single independent variable x which are "intrinsically linear." By this we mean that it is possible to transform one or both of the variables so that the relationship between the new variables is linear. Another class of nonlinear relations is obtained by using polynomial regression functions of the form $\mu_{Y \cdot x} = \beta_0 + \beta_1 x + \beta_2 x^2 + \cdots + \beta_k x^k$; these polynomial models are the subject of Section 13.3. Multiple regression analysis involves building models for relating Y to $k(\geq 2)$ independent variables $x_1, \ldots, x_k$. The focus in Section 13.4 is on interpretation of the parameters of various multiple regression models and on understanding and using the regression output of widely available statistical computer packages. The last section of the chapter presents a matrix algebra formulation of multiple regression models, which facilitates a detailed description of many important results and procedures for such models.

13.1 Aptness of the Model and Model Checking

A plot of the observed pairs (x_i, y_i) is a necessary first step in deciding on the form of a mathematical relationship between x and y. It is possible to fit many functions other than a linear one $y = b_0 + b_1 x$ to the data, using either the principle of least squares or another fitting method. Once a function of the chosen form has been fitted, it is important to check the fit of the model to see whether it is in fact appropriate. One way to study the fit is to superimpose a graph of the best-fit function on the scatter plot of the data. However, any tilt or curvature of the best-fit function may obscure some aspects of the fit that should be investigated. Furthermore, the scale on the vertical axis may make it difficult to assess the extent to which observed values deviate from the best-fit functions.

Residuals and Fitted Values

A more effective approach to assessment of model adequacy is to compute the fitted or predicted values $\hat{y}_i$ and the residuals $e_i = y_i - \hat{y}_i$, and then plot various functions of these computed quantities. We then examine the plots either to confirm our choice of model or for indications that the model is not appropriate. Suppose that the simple linear regression model is correct, and let $y = \hat{\beta}_0 + \hat{\beta}_1 x$ be the equation of the estimated regression line. Then the ith residual is $e_i = y_i - (\hat{\beta}_0 + \hat{\beta}_1 x)$. To derive properties of the residuals, let $e_i = Y_i - \hat{Y}_i$ represent the ith residual as a random variable (before observations are actually made). Then

$$E(Y_i - \hat{Y}_i) = E(Y_i) - E(\hat{\beta}_0 + \hat{\beta}_1 X_i) = \beta_0 + \beta_1 x_i - (\beta_0 + \beta_1 x_i) = 0 \quad (13.1)$$

so each residual has expected value zero. Because $\hat{Y}_i (= \hat{\beta}_0 + \hat{\beta}_1 x_i)$ is a linear function of the Y_j's, so is $Y_i - \hat{Y}_i$ (where the coefficients depend on the x_j's). Thus the normality of the Y_j's implies that each residual is normally distributed. It can also be shown that

$$\text{Var}(Y_i - \hat{Y}_i) = \sigma^2 \cdot \left[1 - \frac{1}{n} - \frac{(x_i - \bar{x})^2}{\Sigma (x_j - \bar{x})^2} \right] \quad (13.2)$$

Since we shall want to know whether any given residual is larger than expected, the **standardized residuals**

$$e_i^* = \frac{y_i - \hat{y}_i}{s \sqrt{1 - \dfrac{1}{n} - \dfrac{(x_i - \bar{x})^2}{\Sigma (x_j - \bar{x})^2}}} \quad (13.3)$$

will be useful. Notice that the variances of the residuals differ from one another. If n is reasonably large, though, the bracketed term in (13.2) will be approximately 1, so some authors use e_i / s as the standardized residual. Computation of the e_i^*'s can be tedious, but several of the most widely used statistical computer packages (such as MINITAB) automatically provide them and (upon request) can construct various plots involving them.

Example 13.1   Exercise 5 in Section 12.1 presented data on $x = $ burner area liberation rate and $y = \text{NO}_x$ emissions. Here we reproduce the data and give the fitted values, residuals, and standardized residuals. The estimated regression line is $y = -45.55 + 1.71x$, and $r^2 = .961$. Notice that the standardized residuals are not a constant multiple of the residuals (that is, $e_i^* \neq e_i/s$).

x_i	y_i	$\hat{y}_i$	e_i	e_i^*
100	150	125.6	24.4	.75
125	140	168.4	−28.4	−.84
125	180	168.4	11.6	.35
150	210	211.1	−1.1	−.03
150	190	211.1	−21.1	−.62
200	320	296.7	23.3	.66
200	280	296.7	−16.7	−.47
250	400	382.3	17.7	.50
250	430	382.3	47.7	1.35
300	440	467.9	−27.9	−.80
300	390	467.9	−77.9	−2.24
350	600	553.4	46.6	1.39
400	610	639.0	−29.0	−.92
400	670	639.0	31.0	.99

Diagnostic Plots

The basic plots which many statisticians recommend for an assessment of model validity and usefulness are

1. e_i^* (or e_i) on the vertical axis versus x_i on the horizontal axis,
2. e_i^* (or e_i) on the vertical axis versus $\hat{y}_i$ on the horizontal axis, and
3. $\hat{y}_i$ on the vertical axis versus y_i on the horizontal axis.

Plots (1) and (2) are called **residual plots** (against the independent variable and fitted values, respectively), while (3) is a plot of fitted against observed values.

 If plot (3) yields points close to the 45° line [slope +1 through (0, 0)], then the estimated regression function gives accurate predictions of the values actually observed. Thus (3) provides a visual assessment of model effectiveness in making predictions. Provided that the model is correct, both residual plots should exhibit no distinct patterns. The residuals should be randomly distributed about zero according to a normal distribution, so all but a very few standardized residuals should lie between −2 and +2 (that is, all but a few residuals within two standard deviations of their expected value zero). The plot of standardized residuals versus $\hat{y}$ is really a combination of the two other plots, showing implicitly both how residuals vary with x and how fitted values compare with observed values. This latter plot is the single one most often recommended for multiple regression analysis.

Example 13.1 (continued) Figure 13.1 presents, in addition to the scatter plot of the data, the three plots recommended in (1), (2), and (3) above. The plot of $\hat{y}$ versus y confirms the impression given by r^2 that x is effective in predicting y, and also indicates that there is no observed y for which the predicted value is terribly far off the mark. Both

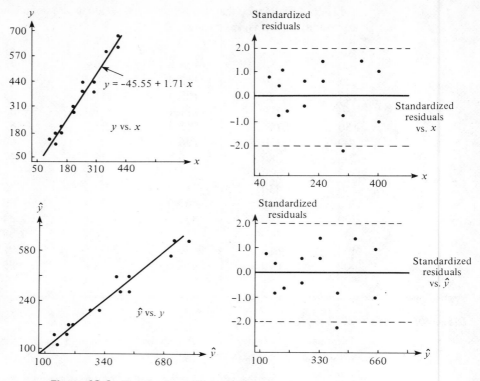

Figure 13.1 Plots for data in Example 13.1

residual plots show no unusual pattern or discrepant values. There is one standard-ized residual slightly outside the interval $(-2, 2)$, but this is not surprising in a sample of size 14. In summary, the plots leave us with no qualms about either the appropriateness of a simple linear relationship or the fit to the given data.

Difficulties and Remedies

While we hope that our analysis will yield plots like those of Figure 1.1, quite frequently the plots will suggest one or more of the following difficulties:

1. A nonlinear probabilistic relationship between x and y is appropriate.
2. The variance of ϵ (and of Y) is not a constant σ^2, but depends on x.
3. The selected model fits the data well except for a very few discrepant or outlying data values, which may have greatly influenced the choice of the best-fit function.
4. The error term ϵ does not have a normal distribution.
5. When the subscript i indicates the time order of the observations, the ϵ_i's exhibit dependence over time.
6. One or more relevant independent variables have been omitted from the model.

Figure 13.2 presents residual plots corresponding to (1)–(3), (5), and (6). To

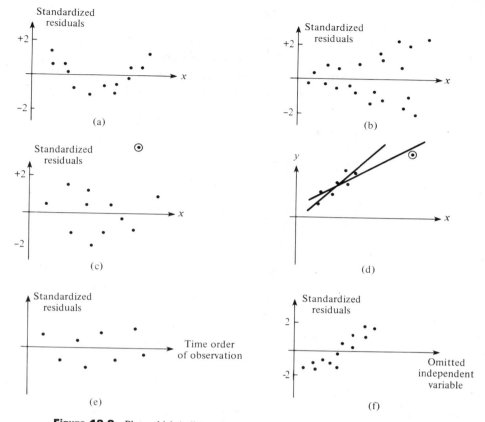

Figure 13.2 Plots which indicate abnormality in data: (a) nonlinear relationship; (b) nonconstant variance; (c) discrepant observation; (d) observation with large influence; (e) dependence in errors; (f) variable omitted

investigate (4), Daniel and Wood (in *Fitting Equations to Data*) recommend the use of normal probability paper, and they give examples of the types of plots which result when the error distribution is normal. Notice that the residuals from the data in Figure 13.2d with the circled point included would not by themselves necessarily suggest further analysis, yet when a new line is fit with that point deleted, the new line differs considerably from the original line. This type of behavior is more difficult to identify in multiple regression. It is most likely to arise when there is a single (or very few) data point(s) with independent variable value(s) far removed from the remainder of the data.

We now indicate briefly what remedies are available for the types of difficulties. For a more comprehensive discussion, one or more of the references on regression analysis should be consulted. If the residual plot looks something like that of Figure 13.2a, exhibiting a curved pattern, then a nonlinear function of x may be fit; in Section 13.2 we show how transformations can be used to this end, while in Section 13.3 we discuss polynomial models.

The residual plot of Figure 13.2b suggests that while a straight-line relationship may be reasonable, the assumption that $\text{Var}(Y_i) = \sigma^2$ for each i is of doubtful validity. When the assumptions of Chapter 12 are valid, it can be shown that among all unbiased estimators of β_0 and β_1, the ordinary least squares estimators have minimum variance. These estimators give equal weight to each (x_i, Y_i). If the variance of Y increases with x, then Y_i's for large x_i should be given less weight than those with small x_i. This suggests that β_0 and β_1 should be estimated by minimizing

$$f_w(b_0, b_1) = \sum w_i [y_i - (b_0 + b_1 x_i)]^2 \tag{13.4}$$

where the w_i's are weights which decrease with increasing x_i. Minimization of (13.4) yields **weighted least squares** estimates. For example, if the standard deviation of Y is proportional to x (for $x > 0$)—that is, $\text{Var}(Y) = kx^2$—then it can be shown that the weights $w_i = 1/x_i^2$ yield best estimators of β_0 and β_1. The books by Neter and Wasserman and by Chaterjee and Price contain more detail. Weighted least squares is used quite frequently by econometricians (economists who use statistical methods) to estimate parameters.

When plots or other evidence suggest that the data set contains outliers or points having large influence on the resulting fit, one possible approach is to omit these outlying points and recompute the estimated regression equation. This would certainly be correct if it was found that the outliers resulted from errors in recording data values or experimental errors. If no assignable cause can be found for the outliers, it is still desirable to report the estimated equation both with and without outliers omitted. Yet another approach is to retain possible outliers but to use an estimation principle which puts relatively less weight on outlying values than does the principle of least squares. One such principle is MAD (minimize absolute deviations), which selects $\hat{\beta}_0$ and $\hat{\beta}_1$ to minimize $\sum |y_i - (b_0 + b_1 x_i)|$. Unlike the estimates of least squares, there are no nice formulas for the MAD estimates: Their values must be found by using an iterative computational procedure. More information about alternative fitting procedures can be found in Chapter 14 of *Data Analysis and Regression* by Mosteller and Tukey. Such procedures are also used when it is suspected that the ϵ_i's have a distribution which is not normal but instead has "heavy tails" (making it much more likely than for the normal distribution that discrepant values will enter the sample); robust regression procedures are those which produce reliable estimates for a wide variety of underlying error distributions. Least squares estimators are not robust in the same way that the sample mean $\bar{X}$ is not a robust estimator for μ.

When a plot suggests time dependence in the error terms, an appropriate analysis may involve a transformation of the y's or else a model explicitly including a time variable. Lastly, a plot such as that of Figure 13.2f, which shows a pattern in the residuals when plotted against an omitted variable, suggests that a multiple regression model which includes the previously omitted variable should be considered. We discuss such models in the last two sections of the chapter.

Robust Regression

In Chapter 6 we discussed robustness in point estimation, and in particular the concept of an M-estimate of the center θ of a symmetric distribution. The estimate is determined by solving $\sum \psi[(x_i - \theta)/cs_1] = 0$, where ψ is selected so that discrepant observations have little influence on the estimate and s_1 is a robust estimate of scale. If in a simple linear regression problem it appears that the distribution of ϵ is heavy-tailed in comparison to the normal distribution, use of the least squares estimates will give too much influence to the few discrepant y's which are likely to appear in the sample. To protect against such eventualities, we can look for robust estimators of β_0 and β_1 which are nearly as good as the least squares estimators when observations are normally distributed and significantly better for a heavy-tailed distribution.

As in the case of a location parameter θ, M-estimates of β_0 and β_1 are obtained by finding b_0 and b_1 which minimize $f(b_0, b_1) = \Sigma \, \rho[(y_i - \beta_0 - \beta_1 x_i)/cs_1]$, where s_1 is a robust estimate of scale (not $\sqrt{SSE/(n - 2)}$. The choice $\rho(u) = u^2$ yields least squares estimates. With $\psi(u) = \rho'(u)$, the estimates are obtained by solving $\partial f/\partial b_0 = \partial f/\partial b_1 = 0$. These equations become

$$\sum_i \psi\left(\frac{y_i - b_0 - b_1 x_i}{cs_1}\right) = 0$$

$$\sum_i \psi\left(\frac{y_i - b_0 - b_1 x_i}{cs_1}\right) x_i = 0$$

(13.5)

The use of the biweight function ψ [expression (6.10) of Section 6.2] is recommended.

The equations (13.5) must in general be solved iteratively. Let $b_0^{(1)}$ and $b_1^{(1)}$ be initial estimates of β_0 and β_1 obtained by minimizing the sum of absolute deviations $\Sigma \, |y_i - b_0 - b_1 x_i|$; there are efficient computer programs available for obtaining $b_0^{(1)}$ and $b_1^{(1)}$. Then let

$$s_1 = \text{median} \frac{|y_i - b_0^{(1)} - b_1^{(1)} x_i|}{.6745}$$

Newton's method can now be used to obtain $b_0^{(2)}$, $b_1^{(2)}$, which can in turn be used to determine $b_0^{(3)}$, $b_1^{(3)}$, and so on. An alternative procedure which does not require ψ' is to use iteratively reweighted least squares. Let

$$w_i = \frac{\psi[y_i - b_0^{(1)} - b_1^{(1)} x_i)/cs_1]}{(y_i - b_0^{(1)} - b_1^{(1)} x_i)/cs_1} \quad i = 1, \ldots, n$$

Then $b_0^{(2)}$ and $b_1^{(2)}$ are found by minimizing $\Sigma \, w_i(y_i - b_0 - b_1 x_i)^2$, the weighted sum of squared deviations. The resulting normal equations are

$$\left(\sum w_i\right) b_0 + \left(\sum w_i x_i\right) b_1 = \sum w_i y_i$$

$$\left(\sum w_i x_i\right) b_0 + \left(\sum w_i x_i^2\right) b_1 = \sum w_i x_i y_i$$

At each iteration the weights are recomputed and the equations resolved using the new weights. This continues until a termination condition is satisfied.

While packaged programs for carrying out this iteratively reweighted least squares analysis are not at present widely available, they are expected to be available in the near future. Although the resulting estimated regression function can be used to obtain point estimates and predictions, information about the distributions of the estimators is limited, so formulas for robust confidence intervals, prediction intervals, and test procedures are also not yet available.

Exercises / Section 13.1

1. Example 12.3 presented the residuals from a simple linear regression of bark lead content y on traffic flow x.
 a. Plot the residuals against x. Does the resulting plot suggest that a straight-line regression function is a reasonable choice of model? Explain your reasoning.
 b. Using $s = 92.19$, compute the values of the standardized residuals. Is $e_i^* \doteq e_i/s$ for $i = 1, \ldots, n$, or are the e_i^*'s not close to being proportional to the e_i's?
 c. Plot the standardized residuals against x. Does the plot differ significantly in general appearance from the plot of (a)?

2. Exercise 5 in Section 12.3 presented data on $y =$ magnetic relaxation time and $x =$ strength of the external magnetic field. A simple linear regression yields the estimated regression function $y = -8.78 + 18.87x$, with $s = 16.61$.
 a. Compute the values of the residuals and plot the residuals against x. Does the plot suggest that a linear regression function is inappropriate?
 b. Compute the values of the standardized residuals and plot them against x. Are there any unusually large (positive or negative) standardized residuals? Does this plot give the same message as the plot of (a) as far as the appropriateness of a linear regression function?
 c. Plot the predicted against the observed values. Does the plot indicate that the model gives reliable predictions?

3. The article "Effects of Gamma Radiation on Juvenile and Mature Cuttings of Quaking Aspen" (*Forest Science*, 1967, pp. 240–245) reported the following data on exposure time to radiation (x, in kr/16 hr) and dry weight of roots (y, in mg $\times 10^{-1}$).

x	0	2	4	6	8
y	110	123	119	86	62

 a. Construct a scatter plot. Does the plot suggest that a linear probabilistic relationship is appropriate?
 b. A linear regression results in the least squares line $y = 127 - 6.65x$, with $s = 16.94$. Compute the residuals and standardized residuals, and then construct residual plots. What do these plots suggest? What type of function should provide a better fit to the data than a straight line does?

4. Exercise 4 in Section 12.4 reported data on $x =$ age of tree and $y =$ annual trunk diameter growth increment. The estimated regression line is $y = 1.78 - .0154x$, with $s = .434$. Does a scatter plot suggest that a particular observation has excessively influenced the choice of a best fit line? Is this impression confirmed by the standardized residuals?

5. Consider the following four (x, y) data sets; the first three have the same x values, so these values are listed only once (from Frank Anscombe, "Graphs in Statistical Analysis," *Amer. Statistician*, 1973, pp. 17–21).

Data set	1–3	1	2	3	4	4
Variable	x	y	y	y	x	y
	10.0	8.04	9.14	7.46	8.0	6.58
	8.0	6.95	8.14	6.77	8.0	5.76
	13.0	7.58	8.74	12.74	8.0	7.71
	9.0	8.81	8.77	7.11	8.0	8.84
	11.0	8.33	9.26	7.81	8.0	8.47
	14.0	9.96	8.10	8.84	8.0	7.04
	6.0	7.24	6.13	6.08	8.0	5.25
	4.0	4.26	3.10	5.39	19.0	12.50
	12.0	10.84	9.13	8.15	8.0	5.56
	7.0	4.82	7.26	6.42	8.0	7.91
	5.0	5.68	4.74	5.73	8.0	6.89

For each of these four data sets, the values of the summary statistics Σx_i, Σx_i^2, Σy_i, Σy_i^2, and Σx_iy_i are identical, so all quantities computed from these five will be identical for the four sets— the least squares line ($y = 3 + .5x$), SSE, s^2, r^2, t intervals, t statistics, and so on. The summary statistics provide no way of distinguishing among the four data sets. Based on a scatter plot and a residual plot for each set, comment on the appropriateness or inappropriateness of fitting a straight-line model; include in your comments any specific suggestions for how a "straight-line analysis" might be modified or qualified.

6. a. Show that $\sum_{i=1}^{n} e_i = 0$ when the e_i's are the residuals from a simple linear regression.
   b. Are the residuals from a simple linear regression independent of one another, positively correlated, or negatively correlated? Explain.
   c. Show that $\sum_{i=1}^{n} x_ie_i = 0$ for the residuals from a simple linear regression. [This result along with (a) shows that there are two linear restrictions on the e_i's, resulting in a loss of 2 d.f. when the squared residuals are used to estimate σ^2].
   d. Is it true that $\sum_{i=1}^{n} e_i^* = 0$? Give a proof or counterexample.

7. a. Express the ith residual $Y_i - \hat{Y}_i$ (where $\hat{Y}_i = \hat{\beta}_0 + \hat{\beta}_1x_i$) in the form Σc_jY_j, a linear function of the Y_j's. Then use rules of variance to verify that $\text{Var}(Y_i - \hat{Y}_i)$ is given by Expression (13.2).
   b. It can be shown that $\hat{Y}_i$ and $Y_i - \hat{Y}_i$ (the ith predicted value and residual) are independent of one another. Use this fact, the relation $Y_i = \hat{Y}_i + (Y_i - \hat{Y}_i)$, and the expression for $\text{Var}(\hat{Y})$ from Section 12.3 to again verify (13.2).
   c. As x_i moves further away from $\bar{x}$, what happens to $\text{Var}(\hat{Y}_i)$ and to $\text{Var}(Y_i - \hat{Y}_i)$?

8. a. Could a linear regression result in residuals 23, -27, 5, 17, -8, 9, and 15? Why or why not?
   b. Could a linear regression result in residuals 23, -27, 5, 17, -8, -12, and 2 corresponding to x values 3, -4, 8, 12, -14, -20, and 25? Why or why not? *Hint:* See Exercise 6.

9. Recall that $\hat{\beta}_0 + \hat{\beta}_1x$ has a normal distribution

with expected value $\beta_0 + \beta_1x$ and variance

$$\sigma^2 \left\{ \frac{1}{n} + \frac{(x - \bar{x})^2}{\Sigma (x_i - \bar{x})^2} \right\}$$

so that

$$Z = \frac{\hat{\beta}_0 + \hat{\beta}_1x - (\beta_0 + \beta_1x)}{\sigma \left(\dfrac{1}{n} + \dfrac{(x - \bar{x})^2}{\Sigma (x_i - \bar{x})^2} \right)^{1/2}}$$

has a standard normal distribution. If $S = \sqrt{SSE/(n - 2)}$ is substituted for σ, the resulting variable has a t distribution with $n - 2$ degrees of freedom. By analogy, what is the distribution of any particular standardized residual? If $n = 25$, what is the probability that a particular standardized residual falls outside the interval $(-2.50, 2.50)$?

10. If there is at least one x value at which more than one Y has been observed, there is a formal test procedure for testing

$H_0: \mu_{Y \cdot x} = \beta_0 + \beta_1x$ for some values β_0, β_1
(the true regression function is linear)

versus

$H_a: H_0$ is not true (the true regression function is not linear)

Suppose that observations are made at x_1, $x_2, \ldots, x_c$, and let $Y_{11}, Y_{12}, \ldots, Y_{1n_1}$ denote the n_1 observations when $x = x_1; \ldots; Y_{c1}$, $Y_{c2}, \ldots, Y_{cn_c}$ denote the n_c observations when $x = x_c$. With $n = \Sigma n_i$ (the total number of observations), SSE has $n - 2$ d.f. We break SSE into two pieces $SSPE$ (pure error) and $SSLF$ (lack of fit) as follows:

$$SSPE = \sum_i \sum_j (Y_{ij} - \bar{Y}_{i.})^2$$

$$= \sum \sum Y_{ij}^2 - \sum n_i\bar{Y}_{i.}^2$$

$$SSLF = SSE - SSPE$$

The n_i observations at x_i contribute $n_i - 1$ d.f. to $SSPE$, so the number of d.f. for $SSPE$ is $\sum_i (n_i - 1) = n - c$ and d.f. for $SSLF$ is $n - 2 - (n - c) = c - 2$. Let $MSPE = SSPE/(n - c)$ and $MSLF = SSLF/(c - 2)$. Then it can be shown that while $E(MSPE) = \sigma^2$ whether or not H_0 is true,

$E(MSLF) = \sigma^2$ if H_0 is true and $E(MSLF) > \sigma^2$ if H_0 is false.

Test statistic: $F = \dfrac{MSLF}{MSPE}$

Rejection region: $F \geq F_{\alpha, c-2, n-c}$

The following data comes from the paper "Changes in Growth Hormone Status Related to Body Weight of Growing Cattle" (*Growth*, 1977, pp. 241–247), with x = body weight and y = metabolic clearance rate/body weight.

x	110	110	110	230	230	230	360
y	235	198	173	174	149	124	115

x	360	360	360	505	505	505	505
y	130	102	95	122	112	98	96

(so $c = 4$, $n_1 = n_2 = 3$, $n_3 = n_4 = 4$).

a. Test H_0 versus H_a at level .05 using the above lack of fit test.

b. Does a scatter plot of the data suggest that the relationship between x and y is linear? How does this compare with the result of (a)? (A nonlinear regression function was used in the paper.)

13.2 Regression with Transformed Variables

The necessity for an alternative model to the linear probabilistic model $Y = \beta_0 + \beta_1 x + \epsilon$ may be suggested either by a theoretical argument or else by examining diagnostic plots from a linear regression analysis. In either case it is desirable to settle on a model whose parameters can be easily estimated. An important class of such models is specified by means of functions which are "intrinsically linear."

Definition: A function relating y to x is **intrinsically linear** if by means of a transformation on x and/or y, the function can be expressed as $y' = \beta_0 + \beta_1 x'$, where x' = the transformed independent variable and y' = the transformed dependent variable.

Four of the most useful intrinsically linear functions are given in Table 13.1 below. In each case the appropriate transformation is either a log transformation—either base 10 or natural logarithm (base e)—or a reciprocal transformation. Representative graphs of the four functions appear in Figure 13.3.

Table 13.1 *Useful Intrinsically Linear Functions**

Function	Transformation(s) to linearize	Linear form
(a) Exponential: $y = \alpha e^{\beta x}$	$y' = \ln(y)$	$y' = \ln(\alpha) + \beta x$
(b) Power: $y = \alpha x^{\beta}$	$y' = \log(y)$, $x' = \log(x)$	$y' = \log(\alpha) + \beta x'$
(c) $y = \alpha + \beta \cdot \log(x)$	$x' = \log(x)$	$y = \alpha + \beta x'$
(d) Reciprocal: $y = \alpha + \beta \cdot \dfrac{1}{x}$	$x' = \dfrac{1}{x}$	$y = \alpha + \beta x'$

*When log ($\cdot$) appears, either a base 10 or base e logarithm can be used.

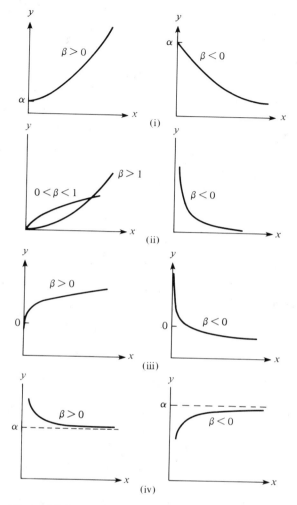

Figure 13.3 Graphs of the intrinsically linear functions given in Table 13.1

Thus for an exponential function relationship, only y is transformed to achieve linearity, while for a power function relationship, both x and y are transformed. Because the variable x is in the exponent in an exponential relationship, y increases (if $\beta > 0$) or decreases (if $\beta < 0$) much more rapidly as x increases than is the case for the power function, though over a short interval of x values it can be difficult to differentiate between the two functions. Examples of functions which are not intrinsically linear are $y = \alpha + \gamma e^{\beta x}$ and $y = \alpha + \gamma x^{\beta}$.

Intrinsically linear functions lead directly to probabilistic models which, though not linear in y as a function, have parameters whose values are easily estimated using ordinary least squares.

> **Definition:** A probabilistic model relating Y to x is **intrinsically linear** if by means of a transformation on Y and/or x, it can be reduced to a linear probabilistic model $Y' = \beta_0 + \beta_1 x' + \epsilon'$.

Corresponding to the four functions of Table 13.1, the intrinsically linear probabilistic models are

a. $Y = \alpha e^{\beta x} \cdot \epsilon$, a multiplicative exponential model, so that $\ln(Y) = Y' = \beta_0 + \beta_1 x' + \epsilon'$ with $x' = x$, $\beta_0 = \ln(\alpha)$, $\beta_1 = \beta$, and $\epsilon' = \ln(\epsilon)$;

b. $Y = \alpha x^{\beta} \cdot \epsilon$, a multiplicative power model, so that $\log(Y) = Y' = \beta_0 + \beta_1 x' + \epsilon'$ with $x' = \log(x)$, $\beta_0 = \log(\alpha)$, $\beta_1 = \beta$, and $\epsilon' = \log(\epsilon)$;

c. $Y = \alpha + \beta \log(x) + \epsilon$, so that $x' = \log(x)$ immediately linearizes the model;

d. $Y = \alpha + \beta \cdot 1/x + \epsilon$, so that $x' = 1/x$ yields a linear model.

The additive exponential and power models, $Y = \alpha e^{\beta x} + \epsilon$ and $Y = \alpha x^{\beta} + \epsilon$, are not intrinsically linear. Notice that both (a) and (b) require a transformation on Y and, as a result, a transformation on the error variable ϵ. In fact if ϵ has a lognormal distribution (see Chapter 4) with $E(\epsilon) = 1$ and $\text{Var}(\epsilon) = \tau^2$ independent of x, then the transformed models for both (a) and (b) will satisfy all the assumptions of Chapter 12 regarding the linear probabilistic model; this in turn implies that all inferences for the parameters of the transformed model based on these assumptions will be valid.

The major advantage of an intrinsically linear model is that the parameters β_0 and β_1 of the transformed model can be immediately estimated using the principle of least squares simply by substituting x' and y' into the estimating formulas:

$$\hat{\beta}_1 = \frac{n \sum x_i' y_i' - \sum x_i' \sum y_i'}{n \sum (x_i')^2 - (\sum x_i')^2}$$

$$\hat{\beta}_0 = \frac{\sum y_i' - \hat{\beta}_1 \sum x_i'}{n}$$

(13.6)

Parameters of the original nonlinear model can then be estimated by transforming back $\hat{\beta}_0$ and/or $\hat{\beta}_1$ if necessary.

Example 13.2   Taylor's equation for tool life y as a function of cutting time x states that $xy^c = k$, or equivalently that $y = \alpha x^{\beta}$. The paper "The Effect of Experimental Error on the Determination of Optimum Metal Cutting Conditions" (*J. Eng. for Industry*, 1967, pp. 315–322) observes that the relationship is not exact (deterministic), and that the parameters α and β must be estimated from data. Thus an appropriate model is the multiplicative power model $Y = \alpha \cdot x^{\beta} \cdot \epsilon$, which the author fit to the accompanying data consisting of 12 carbide tool life observations. In addition to the x, y,

x', and y' values, the predicted transformed values ($\hat{y}'$) and the predicted values on the original scale ($\hat{y}$, after transforming back) are given.

Table 13.2

	x	y	$x' = \ln(x)$	$y' = \ln(y)$	$\hat{y}'$	$\hat{y} = e^{\hat{y}'}$
1	600.	2.3500	6.39693	0.85442	1.12754	3.0881
2	600.	2.6500	6.39693	0.97456	1.12754	3.0881
3	600.	3.0000	6.39693	1.09861	1.12754	3.0881
4	600.	3.6000	6.39693	1.28093	1.12754	3.0881
5	500.	6.4000	6.21461	1.85630	2.11203	8.2650
6	500.	7.8000	6.21461	2.05412	2.11203	8.2650
7	500.	9.8000	6.21461	2.28238	2.11203	8.2650
8	500.	16.5000	6.21461	2.80336	2.11203	8.2650
9	400.	21.5000	5.99146	3.06805	3.31694	27.5760
10	400.	24.5000	5.99146	3.19867	3.31694	27.5760
11	400.	26.0000	5.99146	3.25810	3.31694	27.5760
12	400.	33.0000	5.99146	3.49651	3.31694	27.5760

The summary statistics for fitting a straight line to the transformed data are $\Sigma\, x_i' = 74.41200$, $\Sigma\, y_i' = 26.22601$, $\Sigma\, x_i'^2 = 461.75874$, $\Sigma\, y_i'^2 = 67.74609$, and $\Sigma\, x_i'y_i' = 160.84601$, so

$$\hat{\beta}_1 = \frac{12(160.84601) - (74.41200)(26.22601)}{12(461.75874) - (74.41200)^2} = -5.3996$$

$$\hat{\beta}_0 = \frac{26.22601 - (-5.3996)(74.41200)}{12} = 35.6684$$

The estimated values of α and β, the parameters of the power function model, are $\hat{\beta} = \hat{\beta}_1 = -5.3996$ and $\hat{\alpha} = e^{\hat{\beta}_0} = 3.094491530 \cdot 10^{15}$. Thus the estimated regression function is $\hat{\mu}_{Y \cdot x} = 3.094491530 \cdot 10^{15} \cdot x^{-5.3996}$. To recapture Taylor's (estimated) equation, set $y = 3.094491530 \cdot 10^{15} \cdot x^{-5.3996}$, whence $xy^{.185} = 740$.

Figure 13.4(a) gives a plot of the standardized residuals from the linear regression using transformed variables (for which $r^2 = .922$); there is no apparent pattern in the plot, though one standardized residual is a bit large, and the residuals look as they should for a simple linear regression. Figure 13.4(b) pictures a plot of $\hat{y}$ versus y, which indicates satisfactory predictions on the original scale.

To obtain a confidence interval for true average tool life when cutting time is 500, we transform $x = 500$ to $x' = 6.21461$. Then $\hat{\beta}_0 + \hat{\beta}_1 x' = 2.1120$, and a 95% confidence interval for $\beta_0 + \beta_1(6.21461)$ is (from Section 12.3) $2.1120 \pm (2.228)(.0824) = (1.928, 2.296)$. The 95% confidence interval for $\mu_{Y \cdot 500}$ is then obtained by "exponentiating" each endpoint: $(e^{1.928}, e^{2.296}) = (6.876, 9.930)$. Other inferences and predictions can be made in a similar fashion by working first with transformed values and then transforming back to the original scale.

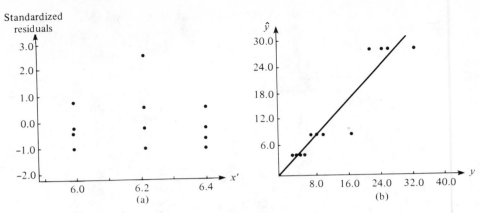

Figure 13.4 (a) Standardized residuals versus x' from Example 13.3; (b) $\hat{y}$ versus y from Example 13.3

Example 13.3 In the article "Ethylene Synthesis in Lettuce Seeds: Its Physiological Significance" (*Plant Physiology*, 1972, pp. 719–722), ethylene content of lettuce seeds (y, in nl/g dry wt) was studied as a function of exposure time (x, in min) to an ethylene absorbant. Figure 13.5 presents both a scatter plot of the data and a plot of the residuals generated from a linear regression of y on x. Both plots show a strong curved pattern, suggesting that a transformation to achieve linearity is appropriate. In addition, a linear regression gives negative predictions for $x = 90$ and $x = 100$.

The author did not give any argument for a theoretical model, but his plot of $y' = \ln(y)$ versus x shows a strong linear relationship, suggesting that an exponential function will provide a good fit to the data. Below we record the data values and other information from a linear regression of y' on x. The estimates of parameters of the linear model are $\hat{\beta}_1 = -.0323$ and $\hat{\beta}_0 = 5.941$, with $r^2 = .995$. The estimated regression function for the exponential model is $\hat{\mu}_{Y \cdot x} = e^{\hat{\beta}_0} \cdot e^{\hat{\beta}_1 x} =$

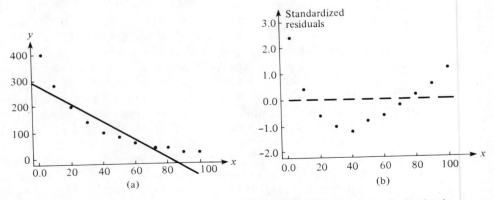

Figure 13.5 (a) Scatter plot; (b) residual plot from linear regression for the data in Example 13.3

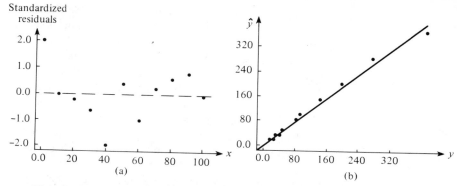

Figure 13.6 Plot of (a) standardized residuals (after transforming) versus x; (b) $\hat{y}$ versus y for data in Example 13.3

$380.32e^{-.0323x}$. The predicted values $\hat{y}_i$ can then be obtained by substitution of $x_i(i = 1, \ldots, n)$ into $\hat{\mu}_{Y \cdot x}$ or else by computing $\hat{y}_i = e^{\hat{y}_i'}$ where the $\hat{y}_i'$'s are the predictions from the transformed straight-line model. Figure 13.6 presents both a plot of $e_i'^*$ versus x (the standardized residuals from a linear regression) and a plot of $\hat{y}$ versus y. These plots support the choice of an exponential model.

Table 13.3

x	y	$y' = \ln(y)$	$\hat{y}'$	$\hat{y} = e^{\hat{y}_i'}$
2	408	6.01	5.876	353.32
10	274	5.61	5.617	275.12
20	196	5.28	5.294	199.12
30	137	4.92	4.971	144.18
40	90	4.50	4.647	104.31
50	78	4.36	4.324	75.50
60	51	3.93	4.001	54.64
70	40	3.69	3.677	39.55
80	30	3.40	3.354	28.62
90	22	3.09	3.031	20.72
100	15	2.71	2.708	15.00

In analyzing transformed data, one should keep in mind the following points:

1. Estimating β_1 and β_0 as in (13.6) above and then transforming back to obtain estimates of the original parameters is not equivalent to using the principle of least squares directly on the original model. Thus for the exponential model, we could estimate α and β by minimizing $\Sigma (y_i - \alpha e^{\beta x_i})^2$. The resulting estimates would not be equal: $\hat{\beta} \neq \hat{\beta}_1$ and $\hat{\alpha} \neq e^{\beta_0}$.

2. If the chosen model is not intrinsically linear, the approach summarized in (13.6) cannot be used. Instead, least squares (or some other fitting procedure) would have to be applied to the untransformed model. Thus for the additive exponential model $Y = \alpha e^{\beta x} + \epsilon$, least squares would involve minimizing $\Sigma (y_i - \alpha e^{\beta x_i})^2$. Taking partial derivatives with respect to α and β results in two

nonlinear normal equations in α and β; these equations must then be solved using an iterative procedure. The book by Draper and Smith gives more details.

3. When the transformed linear model satisfies all the assumptions listed in Chapter 12, least squares yields best estimates of the transformed parameters. However, estimates of the original parameters may not be best in any sense, though they will be reasonable. For example, in the exponential model the estimator $\hat{\alpha} = e^{\hat{\beta}_0}$ will not be unbiased, though it will be the maximum likelihood estimator of α if the error variable ϵ' is normally distributed. It is conceivable that using least squares directly (without transforming) could yield better estimates, though the computations would be quite burdensome.

4. If a transformation on Y has been made and one wishes to use the standard formulas to test hypotheses or construct confidence intervals, ϵ' should be at least approximately normally distributed. To check this, the residuals from the transformed regression should be examined.

5. When Y is transformed, the r^2 value from the resulting regression refers to variation in the y_i''s explained by the transformed regression model. While a high value of r^2 here indicates a good fit of the estimated original nonlinear model to the observed y_i's, r^2 does not refer to these original observations. Perhaps the best way to assess the quality of the fit is to compute the predicted values $\hat{y}_i'$ using the transformed model, transform them back to the original y scale to obtain $\hat{y}_i$, and then plot $\hat{y}$ versus y. A good fit is then evidenced by points close to the 45° line. One could compute $SSE = \Sigma (y_i - \hat{y}_i)^2$ as a numerical measure of the goodness of fit. When the model was linear, we compared this to $SST = \Sigma (y_i - \bar{y})^2$, the total variation about the horizontal line at height $\bar{y}$; this led to r^2. In the nonlinear case, though, it is not necessarily informative to measure total variation in this way, so an r^2 value is not as useful as in the linear case.

Exercises / Section 13.2

1. The accompanying data resulted from an experiment to investigate the relationship between applied stress x and time to rupture y for notched, cold rolled brass specimens exposed to a cracking solution ("On Initiation and Growth of Stress Corrosion Cracks in Tarnished Brass," *J. Electrochemical Soc.*, 1965, pp. 131–138).

x	22.5	25.0	28.0	30.5	38.0	40.5
y	44.0	42.0	33.5	28.0	18.0	13.6

x	42.5	48.0	54.5	55.0	70.0
y	15.0	10.3	9.0	6.3	4.0

A plot on log–log paper shows a pronounced linear pattern, suggesting a power model.

a. Show that fitting $y' = \beta_0 + \beta_1 x' + \epsilon'$, where $y' = \ln(y)$ and $x' = \ln(x)$, results in $\hat{\beta}_1 = -2.16$ and $\hat{\beta}_0 = 10.67$. What are the estimates of the parameters of the power model?

b. Does a residual plot based on the transformed data and analysis confirm the choice of a power model?

c. Does a plot of $\hat{y}$ versus y confirm the efficacy of the power model for prediction of y?

d. Assuming that the power model is correct, test $H_0 : \mu_{Y \cdot x} = \alpha x^{-2}$. $SSE = .128$ for the transformed data.

e. Obtain a 95% confidence interval for the exponent β in the power model.

f. Obtain a 95% confidence interval for expected time to rupture when applied stress $= 40$.

2. The paper "The Luminosity–Spectral Index Relationship for Radio Galaxies" (*Nature*, 1972, pp. 88–89) suggested that for class S galaxies, $\ln(L_{178})$ is linearly related to spectral index, where L_{178} denotes luminosity at 178 MHz. Representative data appears below.

Spectral index	.59	.67	.72	.80	.85	.90
$\ln(L_{178})$	23.6	25.6	26.4	25.7	26.8	26.7

Spectral index	.94	.66	1.00	.86	1.03	.70
$\ln(L_{178})$	27.0	24.9	27.1	27.2	26.9	25.2

a. Estimate the parameters of the exponential model implied by the linear relationship between $\ln(L_{178})$ and spectral index.

b. What value of L_{178} would you predict for a spectral index of .75?

c. Compute a 95% prediction interval for luminosity when the spectral index is .95.

3. The following data on mass rate of burning x and flame length y is representative of that which appeared in the article "Some Burning Characteristics of Filter Paper" (*Combustion Science and Technology*, 1971, pp. 103–120).

x	1.7	2.2	2.3	2.6	2.7	3.0	3.2
y	1.3	1.8	1.6	2.0	2.1	2.2	3.0
x	3.3	4.1	4.3	4.6	5.7	6.1	
y	2.6	4.1	3.7	5.0	5.8	5.3	

a. Estimate the parameters of a power function model.

b. Construct diagnostic plots to check whether a power function is an appropriate model choice.

c. Test $H_0 : \beta = \frac{4}{3}$ versus $H_a : \beta < \frac{4}{3}$, using a level .05 test.

d. Test the null hypothesis which states that expected flame length when burning rate is 5.0 is twice the expected flame length when burning rate is 2.5 against the alternative which states that this is not the case.

4. An investigation of the influence of sodium benzoate concentration on the critical minimum pH necessary for the inhibition of Fe ("Mechanism of the Corrosion Inhibition of Fe by Sodium Benzoate," *Corrosion Science*, 1971, pp. 675–682) yielded the accompanying data, which suggests that expected critical minimum pH is linearly related to the natural logarithm of concentration.

Concentration	.01	.025	.1	.95
pH	5.1	5.5	6.1	7.3

a. What is the implied probabilistic model, and what are the estimates of the model parameters?

b. What critical minimum pH would you predict for a concentration of 1.0? Obtain a 95% prediction interval for critical minimum pH when concentration is 1.0?

5. Thermal endurance tests were performed to study the relationship between temperature and lifetime of polyester enameled wire ("Thermal Endurance of Polyester Enameled Wires Using Twisted Wire Specimens," *IEEE Trans. Insulation*, 1965, pp. 38–44).

Temp.:	200	200	200	200	200	200
Lifetime:	5933	5404	4947	4963	3358	3878
Temp.:	220	220	220	220	220	220
Lifetime:	1561	1494	747	768	609	777
Temp.:	240	240	240	240	240	240
Lifetime:	258	299	209	144	180	184

a. Does a scatter plot of the data suggest a linear probabilistic relationship between lifetime and temperature?

b. What model is implied by a linear relationship between expected ln(lifetime) and 1/temperature? Does a scatter plot of the transformed data appear consistent with this relationship?

c. Estimate the parameters of the model suggested in (b). What lifetime would you predict for a temperature of 220?

d. Because there are multiple observations at each x value, the method in Exercise 13.1.10 can be used to test the null hypothesis which states that the model suggested in (b) is correct. Carry out the test at level .01.

6. Exercise 10 in Section 13.1 presented data on body weight x and MCR/body weight y. Consider the following intrinsically linear functions for specifying the relationship between the two variables: (a) $\ln(y)$ versus x, (b) $\ln(y)$ versus $\ln(x)$, (c) y versus $\ln(x)$, (d) y versus $1/x$, and (e) $\ln(y)$ versus $1/x$. Use any appropriate diagnostic plots and analyses to decide which of these functions you would select to specify a probabilistic model. Explain your reasoning.

7. A plot appearing in the article "Thermal Conductivity of Polyethylene: The Effects of Crystal Size,

Density, and Orientation on the Thermal Conductivity" (*Polymer Eng. and Science*, 1972, pp. 204–208) suggests that the expected value of thermal conductivity y is a linear function of $10^4 \cdot 1/x$ where x is lamellar thickness.

x	240	410	460	490	520	590	745	8300
y	12	14.7	14.7	15.2	15.2	15.6	16.0	18.1

a. Estimate the parameters of the regression function and the regression function itself.
b. Predict the value of thermal conductivity when lamellar thickness is 500 angstroms.

8. In each of the following cases, decide whether the given function is intrinsically linear. If so, identify x' and y', and then explain how a random error term ϵ can be introduced so as to yield an intrinsically linear probabilistic model.
 a. $y = 1/(\alpha + \beta x)$ b. $y = 1/(1 + e^{\alpha + \beta x})$
 c. $y = e^{e^{\alpha + \beta x}}$ (a Gompertz curve)
 d. $y = \alpha + \beta e^{\lambda x}$

9. Suppose that x and y are related according to a probabilistic exponential model $Y = \alpha e^{\beta x} \cdot \epsilon$ with $\text{Var}(\epsilon)$ a constant independent of x (as was the case in the simple linear model $Y = \beta_0 + \beta_1 x + \epsilon$). Is $\text{Var}(Y)$ a constant independent of x [as was the case for $Y = \beta_0 + \beta_1 x + \epsilon$, where $\text{Var}(Y) = \sigma^2$]? Explain your reasoning. Draw a picture of a prototype scatter plot resulting from this model. Answer the same questions for the power model $Y = \alpha x^\beta \cdot \epsilon$.

13.3 Polynomial Regression

The nonlinear yet intrinsically linear models of Section 13.2 involved functions of the independent variable x which were either strictly increasing or strictly decreasing. In many situations either theoretical reasoning or else a scatter plot of the data suggests that the true regression function $\mu_{Y \cdot x}$ has one or more peaks or valleys—that is, at least one relative minimum or maximum. In such cases a polynomial function $y = \beta_0 + \beta_1 x + \ldots + \beta_k x^k$ may provide a satisfactory approximation to the true regression function. The probabilistic model that we shall study in this section is

$$Y = \beta_0 + \beta_1 x + \beta_2 x^2 + \cdots + \beta_k x^k + \epsilon \qquad (13.7)$$

where ϵ is a random error variable with

$$\mu_\epsilon = 0, \qquad \sigma_\epsilon^2 = \sigma^2 \qquad (13.8)$$

From (13.7) and (13.8) it follows immediately that

$$\mu_{Y \cdot x} = \beta_0 + \beta_1 x + \cdots + \beta_k x^k, \quad \sigma_{Y \cdot x}^2 = \sigma^2 \qquad (13.9)$$

In words, the expected value of Y is a kth-degree polynomial function of x, while the variance of Y, which controls the spread of observed values about the regression function, is the same for each value of x. The observed pairs $(x_1, y_1), \ldots, (x_n, y_n)$ are assumed to have been generated independently from the model (13.7). In addition, to obtain confidence intervals for and test hypotheses about the parameters of the model, we shall assume a normal distribution for ϵ, so that the Y_i's will be independent normal variables with mean and variance given by (13.9). Figure 13.7 illustrates both a quadratic and cubic model.

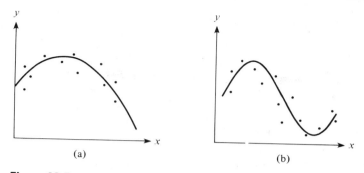

Figure 13.7 (a) Quadratic regression model; (b) cubic regression model

Estimation of Parameters Using Least Squares

In order to estimate $\beta_0, \beta_1, \ldots, \beta_k$, consider a trial regression function $y = b_0 + b_1 x + \cdots + b_k x^k$. Then the "goodness of fit" of this function to the observed data can be measured by computing the sum of squared deviations

$$f(b_0, b_1, \ldots, b_k) = \sum_{i=1}^{n} [y_i - (b_0 + b_1 x_i + b_2 x_i^2 + \cdots + b_k x_i^k)]^2 \quad (13.10)$$

According to the principle of least squares, the estimates $\hat{\beta}_0, \hat{\beta}_1, \ldots, \hat{\beta}_k$ are those values of $b_0, b_1, \ldots, b_k$ which minimize (13.10). It should be noted that when $x_1, x_2, \ldots, x_k$ are all different, there is a polynomial of degree $n - 1$ which fits the data perfectly, so that the minimizing value of (13.10) is zero when $k = n - 1$. However, in virtually all applications the polynomial model (13.7) with k large is quite unrealistic, and in most applications $k = 2$ (quadratic) or $k = 3$ (cubic) is appropriate.

To find the minimizing values in (13.10), we take the $k + 1$ partial derivatives $\partial f/\partial b_0, \partial f/\partial b_1, \ldots, \partial f/\partial b_k$ and equate them to zero, resulting in the system of normal equations for the estimates. Because the trial function $b_0 + b_1 x + \cdots + b_k x^k$ is linear in $b_0, \ldots, b_k$ (though not in x), the $k + 1$ normal equations are linear in these unknowns. The system is

$$
\begin{aligned}
&b_0 n + b_1 \sum x_i + b_2 \sum x_i^2 + \cdots + b_k \sum x_i^k = \sum y_i \\
&b_0 \sum x_i + b_1 \sum x_i^2 + b_2 \sum x_i^3 + \cdots + b_k \sum x_i^{k+1} = \sum x_i y_i \quad (13.11) \\
&\vdots \qquad\qquad\qquad \vdots \qquad\qquad\qquad \vdots \\
&b_0 \sum x_i^k + b_1 \sum x_i^{k+1} + \cdots + b_k \sum x_i^{2k} = \sum x_i^k y_i
\end{aligned}
$$

All standard statistical computer packages will automatically solve the system (13.11) and print out the estimates. To solve by hand, the values of the summary statistics $\Sigma\, x_i$, $\Sigma\, x_i^2$, . . . , $\Sigma\, x_i^{2k}$, $\Sigma\, y_i$, $\Sigma\, x_i y_i$, . . . , and $\Sigma\, x_i^k y_i$ must first be computed. Then after substitution into (13.11), the system can be solved by Gaussian elimination.

For the special case of quadratic regression, explicit formulas for the estimates [the solution of (13.11)] are (with $\overline{x^2} = \Sigma\, x_i^2/n$)*

$$\hat{\beta}_1 = \frac{s_{1y}s_{22} - s_{2y}s_{12}}{s_{11}s_{22} - s_{12}^2}, \; \hat{\beta}_2 = \frac{s_{2y}s_{11} - s_{1y}s_{12}}{s_{11}s_{22} - s_{12}^2}, \; \hat{\beta}_0 = \overline{y} - \hat{\beta}_1\overline{x} - \hat{\beta}_2\overline{x^2} \qquad (13.12)$$

where

$$s_{1y} = \sum x_i y_i - n\overline{x}\,\overline{y}, \; s_{2y} = \sum x_i^2 y_i - n\overline{x^2}\,\overline{y}$$

$$s_{11} = \sum x_i^2 - n\overline{x^2}, \; s_{12} = \sum x_i^3 - n\overline{x}\,\overline{x^2} \qquad (13.13)$$

and

$$s_{22} = \sum x_i^4 - n(\overline{x^2})^2$$

Example 13.4   The article "Determination of Biological Maturity and Effect of Harvesting and Drying Conditions on Milling Quality of Paddy" (*J. Agricultural Eng. Research*, 1975, pp. 353–361) reported the following data on date of harvesting x (number of days after flowering) and yield y (kg/ha) of paddy, a grain farmed in India.

x:	16	18	20	22	24	26	28	30
y:	2508	2518	3304	3423	3057	3190	3500	3883

x:	32	34	36	38	40	42	44	46
y:	3823	3646	3708	3333	3517	3241	3103	2776

Both a scatter plot and a residual plot from a linear regression suggest that the authors' choice of a quadratic model is reasonable. The summary quantities are $\Sigma\, x_i = 496$, $\Sigma\, x_i^2 = 16{,}736$, $\Sigma\, x_i^3 = 603{,}136$, $\Sigma\, x_i^4 = 22{,}825{,}094$, $\Sigma\, y_i = 52{,}530$, $\Sigma\, x_i y_i = 1{,}645{,}108$, and $\Sigma\, x_i^2 y_i = 55{,}565{,}884$, so that $\overline{x} = 31.0000$, $\overline{x^2} = 1046.0000$, and $\overline{y} = 3283.1250$. Substituting into (13.13) gives

$$s_{1y} = 1{,}645{,}108 - 16(31)(3283.125) = 16{,}678$$

$$s_{2y} = 55{,}565{,}884 - 16(1046)(3283.125) = 619{,}504$$

$$s_{11} = 1360, \; s_{12} = 84{,}320, \text{ and } s_{22} = 5{,}319{,}238, \text{ yielding}$$

$$\hat{\beta}_1 = \frac{(16{,}678)(5{,}319{,}238) - (619{,}504)(84{,}320)}{(1360)(5{,}319{,}238) - (84{,}320)^2} = 293.4618$$

* If all s_{ij}'s are multiplied by n, formulas like those of simple linear regression, which involve $\Sigma\, x_i$, $\Sigma\, x_i^2$, and the like rather than $\overline{x}$, $\overline{x^2}$, and so on will result, but then the numbers in the numerator and denominator of β_1 and β_2 may be extremely large.

$$\hat{\beta}_2 = \frac{(619,504)(1360) - (16,678)(84,320)}{(1360)(5,319,238) - (84,320)^2} = -4.5355$$

$$\hat{\beta}_0 = 3283.125 - (293.4618)(31) - (-4.5355)(1046) = -1070.0578$$

The estimated regression function is then $y = -1070.05 + 293.46x - 4.54x^2$, from which predicted values and residuals can be computed. A residual plot indicates that a quadratic regression function is appropriate, though there is somewhat more spread in the residuals for small x than for large x. Weighted least squares can be used to obtain a more satisfactory plot, but the authors did not consider this.

$\hat{\sigma}^2$ and R^2

To make further inferences about the parameters of the regression function, the error variance σ^2 must be estimated. With $\hat{y}_i = \hat{\beta}_0 + \hat{\beta}_1 x_i + \cdots + \hat{\beta}_k x_i^k$, the ith residual is $y_i - \hat{y}_i$ and the sum of squared residuals (error sum of squares) is $SSE = \Sigma (y_i - \hat{y}_i)^2$. A computational formula for SSE is

$$SSE = \sum y_i^2 - \hat{\beta}_0 \sum y_i - \hat{\beta}_1 \sum x_i y_i - \cdots - \hat{\beta}_k \sum x_i^k y_i \qquad (13.14)$$

which can be verified by squaring the residuals and summing the resulting terms. The estimate for σ^2 is then

$$\hat{\sigma}^2 = s^2 = \frac{SSE}{n - (k + 1)} = MSE \qquad (13.15)$$

where the denominator $n - (k + 1)$ is used because $k + 1$ degrees of freedom are lost in estimating $\beta_0, \beta_1, \ldots, \beta_k$.

Example 13.4
(continued)

After computing $\Sigma y_i^2 = 175,087,792$,

$$SSE = 175,087,792 - (-1070.0578)(52,530) - (293.4618)(1,645,108)$$
$$- (-4.5355)(55,565,884) = 540,640.2$$

so

$$s^2 = \frac{540,640.2}{16 - 3} = 41,587.71$$

and

$$\hat{\sigma} = s = 203.93$$

If we again let $SST = \Sigma (y_i - \bar{y})^2$, then SSE/SST is the proportion of the total variation in the observed y_i's which is not explained by the polynomial model. The quantity $1 - SSE/SST$, the proportion of variation explained by the model, is called the **coefficient of multiple determination** and is denoted by R^2. For the data of

Example 13.4 $SST = 2,625,235.7$, so $R^2 = 1 - (540,640.2)/(2,625,235.7) = .794$ and the quadratic model explains 79.4% of the observed variation in y.

Suppose that we consider fitting a cubic model to the data in Example 13.4. Because the cubic model includes the quadratic as a special case, the fit to a cubic will be at least as good as the fit to a quadratic. More generally, with $SSE_k = $ the error sum of squares from a kth degree polynomial, $SSE_{k'} \leq SSE_k$ and $R_{k'}^2 \geq R_k^2$ whenever $k' > k$. Because the objective of regression analysis is to find a model which is both simple (relatively few parameters) and provides a good fit to the data, a higher-degree polynomial may not specify a better model than a lower-degree model in spite of its higher R^2 value. To balance the cost of using more parameters against the gain in R^2, many statisticians use the **adjusted coefficient of multiple determination**

$$\text{adjusted } R^2 = 1 - \frac{n-1}{n-(k+1)} \cdot \frac{SSE}{SST} = \frac{(n-1)R^2 - k}{n-1-k} \tag{13.16}$$

Adjusted R^2 adjusts the proportion of unexplained variation upward (since $(n-1)/(n-k-1) > 1$), which results in adjusted $R^2 < R^2$. Thus if $R_2^2 = .66$, $R_3^2 = .70$, and $n = 10$, then

$$\text{adjusted } R_2^2 = \frac{9(.66) - 2}{10 - 3} = .563$$

$$\text{adjusted } R_3^2 = \frac{9(.70) - 3}{10 - 4} = .550$$

so the small gain in R^2 in going from a quadratic to a cubic model is not enough to offset the cost of adding an extra parameter to the model.

In addition to computing R^2 and adjusted R^2, one should examine the usual diagnostic plots to determine whether model assumptions are valid or whether modification may be appropriate.

Confidence Intervals and Test Procedures

Because the y_i's appear in the normal equations (13.11) only on the right-hand side and in a linear fashion, the resulting estimates $\hat{\beta}_0, \ldots, \hat{\beta}_k$ are themselves linear functions of the y_i's. Thus the estimators are linear functions of the Y_i's, so each $\hat{\beta}_i$ has a normal distribution. It can also be shown that each $\hat{\beta}_i$ is an unbiased estimator of β_i and that

$$T = \frac{\hat{\beta}_i - \beta_i}{S_{\hat{\beta}_i}} \tag{13.17}$$

has a t distribution with $n - (k+1)$ degrees of freedom, where $S_{\hat{\beta}_i}$, the estimated standard deviation of $\hat{\beta}_i$, is obtained by substituting S into $\sigma_{\hat{\beta}_i}$. A confidence interval for β_i is then $\hat{\beta}_i \pm t_{\alpha/2, n-k-1} \cdot s_{\hat{\beta}_i}$, while to test $H_0 : \beta_i = \beta_{i0}$, β_{i0} is substituted into (13.17) for β_i and the appropriate t_α or $t_{\alpha/2}$ value with $n - k - 1$ d.f. is used.

All of this requires $\sigma_{\hat{\beta}_i}$, and here lies the rub. The standard deviation of $\hat{\beta}_i$ involves multiplying σ by a complicated function of the x_i's. In Section 13.5 we use matrix algebra notation to obtain expressions for these standard deviations, as well as for the standard deviation of $\hat{\beta}_0 + \hat{\beta}_1 x + \cdots + \hat{\beta}_k x^k$; without such notation, it is extremely tedious to write down these expressions. Fortunately, the most frequently used statistical computer program packages—MINITAB, BMD, SAS, and SPSS—all compute and print out an estimated standard deviation for each estimated coefficient, and several give also the estimated standard deviation of $\hat{\beta}_0 + \hat{\beta}_1 x_i + \cdots + \hat{\beta}_k x_i^k$ for $i = 1, \ldots, n$, from which confidence and prediction intervals can be constructed and tests regarding $\mu_{Y \cdot x_i}$ can be carried out.

Example 13.4
(continued)

A computer analysis yielded the following (partial) results for the date of harvesting/paddy yield data.

Table 13.4

Coefficient	Estimate $\hat{\beta}_i$	Estimated S.D. $s_{\hat{\beta}_i}$
β_0	-1070.05	617.25
β_1	293.46	42.18
β_2	-4.54	.674

Table 13.5

x_i	y_i	$\hat{y}_i = \hat{\mu}_{Y \cdot x_i}$	Estimated S.D. of $\hat{\mu}_{Y \cdot x_i}$
16	2508	2464	136
18	2518	2743	105
.	.	.	.
.	.	.	.
28	3500	3591	74
30	3883	3652	76
32	3823	3676	76
.	.	.	.
.	.	.	.
46	2776	2832	136

a. For testing $H_0: \beta_2 = 0$ versus $H_a: \beta_2 \neq 0$ (H_0 here says that the quadratic term in the model is unnecessary), the test statistic is $T = \hat{\beta}_2 / S_{\hat{\beta}_2}$, with computed value $-4.54/.674 = -6.74$. The test is based on $n - (k + 1) = 16 - 3 = 13$ d.f.; with $t_{.025, 13} = 2.179$, H_0 is rejected at level .05 if either $t \geq 2.179$ or $t \leq -2.179$. Since $-6.73 \leq -2.179$, H_0 is rejected at level .05, validating the inclusion of the quadratic term.

b. A 95% confidence interval for $\mu_{Y \cdot 30}$ [$= \beta_0 + \beta_1(30) + \beta_2(30)^2$] is $\hat{\mu}_{Y \cdot 30} \pm t_{.025, 13} \cdot$ (estimated S.D. of $\hat{\mu}_{Y \cdot 30}$) $= 3652 \pm (2.179)(76) = (3486.4, 3817.6)$.

c. The form of a prediction interval given in Chapter 12 for a future value Y_f when $x = x_f$ was

$$\hat{y}_f \pm (t_{\alpha/2, \text{d.f.}}) \cdot [\text{estimated S.D. of } (Y_f - \hat{Y}_f)]$$

where $Y_f - \hat{Y}_f$ is the error of prediction. Since a future value Y_f and the prediction $\hat{Y}_f$ are independent of one another, $\text{Var}(Y_f - \hat{Y}_f) = \text{Var}(Y_f) + \text{Var}(\hat{Y}_f) =$

$\sigma^2 + \text{Var}(\hat{Y}_f)$, estimated by $s^2 + [\text{estimated Var}(\hat{Y}_f)]$. The 95% prediction interval for Y_f when $x = 28$ is then (since $s^2 = 41{,}587.71$)

$$3591 \pm (2.179)\sqrt{41{,}587.71 + (76)^2} = 3591 \pm (2.179)(217.6)$$

$$= (3116.8, \ 4065.2)$$

This interval is quite wide because $s = 203.93$ is relatively large.

Centering x Values

For the quadratic model with regression function $\mu_{Y \cdot x} = \beta_0 + \beta_1 x + \beta_2 x^2$, the parameters β_0, β_1, and β_2 characterize the behavior of the function near $x = 0$. For example, β_0 is the height at which the regression function crosses the vertical axis $x = 0$, while β_1 is the first derivative of the function at $x = 0$ (instantaneous rate of change of $\mu_{Y \cdot x}$ at $x = 0$). If the x_i's all lie far from zero, we may not have precise information about the values of these parameters. Let $\bar{x} =$ the average of the x_i's for which observations are to be taken, and consider the model

$$Y = \beta_0^* + \beta_1^*(x - \bar{x}) + \beta_2^*(x - \bar{x})^2 + \epsilon \tag{13.18}$$

In the model (13.18) $\mu_{Y \cdot x} = \beta_0^* + \beta_1^*(x - \bar{x}) + \beta_2^*(x - \bar{x})^2$ and the parameters now describe the behavior of the regression function near the center $\bar{x}$ of the data.

To estimate the parameters of (13.18), we simply subtract $\bar{x}$ off each x_i to obtain $x_i' = x_i - \bar{x}$, and then use the x_i''s in place of the x_i's. An important benefit of this is that the coefficients of $b_0, \ldots, b_k$ in the normal equations (13.11) will be of much smaller magnitude than would be the case were the original x_i's used. When the system is solved by computer, this centering protects against any roundoff error which may result.

Example 13.5   The paper "A Method for Improving the Accuracy of Polynomial Regression Analysis" (*J. Quality Technology*, 1971, pp. 149–155) reported the following data on $x =$ cure temperature (°F) and $y =$ ultimate shear strength of a rubber compound (psi), with $\bar{x} = 297.13$.

x	280	284	292	295	298	305	308	315
x'	-17.13	-13.13	-5.13	-2.13	.87	7.87	10.87	17.87
y	770	800	840	810	735	640	590	560

A computer analysis yielded the following results:

Table 13.6

Parameter	Estimate	Estimated S.D.	Parameter	Estimate	Estimated S.D.
β_0	$-26{,}219.64$	11,912.78	β_0^*	759.36	23.20
β_1	189.21	80.25	β_1^*	-7.61	1.43
β_2	$-.3312$	.1350	β_2^*	$-.3312$	.1350

The estimated regression function using the original model is

$y = -26{,}219.64 + 189.21x - .3312x^2$, while for the centered model the function is $y = 759.36 - 7.61 (x - 297.13) - .3312 (x - 297.13)^2$. These estimated functions are identical; the only difference is that different parameters have been estimated for the two models. The estimated standard deviations indicate clearly that β_0^* and β_1^* have been more accurately estimated than β_0 and β_1. The quadratic parameters are identical ($\beta_2 = \beta_2^*$), as can be seen by comparing the x^2 term in (13.18) with the original model. We emphasize again that a major benefit of centering is the gain in computational accuracy, not only in quadratic but also in higher-degree models.

The book by Neter and Wasserman is again a good source for more information about polynomial regression.

Exercises / Section 13.3

1. The viscosity of an oil (y) was measured by a cone and plate viscometer at six different cone speeds (x). It was assumed that a quadratic regression model was appropriate, and the estimated regression function resulting from the $n = 6$ observations was

$$y = -113.0937 + 3.3684x - .01780x^2$$

a. Estimate $\mu_{Y \cdot 75}$, the expected viscosity when speed is 75 rpm.

b. What viscosity would you predict for a cone speed of 60 rpm?

c. If $\Sigma y_i^2 = 8386.43$, $\Sigma y_i = 210.70$, $\Sigma x_i y_i = 17{,}002.00$, and $\Sigma x_i^2 y_i = 1{,}419{,}780$, compute SSE, s^2, and s.

d. From (c), $SST = 8386.43 - (210.70)^2/6 = 987.35$. Using SSE computed in (c), what is the computed value of the coefficient of multiple determination R^2?

e. If the estimated standard deviation of $\hat{\beta}_2$ is $s_{\hat{\beta}_2} = .00226$, test $H_0: \beta_2 = 0$ versus $H_a: \beta_2 \neq 0$ at level .01.

2. Exercise 3 in Section 13.1 presented the following data on exposure time to radiation x and dry weight of roots y.

x	0	2	4	6	8
y	110	123	119	86	62

a. Show that the estimated quadratic regression function is $y = 111.89 + 8.06x - 1.84x^2$.

b. Compute the predicted values and residuals. Then compute SSE and s^2 using the residuals.

c. Compute SSE using the computational formula,

and then compute s^2.

d. Compute the coefficient of multiple determination R^2.

e. The estimated standard deviation of $\hat{\beta}_2$, the estimator of the quadratic coefficient β_2, is $s_{\hat{\beta}_2} = .480$. Does the quadratic term belong in the model? State and test the appropriate hypotheses at level .05.

f. The estimated standard deviation of $\hat{\beta}_1$ is $s_{\hat{\beta}_1} = 4.01$. Use this and the information in (e) to obtain joint confidence intervals for β_1 and β_2 with joint confidence level (at least) 95%.

g. The estimated standard deviation of $\hat{\mu}_{Y \cdot 4}$ ($= \hat{\beta}_0 + 4\hat{\beta}_1 + 16\hat{\beta}_2$) is 5.01. Compute a 90% confidence interval for $\mu_{Y \cdot 4}$.

h. Estimate the exposure time which maximizes expected dry weight of roots.

3. The article "A Simulation-Based Evaluation of Three Cropping Systems on Cracking-Clay Soils in a Summer Rainfall Environment" (*Agricultural Meteorology*, 1976, pp. 211–229) proposed a quadratic model for the relationship between water supply index (x) and farm wheat yield (y). Representative data and the resulting summary quantities appear below.

x	1.2	1.3	1.5	1.8	2.1	2.3	2.5
y	790	950	740	1230	1000	1465	1370

x	2.9	3.1	3.2	3.3	3.9	4.0	4.3
y	1420	1625	1600	1720	1500	1550	1560

$\Sigma x_i = 37.40$, $\Sigma x_i^2 = 113.42$, $\Sigma x_i^3 = 375.8961$,

$\Sigma x_i^4 = 1322.7388$, $\Sigma y_i = 18,520$, $\Sigma y_i^2 = 25,871,756$, $\Sigma x_i y_i = 53,111$, $\Sigma x_i^2 y_i = 168,248.3$

a. Estimate β_0, β_1, β_2, and the quadratic regression function $\beta_0 + \beta_1 x + \beta_2 x^2$.

b. Compute SSE, estimate σ^2, and compute the coefficient of multiple determination.

c. The estimated standard deviation of $\hat{\beta}_2$ is $s_{\hat{\beta}_2} = 41.97$. Obtain a 95% confidence interval for β_2.

d. The estimated standard deviation of $\hat{\beta}_0 + \hat{\beta}_1 x + \hat{\beta}_2 x^2$ when $x = 2.5$ is 53.5. Test $H_0: \mu_{Y \cdot 2.5} = 1500$ versus $H_a: \mu_{Y \cdot 2.5} < 1500$ using $\alpha = .01$.

e. Obtain a 95% prediction interval for wheat yield when the water supply index is 2.5 by using the information given in (d).

4. The accompanying data was obtained from a study of a certain method for preparing pure alcohol from refinery streams ("Direct Hydration of Olefins," *Industrial and Eng. Chemistry*, 1961, pp. 209–211). The independent variable x is volume hourly space velocity, and the dependent variable y is the amount of conversion of Isobutylene.

x	1	1	2	4	4	4	6
y	23.0	24.5	28.0	30.9	32.0	33.6	20.0

a. Assuming that a quadratic probabilistic model is appropriate, estimate the regression function.

b. Compute the predicted values and residuals, and construct a residual plot. Does the plot look roughly as expected when the quadratic model is correct? Does the plot indicate that any observation has had a great influence on the fit? Does a scatter plot identify a point having large influence? If so, which point?

c. Compute s^2 and R^2. Does the quadratic model provide a good fit to the data?

d. In Exercise 13.1.7 it was noted that the predicted value $\hat{Y}_j$ and the residual $Y_j - \hat{Y}_j$ are independent of one another, so that $\sigma^2 = \text{Var}(Y_j) = \text{Var}(\hat{Y}_j) + \text{Var}(Y_j - \hat{Y}_j)$. A computer printout gives the estimated standard deviations of the predicted values as .955, .955, .712, .777, .777, .777, and 1.407. Use these values along with s^2 to compute the estimated standard deviation of each residual. Then compute the standardized residuals and plot them against x. Does the plot look much like the plot of (b)? Suppose that you had standardized the residuals using just s in the

denominator. Would the resulting values be much different than the correct values?

e. Using information given in (d), compute a 90% prediction interval for Isobutylene conversion when volume hourly space velocity is 4.

5. The following data is a subset of data obtained in an experiment to study the relationship between soil pH x and $y = $ Al. Concentration/EC ("Root Responses of Three Gramineae Species to Soil Acidity in an Oxisol and an Ultisol," *Soil Science*, 1973, pp. 295–302).

x	4.01	4.07	4.08	4.10	4.18
y	1.20	.78	.83	.98	.65

x	4.20	4.23	4.27	4.30	4.41
y	.76	.40	.45	.39	.30

x	4.45	4.50	4.58	4.68	4.70	4.77
y	.20	.24	.10	.13	.07	.04

A cubic model was proposed in the paper, but the version of MINITAB used by the author of the present text refused to include the x^3 term in the model, stating that "x^3 is highly correlated with other predictor variables." To remedy this, $\bar{x} = 4.3456$ was subtracted off each x value to yield $x' = x - \bar{x}$. A cubic regression was then requested to fit the model having regression function

$$y = \beta_0^* + \beta_1^* x' + \beta_2^* (x')^2 + \beta_3^* (x')^3$$

The following computer output resulted:

Parameter	Estimate	Estimated S.D.
β_0^*	.3463	.0366
β_1^*	−1.2933	.2535
β_2^*	2.3964	.5699
β_3^*	−2.3968	2.4590

a. What is the estimated regression function for the "centered" model?

b. What is the estimated value of the coefficient β_3 in the "uncentered" model with regression function $y = \beta_0 + \beta_1 x + \beta_2 x^2 + \beta_3 x^3$? What is the estimate of β_2?

c. Using the cubic model, what value of y would you predict when soil pH is 4.5?

d. Carry out a test to decide whether or not the cubic term should be retained in the model.

6. In many polynomial regression problems, rather than fitting a "centered" regression function using $x' = x - \bar{x}$, computational accuracy can be improved by using a function of the standardized independent variable $x' = (x - \bar{x})/s_x$, where s_x is the standard deviation of the x_i's. Consider fitting the cubic regression function $y = \beta_0^* + \beta_1^* x' + \beta_2^*(x')^2 + \beta_3^*(x')^3$ to the following data resulting from a study of the relation between thrust efficiency y of supersonic propelling rockets and the half-divergence angle x of the rocket nozzle ("More on Correlating Data," *CHEMTECH*, 1976, pp. 266–270).

x	5	10	15	20	25	30	35
y	.985	.996	.988	.962	.940	.915	.878

Parameter	Estimate	Estimated S.D.
β_0^*	.9671	.0026
β_1^*	−.0502	.0051
β_2^*	−.0176	.0023
β_3^*	.0062	.0031

a. What value of y would you predict when the half-divergence angle is 20? When $x = 25$?

b. What is the estimated regression function $\hat{\beta}_0 + \hat{\beta}_1 x + \hat{\beta}_2 x^2 + \hat{\beta}_3 x^3$ for the "unstandardized" model?

c. Use a level .05 test to decide whether or not the cubic term should be deleted from the model.

d. What can you say about the relationship between SSE and R^2 for the standardized and unstandardized models? Explain.

e. SSE for the cubic model is .00006300, while for a quadratic model SSE is .00014367. Compute R^2 for each model. Does the difference between the two suggest that the cubic term can be deleted?

7. The following data resulted from an experiment to assess the potential of unburnt colliery spoil as a medium for plant growth. The variables are $x =$ acid extractable cations and $y =$ exchangeable acidity/total cation exchange capacity ("Exchangeable

Acidity in Unburnt Colliery Spoil," *Nature*, 1969, p. 161).

x	−23	−5	16	26	30	38
y	1.50	1.46	1.32	1.17	.96	.78

x	52	58	67	81	96	100	113
y	.77	.91	.78	.69	.52	.48	.55

Standardizing the independent variable x to obtain $x' = (x - \bar{x})/s_x$ and fitting the regression function $y = \beta_0^* + \beta_1^* x' + \beta_2^*(x')^2$ yielded the accompanying computer output.

Parameter	Estimate	Estimated S.D.
β_0^*	.8733	.0421
β_1^*	−.3255	.0316
β_2^*	.0448	.0319

a. Estimate $\mu_{Y \cdot 50}$.

b. Compute the value of the coefficient of multiple determination.

c. What is the estimated regression function $\hat{\beta}_0 + \hat{\beta}_1 x + \hat{\beta}_2 x^2$ using the unstandardized variable x?

d. What is the estimated standard deviation of $\hat{\beta}_2$ computed in (c)?

e. Carry out a test using the standardized estimates to decide whether or not the quadratic term should be retained in the model. Repeat using the unstandardized estimates. Do your conclusions differ?

8. The paper "The Respiration in Air and in Water of the Limpets Patella Caerulea and Patella Lusitanica" (*Comp. Biochemistry and Physiology*, 1975, pp. 407–411) proposed a simple power model for the relationship between respiration rate y and temperature x for P. caerulea in air. However, a plot of $\ln(y)$ versus x exhibits a curved pattern. Fit the quadratic power model $Y = \alpha e^{\beta x + \gamma x^2} \cdot \epsilon$ to the accompanying data.

x	10	15	20	25	30
y	37.1	70.1	109.7	177.2	222.6

13.4 Multiple Regression Analysis

In multiple regression the objective is to build a probabilistic model which relates a dependent variable Y to more than one independent or predictor variable. The general form of the model to be studied here is

$$Y = \beta_0 + \beta_1 x_1 + \beta_2 x_2 + \cdots + \beta_k x_k + \epsilon \qquad (13.19)$$

where $E(\epsilon) = 0$ and $\text{Var}(\epsilon) = \sigma^2$. This in turn implies that $\sigma^2_{Y \cdot x_1, x_2, \ldots, x_k}$, the variance of Y for given values of $x_1, \ldots, x_k$, equals σ^2 independently of the given values, and that the regression function is

$$\mu_{Y \cdot x_1, \ldots, x_k} = \beta_0 + \beta_1 x_1 + \cdots + \beta_k x_k \qquad (13.20)$$

To construct confidence and prediction intervals and to test hypotheses about the model parameters, we shall also assume that ϵ has a normal distribution.

Each x_i in (13.19) will be called a **carrier,** since it carries information about Y in the model. Thus the model (13.19) contains k carriers. The reason for not using "variable" in place of "carrier" is that we may wish to build a model involving two independent variables x_1 and x_2, with several of the carriers being themselves functions of x_1 and x_2. For example, with $x_3 = x_1^2$ and $x_4 = x_1 x_2$, the model

$$Y = \beta_0 + \beta_1 x_1 + \beta_2 x_2 + \beta_3 x_3 + \beta_4 x_4 + \epsilon$$

has the general form of (13.19) with four carriers, but x_1, x_2, x_3, and x_4 are not functionally independent variables. Thus even a model built from just two independent variables may have a number of carriers.

While the regression function (13.20) will not always be a linear function of the independent variables used to build the model, further analysis depends critically on the fact that it is *a linear function of the unknown parameters* $\beta_0, \beta_1, \ldots, \beta_k$. This property makes it easy to apply the principle of least squares to obtain estimators which can be used for inference and prediction. Before discussing this, we first consider interpretations of different multiple regression models.

Model Interpretation

For the case of two independent variables x_1 and x_2, four useful multiple regression models are

1. the first-order model, with $Y = \beta_0 + \beta_1 x_1 + \beta_2 x_2 + \epsilon$
2. the second-order no-interaction model, with $Y = \beta_0 + \beta_1 x_1 + \beta_2 x_2 + \beta_3 x_1^2 + \beta_4 x_2^2 + \epsilon$
3. the first-order interaction model, with $Y = \beta_0 + \beta_1 x_1 + \beta_2 x_2 + \beta_3 x_1 x_2 + \epsilon$
4. the second-order model with interaction, specified by $Y = \beta_0 + \beta_1 x_1 + \beta_2 x_2 + \beta_3 x_1^2 + \beta_4 x_2^2 + \beta_5 x_1 x_2 + \epsilon$

Understanding the differences between these models is an important first step in building realistic regression models from the independent variables under study.

The first-order model is the most straightforward generalization of simple linear regression. It states that for a fixed value of either variable, the expected value of Y is a linear function of the other variable, and that the expected change in Y for a unit increase in $x_1(x_2)$ is $\beta_1(\beta_2)$ independent of the level of $x_2(x_1)$. Thus if we graph the regression function as a function of x_1 for several different values of x_2, we obtain as contours of the regression function a collection of parallel lines as pictured in Figure 13.8(a). The function $y = \beta_0 + \beta_1 x_1 + \beta_2 x_2$ specifies a plane in three-dimensional space; the first-order model says that each observed value of the dependent variable deviates from this plane by a random amount ϵ.

According to the second-order no-interaction model, if x_2 is fixed, the expected change in Y for a one unit increase in x_1 is

$$\beta_0 + \beta_1(x_1 + 1) + \beta_2 x_2 + \beta_3(x_1 + 1)^2 + \beta_4 x_2^2 -$$

$$(\beta_0 + \beta_1 x_1 + \beta_2 x_2 + \beta_3 x_1^2 + \beta_4 x_2^2) = \beta_1 + 2\beta_3 x_1$$

Because this expected change does not depend on x_2, the contours of the regression function for different values of x_2 are still parallel to one another. However, the dependence of the expected change on the value of x_1 means that the contours are now curves rather than straight lines. This is pictured in Figure 13.8(b). In this case the regression surface is no longer a plane in three-space but is instead a curved surface.

The contours of the regression function for the first-order interaction model are nonparallel straight lines. This is because the expected change in Y when x_1 is increased by 1 is

Figure 13.8 Contours of four different regression functions

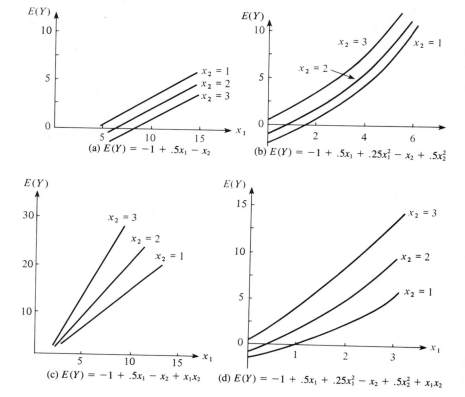

(a) $E(Y) = -1 + .5x_1 - x_2$

(b) $E(Y) = -1 + .5x_1 + .25x_1^2 - x_2 + .5x_2^2$

(c) $E(Y) = -1 + .5x_1 - x_2 + x_1 x_2$

(d) $E(Y) = -1 + .5x_1 + .25x_1^2 - x_2 + .5x_2^2 + x_1 x_2$

$$\beta_0 + \beta_1(x_1 + 1) + \beta_2 x_2 + \beta_3(x_1 + 1)x_2 -$$
$$(\beta_0 + \beta_1 x_1 + \beta_2 x_2 + \beta_3 x_1 x_2) = \beta_0 + \beta_1 + \beta_3 x_2$$

This expected change depends on the value of x_2, so each contour line must have a different slope as in Figure 13.8(c). The word "interaction" reflects the fact that an expected change in Y when one variable increases depends on the value of the other variable.

Finally, for the second-order interaction model, the expected change in Y when x_2 is held fixed while x_1 is increased by one unit is $\beta_1 + 2\beta_3 x_1 + \beta_5 x_2$, which is a function of both x_1 and x_2. This implies that the contours of the regression function are both curved and not parallel to one another, as illustrated in Figure 13.8(d).

Similar considerations apply to models constructed from more than two independent variables. In general, the presence of interaction terms in the model imply that the expected change in Y depends not only on the variable being increased or decreased but also on the values of some of the fixed variables. As in ANOVA, it is possible to have higher-way interaction terms (for example, $x_1 x_2 x_3$), making model interpretation more difficult.

Implicit in the discussion thus far is the assumption that x_1 and x_2 are quantitative variables. It is also possible to build models involving one or more qualitative variables. Suppose that x_1 is a quantitative variable and the other variable of interest is qualitative with three different levels (say, manufacturers 1, 2, and 3). Then two possible models are

$$Y = \beta_0 + \beta_1 x_1 + \beta_2 x_2 + \beta_3 x_3 + \epsilon$$

and

$$Y = \beta_0 + \beta_1 x_1 + \beta_2 x_2 + \beta_3 x_3 + \beta_4 x_1 x_2 + \beta_5 x_1 x_3$$

The carriers x_2 and x_3 here both refer to the qualitative variable, with $x_2 = 0, x_3 = 0$ corresponding to level 1 of the variable, $x_2 = 1$ and $x_3 = 0$ to level 2, and $x_2 = 0$, $x_3 = 1$ to level 3. These carriers are often called dummy or indicator variables. The contour of the regression function for each of the three levels of the qualitative variable is pictured in Figure 13.9 for both these models. Chapter 9 of the book by

Figure 13.9 Regression function contours for a model with one quantitative variable (x_1) and one qualitative variable having three levels

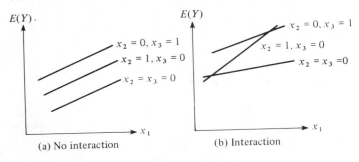

(a) No interaction   (b) Interaction

Neter and Wasserman contains an excellent discussion of models containing qualitative variables.

Estimating Parameters

Instead of n pairs, the data now consists of n $k + 1$-tuples $(x_{11}, x_{21}, \ldots, x_{k1}, y_1)$, $(x_{12}, x_{22}, \ldots, x_{k2}, y_2), \ldots, (x_{1n}, x_{2n}, \ldots, x_{kn}, y_n)$, where x_{ij} is the value of the ith carrier x_i associated with the observed value y_j. The y_j's are assumed to have been observed independently of one another according to the model (13.19). To estimate the parameters $\beta_0, \beta_1, \ldots, \beta_k$ using the principle of least squares, form the sum of squared deviations of the observed y_j's from a trial function $y = b_0 + b_1 x_1 + \cdots + b_k x_k$:

$$f(b_0, b_1, \ldots, b_k) = \sum_j [y_j - (b_0 + b_1 x_{1j} + b_2 x_{2j} + \cdots + b_k x_{kj})]^2 \qquad (13.21)$$

The least squares estimates are those values of $b_0, b_1, \ldots, b_k$ which minimize $f(b_0, \ldots, b_k)$. Upon taking the partial derivative of f with respect to each b_i ($i = 0$, $1, \ldots, k$) and equating all partials to zero, the following system of **normal equations** is obtained:

$$b_0 n + b_1 \sum x_{1j} + b_2 \sum x_{2j} + \cdots + b_k \sum x_{kj} = \sum y_j$$

$$b_0 \sum x_{1j} + b_1 \sum x_{1j}^2 + b_2 \sum x_{1j} x_{2j} + \cdots + b_k \sum x_{1j} x_{kj} = \sum x_{1j} y_j \qquad (13.22)$$

$$\vdots \qquad \qquad \vdots \qquad \qquad \vdots$$

$$b_0 \sum x_{kj} + b_1 \sum x_{1j} x_{kj} + \cdots + b_{k-1} \sum x_{k-1,j} x_{kj} + b_k \sum x_{kj}^2 = \sum x_{kj} y_j$$

That these equations are linear in the unknowns $b_0, b_1, \ldots, b_k$ is a consequence of the regression function being linear in the parameters. Solving (13.22) yields the least squares estimates $\hat{\beta}_0, \hat{\beta}_1, \ldots, \hat{\beta}_k$. In general the system (13.22) can be solved by hand by first computing all coefficients of the b_i's and then using a technique such as Gaussian elimination. This is quite tedious, but fortunately any of the standard statistical regression packages will automatically solve for and print out $\hat{\beta}_0, \ldots, \hat{\beta}_k$.

For the case $k = 2$ (2 carriers) the solution to (13.22) can be written down as was done for quadratic regression. Let

$$s_{1y} = \sum x_{1j} y_j - n \bar{x}_1 \bar{y}, \quad s_{2y} = \sum x_{2j} y_j - n \bar{x}_2 \bar{y}, \quad s_{11} = \sum x_{1j}^2 - n \bar{x}_1^2,$$

$$s_{12} = \sum x_{1j} x_{2j} - n \bar{x}_1 \bar{x}_2, \quad s_{22} = \sum x_{2j}^2 - n \bar{x}_2^2$$

Then

$$\hat{\beta}_1 = \frac{s_{1y}s_{22} - s_{2y}s_{12}}{s_{11}s_{22} - s_{12}^2}, \quad \hat{\beta}_2 = \frac{s_{2y}s_{11} - s_{1y}s_{12}}{s_{11}s_{22} - s_{12}^2}$$

(13.23)

and

$$\hat{\beta}_0 = \bar{y} - \hat{\beta}_1\bar{x}_1 - \hat{\beta}_2\bar{x}_2$$

Example 13.6   In the paper "An Ultracentrifuge Flour Absorption Method" (*Cereal Chemistry*, 1978, pp. 96–101), the authors studied the relationship between water absorption for wheat flour and various characteristics of the flour. In particular the authors used a first-order multiple linear regression model to relate absorption y (%) to flour protein x_1 (%) and starch damage x_2 (Farrand units). The data is shown in Table 13.7.

Table 13.7

x_1	x_2	y	x_1	x_2	y
8.5	2	30.9	12.9	24	47.0
8.9	3	32.7	12.0	25	46.8
10.6	3	36.7	12.9	28	45.9
10.2	20	41.9	13.1	28	48.8
9.8	22	40.9	11.4	32	46.2
10.8	20	42.9	13.2	28	47.8
11.6	31	46.3	11.6	35	49.2
12.0	32	47.6	12.1	34	48.3
12.5	31	47.2	11.3	35	48.6
10.9	28	44.0	11.1	40	50.2
12.2	36	47.7	11.5	45	49.6
11.9	28	43.9	11.6	50	53.2
11.3	30	46.8	11.7	55	54.3
13.0	27	46.2	11.7	57	55.8

The necessary summary quantities are $n = 28$, $\Sigma x_{1j} = 322.3$, $\Sigma x_{2j} = 829.0$, $\Sigma x_{1j}^2 = 3746.4$, $\Sigma x_{2j}^2 = 29{,}327.0$, $\Sigma x_{1j}x_{2j} = 9746.6$, $\Sigma y_j = 1287.4$, $\Sigma x_{1j}y_j = 14{,}940.06$, and $\Sigma x_{2j}y_j = 40{,}016.0$. This gives $s_{11} = 36.49679$, $s_{12} = 204.21786$, $s_{22} = 4782.67857$, $s_{1y} = 121.16643$, and $s_{2y} = 1899.76429$, which in turn yields $\hat{\beta}_1 = 1.44176$, $\hat{\beta}_2 = .33566$, and $\hat{\beta}_0 = 19.44495$. The estimated regression function is $\hat{\mu}_{Y \cdot x_1, x_2} = 19.45 + 1.44x_1 + .34x_2$.

Standardizing Variables

In Section 13.3 we considered transforming x to $x' = x - \bar{x}$ before fitting a polynomial. For multiple regression, especially when values of variables are large in magnitude, it is advantageous to carry this coding one step further. Let $\bar{x}_i$ and s_i be the sample average and sample standard deviation of the x_{ij}'s (j = 1, . . . , n). We now code each variable x_i by $x_i' = (x_i - \bar{x}_i)/s_i$. The coded variable x_i' simply reexpresses any x_i value in units of standard deviation above or below the mean. Thus if $\bar{x}_i = 100$ and $s_i = 20$, $x_i = 130$ becomes $x_i' = 1.5$ because 130 is 1.5 standard deviations above the mean of the values of x_i. For example, the coded full second-order model with two independent variables has regression function

$$E(Y) = \beta_0 + \beta_1\left(\frac{x_1 - \bar{x}_1}{s_1}\right) + \beta_2\left(\frac{x_2 - \bar{x}_2}{s_2}\right) + \beta_3\left(\frac{x_1 - \bar{x}_1}{s_1}\right)^2$$

$$+ \beta_4\left(\frac{x_2 - \bar{x}_2}{s_2}\right)^2 + \beta_5\left(\frac{x_1 - \bar{x}_1}{s_1}\right)\left(\frac{x_2 - \bar{x}_2}{s_2}\right)$$

$$= \beta_0 + \beta_1 x_1' + \beta_2 x_2' + \beta_3 x_3' + \beta_4 x_4' + \beta_5 x_5'$$

The benefits of coding are (a) increased numerical accuracy in all computations (through less computer roundoff error), and (b) more accurate estimation than for the parameters of the uncoded model because the individual parameters of the coded model characterize the behavior of the regression function near the center of the data rather than near the origin.

Example 13.7 The paper "The Value and the Limitations of High-Speed Turbo-Exhausters for the Removal of Tar-Fog from Carburetted Water-Gas" (*Soc. Chemical Industry J.*, 1946, pp. 166–168) presents the accompanying data on y = tar content (grains/100 ft^3) of a gas stream as a function of x_1 = rotor speed (rpm) and x_2 = gas inlet temperature (°F). The data is also considered in the paper "Some Aspects of Nonorthogonal Data Analysis" (*J. Quality Technology*, 1973, pp. 67–79), which suggests using the coded model described above. The means and standard deviations are $\bar{x}_1 = 2991.13$, $s_1 = 387.81$, $\bar{x}_2 = 58.468$,   and $s_2 = 6.944$,   so $x_1' = (x_1 - 2991.13)/387.81$ and $x_2' = (x_2 - 58.468)/6.944$.   With $x_3' = (x_1')^2$, $x_4' = (x_2')^2$, and $x_5' = x_1' \cdot x_2'$, fitting the full second-order model requires solving the system of six normal equations in six unknowns. A computer analysis yielded $\hat{\beta}_0 = 40.2660$, $\hat{\beta}_1 = -13.4041$, $\hat{\beta}_2 = 10.2553$, $\hat{\beta}_3 = 2.3313$, $\hat{\beta}_4 = -2.3405$, and $\hat{\beta}_5 = 2.5978$. The estimated regression equation is then

$$\hat{y} = 40.27 - 13.40x_1' + 10.26x_2' + 2.33x_3' - 2.34x_4' + 2.60x_5'$$

Thus if $x_1 = 3200$ and $x_2 = 57.0$, $x_1' = .539$, $x_2' = -.211$, $x_3' = (.539)^2 = .2901$, $x_4' = (-.211)^2 = .0447$, and $x_5' = (.539)(-.211) = -.1139$, so

$$\hat{y} = 40.27 - (13.40)(.539) + (10.26)(-.211) + (2.33)(.2901)$$

$$- (2.34)(.0447) + (2.60)(-.1139) = 31.16$$

Table 13.8

Run	y	x_1	x_2	x_1'	x_2'
1	60.0000	2400.00	54.5000	−1.52428	−0.57145
2	61.0000	2450.00	56.0000	−1.39535	−0.35543
3	65.0000	2450.00	58.5000	−1.39535	0.00461
4	30.5000	2500.00	43.0000	−1.26642	−2.22763
5	63.5000	2500.00	58.0000	−1.26642	−0.06740
6	65.0000	2500.00	59.0000	−1.26642	0.07662
7	44.0000	2700.00	52.5000	−0.75070	−0.85948
8	52.0000	2700.00	65.5000	−0.75070	1.01272
9	54.5000	2700.00	68.0000	−0.75070	1.37276
10	30.0000	2750.00	45.0000	−0.62177	−1.93960
11	26.0000	2775.00	45.5000	−0.55731	−1.86759
12	23.0000	2800.00	48.0000	−0.49284	−1.50755
13	54.0000	2800.00	63.0000	−0.49284	0.65268
14	36.0000	2900.00	58.5000	−0.23499	0.00461
15	53.5000	2900.00	64.5000	−0.23499	0.86870
16	57.0000	3000.00	66.0000	0.02287	1.08472
17	33.5000	3075.00	57.0000	0.21627	−0.21141
18	34.0000	3100.00	57.5000	0.28073	−0.13941
19	44.0000	3150.00	64.0000	0.40966	0.79669
20	33.0000	3200.00	57.0000	0.53859	−0.21141
21	39.0000	3200.00	64.0000	0.53859	0.79669
22	53.0000	3200.00	69.0000	0.53859	1.51677
23	38.5000	3225.00	68.0000	0.60305	1.37276
24	39.5000	3250.00	62.0000	0.66752	0.50866
25	36.0000	3250.00	64.5000	0.66752	0.86870
26	8.5000	3250.00	48.0000	0.66752	−1.50755
27	30.0000	3500.00	60.0000	1.31216	0.22063
28	29.0000	3500.00	59.0000	1.31216	0.07662
29	26.5000	3500.00	58.0000	1.31216	−0.06740
30	24.5000	3600.00	58.0000	1.57002	−0.06740
31	26.5000	3900.00	61.0000	2.34360	0.36465

$\hat{\sigma}^2$ and R^2

As for both simple linear regression and polynomial regression, the estimate of σ^2 is based on the sum of squared residuals

$$SSE = \sum [y_j - (\hat{\beta}_0 + \hat{\beta}_1 x_{1j} + \cdots + \hat{\beta}_k x_{kj})]^2$$

By squaring out the residuals and carrying the summation through to each term, it can be verified that

$$SSE = \sum y_j^2 - \hat{\beta}_0 \sum y_j - \hat{\beta}_1 \sum x_{1j} y_j - \cdots - \hat{\beta}_k \sum x_{kj} y_j$$

Because $k + 1$ parameters $\beta_0, \beta_1, \ldots, \beta_k$ have been estimated, $k + 1$ degrees of freedom are lost, so $n - (k + 1)$ degrees of freedom are associated with SSE, and

$$\hat{\sigma}^2 = s^2 = \frac{SSE}{n - (k + 1)} = MSE$$

With $SST = \Sigma(y_i - \bar{y})^2$, the proportion of total variation explained by the multiple regression model is $R^2 = 1 - SSE/SST$, the **coefficient of multiple determination.** As in polynomial regression, R^2 is often adjusted for the number of parameters in the model by the formula adjusted $R^2 = [(n - 1)R^2 - k]/[n - (k + 1)]$. The positive square root of the coefficient of multiple determination is called the multiple correlation coefficient R. It can be shown that R is the sample correlation coefficient r between the observed y_j's and the predicted $\hat{y}_j$'s (that is, using $x_j = \hat{y}_j$ in the formula for r results in $r = R$).*

Example 13.6
(continued)

With $\Sigma y_j = 1287.4$ and $\Sigma y_j^2 = 60,035.12$,

$$SSE = 60,035.12 - (19.44495)(1287.4) - (1.44176)(14,940.6)$$

$$- (.33566)(40,016) = 29.496$$

so $\hat{\sigma}^2 = 29.496/[28 - (2 + 1)] = 1.18$ and $\hat{\sigma} = s = 1.09$. Since $SST = 60,035.12 - (1287.4)^2/28 = 842.31$, $R^2 = 1 - (29.496)/(842.31) = .965$, so 96.5% of the variation in the observed y's can be explained by the multiple regression model. Adjusted $R^2 = .962$.

Inferences About Model Parameters

Before testing hypotheses, constructing confidence intervals, and making predictions, one should first examine diagnostic plots to see whether the model needs modification or whether there are outliers in the data. The recommended plots are (standardized) residuals versus each independent variable, residuals versus $\hat{y}$, and $\hat{y}$ versus y. Potential problems are suggested by the same patterns discussed in Section 13.1. Of particular importance is the identification of points which have a large influence on the fit. In the next section we describe several diagnostic tools suitable for this task.

Because each $\hat{\beta}_i$ is a linear function of the y_i's, the standard deviation of each $\hat{\beta}_i$ is the product of σ and a function of the x_{ij}'s, so an estimate $s_{\hat{\beta}_i}$ is obtained by substituting s for σ. Unfortunately the function of the x_{ij}'s is quite complicated—in the next section we obtain these expressions using matrix algebra—but all standard regression computer packages compute and print the $s_{\hat{\beta}_i}$'s. A confidence interval for an individual β_i then has the form $\hat{\beta}_i \pm t_{\alpha/2, n-(k+1)} \cdot s_{\hat{\beta}_i}$, while a test of $H_0 : \beta_i = \beta_{i0}$ is based on the t statistic $T = (\hat{\beta}_i - \beta_{i0})/S_{\hat{\beta}_i}$ having $n - (k + 1)$ d.f. For several simultaneous intervals the Bonferroni technique is recommended.

Example 13.6
(continued)

The parameter estimates and estimated standard deviations for the flour absorption data are

---

* Because each $\hat{\beta}_i$ is a linear function of the observed y_j's, $\hat{y} = \hat{\beta}_0 + \hat{\beta}_1 x_1 + \cdots + \hat{\beta}_k x_k$ is also a linear function of the y_j's. Consider any other such linear function $y' = \Sigma c_j y_j$. Then it can be shown that the correlation between the y_j's and $\hat{y}_j$'s is at least as large as between the y_j's and the y_j''s. Put another way, the estimated regression function yields predicted values which have maximum correlation with the observed values.

Table 13.9

Parameter	Estimate $\hat{\beta}_i$	Estimated S.D. $s_{\hat{\beta}_i}$
β_0	19.440	2.1880
β_1	1.442	.2076
β_2	.336	.0181

If we compute a 99% interval separately for β_1 and for β_2, then the simultaneous confidence level (referring to both intervals containing the true parameter values) is at least 98%. With $t_{.005, 25} = 2.787$, these intervals are $1.442 \pm (2.787)(.2076) = (.85, 2.03)$ for β_1 and $.336 \pm (2.787)(.0181) = (.29, .39)$ for β_2.

Prediction intervals and inferences concerning $\beta_0 + \beta_1 x_1 + \cdots + \beta_k x_k$ for fixed $x_1, \ldots, x_k$ values are also somewhat involved, since the standard deviation of $\hat{\beta}_0 + \hat{\beta}_1 x_1 + \cdots + \hat{\beta}_k x_k$ is a complicated function of the x_{ij}'s. Several regression packages give the estimated standard deviations for the n k-tuples $(x_{1j}, x_{2j}, \ldots, x_{kj})$ for which observations have been made. The package SAS will compute confidence intervals for other values as well.

Conclusions resulting from individually testing hypotheses about the regression parameters can be misleading. For a model with two carriers x_1 and x_2, to decide whether both $\beta_1 = 0$ and $\beta_2 = 0$, it might be tempting to examine the t ratios $\hat{\beta}_1/s_{\hat{\beta}_1}$ and $\hat{\beta}_2/s_{\hat{\beta}_2}$. The difficulty is that the t ratio $\hat{\beta}_1/s_{\hat{\beta}_1}$ for testing H_0: $\beta_1 = 0$ is computed assuming that β_2 belongs in the model, and similarly for testing H_0: $\beta_2 = 0$ using $t = \hat{\beta}_2/s_{\hat{\beta}_2}$. It can happen that both $\hat{\beta}_1/s_{\hat{\beta}_1}$ and $\hat{\beta}_2/s_{\hat{\beta}_2}$ have nonsignificant values, so that neither β_i should be in the model when the other one is, yet this does not imply that both β_i's should be deleted. This is especially likely to occur when the sample values of the x_{1j}'s and x_{2j}'s are highly correlated, so that either carrier can serve as a "proxy" for the other.

The coefficient of multiple determination can be used to test simultaneously whether the parameters associated with all carriers equal zero; intuitively this tells us whether there exists a useful linear relationship between any of the carriers and the dependent variable.

Test of model utility:

Null hypothesis: H_0: $\beta_1 = \beta_2 = \cdots = \beta_k = 0$

Alternative hypothesis: H_a: at least one $\beta_i \neq 0$ $(i = 1, \ldots, k)$

(13.23)

Test statistic: $F = \dfrac{R^2/k}{(1 - R^2)/[n - (k + 1)]}$

Rejection region for a level α test: $F \geq F_{\alpha, k, n-(k+1)}$

Except for a constant multiple, the test statistic here is $R^2/(1 - R^2)$, the ratio of explained to unexplained variation. If the proportion of explained variation is high relative to unexplained, we would naturally want to reject H_0 and confirm the utility of the model.

Example 13.6 (continued)

We previously computed $R^2 = .965$ for the model with $k = 2$ carriers x_1 and x_2. With $n = 28$ and $\alpha = .01$, $F_{.01, 2, 25} = 5.57$. To test $H_0: \beta_1 = \beta_2 = 0$, we compute

$$f = \frac{.965/2}{(1 - .965)/25} = 344.64.$$ Since $344.64 \geq 5.57$, H_0 is rejected in favor of the conclusion that Y is linearly related to at least one of the x_i's.

The above F test was appropriate for testing whether or not all k β_i's associated with carriers were zero. In many problems one first builds a model involving k carriers and then wishes to know whether a particular subset of l carriers provides almost as good a fit as the "full" k-carrier model. Label the parameters associated with carriers as $\beta_1, \ldots, \beta_l, \beta_{l+1}, \ldots, \beta_k$ so that the first l are associated with the carriers of the "reduced" model. We then wish to test

H_0: model is $Y = \beta_0 + \beta_1 x_1 + \cdots + \beta_l x_l + \epsilon$ (reduced model)

versus

H_a: model is $Y = \beta_0 + \beta_1 x_1 + \cdots + \beta_l x_l + \cdots + \beta_k x_k + \epsilon$

(full model)

Since the full model contains not only the parameters of the reduced model but also some extra parameters, it should fit the data at least as well as the reduced model. That is, if we let SSE_k be the sum of squared residuals for the full model and SSE_l be the corresponding sum for the reduced model, then $SSE_k \leq SSE_l$.* Intuitively, if SSE_k is a great deal smaller than SSE_l, the full model provides a much better fit than the reduced model; the appropriate test statistic should then depend on the reduction $SSE_l - SSE_k$ in unexplained variation. The formal procedure is

---

* The estimates $\hat{\beta}_0, \hat{\beta}_1, \ldots, \hat{\beta}_l$ will in general be different for the full and reduced models, so in general two different multiple regressions must be run to obtain SSE_l and SSE_k. If the variables are listed in the above order, though, most computer packages provide an ANOVA table for the full model which can be used to avoid fitting the reduced model.

SSE_k = unexplained variation for the full model

SSE_l = unexplained variation for the reduced model

Test statistic: $F = \dfrac{(SSE_l - SSE_k)/(k - l)}{SSE_k/[n - (k + 1)]}$ (13.24)

Rejection region: $F \geq F_{\alpha, k-l, n-(k+1)}$

Example 13.7
(continued)

We originally fit a (coded) full second-order model to the data. The resulting R^2 is .932, so the F test of (13.23) should certainly indicate that at least one carrier belongs in the model. Suppose that we use (13.24) to decide whether the full second-order model provides a significantly better fit than the first-order no interaction model. That is, we wish to test

$$H_0 : Y = \beta_0 + \beta_1 x_1' + \beta_2 x_2' + \epsilon$$

versus

$$H_a : Y = \beta_0 + \beta_1 x_1' + \beta_2 x_2' + \beta_3 x_3' + \beta_4 x_4' + \beta_5 x_5' + \epsilon$$

A computer analysis yields $SSE_5 = 442.1$ for the full model and $SSE_2 = 982.3$ for the reduced model (the parameter estimates for the reduced model are $\hat{\beta}_0 = 40.98$, $\hat{\beta}_1 = -12.26$, and $\hat{\beta}_2 = 10.32$, which are different from $\hat{\beta}_0$, $\hat{\beta}_1$, and $\hat{\beta}_2$ for the full model). To carry out the test at level .01, we need $F_{.01, 3, 25} = 4.68$. Then

$$f = \frac{(982.3 - 442.1)/3}{442.1/25} = 10.18 \geq 4.68$$

so H_0 is rejected at level .01. Note, however, that if the reduced model had included the quadratic term in x_1 (that is, $\beta_3 x_3'$), then we would have $SSE_3 = 581.0$ replacing 982.8 and 2 replacing 3, yielding $f = 2.62$. Since $F_{.01, 2, 25} = 5.57$, this new H_0 would not be rejected. The R^2 value for this model is .911.

Transformations in Multiple Regression

Often theoretical considerations suggest a nonlinear relation between an independent variable and two or more dependent variables, while on other occasions diagnostic plots indicate that some type of nonlinear function should be used. Frequently a transformation will linearize the model.

Example 13.8

An article in *Lubrication Eng.* ("Accelerated Testing of Solid Film Lubricants," 1972, pp. 365–372) reported on an investigation of wear life for solid film lubricant. Three sets of journal bearing tests were run on a Mil-L-8937 type film at each combination of three loads (3000, 6000, and 10,000 psi) and three speeds (20, 60, and 100 rpm), and the wear life (hours) was recorded for each run.

Table 13.10

s	l(1000's)	w	s	l(1000's)	w
20	3	300.2	60	6	65.9
20	3	310.8	60	10	10.7
20	3	333.0	60	10	34.1
20	6	99.6	60	10	39.1
20	6	136.2	100	3	26.5
20	6	142.4	100	3	22.3
20	10	20.2	100	3	34.8
20	10	28.2	100	6	32.8
20	10	102.7	100	6	25.6
60	3	67.3	100	6	32.7
60	3	77.9	100	10	2.3
60	3	93.9	100	10	4.4
60	6	43.0	100	10	5.8
60	6	44.5			

The article contains the comment that a lognormal distribution is appropriate for W, since ln (W) is known to follow a normal law (recall from Chapter 4 that this is what defines a lognormal distribution). The model that appears is $W = (c/s^a l^b) \cdot \epsilon$, whence $\ln(W) = \ln(c) - a \ln(s) - b \ln(l) + \ln(\epsilon)$, so with $Y = \ln(W)$, $x_1 = \ln(s)$, $x_2 = \ln(l)$, $\beta_0 = \ln(c)$, $\beta_1 = -a$, and $\beta_2 = -b$, we have a multiple linear regression model. After computing $\ln(w_i)$, $\ln(s_i)$, and $\ln(l_i)$ for the above data, a first-order model in the transformed variables yielded the results shown in Table 13.11.

Table 13.11

Parameter β_i	Estimate $\hat{\beta}_i$	Estimated S.D. $s_{\hat{\beta}_i}$	$t = \hat{\beta}_i/s_{\hat{\beta}_i}$
β_0	20.5387	2.1209	9.68
β_1	−1.2060	.1710	−7.05
β_2	−1.3988	.2327	−6.01

The multiple coefficient of determination (for the transformed observations) has value $R^2 = .782$. The estimated regression function for the transformed variables is

$$\ln(w) = 20.54 - 1.21 \ln(s) - 1.40 \ln(l)$$

so that the original regression function is estimated as

$$w = e^{20.54} \cdot s^{-1.21} \cdot l^{-1.40}$$

The Bonferroni approach can be used to obtain simultaneous confidence intervals for β_1 and β_2, and because $\beta_1 = -a$ and $\beta_2 = -b$, intervals for a and b are then immediately available.

Variable Selection

Often an experimenter will have data on a large number of carriers and then wish to build a regression model involving a subset of the carriers. The use of the subset will make the resulting model more manageable, especially if more data is to be subsequently collected, and also result in a model which is easier to interpret and understand than one with many more carriers. Two fundamental questions in connection with variable selection are:

1. If we can examine regressions involving all possible subsets of the carriers for which data is available, what criteria should be used to select a model?
2. If the number of carriers is too large to permit all regressions to be examined, is there a way of examining a reduced number of subsets among which a good model (or models) will be found?

To address (1) first, if the number of carriers is small (≤ 5, say), then it would not be too tedious to examine all possible regression using any one of the readily available statistical computer packages (MINITAB, SAS, BMD, SPSS). If data on at least six carriers is available, all possible regressions involve at least 64 ($= 2^6$) different models. There are efficient computer codes available for examining all regressions for up to 12 carriers; a brief discussion of such codes appears in the review article by Hocking listed in the chapter bibliography. Even when it is not possible to examine all regressions, there is a program called SELECT (discussed in the Hocking article) which for each fixed number of possible carriers will identify the subset with the smallest *SSE*.* These *SSE*'s (or functions of them) can then be compared according to any of the criteria discussed below to decide on a model.

Criteria for Variable Selection

As before we use a subscript k to denote a quantity (say, SSE_k) computed from a model with k carriers (and thus $k + 1$ β_i's, because β_0 will always be included). For a fixed value of k, it is reasonable to identify the best model as the one having minimum SSE_k. The more difficult issue concerns comparison of SSE_k's for different values of k. Three different criteria, each one a simple function of SSE_k, are widely used.

1. R_k^2, the coefficient of multiple determination for a k carrier model. Because R_k^2 will virtually always increase as k does (and can never decrease), it is not the k which maximizes R_k^2 that interests us. Instead we wish to identify a small k for which R_k^2 is nearly as large as R^2 for all carriers in the model.
2. $MSE_k = SSE/(n - k - 1)$, the mean square error for a k carrier model. This is often used in place of R_k^2, since while R_k^2 never decreases with increasing k, a small decrease in SSE_k obtained with one extra carrier can be more than offset by a decrease of one in the denominator of MSE_k. The objective is then to identify the model having minimum MSE_k. Since adjusted $R_k^2 = 1 - MSE_k/MST$ where $MST = SST/(n - 1)$ is constant in k, examination of adjusted R_k^2 is equivalent to consideration of MSE_k.
3. The rationale for the third criterion, C_k, is more difficult to understand, but the criterion is gaining increasing acceptance among data analysts. Suppose that

---

* If the number of carriers is at most 27, the BMDP package will obtain for any specified m between 1 and 10 the best m subsets consisting of 1 carrier, of 2 carriers, of 3 carriers, and so on.

the true regression model is specified by m carriers—that is,

$$Y = \beta_0 + \beta_1 x_1 + \cdots + \beta_m x_m + \epsilon, \; \text{Var}(\epsilon) = \sigma^2$$

so that

$$E(Y) = \beta_0 + \beta_1 x_1 + \cdots + \beta_m x_m$$

Consider fitting a model by using a subset of k of these m carriers; for simplicity of notation, suppose that we use $x_1, x_2, \ldots, x_k$. Then by solving the system of normal equations, estimates $\hat{\beta}_0, \hat{\beta}_1, \ldots, \hat{\beta}_k$ are obtained (but not, of course, estimates of any β's corresponding to carriers not in the fitted model). The true expected value $E(Y)$ can then be estimated by $\hat{Y} = \hat{\beta}_0 + \hat{\beta}_1 x_1 + \cdots + \hat{\beta}_k x_k$. Now consider the **normalized expected total error of estimation**

$$\Gamma_k = \frac{E(\sum_{i=1}^{n} [\hat{Y}_i - E(Y_i)]^2)}{\sigma^2} = \frac{E(SSE_k)}{\sigma^2} + 2(k-1) - n \tag{13.25}$$

The second equality in (13.25) must be taken on faith, because it requires a tricky expected value argument. A particular subset is then appealing if its Γ_k value is small. Unfortunately, though, $E(SSE_k)$ and σ^2 are not known. To remedy this, let s^2 denote the estimate of σ^2 based on the model which includes all carriers for which data is available, and define

$$C_k = \frac{SSE_k}{s^2} + 2(k-1) - n$$

A desirable model is then specified by carriers for which C_k is small.

Example 13.9   The article by Hocking reports on an analysis of data taken from the 1974 issues of *Motor Trend* magazine. The dependent variable y was gas mileage, there were $n = 32$ observations, and the carriers for which data was obtained were x_1 = engine shape (1 = straight and 0 = V), x_2 = number of cylinders, x_3 = transmission type

Table 13.12 *Best Subsets for Gas Mileage Data of Example 13.9*

k = number of carriers	Variables	SSE_k	R_k^2	Adjusted R_k^2	C_k
1	9	247.2	.756	.748	11.6
2	2	169.7	.833	.821	1.2
3	3, 10, −2	150.4	.852	.836	.1
4	6	142.3	.860	.839	.8
5	5	136.2	.866	.840	1.8
6	8	133.3	.869	.837	3.4
7	4	132.0	.870	.832	5.2
8	7	131.3	.871	.826	7.1
9	1	131.1	.871	.818	9.0
10	2	131.0	.871	.809	11.0

(1 = manual and 0 = auto), x_4 = number of transmission speeds, x_5 = engine size, x_6 = horsepower, x_7 = number of carburetor barrels, x_8 = final drive ratio, x_9 = weight, and x_{10} = quarter-mile time. In Table 13.12 we present summary information from the analysis. The table describes for each k the subset having minimum SSE_k; reading down the variables column indicates which variable is added in going from k to $k + 1$ (in going from $k = 2$ to $k = 3$, both x_3 and x_{10} are added, and x_2 is deleted). Figure 13.10 contains plots of R_k^2, adjusted R_k^2, and C_k against k; these plots are an important visual aid in selecting a subset. The estimate of σ^2 is $s^2 = 6.24$, which is MSE_{10}. A simple model which rates highly according to all criteria is the one containing carriers x_3, x_9, and x_{10}.

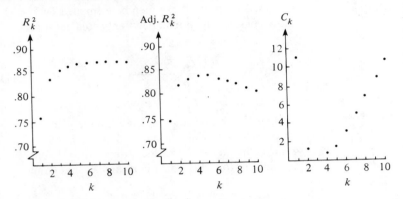

Figure 13.10 R_k^2 and C_k plots for the gas mileage data

Generally speaking, when a subset of k carriers ($k < m$) is used to fit a model, the estimators $\hat{\beta}_0, \hat{\beta}_1, \ldots, \hat{\beta}_k$ will be biased for $\beta_0, \beta_1, \ldots, \beta_k$ and $\hat{Y}$ will also be a biased estimator for the true $E(Y)$ (all this because $m - k$ carriers are missing from the fitted model). However, as measured by the total normalized expected error Γ_k, estimates based on a subset can actually provide more precision than would be obtained using all possible carriers; essentially this greater precision is obtained at the price of introducing a bias in the estimators. A value of k for which $C_k \doteq k + 1$ indicates that the bias associated with this k carrier model would be small.

Stepwise Regression

Because algorithms such as SELECT are not yet readily available, when the number of carriers is too large to allow examination of all possible subsets, there are several alternative selection procedures which generally will identify good models. The simplest such procedure is the backward elimination (BE) method. This method starts with the model in which all carriers under consideration are used. Let the set of all such carriers be $x_1, \ldots, x_m$. Then each t ratio $\hat{\beta}_i / s_{\hat{\beta}_i}$ ($i = 1, \ldots, m$) appropriate for testing H_0: $\beta_i = 0$ versus H_a: $\beta_i \neq 0$ is examined. If the t ratio with the smallest absolute value is less than a prespecified constant t_{out}, that is, if

$$\min_{i=1,\ldots,m} \left| \frac{\hat{\beta}_i}{s_{\hat{\beta}_i}} \right| < t_{out}$$

then the carrier corresponding to the smallest ratio is eliminated from the model. The reduced model is now fit, the $m - 1$ t ratios are again examined, and another carrier is eliminated if it corresponds to the smallest absolute t ratio which is less than t_{out}. In this way the algorithm continues until at some stage, all absolute t ratios are at least t_{out}. The model used is the one containing all carriers which were not eliminated. The value $t_{out} = 2$ is often recommended since most $t_{.05}$ values are near 2.

Example 13.7
(continued)

For the coded full quadratic model in which y = tar content, the five potential carriers are x_1', x_2', $x_3' = x_1'^2$, $x_4' = x_2'^2$, and $x_5' = x_1'x_2'$ (so $m = 5$). Without specifying t_{out}, the carrier with the smallest absolute t ratio (asterisked) was eliminated at each stage, resulting in the sequence of models shown in Table 13.13.

Table 13.13

Step	Carriers	1	2	$\lvert t$ ratio$\rvert$ 3	4	5
1	1, 2, 3, 4, 5	16.0	10.8	2.9	2.8	1.8*
2	1, 2, 3, 4	15.4	10.2	3.7	2.0*	—
3	1, 2, 3	14.5	12.2	4.3*	—	—
4	1, 2	10.9	9.1*	—	—	—
5	1	4.4*	—	—	—	—

Using $t_{out} = 2$, the resulting model would be based on x_1', x_2', and x_3', since at step 3 no carrier could be eliminated. It can be verified that each subset above is actually the best subset of its size, though this is by no means always the case.

An alternative to the BE procedure is forward selection (FS). FS starts with no carriers in the model and considers fitting in turn the model with only x_1, only x_2, . . ., and finally only x_m. The variable which, when fit, yields the largest absolute t ratio enters the model provided that the ratio exceeds the specified constant t_{in}. Suppose that x_1 enters the model. Then models with (x_1, x_2), (x_1, x_3), . . ., (x_1, x_m) are considered in turn. The largest $\lvert \hat{\beta}_j / s_{\hat{\beta}_j} \rvert$ $(j = 2, . . ., m)$ then specifies the entering carrier provided that this maximum also exceeds t_{in}. This continues until at some step no absolute t ratios exceed t_{in}. The carriers which have entered then specify the model. The value $t_{in} = 2$ is often used for the same reason that $t_{out} = 2$ is used in BE. For the tar-content data, FS resulted in the sequence of models given at steps 5, 4, . . ., 1 above, so agreed with BE. This will not always be the case.

The stepwise procedure most widely used is a combination of FS and BE, denoted by FB. This procedure starts off as does forward selection, by adding variables to the model, but after each addition examines those variables previously entered to see if any is a candidate for elimination. For example, if there are eight carriers under consideration and the current set consists of x_2, x_3, x_5, and x_6 with x_5 having just been added, the t ratios $\hat{\beta}_2 / s_{\hat{\beta}_2}$, $\hat{\beta}_3 / s_{\hat{\beta}_3}$, and $\hat{\beta}_6 / s_{\hat{\beta}_6}$ are examined. If the smallest absolute ratio is less than t_{out}, then the corresponding variable is eliminated from the model. The idea behind FB is that with forward selection, a single variable may be more strongly related to y than either of two or more other variables individually, but the combination of these variables may make the single variable subsequently redundant. This actually happened with the gas mileage data of Example 13.9, with x_2 entering and subsequently leaving the model.

The FB procedure is part of several standard computer packages. The BMDP package specifies $t_{in} = 2$ and $t_{out} = \sqrt{3.9}$ (most packages actually use $F = t^2$ rather than t itself).

While in most situations these automatic selection procedures will identify a good model, there is no guarantee that the best or even a nearly best model will result. Close scrutiny should be given to data sets for which there appears to be strong relationships between some of the potential carriers; in the next section we describe an alternative method to least squares for the analysis of such data.

Exercises / Section 13.4

1. The paper "Development of a Model for Use in Maize Replant Decisions" (*Agronomy J.*, 1980, pp. 459–464) reported a summary of regression analyses using as variables $y = \%$ expected maize yield, $x_1 =$ planting date (days after April 20), and $x_2 =$ plant density [.0004047 × (plants/ha)]. The five planting dates were April 20, May 6, May 22, May 31, and June 10, the four plant densities were 30,890, 41,180, 51,480, and 61,780 plants/ha, and n was 180. The model

$$Y = \beta_0 + \beta_1 x_1 + \beta_2 x_2 + \beta_3 x_1^2 + \beta_4 x_2^2 + \epsilon$$

with $R^2 = .820$ gave a good fit to the data. Parameter estimates and estimated SD's were

Parameter	Estimate	Estimated S.D.
β_0	21.09	—
β_1	.653	.14
β_2	5.488	1.407
β_3	−.02059	.0027
β_4	−.10155	.0373

a. Graph the contours of the regression function for each of the given values of x_1 (estimated expected Y versus x_2 for each of the five x_1 values).

b. Assuming that the true model is given as shown above, what is the expected change in yield which results from an increase of one in x_2? Estimate this expected change when plant density is 41,180 plants/ha.

2. Refer back to the maize yield regression discussed in Exercise 1.
   a. Test the hypothesis H_0: $\beta_1 = \beta_2 = \beta_3 = \beta_4 = 0$ (no linear relationship between Y and any of

the carriers) against the alternative H_a: $\beta_i \neq 0$ for at least one i.

b. Compute joint confidence intervals for β_3 and β_4 such that the joint confidence level is at least 95%.

c. Suppose that $SSE = 1275.75$ (not given in the paper). Compute $\hat{\sigma}$ and also SST.

d. The authors also fit the regression function $y = \beta_0 + \beta_1 x_1 + \beta_2 x_2 + \beta_3 x_1^2 + \beta_4 x_2^2 + \beta_5 x_1^3 + \beta_6 x_2^3$. Supposing that the resulting SSE was 1247.30 and using SSE of (c) for the model in Exercise 1, test H_0: $\beta_5 = \beta_6 = 0$ at level .05.

3. The paper "The Influence of Temperature and Sunshine on the Alpha-Acid Contents of Hops (*Agricultural Meteorology*, 1974, pp. 375–382) reported the following data on yield (y), mean temperature over the period between date of coming into hop and date of picking (x_1), and mean percentage of sunshine during the same period (x_2), for the fuggle variety of hop.

x_1:	16.7	17.4	18.4	16.8	18.9	17.1
x_2:	30	42	47	47	43	41
y:	210	110	103	103	91	76

x_1:	17.3	18.2	21.3	21.2	20.7	18.5
x_2:	48	44	43	50	56	60
y:	73	70	68	53	45	31

The summary quantities are $n = 12$, $\Sigma x_{1j} = 222.5$, $\Sigma x_{2j} = 551$, $\Sigma y_j = 1033$, $\Sigma x_{1j}^2 = 4156.47$, $\Sigma x_{2j}^2 = 25,937$, $\Sigma x_{1j}x_{2j} = 10,276$, $\Sigma x_{1j}y_j = 18,680$, $\Sigma x_{2j}y_j = 44,169$, and $\Sigma y_j^2 = 112,123$.

a. Assuming that $\mu_{Y \cdot x_1, x_2} = \beta_0 + \beta_1 x_1 + \beta_2 x_2$ (as was done in the paper), use the summary quantities to verify that the estimated regression function is $\hat{\mu}_{Y \cdot x_1, x_2} = 415.1131 - 6.5982x_1 - 4.5036x_2$.

b. What yield would you predict for a mean temperature of 20 and mean percentage sunshine of 40? What is $\hat{\mu}_{Y \cdot 18.9, 43}$ and what is the residual for these values of x_1 and x_2?

c. Compute SSE, $\hat{\sigma}$, and R^2.

d. Test $H_0: \beta_1 = \beta_2 = 0$ versus H_a: either β_1 or $\beta_2 \neq 0$ at level .05.

e. The estimated standard deviation of $\hat{\beta}_0 + \hat{\beta}_1 x_1 + \hat{\beta}_2 x_2$ when $x_1 = 18.9$ and $x_2 = 43$ is 8.20. Use this to obtain a 95% confidence interval for $\mu_{Y \cdot 18.9, 43}$.

f. Use the information in (e) to obtain a 95% prediction interval for yield in a future experiment when $x_1 = 18.9$ and $x_2 = 43$.

g. The estimated SD of $\hat{\beta}_1$ is $s_{\hat{\beta}_1} = 4.86$. Test to see whether x_1 belongs in the model.

h. When the model $Y = \beta_0 + \beta_2 x_2 + \epsilon$ is fit, the resulting value of R^2 is .721. Verify that the F statistic for testing $H_0: Y = \beta_0 + \beta_2 x_2 + \epsilon$ versus $H_a: Y = \beta_0 + \beta_1 x_1 + \beta_2 x_2 + \epsilon$ satisfies $t^2 = f$, where t is the value of the t statistic computed in part (g).

4. a. When the model $Y = \beta_0 + \beta_1 x_1 + \beta_2 x_2 + \beta_3 x_1^2 + \beta_4 x_2^2 + \beta_5 x_1 x_2 + \epsilon$ is fit to the hops data of Exercise 3, the estimate of β_5 is $\hat{\beta}_5 = .557$ with estimated standard deviation $s_{\hat{\beta}_5} = .94$. Test $H_0: \beta_5 = 0$ versus $H_a: \beta_5 \neq 0$.

b. Each t ratio $\hat{\beta}_i / s_{\hat{\beta}_i}$ $(i = 1, 2, 3, 4, 5)$ for the model of (a) is less than 2 in absolute value, yet $R^2 = .861$ for this model. Would it be cor-

rect to drop each term from the model because of its small t ratio? Explain.

c. Using $R^2 = .861$ for the model of (a), test $H_0: \beta_3 = \beta_4 = \beta_5 = 0$ (which says that all second-order terms can be deleted).

5. The article "The Undrained Strength of Some Thawed Permafrost Soils" (*Canadian Geotechnical J*. 1979, pp. 420–427) contained the accompanying data on undrained shear strength of sandy soil (y, in kPa), depth (x_1, in m), and water content (x_2, in %). The predicted values and residuals were computed by fitting a full quadratic model, which resulted in the estimated regression function

$$y = -151.36 - 16.22x_1 + 13.48x_2 + .094x_1^2 - .253x_2^2 + .492x_1 x_2$$

a. Do plots of e^* versus x_1, e^* versus x_2, and e^* versus $\hat{y}$ suggest that the full quadratic model should be modified? Explain your answer.

b. The value of R^2 for the full quadratic model is .759. Test at level .05 the null hypothesis stating that there is no linear relationship between the dependent variable and any of the five carriers.

c. It can be shown that $\text{Var}(Y) = \sigma^2 = \text{Var}(\hat{Y}) + \text{Var}(Y - \hat{Y})$. The estimate of σ is $\hat{\sigma} = s = 6.99$ (from the full quadratic model). First obtain the estimated SD of $Y - \hat{Y}$, and then estimate the standard deviation of $\hat{Y}$ (that is, of $\hat{\beta}_0 + \hat{\beta}_1 x_1 + \hat{\beta}_2 x_2 + \hat{\beta}_3 x_1^2 + \hat{\beta}_4 x_2^2 + \hat{\beta}_5 x_1 x_2$) when $x_1 = 8.0$ and $x_2 = 33.1$. Finally, compute a 95% confidence interval for $\mu_{Y \cdot 8.0, 33.1}$. *Hint*: What is $(y - \hat{y})/e^*$?

d. Fitting the first-order model with regression

	y	x_1	x_2	$\hat{y}$	$y - \hat{y}$	Standardized residual e^*
1	14.7000	8.9000	31.5000	23.35	-8.65	-1.50
2	48.0000	36.6000	27.0000	46.38	1.62	0.54
3	25.6000	36.8000	25.9000	27.13	-1.53	-0.53
4	10.0000	6.1000	39.1000	10.99	-0.99	-0.17
5	16.0000	6.9000	39.2000	14.10	1.90	0.33
6	16.8000	6.9000	38.3000	16.54	0.26	0.04
7	20.7000	7.3000	33.9000	23.34	-2.64	-0.42
8	38.8000	8.4000	33.8000	25.43	13.37	2.17
9	16.9000	6.5000	27.9000	15.63	1.27	0.23
10	27.0000	8.0000	33.1000	24.29	2.71	0.44
11	16.0000	4.5000	26.3000	15.36	0.64	0.20
12	24.9000	9.9000	37.8000	29.61	-4.71	-0.91
13	7.3000	2.9000	34.6000	15.38	-8.08	-1.53
14	12.8000	2.0000	36.4000	7.96	4.84	1.02

function $\mu_{Y \cdot x_1, x_2} = \beta_0 + \beta_1 x_1 + \beta_2 x_2$ results in $SSE = 894.95$. Test at level .05 the null hypothesis which states that all quadratic terms can be deleted from the model.

6. In an experiment to study factors influencing the wood specific gravity of slash pines ("Anatomical Factors Influencing Wood Specific Gravity of Slash Pines and the Implications for the Development of a High-Quality Pulpwood," *TAPPI*, 1964, pp. 401–404), a sample of 20 mature wood samples was obtained, and measurements were taken on number of fibers/mm^2 in springwood (x_1), number of fibers/mm^2 in summerwood (x_2), % springwood (x_3), light absorption in springwood (x_4), and light absorption in summerwood (x_5).

   a. Fitting the regression function $\mu_{Y \cdot x_1, x_2, x_3, x_4, x_5} = \beta_0 + \beta_1 x_1 + \cdots + \beta_5 x_5$ resulted in $R^2 = .769$. Does the data indicate that there is a linear relationship between specific gravity and at least one of the carriers? Test using $\alpha = .01$.

   b. When x_2 is dropped from the model, the value of R^2 remains at .769. Compute adjusted R^2 for both the full model and the model with x_2 deleted.

   c. When x_1, x_2, and x_4 are all deleted, the resulting value of R^2 is .654. The total sum of squares is $SST = .0196610$. Does the data suggest that all of x_1, x_2, and x_4 have zero coefficients in the true regression model? Test the relevant hypotheses at level .05.

7. a. Referring back to Exercise 6, the mean and standard deviation of x_3 were 52.540 and 5.4447, respectively, while those of x_5 were 89.195 and 3.6660, respectively. When the model involving these two standardized variables was fit, the estimated regression equation was $y = .5255 - .0236 x_3' + .0097 x_5'$. What value of specific gravity would you predict for a wood sample with % springwood = 50 and % light absorption in summerwood = 90?

   b. The estimated standard deviation of the estimated coefficient β_3 of x_3' (that is, $s_{\hat{\beta}_3}$ for β_3 of the standardized model) was .0046. Obtain a 95% confidence interval for β_3.

   c. Using the information in (a) and (b), what is the estimated coefficient of x_3 in the unstandardized model (using only carriers x_3 and x_5), and what is the estimated standard deviation of

the coefficient estimator (that is, $s_{\hat{\beta}_3}$ for $\hat{\beta}_3$ in the unstandardized model)?

   d. The estimate of σ for the two-carrier model is $s = .02001$, while the estimated standard deviation of $\hat{\beta}_0 + \hat{\beta}_3 x_3' + \hat{\beta}_5 x_5'$ when $x_3' = -.3747$ and $x_5' = -.2769$ (that is, when $x_3 = 50.5$ and $x_5 = 88.9$) is .00482. Compute a 95% prediction interval for specific gravity when % springwood = 50.5 and % light absorption in summerwood = 88.9.

8. The accompanying data resulted from a study of the relationship between brightness of finished paper (y) and the variables H_2O_2% by weight (x_1), NaOH% by weight (x_2), silicate % by weight (x_3), and process temperature (x_4) ("Advantages of CE-HDP Bleaching for High Brightness Kraft Pulp Production," *TAPPI*, 1964, pp. 170A–173A). Each independent variable was allowed to assume five different values, and these values were coded for regression analysis as -2, -1, 0, 1, and 2.

   a. When a (coded) model involving all linear terms, all quadratic terms, and all cross product terms was fit, the estimated regression function was

$$y = 84.67 + .650 x_1 - .258 x_2 + .133 x_3$$
$$+ .108 x_4 - .135 x_1^2 + .028 x_2^2 + .028 x_3^2$$
$$- .072 x_4^2 + .038 x_1 x_2 - .075 x_1 x_3$$
$$+ .213 x_1 x_4 + .200 x_2 x_3 - .188 x_2 x_4$$
$$+ .050 x_3 x_4$$

Use this estimated model to predict brightness when H_2O_2 is .4%, NaOH is .4%, silicate is 3.5%, and temperature is 175. What are the values of the residuals for these values of the variables?

   b. Express the estimated regression function in uncoded form.

   c. $SST = 17.2567$, and R^2 for the model of (a) is .885. When a model which includes only the four linear terms is fit, the resulting value of R^2 is .721. State and test at level .05 the null hypothesis which specifies that the coefficients of all quadratic and cross-product terms in the regression function are zero.

   d. The estimated (coded) regression function when only linear terms are included is $\hat{\mu}_{Y \cdot x_1, x_2, x_3, x_4} = 85.5548 + .6500 x_1 - .2583 x_2 + .1333 x_3 + .1083 x_4$. When $x_1 = x_2 = x_3 = x_4 = 0$, the esti-

Variables	−2	−1	0	+1	+2
x_1 Hydrogen peroxide (100%), %wt	0.1	0.2	0.3	0.4	0.5
x_2 NaOH, % wt	0.1	0.2	0.3	0.4	0.5
x_3 Silicate (41° Bé), %wt	0.5	1.5	2.5	3.5	4.5
x_4 Process temp., °F	130	145	160	175	190

Test no.	H_2O_2 conc. (x_1)	NaOH conc. (x_2)	Silicate conc. (x_3)	Temp. (x_4)	Bright. y
1	−1	−1	−1	−1	83.9
2	+1	−1	−1	−1	84.9
3	−1	+1	−1	−1	83.4
4	+1	+1	−1	−1	84.2
5	−1	−1	+1	−1	83.8
6	+1	−1	+1	−1	84.7
7	−1	+1	+1	−1	84.0
8	+1	+1	+1	−1	84.8
9	−1	−1	−1	+1	84.5
10	+1	−1	−1	+1	86.0
11	−1	+1	−1	+1	82.6
12	+1	+1	−1	+1	85.1
13	−1	−1	+1	+1	84.5
14	+1	−1	+1	+1	86.0
15	−1	+1	+1	+1	84.0
16	+1	+1	+1	+1	85.4
17	−2	0	0	0	82.9
18	+2	0	0	0	85.5
19	0	−2	0	0	85.2
20	0	+2	0	0	84.5
21	0	0	−2	0	84.7
22	0	0	+2	0	85.0
23	0	0	0	−2	84.9
24	0	0	0	+2	84.0
25	0	0	0	0	84.5
26	0	0	0	0	84.7
27	0	0	0	0	84.6
28	0	0	0	0	84.9
29	0	0	0	0	84.9
30	0	0	0	0	84.5
31	0	0	0	0	84.6

mated SD of $\hat{\mu}_{Y \cdot 0,0,0,0}$ is .0772. Suppose that it had been believed that expected brightness for these values of the x_i's was at least 85.0. Does the given information contradict this belief? State and test the appropriate hypotheses.

9. The paper "Bank Full Discharge of Rivers" (*Water Resources J*. 1978, pp. 1141–1154) reported data on discharge amount (q, in m^3/sec), flow area (a, in m^2), and slope of the water surface (b, in m/m) obtained at a number of floodplain stations. A subset of the data appears below. The paper proposed a multiplicative power model $Q = \alpha a^\beta b^\gamma \epsilon$.

q	17.6	23.8	5.7	3.0	7.5
a	8.4	31.6	5.7	1.0	3.3
b	.0048	.0073	.0037	.0412	.0416

q	89.2	60.9	27.5	13.2	12.2
a	41.1	26.2	26.4	6.7	9.7
b	.0063	.0061	.0036	.0039	.0025

a. Use an appropriate transformation to make the model linear, and then estimate the regression parameters for the transformed model. Finally, estimate α, β, and γ (the parameters of the original model). What would be your estimate of discharge amount when flow area is 10 and slope is .01?

b. Without actually doing any analysis, how would you fit a multiplicative exponential model $Q = \alpha e^{\beta a} e^{\gamma b} \epsilon$?

c. After the transformation to linearity in (a), a 95% confidence interval for the value of the transformed regression function when $a = 3.3$ and $b = .0046$ was obtained from computer output as (.217, 1.755). Obtain a 95% confidence interval for $\alpha a^\beta b^\gamma$ when $a = 3.3$, $b = .0046$.

10. Below find the smallest SSE for each number of carriers k ($k = 1, 2, 3, 4$) for a regression problem in which y = cumulative heat of hardening in cement, x_1 = % tricalcium aluminate, x_2 = % tricalcium silicate, x_3 = % aluminum ferrate, and x_4 = % dicalcium silicate.

Number of carriers k	Carrier (s)	SSE
1	x_4	880.85
2	x_1, x_2	58.01
3	x_1, x_2, x_3	49.20
4	x_1, x_2, x_3, x_4	47.86

In addition $n = 13$ and $SST = 2715.76$.

a. Use the criteria discussed in the text to recommend the use of a particular regression model.

b. Would forward selection result in the best two carrier models? Explain.

13.5   Regression Using Matrix Algebra

The multiple linear regression model with k carriers $x_1, x_2, \ldots, x_k$ and n observations $Y_1, \ldots, Y_n$ has the form

$$Y_j = \beta_0 + \beta_1 x_{1j} + \beta_2 x_{2j} + \cdots + \beta_k x_{kj} + \epsilon_j \qquad j = 1, \ldots, n \qquad (13.26)$$

To describe the model in matrix form, we introduce a vector of observations **Y**, vector of random errors $\boldsymbol{\epsilon}$, parameter vector $\boldsymbol{\beta}$, and **design matrix X**:

$$\mathbf{Y} = \begin{pmatrix} Y_1 \\ Y_2 \\ \vdots \\ Y_n \end{pmatrix}, \qquad \boldsymbol{\epsilon} = \begin{pmatrix} \epsilon_1 \\ \epsilon_2 \\ \vdots \\ \epsilon_n \end{pmatrix}, \qquad \boldsymbol{\beta} = \begin{pmatrix} \beta_0 \\ \beta_1 \\ \vdots \\ \beta_k \end{pmatrix}, \qquad \mathbf{X} = \begin{pmatrix} 1 & x_{11} x_{21} \cdots x_{k1} \\ 1 & x_{12} x_{22} \cdots x_{k2} \\ & \vdots \\ 1 & x_{1n} x_{2n} \cdots x_{kn} \end{pmatrix}$$

The matrix **X** has n rows and $k + 1$ columns while $\boldsymbol{\beta}$ has $k + 1$ rows, so the product $\mathbf{X}\boldsymbol{\beta}$ is defined. The equations of (13.26) now become $\mathbf{Y} = \mathbf{X}\boldsymbol{\beta} + \boldsymbol{\epsilon}$.

We now wish to estimate the parameter vector $\boldsymbol{\beta}$. Let **b** be a vector of trial estimates. Then the sum of squared deviations is $\Sigma \, (Y_j - b_0 - b_1 x_{1j} - \cdots - b_k x_{kj})^2$, and the least squares estimators are chosen to minimize this sum. It is easily verified that the j deviation is the jth element of the column vector $\mathbf{Y} - \mathbf{Xb}$. With $(\mathbf{Y} - \mathbf{Xb})'$ denoting the transpose of the column vector $\mathbf{Y} - \mathbf{Xb}$ (which is a row vector) the sum of squared deviations can be written as $(\mathbf{Y} - \mathbf{Xb})'(\mathbf{Y} - \mathbf{Xb})$.

Furthermore, the normal equations (13.22) can be written in matrix form. The transpose of the **X** matrix is

$$\mathbf{X}' = \begin{pmatrix} 1 & 1 & 1 & \cdots & 1 \\ x_{11} x_{21} x_{31} & \cdots & x_{k1} \\ & \vdots \\ x_{1n} x_{2n} x_{3n} & \vdots & x_{kn} \end{pmatrix}$$

Because $\mathbf{X}'$ has n columns, it can multiply on the left both **X** and **Y**, since each of these has n rows. The normal equations then take the form

$$(\mathbf{X}'\mathbf{X})\,\mathbf{b} = \mathbf{X}'\mathbf{Y} \qquad\qquad (13.27)$$

It can be shown that if the columns of **X** are linearly independent (that is, no column of **X** can be expressed as a linear combination of other columns of **X**), then the $(k + 1) \times (k + 1)$ matrix $\mathbf{X}'\mathbf{X}$ has an inverse. To solve (13.27), we simply multiply each side of (13.27) on the left by $(\mathbf{X}'\mathbf{X})^{-1}$, yielding

$$\boxed{\mathbf{b} = \hat{\boldsymbol{\beta}} = (\mathbf{X}'\mathbf{X})^{-1}\,\mathbf{X}'\mathbf{Y}} \qquad\qquad (13.28)$$

Example 13.10   For simple linear regression, the **X** matrix has only two columns:

$$\mathbf{X} = \begin{pmatrix} 1 & x_1 \\ 1 & x_2 \\ \vdots & \vdots \\ 1 & x_n \end{pmatrix}, \qquad \mathbf{X}' = \begin{pmatrix} 1 & 1 & \cdots & 1 \\ x_1 & x_2 & \cdots & x_n \end{pmatrix}, \qquad \mathbf{X}'\mathbf{X} = \begin{pmatrix} n & \Sigma\, x_i \\ \Sigma\, x_i & \Sigma\, x_i^2 \end{pmatrix}$$

As long as at least two x_i's differ from one another (and if this is not the case, the slope β_1 cannot be estimated), the two columns of $\mathbf{X}$ are linearly independent, and

$$(\mathbf{X}'\mathbf{X})^{-1} = \frac{1}{n\Sigma\,(x_i - \bar{x})^2}\begin{pmatrix} \Sigma\,x_i^2 & -\Sigma\,x_i \\ -\Sigma\,x_i & n \end{pmatrix}$$

Then

$$\hat{\boldsymbol{\beta}} = (\mathbf{X}'\mathbf{X})^{-1}\,\mathbf{X}'\mathbf{Y} = \frac{1}{n\,\Sigma\,(x_i - \bar{x})^2}\begin{pmatrix} \Sigma\,x_i^2 & -\Sigma\,x_i \\ -\Sigma\,x_i & n \end{pmatrix}\begin{pmatrix} \Sigma\,Y_i \\ \Sigma\,x_iY_i \end{pmatrix}$$

$$= \frac{1}{n\,\Sigma\,(x_i - \bar{x})^2}\begin{pmatrix} (\Sigma\,x_i^2)(\Sigma\,Y_i) - (\Sigma\,x_i)(\Sigma\,x_iY_i) \\ n\,\Sigma\,x_iY_i - (\Sigma\,x_i)(\Sigma\,Y_i) \end{pmatrix}$$

The second component of $\hat{\boldsymbol{\beta}}$ is $\hat{\beta}_1 = [n\,\Sigma\,x_iY_i - (\Sigma\,x_i)(\Sigma\,Y_i)]/n\,\Sigma\,(x_i - \bar{x})^2$ as before, while $\hat{\beta}_0 = (\Sigma\,Y_i - \hat{\beta}_1\,\Sigma\,x_i)/n$ after some algebra.

In problems with more than a single carrier, it is tedious to invert $\mathbf{X}'\mathbf{X}$ by hand. Most regression computer programs will print out $(\mathbf{X}'\mathbf{X})^{-1}$. Because $\mathbf{X}'\mathbf{X}$ is symmetric (the triangle of elements above the main diagonal is the mirror image of the triangle below the main diagonal), so is $(\mathbf{X}'\mathbf{X})^{-1}$, so only the elements on and below the main diagonal are printed.

Example 13.11 For the flour absorption data in Example 13.6,

$$\mathbf{X} = \begin{pmatrix} 1 & 8.5 & 2 \\ 1 & 8.9 & 3 \\ 1 & 10.6 & 3 \\ 1 & 10.2 & 20 \\ \vdots & \vdots & \vdots \\ 1 & 11.7 & 57 \end{pmatrix}, \qquad (\mathbf{X}'\mathbf{X})^{-1} = \begin{pmatrix} 4.00017 & \cdot & \cdot \\ -.36902 & .03601 & \cdot \\ .00957 & -.00154 & .00027 \end{pmatrix}$$

$$\mathbf{X}'\mathbf{y} = \begin{pmatrix} \Sigma\,y_j \\ \Sigma\,x_{1j}y_j \\ \Sigma\,x_{2j}y_j \end{pmatrix} = \begin{pmatrix} 1287.4 \\ 14{,}940.06 \\ 40{,}016.0 \end{pmatrix}$$

Then

$$\hat{\boldsymbol{\beta}} = (\mathbf{X}'\mathbf{X})^{-1}\,\mathbf{X}'\mathbf{y} = \begin{pmatrix} 19.44495 \\ 1.44176 \\ .33566 \end{pmatrix}$$

Once $\hat{\boldsymbol{\beta}}$ is obtained, any predicted value $\hat{y}$ for fixed values $x_1, \ldots, x_k$ of the carriers is

$$\hat{y} = \hat{\beta}_0 + \hat{\beta}_1 x_1 + \cdots + \hat{\beta}_k x_k = \mathbf{x}'\hat{\boldsymbol{\beta}} \qquad (13.29)$$

where $\mathbf{x}' = (1 \quad x_1 \quad x_2 \cdots x_k)$

With $\mathbf{x}'_{(j)} = (1 \quad x_{1j} \quad x_{2j} \; \ldots \; x_{kj})$ denoting the jth row of the $\mathbf{X}$ matrix, the jth predicted value is $\hat{y}_j = \mathbf{x}'_{(j)}\hat{\boldsymbol{\beta}}$. Letting $\hat{\mathbf{y}}$ denote the vector of predicted values (corresponding to values of the carriers for which Y's are observed) and $\mathbf{e}$ the vector of residuals ($e_j = y_j - \hat{y}_j$), (13.28) and (13.29) give

$$\hat{\mathbf{y}} = \mathbf{X}\hat{\boldsymbol{\beta}} = \mathbf{X}(\mathbf{X}'\mathbf{X})^{-1}\,\mathbf{X}'\mathbf{y}, \; \mathbf{e} = \mathbf{y} - \hat{\mathbf{y}} = [\mathbf{I} - \mathbf{X}(\mathbf{X}'\mathbf{X})^{-1}\mathbf{X}']\mathbf{y} \quad (13.30)$$

where $\mathbf{I}$ is the $n \times n$ identity matrix (each diagonal element 1 and all other elements 0).

Identification of Points Having Large Influence

In simple linear regression diagnostic plots can be used to identify both points corresponding to large residuals and points which, because their x values lie far from most sample x_i's, will have a large influence on the resulting fit. In multiple regression plots will often not reveal such information. A basic tool in this investigation is the $n \times n$ matrix $\mathbf{H} = \mathbf{X}(\mathbf{X}'\mathbf{X})^{-1}\mathbf{X}'$, sometimes called the **"hat"** **matrix.** From (13.30) the ith fitted value $\hat{y}_i$ is the product of the ith row of $\mathbf{H}$ with the $\mathbf{y}$ vector:

$$\hat{y}_i = h_{i1}y_1 + h_{i2}y_2 + \cdots + h_{ij}y_j + \cdots + h_{in}y_n$$

Thus the element h_{ij} gives the weight associated with the jth observation in computing the ith predicted value. In particular, h_{ii} measures the influence of y_i on its own predicted value $\hat{y}_i$.

It is therefore of great interest to know whether a particular h_{ii} is relatively large or small (when n is large, so is the $\mathbf{H}$ matrix, so it is easiest to examine the h_{ii}'s). To this end, note first that by using properties of inverses and transposes, it can be shown that $\mathbf{H}^2 = \mathbf{H} \cdot \mathbf{H} = \mathbf{H}$ (a matrix $\mathbf{H}$ is called idempotent if $\mathbf{H} \cdot \mathbf{H} = \mathbf{H}$). This implies that

$$h_{ii} = \sum_{j=1}^{n} h_{ij}^2 = h_{ii}^2 + \sum_{j \neq i} h_{ij}^2 \geq h_{ii}^2$$

so that $0 \leq h_{ii} \leq 1$. An additional argument (beyond the level of this text) shows that $\sum_{i=1}^{n} h_{ii} = k + 1$ when the columns of $\mathbf{X}$ are linearly independent, so that the average of these diagonal elements is $(k + 1)/n$.

Rule of thumb: Any i for which $h_{ii} > 2(k + 1)/n$ is identified as corresponding to a point with large influence on the fit.

This rule of thumb is used in the MINITAB regression command to identify points with large influence.

Example 13.12 The accompanying data appeared in the paper "Testing for the Inclusion of Variables in Linear Regression by a Randomization Technique" (*Technometrics*, 1966, pp.

695–699), and was reanalyzed in David Hoaglin and Roy Welsch, "The Hat Matrix in Regression and ANOVA" (*Amer. Statistician*, 1978, pp. 17–22):

Beam number	Specific gravity (x_1)	Moisture content (x_2)	Strength (y)
1	0.499	11.1	11.14
2	0.558	8.9	12.74
3	0.604	8.8	13.13
4	0.441	8.9	11.51
5	0.550	8.8	12.38
6	0.528	9.9	12.60
7	0.418	10.7	11.13
8	0.480	10.5	11.70
9	0.406	10.5	11.02
10	0.467	10.7	11.41

The **H** matrix (with elements below the diagonal omitted by symmetry) is

i	1	2	3	4	5	6	7	8	9	10
1	.418	−.002	.079	−.274	−.046	.181	.128	.222	.050	.242
2		.242	.292	.136	.243	.128	−.041	.033	−.035	.004
3			.417	−.019	.273	.187	−.126	.044	−.153	.004
4				.604	.197	−.038	.168	−.022	.275	−.028
5					.252	.111	−.030	.019	−.010	−.010
6						.148	.042	.117	.012	.111
7							.262	.145	.277	.174
8								.154	.120	.168
9									.315	.148
10										.187

Here $k = 2$ so $(k + 1)/n = \frac{3}{10} = .3$; since $h_{44} = .604 > 2(.3)$, the fourth data point is identified as exerting large influence.

Another means for deciding whether or not the ith point has large influence is to consider the changes in parameter estimates when the ith data point is deleted from the sample. If the estimates change drastically, this should be reported. Let $\mathbf{X}_{(i)}$ denote the **X** matrix after deleting the ith row, $\mathbf{Y}_{(i)}$ denote the **Y** vector with Y_i deleted, and $\hat{\boldsymbol{\beta}}_{(i)}$ denote the vector of parameter estimates when data point i is removed from the sample. Then

$$\hat{\boldsymbol{\beta}} - \hat{\boldsymbol{\beta}}_{(i)} = (\mathbf{X}'\mathbf{X})^{-1}\mathbf{X}'\mathbf{Y} - (\mathbf{X}'_{(i)}\mathbf{X}_{(i)})^{-1}\mathbf{X}'_{(i)}\mathbf{Y}_{(i)} \tag{13.31}$$

It would seem as though much deletion and recalculation would be required to compute (13.31) for each $i(i = 1, \ldots, n)$. However, it can be shown that

$$\hat{\boldsymbol{\beta}} - \hat{\boldsymbol{\beta}}_{(i)} = \frac{e_i}{1 - h_{ii}}(\mathbf{X}'\mathbf{X})^{-1}\mathbf{x}'_{(i)} \tag{13.32}$$

where e_i is the ith residual from the "no deletions" regression and $\mathbf{x}'_{(i)}$ is the ith row of $\mathbf{X}$. Thus once the h_{ii}'s have been obtained, the computations in (13.32) are straightforward. It is recommended that $\hat{\boldsymbol{\beta}} - \hat{\boldsymbol{\beta}}_{(i)}$ be computed both for i corresponding to large residuals and large h_{ii}.

Example 13.12 Table 13.14
(continued)

Parameter	No deletions estimates	Estimated S.D.	Change when point i is deleted		
			$i = 1$	$i = 4$	$i = 6$
β_0	10.302	1.896	2.710	-2.109	$-.642$
β_1	8.495	1.784	-1.772	1.695	.748
β_2	.2663	.1273	$-.1932$	.1242	.0329
e_i:			-3.25	$-.96$	2.20
h_{ii}:			.418	.604	.148

For deletion of both point 1 and point 4, the change in each estimate is in the range 1–1.5 standard deviations, which is reasonably substantial (this does not tell us what would happen if both points were simultaneously omitted). For point 6, however, the change is roughly .25 standard deviations. Thus points 1 and 4, but not 6, might well be omitted in calculating a regression equation.

Expected Values, Variances, and Covariances of Estimators

Given a vector $\mathbf{Y}$ of random variables, the **mean vector** $\mathbf{E}(\mathbf{Y})$ is defined as the vector of individual expected values $E(Y_1), \ldots, E(Y_n)$. Thus since $E(\epsilon_i) = 0$, $\mathbf{E}(\boldsymbol{\epsilon}) = \mathbf{0}$ (the zero vector), and with $Y = \mathbf{X}\boldsymbol{\beta} + \boldsymbol{\epsilon}$

$$\mathbf{E}(\mathbf{Y}) = \mathbf{E}(\mathbf{X}\boldsymbol{\beta}) + \mathbf{E}(\boldsymbol{\epsilon}) = \mathbf{X}\boldsymbol{\beta} + \mathbf{0} = \mathbf{X}\boldsymbol{\beta}$$

since each component of $\mathbf{X}\boldsymbol{\beta}$ is a constant (nonrandom).

Suppose now that we use a matrix $\mathbf{A}$ to form a new random vector $\mathbf{AY}$. Then since each component of $\mathbf{AY}$ is a linear combination of the Y_i's, it follows that $\mathbf{E}(\mathbf{AY}) = \mathbf{AE}(\mathbf{Y})$. Since $\hat{\boldsymbol{\beta}} = (\mathbf{X}'\mathbf{X})^{-1}\mathbf{X}'\mathbf{Y} = \mathbf{AY}$ with $\mathbf{A} = (\mathbf{X}'\mathbf{X})^{-1}\mathbf{X}'$,

$$\mathbf{E}(\hat{\boldsymbol{\beta}}) = (\mathbf{X}'\mathbf{X})^{-1}\mathbf{X}' \cdot \mathbf{E}(\mathbf{Y}) = (\mathbf{X}'\mathbf{X})^{-1}\mathbf{X}'\mathbf{X}\boldsymbol{\beta} = \boldsymbol{\beta}$$

This verifies the claim made earlier that the least squares estimators are unbiased.

In Chapter 5 we defined the covariance between two random variables Y_1 and Y_2 by $\text{Cov}(Y_1, Y_2) = E[(Y_1 - \mu_1)(Y_2 - \mu_2)]$, where $\mu_1 = E(Y_1)$ and $\mu_2 = E(Y_2)$. Notice that $\text{Cov}(Y_1, Y_1) = \text{Var}(Y_1)$ and similarly for Y_2. Furthermore, we noted that Y_1, Y_2 independent implied that $\text{Cov}(Y_1, Y_2) = 0$. Now given a random vector $\mathbf{Y}$, the **variance-covariance matrix** $\sigma^2(\mathbf{Y})$ is the matrix whose (i, j)th element is $\sigma_{ij} = \text{Cov}(Y_i, Y_j)$. Because $\text{Cov}(Y_i, Y_j) = \text{Cov}(Y_j, Y_i)$, the matrix is symmetric, and the ith diagonal element is $\sigma_{ii} = \text{Var}(Y_i)$. Also $\text{Corr}(Y_i, Y_j) = \sigma_{ij}/\sqrt{\sigma_{ii} \cdot \sigma_{jj}}$. If $\mathbf{Y}$ is the vector of observations in regression, then because the Y_i's are independent

$$\sigma^2(\mathbf{Y}) = \begin{pmatrix} \sigma^2 & 0 & 0 \ldots 0 \\ 0 & \sigma^2 & 0 \ldots \\ \cdot & & & \cdot \\ \cdot & & \cdot & \cdot \\ \cdot & & & \cdot \cdot \\ 0 & & & \sigma^2 \end{pmatrix} = \sigma^2 \cdot \mathbf{I}$$

If we form the new random vector $\mathbf{AY}$, it can be shown that $\sigma^2(\mathbf{AY}) = \mathbf{A}\sigma^2(\mathbf{Y})\mathbf{A}'$. Since $\hat{\boldsymbol{\beta}} = \mathbf{AY}$ with $\mathbf{A} = (\mathbf{X}'\mathbf{X})^{-1}\mathbf{X}'$ and $\mathbf{A}' = \mathbf{X}(\mathbf{X}'\mathbf{X})^{-1}$,

$$\begin{aligned} \sigma^2(\hat{\boldsymbol{\beta}}) &= (\mathbf{X}'\mathbf{X})^{-1}\mathbf{X}' \; \sigma^2(\mathbf{Y}) \; \mathbf{X}(\mathbf{X}'\mathbf{X})^{-1} \\ &= (\mathbf{X}'\mathbf{X})^{-1}\mathbf{X}' \cdot \sigma^2 \cdot \mathbf{I} \cdot \mathbf{X}(\mathbf{X}'\mathbf{X})^{-1} = \sigma^2 \cdot (\mathbf{X}'\mathbf{X})^{-1} \end{aligned} \qquad (13.33)$$

Thus *the variance of each $\hat{\beta}_i$ is the ith diagonal element of* $(\mathbf{X}'\mathbf{X})^{-1}$ *multiplied by* σ^2, while the off-diagonal elements specify $\text{Cov}(\hat{\beta}_i, \hat{\beta}_j)$.

Example 13.11 (continued) The $(\mathbf{X}'\mathbf{X})^{-1}$ matrix is

$$(\mathbf{X}'\mathbf{X})^{-1} = \begin{pmatrix} 4.00017 & -.36902 & .00957 \\ -.36902 & .03601 & -.00154 \\ .00957 & -.00154 & .00027 \end{pmatrix}$$

and $\hat{\sigma}^2 = s^2 = SSE/(n-3) = 1.1972$. The estimated variance of $\hat{\beta}_1$ is then $(1.1972)(.03601) = .0431$, so $s_{\hat{\beta}_1} = .2076$. Similarly, $s_{\hat{\beta}_2} = \sqrt{1.1972} \cdot \sqrt{.00027} = .0180$. These results were reported in Section 13.4. The estimated covariance between $\hat{\beta}_1$ and $\hat{\beta}_2$ is $(1.1972)(-.00154) = -.00184$, so the estimated correlation between $\hat{\beta}_1$ and $\hat{\beta}_2$ is $-.00184/(.2076)(.0180) = -.492$. This indicates that $\hat{\beta}_1$ and $\hat{\beta}_2$ are rather highly correlated.

Suppose now that we wish the variance of $\hat{\beta}_0 + \hat{\beta}_1 x_1 + \cdots + \hat{\beta}_k x_k = \mathbf{x}'\hat{\boldsymbol{\beta}}$ where $\mathbf{x}' = (1 \; x_1 \; \ldots \; x_k)$. Then

$$\sigma^2(\mathbf{x}'\hat{\boldsymbol{\beta}}) = \text{Var}(\mathbf{x}'\hat{\boldsymbol{\beta}}) = \mathbf{x}' \cdot \sigma^2(\hat{\boldsymbol{\beta}}) \cdot \mathbf{x} = \sigma^2 \cdot \mathbf{x}'(\mathbf{X}'\mathbf{X})^{-1}\mathbf{x} \qquad (13.34)$$

In particular, substitution of $\mathbf{x}'_{(i)}$, the ith row of $\mathbf{X}$, into (13.34) gives the variance of the ith fitted value $\hat{Y}_i$:

$$\text{Var}(\hat{Y}_i) = \sigma^2 \cdot \mathbf{x}'_{(i)}(\mathbf{X}'\mathbf{X})^{-1}\mathbf{x}_{(i)}$$

After substituting s^2 into (13.34) in place of σ^2 and taking the square root, the resulting quantity can be used to obtain a confidence interval for $\beta_0 + \beta_1 x_1 + \cdots + \beta_k x_k$ or to test a hypothesis about this quantity (for fixed $x_1, \ldots, x_k$ values).

Example 13.11
(continued)
For $x_1 = 10$, $x_2 = 25$, $\mathbf{x}' = (1 \quad 10 \quad 25)$ and

$$s^2 \cdot \mathbf{x}' (\mathbf{X}'\mathbf{X})^{-1} \mathbf{x} = (1.1972) \cdot (1 \quad 10 \quad 25) \begin{pmatrix} .54922 \\ -.04742 \\ .00092 \end{pmatrix}$$

$$= (1.1972)(.09802) = .11735$$

so the estimated standard deviation of $\hat{\beta}_0 + 10\hat{\beta}_1 + 25\hat{\beta}_2$ is $\sqrt{.11735} = .343$. With $t_{.025,25} = 2.060$, a 95% confidence interval for $\beta_0 + 10\beta_1 + 25\beta_2$ is

$$19.44 + 10(1.44) + 25(.34) \pm (2.060)(.343) = 42.34 \pm .71$$

$$= (41.63, 43.05)$$

Multicollinearity and Biased Estimation

In many multiple regression data sets the carriers $x_1, x_2, \ldots, x_k$ are highly interdependent. Suppose that we consider the usual model

$$Y = \beta_0 + \beta_1 x_1 + \cdots + \beta_k x_k + \epsilon$$

with data $(x_{1j}, \ldots, x_{kj}, y_j)$ $(j = 1, \ldots, n)$ available for fitting. If we use the principle of least squares to regress x_i on the other carriers $x_1, \ldots, x_{i-1}, x_{i+1}, \ldots, x_k$, obtaining

$$\hat{x}_i = a_0 + a_1 x_1 + \cdots + a_{i-1} x_{i-1} + a_{i+1} x_{i+1} + \cdots + a_k x_k$$

it can be shown that

$$\text{Var}(\hat{\beta}_i) = \frac{\sigma^2}{\sum\limits_{j=1}^{n} (x_{ij} - \hat{x}_{ij})^2} \tag{13.35}$$

When the sample x_i values can be predicted very well from the other carrier values, the denominator of (13.35) will be small, so $\text{Var}(\hat{\beta}_i)$ will be quite large. If this is the case for at least one carrier, the data is said to exhibit **multicollinearity.** Multicollinearity is often suggested by a regression computer output in which R^2 is large but some of the t ratios $\hat{\beta}_i / s_{\hat{\beta}_i}$ are small for predictors which, based on prior information and intuition, seem important. Another clue to the presence of multicollinearity lies in a $\hat{\beta}_i$ value which has the opposite sign from that which intuition would suggest, indicating that another carrier or collection of carriers is serving as a "proxy" for x_i.

While one strategy for analysis might be to eliminate carriers from the model, it is not always straightforward to select carriers for elimination when multicollinearity is present. We now describe another strategy which involves abandoning ordinary least squares but retaining all carriers. To see why this is reasonable, consider for the moment estimating a single parameter by an estimator $\hat{\theta}$. Then $E[(\hat{\theta} - \theta)^2]$ is called the **mean square error** of $\hat{\theta}$, and provides a measure of how close $\hat{\theta}$ is to θ on average. With the difference $B(\hat{\theta}) = \theta - E(\hat{\theta})$ denoting the **bias** of $\hat{\theta}$ (so that $B(\hat{\theta}) = 0$ if $\hat{\theta}$ is unbiased), it can be shown that

$$MSE(\hat{\theta}) = E[(\hat{\theta} - \theta)^2] = \text{Var}(\hat{\theta}) + [B(\hat{\theta})]^2 \tag{13.36}$$

If we measure the "goodness" of an estimator by its *MSE*, then (13.36) indicates that a biased estimator with a small variance may be preferred to an unbiased estimator with a large variance; this is pictured in Figure 13.11.

 spans the figure area with two p.d.f. curves.

(a) θ (b) θ

Figure 13.11   (*a*) P.d.f. of an unbiased estimator with large *MSE*, (*b*) p.d.f. of a slightly biased estimator with small *MSE*

In a multiple regression problem, $\sum_i E[(\hat{\beta}_i - \beta_i)^2]$ is the total mean square error. Then

$$\sum_i E[(\hat{\beta}_i - \beta_i)^2] = \sum_i \text{Var}(\hat{\beta}_i) + \sum_i [\beta_i - E(\hat{\beta}_i)]^2 \tag{13.37}$$

Although the least squares estimators are unbiased, so that the second term on the right of (13.37) vanishes, the total mean square error will still be large in the presence of severe multicollinearity because of (13.35). The hope is that by using slightly biased estimators, the total mean square error can be considerably reduced relative to its value for least squares.

Ridge Regression

Ridge regression involves the use of a biasing constant c, with larger c corresponding to greater bias in the estimators. For purposes of selecting an appropriate value of c, we first code all carriers and y. Let $\bar{x}_i$ and s_i be the sample mean and standard deviation of the x_{ij}'s, and let $\bar{y}$ and s_y be the mean and standard deviation of the y_j's. Now code the variables by replacing each x_{ij} by $(x_{ij} - \bar{x}_i)/\sqrt{n-1}\, s_i$ and each y_j by $(y_j - \bar{y})/\sqrt{n-1}\, s_y$. This coding will result in $\mathbf{X'X}$ and $\mathbf{X'Y}$ containing sample correlation coefficients (each diagonal element of $\mathbf{X'X}$ is then 1, and each off-diagonal element is the correlation between the sample values of a pair of carriers).

Because $\bar{y}$ has been subtracted in the coding, the model fit to the coded carriers contains no constant term:

$$Y = \widetilde{\beta}_1 x_1 + \widetilde{\beta}_2 x_2 + \cdots + \widetilde{\beta}_k x_k + \epsilon$$

The estimates of parameters for the coded and uncoded models are related by $\hat{\beta}_i = \hat{\widetilde{\beta}}_i \cdot (s_y/s_i)$, with $\hat{\beta}_0 = \bar{y} - \hat{\beta}_1 \bar{x}_1 - \cdots - \hat{\beta}_k \bar{x}_k$.

Now let $\mathbf{X}$ be the $n \times k$ matrix whose columns consist of the coded carrier values (so there is no column of 1's, since there is no constant term in the coded model), and let $\mathbf{y}$ be the column vector containing the coded values of the dependent variable.

Definition: For any $c > 0$, the **ridge estimates** of $\widetilde{\beta}_1, \ldots, \widetilde{\beta}_k$ are the entries of the column vector $\mathbf{b}_{(c)}$ satisfying the normal equations

$$(\mathbf{X}'\mathbf{X} + c\mathbf{I})\mathbf{b}_{(c)} = \mathbf{X}'\mathbf{y} \tag{13.38}$$

The solution to this system of equations is

$$\mathbf{b}_{(c)} = \widehat{\boldsymbol{\beta}}_{(c)} = (\mathbf{X}'\mathbf{X} + c\mathbf{I})^{-1}\,\mathbf{X}'\mathbf{y} \tag{13.39}$$

Writing the normal equations (13.38) in component form yields

$$(1 + c)b_{1(c)} + r_{12}b_{2(c)} + \cdots + r_{1k}b_{k(c)} = r_{1y}$$

$$r_{12}b_{1(c)} + (1 + c)b_{2(c)} + \cdots + r_{2k}b_{k(c)} = r_{2y}$$

$$\cdot$$
$$\cdot$$
$$\cdot$$

$$r_{1k}b_{1(c)} + \cdots + r_{k-1,k}b_{k-1(c)} + (1 + c)b_{k(c)} = r_{ky}$$

where r_{ij} is the (i, j)th element of $\mathbf{X}'\mathbf{X}$ (the correlation between x_i and x_j values), $r_{ii} = 1$, and r_{iy} is the ith element of $\mathbf{X}'\mathbf{y}$.

If $c = 0$, (13.39) gives the usual unbiased least squares solution, while the bias of the estimators increases as c increases from zero. It can be shown that as c increases from zero, the total *MSE* of the ridge estimators decreases to a minimum at $c = c^*$, say, and then increases. Thus for values of c near c^*, ridge estimators improve on ordinary least squares estimators. Unfortunately c^* depends on the unknown parameters, while the c actually used must depend on the sample. One method for choosing c is to compute $\widehat{\boldsymbol{\beta}}_{(c)}$ for a number of c values between zero and one, and select the smallest value of c for which the ridge estimates have stabilized (stopped changing drastically).* This is conveniently accomplished by plotting each component of $\widetilde{\boldsymbol{\beta}}_{(c)}$ as a function of c; the resulting plot is called the **ridge trace.**

Before giving an example, we note that ridge estimates can easily be obtained from most regression computer packages. Augment both the (coded) $\mathbf{X}$ matrix and $\mathbf{y}$ vector with k additional rows as follows:

$$\mathbf{X}_A = \begin{pmatrix} \mathbf{X} \\ \sqrt{c} \;\; 0 \;\; 0 \ldots 0 \\ 0 \;\; \sqrt{c} \;\; 0 \ldots 0 \\ \vdots \\ 0 \;\; \ldots \;\; 0 \; 0 \sqrt{c} \end{pmatrix} = \begin{pmatrix} \mathbf{X} \\ \sqrt{c}\,\mathbf{I} \end{pmatrix}, \qquad y_A = \begin{pmatrix} \mathbf{y} \\ 0 \\ 0 \\ \vdots \\ 0 \end{pmatrix} = \begin{pmatrix} \mathbf{y} \\ \mathbf{0} \end{pmatrix}$$

Then for this new design matrix and observation vector, the ordinary least squares estimates are exactly the ridge estimates desired:

$$(\mathbf{X}_A'\mathbf{X}_A)^{-1}\,\mathbf{X}_A'y_A = (\mathbf{X}'\mathbf{X} + c\mathbf{I})^{-1}\,\mathbf{X}'\mathbf{y}$$

Thus using a multiple regression command (without a constant term) after inputting $\mathbf{X}_A$ and $\mathbf{y}_A$ will yield $\widehat{\boldsymbol{\beta}}_{(c)}$.

*The method for choosing c is a matter of some controversy among statisticians, and research is continuing in this area.

Example 13.13 The accompanying data was collected in an attempt to relate merchantable volume of subalpine fir trees to various characteristics of the trees. The carriers are $x_1 =$ stand age, $x_2 =$ diameter (at breast height), $x_3 =$ height, and $x_4 =$ total volume. A least squares regression analysis yielded the following output:

Table 13.15

Coefficient	Estimate $\hat{\beta}_i$	Estimated S.D. $s_{\hat{\beta}_i}$	$t = \hat{\beta}_i / s_{\hat{\beta}_i}$
β_0	-1162	448	—
β_1	34.4	30.4	1.13
β_2	-195.5	86.4	-2.26
β_3	37.1	29.4	1.26
β_4	.405	.15	2.70

$R^2 = .858$, adjusted $R^2 = .838$.

Table 13.16

Obs.	x_1	x_2	x_3	x_4	y	c	$\tilde{\beta}_{1(c)}$	$\tilde{\beta}_{2(c)}$	$\tilde{\beta}_{3(c)}$	$\tilde{\beta}_{4(c)}$
1	30.	4.5000	28.	1010.00	340.00	0.00000	1.31500	-1.69800	0.755000	0.410000
2	35.	5.5000	29.	680.00	340.00	0.00100	0.94500	-1.44000	0.851000	0.441000
3	40.	6.8000	38.	1260.00	900.00	0.00200	0.73800	-1.28000	0.887000	0.464000
4	45.	7.1000	41.	1950.00	1060.00	0.00300	0.60700	-1.17000	0.897000	0.481000
5	50.	8.0000	45.	2030.00	1650.00	0.00400	0.51700	-1.08000	0.895000	0.496000
6	55.	8.3000	49.	2620.00	1830.00	0.00500	0.45100	-1.01000	0.886000	0.509000
7	60.	9.1000	53.	2920.00	2530.00	0.00600	0.40000	-0.95600	0.873000	0.520000
8	65.	9.9000	53.	1810.00	1610.00	0.00800	0.32900	-0.86500	0.843000	0.539000
9	70.	11.2000	59.	2510.00	2310.00	0.01000	0.28200	-0.79500	0.812000	0.554000
10	75.	12.2000	62.	3000.00	2870.00	0.01500	0.21300	-0.66900	0.738000	0.583000
11	80.	12.4000	65.	3370.00	3150.00	0.02000	0.17600	-0.58200	0.677000	0.603000
12	85.	12.8000	67.	3860.00	3240.00	0.02500	0.15400	-0.51800	0.626000	0.617000
13	90.	13.6000	70.	4350.00	4120.00	0.03000	0.13900	-0.46800	0.585000	0.627000
14	95.	16.3000	71.	2440.00	2140.00	0.03500	0.12900	-0.42800	0.550000	0.635000
15	100.	17.8000	75.	3000.00	2910.00	0.04000	0.12100	-0.39400	0.520000	0.640000
16	105.	18.2000	76.	3120.00	2960.00	0.05000	0.11100	-0.34200	0.473000	0.647000
17	110.	19.3000	79.	3610.00	3510.00	0.07500	0.10000	-0.25800	0.397000	0.650000
18	115.	20.2000	80.	3740.00	3200.00	0.10000	0.09600	-0.20700	0.352000	0.643000
19	120.	20.8000	82.	4010.00	2830.00	0.20000	0.09600	-0.10800	0.271000	0.597000
20	125.	25.2000	84.	4290.00	2400.00	0.30000	0.10000	-0.06200	0.239000	0.549000
21	130.	27.1000	87.	2050.00	2000.00	0.40000	0.10300	-0.03400	0.220000	0.508000
22	135.	28.3000	89.	2420.00	2010.00	0.50000	0.10600	-0.01400	0.208000	0.473000
23	140.	29.0000	89.	2830.00	2620.00	0.60000	0.10700	0.00300	0.198000	0.443000
24	145.	29.7000	90.	2900.00	2850.00	0.70000	0.10800	0.01100	0.190000	0.416000
25	150.	30.9000	91.	3280.00	3150.00	0.80000	0.10900	0.02000	0.184000	0.393000
						0.90000	0.10900	0.02700	0.180000	0.373000
						1.00000	0.10900	0.03200	0.173000	0.355000

Although R^2 is quite high, none of the individual t ratios is very large. Furthermore, the sign of the estimated coefficient of diameter is negative, which contradicts intuition; a larger diameter should yield on average a higher value of merchantable volume. The catch is that the independent variables are obviously rather strongly interrelated, suggesting that multicollinearity has affected the analysis.

Figure 13.12 pictures the ridge trace for this example (estimated coefficients for the standardized model). As the bias parameter c is increased from $c = 0$ (least

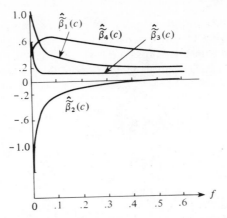

Figure 13.12   Ridge trace for Example 13.13

squares), $\hat{\tilde{\beta}}_{2(c)}$ becomes less and less negative, crossing from a negative value to a positive one between $c = .5$ $[\hat{\tilde{\beta}}_{2(.5)} = -.014]$ and $c = .6$ $[\hat{\tilde{\beta}}_{2(.6)} = .003]$. While $\hat{\beta}_{2(.3)} = -.062$ is still slightly negative, the other coefficients appear to have stabilized at that value, so a reasonable choice of c would be $c = .3$. The fitted model is then

$$y' = .1x_1' - .062x_2' + .239x_3' + .549x_4'$$

where $x_i' = (x_i - \bar{x}_i)/\sqrt{n-1}\,s_i$ and $y' = (y - \bar{y})/\sqrt{n-1})s_y$. Alternatively, the fitted model for $c = .6$ is

$$y' = .107x_1' + .003x_2' + .198x_3' + .443x_4'$$

Some investigators have suggested using the ridge trace to identify carriers which can be deleted from the model. This can be done because, in standardized form, the magnitudes of the estimated coefficients reveal the impact on y (in standard deviations) of changing each x_i by one standard deviation. Thus we might be led here to eliminate x_2 from the model.

Robust Multiple Regression

The least squares estimates of $\beta_0, \beta_1, \ldots, \beta_k$ come from minimizing $\Sigma\,\rho(y_j - b_0 - b_1x_{1j} - \cdots - b_kx_{kj})$, where $\rho(u) = u^2$ (that is, minimizing the sum of squared deviations). As described at the end of Section 13.1 for simple linear regression, robust estimates of $\beta_0, \beta_1, \ldots, \beta_k$ are obtained by replacing $\rho(u) = u^2$ by another function designed to minimize the influence of outlying observations on the resulting fit.

Let $\hat{\beta}_0^{(1)}, \hat{\beta}_1^{(1)}, \ldots, \hat{\beta}_k^{(1)}$ be the estimates of $\beta_0, \ldots, \beta_k$ obtained by minimizing the sum of absolute deviations $\Sigma\,|y_j - b_0 - b_1x_{1j} - \cdots - b_kx_{kj}|$ (there are efficient computer algorithms available for calculating these estimates). Now define a robust estimate of scale by

$$s_1 = \frac{\text{median of}}{\text{(the nonzero deviations)}} \quad \frac{\left|y_j - \hat{\beta}_0^{(1)} - \hat{\beta}_1^{(1)} x_{1j} - \cdots - \hat{\beta}_k^{(1)} x_{kj}\right|}{.6745}$$

The median is taken only over non-zero deviations because with large k, too many deviations can equal zero, yielding an s_1 which is too small.

The M-estimates of $\beta_0, \ldots, \beta_k$ corresponding to a function $\rho(u)$ are those values $b_0, b_1, \ldots, b_k$ which minimize

$$\sum_j \rho \left(\frac{y_j - b_0 - b_1 x_{1j} - \cdots - b_k x_{kj}}{s_1} \right) \tag{13.40}$$

Taking partial derivatives of (13.40) with respect to $b_0, \ldots, b_k$ and letting $\rho'(u) = \psi(u)$ yields the set of normal equations

$$\sum_j \psi \left(\frac{y_j - b_0 x_{0j} - \cdots - b_k x_{kj}}{s_1} \right) x_{ij} = 0 \quad i = 0, \ldots, k \tag{13.41}$$

where $x_{0j} = 1$ for $j = 1, \ldots, n$. A good $\psi(\cdot)$ function to use in (13.41) is the biweight function discussed in Section 6.2. The system (13.41) can be solved by Newton's method, or alternatively by iteratively reweighted least squares. This latter method involves replacing (13.41) by the linear system

$$\sum_j w_j x_{ij} (y_j - b_0 - b_1 x_{1j} - \cdots - b_k x_{kj}) = 0 \quad i = 0, \ldots, k \tag{13.42}$$

where

$$w_i = \frac{\psi[(y_j - \hat{\beta}_0^{(1)} - \hat{\beta}_1^{(1)} x_{1j} - \cdots - \hat{\beta}_k^{(1)} x_{kj})/s_1]}{(y_j - \hat{\beta}_0^{(1)} - \hat{\beta}_1^{(1)} x_{1j} - \cdots - \hat{\beta}_k^{(1)} x_{kj})/s_1} \quad i = 1, \ldots, n \tag{13.43}$$

are the weights in the initial iteration [$\psi(0) = 0$ and the weight associated with a zero residual is $\lim_{u \to 0} \psi(u)/u$]. With $\mathbf{W}$ a diagonal matrix having diagonal elements $w_1, \ldots, w_n$, the solution to (13.42) is

$$\hat{\boldsymbol{\beta}}^{(2)} = (\mathbf{X'WX})^{-1} \mathbf{X'Wy} \tag{13.44}$$

The weights (13.43) are then recomputed using $\hat{\boldsymbol{\beta}}^{(2)}$, resulting in a new solution $\hat{\boldsymbol{\beta}}^{(3)}$ after substitution of the new weights into (13.44). This process continues until there is little change in the estimates on two consecutive iterations.

As an example of the use of robust regression, the article by David Andrews, "A Robust Method for Multiple Linear Regression" (*Technometrics*, 1974, pp. 523–531) reports the result of analyzing a data set consisting of three carriers x_1, x_2, x_3, and 21 observations. The data had previously been analyzed in the book by Daniel and Wood; they relied on their technical knowledge of the underlying scientific problem to discard four observations corresponding to transient states of the system under observation, after which least squares was used on the remaining 17 observations. The *Technometrics* paper reported that esentially the same coefficient estimates were obtained by using a robust $\psi(\cdot)$ function (not the biweight) on all 21 observations. Thus the use of robust regression identified outliers without detailed knowledge of the underlying problem.

Exercises / Section 13.5

1. An analysis of the hop yield data of Exercise 13.4.3 yields the following information for the model $Y = \beta_0 + \beta_1 x_1 + \beta_2 x_2 + \epsilon$

$$(\mathbf{X'X})^{-1} = \begin{pmatrix} 11.3861 & -.5618 & -.0193 \\ -.5618 & .0395 & -.0037 \\ -.0193 & -.0037 & .0019 \end{pmatrix}$$

$$\begin{pmatrix} \hat{\beta}_0 \\ \hat{\beta}_1 \\ \hat{\beta}_2 \end{pmatrix} = \begin{pmatrix} 415.113 \\ -6.593 \\ -4.504 \end{pmatrix}$$

i	1	2	3	4	5	6	7
h_{ii}	.486	.131	.087	.219	.112	.159	.172

i	8	9	10	11	12
h_{ii}	.090	.460	.314	.301	.468

a. According to the rule of thumb given in the text, do any of the data points have a large influence on the fit?

b. The first row of the design matrix is $\mathbf{x'}_{(1)} = (1 \quad 16.7 \quad 30)$, $y_1 = 210$, and the corresponding standardized residual is $e_1^* = 2.29$, suggesting that the effect of deleting this observation should be investigated. Use formula (13.32) to compute the changes in the coefficient estimates resulting from deletion of the first observation.

c. With $s_{\hat{\beta}_0} = 82.5176$, $s_{\hat{\beta}_1} = 4.8593$, and $s_{\hat{\beta}_2} = 1.0712$, are the changes in (b) of a relatively large or small magnitude?

2. a. Using the $(\mathbf{X X})^{-1}$ matrix given in Exercise 1 along with $s = 24.45$, verify that the estimated standard deviations of $\hat{\beta}_0$, $\hat{\beta}_1$, and $\hat{\beta}_2$ are $s_{\hat{\beta}_0} = 82.5176$, $s_{\hat{\beta}_1} = 4.8593$, and $s_{\hat{\beta}_2} = 1.0712$.

b. With $\hat{\beta}_2 = -4.5036$, compute a 95% confidence interval for β_2.

c. When $x_1 = 18.0$ and $x_2 = 43$ [so that $\mathbf{x'} = (1 \quad 18.0 \quad 43)$], what value of hop yield would you predict?

d. What would your estimate of expected hop yield be for a temperature (x_1) of 20 and sunshine percentage (x_2) of 50?

e. Estimate the standard deviation of $\hat{\beta}_0 + \hat{\beta}_1 x_1 + \hat{\beta}_2 x_2$ for the x_1 and x_2 values of (d). Then obtain a 95% confidence interval for expected yield when those values are used.

3. When the water discharge data of Exercise 13.4.9 is transformed by $y = \ln(q)$, $x_1 = \ln(a)$, and $x_2 = \ln(b)$, fitting the regression function $\mu_{Y \cdot x_1, x_2} = \beta_0 + \beta_1 x_1 + \beta_2 x_2$ yields

$$\mathbf{X'} = \begin{pmatrix} 1 & 1 & 1 & 1 \\ 2.128 & 3.453 & 1.741 & 0 \\ -5.339 & -4.920 & -5.599 & -3.189 \\ & & & \\ 1 & 1 & 1 & 1 \\ 1.194 & 3.716 & 3.266 & 2.797 \\ -3.180 & -5.067 & -5.100 & -5.627 \\ & & & \\ 1 & 1 & & \\ 1.902 & 2.272 & & \\ -5.547 & -5.992 & & \end{pmatrix}$$

$$\mathbf{y'} = (2.868 \quad 3.170 \quad 1.740 \quad 1.099 \quad 2.015$$
$$4.491 \quad 4.109 \quad 3.314 \quad 2.580 \quad 2.501)$$

$$(\mathbf{X'X})^{-1} = \begin{pmatrix} 3.07463 & .14863 & .66761 \\ .14863 & .13361 & .09057 \\ .66761 & .09057 & .17577 \end{pmatrix}$$

a. Compute $\hat{\beta}_0$, $\hat{\beta}_1$, and $\hat{\beta}_2$.

c. Using $s = .4179$, compute $s_{\hat{\beta}_0}$, $s_{\hat{\beta}_1}$, and $s_{\hat{\beta}_2}$.

c. Compute the estimated standard deviation of $\hat{\beta}_0 + \hat{\beta}_1 x_1 + \hat{\beta}_2 x_2$ when $x_1 = 2$ and $x_2 = -5$.

d. Compute a 95% prediction interval for a future y when $x_1 = 2$ and $x_2 = -5$. Then transform back to obtain the corresponding prediction interval for q.

4. a. Referring back to Exercise 3, the diagonal elements of the hat matrix (the h_{ii}'s) are .138, .302, .266, .604, .464, .360, .215, .153, .214, and .284. Do any of the data points have a large influence on the fit?

b. The second residual is $e_2 = -.766$ and $e_2^* = 2.19$, while $e_4 = .112$ and $e_4^* = .43$. Compute the change in each estimated coefficient resulting from deleting separately the second data point and the fourth data point.

c. Using the results of part (b) of Exercise 3, comment on whether or not the changes of (b) are substantial.

5. The estimated standard deviation of the ith predicted value $\hat{Y}_i$ (that is, of $\hat{\beta}_0 + \hat{\beta}_1 x_{1i} + \cdots + \hat{\beta}_k x_{ki}$) is $s[\mathbf{x'}_{(i)}(\mathbf{X'X})^{-1} \mathbf{x}_{(i)}]^{1/2}$, while the ith diagonal element of the hat matrix is $h_{ii} = \mathbf{x'}_{(i)}(\mathbf{X'X})^{-1} \mathbf{x}_{(i)}$.

Thus the h_{ii}'s can be calculated once the estimated S.D.'s of the predicted values are available, and they are part of the regression output of the standard statistical packages. When the model $Y = \beta_0 + \beta_1 x_1 + \beta_2 x_2 + \beta_3 x_3 + \beta_4 x_4 + \beta_5 x_5 + \epsilon$ is fit to the specific gravity data referred to in Exercise 13.4.6, $s = .01800$ and the estimated S.D.'s of the

predicted values are .00986, .00626, .00721, .00908, .00734, .00552, .01340, .01289, .01044, .01119, .01275, .01172, .00921, .00732, .00670, .01245, .00944, .00940, .01067, and .00887. Compute the h_{ii}'s and decide whether or not there are any points which have a large influence.

6. The standardized predictor variables $(x_i - \bar{x}_i)/s_i$ for the wood specific gravity data referred to in the previous exercise are $x_1' = (x_1 - 582.70)/74.644$, $x_2' = (x_2 - 1251.80)/173.92$, $x_3' = (x_3 - 52.540)/5.447$, $x_4' = (x_4 - 52.875)/4.6997$, and $x_5' = (x_5 - 89.195)/3.6660$. The estimated coefficients for the standardized model are $\hat{\beta}_0 = .5254$, $\hat{\beta}_1 = .0079$, $\hat{\beta}_2 = .0009$, $\hat{\beta}_3 = -.0286$, $\hat{\beta}_4 = -.0087$, and $\hat{\beta}_5 = .0160$, $s = .01800$, and the (standardized) $(\mathbf{X'X})^{-1}$ matrix is

$$(\mathbf{X'X})^{-1} = \begin{pmatrix} 0.050000 & & & & & \\ 0.000034 & 0.141405 & & & & \\ -0.000052 & -0.119208 & 0.182221 & & & \\ 0.000016 & 0.015313 & -0.056049 & 0.080089 & & \\ -0.000009 & -0.013495 & 0.029823 & 0.004109 & 0.088588 & \\ 0.000011 & 0.033472 & -0.039026 & 0.003591 & -0.052769 & \end{pmatrix}$$

a. What are the estimated S.D.'s $s_{\hat{\beta}_i}$ ($i = 1, \ldots, 5$) of the $\hat{\beta}_i$'s?

b. Compute a 95% confidence interval for $\mu_{Y \cdot x_1, x_2, x_3, x_4, x_5}$ when $x_1 = 600$, $x_2 = 1200$, $x_3 = 50$, $x_4 = 50$, and $x_5 = 90$.

c. Compute a 95% prediction interval for a future y value when the x_i's are as in (b).

d. Among the $\hat{\beta}_i$'s ($i = 1, \ldots, 5$), which pair ap-

pears to be most highly correlated, and what is the estimated correlation coefficient for this pair?

7. In Exercise 13.4.8 we presented data which had been coded so that each x_i ($i = 1, 2, 3, 4$) assumed only the values -2, -1, 0, 1, and 2. For the (coded) model $Y = \beta_0 + \beta_1 x_1 + \beta_2 x_2 + \beta_3 x_3 + \beta_4 x_4 + \epsilon$,

$$(\mathbf{X'X})^{-1} = \begin{pmatrix} .0322581 & 0 & 0 & 0 & 0 \\ 0 & .0416667 & 0 & 0 & 0 \\ 0 & 0 & .0416667 & 0 & 0 \\ 0 & 0 & 0 & .0416667 & 0 \\ 0 & 0 & 0 & 0 & .0416667 \end{pmatrix}$$

a. Compute the $\hat{\beta}_i$'s.

b. Compute the estimated standard deviations and correlation coefficients of the $\hat{\beta}_i$'s.

c. Compute the h_{ii}'s. Do any data points have large influence?

d. Give an expression in component (rather than matrix) form for $\text{Var}(\hat{\beta}_0 + \hat{\beta}_1 x_1 + \cdots + \hat{\beta}_4 x_4)$.

8. The error sum of squares is $SSE = (\mathbf{y} - \mathbf{X}\hat{\boldsymbol{\beta}})'(\mathbf{y} - \mathbf{X}\hat{\boldsymbol{\beta}})$.

a. Show that $SSE = \mathbf{y'y} - \hat{\boldsymbol{\beta}}'\mathbf{X'y}$ (this is the computational formula of Section 13.4 written in matrix notation).

b. Using $\hat{\beta}_0 = 1.5652$, $\hat{\beta}_1 = .9450$, and $\hat{\beta}_2 = .1815$ for the data in Exercise 3, compute SSE and s.

9. From expression (13.30) the residuals $Y_i - \hat{Y}_i$ can be expressed in matrix form as $\mathbf{Y} - \hat{\mathbf{Y}} = [\mathbf{I} - \mathbf{X}(\mathbf{X'X})^{-1}\mathbf{X'}]\mathbf{Y} = \mathbf{AY}$ where $\mathbf{A} = [\mathbf{I} - \mathbf{X}(\mathbf{X'X})^{-1}\mathbf{X'}]$. Use this to show that the expected value of each residual is zero, and also obtain the variance covariance matrix of the residuals. Finally, explain how, once the $(\mathbf{X'X})^{-1}$ matrix has been calculated, the residuals can be standardized.

Bibliography   Chaterjee, S. and Price, Bertram, *Regression Analysis by Example*, John Wiley, New York, 1977. A brief but informative discussion of selected topics, especially multicollinearity and the use of biased estimation methods.

Daniel, Cuthbert and Wood, Fred, *Fitting Equations to Data* (2nd ed.), John Wiley, New York, 1980. Contains many insights and methods which evolved from the authors' extensive consulting experience.

Draper, Norman and Smith, Harry, *Applied Regression Analysis* (2nd ed.), John Wiley, New York, 1981. See the Chapter 12 bibliography.

Hoaglin, David and Welsch, Roy, "The Hat Matrix in Regression and ANOVA," *American Statistician*, 1978, pp. 17–23. Describes methods for detecting influential observations in a regression data set.

Hocking, Ron, "The Analysis and Selection of Variables in Linear Regression," *Biometrics*, 1976, pp. 1–49. An excellent survey of some recent developments.

Mosteller, Frederick and Tukey, John, *Data Analysis and Regression*, Addison-Wesley, Reading, Mass., 1977. Contains many interesting ideas exposited by two pioneers of the methods of exploratory data analysis.

Neter, John and Wasserman, William, *Applied Linear Statistical Models*, Richard D. Irwin, Homewood, Ill., 1974. See the Chapter 12 bibliography.

CHAPTER 14

The Analysis of Categorical Data

Introduction

In the simplest type of situation considered in this chapter, each observation in a sample is classified as belonging to one of a finite number of categories (for example, blood type could be one of the four categories O, A, B, or AB). With p_i denoting the probability that any particular observation belongs in category i (or the proportion of the population belonging to category i), we wish to test a null hypothesis which completely specifies the values of all the p_i's (such as H_0: $p_1 = .45, p_2 = .35$, $p_3 = .15$, $p_4 = .05$, when there are four categories). The test statistic will be a measure of the discrepancy between the observed numbers in the categories and the expected numbers when H_0 is true. Because a decision will be reached by comparing the computed value of the test statistic to a critical value of the chi-squared distribution, the procedure is called a chi-squared goodness-of-fit test.

Sometimes the null hypothesis specifies that the p_i's depend on some smaller number of parameters without specifying the values of these parameters. For example, with three categories the null hypothesis might state that $p_1 = \theta^2$, $p_2 = 2\theta(1 - \theta)$, and $p_3 = (1 - \theta)^2$. For a chi-squared test to be performed, the values of these unspecified parameters must be estimated from the sample data. These problems are discussed in Section 14.2. The methods are then applied to test a null hypothesis which states that the sample comes from a particular family of distributions, such as the Poisson family (with λ estimated from the sample) or the normal family (with μ and σ estimated).

Chi-squared tests for two different situations are presented in Section 14.3. In the first, the null hypothesis states that the p_i's are the same for several different populations. The second type of situation involves taking a sample from a single population and classifying each individual with respect to two different categorical factors (such as religious preference and political party registration). The null hypothesis in this situation is that the two factors are independent within the population. The last section extends this two-factor analysis by introducing and analyzing models for data resulting from classification with respect to three factors.

14.1 Goodness of Fit When Cell Probabilities Are Completely Specified

In Chapter 7 we considered testing the hypothesis H_0: $p = p_0$, where p is the proportion of successes in the population of interest (alternatively, p is the probability that a randomly selected individual is a success) by using the test statistic

$$Z = \frac{\hat{p} - p_0}{\sqrt{p_0 q_0 / n}} \qquad (14.1)$$

with $\hat{p} = X/n$. For the alternative H_a: $p \neq p_0$, H_0 is rejected at (approximately) level α if either $Z \geq z_{\alpha/2}$ or $Z \leq -z_{\alpha/2}$, or equivalently if $Z^2 \geq z_{\alpha/2}^2$. Notice that when H_0 is true, $q = P(\text{failure}) = 1 - p_0 = q_0$.

Rather than continuing to use the labels success and failure, think of each individual as belonging to either category 1 or category 2. With p_1 and p_2 replacing p and q, respectively, the null hypothesis has the form

$$H_0: p_1 = p_{10}, \; p_2 = p_{20} \qquad (14.2)$$

where $p_{20} = 1 - p_{10}$, and H_a states that H_0 is false. Our objective is to develop a procedure for testing a hypothesis analogous to H_0 when each individual in the population falls into exactly one of k possible categories.

χ^2 for the Case $k = 2$

The general test procedure is suggested by reexpressing the above Z test for the case $k = 2$. First replace X and $n - X$, the number of individuals in the sample falling into categories 1 and 2, respectively, by N_1 and N_2 (so that $N_1 + N_2 = n$). The values of N_1 and N_2 can be recorded in a rectangular table consisting of two cells as pictured below; N_1 and N_2 are then called the **observed cell counts** or frequencies. When H_0 is true, the expected cell counts are $E(N_1) = np_{10}$ and $E(N_2) = np_{20}$. The expected cell counts appear below the observed cell counts in Figure 14.1.

	$i = 1$	$i = 2$	Row total
Observed	N_1	N_2	n
Expected	np_{10}	np_{20}	n

Figure 14.1

Observed cell counts which differ greatly from the expected cell counts suggest rejection of H_0 in favor of H_a. We now form a statistic χ^2 which measures the discrepancy between observed and expected counts:

$$\chi^2 = \sum_{\substack{\text{all} \\ \text{cells}}} \frac{(\text{observed} - \text{expected})^2}{\text{expected}} = \sum_{i=1}^{2} \frac{(N_i - np_{i0})^2}{np_{i0}} \tag{14.3}$$

Proposition: $\chi^2 = Z^2$ (where $\hat{p}_1$, p_{10}, and p_{20} replace $\hat{p}$, p_0, and q_0 in Z)

Proof: Because

$$(N_1 - np_{10})^2 = (np_{10} - N_1)^2 = [n - N_1 - n(1 - p_{10})]^2 = (N_2 - np_{20})^2,$$

$$\chi^2 = \frac{(N_1 - np_{10})^2}{np_{10}} + \frac{(N_1 - np_{10})^2}{np_{20}} = \frac{(N_1 - np_{10})^2}{n^2}\left(\frac{n}{p_{10}} + \frac{n}{p_{20}}\right)$$

$$= \left(\frac{N_1}{n} - p_{10}\right)^2 \cdot \frac{n}{p_{10}p_{20}} = \frac{(\hat{p}_1 - p_{10})^2}{p_{10}p_{20}/n} = Z^2$$

The test procedure for H_0 versus H_a can thus be stated as "reject H_0 if $\chi^2 \geq z_{\alpha/2}^2$" with χ^2 defined by (14.3). It is this form of the test procedure which we will generalize to the k-category problem.

In Section 4.4 the chi-squared distribution was introduced, and it was used in Section 7.6 to make inferences about the variance σ^2 of a normal population. The chi-squared distribution has a single parameter ν, called the number of degrees of freedom (d.f.) of the distribution, with possible values 1, 2, 3, ... Analogous to the critical value $t_{\alpha,\nu}$ for the t distribution, $\chi_{\alpha,\nu}^2$ is the value such that α of the area under the χ^2 curve in Figure 14.2 with ν degrees of freedom lies to the right of $\chi_{\alpha,\nu}^2$. Selected values of $\chi_{\alpha,\nu}^2$ are given in Appendix Table A.6.

The next proposition shows how the test procedure can be expressed using the critical value of a chi-squared distribution (see Exercise 12 in Section 4.4).

Proposition: If Z is a standard normal variable, then Z^2 has a chi-squared distribution with $\nu = 1$ degree of freedom. Since Z of (14.1) has approximately a standard normal distribution when $np_{10} \geq 5$ and $nq_{20} \geq 5$, under these conditions the statistic χ^2 of (14.3) has approximately a chi-squared distribution with $\nu = 1$. An approximately level α test for H_0: $p_1 = p_{10}$, $p_2 = p_{20}$ versus H_a: H_0 is not true then rejects H_0 if $\chi^2 \geq \chi_{\alpha,1}^2$.

Figure 14.2   A critical value for a chi-squared distribution

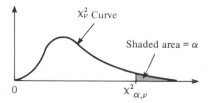

Example 14.1 The developers of an English proficiency exam to be used in a large university system believe that 60% of all incoming freshmen will be able to pass the exam. In a random sample of 200 incoming freshmen, 105 pass the exam. Does this contradict the claim of the developers?

We wish to test $H_0: p_1 = .6, p_2 = .4$ versus $H_a: H_0$ is not true. The expected cell counts are $np_{10} = 200(.6) = 120$ and $np_{20} = 80$, so the computed χ^2 is

$$\chi^2 = \frac{(105 - 120)^2}{120} + \frac{(95 - 80)^2}{80} = 1.88 + 2.81 = 4.69$$

Using $\alpha = .05$, $\chi^2_{.05,1} = 3.843$. Since $4.69 \geq 3.843$, H_0 is rejected at level .05. Notice that $\chi^2_{.025,1} = 5.025$, so the P-value satisfies $.025 < P < .05$. Exactly the same conclusion would have been reached via a two-tailed Z test.

Although there are two cells in the above table of cell counts, the chi-squared statistic is based on only one degree of freedom. Only one of the cell counts is "freely determined" as a result of the sample size n being fixed before the experiment. Once N_1 is determined, so is $N_2 = n - N_1$; thus there is only one degree of freedom in the observed cell counts.

χ^2 for General k

Suppose that each individual in the population of interest can be classified into exactly one of k possible categories. With p_i denoting the proportion of the population falling into the ith category, or equivalently the probability that a randomly selected individual belongs in category i, $\sum_{i=1}^{k} p_i = 1$. We wish to test

$$H_0: p_1 = p_{10}, p_2 = p_{20}, \ldots, p_k = p_{k0}$$

versus \hfill (14.4)

$$H_a: \text{at least one } p_i \text{ does not equal } p_{i0}$$

Let N_i = the number of individuals among the n sampled who belong in category i, so that $\sum_{i=1}^{k} N_i = n$. The expected cell counts when H_0 is true are $E(N_i) = np_{i0}$, yielding the observed and expected cell counts shown in Figure 14.3.

	$i = 1$	$i = 2$	$\ldots$	$i = k$	Row total
Observed	N_1	N_2	$\ldots$	N_k	n
Expected	np_{10}	np_{20}	$\ldots$	np_{k0}	n

Figure 14.3 Observed and expected cell counts

> **Proposition:** When H_0 is true and $np_{i0} \geq 5$ for $i = 1, \ldots, k$, the statistic
>
> $$\chi^2 = \sum_{\substack{\text{all} \\ \text{cells}}} \frac{(\text{observed} - \text{expected})^2}{\text{expected}} = \sum_{i=1}^{k} \frac{(N_i - np_{i0})^2}{np_{i0}} \qquad (14.5)$$
>
> has approximately a chi-squared distribution with $k - 1$ d.f. A test for H_0 versus H_a with approximate level α then rejects H_0 if $\chi^2 \geq \chi^2_{\alpha, k-1}$.

As in the case $k = 2$, because $\Sigma N_i = n$ is fixed, there are only $k - 1$ degrees of freedom among the N_i's, or only $k - 1$ freely determined cell counts in the table.

Example 14.2 If we focus on two different characteristics of an organism, each controlled by a single gene, and cross a pure strain having genotype AABB with a pure strain having genotype aabb (capital letters denoting dominant alleles and small letters recessive alleles), the resulting genotype will be AaBb. If these first-generation organisms are then crossed among themselves (a dihybrid cross), there will be four phenotypes depending on whether or not a dominant allele of either type is present. Mendel's laws of inheritance imply that these four phenotypes should have probabilities $\frac{9}{16}$, $\frac{3}{16}$, $\frac{3}{16}$, and $\frac{1}{16}$ of arising in any given dihybrid cross.

The paper "Linkage Studies of the Tomato" (*Trans. Royal Canadian Institute,* 1931, pp. 1–19) reported the following data on phenotypes from a dihybrid cross of tall cut-leaf tomatoes with dwarf, potato-leaf tomatoes. There are $k = 4$ categories corresponding to the four possible phenotypes, with the null hypothesis being

$$H_o : p_1 = \tfrac{9}{16}, \ p_2 = \tfrac{3}{16}, \ p_3 = \tfrac{3}{16}, \ p_4 = \tfrac{1}{16}$$

The expected cell counts are $9n/16$, $3n/16$, $3n/16$, and $n/16$, and χ^2 is based on $k - 1 = 3$ d.f. The total sample size was $n = 1611$.

	$i = 1$ Tall, cut-leaf	$i = 2$ Tall, potato-leaf	$i = 3$ Dwarf, cut-leaf	$i = 4$ Dwarf, potato-leaf
N_i	926	288	293	104
np_{10}	906.2	302.1	302.1	100.7
$\dfrac{(N_i - np_{i0})^2}{np_{i0}}$	.433	.658	.274	.108

Figure 14.4

Summing the entries in the last row yields $\chi^2 = 1.473$. Since $\chi^2_{.01,3} = 11.344$, H_0 is not rejected at level .01.

When developing a goodness-of-fit test for a k-category problem, intuition might suggest using the test statistic Σ (observed $-$ expected)^2 rather than the χ^2 statistic of (14.5). However, a difference such as (observed $-$ expected) $= 4$ provides stronger evidence against H_0 when the expected cell count is small (say, $np_{i0} = 8$) than when the expected count is large (say, $np_{i0} = 80$). Placing np_{i0} in the denominator results in differences between observed and expected being measured relative to what is expected.

As is the case with all test procedures, one must be careful not to confuse statistical significance with practical significance. A computed χ^2 which exceeds $\chi^2_{\alpha,k-1}$ may be a result of a very large sample size rather than any practical differences between the hypothesized p_{i0}'s and true p_i's. Thus if $p_{10} = p_{20} = p_{30} = \frac{1}{3}$, but the true p_i's have values .330, .340, and .330, a large value of χ^2 is sure to arise with a sufficiently large n. Before rejecting H_0, the $\hat{p}_i$'s should be examined to see if they suggest a model different from that of H_0 from a practical point of view.

Computation of power and choice of sample size for a particular alternative configuration of the p_i's are discussed in the article by Guenther listed in the chapter references.

χ^2 When the p_i's Are Functions of Other Parameters

Frequently the p_i's are hypothesized to depend on a smaller number of parameters $\theta_1, \ldots, \theta_m$ ($m < k$). Then a specific hypothesis involving the θ_i's yields specific p_{i0}'s, which are then used in the χ^2 test.

Example 14.3 In a well-known genetics paper ("The Progeny in Generations F_{12} to F_{17} of a Cross Between a Yellow-Wrinkled and a Green-Round Seeded Pea," *J. Genetics*, 1923, pp. 255–331), the early statistician G. U. Yule analyzed data resulting from crossing garden peas. The dominant alleles in the experiment were Y = yellow color and R = round shape, resulting in the double dominant YR. Yule examined 269 four-seed pods resulting from a dihybrid cross and counted the number of YR seeds in each pod. Letting X denote the number of YR's in a randomly selected pod, possible X values are 0, 1, 2, 3, 4, which we identify with cells 1, 2, 3, 4, and 5 of a rectangular table (so, for example, a pod with $X = 4$ yields an observed count in cell 5).

The hypothesis that the Mendelian laws are operative and that genotypes of individual seeds within a pod are independent of one another implies that X has a binomial distribution with $n = 4$ and $\theta = \frac{9}{16}$. We thus wish to test H_0: $p_1 = p_{10}, \ldots, p_5 = p_{50}$ where

$$p_{i0} = P(i - 1 \text{ YR's among 4 seeds when } H_0 \text{ is true})$$

$$= \binom{4}{i - 1} \theta^{i-1}(1 - \theta)^{4-(i-1)} \quad i = 1, 2, 3, 4, 5; \; \theta = \frac{9}{16}$$

Yule's data and the computations appear in Figure 14.5, with expected cell counts $np_{i0} = 269p_{i0}$.

Cell i	1	2	3	4	5
YR peas/pod	0	1	2	3	4
Observed	16	45	100	82	26
Expected	9.86	50.68	97.75	83.78	26.93
$\dfrac{(\text{observed} - \text{expected})^2}{\text{expected}}$	3.823	.637	.052	.038	.032

Figure 14.5

Thus $\chi^2 = 3.823 + \cdots + .032 = 4.582$. Since $\chi^2_{.01,k-1} = \chi^2_{.01,4} = 13.277$, H_0 is not rejected at level .01.

In Section 14.2 we will discuss goodness-of-fit tests when the values of parameters $\theta_1, \ldots, \theta_m$ are unspecified by H_0, so that we must estimate expected cell counts by first estimating the θ_i's.

χ^2 When the Underlying Distribution Is Continuous

We have so far assumed that the k categories are naturally defined in the context of the experiment under consideration. The χ^2 test can also be used to test whether a sample comes from a specific underlying continuous distribution. Let X denote the variable being sampled, and suppose that the hypothesized p.d.f. of X is $f(x)$. As in the construction of a frequency distribution in Chapter 1, subdivide the measurement scale of X into k intervals $[a_0, a_1), [a_1, a_2), \ldots, [a_{k-1}, a_k)$, where the interval $[a_{i-1}, a_i)$ includes the value a_{i-1} but not a_i. The cell probabilities specified by H_0 are then

$$p_{i0} = P(a_{i-1} \leq X < a_i) = \int_{a_{i-1}}^{a_i} f(x)\, dx$$

The cells should be chosen so that $np_{i0} \geq 5$ for $i = 1, \ldots, k$. Often they are selected so that the np_{i0}'s are equal.

Example 14.4 To see whether the time of onset of labor among expectant mothers is uniformly distributed throughout a 24-hour day, we can divide a day into k periods, each of length $24/k$. The null hypothesis states that $f(x)$ is the uniform p.d.f. on the interval $[0, 24]$, so that $p_{i0} = 1/k$. The article "The Hour of Birth" (*British J. Preventive and Social Medicine*, 1953, pp. 43–59) reported on 1186 onset times which were categorized into $k = 24$ one-hour intervals beginning at midnight, resulting in cell counts of 52, 73, 89, 88, 68, 47, 58, 47, 48, 53, 47, 34, 21, 31, 40, 24, 37, 31, 47, 34, 36, 44, 78, and 59. Each expected cell count is $1186 \cdot \frac{1}{24} = 49.42$, and the resulting value of χ^2 is 162.77. Since $\chi^2_{.01,23} = 41.64$, the computed value is highly significant and the null hypothesis is resoundingly rejected. Generally speaking, it appears that labor is much more likely to commence very late at night than during normal waking hours.

For testing whether or not a sample comes from a specific normal distribution, the fundamental parameters are $\theta_1 = \mu$ and $\theta_2 = \sigma$, and each p_{i0} will be a function of these parameters.

Example 14.5   At a certain university final exams are supposed to last two hours. The psychology department constructed a departmental final for an elementary course which was believed to satisfy the following criteria: (a) actual time taken to complete the exam is normally distributed, (b) $\mu = 100$ min, and (c) exactly 90% of all students will finish within the two-hour period. To see whether this is actually the case, 120 students were randomly selected and their completion times recorded. It was decided that $k = 8$ intervals should be used. The criteria imply that the ninetieth percentile of the completion time distribution is $\mu + 1.28\sigma = 120$. Since $\mu = 100$, this implies that $\sigma = 15.63$.

The eight intervals which divide the standard normal scale into eight equally likely segments are $[0, .32)$, $[.32, .675)$, $[.675, 1.15)$, $[1.15, \infty)$, and their four counterparts on the other side of 0. For $\mu = 100$ and $\sigma = 15.63$, these intervals become $[100, 105)$, $[105, 110.55)$, $[110.55, 117.97)$, and $[117.97, \infty)$. Thus $p_{i0} = \frac{1}{8} = .125$ ($i = 1, \ldots, 8$), so each expected cell count is $np_{i0} = 120(.125) = 15$. The observed cell counts were 21, 17, 12, 16, 10, 15, 19, and 10, resulting in a χ^2 of 7.73. Since $\chi^2_{.10,7} = 12.017$ and 7.73 is not ≥ 12.017, there is no evidence for concluding that the criteria have not been met.

Exercises / Section 14.1

1. Student consultants at a particular computer center have questions about programs written in FORTRAN (1), BASIC (2), PASCAL (3), and PL1 (4). The consultants have been hired based on the assumption that 40% of all questions concern FORTRAN programs, 25% concern BASIC programs, 25% concern PASCAL programs, and 10% concern PL1 programs. Let p_i denote the probability that a randomly selected question concerns a program written in language i ($i = 1, 2, 3, 4$). Use the accompanying data to test

H_0: $p_1 = .4$, $p_2 = .25$, $p_3 = .25$, $p_4 = .10$

versus

H_a: H_0 is not true

using a level .05 chi-squared test.

Cell	1	2	3	4
Frequency	52	38	21	9

2. It is hypothesized that when homing pigeons are disoriented in a certain manner, they will exhibit no preference for any direction of flight after takeoff (so that the direction X should be uniformly distributed on the interval from $0°$ to $360°$). To test this, 120 pigeons are disoriented, let loose, and the direction of flight of each is recorded; the resulting data appears below. Use the chi-squared test at level .10 to see if the data supports the hypothesis.

Direction	0– < 45°	45– < 90°	90– < 135°
Frequency	12	16	17

Direction	135– < 180°	180– < 225°	225– < 270°
Frequency	15	13	20

Direction	270– < 315°	315– < 360°
Frequency	17	10

3. Information has been loaded onto a single-disk storage device with 10 concentric tracks in such a way that the probability of the access arm next being required to access track i is $p_i = (5.5 - |i - 5.5|)/30$ for $i = 1, \ldots, 10$ (that is, the p_i's are $\frac{1}{30}, \frac{2}{30}, \frac{3}{30}, \frac{4}{30}, \frac{5}{30}, \frac{5}{30}, \frac{4}{30}, \frac{3}{30}, \frac{2}{30}, \frac{1}{30}$). A sample of 200 track numbers accessed resulted in the following data. Use the chi-squared test at level .10 to decide whether the data is consistent with the above access probabilities.

Track number	1	2	3	4	5	6	7	8	9	10
Frequency	4	15	23	25	38	31	32	14	10	8

4. Sorghum is an important cereal crop whose quality and appearance could be affected by the presence of pigments in the pericarp (the walls of the plant ovary). The paper "A Genetic and Biochemical Study on Pericarp Pigments in a Cross Between Two Cultivars of Grain Sorghum, Sorghum Bicolor" (*Heridity*, 1976, pp. 413–416) reported on an experiment which involved an initial cross between CK60 sorghum (an American variety with white seeds) and Abu Taima (an Ethiopian variety with yellow seeds) to produce plants with red seeds, and then a self-cross of the red-seeded plants. According to genetic theory, this F_2 cross should produce plants with red, yellow, or white seeds in the ratio 9 : 3 : 4. The data from the experiment appears below; does the data confirm or contradict the genetic theory? Test at level .05?

Seed color	Red	Yellow	White
Observed frequency	195	73	100

5. The response time of a computer system to a request for a certain type of information is hypothesized to have an exponential distribution with parameter $\lambda = 1$ sec [so if $X =$ response time, the p.d.f. of X under H_0 is $f(x) = e^{-x}$ for $x \geq 0$].
   a. If you had observed $X_1, X_2, \ldots, X_n$ and wanted to use the chi-squared test with five class intervals having equal probability under H_0, what would the resulting class intervals be?
   b. Carry out the chi-squared test using the following data resulting from a random sample of 40 response times:

   .10, .99, 1.14, 1.26, 3.24, .12, .26, .80, .79, 1.16, 1.76, .41, .59, .27, 2.22, .66, .71, 2.21, .68, .43, .11, .46, .69, .38, .91, .55, .81, 2.51, 2.77, .16, 1.11, .02, 2.13, .19, 1.21, 1.13, 2.93, 2.14, .34, .44

6. a. Show that another expression for the chi-squared statistic is

$$\chi^2 = \sum_{i=1}^{k} \frac{N_i^2}{np_{i0}} - n$$

   Why is it more efficient to compute χ^2 using this formula?
   b. When the null hypothesis is H_0: $p_1 = p_2 = \cdots = p_k = 1/k$ (that is, $p_{i0} = 1/k$ for all i), how does the formula of (a) simplify? Use the simplified expression to calculate χ^2 for the pigeon/direction data in Exercise 2.

7. a. Having obtained a random sample from a population, you wish to use a chi-squared test to decide whether or not the population distribution is standard normal. If you base the test on six class intervals having equal probability under H_0, what should the class intervals be?
   b. If you wish to use a chi-squared test to test H_0: the population distribution is normal with $\mu = .5$, $\sigma = .002$, and the test is to be based on six equiprobable (under H_0) class intervals, what should these intervals be?
   c. Use the chi-squared test with the intervals of (b) to decide, based on the following 45 bolt diameters, whether bolt diameter is a normally distributed variable with $\mu = .5$ in., $\sigma = .002$ in.

   .4974, .4976, .4991, .5014, .5008, .4993, .4994, .5010, .4997, .4993, .5013, .5000, .5017, .4984, .4967, .5028, .4975, .5013, .4972, .5047, .5069, .4977, .4961, .4987, .4990, .4974, .5008, .5000, .4967, .4977, .4992, .5007, .4975, .4998, .5000, .5008, .5021, .4959, .5015, .5012, .5056, .4991, .5006, .4987, .4968

14.2 Goodness of Fit for Composite Hypotheses

In the previous section we presented a goodness-of-fit test based on a χ^2 statistic for deciding between H_0: $p_i = p_{i0}$, . . ., $p_k = p_{k0}$ and the alternative H_a stating that H_0 is not true. The null hypothesis was a **simple hypothesis** in the sense that each p_{i0} was a specified number, so that the expected cell counts when H_0 was true were uniquely determined numbers.

In many situations there are k naturally occurring categories, but H_0 states only that the p_i's are functions of other parameters θ_1, . . ., θ_m without specifying the values of these θ's. For example, a population may be in equilibrium with respect to proportions of the three genotypes AA, Aa, and aa. With p_1, p_2, and p_3 denoting these proportions (probabilities), one may wish to test

$$H_0: p_1 = \theta^2, \; p_2 = 2\theta(1 - \theta), \; p_3 = (1 - \theta)^2 \tag{14.6}$$

where θ represents the proportion of gene A in the population. This hypothesis is **composite** because knowing that H_0 is true does not uniquely determine the cell probabilities and expected cell counts, but only their general form. To carry out a χ^2 test, the unknown θ_i's must first be estimated.

Similarly, we may be interested in testing to see whether a sample came from a particular family of distributions without specifying any particular member of the family. To use the χ^2 test to see whether the distribution is Poisson, for example, the parameter λ must be estimated. In addition, because there are actually an infinite number of possible values of a Poisson variable, these values must be grouped so that there are a finite number of cells. If H_0 states that the underlying distribution is normal, use of a χ^2 test must be preceded by a choice of cells and estimation of μ and σ.

χ^2 When Parameters Are Estimated

As before, k will denote the number of categories or cells and p_i will denote the probability of an observation falling in the ith cell. The null hypothesis now states that each p_i is a function of a small number of parameters θ_1, . . ., θ_m with the θ_i's otherwise unspecified:

$$H_0: p_1 = \pi_1(\boldsymbol{\theta}), \; . . ., \; p_k = \pi_k(\boldsymbol{\theta}) \quad \text{where} \quad \boldsymbol{\theta} = (\theta_1, \; . . ., \; \theta_m)$$

$$H_a: \text{the hypothesis } H_0 \text{ is not true} \tag{14.7}$$

For example, for H_0 of (14.6), $m = 1$ (there is only one θ), $\pi_1(\theta) = \theta^2$, $\pi_2(\theta) = 2\theta(1 - \theta)$, and $\pi_3(\theta) = (1 - \theta)^2$.

In the case $k = 2$ there is really only a single random variable N_1 (since $N_1 + N_2 = n$), which has a binomial distribution. The joint probability that $N_1 = n_1$ and $N_2 = n_2$ is then

$$P(N_1 = n_1, N_2 = n_2) = \binom{n}{n_1} p_1^{n_1} \cdot p_2^{n_2} \propto p_1^{n_1} \cdot p_2^{n_2}$$

where $p_1 + p_2 = 1$ and $n_1 + n_2 = n$. For general k, the joint distribution of N_1, . . ., N_k is the multinomial distribution (Section 5.1) with

$$P(N_1 = n_1, \; . . ., \; N_k = n_k) \propto p_1^{n_1} \cdot p_2^{n_2} \ldots p_k^{n_k} \tag{14.8}$$

When H_0 is true, (14.8) becomes

$$P(N_1 = n_1, \ldots, N_k = n_k) \propto [\pi_1(\boldsymbol{\theta})]^{n_1} \ldots [\pi_k(\boldsymbol{\theta})]^{n_k} \tag{14.9}$$

To apply a chi-squared test, $\boldsymbol{\theta} = (\theta_1, \ldots, \theta_m)$ must be estimated.

Method of estimation: Let $n_1, n_2, \ldots, n_k$ denote the observed values of $N_1, \ldots, N_k$. Then $\hat{\theta}_1, \ldots, \hat{\theta}_m$ are those values of the θ_i's which maximize (14.9).

The resulting estimators $\hat{\theta}_1, \ldots, \hat{\theta}_m$ are the **maximum likelihood estimators** of $\theta_1, \ldots, \theta_m$; this principle of estimation was discussed in Section 6.2.

Example 14.6   In humans there is a blood group, the MN group, which is composed of individuals having one of the three blood types M, MN, and N. Type is determined by two alleles and there is no dominance, so the three possible genotypes give rise to three phenotypes. A population consisting of individuals in the MN group is in equilibrium if

$$P(M) = p_1 = \theta^2$$
$$P(MN) = p_2 = 2\theta(1 - \theta)$$
$$P(N) = p_3 = (1 - \theta)^2$$

for some θ. Suppose that a sample from such a population yielded the following results:

Type	M	MN	N	
Observed	125	225	150	$n = 500$

Figure 14.6

Then $[\pi_1(\theta)]^{n_1}[\pi_2(\theta)]^{n_2}[\pi_3(\theta)]^{n_3} = [\theta^2]^{n_1}[2\theta(1 - \theta)]^{n_2}[(1 - \theta)^2]^{n_3}$

$$= 2^{n_2} \cdot \theta^{2n_1 + n_2} \cdot (1 - \theta)^{n_2 + 2n_3}$$

Maximizing this with respect to θ (or, equivalently, maximizing the ln of this quantity, which is easier to differentiate) yields

$$\hat{\theta} = \frac{2n_1 + n_2}{[(2n_1 + n_2) + (n_2 + 2n_3)]} = \frac{2n_1 + n_2}{2n}$$

With $n_1 = 125$ and $n_2 = 225$, $\hat{\theta} = \frac{475}{1000} = .475$.

Once $\boldsymbol{\theta} = (\theta_1, \ldots, \theta_m)$ has been estimated by $\hat{\boldsymbol{\theta}} = (\hat{\theta}_1, \ldots, \hat{\theta}_m)$, the estimated expected cell counts are the $n\pi_i(\hat{\boldsymbol{\theta}})$'s. These are now used in place of the np_{i0}'s of Section 14.1 to specify a χ^2 statistic.

> **Theorem:** Under general "regularity" conditions on $\theta_1, \ldots, \theta_m$ and the $\pi_i(\boldsymbol{\theta})$'s, if $\theta_1, \ldots, \theta_m$ are estimated by the method of maximum likelihood as described above and n is large,
>
> $$\chi^2 = \sum_{\substack{\text{all} \\ \text{cells}}} \frac{(\text{observed} - \text{estimated expected})^2}{\text{estimated expected}} = \sum_{i=1}^{k} \frac{[N_i - n\pi_i(\hat{\boldsymbol{\theta}})]^2}{n\pi_i(\hat{\boldsymbol{\theta}})} \quad (14.10)$$
>
> has approximately a chi-squared distribution with $k - 1 - m$ degrees of freedom when H_0 of (14.7) is true. An approximately level α test of H_0 versus H_a is then to reject H_0 if $\chi^2 \geq \chi^2_{\alpha,k-1-m}$.

In practice, if $n\pi_i(\hat{\boldsymbol{\theta}}) \geq 5$ for every i, then the test can be used. Notice that *the number of degrees of freedom is reduced by the number of θ_i's estimated.*

Example 14.6 (continued)   With $\hat{\theta} = .475$ and $n = 500$, the estimated expected cell counts are $n\pi_1(\hat{\theta}) = 500(\hat{\theta})^2 = 112.81$, $n\pi_2(\hat{\theta}) = (500)(2)(.475)(1 - .475) = 249.98$, and $n\pi_3(\hat{\theta}) = 500 - 112.81 - 249.98 = 137.82$. Then

$$\chi^2 = \frac{(125 - 112.81)^2}{112.81} + \frac{(225 - 249.98)^2}{249.98} + \frac{(150 - 137.82)^2}{137.82} = 4.89$$

Since $\chi^2_{.05,k-1-m} = \chi^2_{.05,3-1-1} = \chi^2_{.05,1} = 3.843$ and $4.89 \geq 3.843$, H_0 is rejected.

Example 14.7   Consider a series of games between two teams, I and II, which terminates as soon as one team has won four games (with no possibility of a tie). A simple probability model for such a series assumes that outcomes of successive games are independent and that the probability of team I winning any particular game is a constant θ. We arbitrarily designate I the better team, so that $\theta \geq .5$. Any particular series can then terminate after 4, 5, 6, or 7 games. Let $\pi_1(\theta), \pi_2(\theta), \pi_3(\theta), \pi_4(\theta)$ denote the probability of termination in 4, 5, 6, and 7 games, respectively. Then

$$\pi_1(\theta) = P(\text{I wins in 4 games}) + P(\text{II wins in 4 games})$$
$$= \theta^4 + (1 - \theta)^4$$
$$\pi_2(\theta) = P(\text{I wins 3 of the first 4 and the fifth})$$
$$+ P(\text{I loses 3 of the first 4 and the fifth})$$
$$= \binom{4}{3}\theta^3(1 - \theta) \cdot \theta + \binom{4}{1}\theta(1 - \theta)^3 \cdot (1 - \theta)$$
$$= 4\theta(1 - \theta)[\theta^3 + (1 - \theta)^3]$$
$$\pi_3(\theta) = 10\theta^2(1 - \theta)^2[\theta^2 + (1 - \theta)^2]$$
$$\pi_4(\theta) = 20\theta^3(1 - \theta)^3$$

The paper "Seven Game Series in Sports" by Groeneveld and Meeden (*Mathematics Magazine*, 1975, pp. 187–192) tested the fit of this model to results of National Hockey League playoffs during the period 1943–1967 (when league membership was stable). The data appears in Figure 14.7.

Cell	1	2	3	4	
Number of games played	4	5	6	7	
Observed frequency	15	26	24	18	$n = 83$
Estimated expected frequency	16.351	24.153	23.240	19.256	

Figure 14.7

The estimated expected cell counts are $83\pi_i(\hat{\theta})$, where $\hat{\theta}$ is the value of θ which maximizes

$$\{\theta^4 + (1 - \theta)^4\}^{15} \cdot \{4\theta(1 - \theta)[\theta^3 + (1 - \theta)^3]\}^{26}$$
$$\cdot \{10\theta^2(1 - \theta)^2[\theta^2 + (1 - \theta)^2]\}^{24} \cdot \{20\theta^3(1 - \theta)^3\}^{18} \qquad (14.11)$$

Standard calculus methods fail to yield a nice formula for the maximizing value $\hat{\theta}$, so it must be computed using numerical methods. The result is $\hat{\theta} = .654$, from which $\pi_i(\hat{\theta})$ and the estimated expected cell counts are computed. The computed value of χ^2 is .360, while (since $k - 1 - m = 4 - 1 - 1 = 2$) $\chi^2_{.10,2} = 4.605$. There is thus no reason to reject the simple model as applied to NHL playoff series.

The cited paper also considered World Series data for the period 1903–1973. For the simple model, $\chi^2 = 5.97$, so the model does not seem appropriate. The suggested reason for this is that for the simple model

$$P(\text{series lasts six games} \mid \text{series lasts at least six games}) \geq .5 \qquad (14.12)$$

while of the 38 series which actually lasted at least six games, only 13 lasted exactly six. The following alternative model is then introduced:

$$\pi_1(\theta_1, \theta_2) = \theta_1^4 + (1 - \theta_1)^4, \quad \pi_2(\theta_1, \theta_2) = 4\theta_1(1 - \theta_1)[\theta_1^3 + (1 - \theta_1)^3],$$
$$\pi_3(\theta_1, \theta_2) = 10\theta_1^2(1 - \theta_1)^2\theta_2, \quad \pi_4(\theta_1, \theta_2) = 10\theta_1^2(1 - \theta_1)^2(1 - \theta_2)$$

The first two π_i's are identical to the simple model, while θ_2 is the conditional probability of (14.12) (which can now be any number between zero and one). The values of $\hat{\theta}_1$ and $\hat{\theta}_2$ which maximize the expression analogous to (14.11) are determined numerically as $\hat{\theta}_1 = .614$, $\hat{\theta}_2 = .342$. A summary appears in Figure 14.8, with $\chi^2 = .384$. Because two parameters are estimated, d.f. $= k - 1 - m = 1$ with $\chi^2_{.10,1} = 2.706$, indicating a good fit of the data to this new model.

Number of games played	4	5	6	7
Observed frequency	12	16	13	25
Estimated expected frequency	10.85	18.08	12.68	24.39

Figure 14.8

One of the regularity conditions on the θ_i's in the theorem is that they be functionally independent of one another. That is, no single θ_i can be determined from the values of other θ_i's, so that m is the number of functionally independent parameters estimated. A general rule of thumb for d.f. in a chi-squared test is

$$\chi^2 \text{ d.f.} = \text{number of freely determined cell counts}$$
$$- \text{number of independent parameters estimated} \qquad (14.13)$$

This rule will be used in connection with several different chi-squared tests in the next section.

Goodness of Fit for Discrete Distributions

Many experiments involve observing a random sample $X_1, X_2, \ldots, X_n$ from some discrete distribution. One may then wish to investigate whether the underlying distribution is a member of a particular family, such as the Poisson or negative binomial family. In the case of both a Poisson and a negative binomial distribution, the set of possible values is infinite, so the values must be grouped into k subsets before a chi-squared test can be used. The groupings should be done so that the expected frequency in each cell (group) is at least five. The last cell will then correspond to X values of $c, c + 1, c + 2, \ldots$ for some value c.

This grouping can considerably complicate the computation of the $\hat{\theta}_i$'s and estimated expected cell counts. This is because the theorem requires that the $\hat{\theta}_i$'s be obtained from the cell counts $N_1, \ldots, N_k$ rather than the sample values $X_1, \ldots, X_n$.

Example 14.8 The article "Some Sampling Characteristics of Plants and Arthropods of the Arizona Desert" (*Ecology*, 1962, pp. 567–571) reported the following count data on the number of Larrea divaricata plants found in each of 48 sampling quadrats.

Cell	1	2	3	4	5
Number of plants	0	1	2	3	≥ 4
Frequency	9	9	10	14	6

Figure 14.9

The author fit a Poisson distribution to the data. Let λ denote the Poisson parameter, and suppose for the moment that the six counts in cell 5 were actually 4, 4, 5, 5, 6, 6. Then denoting sample values by $x_1, \ldots, x_{48}$, nine of the x_i's were 0, nine were 1, and so on. The likelihood of the observed sample is

$$\frac{e^{-\lambda}\lambda^{x_1}}{x_1!} \cdots \frac{e^{-\lambda}\lambda^{x_{48}}}{x_{48}!} = \frac{e^{-48\lambda}\lambda^{\Sigma x_i}}{x_1! \ldots x_{48}!} = \frac{e^{-48\lambda}\lambda^{101}}{x_1! \ldots x_{48}!}$$

The value of λ for which this is maximized is $\hat{\lambda} = \Sigma\, x_i/n = 101/48 = 2.10$ (the value reported in the paper).

However, the $\hat{\lambda}$ required for χ^2 is obtained by maximizing Expression (14.9) rather than the likelihood of the full sample. The cell probabilities are

$$\pi_i(\lambda) = \frac{e^{-\lambda}\lambda^{i-1}}{(i-1)!} \quad i = 1, 2, 3, 4; \quad \pi_5(\lambda) = 1 - \sum_{i=0}^{3} \frac{e^{-\lambda}\lambda^i}{i!}$$

so the right-hand side of (14.9) becomes

$$\left[\frac{e^{-\lambda}\lambda^0}{0!}\right]^9 \left[\frac{e^{-\lambda}\lambda^1}{1!}\right]^9 \left[\frac{e^{-\lambda}\lambda^2}{2!}\right]^{10} \left[\frac{e^{-\lambda}\lambda^3}{3!}\right]^{14} \left[1 - \sum_{i=0}^{3} \frac{e^{-\lambda}\lambda^i}{i!}\right]^6$$

There is no nice formula for $\hat{\lambda}$, the maximizing value of λ, in this latter expression, so it must be obtained numerically.

Because the parameter estimates are usually much more difficult to compute from the grouped data than from the full sample, they are virtually always computed using this latter method. When these "full" estimators are used in the chi-squared statistic, the distribution of the statistic is altered and a level α test is no longer specified by the critical value $\chi^2_{\alpha,k-1-m}$.

Theorem: Let $\hat{\theta}_1, \ldots, \hat{\theta}_m$ be the maximum likelihood estimators of $\theta_1, \ldots, \theta_m$ based on the full sample $X_1, \ldots, X_n$, and let χ^2 denote the statistic (14.10) based on these estimators. Then the critical value c_α which specifies a level α upper-tailed test satisfies

$$\chi^2_{\alpha,k-1-m} \leq c_\alpha \leq \chi^2_{\alpha,k-1} \qquad (14.14)$$

The test procedure implied by this theorem is:

$$\begin{aligned}
&\text{if } \chi^2 \geq \chi^2_{\alpha,k-1}, \quad \text{reject } H_0 \\
&\text{if } \chi^2 \leq \chi^2_{\alpha,k-1-m}, \quad \text{don't reject } H_0 \\
&\text{if } \chi^2_{\alpha,k-1-m} < \chi^2 < \chi^2_{\alpha,k-1}, \quad \text{withhold judgment}
\end{aligned} \qquad (14.15)$$

Example 14.8
(continued)

Using $\hat{\lambda} = 2.10$, the estimated expected cell counts are computed from $n\pi_i(\hat{\lambda})$ where $n = 48$. For example

$$n\pi_1(\hat{\lambda}) = 48 \cdot \frac{e^{-2.1}(2.1)^0}{0!} = (48)(e^{-2.1}) = 5.88$$

Similarly, $n\pi_2(\hat{\lambda}) = 12.34$, $n\pi_3(\hat{\lambda}) = 12.96$, $n\pi_4(\hat{\lambda}) = 9.07$, and $n\pi_5(\hat{\lambda}) = 48 - 5.88 - \cdots - 9.07 = 7.75$. Then

$$\chi^2 = \frac{(9 - 5.88)^2}{5.88} + \cdots + \frac{(6 - 7.75)^2}{7.75} = 6.31$$

Since $m = 1$ and $k = 5$, at level .05 we need $\chi^2_{.05,3} = 7.815$ and $\chi^2_{.05,4} = 9.488$. Because $6.31 \le 7.815$, we do not reject H_0; at the 5% level, the Poisson distribution provides a reasonable fit to the data. Notice that $\chi^2_{.10,3} = 6.251$ and $\chi^2_{.10,4} = 7.779$, so at level .10 we would have to withhold judgment on whether or not the Poisson distribution was appropriate.

Sometimes even the maximum likelihood estimates based on the full sample are quite difficult to compute. This is the case, for example, for the two-parameter (generalized) negative binomial distribution. In such situations method-of-moments estimates are often used and the resulting χ^2 compared to $\chi^2_{\alpha,k-1-m}$, though it is not known to what extent the use of moments estimators affects the true critical value.

Goodness of Fit for Continuous Distributions

The chi-squared test can also be used to test whether or not the sample comes from a specified family of continuous distributions, such as the exponential family or the normal family. The choice of cells (class intervals) is even more arbitrary in the continuous case than in the discrete case. To ensure that the chi-squared test is valid, the cells should be chosen independently of the sample observations. Once the cells are chosen, it is almost always quite difficult to estimate unspecified parameters (such as μ and σ in the normal case) from the observed cell counts, so instead maximum likelihood estimates based on the full sample are computed. The critical value c_α again satisfies (14.14), and the test procedure is given by (14.15).

Example 14.9

In Example 9.1 we presented 49 serum cholesterol measurements. We now wish to see whether or not the underlying distribution is normal. Suppose that before observing $X_1, \ldots, X_{49}$, it was felt that plausible values for μ and σ were 150 and 30, respectively. The seven equiprobable class intervals for the standard normal distribution are $(-\infty, -1.07)$, $(-1.07, -.57)$, $(-.57, -.18)$, $(-.18, .18)$, $(.18, .57)$, $(.57, 1.07)$, and $(1.07, \infty)$, with each endpoint also giving the distance in standard deviations from the mean for any other normal distribution. For $\mu = 150$ and $\sigma = 30$, these intervals become $(-\infty, 117.9)$, $(117.9, 132.9)$, $(132.9, 144.6)$, $(144.6, 155.4)$, $(155.4, 167.1)$, $(167.1, 182.1)$, and $(182.1, \infty)$.

To obtain the estimated cell probabilities $\pi_1(\hat{\mu}, \hat{\sigma}), \ldots, \pi_7(\hat{\mu}, \hat{\sigma})$, we first need the maximum likelihood estimates $\hat{\mu}$ and $\hat{\sigma}$. In Chapter 6 the maximum likelihood estimate of σ was shown to be $[\Sigma\,(x_i - \bar{x})^2/n]^{1/2}$ (rather than s), so with $s = 31.75$

$$\hat{\mu} = \bar{x} = 157.02, \quad \hat{\sigma} = \left[\frac{\Sigma\,(x_i - \bar{x})^2}{n}\right]^{1/2} = \left[\frac{(n-1)s^2}{n}\right]^{1/2} = 31.42$$

Each $\pi_i(\hat{\mu}, \hat{\sigma})$ is then the probability that a normal random variable X with mean 157.02 and standard deviation 31.42 falls in the ith class interval. For example,

$$\pi_2(\hat{\mu}, \hat{\sigma}) = P(117.9 \le X \le 132.9) = P(-1.25 \le Z \le -.77) = .1150$$

so $n\pi_2(\hat{\mu}, \hat{\sigma}) = 49(.1150) = 5.64$. Observed and estimated expected cell counts are shown in Figure 14.10.

Cell	$(-\infty, 117.9)$	$(117.9, 132.9)$	$(132.9, 144.6)$	$(144.6, 155.4)$
Observed	5	5	11	6
Estimated expected	5.17	5.64	6.08	6.64

Cell	$(155.4, 167.1)$	$(167.1, 182.1)$	$(182.1, \infty)$
Observed	6	7	9
Estimated expected	7.12	7.98	10.38

Figure 14.10

The computed χ^2 is 4.60. With $k = 7$ cells and $m = 2$ parameters estimated, $\chi^2_{.05,k-1} = \chi^2_{.05,6} = 12.592$ and $\chi^2_{.05,k-1-m} = \chi^2_{.05,4} = 9.488$. Since $4.60 \le 9.488$, a normal distribution provides quite a good fit to the data.

Example 14.10 The paper "Some Studies on Tuft Weight Distribution in the Opening Room" (*Textile Research J.*, 1976, pp. 567–573) reported the accompanying data on the distribution of output tuft weight X (mg) of cotton fibers for the input weight $x_0 = 70$.

Interval	0–8	8–16	16–24	24–32	32–40	40–48	48–56	56–64	64–70
Observed frequency	20	8	7	1	2	1	0	1	0
Expected frequency	18.0	9.9	5.5	3.0	1.8	.9	.5	.3	.1

Figure 14.11

The authors postulated a truncated exponential distribution:

$$H_0: f(x) = \frac{e^{-\lambda x}}{1 - e^{-\lambda x_0}} \quad 0 \leq X \leq x_0$$

The mean of this distribution is

$$\mu = \int_0^{x_0} x f(x) \, dx = \frac{1}{\lambda} - \frac{x_0 e^{-\lambda x_0}}{1 - e^{-\lambda x_0}}$$

The parameter λ was estimated by replacing μ by $\bar{x} = 13.086$ and solving the resulting equation to obtain $\hat{\lambda} = .0742$ (so $\hat{\lambda}$ is a method-of-moments estimate and not a maximum likelihood estimate). Then with $\hat{\lambda}$ replacing λ in $f(x)$, the estimated expected cell frequencies as displayed above are computed as

$$40\pi_i(\hat{\lambda}) = 40P(a_{i-1} \leq X < a_i) = 40 \int_{a_{i-1}}^{a_i} f(x) \, dx = \frac{40(e^{-\hat{\lambda} a_{i-1}} - e^{-\hat{\lambda} a_i})}{1 - e^{-\hat{\lambda} x_0}}$$

where $[a_{i-1}, a_i)$ is the ith class interval. To obtain expected cell counts of at least 5, the last six cells are combined to yield observed counts of 20, 8, 7, 5 and expected counts of 18.0, 9.9, 5.5, 6.6. The computed value of chi-squared is then $\chi^2 = 1.34$. Because $\chi^2_{.05,2} = 5.992$, H_0 is not rejected, so the truncated exponential model provides a good fit.

A Special Test for Normality

Because the chi-squared test can be used to test the fit for any given family, it is plausible that if H_0 specifies some particular family of interest, we should be able to design a test which is better able to detect a departure from H_0 than is the chi-squared test. That is, for any specific hypothesized family, there should exist a more powerful test than the chi-squared test. There is a large body of literature on such test procedures. Here we focus on a test for normality which is both more powerful than the chi-squared test and requires considerably less computational effort.

The test procedure, called **Geary's test of normality,** is based on the ratio of the average absolute deviation to the square root of the average squared deviation. Define a random variable U by

$$U = \frac{\sqrt{\pi/2}\ \Sigma |X_i - \bar{X}|/n}{\sqrt{\Sigma (X_i - \bar{X})^2/n}} = \frac{1.2533\ \Sigma |X_i - \bar{X}|/n}{\sqrt{\Sigma (X_i - \bar{X})^2/n}}$$

When the underlying distribution is normal, both the numerator and denominator of U estimate σ, and the expected value of U is approximately 1. A departure from normality is indicated by a value of U which differs considerably from 1.

Although there are tables of the critical values of U for various values of n, we describe here only a large-sample test ($n > 20$) based on standardizing U.

H_0: the underlying distribution is normal

H_a: the underlying distribution is not normal

Test statistic: $Z = \dfrac{U - 1}{.2661/\sqrt{n}}$

Rejection region: either $Z \geq z_{\alpha/2}$ or $Z \leq -z_{\alpha/2}$

To compute U, note that $\Sigma |X_i - \overline{X}| = 2(\Sigma' X_i - n'\overline{X})$, where n' denotes the number of X_i's which exceed $\overline{X}$ and Σ' is the sum of those n' observations.

Example 14.9
(continued)

For the serum cholesterol data, $n = 49$, $\overline{x} = 157.02$, and $[\Sigma (x_i - \overline{x})^2/n]^{1/2} = 31.42$. Also $n' = 21$ and $\Sigma' x_i = 3911$, so $\Sigma |x_i - \overline{x}|/n = 2[3911 - 21(157.02)]/49 = 25.04$. Thus

$$u = \frac{(1.2533)(25.04)}{31.42} = .9988, \quad z = \frac{.9988 - 1}{.2661/\sqrt{49}} = -.03$$

This computed value of Z is quite insignificant, again indicating that normality is very plausible.

For more information on Geary's test, the expository paper "Simple Compact Portable Test of Normality: Geary's Test Revisited" by Ralph D'Agostino (*Psychological Bulletin*, 1970, pp. 138–140) can be consulted.

Exercises / Section 14.2

1. A study of sterility in the fruit fly ("Hybrid Dysgensis in Drosophila Melanogaster: The Biology of Female and Male Sterility," *Genetics*, 1979, pp. 161–174) reported the following data on the number of ovaries developed for each female fly in a sample of size 1388. One model for unilateral sterility states that each ovary develops with some probability p independently of the other ovary. Test the fit of this model using χ^2 (first p must be estimated, as in Example 14.6).

x = number of ovaries developed	0	1	2
Observed count	1212	118	58

2. The article "Feeding Ecology of the Red-Eyed Vireo and Associated Foliage-Gleaning Birds" (*Ecological Monographs*, 1971, pp. 129–152) presented the accompanying data on the variable X = the number of hops before the first flight and preceded by a flight. The author then proposed and fit a geometric probability distribution $[p(x) = P(X = x) = p^{x-1} \cdot q$ for $x = 1, 2, \ldots$ where $q = 1 - p]$ to the data. The total sample size was $n = 130$.

x	1	2	4	5	6	7	8	9	10	11	12
Number of times x observed	48	31	20	9	6	5	4	2	1	1	2	1

a. The likelihood is $(p^{x_1} \cdot q) \ldots (p^{x_n} \cdot q) = p^{\Sigma x_i} \cdot q^n$. Show that the maximum likelihood estimate of p is $\hat{p} = (\Sigma x_i - n)/\Sigma x_i$, and compute $\hat{p}$ for the given data.

b. Estimate the expected cell counts using $\hat{p}$ of (a) [expected cell counts = $n \cdot (\hat{p})^x \cdot \hat{q}$ for

$x = 1, 2, \ldots]$ and test the fit of the model using a χ^2 test by combining the counts for $x = 7, 8, \ldots,$ and 12 into one cell ($x \geq 7$).

3. A certain type of flashlight is sold with the four batteries included. A random sample of 150 flashlights is obtained and the number of defective batteries in each is determined, resulting in the following data:

Number defective	0	1	2	3	4
Frequency	26	51	47	16	10

Let X be the number of defective batteries in a randomly selected flashlight. Test the null hypothesis that the distribution of X is Bin $(4, \theta)$. That is, with $p_i = P(i \text{ defectives})$, test

$$H_0: p_i = \binom{4}{i} \theta^i (1 - \theta)^{4-i} \quad i = 0, 1, 2, 3, 4$$

Hint: To obtain the maximum likelihood estimate of θ, write the likelihood (the function to be maximized) as $\theta^u (1 - \theta)^v$ where the exponents u and v are linear functions of the cell counts. Then take the natural log, differentiate with respect to θ, equate the result to zero, and solve for $\hat{\theta}$.

4. In a genetic experiment, investigators looked at 300 chromosomes of a particular type and counted the number of sister-chromatid exchanges on each ("On the Nature of Sister-Chromatid Exchanges in 5-Bromodeoxyuridine-Substituted Chromosomes," *Genetics,* 1979, pp. 1251–1264). A Poisson model was hypothesized for the distribution of the number of exchanges. Test the fit of a Poisson distribution to the data by first estimating λ and then combining the counts for $x = 8$ and $x = 9$ into one cell.

X = number of exchanges	0	1	2	3	4	5	6	7	8	9
Observed counts	6	24	42	59	62	44	41	14	6	2

5. A paper in *Annals of Mathematical Statistics* reported the following data on the number of borers in each of 120 groups of borers. Does the Poisson p.m.f. provide a plausible model for the distribution of the number of borers in a group? *Hint:* Add the frequencies for 7, 8, ..., 12 to establish a single category "≥ 7."

Number of borers	0	1	2	3	4	5	6	7	8	9	10	11	12
Frequency	24	16	16	18	15	9	6	5	3	4	3	0	1

6. The article "A Probabilistic Analysis of Dissolved Oxygen–Biochemical Oxygen Demand Relationship in Streams" (*J. Water Resources Control Fed.,* 1969, pp. 73–90) reported data on the rate of oxygenation in streams at 20 °C in a certain region. The sample mean and standard deviation were computed as $\bar{x} = .173$ and $s = .066$. Based on the accompanying frequency distribution, can it be concluded that oxygenation rate is a normally distributed variable? Use the chi-squared test with $\alpha = .05$.

Rate (*per day*)	Frequency
below .100	12
.100–below .150	20
.150–below .200	23
.200–below .250	15
.250 or more	13

7. Each headlight on an automobile undergoing an annual vehicle inspection can be focused either too high (H), too low (L), or properly (N). Checking the two headlights simultaneously (and not distinguishing between left and right) results in the six possible outcomes HH, LL, NN, HL, HN, and LN. If the probabilities (population proportions) for the single headlight focus direction are $P(H) = \theta_1$, $P(L) = \theta_2$, $P(N) = 1 - \theta_1 - \theta_2$ and the two headlights are focused independently of one another, the probabilities of the six outcomes for a randomly selected car are

$$p_1 = \theta_1^2, \; p_2 = \theta_2^2, \; p_3 = (1 - \theta_1 - \theta_2)^2$$
$$p_4 = 2\theta_1\theta_2, \; p_5 = 2\theta_1(1 - \theta_1 - \theta_2),$$
$$p_6 = 2\theta_2(1 - \theta_1 - \theta_2)$$

Use the data given below to test the null hypothesis

$$H_0: p_1 = \pi_1(\theta_1, \theta_2), \ldots, p_6 = \pi_6(\theta_1, \theta_2)$$

where the $\pi_i(\theta_1, \theta_2)$'s are given above.

Outcome	HH	LL	NN	HL	HN	LN
Frequency	49	26	14	20	53	38

Hint: Write the likelihood as a function of θ_1 and θ_2, take the natural log, then compute $\partial/\partial\theta_1$ and $\partial/\partial\theta_2$, equate them to zero, and solve for $\hat{\theta}_1, \hat{\theta}_2$.

8. The following data set consists of 26 observations on fracture toughness of base plate of 18% nickel maroging steel (from "Fracture Testing of Weldments," *ASTM Special Publ. No. 381,* 1965, pp. 328–356). Use Geary's test to decide whether a

normal distribution provides a plausible model for fracture toughness.

69.5, 71.9, 72.6, 73.1, 73.3, 73.5, 74.1,
74.2, 75.3, 75.5, 75.7, 75.8, 76.1, 76.2,
76.9, 77.0, 77.9, 78.1, 79.6, 79.7, 79.9,
80.1, 82.2, 83.7, 93.7

9. The paper "Nonbloated Burned Clay Aggregate Concrete" (*J. Materials*, 1972, pp. 555–563) reported the following data on seven-day flexural

strength of nonbloated burned clay aggregate concrete samples (psi):

257, 327, 317, 300, 340, 340, 343, 374, 377, 386,
383, 393, 407, 407, 434, 427, 440, 407, 450, 440,
456, 460, 456, 476, 480, 490, 497, 526, 546, 700

Use Geary's test at level .10 to decide whether or not flexural strength is a normally distributed variable.

14.3 Two-Way Contingency Tables

In the previous two sections we discussed inferential problems in which the count data was displayed in a rectangular table of cells. Each table consisted of one row and a specified number of columns, where the columns corresponded to categories into which the population had been divided. We now study problems in which the data also consists of counts or frequencies, but the data table will now have I rows ($I \geq 2$) and J columns, so IJ cells. There are two commonly encountered situations in which such data arises:

1. There are I populations of interest, each corresponding to a different row of the table, and each population is divided into the same J categories. A sample is taken from the ith population ($i = 1, \ldots, I$) and the counts are entered in the cells in the ith row of the table.

2. There is a single population of interest, with each individual in the population categorized with respect to two different factors. There are I categories associated with the first factor and J categories associated with the second factor. A single sample is taken, and the number of individuals belonging in category i of factor 1 and category j of factor 2 is entered in the cell in row i, column j ($i = 1, \ldots, I; j = 1, \ldots, J$).

Let N_{ij} denote the number of individuals in the sample(s) falling in the (i, j)th cell (row i, column j) of the table—that is, the (i, j)th cell count. The table displaying the observed N_{ij}'s is called a **two-way contingency table.**

	1	2	...	j	...	J
1	N_{11}	N_{12}	...	N_{1j}	...	N_{1J}
2	N_{21}					⋮
⋮	⋮					
i	N_{i1}	...		N_{ij}	...	
⋮	⋮					
I	N_{I1}	...				N_{IJ}

Figure 14.12 A two-way contingency table

In situations of type 1, we want to investigate whether or not the proportions in the different categories are the same for all populations. The null hypothesis states that the populations are **homogeneous** with respect to these categories. In type-2 situations, we investigate whether or not the categories of the two factors occur independently of one another in the population.

Homogeneity for Two Dichotomous Populations

In Chapter 8 we considered two success/failure populations, with p_1 and p_2 denoting the proportion of successes in the first and second populations, respectively. We then tested $H_0: p_1 = p_2$ versus $H_a: p_1 \neq p_2$ using a two-tailed Z test with test statistic

$$Z = \frac{\hat{p}_1 - \hat{p}_2}{\sqrt{\hat{p}(1 - \hat{p})\left(\frac{1}{n_1} + \frac{1}{n_2}\right)}} \tag{14.16}$$

Before generalizing to more than two categories and more than two populations, we first show that this Z test is equivalent to a chi-squared test based on data from a 2×2 contingency table.

Instead of using X_1 and X_2 to denote the observed numbers of successes from the two samples, we now use notation displayed in Figure 14.13.

	Successes	Failures	Sample sizes
Sample from first population	N_{11}	N_{12}	n_1
Sample from second population	N_{21}	N_{22}	n_2
			$n_1 + n_2 = n$

Figure 14.13   A 2×2 contingency table

The first subscript in N_{ij} refers to the population sampled and the second subscript to the category ($j = 1$ corresponds to success and $j = 2$ to failure). Now $\hat{p}_1 = N_{11}/n_1$ and $\hat{p}_2 = N_{21}/n_2$.

Now let E_{ij} denote the expected count in cell (i, j)—that is, $E_{ij} = E(N_{ij})$. Then if $H_0: p_1 = p_2$ is true and p denotes the common value of the p_i's

$$E_{11} = n_1 p, \; E_{12} = n_1(1 - p), \; E_{21} = n_2 p, \; E_{22} = n_2(1 - p) \tag{14.17}$$

To form a chi-squared statistic, the E_{ij}'s must first be estimated:

$$\hat{E}_{11} = n_1\hat{p}, \; \hat{E}_{12} = n_1(1 - \hat{p}), \; \hat{E}_{21} = n_2\hat{p}, \; \hat{E}_{22} = n_2(1 - \hat{p}) \tag{14.18}$$

where

$$\hat{p} = \frac{N_{11} + N_{21}}{n}$$

Because $\hat{p}$, the total number of successes divided by the total sample size, is the maximum likelihood estimator of p, we are using the same principle of estimation used in Section 14.2. Observed N_{ij}'s which differ greatly from the $\hat{E}_{ij}$'s suggest that H_0 is false. Now let

$$\chi^2 = \sum_{\substack{\text{all} \\ \text{cells}}} \frac{(\text{observed} - \text{estimated expected})^2}{\text{estimated expected}} = \sum_{i=1}^{2} \sum_{j=1}^{2} \frac{(N_{ij} - \hat{E}_{ij})^2}{\hat{E}_{ij}} \qquad (14.19)$$

Proposition: When H_0: $p_1 = p_2$ is true and sample sizes are such that $\hat{E}_{ij} \geq 5$ for each i, j, χ^2 of (14.19) has approximately a chi-squared distribution with one degree of freedom. A test of H_0 versus H_a: $p_1 \neq p_2$ with approximate level α then rejects H_0 if $\chi^2 \geq \chi^2_{\alpha,1}$. This test is identical to the two-tailed Z test because $Z^2 = \chi^2$ and $z^2_{\alpha/2} = \chi^2_{\alpha,1}$.

The proof of $Z^2 = \chi^2$ involves some straightforward but somewhat messy algebra, so we omit it.

Example 14.11 The paper "Impulsive and Premeditated Homicide: An Analysis of the Subsequent Parole Risk of the Murderer" (*J. Criminal Law and Criminology*, 1978, pp. 108–114) reported the accompanying data on successes and failures in parole for two different types of murderer. Here $\hat{p} = 35/82 = .427$, so $\hat{E}_{11} = 42(.427) = 17.93$, and similarly for the other $\hat{E}_{ij}$'s.

Observed	S	F	
Impulsive	13	29	$n_1 = 42$
Premeditated	22	18	$n_2 = 40$
	35	47	$n = 82$

Expected	S	F
Impulsive	17.93	24.07
Premeditated	17.08	24.92

Then

$$\chi^2 = \frac{(13 - 17.93)^2}{17.93} + \cdots + \frac{(18 - 24.92)^2}{24.92} = 4.84$$

Since $\chi^2_{.05,1} = 3.843$ and $\chi^2_{.025,1} = 5.025$, the P-value satisfies $.025 < P < .05$, so H_0 is rejected at level .05 but not at level .025.

The fact that d.f. $= 1$ for this chi-squared test is a consequence of our general rule of thumb (14.13) of Section 14.2. There is one freely determined cell count in each row (since n_1 and n_2 are fixed), so two freely determined counts in the table, and one estimated parameter (p), implying that d.f. $= 2 - 1 = 1$.

Homogeneity for *I* Dichotomous Populations

We suppose now that there are I different populations of interest, with $p_i =$ the proportion of successes in the ith population for $i = 1, \ldots, I$. The hypotheses of interest are

---

$H_0: p_1 = p_2 = \cdots = p_I$

$H_a:$ at least two of the p_i's are unequal

To test H_0 against H_a, we obtain a random sample of n_i individuals from the ith population $(i = 1, \ldots, I)$. Let N_{i1} denote the number of successes in the ith sample and N_{i2} denote the number of failures. The observations can be displayed in an $I \times 2$ contingency table. The N_{ij}'s and column totals $N_{.1}$ and $N_{.2}$ are random, while the row totals are fixed in advance of the experiment.

	Successes	Failures	Sample sizes
Sample from population 1	N_{11}	N_{12}	n_1
Sample from population 2	N_{21}	N_{22}	n_2
$\vdots$	$\vdots$	$\vdots$	$\vdots$
Sample from population I	N_{I1}	N_{I2}	n_I
	$N_{.1}$	$N_{.2}$	n

Figure 14.14   An I $\times$ 2 contingency table

When H_0 is true, let p denote the common value of the p_i's. Then the appropriate estimator of p is

$$\hat{p} = \frac{N_{11} + N_{21} + \cdots + N_{I1}}{n_1 + \cdots + n_I} = \frac{N_{.1}}{n} \tag{14.20}$$

the ratio of the total number of successes in the I samples to the total sample size. The expected cell counts are

$$E_{ij} = \begin{cases} n_i p & j = 1 \quad (successes) \\ n_i(1 - p) & j = 2 \quad (failures) \end{cases} \quad i = 1, \ldots, I$$

Inserting $\hat{p}$ from (14.20) in place of p then yields the estimated expected cell counts

$$\hat{E}_{ij} = \begin{cases} \dfrac{n_i \cdot N_{\cdot 1}}{n} & j = 1 \\ n_i\left(1 - \dfrac{N_{\cdot 1}}{n}\right) = \dfrac{n_i \cdot N_{\cdot 2}}{n} & j = 2 \end{cases} \qquad (14.21)$$

The formula for $\hat{E}_{ij}$ can be rewritten as

$$\hat{E}_{ij} = \frac{(i\text{th row total}) \cdot (j\text{th column total})}{n} \qquad (14.22)$$

so that the $\hat{E}_{ij}$'s are easily computed by first computing the two column totals. There are I freely determined cells in the table (one in each row) and a single parameter p is estimated, so our rule of thumb for degrees of freedom yields d.f. $= I - 1$.

> **Proposition:** Let χ^2 be defined by
>
> $$\chi^2 = \sum_{\text{all cells}} \frac{(\text{observed} - \text{estimated expected})^2}{\text{estimated expected}} = \sum_{i=1}^{I} \sum_{j=1}^{2} \frac{(N_{ij} - \hat{E}_{ij})^2}{\hat{E}_{ij}}$$
>
> where $\hat{E}_{ij}$ is given by (14.21) or (14.22) above. If $H_0: p_1 = \cdots = p_I$ is true and the sample sizes are such that $\hat{E}_{ij} \geq 5$ for every i, j, then χ^2 has approximately a chi-squared distribution with $I - 1$ degrees of freedom. An approximately level α test for H_0 versus the alternative that at least two p_i's are unequal consists of rejecting H_0 if $\chi^2 \geq \chi^2_{\alpha, I-1}$.

Example 14.12 Three different sites off the coast of California are being considered as possible locations for a liquid natural gas terminal. To get information on public opinion regarding the LNG project, a separate random sample of individuals in each region is obtained. Each individual is then asked whether he/she favors building an LNG terminal at the proposed location in the region in which the individual resides. The observed cell counts appear below.

The hypothesis of interest is $H_0: p_1 = p_2 = p_3$, where p_i is the true proportion of individuals in region i who favor an LNG facility in that region. To compute the estimated expected cell counts, the column totals are computed (the row totals n_1, n_2, and n_3 are fixed in advance, since three different regions are sampled), and formula (14.22) is then used.

Observed

	Favor	Oppose	
Region 1	198	202	$n_1 = 400$
Region 2	140	210	$n_2 = 350$
Region 3	133	217	$n_3 = 350$

$$N._1 = 471 \quad N._2 = 629$$

Estimated

	Favor	Oppose	
Region 1	$\dfrac{(400)\,(471)}{1100} = 171.27$	$\dfrac{(400)\,(629)}{1100} = 228.73$	400
Region 2	$\dfrac{(350)\,(471)}{1100} = 149.86$	$\dfrac{(350)\,(629)}{1100} = 200.14$	350
Region 3	$\dfrac{(350)\,(471)}{1100} = 149.86$	$\dfrac{(350)\,(629)}{1100} = 200.14$	350

| 471 | 629 |

Figure 14.15

$$\chi^2 = \frac{(198 - 171.27)^2}{171.27} + \cdots + \frac{(217 - 200.14)^2}{200.14} = 11.75, \quad \chi^2_{.05,2} = 5.992$$

Since $11.75 \geq 5.992$, the hypothesis of equal p_i's is rejected at level .05 in favor of the alternative that at least two p_i's are unequal.

In the analysis of variance, once the null hypothesis H_0: $\mu_1 = \cdots = \mu_I$ was rejected, we used multiple comparison methods to identify significant differences among the μ_i's. Similarly, once the null hypothesis of homogeneity is rejected, one would like to investigate further differences among the p_i's. One method for doing this is based on Bonferroni confidence intervals, which were discussed in both Chapter 10 and Chapter 12. The method involves constructing a confidence interval for each of the $\binom{I}{2}$ $[=I\,(I-1)/2]$ pairwise differences $p_i - p_j\,(i < j)$. The estimator of $p_i - p_j$ and its variance are

$$(p_i \overset{\wedge}{-} p_j) = \hat{p}_i - \hat{p}_j = \frac{N_{i1}}{n_i} - \frac{N_{j1}}{n_j}, \quad \mathrm{Var}(\hat{p}_i - \hat{p}_j) = \frac{p_i q_i}{n_i} + \frac{p_j q_j}{n_j}$$

Proposition: Consider for each pair i, j with $i < j$ the interval

$$\hat{p}_i - \hat{p}_j \pm z_{\alpha/2k} \sqrt{\frac{\hat{p}_i \hat{q}_i}{n_i} + \frac{\hat{p}_j \hat{q}_j}{n_j}} \tag{14.23}$$

where $k = \binom{I}{2}$, the number of pairwise differences $p_i - p_j$ with $i < j$. Then with confidence coefficient at least $100(1 - \alpha)\%$, each of the intervals (14.23) will include the true value of the corresponding $p_i - p_j$.

To identify significant differences among the p_i's, construct every interval of the form (14.23). Those intervals which do not include zero correspond to proportions p_i and p_j which are said to differ significantly from one another. To display the conclusions, the sample $\hat{p}_i$'s can be computed and written down in increasing order of magnitude. Then any underscored pair of $\hat{p}_i$'s corresponds to p_i's which do not differ significantly.

Example 14.12
(continued)

The estimates of the p_i's are $\hat{p}_1 = 198/400 = .495$, $\hat{p}_2 = .400$, and $\hat{p}_3 = .380$. The number of pairs is $k = 3(2)/2 = 3$, so for a simultaneous confidence coefficient of at least 95%, $\alpha = .05$ and $z_{\alpha/2k} = z_{.0083} = 2.495$ (look for .9917 inside the standard normal table). The intervals are

Difference	*Interval*
$p_1 - p_2$	$.095 \pm 2.495 \sqrt{\dfrac{(.495)(.505)}{400} + \dfrac{(.400)(.600)}{350}} = (.005, .185)$
$p_1 - p_3$	$.115 \pm 2.495 \sqrt{\dfrac{(.495)(.505)}{400} + \dfrac{(.380)(.620)}{350}} = (.025, .205)$
$p_2 - p_3$	$.020 \pm 2.495 \sqrt{\dfrac{(.400)(.600)}{350} + \dfrac{(.380)(.620)}{350}} = (-.072, .112)$

Since the intervals for $p_1 - p_2$ and $p_1 - p_3$ do not include zero, p_1 and p_2 differ significantly as do p_1 and p_3.

$\hat{p}_3$	$\hat{p}_2$	$\hat{p}_1$
.380	.400	.495

Homogeneity for *I* Populations, *J* Categories

We now assume that each individual in every one of the I populations belongs in exactly one of J categories. A sample of n_i individuals is taken from the ith population; let $n = \Sigma\, n_i$ and

N_{ij} = the number of individuals in the ith sample who fall into category j

$N_{\cdot j} = \sum_{i=1}^{I} N_{ij}$ = the total number of individuals among the n sampled who fall into category j

The observed N_{ij}'s are recorded in a two-way contingency table with I rows and J columns. The sum of the N_{ij}'s in the ith row is n_i, while the sum of entries in the jth column is $N_{\cdot j}$.

With $J > 2$, the notation necessary to state the hypotheses is more cumbersome than in the case $J = 2$ just considered. Let

p_{ij} = the proportion of the individuals in population i who fall into category j

Thus for population 1 the J proportions are $p_{11}, p_{12}, \ldots, p_{1J}$ (which sum to 1), and similarly for the other populations. The **null hypothesis of homogeneity** states that the proportion of individuals in category j is the same for each population, and that this is true for every category:

$H_0: p_{1j} = p_{2j} = \cdots = p_{Ij} \quad j = 1, 2, \ldots, J$

$H_a: H_0$ is not true

When H_0 is true, we can use $p_1, p_2, \ldots, p_J$ to denote the population proportions in the J different categories; these proportions are common to all I populations. The expected number of individuals in the ith sample who fall in the jth category when H_0 is true is then $E(N_{ij}) = n_i \cdot p_j$. To estimate $E(N_{ij})$, we must first estimate p_j, the proportion in category j. Among the total sample of n individuals, $N_{\cdot j}$ fall into category j, so we use $\hat{p}_j = N_{\cdot j}/n$ as the estimator (this can be shown to be the maximum likelihood estimator of p_j). Then

$$\hat{E}_{ij} = \text{estimated expected count in cell } (i, j) = n_i \cdot \frac{N_{\cdot j}}{n}$$

$$= \frac{(i\text{th row total})(j\text{th column total})}{n} \tag{14.24}$$

Proposition: When H_0 is true and $\hat{E}_{ij} \geq 5$ for all i, j, the statistic

$$\chi^2 = \sum_{i=1}^{I} \sum_{j=1}^{J} \frac{(N_{ij} - \hat{E}_{ij})^2}{\hat{E}_{ij}}$$

has approximately a chi-squared distribution with $(I - 1)(J - 1)$ degrees of freedom. A test for H_0 versus H_a which has approximate level α then rejects H_0 if $\chi^2 \geq \chi^2_{\alpha,(I-1)(J-1)}$.

The number of degrees of freedom comes from the general rule of thumb. In each row of the table there are $J - 1$ freely determined cell counts (the sample size n_i is fixed), so there are a total of $I(J - 1)$ freely determined cells. Parameters $p_1, \ldots, p_J$ are estimated, but because $\Sigma p_i = 1$, only $J - 1$ of these are independent. Therefore d.f. $= I(J - 1) - (J - 1) = (J - 1)(I - 1)$ as stated.

Example 14.13   In a study to investigate the extent to which individuals are aware of industrial odors in a certain region ("Annoyance and Health Reactions to Odor from Refineries and Other Industries in Carson, California," *Environmental Research*, 1978, pp. 119–132), a sample of individuals was obtained from each of three different areas near industrial facilities. Each individual was asked whether he/she noticed odors (1) every day, (2) at least once/week, (3) at least once/month, (4) less often than once/month, or (5) not at all, resulting in the data given in Figure 14.16, with the estimated expected cell counts computed from (14.24).

Category

Observed	1	2	3	4	5	n_i
Area I	20	28	23	14	12	97
Area II	14	34	21	14	12	95
Area III	4	12	10	20	53	99
$N_{\cdot j}$	38	74	54	48	77	291

Expected	1	2	3	4	5	
Area I	12.67	24.67	18.00	16.00	25.67	97
Area II	12.41	24.16	17.63	15.67	25.14	95
Area III	12.93	25.18	18.37	16.33	26.20	99
	38	74	54	48	77	

Figure 14.16

Then

$$\chi^2 = \frac{(20 - 12.67)^2}{12.67} + \cdots + \frac{(53 - 26.20)^2}{26.20} = 71.36$$

With $(I - 1)(J - 1) = (2)(4) = 8$, $\chi^2_{.005,8} = 21.954$. Because $71.36 \geq 21.954$, the hypothesis of homogeneity is rejected at level .005 in favor of the conclusion that there are differences in perception of odors among the three areas.

One way to extend the chi-squared analysis described here is to partition the overall χ^2 statistic into independent components. In Example 14.13 χ^2 can be broken down into χ^2 for area I versus area II with d.f. = 4 and areas I and II versus area III with d.f. = 4. The conclusion is that areas I and II don't differ significantly from one another while in combination they differ significantly from area III. This type of partitioning should be planned before the data is actually obtained. For unplanned comparisons, there are methods of multiple comparisons which can be employed. The book by Everitt contains a good exposition of these topics.

Independence of Factors in a Single Population

We focus now on the relationship between two different factors in a single population. The number of categories of the first factor will be denoted by I and the number of categories of the second factor by J. Each individual in the population is assumed to belong in exactly one of the I categories associated with the first factor and exactly one of the J categories associated with the second factor.

For a sample of n individuals taken from the population, let N_{ij} denote the number among the n who fall both in category i of the first factor and category j of the second factor. The N_{ij}'s can be displayed in a two-way contingency table with I rows and J columns. In the case of homogeneity for I populations, the row totals were fixed in advance and only the J column totals were random. Now only the total sample size is fixed, and the row totals $N_i.$ $(i = 1, \ldots, I)$ and column totals $N_{\cdot j}$ $(j = 1, \ldots, J)$ are all random. To state the hypotheses of interest, let

p_{ij} = the proportion of individuals in the population who belong in category i of factor 1 and category j of factor 2

= P(a randomly selected individual falls in both category i of factor 1 and category j of factor 2)

Then

$$p_i. = \sum_j p_{ij} = P(\text{a randomly selected individual falls in category } i \text{ of factor 1})$$

$$p_{\cdot j} = \sum_i p_{ij} = P(\text{a randomly selected individual falls in category } j \text{ of factor 2})$$

Recall from probability that two events A and B are independent if $P(A \cap B) = P(A) \cdot P(B)$. The *null hypothesis here says that an individual's category with respect to factor 1 is independent of the category with respect to factor 2*. In symbols

H_0: $p_{ij} = p_i. \cdot p_{\cdot j}$ $i = 1, \ldots, I$ and $j = 1, \ldots, J$

H_a: H_0 is not true

The expected count in cell (i, j) is $n \cdot p_{ij}$, so when H_0 is true, $E(N_{ij}) = n \cdot p_i. \cdot p_{\cdot j}$. To obtain a chi-squared statistic, we must therefore estimate the $p_i.$'s $(i = 1, \ldots, I)$ and $p_{\cdot j}$'s $(j = 1, \ldots, J)$. The (maximum likelihood) estimators are

$$\hat{p}_i. = \frac{N_i.}{n} = \text{sample proportion for category } i \text{ of factor 1}$$

and

$$\hat{p}_{\cdot j} = \frac{N_{\cdot j}}{n} = \text{sample proportion for category } j \text{ of factor 2}$$

This gives as estimated expected cell counts,

$$\hat{E}_{ij} = n \cdot \hat{p}_i. \cdot \hat{p}_{\cdot j} = n \cdot \frac{N_i.}{n} \cdot \frac{N_{\cdot j}}{n} = \frac{N_i. \cdot N_{\cdot j}}{n}$$

$$= \frac{(i\text{th row total})(j\text{th column total})}{n}$$

Thus the $\hat{E}_{ij}$'s for independence are computed exactly as were the $\hat{E}_{ij}$'s for homogeneity.

The number of degrees of freedom is also identical for the two situations. The number of freely determined cell counts is $IJ - 1$, since only the total n is fixed in advance. There are I $p_{i\cdot}$'s estimated, but only $I - 1$ are independently estimated since $\Sigma\, p_{i\cdot} = 1$, and similarly $J - 1$ $p_{\cdot j}$'s are independently estimated, so $I + J - 2$ parameters are independently estimated. The rule of thumb for degrees of freedom now yields d.f. $= IJ - 1 - (I + J - 2) = IJ - I - J + 1 = (I - 1) \cdot (J - 1)$.

Proposition: When the null hypothesis of independence of factors is true and $\hat{E}_{ij} \geq 5$ for all i, j, the statistic

$$\chi^2 = \sum_{i=1}^{I} \sum_{j=1}^{J} \frac{(N_{ij} - \hat{E}_{ij})^2}{\hat{E}_{ij}}$$

has approximately a chi-squared distribution with $(I - 1)(J - 1)$ degrees of freedom. A test for H_0 versus H_a which has approximate level α consists of rejecting H_0 if $\chi^2 \geq \chi^2_{\alpha,(I-1)(J-1)}$.

Example 14.14 A study of the relationship between facility conditions at gasoline stations and aggressiveness in the pricing of gasoline ("An Analysis of Price Aggressiveness in Gasoline Marketing," *J. Marketing Research*, 1970, pp. 36–42) reported the accompanying data based on a sample of $n = 441$ stations. At level .01, does the data suggest that facility conditions and pricing policy are independent of one another?

	Observed				**Expected**			
	Pricing Policy					**Pricing Policy**		
	Aggressive	Neutral	Nonaggressive	$N_{i\cdot}$				
Substandard	24	15	17	56	17.02	22.10	16.89	56
Standard	52	73	80	205	62.29	80.88	61.83	205
Modern	58	86	36	180	54.69	71.02	54.29	180
$N_{\cdot j}$	134	174	133	441	134	174	133	441

Condition labels the rows (Substandard, Standard, Modern).

Figure 14.16

Thus

$$\chi^2 = \frac{(24 - 17.02)^2}{17.02} + \cdots + \frac{(36 - 54.29)^2}{54.29} = 22.47$$

and because $\chi^2_{.01,4} = 13.277$, the hypothesis of independence is rejected.

We conclude that knowledge of a station's pricing policy does give information about the condition of facilities at the station. In particular, stations with an aggressive pricing policy appear more likely to have substandard facilities than stations with a neutral or nonaggressive policy.

Exercises / Section 14.3

1. The accompanying data refers to leaf marks found on white clover samples selected from both long grass areas and short grass areas ("The Biology of the Leaf Mark Polymorphism in Trifolium Repens L.," *Heredity*, 1976, pp. 306–325). Use a χ^2 test to decide whether the true proportions of different marks are identical for the two types of regions.

Type of mark

	L	LL	Y + YL	O	Others	Sample size
Long-grass areas	409	11	22	7	277	726
Short-grass areas	512	4	14	11	220	761

2. The following data resulted from an experiment to study the effects of leaf removal on the ability of fruit of a certain type to mature ("Fruit Set, Herbivory, Fruit Reproduction, and the Fruiting Strategy of Catalpa Speciosa," *Ecology*, 1980, pp. 57–64).

Treatment	Number of fruits matured	Number of fruits aborted
Control	141	206
Two leaves removed	28	69
Four leaves removed	25	73
Six leaves removed	24	78
Eight leaves removed	20	82

a. Does the data suggest that the chance of a fruit maturing is affected by the number of leaves removed? State and test the appropriate hypotheses at level .01.

b. Let $p_i = P$(fruit matures when given treatment i). Compute simultaneous confidence intervals to identify significant differences among the p_i's.

3. The article "Human Lateralization from Head to Foot: Sex Related Factors" (*Science*, 1978, pp. 1291–1292) reported for both a sample of right-handed males and a sample of right-handed females the number of individuals whose feet were the same size, had a bigger left than right foot (a difference of half a shoe size or more), or had a bigger right than left foot.

	L > R	L = R	L < R	Sample size
Males	2	10	28	40
Females	55	18	14	87

Does the data indicate that sex has a strong effect on the development of foot asymmetry? State the appropriate null and alternative hypotheses, compute the value of χ^2, and place a bound or bounds on the P-value.

4. The paper "Susceptibility of Mice to Audiogenic Seizure Is Increased by Handling Their Dams During Gestation" (*Science*, 1976, pp. 427–428) reported on research into the effect of different injection treatments on the frequencies of audiogenic seizures.

Treatment	No response	Wild running	Clonic seizure	Tonic seizure
Thienylalanine	21	7	24	44
Solvent	15	14	20	54
Sham	23	10	23	48
Unhandled	47	13	28	32

Does the data suggest that the true percentages in the different response categories depend on the nature of the injection treatment? State and test the appropriate hypotheses using $\alpha = .005$.

5. The accompanying data on sex combinations of two recombinants resulting from six different male genotypes appeared in the article "A New Method for Distinguishing Between Meiotic and Premeiotic Recombinational Events in Drosophila Melonogaster" (*Genetics*, 1979, pp. 543–554). Does the data support the hypothesis that the frequency distribution among the three sex combinations is homogeneous with respect to the different genotypes? Define the parameters of interest, state the appropriate H_0 and H_a, and perform the analysis.

		Sex combination		
		M/M	M/F	F/F
Male genotype	1.	35	80	39
	2.	41	84	45
	3.	33	87	31
	4.	8	26	8
	5.	5	11	6
	6.	30	65	20

6. Each of 325 individuals participating in a certain drug program was categorized both with respect to the presence or absence of hypoglycemia and with respect to mean daily dosage of insulin ("Relation of Body Weight and Insulin Dose to the Frequency of Hypoglycemia," *J. Amer. Medical Assn.*, 1974, pp. 192–194). Does the accompanying data support the claim that the presence/absence of hypoglycemia is independent of insulin dosage? Test using $\alpha = .05$.

		Mean daily insulin dose				
		<.25	.25–.49	.50–.74	.75–.99	≥1.0
Hypoglycemia condition	Present	4	21	28	15	12
	Absent	40	74	59	26	46

7. A random sample of individuals who drive alone to work in a large metropolitan area was obtained, and each individual was categorized with respect to both size of car and commuting distance. Does the accompanying data suggest that commuting distance and size of car are related in the population sampled? State the appropriate hypotheses and use a level .05 chi-squared test.

		Commuting distance		
		0–<10	10–<20	≥20
Size of car	Subcompact	6	27	19
	Compact	8	36	17
	Midsize	21	45	33
	Full-size	14	18	6

8. Each individual in a random sample of high school and college students was cross classified with respect to both political views and marijuana usage, resulting in the data displayed in the accompanying two-way table ("Attitudes About Marijuana and Political Views," *Psychological Reports*, 1973, pp. 1051–1054). Does the data support the hypothesis that political views and marijuana usage level are independent within the population? Test the appropriate hypotheses using level of significance .01.

		Usage level		
		Never	Rarely	Frequently
Political views	Liberal	479	173	119
	Conservative	214	47	15
	Other	172	45	85

9. Show that the chi-squared statistic for the test of independence can be written in the form

$$\chi^2 = \sum_{i=1}^{I} \sum_{j=1}^{J} \left(\frac{N_{ij}^2}{\hat{E}_{ij}} \right) - n$$

Why is this formula more efficient computationally than the defining formula for χ^2?

10. Suppose that in Exercise 8 each student had been categorized with respect to political views, marijuana usage, and religious preference, with the categories of this latter factor being Protestant, Catholic, and other. The data could be displayed in three different two-way tables, one corresponding to each category of the third factor. With $p_{ijk} = P$ (political category i, marijuana category j, and religious category k), the null hypothesis of independence of all three factors states that $p_{ijk} = p_{i..} \cdot p_{.j.} \cdot p_{..k}$. Let N_{ijk} denote the observed frequency in cell (i, j, k). Show how to estimate the expected cell counts assuming that H_0 is true $(\hat{E}_{ijk} = n\hat{p}_{ijk}$, so the $\hat{p}_{ijk}$'s must be determined). Then use the general rule of thumb to determine the number of degrees of freedom for the chi-squared statistic.

11. Suppose that in a particular state which consists of four distinct regions, a random sample of n_k voters is obtained from the kth region for $k = 1, 2, 3, 4$. Each voter is then classified according to which of candidates 1, 2, or 3 he/she prefers and according to voter registration (1 = Dem., 2 = Rep., 3 = Indep.). Let p_{ijk} denote the proportion of voters in region k who belong in candidate category i and registration category j. The null hypothesis of homogeneous regions is H_0: $p_{ij1} = p_{ij2} = p_{ij3} = p_{ij4}$ for all i, j (that is, the proportion within each candidate-registration combination is the same for all four regions). Assuming that H_0 is true, determine $\hat{p}_{ijk}$ and $\hat{E}_{ijk}$ as functions of the observed N_{ijk}'s, and use the general rule of thumb to obtain the number of degrees of freedom for the chi-squared test.

14.4 Multiway Contingency Tables and Log-Linear Models

In Section 14.3 we discussed the independence of two factors, one consisting of I categories and the other of J categories. An investigator will often wish to study the interrelationships between three or more different factors. Here we discuss models and analyses for three-factor problems. In such situations there is the potential for relationships between factors which are more complex than is the case for just two factors. Once three-factor models and analyses are understood, the extension to more than three factors is relatively straightforward.

The three factors of interest will be denoted by A, B, and C. Each individual in the population is assumed to belong to exactly one of the I categories of A, one of the J categories of B, and one of the K categories of C. We shall use A_i to represent the event that a randomly selected individual belongs in category i of factor A, and similarly for B_j and C_k. The event that an individual belongs simultaneously in category i of A, category j of B, and category k of C, is then denoted by $A_iB_jC_k$, with $p_{ijk} = P(A_iB_jC_k)$ being the joint probability (or population proportion) associated with categories i, j, and k of factors A, B, and C, respectively.

To obtain information about the p_{ijk}'s, we shall have available a random sample of n individuals from the population. Let N_{ijk} be the number among the n who are simultaneously in category i of factor A, category j of factor B, and category k of factor C. The N_{ijk}'s are the observed frequencies or cell counts. As before, a dot replacing a particular subscript indicates summation over all values of the subscript, so that $N_{ij\cdot} = \sum_k N_{ijk}$, $n = \sum_i \sum_j \sum_k N_{ijk} = N\ldots$, and so on. To display the observed N_{ijk}'s we use a collection of $I \times J$ two-way tables; there will be one such table for each level of the third factor C, so a total of K such tables.

Example 14.15 The article "Human Lateralization from Head to Foot: Sex-Related Factors" (*Science*, 1979, pp. 1291–1292) reported the accompanying data in an investigation of relationships between relative foot size (factor A), sex (factor B), and "handedness" (factor C). For relative foot size, the size of the left foot was compared to that of the right foot; L = R corresponds to a difference of at most one-half a shoe size between the two feet.

	Right-handed				Not right-handed		
	M	F	$N_{i\cdot 1}$		M	F	$N_{i\cdot 2}$
L > R	2	55	57		6	0	6
L = R	10	18	28		6	2	8
L < R	28	14	42		0	9	9
$N_{\cdot j1}$	40	87	127	$N_{\cdot j2}$	12	11	23

Figure 14.17

The total sample size is $n = N_{\cdot\cdot 1} + N_{\cdot\cdot 2} = 127 + 23 = 150$.

A Look Back at Two Factors

The basic approach to three-factor problems is to express the cell probabilities p_{ijk} as functions of other parameters in much the same way that in ANOVA we expressed μ_{ijk} in terms of main effect and interaction parameters. To see how this is done, first consider the case of two factors which are independent. Then $p_{ij} = p_{i\cdot} \cdot p_{\cdot j}$, so

$$E_{ij} = np_{ij} = n \cdot p_{i\cdot} \cdot p_{\cdot j} = \frac{(np_{i\cdot})(np_{\cdot j})}{n} = \frac{E_{i\cdot} \cdot E_{\cdot j}}{n}$$

where $E_{i\cdot} = \sum_j E_{ij}$ and $E_{\cdot j} = \sum_i E_{ij}$. If we now take logarithms (to the base e), the log of the expected (i, j)th cell count is

$$\ln(E_{ij}) = -\ln(n) + \ln(E_{i\cdot}) + \ln(E_{\cdot j}) \tag{14.25}$$

Thus $\ln(E_{ij})$ is expressed in the form $\lambda + \alpha_i + \beta_j$, but α_i and β_j do not satisfy the "ANOVA-like" side conditions $\alpha_\cdot = \beta_\cdot = 0$. To achieve this, let

$$\lambda = \frac{\sum_i \sum_j \ln(E_{ij})}{IJ}, \quad \alpha_i = \frac{\sum_j \ln(E_{ij})}{J} - \lambda, \quad \beta_j = \frac{\sum_i \ln(E_{ij})}{I} - \lambda \tag{14.26}$$

Then α_i is the deviation of the average $\ln(E_{ij})$ for the ith row of the table from the overall average λ, so that $\sum_i \alpha_i = \alpha_\cdot = 0$. Similarly, $\beta_\cdot = 0$. If $\ln(E_{ij})$ from (14.25) is now substituted into the quantities in (14.26), some algebra yields

$$\ln(E_{ij}) = \lambda + \alpha_i + \beta_j \quad i = 1, \ldots, I; \quad j = 1, \ldots, J \tag{14.27}$$

Thus *when the two factors are independent, the log expected cell count can be expressed in an additive form* as in (14.27). The model of (14.27) is called a **log-linear** model, since $\ln(E_{ij})$ is written in a linear model format.

If the factors are not independent, then the model (14.27) is not correct for all $\ln(E_{ij})$'s. Denoting the difference between the left- and right-hand sides of (14.27) by γ_{ij}, the model is

$$\ln(E_{ij}) = \lambda + \alpha_i + \beta_j + \gamma_{ij} \tag{14.28}$$

If $\gamma_{ij} = 0$ for all i, j, the factors are independent; if $\gamma_{ij} \neq 0$ for at least one pair i, j, then the two factors interact, so the γ_{ij}'s are the interaction parameters. To test for independence of factors, the parameters of the model (14.27) can be estimated and the estimated $\ln(E_{ij})$'s compared with the observed $\ln(N_{ij})$'s to see if the independence model provides a good fit. Since the chi-squared test for independence of two factors has already been discussed, we now turn attention to models analogous to (14.27) and (14.28) for the case of three factors.

Log-Linear Models for Three Factors

The most general model for the $\ln(E_{ijk})$'s, usually referred to as the **saturated model**, involves a constant term, main-effect terms, and two- and three-factor interactions:

$$\ln(E_{ijk}) = \lambda + \alpha_i + \beta_j + \delta_k + \gamma_{ij}^{AB} + \gamma_{ik}^{AC} + \gamma_{jk}^{BC} + \gamma_{ijk} \tag{14.29}$$

where the side conditions on the parameters are $\alpha. = \beta. = \delta. = \gamma_{i\cdot}^{AB} = \gamma_{\cdot j}^{AB} = \cdots = \gamma_{i\cdot k} = \gamma_{\cdot jk} = 0$. With these side conditions, there are exactly IJK independently determined parameters in the model (14.29).

Various relationships among the factors can be described by models in which some of the parameters in (14.29) are deleted. Consider the following hierarchy of models:

$$M_1 : \ln(E_{ijk}) = \lambda + \alpha_i + \beta_j + \delta_k$$

$$M_2 : \ln(E_{ijk}) = \lambda + \alpha_i + \beta_j + \delta_k + \gamma_{ij}^{AB}$$

$$M_3 : \ln(E_{ijk}) = \lambda + \alpha_i + \beta_j + \delta_k + \gamma_{ij}^{AB} + \gamma_{ik}^{AC}$$

$$M_4 : \ln(E_{ijk}) = \lambda + \alpha_i + \beta_j + \delta_k + \gamma_{ij}^{AB} + \gamma_{ik}^{AC} + \gamma_{jk}^{BC}$$

The models are listed in order of increasing complexity. Taking the antilog in M_1, we have

$$E_{ijk} = np_{ijk} = e^\lambda \cdot e^{\alpha_i} \cdot e^{\beta_j} \cdot e^{\delta_k}$$

so that p_{ijk} is the product of terms involving separately i, j, and k. This is exactly what is meant by saying that factors A, B, and C are mutually independent $[p_{ijk} = P(A_iB_jC_k) = P(A_i)P(B_j)P(C_k)]$, so that the model M_1 specifies **mutual independence** of the three factors.

Model M_2 involves the main-effect term and a single interaction term. Taking antilogs shows that for M_2, p_{ijk} is a product of a term involving k alone and a term involving i and j jointly $[P(A_iB_jC_k) = P(A_iB_j)P(C_k)]$. Thus M_2 specifies independence of the third factor and the first two factors, usually referred to as **partial independence.**

For model M_3, which has all main effect and two two-factor interaction terms, taking antilogs and manipulating the result yields $p_{ijk} = (p_{ij\cdot})(p_{i\cdot k})/p_{i\cdot\cdot}$. This in turn implies that the conditional probability of (B_j and C_k) given A_i is

$$P(B_jC_k|A_i) = \frac{P(A_iB_jC_k)}{P(A_i)} = \frac{p_{ijk}}{p_{i\cdot\cdot}} = \frac{(p_{ij\cdot})(p_{i\cdot k})}{(p_{i\cdot\cdot})(p_{i\cdot\cdot})} = P(B_j|A_i) \cdot P(C_k|A_i)$$

That is, model M_3 says that factors B and C are **conditionally independent** in that for each fixed level of the first factor, they are independent. Thus once we restrict ourselves to any particular category of factor A, knowledge of the factor-B category provides no information about the category of factor C and vice versa.

Model M_4, with all three two-factor interactions present but three-factor interactions absent, is the most difficult model to interpret. It essentially states that the degree of association between any two of the factors is the same for all categories of the other factor. We shall call this model one of **constant association.** The distinction between M_4 and the saturated model is roughly the same as that between three-factor ANOVA models without and with the three-factor interaction terms.

Models other than $M_1 - M_4$ can be obtained by permuting subscripts. For example, there is a model analogous to M_3 for the conditional independence of A and

C given B. We do, however, restrict ourselves to **hierarchical models** in which the presence of higher-order terms (say, γ_{ij}^{AB}) automatically imply that all corresponding lower-order terms (α_i and β_j) are also included.

Fitting Models

To investigate whether or not any particular model provides a good explanation for the observed data, we must first estimate the E_{ijk}'s for that model and then compare them to the observed N_{ijk}'s. Once the $\hat{E}_{ijk}$'s have been computed, a chi-squared statistic will be used to assess the fit—but one which is different from Pearson's chi-squared statistic discussed in Sections 14.1–14.3.

As before, the $\hat{E}_{ijk}$'s (equivalently $\hat{p}_{ijk}$'s) are maximum likelihood estimates. For any particular model the $\hat{E}_{ijk}$'s are obtained directly rather than by first estimating the parameters λ, α_i, . . . , γ_{ij}^{AB}, . . . which define the model. For models M_1, M_2, and M_3 there are simple formulas for the $\hat{E}_{ijk}$'s, but for model M_4 the estimates must be obtained by an iterative fitting procedure. We will shortly describe one such method. Many models involving more than three factors require such an iterative procedure for obtaining the estimated expected cell counts.*

Model	Description	$\hat{E}_{ijk}\ (=n\hat{p}_{ijk})$
M_1	complete independence	$\dfrac{N_{i..} \cdot N_{.j.} \cdot N_{..k}}{n^2}$
M_2	partial independence $[(A,\ B)$ and $C]$	$\dfrac{N_{ij.} \cdot N_{..k}}{n}$
M_3	conditional independence $[(B$ and $C)$ given $A]$	$\dfrac{N_{ij.} \cdot N_{i \cdot k}}{N_{i..}}$
M_4	constant association	obtained iteratively

The general rule of thumb can now be used to determine the appropriate number of degrees of freedom for each model. The number of freely determined cell counts is $IJK - 1$, since only the total sample size n is fixed in advance. Although the model parameters (λ, main effect parameters, and interaction parameters) are not estimated directly but only implicitly through calculation of the $\hat{E}_{ijk}$'s, we can still determine degrees of freedom by counting the number of parameters independently estimated. Because $\Sigma \Sigma \Sigma\, p_{ijk} = p\ldots = 1$, once all other parameters are estimated, the estimate of λ is determined, so we do not count λ among the parameters estimated. Considering model M_2 as an example, there are $I - 1$ α_i's, $J - 1$ β_j's, $K - 1$ δ_k's, and $(I - 1)(J - 1)$ γ_{ij}^{AB}'s independently estimated (recall the side conditions $\alpha. = \beta. = \delta. = \gamma_{i.}^{AB} = \gamma_{.j}^{AB} = 0$), giving a total of $IJ + K - 2$ independent estimates. Thus d.f. for M_2 is $IJK - 1 - (IJ + K - 2) = (IJ - 1) \cdot (K - 1)$. Proceeding analogously for other models yields the following results:

---

*The BMD package of statistical computer programs, as well as some other statistical packages, can be used to compute $\hat{E}_{ijk}$'s for various models and to assess the fit of these models.

Model	Model Degrees of Freedom
M_1	$IJK - I - J - K + 2$
M_2	$(IJ - 1)(K - 1)$
M_3	$I(J - 1)(K - 1)$
M_4	$(I - 1)(J - 1)(K - 1)$

To assess the goodness of fit of a particular model to the data, form the statistic

$$
\begin{aligned}
G^2 &= 2 \sum_i \sum_j \sum_k N_{ijk} \cdot \ln\left(\frac{N_{ijk}}{\hat{E}_{ijk}}\right) \\
&= 2 \sum_i \sum_j \sum_k N_{ijk} \cdot \ln(N_{ijk}) - 2 \sum_i \sum_j \sum_k N_{ijk} \cdot \ln(\hat{E}_{ijk})
\end{aligned}
\tag{14.30}
$$

The plausibility of G^2 as a measure of fit comes from consideration of the quantity $\sum \sum \sum N_{ijk} \cdot \ln(F_{ijk})$ where it is required that $N... = F... = n$. It can then be shown that the choice $F_{ijk} = N_{ijk}$ maximizes the quantity under consideration, so that $\sum \sum \sum N_{ijk} \cdot \ln(N_{ijk})$ provides a standard against which models can be judged. If a model provides a poor fit, $\sum \sum \sum N_{ijk} \cdot \ln(\hat{E}_{ijk})$ will be much smaller than $\sum \sum \sum N_{ijk} \cdot \ln(N_{ijk})$, resulting in a large positive value of G^2. A small positive value of G^2 indicates a model which fits the data quite well. The reason for using G^2 rather than our earlier χ^2 is that it will be straightforward to test H_0 versus H_a where H_0 specifies a model which is a special case of the model specified by H_a (that is, the model in H_0 is lower in the hierarchy than the model in H_a, such as M_2 versus M_3). In such a testing situation, G^2 can be derived as the test statistic by applying the likelihood ratio principle mentioned in Chapter 7.

Proposition: When model M_t is the correct model, the $\hat{E}_{ijk}$'s are the estimated expected cell counts associated with M_t, and $\hat{E}_{ijk} \geq 5$ for all i, j, k, then the statistic G^2 has approximately a chi-squared distribution with number of degrees of freedom corresponding to M_t.

Example 14.16 The paper "The Anolis Lizards of Bimini: Resource Partitioning in a Complex Fauna" (*Ecology*, 1968, pp. 704–726) presents the following data on perch height (factor A) and perch diameter (factor B) for two different species of lizards.

To fit the model of conditional independence of factors A and B for each level of C (that is, that for each species separately perch height and perch diameter are independent of one another), we use $\hat{E}_{ijk} = N_{i \cdot k} N_{\cdot jk} / N_{\cdot \cdot k}$. The model is identical to M_3 except for changes in the subscripts, since we are conditioning on factor C here

		Sagrei species Perch diameter				Distichus species Perch diameter		
		Low	High	$N_{i \cdot 1}$		Low	High	$N_{i \cdot 2}$
Perch	High	32	11	43	High	61	41	102
height	Low	86	35	121	Low	73	70	143
	$N_{\cdot j1}$	118	46	164	$N_{\cdot j2}$	134	111	245

Figure 14.18

rather than factor A. The estimating formula states that $\hat{E}_{ijk}$ = (row total) · (column total)/(table total) separately for each table, just as was the case for independence in a two-way table.

$\hat{E}_{ijk}$:	30.94	12.06	43	55.79	46.21	102
	87.06	33.94	121	78.21	64.79	143
	118	46	164	134	111	245

Figure 14.19

Thus

$$G^2 = 2\left[32 \ln\left(\frac{32}{30.94}\right) + 11 \ln\left(\frac{11}{12.06}\right) + \cdots + 70 \ln\left(\frac{70}{64.79}\right)\right] = 2.02$$

For this model, d.f. $= (I - 1)(J - 1)K = (1)(1)(2) = 2$ and $\chi^2_{.10,2} = 4.605$. In fact, the seventieth percentile of the chi-squared distribution with two degrees of freedom is 2.41, so the model of conditional independence would yield a chi-square of at least 2.02 more than 30% of the time. This shows that $\chi^2 = 2.02$ is quite a reasonable value if the conditional independence model is correct, so this model provides a good fit to the data.

Estimates for Model M_4

There is no general formula for the estimates $\hat{E}_{ijk}$ corresponding to the constant association model (in which the two-factor interactions γ_{ij}^{AB}, γ_{ik}^{AB}, and γ_{jk}^{BC} are present but there are no three-factor interaction terms), so these estimates must be obtained numerically for each different data set. The fact that the model contains the three two-factor interaction terms implies that the corresponding marginal totals of observed counts and marginal totals of estimated expected counts must be equal:

$$\hat{E}_{ij \cdot} = N_{ij \cdot}, \ \hat{E}_{i \cdot k} = N_{i \cdot k}, \ \hat{E}_{\cdot jk} = N_{\cdot jk} \quad \begin{array}{l} i = 1, \ldots, I; \\ j = 1, \ldots, J; \\ k = 1, \ldots, K \end{array} \quad (14.31)$$

In general, the presence of the highest-order terms involving any particular factor implies the equality of the corresponding observed and estimated expected marginal totals. Thus in model M_2 specifying partial independence of (A, B) and C, the highest-order terms are γ_{ij}^{AB} and δ_k, so $\hat{E}_{ij\cdot} = N_{ij\cdot}$ and $\hat{E}_{\cdot\cdot k} = N_{\cdot\cdot k}$.

To compute the estimates, we make an initial choice $\hat{E}_{ijk}^{(0)}$ and then generate a sequence of estimates by successively readjusting so that the estimates satisfy the first condition in (14.31), then the second condition, then the third condition, then the first condition again, and so on. The procedure continues until, to a very good approximation, all three conditions are simultaneously satisfied.

The starting values usually recommended are $\hat{E}_{ijk}^{(0)} = 1$ for all i, j, k. Then

$$\hat{E}_{ijk}^{(1)} = \frac{N_{ij\cdot}}{\hat{E}_{ij\cdot}^{(0)}} \cdot \hat{E}_{ijk}^{(0)}, \quad \hat{E}_{ijk}^{(2)} = \frac{N_{i\cdot k}}{\hat{E}_{i\cdot k}^{(1)}} \cdot \hat{E}_{ijk}^{(1)}, \quad \hat{E}_{ijk}^{(3)} = \frac{N_{\cdot jk}}{\hat{E}_{\cdot jk}^{(2)}} \cdot \hat{E}_{ijk}^{(2)} \tag{14.32}$$

complete the first iteration. Notice that each successive $\hat{E}_{ijk}$ is readjusted by a constant of proportionality to satisfy first the first condition in (14.31), then the second condition, and then the third condition. In the next iteration $\hat{E}_{ijk}^{(4)}$ is computed by first replacing $\hat{E}_{ij\cdot}^{(0)}$ by $\hat{E}_{ijk}^{(3)}$ and $\hat{E}_{ij\cdot}^{(0)}$ by $\hat{E}_{ij\cdot}^{(3)}$, after which $\hat{E}_{ijk}^{(5)}$ and $\hat{E}_{ijk}^{(6)}$ are computed to finish the second iteration. The process continues to cycle in this manner, with each iteration consisting of three adjustments analogous to those of (14.32).

Example 14.17 The accompanying data was extracted from the paper "Social Class and Corporal Punishment in Childrearing: A Reassessment" (*Amer. Sociological Review*, 1974, pp. 68–85). A sample of 851 adults was obtained and each one was cross-classified according to frequency of spanking as a child (factor A), age group (factor B), and social class (factor C).

	Middle class 18–50	Middle class 51+	Working class 18–50	Working class 51+
Infrequently	145	67	234	140
Frequently	39	28	105	93

Figure 14.20

To fit the constant association model, we need the marginal totals $N_{ij\cdot}$, $N_{i\cdot k}$, and $N_{\cdot jk}$. For example, $N_{11\cdot} = N_{111} + N_{112} = 145 + 234 = 379$, while $N_{2\cdot 1} = N_{211} + N_{221} = 39 + 28 = 67$. All such marginal totals are displayed below.

$N_{ij\cdot}$	$j = 1$	$j = 2$
$i = 1$	379	207
$i = 2$	144	121

$N_{i\cdot k}$	$k = 1$	$k = 2$
$i = 1$	212	374
$i = 2$	67	198

$N_{\cdot jk}$	$k = 1$	$k = 2$
$j = 1$	184	339
$j = 2$	95	233

Figure 14.21

Now with $\hat{E}_{ijk}^{(0)} = 1$, $\hat{E}_{ij\cdot}^{(0)} = 1 + 1 = 2$, so

$$\hat{E}_{ijk}^{(1)} = \frac{N_{ij\cdot}}{\hat{E}_{ij\cdot}^{(0)}} \cdot \hat{E}_{ijk}^{(0)} = \frac{1}{2} N_{ij\cdot}.$$

The $\hat{E}_{ijk}^{(1)}$'s are computed from the first table of marginal totals; their values are recorded in the table below. For example,

$$\hat{E}_{211}^{(1)} = \tfrac{1}{2} N_{21\cdot} = \tfrac{1}{2}(144) = 72$$

Then the $\hat{E}_{ijk}^{(2)}$'s are computed using the second marginal table according to (14.32). Thus, for example,

$$\hat{E}_{211}^{(2)} = \frac{N_{2\cdot 1}}{\hat{E}_{2\cdot 1}^{(1)}} \cdot \hat{E}_{211}^{(1)} = \frac{67}{72 + 60.5} \cdot 72 = 36.4$$

$$\hat{E}_{221}^{(2)} = \frac{67}{132.5} \cdot 60.5 = 30.6$$

Table 14.1 $E_{ijk}^{(1)}$'s for the Data in Example 14.17

	Cell							
	111	112	121	122	211	212	221	222
N_{ijk}	145	234	67	140	39	105	28	93
$E_{ijk}^{(0)}$	1	1	1	1	1	1	1	1
$\hat{E}_{ijk}^{(1)}$	189.5	189.5	103.5	103.5	72	72	60.5	60.5
$E_{ijk}^{(2)}$	137.1	241.9	74.9	132.1	36.4	107.6	30.6	90.4
$E_{ijk}^{(3)}$	145.4	234.6	67.5	138.3	38.6	104.4	27.6	94.7
$\hat{E}_{ijk}^{(4)}$	145.0	234.0	67.9	139.1	38.9	105.1	27.3	93.7
$E_{ijk}^{(5)}$	144.4	234.6	67.6	139.4	39.4	104.7	27.6	93.3
$E_{ijk}^{(6)}$	144.6	234.4	67.5	139.6	39.4	104.6	27.5	93.4
$\hat{E}_{ijk}^{(7)}$	144.6	234.4	67.5	139.5	39.4	104.6	27.5	93.5
$E_{ijk}^{(8)}$	144.5	234.5	67.5	139.5	39.5	104.6	27.5	93.5
$E_{ijk}^{(9)}$	144.5	234.4	67.5	139.5	39.5	104.6	27.5	93.5
$\left(\dfrac{N_{ijk}}{\hat{E}_{ijk}} \right)$	1.003	.998	.993	.996	.987	1.004	1.108	.995

From Table 14.1 we see that the iterated values have stabilized at the end of three complete iterations (the estimates differ from those at the end of the previous iteration by at most .1), so the estimated expected cell counts are $\hat{E}_{ijk}^{(9)}$. The ratios $(N_{ijk}/\hat{E}_{ijk})$ appear in the last row, from which

$$G^2 = 2[145 \ln(1.003) + \cdots + 93 \ln(.995)] = 2.497$$

The constant association model specifies one degree of freedom, and $\chi^2_{.1,1} = 2.706$. Thus when the model of constant association is correct, a value exceeding the computed G^2 would be observed more than 10% of the time, indicating that the model fits the data reasonably well.

Model Selection

The selection of an appropriate log-linear model involves the same sort of subjective judgments as did the selection of a set of carriers in multiple regression. When one is trying to decide between different models in a given hierarchy, such as the hierarchy M_1, M_2, M_3, and M_4 discussed earlier, examination of differences between corresponding G^2 statistics is recommended.

Proposition: Let $G^2(M_t)$ denote the G^2 statistic for the log-linear model M_t in the hierarchy M_1, M_2, M_3 and M_4, so that M_t is a special case of M_{t+1}. Then $G^2(M_t) \geq G^2(M_{t+1})$, and when model M_t is correct, the difference $G^2(M_t) - G^2(M_{t+1})$ has approximately a chi-squared distribution with number of degrees of freedom equal to the difference between d.f. for M_t and d.f. for M_{t+1}.

The strategy suggested by this result is to examine first the fit to M_4, the most complicated unsaturated model. If $G^2(M_4)$, when compared to a chi-squared critical value with d.f. for M_4 ($\alpha = .05$ or $.01$ are often used), suggests that M_4 provides a reasonable fit, we examine both $G^2(M_3)$ and the difference $G^2(M_3) - G^2(M_4)$ to see by how much the goodness of fit deteriorates in going from M_4 to M_3. If the deterioration is not significant, $G^2(M_2)$ and $G^2(M_2) - G^2(M_3)$ are next considered, and so on. A major reason for not measuring goodness of fit by the χ^2 statistic is that $\chi^2(M_t)$ may actually be smaller than $\chi^2(M_{t+1})$, indicating that the simpler model provides a better fit than the one having more parameters. With G^2, a model lower in the hierarchy will always provide a worse fit than one higher; the investigator must then judge if the fit is "significantly worse."

All this presumes that the hierarchy of interest has been identified. With just three factors, all hierarchical models can be fit and all hierarchies examined, so that the investigator can "snoop" for a good model.

Example 14.17 (continued) We have already seen that the log-linear model M_4 with no three-factor interaction terms provides a reasonable fit to the data on frequency of spanking, age, and social class. The next type of model M_3 in the hierarchy postulates conditional independence of two factors for each level of the third factor. The model which states that frequency of spanking and age are independent within each social class is

$$\ln(E_{ijk}) = \lambda + \alpha_i + \beta_j + \delta_k + \gamma_{ik}^{AC} + \gamma_{jk}^{BC}$$

with

$$\hat{E}_{ijk} = \frac{N_{i \cdot k} \cdot N_{\cdot jk}}{N_{\cdot \cdot k}}$$

		$k = 1$			$k = 2$	
		$j = 1$	$j = 2$		$j = 1$	$j = 2$
$\hat{E}_{ijk}$:	$i = 1$	139.81	72.19		221.65	152.35
	$i = 2$	44.19	22.81		117.35	80.65

Figure 14.22

Notice that because γ_{ik}^{AC} is the highest-order term in the model involving i and k, $N_{i \cdot k} = \hat{E}_{i \cdot k}$, and similarly $N_{\cdot jk} = \hat{E}_{\cdot jk}$.

The computed value of G^2 for this model is $G^2(M_3) = 7.162$, so $G^2(M_3) - G^2(M_4) = 7.162 - 2.497 = 3.665$. Model M_3 and M_4 have d.f. 2 and 1, respectively, so we compare $G^2(M_3)$ to $\chi^2_{.05,2} = 5.992$ and $G^2(M_3) - G^2(M_4)$ to $\chi^2_{.05,1} = 3.843$. While the difference in G^2 values does not exceed 3.843, the value of $G^2(M_3)$ itself does exceed 5.992, suggesting that the conditional independence model does not provide a good fit to the data.

Similarly, the model which states that frequency of spanking and class are independent conditional on age yields $G^2(M_3) = 9.10$, so this model is also unsatisfactory.

With more than three factors, the examination of all models can be quite tedious. The introductory book by Fienberg listed in the chapter bibliography discusses stepwise procedures analogous to those of regression analysis for selecting a model. These procedures have been implemented in several computer programs for performing log-linear analyses.

Other Topics

Until now we have assumed that only the total sample size $n = N \ldots$ was fixed and that any hierarchical model could be fit. Many times some of the marginal totals are fixed by the design of the experiment. For example, the first factor might refer to different times at which samples were obtained, in which case $N_{i \cdot \cdot}$ would be fixed. If the second factor was the region sampled and the first was time period, then the $N_{ij \cdot}$ marginals would be fixed. Log-linear models can be used in such situations provided that model parameters associated with the highest-level fixed marginals are included in the model being fit. For example, if the $N_{ij \cdot}$'s are fixed, then any model fit should include γ_{ij}^{AB} (and of course α_i and β_j). If only $N_{i \cdot \cdot}$ is fixed for each i, then α_i should be included in any model under consideration.

In some situations a distinction is made between explanatory factors and a response factor or variable. For example, in studying the incidence of cancer the responses would be cancer/no cancer, and explanatory variables might include frequency of smoking, proximity to industrial pollution, and the like. With m_{ijkl} representing the expected number of individuals in category $l(l = 1, 2)$ of the response factor when the explanatory factors are at levels i, j, and k, respectively, models called **logit models** have been developed which postulate a log-linear structure for $\ln(m_{ijk1}/m_{ijk2})$. The book by Fienberg or the one by Everitt contain an introduction to these models, as well as much other material on log-linear models.

Exercises / Section 14.4

1. a. For the situation described in Example 14.15, which model for $\ln(E_{ijk})$ corresponds to relative foot size and handedness being dependent on one another but jointly independent of sex?

b. Again referring to Example 14.15, what model corresponds to independence of handedness and relative foot size for each sex separately?

2. Suppose that each gas station is categorized with respect to four different factors: (A) condition (substandard, standard, or modern), (B) pricing policy (aggressive, neutral, nonaggressive), (C) brand of gasoline (major, independent), and (D) location (on a highway or major thoroughfare, or not). To describe a log-linear model for $\ln(E_{ijkl})$, let α_i^A, α_j^B, α_k^C, and α_l^D denote the main-effect (single-factor) parameters, let β_{ij}^{AB}, . . . , β_{kl}^{CD} denote the two-factor interaction parameters, let δ_{ijk}^{ABC}, . . . , δ_{jkl}^{BCD} denote the three-factor interaction parameters, and let γ_{ijkl} denote a four-factor interaction parameter.

a. What are the side conditions on the parameters which are analogous to the conditions for a three-factor model?

b. How many independently determined main-effect parameters are there (both in general and for the specific example under consideration)? How many independently determined second-order interaction parameters are there?

c. Which model corresponds to mutual independence of the four factors?

d. How many degrees of freedom are associated with the model of mutual four-factor independence?

e. Which model corresponds to independence of condition, location, and the pair (pricing policy, brand) (with possible dependence between pricing policy and brand)?

f. Which model corresponds to independence of the two pairs (condition, location) and (pricing policy, brand) while allowing for dependence of the two factors within each pair?

g. Which model corresponds to the independence of condition, pricing policy, and brand for each type of location (a conditional independence model)?

h. Which model corresponds to independence of

the pair (pricing policy, brand) and the factor condition for each type of location (also a conditional independence model, but one in which pricing policy and brand are not necessarily conditionally independent given the type of location)?

3. a. Compute the $\hat{E}_{ijk}$'s for the data in Example 14.16 assuming that the three factors are mutually independent.

b. Compute the value of the statistic G^2 for the mutual independence model. Does this model provide a good fit to the data?

4. a. Compute the $\hat{E}_{ijk}$'s for the data in Example 14.16 assuming that the pair (perch height, perch diameter) is independent of species (a partial independence model).

b. Compute the value of the statistic G^2 for the model of (a). Does this model provide a reasonable fit to the data?

5. Following an outbreak of food poisoning at an outing, participants were surveyed and the following data was obtained ("A Coordinated Investigation of a Food Poisoning Outbreak," *Public Health Reports*, 1952, pp. 909–913):

		Ate crabmeat (C)			
		Yes		No	
		Ate potato salad (B)		Ate potato salad (B)	
		yes	no	yes	no
Illness (A)	yes	120	4	22	0
	no	80	31	24	23

a. Consider the model which specifies no third-order interaction and no interaction between illness and potato salad consumption—that is,

$$\ln(E_{ijk}) = \lambda + \alpha_i + \beta_j + \delta_k + \gamma_{ik}^{AC} + \gamma_{jk}^{BC}$$

(This is a conditional independence model which says that at each level of crabmeat consumption, illness and potato salad consumption are independent.) Calculate G^2 for this model [be careful—the model looks like M_3, except that γ_{jk}^{BC} appears in place of γ_{ij}^{AB}; use $0 \cdot \ln(0) = 0$].

b. It can be shown that $G^2 = 2.74$ for the constant association model M_4. Based on both this G^2 and the G^2 of (a), does the conditional independence model of (a) provide a reasonable fit to the data? Why or why not?

c. Use the iterative method described in Example 14.17 to compute the $\hat{E}_{ijk}$'s for the constant association model, and then verify that $G^2 = 2.74$ (even though several $\hat{E}_{ijk}$'s are small, Fienberg and others have used the chi-squared approximation to assess the fit of this model).

6. A random sample of 300 cars which had recently been checked at a regional vehicle inspection station yielded the following data on the results of a brake test (A), a headlight test (B), and an emissions test (C).

		\multicolumn{2}{c}{Emissions test}			
		\multicolumn{2}{c}{Pass}	\multicolumn{2}{c}{Fail}	
		\multicolumn{2}{c}{Headlight test}	\multicolumn{2}{c}{Headlight test}	
		pass	fail	pass	fail
Brake test	pass	54	63	12	42
	fail	49	14	35	31

a. Use the iterative method illustrated in Example 14.17 to compute the $\hat{E}_{ijk}$'s for the constant association model M_4.

b. Do any of the three conditional independence models provide a reasonable fit to the data? Answer by computing $G^2(M_3)$ and $G^2(M_3) - G^2(M_4)$ for each possible model of the form M_3.

Bibliography

Everitt, B. S., *The Analysis of Contingency Tables,* Halsted Press, New York, 1977. A compact but informative survey of methods for analyzing categorical data, exposited with a minimum of mathematics.

Fienberg, Stephen, *The Analysis of Cross-Classified Categorical Data,* MIT Press, Cambridge, Mass., 1977. A good introduction to log-linear models for contingency table data.

Guenther, William, "Power and Sample Size for Approximate Chi-Squared Tests," *Amer. Statistician,* 1977, pp. 83–85.

Mosteller, Frederick and Rourke, Richard, *Sturdy Statistics,* Addison-Wesley, Reading, Mass., 1973. Contains several very readable chapters on the varied uses of chi-square.

CHAPTER 15

Distribution-Free Procedures

Introduction

When the underlying population or populations are nonnormal, the t and F tests and t confidence intervals of Chapters 7–13 will in general have actual levels of significance or confidence levels which differ from the nominal levels (those prescribed by the experimenter through the choice of, say, $t_{.025}$, $F_{.01}$, and so on) α and $100(1 - \alpha)\%$, although the difference between actual and nominal levels may not be large when the departure from normality is not too severe. Because the t and F procedures require the distributional assumption of normality, they are not "distribution-free" procedures—alternatively, because they are based on a particular parametric family of distributions (normal), they are not "nonparametric" procedures.

In this chapter we describe procedures which are valid [actual level α or confidence level $100(1 - \alpha)\%$] simultaneously for many different types of underlying distributions. Such procedures are called **distribution-free** or **nonparametric.** Sections 15.1 and 15.2 discuss two test procedures for analyzing a single sample of data; Section 15.3 presents a test procedure for use in two-sample problems. In Section 15.4 we develop distribution-free confidence intervals for μ and $\mu_1 - \mu_2$, while Section 15.5 describes distribution-free ANOVA procedures. These procedures are all competitors of the parametric (t and F) procedures described in earlier chapters, so it is important to compare the performance of the two types of procedures under both normal and nonnormal population models. Generally speaking, the distribution-free procedures perform almost as well as their t and F counterparts on the "homeground" of the normal distribution and will often yield a considerable improvement under nonnormal conditions.

15.1 The Sign Test

The sign test is a procedure for testing hypotheses about the median of a continuous distribution. Recall from Chapter 4 that the median $\tilde{\mu}$ of such a distribution is a number such that half the area under the density curve lies to the left of $\tilde{\mu}$ and half lies to the right of $\tilde{\mu}$. If X denotes the random variable whose distribution is under investigation, then $P(X \le \tilde{\mu}) = P(X \ge \tilde{\mu}) = .5$. The general null hypothesis will have the form $H_0: \tilde{\mu} = \tilde{\mu}_0$. To test H_0 against one of the three commonly encountered alternative hypotheses, we will have available a random sample $X_1, \ldots, X_n$ from the distribution of interest.

In Chapter 7 the t test was used to test hypotheses about the mean of a normal population. Because any normal distribution is symmetric, $\tilde{\mu} = \mu$, so the sign test can also be used to test hypotheses about a normal mean. We shall comment shortly on the relative merits of the two test procedures. For now, we note that the sign test is valid whenever the distribution is continuous, whereas the t test was designed specifically for the normal distribution. Because the sign test can be used for a wide variety of underlying distributions, it is often referred to as a distribution-free test.

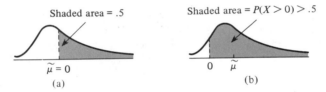

Figure 15.1 (a) $\tilde{\mu} = 0$, (b) $\tilde{\mu} > 0$

Testing $H_0: \tilde{\mu} = 0$

When $\tilde{\mu} = 0$, any X_i is equally likely to be positive or negative. If, however, the true value of $\tilde{\mu}$ is much larger than zero, X_i is much more likely to be positive than negative, so we would expect most of the observed X_i's to be positive in this case. When most sample X_i's are positive, we would suspect that $\tilde{\mu} > 0$ rather than $\tilde{\mu} = 0$.

Define a test statistic Y by

$$Y = \text{the number of } X_i\text{'s for which } X_i > 0$$

For testing $H_0: \tilde{\mu} = 0$ versus $H_a: \tilde{\mu} > 0$, the sign test rejects H_0 when $Y \ge c$ (when a large number of the observed X_i's are positive). The constant c should now be chosen to control the probability of a type I error (rejecting H_0 when H_0 is true). Regard the observation of each X_i as constituting a trial, so that the experiment consists of n identical trials. If we now identify a positive X_i with a success and a nonpositive X_i with a failure, then when H_0 is true,

$$p = P(\text{success}) = P(X_i > 0) = P(X_i > \tilde{\mu}) = .5$$

Since the trials are independent, we have

Proposition: When H_0: $\widetilde{\mu} = 0$ is true, the statistic Y has a binomial distribution with parameters n and $p = .5$.

According to this proposition, a critical value c which ensures that P (reject H_0 when H_0 is true) $= \alpha$ for a specified α can be found from the binomial distribution with appropriate n and $p = .5$.

Example 15.1   The use of a particular cloud-seeding technique for inducing rain in a series of 15 independent experiments performed under similar climatic conditions yielded the following observations on the net gain in inches of rain after the seeding: $-.30$, 1.50, .41, $-.11$, 2.03, .28, .76, $-.29$, -1.14, .09, $-.92$, .71, .36, .25, $-.50$. Does this data indicate that the cloud-seeding technique is successful in its objective under the given conditions?

To answer this question using the sign test, let X represent the net gain in inches resulting from one particular seeding experiment, and let $\widetilde{\mu}$ denote the median of the distribution of X (assumed continuous). Then we shall say that seeding is successful if $\widetilde{\mu} > 0$, so that more than 50% of all seeding experiments result in a positive gain in rainfall. Because seeding is expensive, we will be convinced of its validity only if the data strongly indicates that $\widetilde{\mu} > 0$, so we test H_0: $\widetilde{\mu} = 0$ versus H_a: $\widetilde{\mu} > 0$.

The test statistic Y has a binomial distribution with $n = 15$ and $p = .5$ when H_0 is true. From the binomial tables, $P(Y \geq 12) = 1 - B(11; 15, .5) = .018$ while $P(Y \geq 13) = .004$. Thus a test with level of significance approximately .02 rejects H_0 if $Y \geq 12$. Since only nine of the 15 x_i's in the sample are positive, the observed value of Y is $y = 9$, which is not in the rejection region. At the chosen level of significance, H_0 cannot be rejected, so we conclude that seeding is not successful.

The test statistic Y has a discrete distribution (binomial), so for any given level α there will not in general be a critical value which yields a test with exactly the desired level. In Example 15.1, there was no value of c yielding exactly level .01, so we used a test with a slightly higher level of significance. When the distribution of a test statistic is discrete, as is usually the case with a distribution-free procedure, the rejection region is chosen to yield an α as close to the desired level as possible.

The Sign Test for H_0: $\widetilde{\mu} = \widetilde{\mu}_0$

A minor modification of the test procedure for $\widetilde{\mu}_0 = 0$ yields a procedure appropriate for any other null value. The new test statistic is the number of positive $(X_i - \widetilde{\mu}_0)$'s, or equivalently the number of sample observations which exceed $\widetilde{\mu}_0$.

Null hypothesis: H_0: $\widetilde{\mu} = \widetilde{\mu}_0$

Test statistic: $Y = $ the number of X_i's which exceed $\widetilde{\mu}_0$

Alternative Hypothesis		Rejection Region
$H_a: \widetilde{\mu} > \widetilde{\mu}_0$		$Y \geq c_1$ (large Y)
$H_a: \widetilde{\mu} < \widetilde{\mu}_0$		$Y \leq c_2$ (small Y)
$H_a: \widetilde{\mu} \neq \widetilde{\mu}_0$	either	$Y \geq c$ or $Y \leq n - c$ (small or large Y)

where the critical values c_1, c_2, and c are obtained from the binomial distribution Bin $(n, .5)$ to yield the desired α.

Example 15.2 An electronics company will shortly begin to market a kit for a CB radio. The company wishes to claim that the kit can be assembled quite easily even by someone with no previous electronics experience. Let X be the assembly time for a randomly selected person of this type, with $f(x)$ denoting the distribution of X and $\widetilde{\mu}$ the median of this distribution. It is decided that the claim will be validated only if experimental evidence strongly suggests that at least half of all such individuals can assemble the kit in less than 20 hours. The null hypothesis is then $H_0: \widetilde{\mu} = 20$, while the appropriate alternative is $H_a: \widetilde{\mu} < 20$; this choice of H_a ensures that the burden of proof lies with the claim of easy assembly.

With a sample size of $n = 15$, the test that rejects H_0 when Y (the number among the 15 assembly times which exceed 20) is 0, 1, 2, or 3 has $\alpha = .018$. For the sample $x_1 = 18.5$, $x_2 = 19.1$, $x_3 = 22.2$, $x_4 = 16.9$, $x_5 = 19.6$, $x_6 = 18.8$, $x_7 = 19.8$, $x_8 = 19.0$, $x_9 = 18.3$, $x_{10} = 19.2$, $x_{11} = 21.6$, $x_{12} = 17.5$, $x_{13} = 18.9$, $x_{14} = 18.0$, and $x_{15} = 19.4$, the test statistic has value $y = 2$. Since 2 falls in the rejection region for a level .018 test, H_0 is rejected at this level in favor of the conclusion that $\widetilde{\mu} < 20$.

The Normal Approximation for Large *n*

When $p = .5$, the binomial distribution can be approximated by a normal distribution for n as small as 10. Since $\mu_Y = np = .5n$ and $\sigma_Y^2 = np(1 - p) = .25n$, the test statistic

$$Z = \frac{Y - .5n}{.5\sqrt{n}}$$

has approximately a standard normal distribution when H_0 is true and $n > 10$. The upper-tailed test then rejects H_0 in favor of $H_a: \widetilde{\mu} > \widetilde{\mu}_0$ when $Z \geq z_\alpha$, and rejection regions for the other two standard alternatives are identical to those of previous lower and two-tailed Z tests.

Example 15.3 Each of 20 randomly selected homeowners in a particular city was asked to reveal the amount by which his or her property tax bill had increased as a result of the most recent reassessment. The sample (dollar) amounts were

342, 176, 517, 296, 312, 143, 279, 412, 228, 209, 195, 241, 211, 285, 329, 137, 188, 260, 233, 357

The county assessor has stated that the median increase for all county homeowners was \$300. Does the sample data strongly indicate that the true median increase for residents of this city was less than that for the county as a whole? Use the sign test with $\alpha = .05$.

The null hypothesis here is H_0: $\widetilde{\mu} = 300$ and the alternative is H_a: $\widetilde{\mu} < 300$, where $\widetilde{\mu}$ is the median increase for city homeowners. The number of sample values which exceed 300 is $y = 6$, so the computed value of Z is

$$z = \frac{y - .5n}{.5\sqrt{n}} = \frac{6 - (.5)(20)}{.5\sqrt{20}} = -1.79$$

Using a level .05 test, H_0 is rejected since -1.79 is ≤ -1.645.

Power of the Sign Test

For testing H_0: $\widetilde{\mu} = \widetilde{\mu}_0$, as long as the underlying distribution is continuous the sign test will control the probability of a type I error at the desired level α. To assess the ability of the test procedure to draw correct conclusions, we should also examine probabilities of type II errors, or equivalently the power of the test [since power $= 1 - P$ (type II error)]. The test should be able to detect departures from the null hypothesis as evidenced by high power when the alternative hypothesis is true.

The difficulty with power calculations for the sign test is that we must assume not only a particular alternative value of $\widetilde{\mu}$, but also a specific shape for the underlying density function. For example, if $\widetilde{\mu} = \widetilde{\mu}_0 + 1$ rather than $\widetilde{\mu}_0$, the power of the test will also depend on whether the distribution assumed for the calculation is normal, uniform, exponential, or whatever. However, *for any particular underlying distribution, the test statistic Y is a binomial random variable,* but if $\widetilde{\mu} \neq \widetilde{\mu}_0$ then p will be something other than .5. Thus all power calculations reduce to binomial probability computations.

Example 15.1
(continued)

Suppose that H_0 is false because $\widetilde{\mu} = 1$, and consider the two possible distributions of Figure 15.2.

For the normal distribution, the probability of a positive net change is

$$p = P(X_i > 0) = P(Z > -1) = 1 - \Phi(-1) = .8413$$

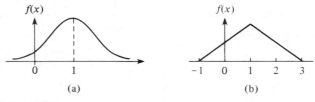

(a)   (b)

Figure 15.2   Two distributions for which H_0: $\widetilde{\mu} = 0$ is false: (a) Normal, $\widetilde{\mu} = 1$, $\sigma = 1$, (b) Symmetric triangular, $\widetilde{\mu} = 1$

Thus Y has a binomial distribution with $n = 15$ and $p = .8413$. Using the rejection region $\{Y \geq 12\} = \{12, 13, 14, 15\}$, the probability of rejecting H_0 is

$$\pi(1) = P(Y \geq 12 \quad \text{when} \quad Y \sim \text{Bin}(15, .8413))$$

$$= \sum_{y=12}^{15} \binom{15}{y} (.8413)^y (.1587)^{15-y} = .7950$$

If the underlying distribution is triangular as in Figure 15.2b, then a geometrical argument yields $p = .875$, so that

$$\pi(1) = P(Y \geq 12 \quad \text{when} \quad Y \sim \text{Bin}(15, .875)) = .8921$$

The above calculations show that the power of the sign test depends not just on the alternative value of $\widetilde{\mu}$ but on the area under the density curve to the right of $\widetilde{\mu}_0$. This area in turn depends on the particular shape of the density function.

Comparison of the Sign and *t* Tests

If it is assumed that the underlying distribution is normal, then either the sign test or the *t* test can be used to test H_0: $\widetilde{\mu} = \widetilde{\mu}_0$ against one of the usual alternative hypotheses. In this case the *t* test is known to be most powerful (minimum type II error probabilities) among all level α tests, so is certainly more powerful than the sign test. Furthermore, although the power for any particular alternative value of $\widetilde{\mu}$ depends both on that value and on the sample size n, in general the power of the sign test is substantially less than that of the *t* test. The sign test is thus better regarded as a procedure for drawing conclusions about a population median than as a serious competitor to the *t* test. In the next section we describe another distribution-free test procedure which compares quite favorably to the *t* test for both normal and non-normal populations.

Exercises / Section 15.1

1. Compute the median of
   a. a uniform distribution with p.d.f.

$$f(x) = \begin{cases} \dfrac{1}{b-a} & a < x < b \\ 0 & \text{otherwise} \end{cases}$$

   b. an exponential distribution with p.d.f.

$$f(x) = \begin{cases} \lambda e^{-\lambda x} & x > 0 \\ 0 & x \leq 0 \end{cases}$$

2. The July 3, 1978, *Los Angeles Times* reported that under conditions of the desegregation plan to be implemented, any students who travel more than 45 minutes one way on a bus are required to have only three years of integrated schooling, while those traveling at most 45 minutes face a five-year requirement. Let $\widetilde{\mu}$ be the median of the distribution of travel time between two particular locations, and suppose that a random sample of 20 travel times consists of

47.3, 44.1, 44.5, 43.7, 46.2, 45.3, 47.0, 42.5, 43.9, 45.6, 46.0, 44.8, 45.7, 47.8, 46.6, 44.2, 47.2, 46.5, 43.6, 45.4

Use the sign test at level .10, with Y = the number of times which exceed 45 minutes, to test H_0: $\widetilde{\mu} = 45$ versus H_a: $\widetilde{\mu} > 45$.

3. Ten soil samples taken from a particular region were subjected to chemical analysis to determine the pH of each sample. The sample pH's were found to be

   5.93, 6.08, 5.86, 5.91, 6.12, 5.90, 5.95, 5.89, 5.98, 5.96

   It had previously been believed that the median soil pH in this region was 6.0. Does this sample data strongly indicate that the true median pH is something other than 6.0? Let Y = the number of pH's in the sample which exceed 6.0, and use the sign test at level .05 to test the appropriate hypotheses.

4. Suppose that the distribution of times between successive alarms at a fire station is exponential, and we wish to test $H_0: \widetilde{\mu} \leq 3$ hrs versus $H_a: \widetilde{\mu} > 3$, where $\widetilde{\mu}$ is the median time between successive alarms. Suppose that a sample of 10 such times is obtained, let Y = the number of times which exceed 3 hours, and consider using the sign test with rejection region $\{8, 9, 10\}$.

   a. What is the probability of a type I error when $\lambda = .231$ (λ is the parameter of the exponential distribution, which is related to $\widetilde{\mu}$ by the result of Exercise 1, part b)?

   b. What is the probability of a type II error if $\lambda = .583$?

5. In the manufacture of a certain type of cylindrical rod, a lathe is periodically reset to the desired diameter of the rod. A production supervisor has recently become concerned over a possible tendency for the lathes to produce rods with larger and larger diameters immediately subsequent to setting (an upward trend in observed diameters). A sample of 12 rods is obtained and the measured diameters are

   $x_1 = 4.935$, $x_2 = 4.967$, $x_3 = 4.989$, $x_4 = 5.012$, $x_5 = 5.020$, $x_6 = 4.997$, $x_7 = 5.025$, $x_8 = 5.041$, $x_9 = 5.008$, $x_{10} = 5.035$, $x_{11} = 5.016$, and $x_{12} = 5.022$.

   a. Let p = the probability that the $(i + 6)$th diameter X_{i+6} exceeds the ith diameter X_i. If there is actually no upward or downward trend in diameters, so that X_{i+6} is no more or less likely to exceed X_i than it is to fall below X_i, what is the value of p?

   b. If Y = the number of values of i for which $X_{i+6} > X_i$, and there is no upward or downward

trend in diameters, what probability distribution does Y have?

   c. Use the result of (b) to test H_0: there is no trend versus H_a: there is an upward trend. Your test statistic should be Y, and you should use a level .1 (approximately) test. This simple test for trends, a modification of the sign test, is due to Cox and Stuart. A good reference is the book by Daniel.

6. Suppose that the underlying population is normal with $\sigma = 1$. For testing $H_0: \mu = 0$ versus $H_a: \mu > 0$ based on a sample of size $n = 15$, consider the Z test that rejects H_0 if $Z \geq 2.10$ (where $Z = (\overline{X} - \mu_0)/(\sigma/\sqrt{n}) = \sqrt{n}\,\overline{X}$).

   a. What is α for this test?

   b. When $\mu = 1$, how does the power of this test compare to the power of the sign test which rejects H_0 if $Y \geq 12$?

7. A process for producing stainless steel bars was originally designed so that the nickel content of the bars was 10% by weight. A potential customer feels that the bars will be suitable for a particular application only if more than 75% of all bars produced contain at least 10% nickel (that is, that fewer than 25% of all bars contain less than 10% nickel). The customer will not agree to purchase the bars unless experimental evidence strongly indicates that this requirement is satisfied.

   If $f(x)$ denotes the density function of nickel content in a randomly selected bar, then the requirement is met only if the area under the graph of $f(x)$ to the right of 10 is more than .75. Equivalently, if θ denotes the twenty-fifth percentile of the nickel-content distribution, then we wish to test $H_0: \theta = 10$ versus $H_a: \theta > 10$ (so H_a states that more than 75% of the area under the density curve lies to the right of 10). Suppose that 20 bars are sampled and the observed values of nickel content are

   10.23, 10.95, 11.02, 10.14, 9.83, 10.30, 12.05, 11.26, 10.58, 10.18, 10.38, 11.49, 9.52, 10.86, 11.00, 10.19, 10.94, 11.13, 10.25, 11.02

   Develop a test procedure for H_0 versus H_a and carry out the test for this data. Your α should be as close to .10 as possible. *Hint:* Let Y = the number of X_i's which exceed 10. If H_0 is true, what is the distribution of Y?

15.2 The Wilcoxon Signed-Rank Test

Consider again the problem of testing a hypothesis about the median $\tilde{\mu}$ of a continuous distribution (that is, population). If we wish to test H_0: $\tilde{\mu} = 0$ versus H_a: $\tilde{\mu} > 0$ using the sign test on the basis of ten observations from the population, then the rejection region $R = \{8, 9, 10\}$ has type I error probability $\alpha = .055$. Suppose that our sample consists of observations $x_1 = -0.19$, $x_2 = 2.17$, $x_3 = 2.68$, $x_4 = -.57$, $x_5 = 2.02$, $x_6 = 2.46$, $x_7 = 3.02$, $x_8 = -.05$, $x_9 = .76$, and $x_{10} = 1.30$. Then the observed value of the test statistic Y, the number of positive observations, is $y = 7$, so at level $\alpha = .055$ we would not reject H_0 in favor of H_a.

A closer look at the data reveals that while there are indeed three negative observations, these three observations are quite small in magnitude compared with the seven positive observations. That is, x_1, x_4, and x_8 are barely negative, while the positive observations are relatively distant from zero. The nonrejection of H_0 for this data is due to the way in which the sign test extracts information from the sample; magnitudes of the observations are discarded and only the sign of each observation is retained. In a sense this is the price that must be paid for having a test which is valid for a very large class of populations (*all* continuous populations).

Symmetric Distributions and a New Test Statistic

Frequently an experimenter may be willing to assume more about the population of interest than just that it is continuous, and in particular may be willing to assume that the population distribution has a symmetrical shape. A population distribution is **symmetric** if the graph of the p.d.f. to the left of $\tilde{\mu}$ is a reflection or mirror image of the graph to the right of $\tilde{\mu}$. The graphs of several symmetric p.d.f.'s are pictured in Figure 15.3.

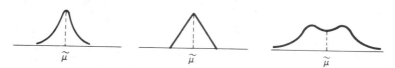

Figure 15.3 Symmetric p.d.f.'s

If we assume that the population which yielded the above sample is symmetric, then the conclusion that $\tilde{\mu} = 0$ (actually ≤ 0) seems quite suspect, for the data suggests strongly that the point of symmetry lies to the right of zero ($\bar{x} = 1.66$, which is much larger in magnitude than any of the negative observations). We now assume that *our sample comes from a continuous symmetric probability distribution,* and develop a test for making an inference about $\tilde{\mu}$ which is different from the sign test. When this new test is applied to the above sample, the data will argue strongly for the rejection of H_0 in favor of H_a. This is because the new test will take into account the magnitudes of the observations as well as the signs.

Figure 15.4 shows two different symmetric p.d.f.'s, one for which H_0 is true and one for which H_a is true. When H_0 is true then we expect that the negative

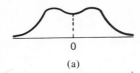

Figure 15.4

observations in the sample should be comparable in magnitude to the magnitudes of the positive observations. If, however, H_0 is "grossly" untrue (Figure 15.4b), then observations of large absolute magnitude will tend to be positive rather than negative. For the ten observations in our sample, suppose we proceed as follows:

1. Disregarding the signs of the observations, rank the collection of 10 numbers in order of increasing magnitude.
2. Use as a test statistic S_+ = sum of the ranks associated with the positive observations.

Absolute magnitude:	.05	.19	.57	.76	1.30	2.02	2.17	2.46	2.68	3.02
Rank:	1	2	3	4	5	6	7	8	9	10
Signed rank:	-1	-2	-3	4	5	6	7	8	9	10

The observed value of S_+ is $s_+ = 4 + 5 + 6 + 7 + 8 + 9 + 10 = 49$. H_0 is now rejected in favor of H_a: $\tilde{\mu} > 0$ when s_+ is too large because a large value of s_+ indicates that most of the observations with large absolute magnitude are positive, which in turn indicates a median greater than zero. If on the other hand, observations with negative signs are intermingled with positive observations when signs are disregarded, then the observed s_+ will not be very large.

The Distribution of S_+

To decide between H_0 and H_a, we must know whether 49 is a "surprisingly large" value of S_+ when the null hypothesis is true. This entails obtaining the probability distribution of S_+ when H_0 is true, and selecting as the rejection region a set of large S_+ values with small probability under H_0 (small α). If our observed s_+ is in the rejection region, we can then reject H_0 at the desired level α, and otherwise not reject H_0.

The probability distribution of S_+ is more difficult to obtain than was the distribution for the sign test. To illustrate, consider the case $n = 5$. The key observation in the derivation is this: When H_0 is true, *any* collection of signed ranks is *just as* likely as any other collection. Worded differently, the smallest observation in absolute magnitude is just as likely to be positive as negative, the second smallest observation in absolute magnitude is just as likely to be positive as negative, and similarly for the other observations. Thus for $n = 5$ the sequence $-1, +2, +3, -4, +5$ of signed ranks is just as likely to be observed (when H_0 is true) as is the sequence $-1, -2, -3, -4, +5$, or any other sequence. Table 15.1 lists the possible signed rank sequences for $n = 5$ and the s_+ value associated with each sequence. Since there are 32 such sequences (two possible signs for the smallest observation, two for the second smallest, and so on, or $2 \cdot 2 \cdot 2 \cdot 2 \cdot 2 = 32$ sequences), each sequence has probability 1/32 of being observed when H_0 is true. From this the p.m.f. of S_+ under H_0 can immediately be obtained as in Table 15.2. Note that for $n = 10$ observations there are $2^{10} = 1024$ possible signed rank sequences, and to list these sequences would be very tedious. Each sequence, though, would receive probability 1/1024 under H_0, from which the null distribution (distribution when H_0 is true) of S_+ can be easily obtained.

Table 15.1 *Possible Signed Rank Sequences for n = 5*

Sequence	s_+	Sequence	s_+
-1 -2 -3 -4 -5	0	-1 -2 -3 $+4$ -5	4
$+1$ -2 -3 $''$ $''$	1	$+1$ -2 -3 $''$ $''$	5
-1 $+2$ -3 $''$ $''$	2	-1 $+2$ -3 $''$ $''$	6
-1 -2 $+3$ $''$ $''$	3	-1 -2 $+3$ $''$ $''$	7
$+1$ $+2$ -3 $''$ $''$	3	$+1$ $+2$ -3 $''$ $''$	7
$+1$ -2 $+3$ $''$ $''$	4	$+1$ -2 $+3$ $''$ $''$	8
-1 $+2$ $+3$ $''$ $''$	5	-1 $+2$ $+3$ $''$ $''$	9
$+1$ $+2$ $+3$ $''$ $''$	6	$+1$ $+2$ $+3$ $''$ $''$	10
-1 -2 -3 -4 $+5$	5	-1 -2 -3 $+4$ $+5$	9
$+1$ -2 -3 $''$ $''$	6	$+1$ -2 -3 $''$ $''$	10
-1 $+2$ -3 $''$ $''$	7	-1 $+2$ -3 $''$ $''$	11
-1 -2 $+3$ $''$ $''$	8	-1 -2 $+3$ $''$ $''$	12
$+1$ $+2$ -3 $''$ $''$	8	$+1$ $+2$ -3 $''$ $''$	12
$+1$ -2 $+3$ $''$ $''$	9	$+1$ -2 $+3$ $''$ $''$	13
-1 $+2$ $+3$ $''$ $''$	10	-1 $+2$ $+3$ $''$ $''$	14
$+1$ $+2$ $+3$ $''$ $''$	11	$+1$ $+2$ $+3$ $''$ $''$	15

Table 15.2 *Null Distribution of* S_+ *When n = 5*

s_+	0	1	2	3	4	5	6	7
$p(s_+)$	1/32	1/32	1/32	2/32	2/32	3/32	3/32	3/32
s_+	8	9	10	11	12	13	14	15
$p(s_+)$	3/32	3/32	3/32	2/32	2/32	1/32	1/32	1/32

As an example of how the probabilities in Table 15.2 were obtained, consider the S_+ value 7; there are three signed rank sequences for which $S_+ = 7$, and each receives probability 1/32 under H_0, so that $P(S_+ = 7$ when H_0 is true$) = 3/32$.

Table 15.2 can now be used to specify a rejection region for H_0: $\tilde{\mu} = 0$ versus H_a: $\tilde{\mu} > 0$ for which α can be calculated. Consider the rejection region $R = \{13, 14, 15\}$. Then

$$\alpha = P(\text{reject } H_0 \text{ when } H_0 \text{ is true}) = P(S_+ = 13, 14, \text{ or } 15 \text{ when } H_0 \text{ is true})$$
$$= 1/32 + 1/32 + 1/32 = 3/32 = .094 \text{ from Table 15.2}$$

so that $R = \{13, 14, 15\}$ specifies a test with approximate level .1. For the rejection region $\{14, 15\}$, $\alpha = 2/32 = .063$. For the data set $x_1 = .58$, $x_2 = 2.50$, $x_3 = -.21$, $x_4 = 1.23$, $x_5 = .97$, the signed rank sequence is -1, $+2$, $+3$, $+4$, $+5$, so $s_+ = 14$ and at level .063 H_0 would be rejected.

A General Description of the Test

Because the underlying distribution is assumed symmetric, $\mu = \tilde{\mu}$, so we shall state the hypotheses of interest in terms of μ rather than $\tilde{\mu}$.*

Assumption: $X_1, X_2, \ldots, X_n$ is a random sample from a continuous and symmetric probability distribution with mean (and median) μ.

When the hypothesized value of μ is μ_0, the absolute differences $|X_1 - \mu_0|, \ldots, |X_n - \mu_0|$ must be ranked from smallest to largest.

Null hypothesis: H_0: $\mu = \mu_0$

Test statistic: $S_+ =$ the sum of the ranks associated with positive $(X_i - \mu_0)$'s

Alternative Hypothesis	*Rejection Region for a Level α Test*
H_a: $\mu > \mu_0$	$S_+ \geq c_1$
H_a: $\mu < \mu_0$	$S_+ \leq c_2$ [where $c_2 = n(n+1)/2 - c_1$]
H_a: $\mu \neq \mu_0$	either $\quad S_+ \geq c$ or $S_+ \leq n(n+1)/2 - c$

where the critical values c_1 and c are obtained from Appendix Table A.9 and satisfy $P(S_+ \geq c_1) \doteq \alpha$ and $P(S_+ \geq c) \doteq \alpha/2$.

Example 15.4

A manufacturer of electric irons, wishing to test the accuracy of the thermostat control at the 500 °F setting, instructs a test engineer to obtain actual temperatures at that setting for 15 irons using a thermocouple. The resulting measurements are

494.6, 510.8, 487.5, 493.2, 502.6, 485.0, 495.9, 498.2, 501.6, 497.3, 492.0, 504.3, 499.2, 493.5, 505.8

The engineer feels that it is reasonable to assume that a temperature deviation from 500° of any particular magnitude is just as likely to be positive as negative (the assumption of symmetry), but wants to protect against possible nonnormality of the actual temperature distribution, so decides to use the Wilcoxon signed-rank test to see whether the data strongly suggests incorrect calibration of the iron.

The hypotheses are H_0: $\mu = 500$ versus H_a: $\mu \neq 500$, where $\mu =$ the true average actual temperature at the 500 °F setting. Subtracting 500 from each x_i gives

−5.6, 10.8, −12.5, −6.8, 2.6, −15.0, −4.1, −1.8, 1.6, −2.7, −8.0, 4.3, −.8, −6.5, 5.8

The ranks are obtained by ordering these from smallest to largest without regard to sign:

---

*If the tails of the distribution are "too heavy," as was the case with the Cauchy distribution mentioned in Chapter 6, then μ will not exist. In such cases the Wilcoxon test will still be valid for tests concerning $\tilde{\mu}$.

Absolute magnitude:	.8	1.6	1.8	2.6	2.7	4.1	4.3	5.6	5.8	6.5	6.8	8.0	10.8	12.5	15.0
Rank:	1	2	3	4	5	6	7	8	9	10	11	12	13	14	15
Sign:	−	+	−	+	−	−	+	−	+	−	−	−	+	−	−

The computed value of S_+ for the data is $s_+ = 2 + 4 + 7 + 9 + 13 = 35$. From Appendix Table A.9, $P(S_+ \geq 95) = P(S_+ \leq 25) = .026$ when H_0 is true, so the approximately level .05 two-tailed test rejects when either S_+ is ≥ 95 or ≤ 25 (the exact α is $2(.026) = .052$). Since $s_+ = 35$ is not in the rejection region, it cannot be concluded at level .05 that μ is anything other than 500. Even at level .094 (approximately .1), H_0 is not rejected, since $P(S_+ \leq 30) = .047$ implies that s_+ values between 30 and 90 are not significant at that level. The P-value of the data is thus greater than .1.

Paired Observations

When the data consisted of pairs $(X_1, Y_1), \ldots, (X_n, Y_n)$ and the X_i's and Y_i's each had normal distributions, in Chapter 8 we formed the differences $D_i = X_i - Y_i$ $(i = 1, \ldots, n)$ and used a one-sample t test to test hypotheses about the expected difference μ_D. If normality is not assumed, hypotheses about μ_D can be tested by using the Wilcoxon signed-rank test on the D_i's provided that the distribution of the differences is continuous and symmetric. *If X_i and Y_i both have continuous distributions which differ only with respect to their means* (so the Y distribution is the X distribution shifted by $\mu_1 - \mu_2 = \mu_D$), *then D_i will have a continuous symmetric distribution* (it is *not* necessary for the X and Y distributions to be symmetric individually). The null hypothesis is H_0: $\mu_D = \Delta_0$, and the test statistic S_+ is the sum of the ranks associated with the positive $(D_i - \Delta_0)$'s.

Example 15.5 In Example 8.12 we presented paired data in which $X(Y)$ referred to seconds of useful consciousness before (after) a blood transfusion. Suppose that it is believed before the experiment that a transfusion will increase time of consciousness by at most five seconds for the "average individual." Does the data contradict this prior belief?

Since a five-second increase corresponds to $\mu_D = -5$, we wish to test H_0: $\mu_D = -5$ versus H_a: $\mu_D < -5$.

d_i:	−7	−12	−13	−11	−6
$d_i - (-5)$:	−2	−7	−8	−6	−1
Rank of $|d_i - (-5)|$:	2	4	5	3	1
Sign:	−	−	−	−	−

Since all signs are negative, the computed value of S_+ is $s_+ = 0$. From Appendix Table A.9 or Table 15.2, the test which rejects H_0 when $S_+ = 0$ has level .031, so the value 0 is significant at level .05. H_0 is thus rejected in favor of the conclusion that a transfusion increases time of useful consciousness by more than five seconds on average.

Ties in the Wilcoxon Test

Although a theoretical implication of the continuity of the underlying distribution is that ties will not occur, in practice they often do because of the discreteness of measuring instruments. If there are several data values with the same absolute magnitude, then they would be assigned the average of the ranks they would receive if they differed very slightly from one another. For example, if in Example 15.4 $x_8 = 498.2$ is changed to 498.4, then two different $(x_i - 500)$'s would have absolute magnitude 1.6. The ranks to be averaged would be 2 and 3, so each would be assigned rank 2.5.

A Large-Sample Approximation

Appendix Table A.9 provides critical values for level α tests only when $n \leq 20$. For $n > 20$, it can be shown that S_+ has approximately a normal distribution with

$$
\mu_{S_+} = \frac{n(n + 1)}{4}, \quad \sigma^2_{S_+} = \frac{n(n + 1)(2n + 1)}{24} \quad \text{when } H_0 \text{ is true}
$$

The mean and variance result from noting that when H_0 is true (the symmetric distribution is centered at μ_0), then the rank i is just as likely to receive a $+$ sign as it is to receive a $-$ sign. Thus

$$
S_+ = W_1 + W_2 + W_3 + \cdots + W_n
$$

where

$$
W_1 = \begin{cases} +1 & \text{with probability } \frac{1}{2} \\ 0 & \text{with probability } \frac{1}{2} \end{cases}
$$

$$
\cdot
$$
$$
\cdot
$$
$$
\cdot
$$

$$
W_n = \begin{cases} +n & \text{with probability } \frac{1}{2} \\ 0 & \text{with probability } \frac{1}{2} \end{cases}
$$

($W_i = 0$ is equivalent to rank i being associated with a $-$, so i does not contribute to S_+).

S_+ is then a sum of random variables, and when H_0 is true, these W_i's can be shown to be independent. Application of the rules of expected value and variance gives the mean and variance of S_+. Because the W_i's are not identically distributed, our version of the Central Limit Theorem cannot be applied, but there is a more general version of the theorem which can be used to justify the normality conclusion.

The large-sample test statistic is now given by

$$
Z = \frac{S_+ - n(n + 1)/4}{\sqrt{n(n + 1)(2n + 1)/24}} \tag{15.1}
$$

For the three standard alternatives, the critical values for level α tests are the usual standard normal values z_α, $-z_\alpha$, and $\pm z_{\alpha/2}$.

Example 15.6

A particular type of steel beam has been designed to have a compressive strength (pounds per square inch) of at least 50,000. For each beam in a sample of 25 beams, the compressive strength was determined and is given below. Assuming that actual compressive strength is distributed symmetrically about the true average value, use the Wilcoxon test to decide if the true average compressive strength is less than the specified value. That is, test H_0: $\mu = 50,000$ versus H_a: $\mu < 50,000$ (favoring the claim that average compressive strength is at least 50,000).

Table 15.3

$x_i - 50,000$	Signed rank	$x_i - 50,000$	Signed rank	$x_i - 50,000$	Signed rank
-10	-1	-99	-10	165	$+18$
-27	-2	113	$+11$	-178	-19
36	$+3$	-127	-12	-183	-20
-55	-4	-129	-13	-192	-21
73	$+5$	136	$+14$	-199	-22
-77	-6	-150	-15	-212	-23
-81	-7	-155	-16	-217	-24
90	$+8$	-159	-17	-229	-25
-95	-9				

The sum of the positively signed ranks is $3 + 5 + 8 + 11 + 14 + 18 = 59$, $n(n + 1)/4 = 162.5$ and $n(n + 1)(2n + 1)/24 = 1381.25$, so

$$z = \frac{59 - 162.5}{\sqrt{1381.25}} = -2.78$$

The lower-tailed level .01 test rejects H_0 if $Z \leq -2.33$. Since -2.78 is ≤ -2.33, H_0 is rejected in favor of the conclusion that true average compressive strength is less than 50,000.

When there are ties in the absolute magnitudes, so that average ranks must be used, it is still correct to standardize S_+ by subtracting $n(n + 1)/4$, but the following corrected formula for variance should be used:

$$\sigma^2_{S_+} = \tfrac{1}{24} n(n + 1)(2n + 1) - \tfrac{1}{48} \sum (\tau_i - 1)(\tau_i)(\tau_i + 1) \qquad (15.2)$$

where τ_i is the number of ties in the ith set of tied values and the sum is over all sets of tied values. If, for example, $n = 10$ and the signed ranks are $1, 2, -4, -4, 4$, $6, 7, 8.5, 8.5, 9$, and 10, then there are two tied sets with $\tau_1 = 3$ and $\tau_2 = 2$, so the summation is $(2)(3)(4) + (1)(2)(3) = 30$ and (15.2) becomes $96.25 - 30/48 = 95.62$. The denominator in (15.1) should be replaced by the square root of (15.2), though as this example shows, the correction is usually insignificant.

Efficiency of the Wilcoxon Test

When the underlying distribution being sampled is normal, either the $t(Z)$ test or the signed-rank test can be used to test a hypothesis about μ. The t test is the best test in such a situation because among all level α tests it is the one having maximum power. Since it is generally agreed that there are many experimental situations in which normality can be reasonably assumed, as well as some in which it should not be, there are two questions which must be addressed in an attempt to compare the two tests:

1. When the underlying distribution is normal (the "home ground" of the t test), how much is lost by using the signed-rank test?
2. When the underlying distribution is not normal, can a significant improvement be achieved by using the signed-rank test?

If the Wilcoxon test does not suffer much with respect to the t test on the "home ground" of the latter, and performs significantly better than the t test for a large number of other distributions, then there will be a strong case for using the Wilcoxon test.

Unfortunately there are no simple answers to the two questions. Upon reflection, it is not surprising that the t test can perform poorly when the underlying distribution has "heavy tails" (that is, when observed values lying far from μ are relatively more likely than they are when the distribution is normal). This is because the t test depends in its behavior on the sample mean, which can be very unstable in the presence of heavy tails. The difficulty in producing answers to the two questions is that the power function of the Wilcoxon test is very difficult to obtain and study for *any* underlying distribution, and the same can be said for the t test when the distribution is not normal. Even if the power functions were easily obtained, any measure of efficiency would clearly depend on which underlying distribution was postulated. A number of different efficiency measures have been proposed by statisticians; one which many statisticians regard as credible is called **asymptotic relative efficiency** (ARE). The ARE of one test with respect to another is essentially the limiting ratio of sample sizes necessary to obtain identical power functions (error probabilities) for the two tests. Thus if the ARE of one test with respect to a second equals .5, then when sample sizes are large, twice as large a sample size will be required of the first test in order that it perform as well as the second test. Although the ARE does not characterize test performance for small sample sizes, the following results can be shown to hold:

1. When the underlying distribution is normal, the ARE of the Wilcoxon test with respect to the t test is approximately .95.
2. For any distribution, the ARE will be at least .86, and for many distributions will be much greater than 1.

We can summarize these results by saying that in large-sample problems, the Wilcoxon test is never very much less efficient than the t test and may be much more efficient if the underlying distribution is far from normal. Though the issue is far from resolved in the case of sample sizes obtained in most practical problems, studies have shown that the Wilcoxon test performs reasonably and is thus a viable alternative to the *t* test. For a more thorough discussion of these issues, the reader is referred either to the book by Mosteller and Rourke or the book by Lehmann, both of which are listed in the bibliography.

Exercises / Section 15.2

1. Reconsider the problem described in Example 15.2 and use the Wilcoxon test with $\alpha = .01$ to test the given hypotheses.

2. Use the Wilcoxon test to analyze the data given in Exercise 3, Section 15.1.

3. The accompanying data is a subset of the data reported in the paper "Synovial Fluid pH, Lactate, Oxygen and Carbon Dioxide Partial Pressure in Various Joint Diseases" (*Arthritis and Rheumatism*, 1971, pp. 476–477). The observations are pH values of synovial fluid (which lubricates joints and tendons) taken from the knees of individuals suffering from arthritis. Assuming that true average pH for nonarthritic individuals is 7.39, test at level .05 to see whether the data indicates a difference between average pH values for arthritic and nonarthritic individuals.

7.02, 7.35, 7.34, 7.17, 7.28, 7.77, 7.09, 7.22, 7.45, 6.95, 7.40, 7.10, 7.32, 7.14

4. Both a gravimetric and a spectrophotometric method are under consideration for determining phosphate content of a particular material. Twelve samples of the material are obtained, each is split in half, and a determination is made on each half using one of the two methods, resulting in the following data.

Sample:	1	2	3	4	5	6
Gravimetric:	54.7	58.5	66.8	46.1	52.3	74.3
Spectrophotometric:	55.0	55.7	62.9	45.5	51.1	75.4

Sample:	7	8	9	10	11	12
Gravimetric:	92.5	40.2	87.3	74.8	63.2	68.5
Spectrophotometric:	89.6	38.4	86.8	72.5	62.3	66.0

Use the Wilcoxon test to decide whether or not one technique gives on average a different value than the other technique for this type of material.

5. Analyze the data in Exercise 15.1.2 using the large-sample version of the Wilcoxon procedure, and compare the result with an analysis using a critical value from Appendix Table A.9.

6. The accompanying 26 observations on fracture toughness of base plate of 18% nickel maraging steel were reported in the paper "Fracture Testing of Weldments," *ASTM Special Publ. No. 381*, 1965, pp. 328–356. Suppose that a company will agree to purchase this steel for a particular application only if it can be strongly demonstrated from experimental evidence that true average toughness exceeds 75. Assuming that the fracture toughness distribution is symmetric, state and test the appropriate hypotheses at level .05, and also compute a *P*-value.

69.5, 71.9, 72.6, 73.1, 73.3, 73.5, 74.1, 74.2, 75.3, 75.5, 75.7, 75.8, 76.1, 76.2, 76.2, 76.9, 77.0, 77.9, 78.1, 79.6, 79.7, 80.1, 82.2, 83.7, 93.7

7. Suppose that observations $X_1, X_2, \ldots, X_n$ are made on a process at times 1, 2, ..., *n*. On the basis of this data, we wish to test

H_0: the X_i's constitute an independent and identically distributed sequence

versus

H_a: X_{i+1} tends to be larger than X_i for $i = 1, \ldots, n$ (an increasing trend)

Suppose that the X_i's are ranked from 1 to *n*. Then when H_a is true, larger ranks tend to occur later in the sequence, while if H_0 is true large and small ranks tend to be mixed together. Let R_i be the rank of X_i and consider the test statistic $D = \sum_{i=1}^{n} (R_i - i)^2$. Then small values of *D* give support to H_a (for example, the smallest value is 0 for $R_1 = 1, R_2 = 2, \ldots, R_n = n$), so H_0 should be rejected in favor of H_a if $D \leq c$. When H_0 is true, any sequence of ranks has probability $1/n!$. Use this to find *c* for which the test has level as close to .10 as possible in the case $n = 4$. *Hint:* List the 4! rank sequences and compute *D* for each one, and then obtain the null distribution of *D*. See the Lehmann book, p. 290, for more information.

15.3 The Wilcoxon Rank-Sum Test

When at least one of the sample sizes in a two-sample problem is small, the *t* test requires the assumption of normality (at least approximately). While the test is the best test in the sense of maximizing power for fixed α, there are situations in which an investigator would want to use a test which is valid even if the underlying distributions are quite nonnormal. We now describe such a test, called the Wilcoxon rank-sum test. An alternative name for the procedure is the Mann-Whitney test, though the Mann-Whitney test statistic is sometimes expressed in a slightly different form from that of the Wilcoxon test. The Wilcoxon test procedure is distribution-free because it will have the desired level of significance for a very large class of underlying distributions.

> **Assumptions:** $X_1, \ldots, X_m$ and $Y_1, \ldots, Y_n$ are two independent random samples from continuous distributions with means μ_1 and μ_2, respectively. The *X* and *Y* distributions have exactly the same shape and spread, the only possible difference between the two being in the values of μ_1 and μ_2.

When $H_0: \mu_1 - \mu_2 = \Delta_0$ is true, the *X* distribution is shifted by the amount Δ_0 to the right of the *Y* distribution, while when H_0 is false, the shift is by an amount other than Δ_0.

Development of the Test When $m = 3$, $n = 4$

Consider first testing $H_0: \mu_1 - \mu_2 = 0$. If μ_1 is actually much larger than μ_2, then most of the observed *x*'s will fall to the right of the observed *y*'s, while if H_0 is true then the observed values from the two samples should be intermingled. The test statistic will provide a quantification of how much intermingling there is in the two samples.

For concreteness, let $m = 3$ and $n = 4$. Then if all three observed *x*'s were to the right of all four observed *y*'s, this would provide strong evidence for rejecting H_0 in favor of $H_a: \mu_1 - \mu_2 \neq 0$, with a similar conclusion being appropriate if all three *x*'s fall below all four of the *y*'s. Suppose that we pool the *X*'s and *Y*'s into a combined sample of size $m + n = 7$ and rank these observations from smallest to largest, with the smallest receiving rank 1 and the largest, rank 7. If either most of the largest ranks or most of the smallest ranks were associated with *X* observations, we would begin to suspect H_0. This suggests the test statistic

$$W = \text{the sum of the ranks in the combined sample associated with } X \text{ observations} \tag{15.3}$$

For the values of *m* and *n* under consideration, the smallest possible value of *W* is $w = 1 + 2 + 3 = 6$ (if all three *x*'s are smaller than all four *y*'s) and the largest possible value is $w = 5 + 6 + 7 = 18$ (if all three *x*'s are larger than all four *y*'s).

As an example of the calculation of the observed *W*, suppose that $x_1 = -3.10$, $x_2 = 1.67$, $x_3 = 2.01$, $y_1 = 5.27$, $y_2 = 1.89$, $y_3 = 3.86$, and $y_4 = 0.19$. Then the pooled ordered sample is $-3.10, 0.19, 1.67, 1.89, 2.01, 3.86,$ and 5.27. The *X* ranks for this sample are 1 (for -3.10), 3 (for 1.67), and 5 (for 2.01), so the computed value of W is $w = 1 + 3 + 5 = 9$.

The test procedure based on the statistic (15.3) is to reject H_0 if the computed w is "too extreme"—that is, $\geq c$ for an upper-tailed test, $\leq c$ for a lower-tailed test, and either $\geq c_1$ or $\leq c_2$ for a two-tailed test. The critical constant(s) c (c_1, c_2) should be chosen so that the test has the desired level of significance α. To see how this should be done, recall that when H_0 is true, all seven observations come from the same population. This means that under H_0, any possible triple of ranks associated with the three x's—such as $(1, 4, 5)$, $(3, 5, 6)$, or $(5, 6, 7)$—has the same probability as any other possible rank triple. Since there are $\binom{7}{3} = 35$ possible rank triples, under H_0 each rank triple has probability $\frac{1}{35}$. From a list of all 35 rank triples and the w value associated with each, the probability distribution of W can immediately be determined. For example, there are four rank triples which have w value 11—$(1, 3, 7)$, $(1, 4, 6)$, $(2, 3, 6)$, and $(2, 4, 5)$—so $P(W = 11) = \frac{4}{35}$. The summary of the listing and computations appears in Table 15.4.

Table 15.4 *Probability Distribution of W ($m = 3$, $n = 4$) When H_0 Is True*

w	6	7	8	9	10	11	12	13	14	15	16	17	18
$P(W = w)$	$\frac{1}{35}$	$\frac{1}{35}$	$\frac{2}{35}$	$\frac{3}{35}$	$\frac{4}{35}$	$\frac{4}{35}$	$\frac{5}{35}$	$\frac{4}{35}$	$\frac{4}{35}$	$\frac{3}{35}$	$\frac{2}{35}$	$\frac{1}{35}$	$\frac{1}{35}$

The distribution of Table 15.4 is symmetric about the value $w = (6 + 18)/2 = 12$, which is the middle value in the ordered list of possible W values. This is because the two rank triples (r, s, t) (with $r < s < t$) and $(8 - t, 8 - s, 8 - r)$ have values of w which are symmetric about 12, so for each triple with w value below 12, there is a triple with w value above 12 by the same amount.

If the alternative hypothesis is $H_a: \mu_1 - \mu_2 > 0$, then H_0 should be rejected in favor of H_a for large W values. Choosing as the rejection region the set of W values $\{17, 18\}$, $\alpha = P(\text{type I error}) = P(\text{reject } H_0 \text{ when } H_0 \text{ is true}) = P(W = 17 \text{ or } 18$ when H_0 is true) $= \frac{1}{35} + \frac{1}{35} = \frac{2}{35} = .057$; the region $\{17, 18\}$ therefore specifies a test with level of significance approximately .05. Similarly, the region $\{6, 7\}$, which is appropriate for $H_a: \mu_1 - \mu_2 < 0$ has $\alpha = .057 \doteq .05$. The region $\{0, 1, 17, 18\}$, which is appropriate for the two-sided alternative, has $\alpha = \frac{4}{35} = .114$. The W value for the data given several paragraphs earlier was $w = 9$ which is rather close to the middle value 12, so H_0 would not be rejected at any reasonable level α for any one of the three H_a's.

General Description of the Wilcoxon Rank-Sum Test

The null hypothesis $H_0: \mu_1 - \mu_2 = \Delta_0$ is handled by subtracting Δ_0 off each X_i and using the $(X_i - \Delta_0)$'s as the X_i's were previously used. Recalling that for any positive integer K, the sum of the first K integers is $K(K + 1)/2$, the smallest possible value of the statistic W is $m(m + 1)/2$, which occurs when the $(X_i - \Delta_0)$'s are all to the left of the Y sample. The largest possible value of W occurs when the $(X_i - \Delta_0)$'s lie entirely to the right of the Y's; in this case $W = (m + 1) + \cdots + (m + n) = (\text{sum of first } m + n \text{ integers}) - (\text{sum of first } m \text{ integers}) = m(m + 2n + 1)/2$. As with the special case $m = 3$, $n = 4$, the distribution of W is symmetric about the value which is halfway between the smallest and largest values; this middle value is $m(m + n + 1)/2$. Because of this symmetry, probabilities involving lower-tail critical values can be obtained from corresponding upper-tail values.

Null hypothesis: H_0: $\mu_1 - \mu_2 = \Delta_0$

Test statistic: $W = \sum_{i=1}^{m} R_i$ where R_i = rank of $(X_i - \Delta_0)$ in the combined sample of $m + n$ X's and Y's

Alternative Hypothesis	*Rejection Region*
H_a: $\mu_1 - \mu_2 > \Delta_0$	$W \geq c_1$
H_a: $\mu_1 - \mu_2 < \Delta_0$	$W \leq m(m + n + 1) - c_1$
H_a: $\mu_1 - \mu_2 \neq \Delta_0$	either $W \geq c$ or $W \leq m(m + n + 1) - c$

where $P(W \geq c_1$ when H_0 is true$) = \alpha$, $P(W \geq c$ when H_0 is true$) = \alpha/2$.

Because W has a discrete probability distribution, there will not always exist a critical value corresponding exactly to one of the usual levels of significance. Appendix Table A.10 gives upper-tail critical values for probabilities closest to .05, .025, .01, and .005, from which level .05 or .01 one- and two-tailed tests can be obtained. The table gives information only for $m = 3, 4, \ldots, 8$ and $n = m$, $m + 1, \ldots, 8$ (that is, $3 \leq m \leq n \leq 8$). For values of m and n which exceed 8, a normal approximation can be used. To use the table for small m and n, though, *the X and Y samples should be labeled so that $m \leq n$.*

Example 15.7 The urinary fluoride concentration (parts per million) was measured both for a sample of livestock which had been grazing in an area previously exposed to fluoride pollution and for a similar sample which had grazed in an unpolluted region.

  polluted: 21.3, 18.7, 23.0, 17.1, 16.8, 20.9, 19.7
unpolluted: 14.2, 18.3, 17.2, 18.4, 20.0

Does the data indicate strongly that the true average fluoride concentration for livestock grazing in the polluted region is larger than for the unpolluted region? Use the Wilcoxon rank-sum test at level $\alpha = .05$.

  The sample sizes here are 7 and 5. To obtain $m \leq n$, label the unpolluted observations as the x's ($x_1 = 14.2, \ldots, x_5 = 20.0$) and the polluted observations as the y's. Thus μ_1 is the true average fluoride concentration without pollution and μ_2 the true average concentration with pollution. The alternative hypothesis is H_a: $\mu_1 - \mu_2 < 0$ (pollution causes an increase in concentration), so a lower-tailed test is appropriate. Entering Appendix Table A.10 with $m = 5$ and $n = 7$, $P(W \geq 47$ when H_0 is true$) \doteq .01$. The critical value for the lower-tailed test is $m(m + n + 1) - 47 = 5(13) - 47 = 18$; H_0 will now be rejected if $W \leq 18$. The pooled ordered sample appears below; the computed W is $w = r_1 + r_2 + \cdots + r_5$ (where r_i is the rank of x_i) = $1 + 5 + 4 + 6 + 9 = 25$. Since 25 is not ≤ 18, H_0 is not rejected at (approximately) level .01.

x	y	y	x	x	x	y	y	x	y	y	y
14.2	16.8	17.1	17.2	18.3	18.4	18.7	19.7	20.0	20.9	21.3	23.0
1	2	3	4	5	6	7	8	9	10	11	12

A Normal Approximation for *W*

When both m and n exceed 8, the distribution of W can be approximated by an appropriate normal curve, and this approximation can be used in place of Appendix Table A.10 (or the more extensive tables which can be found in the book by Hollander and Wolfe). To obtain the approximation, we need μ_W and σ_W^2 when H_0 is true. In this case the rank R_i of $X_i - \Delta_0$ is equally likely to be any one of the possible values $1, 2, 3, \ldots, m + n$ (R_i has a discrete uniform distribution on the first $m + n$ positive integers), so $\mu_{R_i} = (m + n + 1)/2$. This gives, since $W = \Sigma R_i$,

$$\mu_W = \mu_{R_1} + \mu_{R_2} + \cdots + \mu_{R_m} = \frac{m(m + n + 1)}{2} \tag{15.4}$$

The variance of R_i is also easily computed to be $(m + n + 1)(m + n - 1)/12$. However, because the R_i's are not independent, $\mathrm{Var}(W) \neq m\mathrm{Var}(R_i)$. Using the fact that for any two distinct integers a and b between 1 and $m + n$ inclusive, $P(R_i = a, R_j = b) = 1/(m + n)(m + n - 1)$ (two integers are being sampled without replacement), $\mathrm{Cov}(R_i, R_j) = -(m + n + 1)/12$, yielding

$$\sigma_W^2 = \sum_{i=1}^{m} \mathrm{Var}(R_i) + \sum\sum_{i \neq j} \mathrm{Cov}(R_i, R_j) = \frac{mn(m + n + 1)}{12} \tag{15.5}$$

A Central Limit Theorem can then be used to conclude that when H_0 is true, the **test statistic**

$$Z = \frac{W - m(m + n + 1)/2}{\sqrt{mn(m + n + 1)/12}}$$

has approximately a standard normal distribution. This Z statistic is used in conjunction with the critical values z_α, $-z_\alpha$, and $\pm z_{\alpha/2}$ for upper-, lower-, and two-tailed tests, respectively.

Example 15.8 A paper which appeared in the *Journal of Applied Physiology* ("Histamine Content in Sputum from Allergic and Non-Allergic Individuals," 1969, pp. 535–539) reported the following data on sputum histamine level (micrograms per gram dry weight of sputum) for a sample of nine individuals classified as allergics and another sample of 13 individuals classified as nonallergics.

Allergics: 67.6, 39.6, 1651.0, 100.0, 65.9, 1112.0, 31.0, 102.4, 64.7
Nonallergics: 34.3, 27.3, 35.4, 48.1, 5.2, 29.1, 4.7, 41.7, 48.0, 6.6, 18.9, 32.4, 45.5

Does the data indicate that there is a difference in true average sputum histamine level between allergics and nonallergics?

Since both sample sizes exceed eight, we use the normal approximation. The null hypothesis is $H_0: \mu_1 - \mu_2 = 0$, and observed ranks of the x_i's are $r_1 = 18$, $r_2 = 11$, $r_3 = 22$, $r_4 = 19$, $r_5 = 17$, $r_6 = 21$, $r_7 = 7$, $r_8 = 20$, and $r_9 = 16$, so $w = \Sigma\, r_i = 151$. The mean and variance of W are given by $\mu_W = 9(23)/2 = 103.5$ and $\sigma_W^2 = 9(13)(23)/12 = 224.25$, so

$$z = \frac{151 - 103.5}{\sqrt{224.25}} = 3.17$$

The alternative hypothesis is $H_a: \mu_1 - \mu_2 \neq 0$, so at level .01 H_0 is rejected if either $Z \geq 2.58$ or $Z \leq -2.58$. Because $3.17 \geq 2.58$, H_0 is rejected, and we conclude that there is a difference in true average sputum histamine levels (the paper also used the Wilcoxon test).

Ties in the Wilcoxon Rank-Sum Test

Theoretically the assumption of continuity of the two distributions ensures that all $m + n$ observed x's and y's will have different values. In practice, though, there will often be ties in the observed values. As with the Wilcoxon signed-rank test, the common practice in dealing with ties is to assign each of the tied observations in a particular set of ties the average of the ranks they would receive if they differed very slightly from one another.

If the sample sizes both exceed eight, the numerator of Z is still appropriate but the denominator should be replaced by the square root of the adjusted variance

$$\sigma_W^2 = \frac{mn(m + n + 1)}{12} - \frac{mn}{12(m + n)(m + n - 1)} \sum (\tau_i - 1)(\tau_i)(\tau_i + 1) \tag{15.6}$$

where τ_i is the number of tied observations in the ith set of ties and the sum is over all sets of ties. Unless there are a great many ties, there is little difference between (15.6) and (15.5).

Efficiency of the Rank-Sum Test

When the distributions being sampled are both normal with $\sigma_1^2 = \sigma_2^2$, both the two-sample t test and the Wilcoxon rank-sum test can be used, but the t test is known to be the best among all possible tests in the sense of maximizing power for a fixed α. Because the Wilcoxon test is valid for a much broader class of underlying distributions than just the normal, it might be used if the validity of the normality assumption is open to question. One aspect of efficiency relates to how much is lost by using W instead of T when normality is correct. The other aspect of efficiency is concerned with the performance of W relative to T in nonnormal situations.

In the previous section the notion of efficiency of test procedures was discussed in connection with the one-sample t test and Wilcoxon signed-rank test. The results for two-sample tests are the same as for the one-sample tests. When the normailty assumption is correct, the rank-sum test is approximately 95% as efficient as the t test in large samples—meaning that the t test can give the same power as the

Wilcoxon test using slightly smaller sample sizes. On the other hand, the Wilcoxon test will always be at least 86% as efficient as the t test and may be many times as efficient if the underlying distributions are very nonnormal.

In summary, the Wilcoxon test never gives up very much to the t test and may be a distinct improvement on it, so if there is much doubt about the assumption of normality, the Wilcoxon test provides a good alternative means of analysis.

Lastly we note that power calculations for the Wilcoxon test are quite difficult, because the distribution of W when H_0 is false depends not only on $\mu_1 - \mu_2$ but also on the shapes of the two distributions. For most underlying distributions, the nonnull distribution of W is virtually intractable. This is why statisticians have developed large-sample (asymptotic relative) efficiency as a means of comparing tests.

Exercises / Section 15.3

1. In an experiment to compare the bond strength of two different adhesives, each adhesive was used in five bondings of two surfaces, and the force necessary to separate the surfaces was determined for each bonding. For adhesive 1 the resulting values were 229, 286, 245, 299, and 250, while the adhesive 2 observations were 213, 179, 163, 247, and 225. Let μ_i denote the true average bond strength of adhesive type i. Use the Wilcoxon rank-sum test at level .05 to test H_0: $\mu_1 = \mu_2$ versus H_a: $\mu_1 > \mu_2$.

2. The article "A Study of Wood Stove Particulate Emissions" (*J. Air Pollution Control Assn.*, 1979, pp. 724–728) reported the following data on burn time (hours) for samples of oak and pine. Test at level .05 to see whether or not there is any difference in true average burn time for the two types of wood.

   Oak: 1.72, .67, 1.55, 1.56, 1.42, 1.23, 1.77, .48
   Pine: .98, 1.40, 1.33, 1.52, .73, 1.20

3. A modification has been made to the process for producing a certain type of "time-zero" film (film that begins to develop as soon as a picture is taken). Because the modification involves extra cost, it will be incorporated only if sample data strongly indicates that the modification has decreased true average developing time by more than one second. Assuming that both developing time distributions differ only with respect to location if at all, use the Wilcoxon rank-sum test at level .05 on the accompanying data to test the appropriate hypotheses.

   Original process: 8.6, 5.1, 4.5, 5.4, 6.3, 6.6, 5.7, 8.5
   Modified process: 5.5, 4.0, 3.8, 6.0, 5.8, 4.9, 7.0, 5.7

4. The accompanying data resulted from an experiment to compare the effects of vitamin C in orange juice and in synthetic ascorbic acid on the length of odontoblasts in guinea pigs over a six-week period ("The Growth of the Odontoblasts of the Incisor Tooth as a Criterion of the Vitamin C Intake of the Guinea Pig," *J. Nutrition*, 1947, pp. 491–504). Use the Wilcoxon rank-sum test at level .01 to decide whether or not true average length differs for the two types of vitamin C intake. Compute also an approximate P-value.

   Orange juice: 8.2, 9.4, 9.6, 9.7, 10.0, 14.5, 15.2, 16.1, 17.6, 21.5
   Ascorbic acid: 4.2, 5.2, 5.8, 6.4, 7.0, 7.3, 10.1, 11.2, 11.3, 11.5

5. Test the hypotheses suggested in Exercise 4 using the following data:

   Orange juice: 8.2, 9.5, 9.5, 9.7, 10.0, 14.5, 15.2, 16.1, 17.6, 21.5
   Ascorbic acid: 4.2, 5.2, 5.8, 6.4, 7.0, 7.3, 9.5, 10.0, 11.5, 11.5

6. Suppose that we wish to test

   H_0: the X and Y distributions are identical

   versus

   H_a: the X distribution is less spread out than the Y distribution

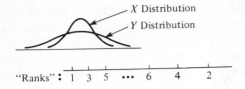

"Ranks": 1   3   5 ··· 6   4   2

The accompanying figure pictures X and Y distributions for which H_a is true. The Wilcoxon rank-sum test is not appropriate in this situation because when H_a is true as pictured, the Y's will tend to be at the extreme ends of the combined sample (resulting in small and large Y ranks), so the sum of X ranks will result in a W value which is neither large nor small.

Consider modifying the procedure for assigning ranks as follows: After ordering the combined sample of $m + n$ observations, the smallest observation is given rank 1, the largest observation is given rank 2, the second smallest is given rank 3, the second largest is given rank 4, and so on. Then if H_a is true as pictured, the X values will tend to be in the middle of the sample and thus receive large ranks. Let W' denote the sum of the X ranks, and consider rejecting H_0 in favor of H_a when $W' \geq c$. When H_0 is true, any possible set of X ranks has the same probability, so W' *has the same distribution as does W when H_0 is true.* Thus c can be chosen from Appendix Table A.10 to yield a level

α test. The data below refers to medial muscle thickness for arterioles from the lungs of both children who died from sudden infant death syndrome (x's) and a control group of children (y's). Carry out the test of H_0 versus H_a at level .05.

SIDS: 4.0, 4.4, 4.8, 4.9
Control: 3.7, 4.1, 4.3, 5.1, 5.6

See Lehmann for more information on this, called the Siegel-Tukey test.

7. The ranking procedure described in Exercise 6 is somewhat asymmetrical, since the smallest observation receives rank 1 while the largest receives rank 2, and so on. Suppose that both the smallest and the largest receive rank 1, the second smallest and second largest receive rank 2, and so on, and let W'' be the sum of the X ranks. The null distribution of W'' is no longer identical to the null distribution of W, so different tables are needed. Consider the case $m = 3$, $n = 4$. List all 35 possible orderings of the three X values among the seven observations (say, 1, 3, 7 or 4, 5, 6), assign ranks in the manner described, compute the value of W'' for each possibility, and then tabulate the null distribution of W''. For the test which rejects if $W'' \geq c$, what value of c prescribes approximately a level .10 test? This is the Ansari-Bradley test; see the book by Hollander and Wolfe.

15.4 Distribution-Free Confidence Intervals

The method we have used so far to construct a confidence interval can be described as follows: Start with a random variable (Z, T, χ^2, F, or the like) that depends on the parameter of interest and a probability statement involving the variable, manipulate the inequalities of the statement to isolate the parameter between random endpoints, and finally substitute computed values for random variables. Another general method for obtaining confidence intervals takes advantage of a relationship between test procedures and confidence intervals; a $100(1 - \alpha)\%$ confidence interval for a parameter θ can be obtained from a level α test for H_0: $\theta = \theta_0$ versus H_a: $\theta \neq \theta_0$. This method will be used to derive intervals associated with the sign test, the Wilcoxon signed-rank test, and the Wilcoxon rank-sum test.

Before using the method to derive new intervals, reconsider the t test and the t interval. Suppose that a random sample of $n = 25$ observations from a normal population yields summary statistics $\bar{x} = 100$, $s = 20$. Then a 90% confidence interval for μ is

$$\left(\bar{x} - t_{.05,24} \cdot \frac{s}{\sqrt{25}}, \bar{x} + t_{.05,24} \cdot \frac{s}{\sqrt{25}} \right) = (93.16, \ 106.84) \tag{15.7}$$

Suppose that instead of a confidence interval, we had wished to test a hypothesis about μ. For $H_0: \mu = \mu_0$ versus $H_a: \mu \neq \mu_0$, the t test at level .10 specifies that H_0 should be rejected if T is either ≥ 1.711 or ≤ -1.711, where the computed T is

$$t = \frac{\bar{x} - \mu_0}{s/\sqrt{25}} = \frac{100 - \mu_0}{20/\sqrt{25}} = \frac{100 - \mu_0}{4} \tag{15.8}$$

Consider now the null value $\mu_0 = 95$. Then $t = 1.25$, so H_0 is not rejected. Similarly, if $\mu_0 = 104$, then $t = -1$, so again H_0 is not rejected. However, if $\mu_0 = 90$, then $t = 2.5$, so H_0 is rejected, and if $\mu_0 = 108$, then $t = -2$, so H_0 is again rejected. By considering other values of μ_0 and the decision resulting from each one, the following general fact emerges: *Every number inside the interval (15.7) specifies a value of μ_0 for which t of (15.8) leads to acceptance of H_0, while every number outside (15.7) corresponds to a t for which H_0 is rejected.* That is, for the fixed values of n, $\bar{x}$, and s, the set of all μ_0 values for which $H_0: \mu = \mu_0$ versus $H_a: \mu \neq \mu_0$ is accepted is precisely the interval (15.7).

> **Proposition:** Suppose that we have a level α test procedure for testing $H_0: \theta = \theta_0$ versus $H_a: \theta \neq \theta_0$. For fixed sample values, let A denote the set of all values θ_0 for which H_0 is accepted. Then A is a $100(1 - \alpha)\%$ confidence interval for θ.

There are actually pathological examples in which the set A defined in the proposition is not an interval of θ values, but instead the complement of an interval or something even stranger. To be more precise, we should really replace the notion of a confidence interval with that of a confidence set. In the three cases of interest dealt with below, the set A does turn out to be an interval.

The Sign Interval for $\tilde{\mu}$

Let the sample be denoted by $X_1, \ldots, X_n$. Then for testing $H_0: \tilde{\mu} = \tilde{\mu}_0$ versus $H_a: \tilde{\mu} \neq \tilde{\mu}_0$, the sign test statistic is Y = number of X_i's which are $> \tilde{\mu}_0$ [the number of $(X_i - \tilde{\mu}_0)$'s which have positive signs]. H_0 is rejected if $Y \geq c$ or $\leq n - c$; under H_0, Y has a binomial distribution with $p = .5$ and n, and c is determined by $P(Y \geq c) + P(Y \leq n - c) = \alpha$.

Example 15.9 The daily dietetic intake of nicotinic acid-equivalents (mg) was determined for a sample of 12 healthy individuals ("Assessment of Nicotinic Acid Status of Population Groups," *Amer. J. Clinical Nutrition*, 1964, pp. 169–174). The sample observations, listed in order of increasing magnitude, were

16.6, 19.4, 21.7, 23.4, 24.2, 26.9, 27.3, 27.4, 28.0, 29.5, 30.7, and 58.8

To obtain a $100(1 - \alpha)\%$ confidence interval for $\tilde{\mu}$ = the true median daily nicotinic acid-equivalent intake, we must use the sign test at level α to test $H_0: \tilde{\mu} = \tilde{\mu}_0$ versus $H_a: \tilde{\mu} \neq \tilde{\mu}_0$ for various values of $\tilde{\mu}_0$. The confidence interval is then the set $A = \{$all $\tilde{\mu}_0: H_0$ is accepted$\}$.

For $n = 12$ and $p = .5$, $P(Y \le 2) + P(Y \ge 10) = .0193 + .0193 = .0396 \doteq .04$. If the rejection region for the sign test is $R = \{0, 1, 2, 10, 11, 12\}$, the test has (approximately) level .04, so the confidence level of the resulting interval will be 96%. The confidence interval is most easily identified from the plot of the data in Figure 15.5.

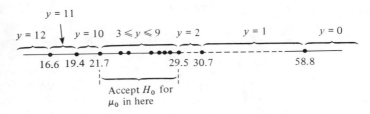

Figure 15.5 Plot of the data for Example 15.9

If $\widetilde{\mu}_0 < 21.7$, then $y =$ number of x_i's $> \widetilde{\mu}_0$ is at least 10 (10, 11, or 12), so H_0 would be rejected. Similarly, if $\widetilde{\mu}_0 > 29.5$, then y is at most 2 (0, 1, or 2), so again H_0 is rejected. However, as Figure 15.5 indicates, if $\widetilde{\mu}_0$ is between 21.7 and 29.5, then $3 \le y \le 9$ so H_0 is accepted. The set A is then $\{\widetilde{\mu}_0: 21.7 < \widetilde{\mu}_0 < 29.5\}$, so our 96% confidence interval for $\widetilde{\mu}$ is (21.7, 29.5).

Example 15.9 shows that after the data values are ordered from smallest to largest, the interval for $\widetilde{\mu}$ includes all numbers between two of these ordered values.

Proposition: Denote the ordered sample values by $x_{(1)}, x_{(2)}, \ldots x_{(n)}$, where $x_{(1)}$ is the smallest, $x_{(2)}$ the next smallest, and so on. Suppose that the level α sign test rejects H_0: $\widetilde{\mu} = \widetilde{\mu}_0$ if either $Y \ge c$ or $\le n - c$. Then a $100(1 - \alpha)\%$ confidence interval for $\widetilde{\mu}$ is $(x_{(n-c+1)}, x_{(c)})$.

Example 15.9 (continued)

For $n = 12$ and $c = 10$, the level of significance is .04. Therefore the $100(1 - .04) = 96\%$ confidence interval is $(x_{(12-10+1)}, x_{(10)}) = (x_{(3)}, x_{(10)}) = (21.7, 29.5)$. The sample mean and sample standard deviation are easily computed as $\bar{x} = 27.83$ and $s = 10.61$, from which the 95% t interval for μ, assuming the distribution to be normal, is (21.09, 34.57). That this t interval is much longer than the sign interval is due primarily to the single outlying value 58.8.

Notice that in contrast to the t interval, the sign interval is relatively insensitive to a few outlying values. In Example 15.9 the two extreme values on each end of the sample can be moved further from the middle without affecting the interval. When the population is normal, $\widetilde{\mu} = \mu$, so the sign interval is a competitor to the t interval. In this case the sign interval will usually be longer than the t interval, since outlying values in a normal sample are very unusual. The Wilcoxon interval, to be described next, is a better competitor of the t interval.

Finally, because of the discreteness of the binomial distribution, the usual confidence levels are not achievable. In Example 15.9 we settled for 96% rather than 95% confidence.

The Wilcoxon Signed-Rank Interval

To test H_0: $\mu = \mu_0$ versus H_a: $\mu \neq \mu_0$ using the Wilcoxon signed-rank test, where μ is the mean of a continuous symmetric distribution, the absolute values $|x_1 - \mu_0|, \ldots, |x_n - \mu_0|$ are ordered from smallest to largest, with the smallest receiving rank 1 and the largest rank n. Each rank is then given the sign of its associated $x_i - \mu_0$, and the test statistic S_+ is the sum of the positively signed ranks. The two-tailed test rejects H_0 if S_+ is either $\geq c$ or $\leq n(n + 1)/2 - c$, where c is obtained from Appendix Table A.9 once the desired level of significance α is specified.

Just as was the case for the sign interval, for fixed $x_1, \ldots, x_n$ the $100(1 - \alpha)\%$ signed-rank interval will consist of all μ_0 for which H_0: $\mu = \mu_0$ is accepted at level α. To identify this interval, it is convenient to express the test statistic S_+ in another form:

$$S_+ = \text{the number of pairwise averages } (X_i + X_j)/2 \text{ with } i \leq j \text{ which are } \geq \mu_0$$
$$(15.9)$$

That is, if we average each X_j in the list with each X_i to its left, including $(X_j + X_j)/2$ (which is just X_j), and count the number of these averages which are $\geq \mu_0$, S_+ results. In moving from left to right in the list of sample values, we are simply averaging every pair of observations in the sample [again including $(X_j + X_j)/2$] exactly once, so the order in which the observations are listed before averaging is not important. The equivalence of the two methods for computing S_+ is not difficult to verify. The number of pairwise averages is $\binom{n}{2} + n$ (the first term due to averaging of different observations and the second due to averaging each x_i with itself), which equals $n(n + 1)/2$. If either too many or two few of these pairwise averages are $\geq \mu_0$, H_0 is rejected.

Example 15.10 The following data was introduced in Example 1.12 and refers to cerebral metabolic rate for rhesus monkeys: $x_1 = 4.51$, $x_2 = 4.59$, $x_3 = 4.90$, $x_4 = 4.93$, $x_5 = 6.80$, $x_6 = 5.08$, $x_7 = 5.67$. There are 28 pairwise averages which are, in increasing order

4.51, 4.55, 4.59, 4.705, 4.72, 4.745, 4.76, 4.795, 4.835, 4.90, 4.915, 4.93, 4.99, 5.005, 5.08, 5.09, 5.13, 5.285, 5.30, 5.375, 5.655, 5.67, 5.695, 5.85, 5.865, 5.94, 6.235, 6.80

The first few and the last few of these are pictured on a measurement axis in Figure 15.6.

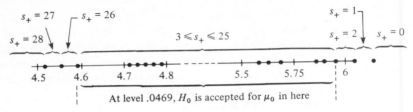

Figure 15.6 Plot of the data for Example 15.10

Because of the discreteness of the distribution of S_+, $\alpha = .05$ cannot be obtained exactly. The rejection region $\{0, 1, 2, 26, 27, 28\}$ has $\alpha = .0469$, which is as close as possible to .05, so the level is approximately .05. Thus if the number of pairwise averages $\geq \mu_0$ is between 3 and 25, inclusive, H_0 is accepted. From Figure 15.6 the (approximate) 95% confidence interval for μ is (4.59, 5.94).

In general, once the pairwise averages are ordered from smallest to largest, the endpoints of the Wilcoxon interval are two of the "extreme" averages. To express this precisely, let the smallest pairwise average be denoted by $\overline{x}_{(1)}$, the next smallest by $\overline{x}_{(2)}, \ldots$, and the largest by $\overline{x}_{(n(n+1)/2)}$.

> **Proposition:** If the level α Wilcoxon signed rank test for H_0: $\mu = \mu_0$ versus H_a: $\mu \neq \mu_0$ is to reject H_0 if either $S_+ \geq c$ or $S_+ \leq n(n + 1)/2 - c$, then a $100(1 - \alpha)\%$ confidence interval for μ is
>
> $$(\overline{x}_{(n(n+1)/2-c+1)}, \overline{x}_{(c)}) \tag{15.10}$$

In words, the interval extends from the dth smallest pairwise average to the dth largest average, where $d = n(n + 1)/2 - c + 1$. Appendix Table A.11 gives for $n = 5, 6, \ldots, 25$ the values of c which correspond to the usual confidence levels.

Example 15.10 (continued) For $n = 7$, an 89.1% interval (approximately 90%) is obtained by using $c = 24$ (since the rejection region $\{0, 1, 2, 3, 4, 24, 25, 26, 27, 28\}$ has $\alpha = .109$). The interval is $(\overline{x}_{(28-24+1)}, \overline{x}_{(24)}) = (\overline{x}_{(5)}, \overline{x}_{(24)}) = (4.72, 5.85)$, which extends from the fifth smallest to the fifth largest pairwise average.

Other Uses and Properties of the Signed-Rank Interval

While the derivation of the interval depended on having a single sample from a continuous symmetric distribution with mean (median) μ, when the data is paired, the interval constructed from the differences $d_1, d_2, \ldots, d_n$ is a confidence interval for the mean (median) difference μ_D. In this case the symmetry of X and Y distributions need not be assumed; as long as the X and Y distributions have the same shape, the $X - Y$ distribution will be symmetric, so only continuity is required.

For $n > 20$, the large-sample approximation to the Wilcoxon test based on standardizing S_+ gives an approximation to c in (15.10). The result [for a $100(1 - \alpha)\%$ interval] is

$$c \doteq \frac{n(n + 1)}{4} + z_{\alpha/2} \sqrt{\frac{n(n + 1)(2n + 1)}{24}}$$

The efficiency of the Wilcoxon interval relative to the t interval is roughly the same as that for the Wilcoxon test relative to the t test. In particular, for large samples when the underlying population is normal the Wilcoxon interval will tend to be slightly longer than the t interval, but if the population is quite nonnormal (symmetric, but with heavy tails), then the Wilcoxon interval will tend to be much shorter than the t interval. Again the books by Hollander and Wolfe and by Daniel provide more information on this and other distribution-free procedures.

The Wilcoxon Rank-Sum Interval

The Wilcoxon rank-sum test for testing $H_0: \mu_1 - \mu_2 = \Delta_0$ is carried out by first combining the $(X_i - \Delta_0)$'s and Y_j's into one sample of size $m + n$ and ranking them from smallest (rank 1) to largest (rank $m + n$). The test statistic W is then the sum of the ranks of the $(X_i - \Delta_0)$'s. For the two-sided alternative, H_0 is rejected if W is either too small or too large.

To obtain the associated confidence interval, for fixed x_i's and y_j's we must determine the set of all Δ_0 values for which H_0 is accepted. This is easiest to do if we first express the test statistic in a slightly different form. The smallest possible value of W is $m(m + 1)/2$, corresponding to every $(X_i - \Delta_0)$ less than every Y_j, and there are mn differences of the form $(X_i - \Delta_0) - Y_j$. A bit of manipulation gives

$$W = [\text{number of } (X_i - Y_j - \Delta_0)\text{'s} \geq 0] + \frac{m(m + 1)}{2}$$

$$= [\text{number of } (X_i - Y_j)\text{'s} \geq \Delta_0] + \frac{m(m + 1)}{2}$$

(15.11)

Thus rejecting H_0 if the number of $(X_i - Y_j)$'s $\geq \Delta_0$ is either too small or too large is equivalent to rejecting H_0 for small or large W.

Expression (15.11) suggests that we compute $x_i - y_j$ for each i and j, and order these mn differences from smallest to largest. Then if the null value Δ_0 is not either smaller than most of these differences or larger than most, $H_0: \mu_1 - \mu_2 = \Delta_0$ is accepted. Varying Δ_0 now shows that a confidence interval for $\mu_1 - \mu_2$ will have as its lower endpoint one of the ordered $(x_i - y_j)$'s, and similarly for the upper endpoint.

Proposition: Let $x_1, \ldots, x_m$ and $y_1, \ldots, y_n$ be the observed values in two independent samples from continuous distributions which differ only in location (and not in shape). With $d_{ij} = x_i - y_j$ and the ordered differences denoted by $d_{ij(1)}, d_{ij(2)}, \ldots, d_{ij(mn)}$, the general form of a $100(1 - \alpha)\%$ confidence interval for $\mu_1 - \mu_2$ is

$$(d_{ij(mn-c+1)}, d_{ij(c)}) \tag{15.12}$$

where c is the critical constant for the two-tailed level α Wilcoxon rank-sum test.

Notice that the form of the Wilcoxon rank-sum interval (15.12) is very similar to the Wilcoxon signed rank interval (15.10); (15.10) uses pairwise averages from a single sample, while (15.12) uses pairwise differences from two samples. Appendix Table A.12 gives values of c for selected values of m and n.

Example 15.11 The paper "Some Mechanical Properties of Impregnated Bark Board (*Forest Products J.*, 1977, pp. 31–38) reported the following data on maximum crushing strength (psi) for a sample of epoxy-impregnated barkboard and for a sample of barkboard impregnated with another polymer:

Epoxy (x's): 10,860, 11,120, 11,340, 12,130, 14,380, 13,070
Other (y's): 4,590, 4,850, 6,510, 5,640, 6,390

Obtain a 95% confidence interval for the true average difference in crushing strength between the epoxy-impregnated board and the other type of board.

From Appendix Table A.12, since the smaller sample size is five and the larger sample size is six, $c = 26$ for a confidence level of approximately 95%. The d_{ij}'s appear in Figure 15.7.

d_{ij}	4590	4850	y_j 5640	6390	6510
10,860	6270	6010	5220	4470	4350
11,120	6530	6270	5480	4730	4610
11,340	6750	6490	5700	4950	4830
x_i 12,130	7540	7280	6490	5740	5620
13,070	8480	8220	7430	6680	6560
14,380	9790	9530	8740	7990	7870

Figure 15.7

The five smallest d_{ij}'s $[d_{ij(1)}, \ldots, d_{ij(5)}]$ are 4350, 4470, 4610, 4730, and 4830, while the five largest d_{ij}'s are (in descending order) 9790, 9530, 8740, 8480, and 8220. Thus the confidence interval is $(d_{ij(5)}, d_{ij(26)}) = (4830, 8220)$.

Other Properties of the Rank-Sum Interval

When m and n are both large, the Wilcoxon test statistic has approximately a normal distribution. This can be used to derive a large-sample approximation for the value c in (15.12). The result is

$$c \doteq \frac{mn}{2} + z_{\alpha/2} \sqrt{\frac{mn(m + n + 1)}{12}} \tag{15.13}$$

As with the signed-rank interval, the rank-sum interval (15.12) is quite efficient with respect to the t interval; in large samples (15.12) will tend to be only a bit longer than the t interval when the underlying populations are normal, and may be considerably shorter than the t interval if the underlying populations have heavier tails than do normal populations.

Exercises / Section 15.4

1. Use the interval associated with the sign test to compute a 99% confidence interval for the true median travel time $\tilde{\mu}$ using the data in Exercise 15.1.2.

2. Compute a 98% confidence interval for the true median soil pH $\tilde{\mu}$ using the data in Exercise 15.1.3. For the given sample size, can a 95% confidence interval be obtained? Explain.

3. The paper "The Lead Content and Acidity of Christchurch Precipitation" (*New Zealand J. Science*, 1980, pp. 311–312) reported the accompanying data on lead concentration ($\mu g/1$) in samples gathered during eight different summer rainfalls. Assuming that the lead content distribution is symmetric, use the Wilcoxon signed-rank interval to obtain a 95% confidence interval for μ.

   17.0, 21.4, 30.6, 5.0, 12.2, 11.8, 17.3, 18.8

4. Compute the 99% signed-rank interval for true average pH μ (assuming symmetry) using the data in Exercise 15.2.3. *Hint:* Try to compute only those pairwise averages having relatively small or large

   values (rather than all 105 averages).

5. Compute a 94% confidence interval for μ_D of Example 15.5 using the data given there.

6. The following observations are amounts of hydrocarbon emissions resulting from road wear of bias-belted tires under a 522 kg load inflated at 228 kPa and driven at 64 km/hr for six hours ("Characterization of Tire Emissions Using an Indoor Test Facility," *Rubber Chemistry and Technology*, 1978, pp. 7–25):

   .045, .117, .062, .072

   What confidence levels are achievable for this sample size using the signed-rank interval? Select an appropriate confidence level and compute the interval.

7. Compute the 90% rank-sum confidence interval for $\mu_1 - \mu_2$ using the data in Exercise 15.3.1.

8. Compute a 99% confidence interval for $\mu_1 - \mu_2$ using the data in Exercise 15.3.2.

15.5 Distribution-Free Analysis of Variance

The single-factor ANOVA model of Chapter 10 for comparing I population or treatment means assumed that for $i = 1, 2, \ldots, I$, a random sample of size J_i was drawn from a normal population with mean μ_i and variance σ^2. This can be written as

$$X_{ij} = \mu_i + \epsilon_{ij} \quad j = 1, \ldots, J_i; \quad i = 1, \ldots, I \qquad (15.14)$$

where the ϵ_{ij}'s are independent and normally distributed with mean zero and variance σ^2. While the normality assumption was required for the validity of the F test described in Chapter 10, the validity of the Kruskal-Wallis test for testing equality of the μ_i's depends only on the ϵ_{ij}'s having the same continuous distribution.

The Kruskal-Wallis Test

Let $N = \Sigma J_i$, the total number of observations in the data set, and suppose that we rank all N observations from 1 (the smallest X_{ij}) to N (the largest X_{ij}). When $H_0: \mu_1 = \mu_2 = \cdots = \mu_i$ is true, the N observations all come from the same distribution, in which case all possible assignments of the ranks 1, 2, . . ., N to the I samples are equally likely and we expect ranks to be intermingled in these samples. If, however, H_0 is false, then some samples will consist mostly of observations having small ranks in the combined sample while others will consist mostly of observations having large ranks. More specifically, if R_{ij} denotes the rank of X_{ij} among the N observations and $R_i.$ and $\overline{R}_i.$ denote respectively the total and average of the ranks in the ith sample, then when H_0 is true

$$E(R_{ij}) = \frac{N + 1}{2}$$

$$E(\overline{R}_i.) = \frac{1}{J_i} \sum_J E(R_{ij}) = \frac{N + 1}{2}$$

The K-W test statistic is a measure of the extent to which the $\overline{R}_i.$'s deviate from their common expected value $(N + 1)/2$, and H_0 is rejected if the computed value of the statistic indicates too great a discrepancy between observed and expected rank averages.

$$
\text{Test statistic:} \quad K = \frac{12}{N(N + 1)} \sum_{j=1}^{I} J_i \left(\overline{R}_i. - \frac{N + 1}{2} \right)^2
$$
$$
= \frac{12}{N(N + 1)} \sum_{i=1}^{I} \frac{R_i.^2}{J_i} - 3(N + 1)
$$

(15.15)

The second expression for K is the computational formula; it involves the rank totals ($R_i.$'s) rather than the averages and requires only one subtraction, so is analogous to computational formulas for ANOVA sums of squares.

If H_0 is rejected when $K \geq c$, then c should be chosen so that the test has level α. That is, c should be the upper-tail critical value of the distribution of K when H_0 is true. Under H_0, each possible assignment of the ranks to the I samples is equally likely, so in theory all such assignments can be enumerated, the value of K determined for each one, and the null distribution obtained by counting the number of times each value of K occurs. Clearly this computation is tedious, so while there are tables of the exact null distribution and critical values for small values of the J_i's, we shall use the following "large-sample" approximation.

> **Proposition:** When H_0 is true and either
>
> $$I = 3, J_i \geq 6 \quad i = 1, 2, 3, \text{ or}$$
>
> $$I > 3, J_i \geq 5 \quad i = 1, \ldots, I$$
>
> then K has approximately a chi-squared distribution with $I - 1$ d.f.

This result implies that an approximate level α test is given by

> reject H_0 if $K \geq \chi^2_{\alpha, I-1}$ (15.16)

Example 15.12 The accompanying observations on axial stiffness index resulted from a study of metal plate connected trusses in which five different plate lengths—4 in., 6 in., 8 in., 10 in., and 12 in.—were used ("Modeling Joints Made with Light-Gauge Metal Connector Plates," *Forest Products J.*, 1979, pp. 39–44).

$$
\begin{array}{llllllll}
i = 1 \ (4''): & 309.2 & 309.7 & 311.0 & 316.8 & 326.5 & 349.8 & 409.5 \\
i = 2 \ (6''): & 331.0 & 347.2 & 348.9 & 361.0 & 381.7 & 402.1 & 404.5 \\
i = 3 \ (8''): & 351.0 & 357.1 & 366.2 & 367.3 & 382.0 & 392.4 & 409.9 \\
i = 4 \ (10''): & 346.7 & 362.6 & 384.2 & 410.6 & 433.1 & 452.9 & 461.4 \\
i = 5 \ (12''): & 407.4 & 410.7 & 419.9 & 441.2 & 441.8 & 465.8 & 473.4 \\
\end{array}
$$

									$R_{i\cdot}$	$\overline{R}_{i\cdot}$
	$i = 1$:	1	2	3	4	5	10	24	49	7.00
	$i = 2$:	6	8	9	13	17	21	22	96	13.71
Ranks	$i = 3$:	11	12	15	16	18	20	25	117	16.71
	$i = 4$:	7	14	19	26	29	32	33	160	22.86
	$i = 5$:	23	27	28	30	31	34	35	208	29.71

Figure 15.8

The computed value of K is

$$
k = \frac{12}{35(36)} \left[\frac{(49)^2}{7} + \frac{(96)^2}{7} + \frac{(117)^2}{7} + \frac{(160)^2}{7} + \frac{(208)^2}{7} \right] - 3(36) = 20.12
$$

At level .01, $\chi^2_{.01,4} = 13.28$, and since 20.12 is ≥ 13.28, H_0 is rejected and we conclude that expected axial stiffness does depend on plate length.

When the data set contains ties, ranks are given just as in the Wilcoxon rank-sum test—by assigning to each tied value the average of the ranks that the values would receive if they differed slightly from one another. The test statistic is then adjusted for ties by dividing by the adjustment factor $1 - [\Sigma_i(\tau_i - 1)\tau_i(\tau_i + 1)]/(N^3 - N)$, where τ_i is the number of ties in the ith group of ties and the sum is over all groups of ties.

Multiple Comparisons

When the K-W test rejects H_0, one would usually wish to further analyze the data to identify treatments or populations which differ significantly from one another. One procedure is based on the fact that when H_0 is true, $\overline{R}_{i\cdot} - \overline{R}_{i'\cdot}$ has approximately a normal distribution with mean zero and variance $N(N + 1)[(1/J_i) + (1/J_{i'})]/12$, so that

$$Z = \frac{\overline{R}_{i\cdot} - \overline{R}_{i'\cdot}}{\sqrt{\dfrac{N(N + 1)}{12}\left(\dfrac{1}{J_i} + \dfrac{1}{J_{i'}}\right)}}$$

has approximately a standard normal distribution. It then follows that when H_0 is true,

$$\left|\overline{R}_{i\cdot} - \overline{R}_{i'\cdot}\right| \le z_{\alpha/2m} \sqrt{\frac{N(N + 1)}{12}\left(\frac{1}{J_i} + \frac{1}{J_{i'}}\right)} \tag{15.17}$$

for every pair i, i' with probability at least $1 - \alpha$, where $m = I(I - 1)/2$ is the number of pairs to be compared. Every pair i, i' for which the inequality (15.17) is *not* satisfied identifies a pair μ_i, $\mu_{i'}$ which are said to differ significantly from one another. As with other multiple comparison procedures, α here refers to an experimentwise, as opposed to per comparison, error rate, so α is often taken to be .10 if many comparisons are involved. Also, if all J_i's are equal, the square root in (15.17) need be computed only once.

Example 15.12 (continued) With $I = 5$, $m = 10$, so for $\alpha = .10$, $\alpha/2m = .10/20 = .005$, and $z_{.005} = 2.58$. Since $J_i = 7$ for $i = 1, \ldots, 5$,

$$z_{\alpha/2m} \sqrt{\frac{N(N + 1)}{12}\left(\frac{1}{J_i} + \frac{1}{J_{i'}}\right)} = 2.58 \sqrt{\frac{(35)(36)}{12}\left(\frac{1}{7} + \frac{1}{7}\right)} = 14.13$$

The rank averages are $\overline{R}_{1\cdot} = 7.00$, $\overline{R}_{2\cdot} = 13.71$, $\overline{R}_{3\cdot} = 16.71$, $\overline{R}_{4\cdot} = 22.86$, $\overline{R}_{5\cdot} = 29.71$, so $|\overline{R}_{i\cdot} - \overline{R}_{i'\cdot}|$ exceeds 14.13 for the (i, i') pairs $(1, 4)$, $(1, 5)$, and $(2, 5)$. Thus μ_1 differs significantly from μ_4 and μ_5, while μ_2 and μ_5 also differ significantly from one another.

Friedman's Test for a Randomized Block Experiment

Suppose that $X_{ij} = \mu + \alpha_i + \beta_j + \epsilon_{ij}$ where α_i is the ith treatment effect, β_j is the jth block effect, and the ϵ_{ij}'s are drawn independently from the same continuous (but not necessarily normal) distribution. Then to test $H_0: \alpha_1 = \alpha_2 = \cdots = \alpha_I = 0$, the null hypothesis of no treatment effects, the observations are first ranked separately from 1 to I within each block, and then the rank average $\overline{R}_{i\cdot}$ is computed for each of the I treatments. When H_0 is true, the $\overline{R}_{i\cdot}$'s should be close to one another, since within each block all $I!$ assignments of ranks to treatments are equally likely. Friedman's test statistic measures the discrepancy between the expected value $(I + 1)/2$ of each rank average and the observed $\overline{R}_{i\cdot}$'s.

$$\text{Test statistic: } F_r = \frac{12J}{I(I + 1)} \sum_{i=1}^{I} \left(\overline{R}_{i\cdot} - \frac{I + 1}{2}\right)^2$$

$$= \frac{12}{IJ(I + 1)} \sum R_{i\cdot}^2 - 3J(I + 1)$$

As with the K-W test, Friedman's test rejects H_0 when the computed value of the test statistic is too large. For the cases $I = 3$, $J = 2, \ldots, 15$ and $I = 4$, $J = 2, \ldots, 8$, Lehmann's book gives the upper-tail critical values for the test. Alternatively, for even moderate values of J, the test statistic F_r has approximately a chi-squared distribution with $I - 1$ d.f. when H_0 is true, so H_0 can be rejected if $F_r \geq \chi^2_{\alpha, I-1}$.

Example 15.13 The paper "Physiological Effects During Hypnotically Requested Emotions" (*Psychosomatic Med.*, 1963, pp. 334–343) reported the following data on skin potential (millivolts) when the emotions of fear, happiness, depression, and calmness were requested from each of eight subjects.

Blocks (Subjects)

x_{ij}	1	2	3	4	5	6	7	8		
Fear	23.1	57.6	10.5	23.6	11.9	54.6	21.0	20.3		
Happiness	22.7	53.2	9.7	19.6	13.8	47.1	13.6	23.6		
Depression	22.5	53.7	10.8	21.1	13.7	39.2	13.7	16.3		
Calmness	22.6	53.1	8.3	21.6	13.3	37.0	14.8	14.8		
Ranks	1	2	3	4	5	6	7	8	$R_{i\cdot}$	$R_{i\cdot}^2$
Fear	4	4	3	4	1	4	4	3	27	729
Happiness	3	2	2	1	4	3	1	4	20	400
Depression	1	3	4	2	3	2	2	2	19	361
Calmness	2	1	1	3	2	1	3	1	14	196
										1686

Figure 15.9

Thus

$$F_r = \frac{12}{4(8)(5)}(1686) - 3(8)(5) = 6.45$$

At level .05, $\chi^2_{.05,3} = 7.815$, and because 6.45 is not ≥ 7.815, H_0 is not rejected. We conclude that average skin potential does not depend on which emotion is requested.

The book by Hollander and Wolfe discusses multiple comparisons when H_0 is rejected by Friedman's test, as well as other aspects of distribution-free ANOVA.

Exercises / Section 15.5

1. The accompanying data refers to concentration of the radioactive isotope strontium-90 in milk samples obtained from five randomly selected dairies in each of four different regions.

Region
1. 6.4, 5.8, 6.5, 7.7, 6.1
2. 7.1, 9.9, 11.2, 10.5, 8.8
3. 5.7, 5.9, 8.2, 6.6, 5.1
4. 9.5, 12.1, 10.3, 12.4, 11.7

   a. Test at level .10 to see whether or not true average strontium-90 concentration differs for at least two of the regions.
   b. Use the multiple comparison procedure with $\alpha = .10$ to identify significantly different regions.

2. The paper "Production of Gaseous Nitrogen in Human Steady-State Conditions" (*J. Applied Physiology*, 1972, pp. 155–159) reported the following observations on the amount of nitrogen expired (in liters) under four dietary regimens: fasting (1), 23% protein (2), 32% protein (3), and 67% protein (4). Use the K-W test at level .05 to test equality of the corresponding μ_i's.

1.	4.079	4.859	3.540	5.047	3.298
2.	4.368	5.668	3.752	5.848	3.802
3.	4.169	5.709	4.416	5.666	4.123
4.	4.928	5.608	4.940	5.291	4.674

1.	4.679	2.870	4.648	3.847
2.	4.844	3.578	5.393	4.374
3.	5.059	4.403	4.496	4.688
4.	5.038	4.905	5.208	4.806

3. The accompanying data on cortisol level was reported in the paper "Cortisol, Cortisone, and 11-Deoxycortisol Levels in Human Umbilical and Maternal Plasma in Relation to the Onset of Labor" (*J. Obstetric Gynaecology British Commonwealth*, 1974, pp. 737–745). Experimental subjects were pregnant women who were delivered between 38 and 42 weeks gestation. Group 1 individuals elected to deliver by Caesarean section before labor onset, group 2 delivered by emergency Caesarean during induced labor, and group 3 individuals experienced spontaneous labor.

   a. Use the K-W test at level .05 to test for equality of the three population means.
   b. Use the multiple comparisons procedure with $\alpha = .05$ to identify significant differences.

Group 1: 262 307 211 323 454 339 304 154 287 356
Group 2: 465 501 455 355 468 362
Group 3: 343 772 207 1048 838 687

4. In a test to determine if soil pretreated with small amounts of Basic-H makes the soil more permeable to water, soil samples were divided into blocks and each block received each of the four treatments under study. The treatments were (A) water with .001% Basic-H flooded on control soil, (B) water without Basic-H on control, (C) water with Basic-H flooded on soil pretreated with Basic-H, and (D) water without Basic-H on soil pretreated with Basic-H. Test at level .01 to see if there are any effects due to the different treatments.

| | \multicolumn{5}{c}{Blocks} |
	1	2	3	4	5
A	37.1	31.8	28.0	25.9	25.5
B	33.2	25.3	20.2	20.3	18.3
C	58.9	54.2	49.2	47.9	38.2
D	56.7	49.6	46.4	40.9	39.4

	6	7	8	9	10
A	25.3	23.7	24.4	21.7	26.2
B	19.3	17.3	17.0	16.7	18.3
C	48.8	47.8	40.2	44.0	46.4
D	37.1	37.5	39.6	35.1	36.5

5. In an experiment to study the way in which different anesthetics affected plasma epinephrine concentration, 10 dogs were selected and concentration was measured while under the influence of the anesthetics isoflurane, halothane, and cyclopropane ("Sympathoadrenal and Hemodynamic Effects of Isoflurane, Halothane, and Cyclopropane in Dogs," *Anesthesiology*, 1974, pp. 465–470). Test at level .05 to see whether there is an anesthetic effect on concentration.

	Dog				
	1	2	3	4	5
Isoflurane:	.28	.51	1.00	.39	.29
Halothane:	.30	.39	.63	.38	.21
Cyclopropane:	1.07	1.35	.69	.28	1.24
	6	7	8	9	10
Isoflurane:	.36	.32	.69	.17	.33
Halothane:	.88	.39	.51	.32	.42
Cyclopropane:	1.53	.49	.56	1.02	.30

Bibliography

Daniel, Wayne, *Applied Nonparametric Statistics,* Houghton, Mifflin, Boston, 1978. Nonmathematical, with emphasis on how methods work rather than why they work.

Hollander, Myles and Wolfe, Douglas, *Nonparametric Statistical Methods,* John Wiley, New York, 1973. A very good reference on distribution-free methods with an excellent collection of tables.

Lehmann, Erich, *Nonparametrics: Statistical Methods Based on Ranks,* Holden-Day, San Francisco, 1975. An excellent discussion of the most important distribution-free methods, presented with a great deal of insightful commentary.

Marascuilo, Leonard and McSweeney, Maryellen, *Nonparametric and Distribution-Free Methods for the Social Sciences,* Brooks/Cole, Monterey, Calif., 1977. A good survey, with particular attention given to multiple comparison methods.

Mosteller, Frederick and Rourke, Richard, *Sturdy Statistics,* Addison-Wesley, Reading, Mass., 1973. An excellent elementary exposition of nonparametric methods.

Answers to Odd-Numbered Exercises

Chapter 1

Section 1.2

1.b. The eight frequencies are 2, 8, 12, 12, 3, 1, 0, and 2. **3.** The eight classes 1100–1199, . . ., 1800–1899 have frequencies 9, 5, 7, 9, 7, 3, 5, 5, and relative frequencies .18, .10, .14, .18, .14, .06, .10, .10. **5.** The seven classes 0.0–under 4.0, . . ., 24.0–28.0 have frequencies 2, 14, 11, 8, 4, 0, 1 and relative frequencies .050, .350, .275, .200, .100, .000, .025. **7.** Cumulative frequencies are .050, .400, .675, .875, .975, .975, and 1.000.

Section 1.3

1.a. 3.912. **b.** 3.90. **3.a.** $\bar{x} = 327.2$. **b.** $\tilde{x} = 341$. **c.** $\bar{x}_{\text{tr}(10)}$ $= 334.9$. **5.a.** .613, .506. **b.** .577. **7.a.** $\bar{y} = \bar{x} + c$. **b.** $\bar{y} = c\bar{x}$.

Section 1.4

1.a. 11.2. **b.** 24.12. **c.** 4.91. **3.a–c.** $s^2 = .095$. **5.** .0558, .236.

Supplementary Exercises

1.a. 5. **b.** 84. **3.** 10.65. **5.a.** $\bar{y} = a\bar{x} + b$, $s_y^2 = a^2 s_x^2$.

Chapter 2

Section 2.1

1.a. $\{(A, A), (A, E), (A, J), (E, A), (E, E), (E, J), (J, A), (J, E), (J, J)\}$.
b. $\{(A, E), (A, J), (E, A), (E, J)\}$. **c.** $\{(A, E), (A, J), (E, A), (E, E), (E, J), (J, A), (J, E), (J, J)\}$; complement $= \{(A, A)\}$, a simple event. **3.a.** $\mathcal{S} = \{(1, 1, 1), (1, 1, 2), (1, 1, 3), (1, 2, 1), (1, 2, 2), (1, 2, 3), (1, 3, 1), (1, 3, 2), (1, 3, 3), (2, 1, 1), (2, 1, 2), (2, 1, 3), (2, 2, 1), (2, 2, 2), (2, 2, 3), (2, 3, 1), (2, 3, 2), (2, 3, 3), (3, 1, 1), (3, 1, 2), (3, 1, 3), (3, 2, 1), (3, 2, 2), (3, 2, 3), (3, 3, 1), (3, 3, 2), (3, 3, 3)\}$. **b.** $\{(1, 1, 1), (2, 2, 2), (3, 3, 3)\}$. **c.** $\{(1, 2, 3), (1, 3, 2), (2, 1, 3), (2, 3, 1), (3, 1, 2), (3, 2, 1)\}$. **d.** $\{(1, 1, 1), (1, 1, 3), (1, 3, 1), (1, 3, 3), (3, 1, 1), (3, 1, 3), (3, 3, 1), (3, 3, 3)\}$. **5.a.** $\{11, 22, 33\}$.
b. $\{1213, 1312, 1231, 1321, 2123, 2132, 2312, 2321, 3123, 3132, 3213, 3231\}$.

Section 2.2   **1.a.** .8.   **b.** .2.   **c.** .4.   **3.a.** .6.   **b.** .3.   **5.a.** .4.   **b.** .75.
c. .25.   **7.a.** .2.   **b.** .5.   **c.** .3.   **9.a.** 1/4, 1/2, 3/13,
5/13.   **b.** 3/4, 8/13.   **c.** 35/52, 9/13.   **d.** 47/52.   **11.a.** .95.
b. .05.   **c.** .15.   **d.** .3.   **13.a.** 1/9.   **b.** 8/9.   **c.** 2/9.

Section 2.3   **1.a.** 20.   **b.** 60.   **c.** 10.   **3.a.** 243.   **b.** Approximately 10 years.
5.a. 2401.   **b.** .050.   **c.** .713.   **7.a.** 15,504.   **c.** 56.   **c.** .0578.
9.a. .929.   **b.** .0714.   **c.** .99997520.   **11.** .003940, .00001539.

Section 2.4   **1.a.** .4.   **b.** .429.   **c.** .675   **d.** .360.   **e.** .360.   **f.** .714.
g. 430.   **3.** .027   **5.a** .5.   **b.** .0455.   **c.** .682.   **d.** .0189.
9.a. .24.   **b.** .34.   **c.** .706, .206. **11.a.** .0588.   **b.** .6735.
c. .9996.   **13.** .8182.

Section 2.5   **3.a.** No.   **b.** .2, .5.   **5.** .248.   **7.a.** .2963.   **b.** .7037.   **c.** .2222.
d. .2593.   **e.** .3077.   **9.a.** .2.   **b.** .4.   **11.a.** For route 1,
$P(\text{late}) = .0523$, and for route 2 $P(\text{late}) = .19$, so take 1.   **b.** .4317.
13.a. $2\pi(1 - \pi)/(1 - \pi^2)$.   **b.** The estimate of π is $m_1/(m_1 + 2m_2)$.

Supplementary Exercises

1.a. 1140.   **b.** 969.   **c.** 1020.   **d.** .85.   **3.** $P(\text{hire 1}) = 6/24, 11/24$,
10/24, and 6/24 for $s = 0, 1, 2, 3$, respectively, so $s = 1$ is best.
5. $P(A) = .0204, P(B) = .0098$.   **7.a.** .2.   **b.** .2.   **9.** .1071, .5357.
11. .5.   **13.** $1 - (1 - p_1)(1 - p_2) \cdots (1 - p_n)$.   **15.a.** .0417.   **b.** .375

Chapter 3

Section 3.1   **1.** $x = 0$ for *FFF*; $x = 1$ for *SFF, FSF*, and *FFS*; $x = 2$ for *SSF, SFS*, and *FSS*, and
$x = 3$ for *SSS*.   **3.** Z = average of the two numbers, with possible values 2/2,
3/2, . . ., 12/2; W = absolute value of the difference, with possible values 0, 1, 2,
3, 4, 5.   **5.** No. In Example 3.4, let $Y = 1$ if at most three batteries are examined and let $Y = 0$ otherwise. Then Y has only two values.
7.a. $\{0, 1, \ldots, 12\}$, discrete. **c.** $\{1, 2, 3, \ldots\}$, discrete.   **e.** $\{0, c, 2c, \ldots,$
$10,000c\}$ where c is the royalty per book, discrete.   **g.** $\{x: m \leq x \leq M\}$ where
$m(M)$ is the minimum (maximum) possible tension, continuous.
9.a. $\{2, \ldots, 10\}$.   **c.** $\{0, 1, 2\}$.

Section 3.2   **1.a.** $p(x) = .25, .4, .35$ for $x = 2, 4, 6$.   **c.** $F(x) = 0$ for $x < 2$, .25 for
$2 \leq x < 4$, .65 for $4 \leq x < 6$, 1 for $6 \leq x$.   **3.a** 1/15.   **b.** .4.
c. .6.   **d.** No.   **5.** $p(y) = 1/4$ for $y = 1, 2, 3, 4$.   **7.** $p(144.9) =$
$.3, p(145.9) = .24, p(147.6) = .18, p(149.9) = .28$.   **9.** $p(y) = (1 - p)^y \cdot p$
for $y = 1, 2, 3, \ldots$.

Section 3.3   **1.a.** 2.06.   **b.** .9364.   **c.** .9677.   **3.a.** 16.38, 272.30, 3.9936.
b. 401.   **c.** 2496.   **d.** 13.657.   **5.** Yes.   **7.** For A, expected
return = .5, and for B, expected return = .97.   **9.** $E(1/X) = .4083$, so gamble.   **13.a.** 32.5.   **b.** 7.5.

Section 3.4   **1.a.** .124.   **b.** .279.   **c.** .635.   **d.** .718.   **3.a.** .585.   **b.** .202.
c. .909.   **d.** .959.   **e.** .762.   **f.** .004.   **g.** 15, 3.75.
5.a. .944.   **b.** .449.   **c.** .816.   **7.a.** .007.   **b.** .098.   **c.** .579.
9.a. Yes.   **b.** $p = .5$.

Section 3.5   **1.a.** .050.   **b.** .500, .050.   **c.** Don't reject H_0; type II error.   **3.a.** $\{15,$
$16, \ldots, 20\}$.   **b.** .021; yes; yes.   **c.** .874, .196.   **d.** .126, .804.

e. No. **5.** $n = 25$ is necessary, with $R = \{5, 6, \ldots, 25\}$, $\alpha = .098$, $\beta(.3) = .090$.

Section 3.6

1.a. .420. **b.** .714. **c.** .958. **d.** 2, .857. **3.a.** .087.

b. .400. **c.** 3.333, .794. **7.a.** $\binom{10}{x}\binom{10}{10-x}\Big/\binom{20}{10}$.

b. .033. **9.a.** $nb(x; 2, .5)$. **b.** .188. **c.** .376. **d.** 2; 4.

11. $nb(x; 6, .5)$; 6, which is $3 \cdot$ (expected number born to any particular brother).

Section 3.7

1.a. .932. **b.** .065. **c.** .068. **d.** .492. **e.** .251. **3.a.** .583.

b. .493. **c.** 10, 3.162. **5.a.** .176. **b.** .875. **c.** 3.75.

7. .033. **9.a.** .221. **b.** 6,800,000. **c.** $p(x; 20.11)$.

Supplementary Exercises

1.

w	6	7	8	9	10	11	12	13	14	15	16	17	18
$p(w)$	$\frac{1}{35}$	$\frac{1}{35}$	$\frac{2}{35}$	$\frac{3}{35}$	$\frac{4}{35}$	$\frac{4}{35}$	$\frac{5}{35}$	$\frac{4}{35}$	$\frac{4}{35}$	$\frac{3}{35}$	$\frac{2}{35}$	$\frac{1}{35}$	$\frac{1}{35}$

$\mu = 12$, $\sigma^2 = 8$ **5.a.** $p(x; .5)$. **b.** .076. **c.** .090. **7.a.** 16.

b. 4. **c.** .804. **d.** A hypergeometric p.m.f. **9.a.** .135.

b. .00144. **11.** 3.590. **13.a.** $p_1^{10} + p_2^{10} - (p_1 p_2)^{10}$.

b. $\left[\dfrac{p_1 p_2}{(1 - p_1)(1 - p_2)}\right] \displaystyle\sum_{x=10}^{\infty} \binom{x-1}{9}(1 - p_1)^x (1 - p_2)^x$. **15.b.** $.6 p(x; \lambda) +$ $.4\, p(x; \mu)$. **c.** $(\lambda + \mu)/2$. **d.** $(\lambda - \mu)^2/4 + (\lambda + \mu)/2$.

Chapter 4

Section 4.1

1.a. $1/4$. **b.** $1/2$. **c.** $7/16$. **3.b.** $1/2$. **c.** $33/48$. **d.** $81/128$.

5.b. $k = 3/4$. **c.** $1/2$. **d.** $47/96$. **e.** $5/12$. **7.c.** $9/50$.

d. $23/25$. **e.** $37/50$. **f.** $2/5$. **9.c.** $1 - (\theta/b)^k$.

d. $(\theta/a)^k - (\theta/b)^k$. **e.** $c = \dfrac{k}{\theta_1^{k+1}}[(1/\theta_1^k) - (1/\theta_2^k)]^{-1}$

Section 4.2

1.a. $1/4$. **b.** $3/16$. **c.** $15/16$. **d.** 1.414. **e.** $f(x) = x/2$ for $0 < x < 2$. **3.a.** $4/3$. **b.** $\sigma^2 = 2/9$. **c.** $E(X^2) = 2$.

5.a. $F(x) = 0$ if $x < 0$, x^3 if $0 \le x \le 1$, and 1 if $x > 1$. **b.** $1/8$. **c.** $7/64$, $7/64$. **d.** .9086. **e.** .7937. **f.** $E(X) = 3/4$, $\text{Var}(X) = 3/80$.

7.a. $A + p(B - A)$. **b.** $E(X) = (A + B)/2$, $\text{Var}(X) = (B - A)^2/12$.

c. $[B^{n+1} - A^{n+1}]/(n + 1)(B - A)$. **9.** 100.07π. **11.** $248°, 3.60°$.

Section 4.3

1.a. .4850. **b.** .3413. **c.** .4938. **d.** .9876. **e.** .9147.

f. .9599. **g.** .9104. **h.** .0791. **i.** .0668. **j.** .9876.

3.a. 1.34. **b.** -1.34. **c.** .675. **d.** $-.675$. **e.** -1.55.

5.a. .9772. **b.** .5. **c.** .9104. **d.** .8413. **e.** .2417. **f.** .6826.

7.a. .1251. **b.** .9382. **c.** .0548. **11.a.** 31.7. **b.** 17.25.

c. .2533. **d.** .2526. **13.a.** .9050. **b.** 5.88. **c.** 9.05.

d. .264. **15.** 100 min. **17.a.** .7580. **b.** .6727. **19.** No.

Section 4.4

1.a. 120. **b.** 1.329. **c.** .371. **d.** .735. **e.** 0. **3.a.** .594.

b. .092. **c.** .537. **5.a.** .424. **b.** .567; yes. **c.** 60. **d.** 66.

7.a. $1 - e^{-1.5} = .777$. **b.** $e^{-1} = .368$. **c.** $e^{-1} - e^{-1.5} = .145$.

d. 13.863. **11.a.** $A_1 \cap A_2 \cap A_3 \cap A_4 \cap A_5$. **b.** $e^{-.05t}$; $F(t) = 1 - e^{-.05t}$, exponential with $\lambda = .05$. **c.** Exponential with parameter $\lambda = .01n$.

Section 4.5

1.a. $\mu = 2.660$, $\sigma^2 = 1.926$. **b.** $1 - e^{-4} = .982$.

c. $e^{-.25} - e^{-4} = .760$. **5.a.** .8212. **b.** .1120.

7.a. $e^{5.005} = 149.157$, $e^{10.01}(e^{.01} - 1) = 223.595$.   **b.** .9830.   **c.** .0921.
d. 148.41.   **e.** 9.83.   **f.** 145.99.   **9.a.** 5/7, 5/196.   **b.** .0016.
c. .03936.   **d.** 2/7.   **11.a.** $\alpha = \beta = 3$.   **b.** .365.   **c.** .635.

Supplementary Exercises

1.a. .4.   **b.** .6.   **c.** $F(x) = x/25$ for $0 \le x \le 25$.   **d.** 12.5, 52.083.
3.a. $f(x) = x^2$ for $0 \le x < 1$, $7/4 - (3/4)x$ for $1 \le x \le 7/3$, 0 otherwise.
b. 11/12.   **c.** 131/108.   **5.a.** c/λ.   **b.** $(.5\lambda - a)/(\lambda - a)$.
7.b. $F(x) = (1/2)\,e^{.2x}$ for $x < 0$ and $F(x) = 1 - (1/2)\,e^{-.2x}$ for $x \ge 0$.   **c.** 1/2,
.665, .256, .670.   **9.a.** $k = (\alpha - 1)5^{\alpha - 1}$; $\alpha > 1$.   **b.** $F(x) = [k/(\alpha - 1)] \cdot$
$[1/5^{\alpha - 1} - 1/x^{\alpha - 1}]$ for $x \ge 5$.   **c.** $k/(\alpha - 2)\,5^{\alpha - 2}$; $\alpha > 2$.

Chapter 5

Section 5.1   **1.a.** .2.   **b.** .42.   **c.** At least one hose is in use at each pump; .70.
d. $p_X(x) = .16, .34, .5$ for $x = 0, 1, 2$, respectively; $p_Y(y) = .24, .38, .38$ for
$y = 0, 1, 2$, respectively; $P(X \le 1) = .5$.   **e.** No; $p(0, 0) \ne p_X(0) \cdot p_Y(0)$.

3.a.

		y		
		0	1	2
	0	1/4	1/4	1/16
x	1	1/4	1/8	0
	2	1/16	0	0

b.

		y			
		1	2	3	4
	1	1/16	1/16	1/16	1/16
x	2	1/16	1/16	1/16	1/16
	3	1/16	1/16	1/16	1/16
	4	1/16	1/16	1/16	1/16

5.a. $k = 3/380,000$.   **b.** 3024.   **c.** .3593.   **d.** $f_X(x) = 10kx^2 + .05$ for
$20 \le x \le 30$.   **e.** No.   **7.a.** $p(x, y) = [e^{-(\lambda + \mu)}\lambda^x \mu^y]/[x!\,y!]$ for $x = 0$,
1, ... and $y = 0, 1, \ldots$ **b.** $e^{-(\lambda + \mu)}(1 + \lambda + \mu)$.
c. $e^{-(\lambda + \mu)}(\lambda + \mu)^m/m!$.   **9.a.** $(1 - e^{-\lambda t})^{10}$.   **b.** $\binom{10}{k}(1 - e^{-\lambda t})^k (e^{-\lambda t})^{10 - k}$.
c. $\binom{9}{5}(1 - e^{-\lambda t})^5 (e^{-\lambda t})^4 (e^{-\mu t}) + \binom{9}{4}(1 - e^{-\lambda t})^4 (e^{-\lambda t})^5 (1 - e^{-\mu t})$.
11.a. $f(x_1, x_3) = 72\,x_1(1 - x_3)(1 - x_1 - x_3)^2$ for $0 \le x_1$, $0 \le x_3$, $x_1 + x_3 \le 1$.
b. .53125.   **c.** $f(x_1) = 18x_1 - 48x_1^2 + 36x_1^3 - 6x_1^5$ for $0 \le x_1 \le 1$.

Section 5.2   **1.a.** 14.10.   **b.** 9.60.   **3.** L^2.   **7.a.** -3.20.   **b.** $-.21$.

Section 5.3   **1.a.**

t	2	3	4	5	6	7	8	$\bar{x}$	1	1.5	2	2.5	3	3.5	4
$p_{T_0}(t)$	.04	.12	.25	.28	.22	.08	.01	$p_{\bar{x}}(\bar{x})$	.04	.12	.25	.28	.22	.08	.01

c. $E(T_0) = 4.8$, $E(\bar{X}) = 2.4$.   **d.** $\text{Var}(T_o) = 1.68$, $\text{Var}(\bar{X}) = .42$.

3.a.

t	320	360	400	440	480	$\bar{x}$	160	180	200	220	240
$p_{T_0}(t)$	.16	.32	.32	.16	.04	$p_{\bar{x}}(\bar{x})$	.16	.32	.32	.16	.04

(X_1, X_2) is a random sample.

b.

t	320	360	400	440	480
$p_{T_0}(t)$	.16	.24	.32	.20	.08

X_1 and X_2 do not have the same distribution.

c. $\mu_{T_0} = 392$, $\sigma_{T_0}^2 = 2176$.   **d.** X_1 and X_2 are not independent.

t	360	400	440
$p_{T_o}(t)$	16/30	7/30	7/30

e. $\mu_{T_0} = 388$.   **5.a.** 37.5.   **b.** 52.083.
c. -2.5, 10.417.   **d.** -7.5, 52.083.   **7.a.** 1300, 100, 10.
b. 52, .16, .4.   **9a.** 2400.   **b.** Assuming independence, $\sigma_{T_0}^2 = 1205$.
c. 2400, 36.37.

Section 5.4 **1.a.** .9332, .9544. **b.** .0062, .6826. **c.** .8719. **d.** .8413, .6826. **3.a.** .9876. **b.** .6915. **5.a.** .7620. **b.** .8414. **7.a.** .0968. **b.** .0882. **9.a.** 158, 430.25. **b.** .9788. **11.a.** Both approximately normal. **b.** Approximately normal with mean 2 and variance 2.629. **c.** .2354. **d.** .0068; yes. **13.a.** .9838. **b.** .8926.

Supplementary Exercises

1.a. $p_X(x) = .2, .5, .3$ for $x = 7, 9, 10$; $p_Y(y) = .1, .35, .55$ for $y = 7, 9, 10$. **b.** .25. **c.** No; $p(10, 7) = 0 \neq (.3)(.1)$. **d.** 17.75. **e.** 1.05. **3.** $f_Y(y) = [n/(100)^n](y - 100)^{n-1}$ for $100 \leq y \leq 200$, so $E(Y) = 100 [(2n + 1)/(n + 1)]$. **5.** .9686. **7.** .9099. **11.b,c.** Chi-squared with n d.f.

Chapter 6

Section 6.1 **1.a.** $\hat{\mu} = 65.72$, $\hat{\sigma} = 4.53$. **b.** $\tilde{\hat{\mu}} = 65.10$. **c.** $\hat{\mu} = \bar{x}_{tr(10)} = 66.05$. **d.** $\hat{p} = .7$. **3.a.** $\hat{\mu} = 76.818; \bar{X}$. **b.** $\tilde{\hat{\mu}} = 76.88$. **c.** $\hat{p} = .083$. **d.** $s/\sqrt{n} = .216$. **5.a.** $\hat{\lambda} = \bar{x} = 2.11$. **b.** $\sqrt{\hat{\lambda}}/\sqrt{n} = .119$. **7.b.** $[(p_1 q_1/n_1) + (p_2 q_2/n_2)]^{1/2}$. **c.** Substitute $\hat{p}_i = x_i/n_i$, $\hat{q}_i = 1 - \hat{p}_i$ $(i = 1, 2)$ into the standard error formula. **d.** $-.245$. **e.** .041. **9.** $\hat{\theta} = 6.3$; no; it is not unbiased. **11.b.** $\hat{p} = .444$. **13.a.** $\hat{p} = 2\hat{\lambda} - .3 = .20$. **c.** $\hat{p} = (100\hat{\lambda} - 63)/70$.

Section 6.2 **1.a.** $\hat{p} = x/n = .15$. **b.** Yes. **c.** $(1 - \hat{p})^5 = .444$. **3.a.** $E(X) = 1 - 1/(\theta + 2)$ so $\theta = 1/[1 - E(X)] - 2$ and $\hat{\theta} = 1/(1 - \bar{x}) - 2 = 3.00$. **b.** $\hat{\theta} = -n/[\Sigma \ln(x_i)] - 1 = 3.12$. **5.** $\hat{p} = r/(r + x)$; yes; no. **7.** M.l.e.'s of μ and σ are $\hat{\mu} = \bar{x} = 384.40$, $\hat{\sigma} = [(1/n) \Sigma (x_i - \bar{x})^2]^{1/2} = 18.86$ (not s), so the m.l.e. of $P(X \leq 400)$ is $\Phi((400 - \hat{\mu})/\hat{\sigma}) = .7881$. **9.** $p = e^{-\lambda(24)}$, so $\lambda = -\ln(p)/24$, $\hat{\lambda} = -\ln(\hat{p})/24 = .0120$.

Supplementary Exercises

3. With x_i = time between birth $i - 1$ and birth i, $\hat{\lambda} = 6/(x_1 + 2x_2 + \cdots + 6x_6) = .0436$. **5.** $\hat{\mu} = 9.55$.

Chapter 7

Section 7.1 **1.a.** $z = 2.51$; reject H_0. **b.** .1587. **c.** $n = 143$. **d.** .006. **e.** .0038. **3.a.** Either $Z \geq 2.58$ or $Z \leq -2.58$. **b.** Don't reject H_0. **c.** .005. **d.** .9649. **e.** .8729. **f.** .0124. **g.** Reject H_0. **h.** 131.135 or 128.865. **5.a.** $z = 2.18$; reject H_0. **b.** .0146. **c.** .9099, .0901. **d.** A different choice of H_0 and H_a. **7.a.** $z = -3.33$; at level .01, reject H_0. **b.** .1056. **c.** .0008. **9.** At level .05, recalibrate.

Section 7.2 **1.a.** $z = -2.34$; don't reject H_0; no. **b.** .0192. **3.** $z = 1.23$; don't reject H_a. **5.** $z = -1.83$; don't reject H_0. **7.** $z = -1.83$, an upper-tailed test is appropriate, the data strongly supports the claim. **b.** $P = .9664$. **9.** $H_a: \mu > 15$, $z = 2.72$, reject H_0 and retain the service.

Section 7.3 **1.** $z = 1.40$; don't reject H_0. **3.** $z = 3.55$; reject H_0 and proceed with underground installation. **5.a.** $z = -2.61$; yes. **b.** .0045. **c.** 89. **7.a.** $P = .1660$; no modification is necessary. **b.** .9974

Section 7.4 **1.a.** 1.345. **b.** 1.761. **c.** 1.725. **d.** 1.684. **e.** 2.528. **f.** -1.711. **g.** .001. **h.** .98. **3.** $t = -.21$; don't reject H_0.

5.a. $t = 1.88$; reject H_0; approximately normal.   **b.** $.025 < P < .05$.
c. $n = 16$.
7. $t = .87$; no.   **9.a.** $t = 2.58$; start refusing service; $P \doteq .008$.   **b.** No.

Section 7.5   **1.a.** 22.307.   **b.** 34.381.   **c.** 44.313.   **d.** 46.925.   **e.** 11.523.
f. 10.519.   **g.** 18.307.   **h.** 3.940.   **j.** .95.   **3.** $\chi^2 = 6.74$; yes.
5. $z = 2.56$; no.   **7.** $\chi^2 = 21.31$; don't use the steel.

Section 7.6   **1.a.** .1020, .8051, .9986.   **b.** $P \ll .0002$; yes.   **c.** No.

Supplementary Exercises
1.a. $z = -3.11$; yes.   **b.** $P = .0009$.   **3.** Yes.   **5.a.** $t = 1.70$; no.
b. $P \doteq .10$.   **7.** $z = -3.32$; yes.   **9.** $H_a: \lambda > 4$, $z = 1.33$, so don't reject H_0.   **11.a.** $\chi^2 = 9.14$; no.   **b.** $.05 < P < .10$.
13.a. $\bar{X} + 2.33S$.   **b.** $\sigma_\theta = 1.927\sigma/\sqrt{n}$, $\hat{\sigma}_\theta = 1927S/\sqrt{n}$;
$Z = (\bar{X} + 2.33S - \theta_0)/(1.927S/\sqrt{n})$, $z = -1.23$; no.

Chapter 8

Section 8.1   **1.a.** $z = -2.90$; reject H_0.   **b.** $P = .0019$.   **c.** .8186.   **d.** $m = n = 66$.
3. $z = -3.67$; yes.   **5.a.** $z = 2.89$; don't use the high-purity steel.   **b.** .7019,
.2981.   **c.** s_1 and s_2 replace σ_1 and σ_2; same conclusion.   **7.a.** $z = .84$; no.
b. .3577.   **c.** 274.   **9.** 50.   **13.** $P = .0793$ for $n = 100$ and .0023
for $n = 400$.

Section 8.2   **1.a.** .1760, .420.   **b.** $t = 3.55$; no.   **c.** $.002 < P < .01$.   **3.** $t = -3.36$;
yes; $P \doteq .001$.   **5.** $t = 4.25$; $P < .001$; yes.   **7.** $t = .92$; $P > .10$; no.
9. 26.

Section 8.3   **1.** $t = .05$; don't reject H_0.   **3.** $t = .36$; no.   **5.** $t = 1.22$; no.   **7.** (x_1, y_1)
$= (6,6)$, $(x_2, y_2) = (15, 15)$, $(x_3, y_3) = (1, 1)$, $(x_4, y_4) = (21, 21)$, so $t_{\text{paired}} = \infty$,
$t = 0$.

Section 8.4   **1.a.** $z = -4.84$, so reject H_0.   **b.** .9988.   **3.a.** $z = 1.48$; conclude that no difference existed.   **b.** $n = 6582$.   **5.** $z = -2.84$; conclude that there is a difference.   **7.** $P(X = 0, 1, 6,$ or $7 \mid X + Y = 7) = .0743$, so $P > .0743$; conclude that no difference exists.

Section 8.5   **1.a.** 3.69.   **b.** 4.82.   **c.** .207.   **d.** .271.   **e.** 4.30.   **f.** .212.
g. .95.   **h.** .94.   **3.** $f = .585$; no.   **5.** $z = .68$; don't reject
$H_0: \sigma_1 = \sigma_2$.

Supplementary Exercises
1. $t = .65$; no.   **3.** $t_{\text{paired}} = -1.65$; no.   **5.** $f = .228$, so don't reject
$H_0: \sigma_1 = \sigma_2$.   **7.** $\beta = .9015, .8264, .0294, .0000$.   **9.** $z = .63$; no.
11. $z = -1.34$; since $P > .05$, erogotamine does not seem effective.

Chapter 9

Section 9.1   **1.a.** (57.1, 59.5).   **b.** (57.7, 58.9).   **c.** (57.5, 59.1).   **d.** (57.9, 58.7).
e. 240.   **3.a.** (8406.1, 8471.9).   **b.** Use $z_{.04} = 1.75$.
5.a. $(\bar{x} - z_{\alpha_2} \cdot \sigma/\sqrt{n}, \bar{x} + z_{\alpha_1} \cdot \sigma/\sqrt{n})$.   **b.** Longer.
7.a. $\left(\dfrac{(n-1)S^2}{\chi^2_{\alpha/2,n-1}}, \dfrac{(n-1)S^2}{\chi^2_{1-\alpha/2,n-1}} \right)$.   **c.** (3.60, 28.98) for σ^2, (1.90, 5.38) for σ.

Section 9.2
1. $(37.5, 39.9)$; none. 3. $(150.2, 154.4)$. 5. $(-9.6, 0.0)$. 7. $(-63.05, -13.03)$; no; reject H_0.

Section 9.3
1. $(.182, .316)$. 3.a. 385. b. 342. 5. $(.021, .107)$. 7. $(-.055, .215)$. 9. 769.

Section 9.4
1. $(8.19, 10.85)$. 3. $(33.53, 43.79)$. 5. $(-1.81, 1.15)$. 7. $(-11.22, -10.32)$. 9. $\left(\dfrac{s_2^2}{s_1^2} \cdot \dfrac{1}{F_{\alpha/2,n-1,m-1}}, \dfrac{s_2^2}{s_1^2} \cdot F_{\alpha/2,m-1,n-1}\right)$.

Supplementary Exercises
1. $(490.0, 786.2)$, normal distribution of histamine content. 3.a. $(.626, .734)$. b. 1083. c. $(.636, .724)$. 5. $(.185, .443)$. 7. $(.198, .230)$. 9. $(12.11, 164.55)$. 11.a. $m = 141$, $n = 47$. b. $m = 240$, $n = 160$. 13.a. $\left(\dfrac{\chi^2_{1-\alpha/2,2n}}{2\sum x_i}, \dfrac{\chi^2_{\alpha/2,2n}}{2\sum x_i}\right)$ for λ; $\left(\dfrac{2\sum x_i}{\chi^2_{\alpha/2,2n}}, \dfrac{2\sum x_i}{\chi^2_{1-\alpha/2,2n}}\right)$ for μ. b. $(22.48, 80.88)$. c. $(15.58, 23.68)$. 15.a. $\left[\dfrac{\max(x_i)}{(1-\alpha/2)^{1/n}}, \dfrac{\max(x_i)}{(\alpha/2)^{1/n}}\right]$. b. $\left[\max(x_i), \dfrac{\max(x_i)}{\alpha^{1/n}}\right]$. c. The interval of (b); $(4.2, 7.65)$.

Chapter 10

Section 10.1
1. $f=1.85$, $F_{.05,4,15} = 3.06$; so don't reject H_0. 3. $f=6.43$; reject H_0 and conclude that there are differences. 5. $f=1.70$; brands don't appear to differ significantly from one another. 7. $f=10.48$, $F_{.01,4,30} = 4.02$, so axial stiffness appears to depend on plate length. 9.a. $MSE = 1,681.83$. b. $MSTr = 37,990.25$. c. $f=22.59$, so reject H_0.

Section 10.2
1. $w = 36.09$; $\underline{437.5 \quad 462.0 \quad 469.3} \quad \underline{512.8 \quad 532.1}$; brands 1, 3, 4 do not differ significantly from one another, but do differ significantly from both brand 2 and brand 5, while the latter two brands do not differ significantly from one another.

3.
$$\begin{array}{ccccc} 3 & 1 & 4 & 2 & 5 \\ \underline{427.5} & \underline{462.0} & \underline{469.3} & \underline{502.8} & \underline{532.1} \end{array}$$

1 does not differ significantly from 3 or 4; 2 does not differ significantly from 4 or 5; 3 does not differ significantly from 1; 4 does not differ significantly from 1 or 2; 5 does not differ significantly from 2; all other differences appear significant.

5. 1 and 2 differ significantly from 4, and there are no other significant differences (using $\alpha = .01$).

7.
$$\begin{array}{cccccc} 1 & 6 & 5 & 4 & 3 & 2 \\ \underline{166} & \underline{184} & \underline{202} & \underline{212} & \underline{266} & 303 \end{array}$$

9. $w = 6.35$, resulting in the same underscoring as in exercise 5 using Tukey's procedure. 11. $MSE = .108$, $\sum c_i^2 = 1.25$, so a 95% C.I. is $.165 \pm .309 = (-.144, .474)$. 13. No.

Section 10.3
1. $SSTr = 456.50$, $MSTr = 152.17$, $SSE = 124.50$, $MSE = 8.89$, $f=17.12$, so reject H_0 and conclude that salinity level affects average yield.

3.
Source	$d.f.$	SS	MS	F
Groups	2	152.18	76.09	5.56
Error	71	970.96	13.68	
Total	73	1123.14		

5. $f=.39$; no. 9. The computed value doesn't change. 11. For $\sigma = 1$ (the answer is independent of this choice), $\alpha_1 = \alpha_2 = \alpha_3 = 2/5$, $\alpha_4 = \alpha_5 = -3/5$, so

$\phi = 1.2$ and power $\doteq .45$.   **13.** Taking square roots and applying ANOVA yields $f = 3.23$, so conclude that brands do not differ significantly at level .01.

Chapter 11

Section 11.1   **1.a.** $f_A = 1.55$, so don't reject H_{0A}.   **b.** $f_B = 2.98$, so don't reject H_{0B}.
3.a. $f_A = 12.98$, $F_{01,3,9}$, so conclude that there is a gas rate effect; $f_B = 105.31$, so conclude that there is a liquid rate effect.   **b.** $w = 95.44$; $\underline{231.75 \quad 325.25} \quad 441.0$
613.25, so only the lowest two rates do not differ significantly from one another.
c. $\underline{336.75 \quad 382.25 \quad 419.25} \quad 473$ so only the lowest and highest rates appear to differ significantly from one another.   **5.** $f_A = 2.65$, $F_{01,3,12} = 5.95$, so there appears to be no effect due to angle of pull.

7.a.

Source	d.f.	SS	MS	F
Treatments	2	28.78	14.39	1.04
Blocks	17	2977.67	175.16	12.68
Error	34	469.55	13.81	
Total	53	3476.00		

True average adaptation score does not appear to depend on which treatment is given.   **b.** Yes; f_B is quite large, suggesting great variability between subjects.
9. $f_B = 8.87$, $F_{01,4,8} = 7.01$, so conclude that $\sigma_B^2 > 0$.

Section 11.2   **1.a.**

Source	d.f.	SS	MS	F
A	2	30,763.0	15,381.50	3.79
B	3	34,185.6	11,395.20	2.81
AB	6	43,581.2	7263.53	1.79
Error	24	97,436.8	4059.87	
Total	35	205,966.6		

b. $F_{05,6,24} = 2.51$, $f_{AB} = 1.79$, so do not reject H_{0AB}.   **c.** $F_{05,2,24} = 3.40$, $f_A = 3.79$, so reject H_{0A}.   **d.** Don't reject H_{0B}.   **e.** $w = 64.93$; only times 2 and 3 differ significantly from one another.   **3.** $f_{AB} = .32$, $f_A = 192.09$, $f_B = 8.96$, so interactions are not significant, but both main effects are significant.
5. $F_{01,8,30} = 3.17$, $f_{AB} = 1.38$, so don't reject H_{0G}; $f_A = MSA/MSAB = 26.70$, $F_{.01,2,8} = 8.65$, so reject H_{0A} and conclude that at least one $\alpha_i \neq 0$; $f_B = 28.51$, so reject H_{0B} and conclude that $\sigma_B^2 > 0$.   **9.a.** $MSAB/MSE$.   **b.** $MSA/MSAB$ for testing H_{0A}; $MSB/MSAB$ for testing H_{0B}.

Section 11.3   **1.a.**

Source	d.f.	SS	MS	F
A	2	14,144.44	7,072.22	61.06
B	2	5,511.27	2,755.64	23.79
C	2	244,696.39	122,348.20	1,056.27
AB	4	1,069.62	267.41	2.31
AC	4	62.67	15.67	.14
BC	4	331.67	82.92	.72
ABC	8	1,080.77	135.10	1.17
Error	27	3,127.50	115.83	
Total	53	270,024.33		

d. $Q_{.05,3,27} = 3.51$, $w = 8.90$, and all three of the levels differ significantly from one another.

3.

Source	d.f.	SS	MS	F
A	2	12.896	6.448	1.04
B	1	100.041	100.041	16.10

C	3	393.416	131.139	21.10
AB	2	1.646	.823	< 1
AC	6	71.021	11.837	1.905
BC	3	1.542	.514	< 1
ABC	6	9.771	1.629	< 1
Error	72	447.500	6.215	
Total	95	1037.833		

a. No interaction effects are significant. **b.** Factor B and factor C main effects are significant. **c.** $w = 1.89$; only machines 2 and 4 do not differ significantly from one another.

5.

Source	d.f.	SS	MS	F
A	2	42.12	21.06	8.78
B	2	110.74	55.37	23.07
C	2	68.14	34.07	14.20
AB	4	67.76	16.94	7.06
AC	4	35.18	8.80	3.67
BC	4	136.43	34.11	14.21
Error	8	19.22	2.40	
Total	26	479.59		

At level .01, several interaction effects are significant.

7.

Source	d.f.	SS	MS	F
A	6	67.32	11.02	
B	6	51.06	8.51	
C	6	5.43	.91	.61
Error	30	44.26	1.48	
Total	48	168.07		

$F_{.05,6,30} = 2.42$, $f_C = .61$, so H_{0C} is not rejected.

9.

Source	d.f.	SS	MS	F
A	4	28.88	7.22	10.7
B	4	23.70	5.93	8.79
C	4	.62	.155	< 1
Error	12	8.10	.675	
Total	24	61.30		

Since $F_{05,4,12} = 3.26$, both A and B are significant.

Section 11.4

1.a. $\hat{\beta}_1 = 54.38$, $\hat{\gamma}_{11}^{AC} = -2.21$, $\hat{\gamma}_{21}^{AC} = 2.21$.

b.

Source	Effect Contrast	MS	F
A	1307	71,177.04	436.7
B	1305	70,959.34	435.4
C	529	11,660.04	71.54
AB	199	1,650.04	10.12
AC	−53	117.04	< 1
BC	57	135.38	< 1
ABC	27	30.38	< 1
Error		162.98	

3.

Source	SS	F
A	136,640.02	999.3
B	139,644.19	1021.3
C	24,616.02	180.0
D	20,377.52	149.0

AB	2173.52	15.90
AC	2.52	< 1
AD	58.52	< 1
BC	165.02	1.21
BD	9.19	< 1
CD	17.52	< 1
ABC	42.19	< 1
ABD	117.19	< 1
ACD	188.02	1.38
BCD	13.02	< 1
ABCD	204.19	1.49
Error	4375.33	
Total	328,607.98	

$F_{.05,1,32} \doteq 4.15$, so only the four main effects and the *AB* interaction appear significant.

5.
Source	d.f.	SS	F
A	1	.436	1.12
B	1	.099	< 1
C	1	.109	< 1
D	1	414.12	1067
AB	1	.497	1.28
AC	1	.078	< 1
AD	1	.017	< 1
BC	1	1.404	3.62
BD	1	.456	1.18
CD	1	2.190	5.64
Error	5	.388	

$F_{05,1,5} = 6.61$, so only the factor *D* main effect is judged significant.

7.a. 1: (1), *ab*, *cd*, *abcd*, 2: *a*, *b*, *acd*, *bcd*, 3: *c*, *d*, *abc*, *abd*, 4: *ac*, *bc*, *ad*, *bd*.

b.
Source	d.f.	SS	F
A	1	14,028.13	53.89
B	1	92,235.13	345.33
C	1	3.13	< 1
D	1	18.00	< 1
AC	1	105.13	< 1
AD	1	200.00	< 1
BC	1	91.13	< 1
BD	1	420.50	1.62
ABC	1	276.13	1.06
ABD	1	2.00	< 1
ACD	1	450.00	1.73
BCD	1	2.00	< 1
Blocks	7	898.88	< 1
Error	12	3,123.72	
Total	31	111,853.88	

$F_{01,1,12} = 9.33$, so only the *A* and *B* main effects are significant.   **9.a.**   *ABFG*; (1), *ab*, *cd*, *ce*, *de*, *fg*, *acf*, *adf*, *adg*, *aef*, *acg*, *aeg*, *bcg*, *bcf*, *bdf*, *bdg*, *bef*, *beg*, *abcd*, *abce*, *abde*, *abfg*, *cdfg*, *cefg*, *defg*, *acdef*, *acdeg*, *bcdef*, *bcdeg*, *abcdfg*, *abcefg*, *abdefg*. {*A*, *BCDE*, *ACDEFG*, *BFG*}, {*B*, *ACDE*, *BCDEFG*, *AFG*}, {*C*, *ABDE*, *DEFG*, *ABCFG*}, {*D*, *ABCE*, *CEFG*, *ABDFG*}, {*E*, *ABCD*, *CDFG*, *ABEFG*}, {*F*, *ABCDEF*, *CDEG*, *ABG*}, {*G*, *ABCDEG*, *CDEF*, *ABF*}.

b. 1: (1), aef, beg, $abcd$, $abfg$, $cdfg$, $acdeg$, $bcdef$; 2: ab, cd, fg, aeg, bef, $acdef$, $bcdeg$, $abcdfg$; 3: de, acg, adf, bcf, bdg, $abce$, $cefg$, $abdefg$; 4: ce, acf, adg, bcg, bdf, $abde$, $defg$, $abcefg$.

Chapter 12

Section 12.1

1.a. 5050. **b.** 1.3. **c.** 130. **d.** −130. **e.** .0436. **3.a.** $10b_0 + 1269b_1 = 475$, $1269b_0 + 172,809b_1 = 62,631$. **b.** $\hat{\beta}_1 = .19990826 \doteq .200$, $\hat{\beta}_0 = 22.13164131 \doteq 22.13$. **c.** .20 taps/sec. **d.** 2.40. **e.** 46.13. **f.** 218.02. **5.a.** $y = -45.5519 + 1.7114x$. **b.** 339.51. **c.** −85.57. **d.** $\hat{y}_i$'s are 125.6, 168.4, 168.4, 211.1, 211.1, 296.7, 296.7, 382.3, 382.3, 467.9, 467.9, 553.4, 639.0, 639.0; a 45° line. **7.b.** $y = -2.182 + .660x$. **c.** 7.72. **d.** 7.72. **e.** $Y_{new} - [\hat{\beta}_0 + \hat{\beta}_1(15)]$. **f.** $\beta_0 + \beta_1(15) - [\hat{\beta}_0 + \hat{\beta}_1(15)]$; prediction. **11.a,b.** $\hat{\beta}_1$ for "transformed" data = $\hat{\beta}_1$ for original data, $\hat{\beta}_0$ for "transformed" data = $\bar{y}$.

Section 12.2

1.a. Residuals are 71.08, −64.49, −140.07, 4.36, 38.78, 73.21, 97.63, 32.06, −13.52, −99.09. **b.** $SSE = 56,274.6125$, $s^2 = 7,034.33$, $s = 83.87$. **c.** $s = 83.86$; $s = 83.79$. **3.a.** $s^2 = .0037813$, $s = .0616$. **b.** .00000103, .001017. **5.a.** .1101, .000262. **b.** $t_{.05,8} = 1.860$, $t = 3.06$, so reject H_0. **7.** (−.0332, −.0170). **9.** H_0: $\beta_1 = 0$ versus H_a: $\beta_1 \neq 0$, $t = 4.24$, so reject H_0.

Section 12.3

1.a. (48.97, 52.63). **b.** $t = .78$; don't reject H_0. **3.** $t = .032$, $P > .2$. **5.a.** (318.70, 343.18). **b.** (291.40, 370.48). **c.** (258.76, 289.88) when $x = 15$; (357.35, 380.03) when $x = 20$. **7.** Using $t_{.01,5}$ (98% for each interval) yields simultaneous confidence of at least 94%; (326.96, 373.64) for $x = 1200$, (359.86, 404.74) for $x = 1250$, (390.27, 438.33) for $x = 1300$.

Section 12.4

1.a. .743. **b.** .552. **3.a.** .9963. **b.** .993. **5.** H_0: $\rho = .8$ versus H_a: $\rho > .8$, $r = .9208$, $z = 1.85$, so reject H_0. **7.** $SST = 178.23$, $SSE = 29.23$, $r^2 = .836$, so 83.6% is explained.

Chapter 13*

Section 13.1

1.a. Yes. **b.** e_i^*'s are −.75, .31, −.74, 1.13, .43, −.72, 1.43, .93, −1.51, −1.27, .90. **c.** No. **3.a.** No. **b.** e_i's are −16.60, 9.70, 19.00, −.70, −11.40; e_i^*'s are −1.55, .68, 1.25, −.05, −1.06; a quadratic function. **5.** For set 1, a simple linear regression is appropriate; for set 2, a quadratic regression is appropriate; for set 3, the single point (13, 12.74) appears to be quite inconsistent with the remaining data; for set 4, the single point (19, 12.5) determines $\hat{\beta}_1$, and the evidence for a linear relationship is not compelling. **7.c.** $\text{Var}(\hat{Y}_i)$ increases and $\text{Var}(Y_i - \hat{Y}_i)$ decreases. **9.** A t distribution with $n - 2$ d.f.; .02.

Section 13.2

1.a. $\hat{\alpha} = 43,044.94$; $\hat{\beta} = -2.16$. **b.** Yes. **c.** Yes. **d.** H_0: $\beta_1 = -2$ versus H_a: $\beta_1 \neq -2$, $t = -1.53$, so don't reject H_0. **e.** (−2.40, −1.92). **f.** $x = 40 = $; $x' = 3.6889$; the interval for $\beta_0 + \beta_1 x'$ is (2.6203, 2.7833), so the desired interval is (13.74, 16.17). **3.a.** $\Sigma x_i' = 15.501$, $\Sigma y_i' = 13.352$, $\Sigma x_i'^2 = 20.228$, $\Sigma y_i'^2 = 16.572$, $\Sigma x_i' y_i' = 18.109$, $\hat{\beta}_1 = 1.254$, $\hat{\beta}_0 = -.468$, so $\hat{\alpha} = .626$, $\hat{\beta} = 1.254$. **c.** $t = -1.07$, so don't reject

---

*Some answers have been obtained from computer output, so may differ slightly from hand-calculated answers.

H_0. **d.** H_0: $\beta = 1$, $t = -4.30$, so reject H_0. **5.a.** No. **b.** $Y' = \beta_0 + \beta_1 \cdot 1/t + \epsilon'$ where $Y' = \ln(Y)$, so $Y = \alpha e^{\beta/t} \cdot \epsilon$. **c.** $\hat{\beta} = \hat{\beta}_1 = 3735.45$, $\hat{\beta}_0 = -10.2045$, $\hat{\alpha} = 3.70034 \cdot 10^{-5}$, $\hat{y}' = 6.7748$, $\hat{y} = 875.5$. **d.** $SSE = 1.39587$, $SSPE = 1.36594$, $SSLF = .02993$ (all from transformed values), $f = .26$, $F_{.01,1,12} = 3.80$, so don't reject H_0. **7.a.** $\hat{\mu}_{Y \cdot x} = 18.14 - 1485/x$. **b.** $\hat{y} = 15.17$. **9.** For the exponential model, $\text{Var}(Y) = \alpha^2 e^{2\beta x} \sigma^2$, which does depend on x; a similar result holds for the power model.

Section 13.3　**1.a.** 39.41. **b.** 24.93. **c.** 217.82; 72.61; 8.52. **d.** .779. **e.** $t = -7.88$, so reject H_0. **3.a.** $y = -251.719114 + 1000.202871x - 135.456884x^2$. **b.** $SSE = 202,209.77$, $s^2 = 18,382.71$, $R^2 = .853$. **c.** $(-227.84, -43.08)$. **d.** $t = -1.83$, $-t_{.01,11} = -2.718$, so don't reject H_0. **e.** $(1081.37, 1722.99)$. **5.a.** $.3463 - 1.2933 \cdot (x - \bar{x}) + 2.3964 \cdot (x - \bar{x})^2 - 2.3968(x - \bar{x})^3$. **b.** $\hat{\beta}_3 = -2.3968$, $\hat{\beta}_2 = 33.6430$. **c.** 2.411. **d.** $t = -.97$, so the cubic term can be deleted. **7.a.** .873. **b.** .919. **c.** $1.200887 - .01048314x + .00002618x^2$. **d.** .00077118. **e.** $t = 1.40$, so the quadratic term should not be retained; no.

Section 13.4　**1.b.** $\beta_2 + \beta_4(2x_2 + 1)$; 2.00. **3.b.** 103.11; 96.85; -5.85. **c.** 5384.18; 24.46; .768. **d.** $f = 14.89$, $F_{.05,2,9} = 3.89$, so reject H_0. **e.** $(78.30, 115.40)$. **f.** $(38.50, 155.20)$. **g.** $t = -1.36$, so x_1 can be eliminated. **5.a.** No. **b.** $f = 5.04$, $F_{.05,5,8} = 3.69$, so reject H_0. **c.** $(16.67, 31.91)$. **d.** $f = 3.44$, $F_{.05,3,8} = 4.07$, so the simpler model appears to be adequate. **7.a.** .5386. **b.** $(-.0333, -.0139)$. **c.** .004334; .000845. **d.** $(.4336, .5804)$. **9.a.** $y = 1.5652 + .9450x_1 + .1815x_2$ where $y = \ln(q)$, $x_1 = \ln(a)$, $x_2 = \ln(b)$; $\hat{q} = 18.271$. **b.** Regress $\ln(q)$ on a and b. **c.** $(1.242, 5.783)$.

Section 13.5　**1.a.** No. **b.** 117.56 for $\hat{\beta}_0$, -1.09 for $\hat{\beta}_1$, -1.99 for $\hat{\beta}_2$. **c.** In S.D.'s changes are 1.42, $-.224$, -1.86; the first and last are substantial. **3.a.** 1.5652; .9450; .1815. **b.** .7328; .1528; .1752. **c.** .1388. **d.** $(4.53, 36.23)$. **5.** h_{ii}'s are .3001, .1209, .1604, .2545, .1663, .0940, .5542, .5128, .3364, .3865, .5017, .4239, .2618, .1654, .1385, .4784, .2750, .2727, .3514, .2428; $2(k + 1)/n = .6$, so none is large. **7.a.** 84.5549; .6500; $-.2583$; .1333; .1083. **b.** $s_{\hat{\beta}_0} = .077430$, $s_{\hat{\beta}_i} = .056800$ ($i = 1, 2, 3, 4$), $\text{Corr}(\hat{\beta}_i, \hat{\beta}_j) = 0$ ($i \neq j$). **c.** $h_{ii} = a + 4b$ ($i = 1, \ldots, 24$), $= a$ ($i = 25, \ldots, 31$), where $a = .0322581$, $b = .0416667$, so no data points have large influence. **d.** $\sigma^2[a + bx_1^2 + bx_2^2 + bx_3^2 + bx_4^2]$ where $a = .0322581$, $b = .0416667$.

Chapter 14

Section 14.1　**1.** $\chi^2 = 7.073$, $\chi^2_{.05,3} = 7.815$, so don't reject H_0. **3.** $\chi^2 = 6.612$, $\chi^2_{.10,9} = 14.684$, so don't reject H_0. **5.a.** $[0, .2231)$, $[.2231, .5108)$, $[.5108, .9163)$, $[.9163, 1.0694)$, $[1.0694, \infty)$. **b.** $\chi^2 = 1.25$. **7.a.** $(-\infty, -.97)$, $[-.97, -.43)$, $[-.43, 0)$, $[0, .43)$, $[.43, .97)$, $[.97, \infty)$. **b.** $(-\infty, .49806)$, $[.49806, .49914)$, $[.49914, .5)$, $[.5, .50086)$, $[.50086, .50194)$, $[.50194, \infty)$. **c.** $\chi^2 = 6.11$.

Section 14.2　**1.** $\hat{p} = .0843$, $\chi^2 = 280.3$, so the model gives a poor fit. **3.** Likelihood $\propto \theta^{233}(1 - \theta)^{367}$, $\hat{\theta} = .3883$, estimated expected frequencies are 21.00, 53.33, 50.78, 21.50, 3.41; combining cells 4 and 5, $\chi^2 = 1.62$, so don't reject H_0. **5.** $\hat{\lambda} = 3.167$, $\chi^2 = 103.98$, so the Poisson distribution does not provide a good fit. **7.** $\hat{\theta}_1 = (2n_1 + n_3 + n_5)/2n = .4125$, $\hat{\theta}_2 = .2750$, $\chi^2 = 30.58$, $\chi^2_{.01,3} = 11.344$, so reject H_0. **9.** $u = .9457$, $z = -1.12$, so the normal distribution provides a good model.

Section 14.3 **1.** $\chi^2 = 23.18$, $\chi^2_{.01,5} = 15.085$, so reject the hypothesis of homogeneity and conclude that proportions differ. **3.** $\chi^2 = 44.98$, so conclude that sex and the nature of foot asymmetry are related. **5.** $\chi^2 = 6.49$, $\chi^2_{.10,10} = 15.987$, so don't reject H_0.
7. $\chi^2 = 14.15$, $\chi^2_{.05,6} = 12.592$, so at level .05 conclude that the two factors are related.
11. $\hat{p}_{ij} = N_{ij\cdot}/n$, $\hat{E}_{ijk} = N_{ij\cdot} \cdot n_k/n$, d.f. $= 32 - 8 = 24$.

Section 14.4 **1.a.** $\ln E_{ijk} = \lambda + \alpha_i + \beta_j + \delta_k + \gamma_{ik}^{AC}$.
b. $\ln E_{ijk} = \lambda + \alpha_i + \beta_j + \delta_k + \gamma_{ij}^{AB} + \gamma_{jk}^{BC}$.
3.a. $\hat{E}_{111} = 35.82$, $\hat{E}_{121} = 22.32$, $\hat{E}_{211} = 65.22$, $\hat{E}_{221} = 40.63$, $\hat{E}_{112} = 53.51$, $\hat{E}_{122} = 33.34$, $\hat{E}_{212} = 97.44$, $\hat{E}_{222} = 60.70$. **b.** $G^2 = 25.084$, $\chi^2_{.01,4} = 13.277$, so the complete independence model does not provide a good fit. **5.a.** $\hat{E}_{111} = 105.53$, $\hat{E}_{121} = 18.47$, $\hat{E}_{211} = 94.47$, $\hat{E}_{221} = 16.53$, $\hat{E}_{112} = 14.67$, $\hat{E}_{122} = 7.33$, $\hat{E}_{212} = 31.33$, $\hat{E}_{222} = 15.67$; $G^2 = 53.68$. **b.** No, since $G^2(M_3)$ and $G^2(M_3) - G^2(M_4)$ are both large. **c.** $\hat{E}_{111} = 121.08$, $\hat{E}_{121} = 2.29$, $\hat{E}_{211} = 78.92$, $\hat{E}_{221} = 32.08$, $\hat{E}_{112} = 20.92$, $\hat{E}_{122} = 1.08$, $\hat{E}_{212} = 25.09$, $\hat{E}_{221} = 21.92$.

Chapter 15

Section 15.1 **1.a.** $(a + b)/2$. **b.** $\ln(2)/\lambda$. **3.** At level .05, reject H_0 if $Y \le 1$ or ≥ 9; since $y = 2$, don't reject H_0. **5.a.** $p = .5$. **b.** Bin$(6, .5)$. **c.** The region $\{5, 6\}$ gives $\alpha = .1094$; since $y = 5$, reject H_0. **7.** When H_0 is true, $Y \sim$ Bin$(20, .75)$, so $\{18, 19, 20\}$ yields $\alpha = .091$; since $y = 18$, reject H_0.

Section 15.2 **1.** Reject H_0 if $S_+ \le 20$; $s_+ = 23$, so don't reject H_0. **3.** Reject H_0 if $S_+ \ge 84$ or ≤ 21; $s_+ = 18$, so reject H_0. **5.** $s_+ = 136$, $z = 1.16$; since 136 is not ≥ 140 and 1.16 is not ≥ 1.28, neither test rejects H_0. **7.** The null distribution of D is

d	0	2	4	6	8	10	12	14	16	18	20
$p(d)$	1/24	3/24	1/24	4/24	2/24	2/24	2/24	4/24	1/24	3/24	1/24

so $c = 0$ gives $\alpha = .042$ and $c = 2$ gives $\alpha = .167$.

Section 15.3 **1.** $w = 38$ is ≥ 36, so reject H_0. **3.** $w = 65$ is not ≥ 84, so don't reject H_0.
5. $z = 2.54$, so don't reject H_0.
7.

w''	4	5	6	7	8	9	10
$p(w'')$	2/35	4/35	9/35	8/35	7/35	4/35	1/35

$c = 10$ yields $\alpha = .029$, $c = 9$ yields $\alpha = .143$.

Section 15.4 **1.** $(x_{(5)}, x_{(16)}) = (44.1, 46.6)$. **3.** $(\bar{x}_{(5)}, \bar{x}_{(32)}) = (11.15, 23.80)$. **5.** $(-13.0, -6.0)$. **7.** $(d_{ij(5)}, d_{ij(21)}) = (16, 87)$.

Section 15.5 **1.a** $k = 14.06$, $\chi^2_{.10,3} = 6.251$, so reject H_0. **b.** $\underline{5.20}$ $\underline{6.20}$ $\underline{13.60}$ $\underline{17.00}$, so 3, 4 differ and 1, 2 differ. **3.a.** $k = 9.23$, $\chi^2_{.05,2} = 5.992$, so reject H_0. **b.** 1, 2 differ and 1, 3 differ. **5.** $f_r = 2.60$, $\chi^2_{.05,2} = 5.992$, so don't reject H_0.

Appendix Tables

Table A.1 Cumulative Binomial Probabilities

$$B(x; n, p) = \sum_{y=0}^{x} b(y; n, p)$$

a. $n = 5$

| | | | | | | | p | | | | | | | | |
	0.01	0.05	0.10	0.20	0.25	0.30	0.40	0.50	0.60	0.70	0.75	0.80	0.90	0.95	0.99
0	.951	.774	.590	.328	.237	.168	.078	.031	.010	.002	.001	.000	.000	.000	.000
1	.999	.977	.919	.737	.633	.528	.337	.188	.087	.031	.016	.007	.000	.000	.000
x 2	1.000	.999	.991	.942	.896	.837	.683	.500	.317	.163	.104	.058	.009	.001	.000
3	1.000	1.000	1.000	.993	.984	.969	.913	.812	.663	.472	.367	.263	.081	.023	.001
4	1.000	1.000	1.000	1.000	.999	.998	.990	.969	.922	.832	.763	.672	.410	.226	.049

b. $n = 10$

| | | | | | | | p | | | | | | | | |
	0.01	0.05	0.10	0.20	0.25	0.30	0.40	0.50	0.60	0.70	0.75	0.80	0.90	0.95	0.99
0	.904	.599	.349	.107	.056	.028	.006	.001	.000	.000	.000	.000	.000	.000	.000
1	.996	.914	.736	.376	.244	.149	.046	.011	.002	.000	.000	.000	.000	.000	.000
2	1.000	.988	.930	.678	.526	.383	.167	.055	.012	.002	.000	.000	.000	.000	.000
3	1.000	.999	.987	.879	.776	.650	.382	.172	.055	.011	.004	.001	.000	.000	.000
4	1.000	1.000	.998	.967	.922	.850	.633	.377	.166	.047	.020	.006	.000	.000	.000
x 5	1.000	1.000	1.000	.994	.980	.953	.834	.623	.367	.150	.078	.033	.002	.000	.000
6	1.000	1.000	1.000	.999	.996	.989	.945	.828	.618	.350	.224	.121	.013	.001	.000
7	1.000	1.000	1.000	1.000	1.000	.998	.988	.945	.833	.617	.474	.322	.070	.012	.000
8	1.000	1.000	1.000	1.000	1.000	1.000	.998	.989	.954	.851	.756	.624	.264	.086	.004
9	1.000	1.000	1.000	1.000	1.000	1.000	1.000	.999	.994	.972	.944	.893	.651	.401	.096

c. $n = 15$

| | | | | | | | p | | | | | | | | |
	0.01	0.05	0.10	0.20	0.25	0.30	0.40	0.50	0.60	0.70	0.75	0.80	0.90	0.95	0.99
0	.860	.463	.206	.035	.013	.005	.000	.000	.000	.000	.000	.000	.000	.000	.000
1	.990	.829	.549	.167	.080	.035	.005	.000	.000	.000	.000	.000	.000	.000	.000
2	1.000	.964	.816	.398	.236	.127	.027	.004	.000	.000	.000	.000	.000	.000	.000
3	1.000	.995	.944	.648	.461	.297	.091	.018	.002	.000	.000	.000	.000	.000	.000
4	1.000	.999	.987	.836	.686	.515	.217	.059	.009	.001	.000	.000	.000	.000	.000
5	1.000	1.000	.998	.939	.852	.722	.403	.151	.034	.004	.001	.000	.000	.000	.000
6	1.000	1.000	1.000	.982	.943	.869	.610	.304	.095	.015	.004	.001	.000	.000	.000
x 7	1.000	1.000	1.000	.996	.983	.950	.787	.500	.213	.050	.017	.004	.000	.000	.000
8	1.000	1.000	1.000	.999	.996	.985	.905	.696	.390	.131	.057	.018	.000	.000	.000
9	1.000	1.000	1.000	1.000	.999	.996	.966	.849	.597	.278	.148	.061	.002	.000	.000
10	1.000	1.000	1.000	1.000	1.000	.999	.991	.941	.783	.485	.314	.164	.013	.001	.000
11	1.000	1.000	1.000	1.000	1.000	1.000	.998	.982	.909	.703	.539	.352	.056~	.005	.000
12	1.000	1.000	1.000	1.000	1.000	1.000	1.000	.996	.973	.873	.764	.602	.184	.036	.000
13	1.000	1.000	1.000	1.000	1.000	1.000	1.000	1.000	.995	.965	.920	.833	.451	.171	.010
14	1.000	1.000	1.000	1.000	1.000	1.000	1.000	1.000	1.000	.995	.987	.965	.794	.537	.140

d. $n = 20$

| | | | | | | | p | | | | | | | | |
	0.01	0.05	0.10	0.20	0.25	0.30	0.40	0.50	0.60	0.70	0.75	0.80	0.90	0.95	0.99
0	.818	.358	.122	.002	.003	.001	.000	.000	.000	.000	.000	.000	.000	.000	.000
1	.983	.736	.392	.069	.024	.008	.001	.000	.000	.000	.000	.000	.000	.000	.000
2	.999	.925	.677	.206	.091	.035	.004	.000	.000	.000	.000	.000	.000	.000	.000
3	1.000	.984	.867	.411	.225	.107	.016	.001	.000	.000	.000	.000	.000	.000	.000
4	1.000	.997	.957	.630	.415	.238	.051	.006	.000	.000	.000	.000	.000	.000	.000

(*continued*)

Table A.1 Cumulative Binomial Probabilities (*cont.*) $\qquad B(x; n, p) = \sum_{y=0}^{x} b(y; n, p)$

d. $n = 20$ (*continued*)

	0.01	0.05	0.10	0.20	0.25	0.30	*p* 0.40	0.50	0.60	0.70	0.75	0.80	0.90	0.95	0.99
5	1.000	1.000	.989	.804	.617	.416	.126	.021	.002	.000	.000	.000	.000	.000	.000
6	1.000	1.000	.998	.913	.786	.608	.250	.058	.006	.000	.000	.000	.000	.000	.000
7	1.000	1.000	1.000	.968	.898	.772	.416	.132	.021	.001	.000	.000	.000	.000	.000
8	1.000	1.000	1.000	.990	.959	.887	.596	.252	.057	.005	.001	.000	.000	.000	.000
9	1.000	1.000	1.000	.997	.986	.952	.755	.412	.128	.017	.004	.001	.000	.000	.000
10	1.000	1.000	1.000	.999	.996	.983	.872	.588	.245	.048	.014	.003	.000	.000	.000
11	1.000	1.000	1.000	1.000	.999	.995	.943	.748	.404	.113	.041	.010	.000	.000	.000
12	1.000	1.000	1.000	1.000	1.000	.999	.979	.868	.584	.228	.102	.032	.000	.000	.000
13	1.000	1.000	1.000	1.000	1.000	1.000	.994	.942	.750	.392	.214	.087	.002	.000	.000
14	1.000	1.000	1.000	1.000	1.000	1.000	.998	.979	.874	.584	.383	.196	.011	.000	.000
15	1.000	1.000	1.000	1.000	1.000	1.000	1.000	.994	.949	.762	.585	.370	.043	.003	.000
16	1.000	1.000	1.000	1.000	1.000	1.000	1.000	.999	.984	.893	.775	.589	.133	.016	.000
17	1.000	1.000	1.000	1.000	1.000	1.000	1.000	1.000	.996	.965	.909	.794	.323	.075	.001
18	1.000	1.000	1.000	1.000	1.000	1.000	1.000	1.000	.999	.992	.976	.931	.608	.264	.017
19	1.000	1.000	1.000	1.000	1.000	1.000	1.000	1.000	1.000	.999	.997	.988	.878	.642	.182

x labels rows 10–14 region.

e. $n = 25$

	0.01	0.05	0.10	0.20	0.25	0.30	*p* 0.40	0.50	0.60	0.70	0.75	0.80	0.90	0.95	0.99
0	.778	.277	.072	.004	.001	.000	.000	.000	.000	.000	.000	.000	.000	.000	.000
1	.974	.642	.271	.027	.007	.002	.000	.000	.000	.000	.000	.000	.000	.000	.000
2	.998	.873	.537	.098	.032	.009	.000	.000	.000	.000	.000	.000	.000	.000	.000
3	1.000	.966	.764	.234	.096	.033	.002	.000	.000	.000	.000	.000	.000	.000	.000
4	1.000	.993	.902	.421	.214	.090	.009	.000	.000	.000	.000	.000	.000	.000	.000
5	1.000	.999	.967	.617	.378	.193	.029	.002	.000	.000	.000	.000	.000	.000	.000
6	1.000	1.000	.991	.780	.561	.341	.074	.007	.000	.000	.000	.000	.000	.000	.000
7	1.000	1.000	.998	.891	.727	.512	.154	.022	.001	.000	.000	.000	.000	.000	.000
8	1.000	1.000	1.000	.953	.851	.677	.274	.054	.004	.000	.000	.000	.000	.000	.000
9	1.000	1.000	1.000	.983	.929	.811	.425	.115	.013	.000	.000	.000	.000	.000	.000
10	1.000	1.000	1.000	.994	.970	.902	.586	.212	.034	.002	.000	.000	.000	.000	.000
11	1.000	1.000	1.000	.998	.980	.956	.732	.345	.078	.006	.001	.000	.000	.000	.000
12	1.000	1.000	1.000	1.000	.997	.983	.846	.500	.154	.017	.003	.000	.000	.000	.000
13	1.000	1.000	1.000	1.000	.999	.994	.922	.655	.268	.044	.020	.002	.000	.000	.000
14	1.000	1.000	1.000	1.000	1.000	.998	.966	.788	.414	.098	.030	.006	.000	.000	.000
15	1.000	1.000	1.000	1.000	1.000	1.000	.987	.885	.575	.189	.071	.017	.000	.000	.000
16	1.000	1.000	1.000	1.000	1.000	1.000	.996	.946	.726	.323	.149	.047	.000	.000	.000
17	1.000	1.000	1.000	1.000	1.000	1.000	.999	.978	.846	.488	.273	.109	.002	.000	.000
18	1.000	1.000	1.000	1.000	1.000	1.000	1.000	.993	.926	.659	.439	.220	.009	.000	.000
19	1.000	1.000	1.000	1.000	1.000	1.000	1.000	.998	.971	.807	.622	.383	.033	.001	.000
20	1.000	1.000	1.000	1.000	1.000	1.000	1.000	1.000	.991	.910	.786	.579	.098	.007	.000
21	1.000	1.000	1.000	1.000	1.000	1.000	1.000	1.000	.998	.967	.904	.766	.236	.034	.000
22	1.000	1.000	1.000	1.000	1.000	1.000	1.000	1.000	1.000	.991	.968	.902	.463	.127	.002
23	1.000	1.000	1.000	1.000	1.000	1.000	1.000	1.000	1.000	.998	.993	.973	.729	.358	.026
24	1.000	1.000	1.000	1.000	1.000	1.000	1.000	1.000	1.000	1.000	.999	.996	.928	.723	.222

x labels row 12.

Table A.2 Cumulative Poisson Probabilities

$$F(x; \lambda) = \sum_{y=0}^{x} \frac{e^{-\lambda}\lambda^{y}}{y!}$$

					λ					
	.1	.2	.3	.4	.5	.6	.7	.8	.9	1.0
0	.905	.819	.741	.670	.607	.549	.497	.449	.407	.368
1	.995	.982	.963	.938	.910	.878	.844	.809	.772	.736
2	1.000	.999	.996	.992	.986	.977	.966	.953	.937	.920
x 3		1.000	1.000	.999	.998	.997	.994	.991	.945	.981
4				1.000	1.000	1.000	.999	.999	.989	.996
5							1.000	1.000	.998	.999
6									1.000	1.000

					λ						
	2.0	3.0	4.0	5.0	6.0	7.0	8.0	9.0	10.0	15.0	20.0
0	.135	.050	.018	.007	.002	.001	.000	.000	.000	.000	.000
1	.406	.199	.092	.040	.017	.007	.003	.001	.000	.000	.000
2	.677	.423	.238	.125	.062	.030	.014	.006	.003	.000	.000
3	.857	.647	.433	.265	.151	.082	.042	.021	.010	.000	.000
4	.947	.815	.629	.440	.285	.173	.100	.055	.029	.001	.000
5	.983	.916	.785	.616	.446	.301	.191	.116	.067	.003	.000
6	.995	.966	.889	.762	.606	.456	.313	.207	.130	.008	.000
7	.999	.988	.949	.867	.744	.599	.453	.324	.220	.018	.001
8	1.000	.996	.979	.932	.847	.729	.593	.456	.333	.037	.002
9		.999	.992	.968	.916	.830	.717	.587	.458	.070	.005
10		1.000	.997	.986	.957	.901	.816	.706	.583	.118	.011
11			.999	.995	.980	.947	.888	.803	.697	.185	.021
12			1.000	.998	.991	.973	.936	.876	.792	.268	.039
13				.999	.996	.987	.966	.926	.864	.363	.066
14				1.000	.999	.994	.983	.959	.917	.466	.105
15					.999	.998	.992	.978	.951	.568	.157
16					1.000	.999	.996	.989	.973	.664	.221
17						1.000	.998	.995	.986	.794	.297
x 18							1.000	.999	.993	.819	.381
19								1.000	.997	.875	.470
20									.998	.917	.559
21									.999	.947	.644
22									1.000	.967	.721
23										.981	.787
24										.989	.843
25										.994	.888
26										.997	.922
27										.998	.948
28										.999	.966
29										1.000	.978
30											.987
31											.992
32											.995
33											.997
34											.999
35											.999
36											1.000

Source: Lincoln L. Chao, *Statistics: Methods and Analysis*, (2nd ed.), New York: McGraw-Hill Book Company, Copyright © 1974. Reprinted by permission.

Table A.3   Standard Normal Curve Areas

$\Phi(z) = P(Z \leq z)$

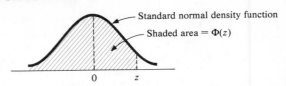

Standard normal density function

Shaded area = $\Phi(z)$

z	0.00	0.01	0.02	0.03	0.04	0.05	0.06	0.07	0.08	0.09
−3.4	0.0003	0.0003	0.0003	0.0003	0.0003	0.0003	0.0003	0.0003	0.0003	0.0002
−3.3	0.0005	0.0005	0.0005	0.0004	0.0004	0.0004	0.0004	0.0004	0.0004	0.0003
−3.2	0.0007	0.0007	0.0006	0.0006	0.0006	0.0006	0.0006	0.0005	0.0005	0.0005
−3.1	0.0010	0.0009	0.0009	0.0009	0.0008	0.0008	0.0008	0.0008	0.0007	0.0007
−3.0	0.0013	0.0013	0.0013	0.0012	0.0012	0.0011	0.0011	0.0011	0.0010	0.0010
−2.9	0.0019	0.0018	0.0017	0.0017	0.0016	0.0016	0.0015	0.0015	0.0014	0.0014
−2.8	0.0026	0.0025	0.0024	0.0023	0.0023	0.0022	0.0021	0.0021	0.0020	0.0019
−2.7	0.0035	0.0034	0.0033	0.0032	0.0031	0.0030	0.0029	0.0028	0.0027	0.0026
−2.6	0.0047	0.0045	0.0044	0.0043	0.0041	0.0040	0.0039	0.0038	0.0037	0.0036
−2.5	0.0062	0.0060	0.0059	0.0057	0.0055	0.0054	0.0052	0.0051	0.0049	0.0048
−2.4	0.0082	0.0080	0.0078	0.0075	0.0073	0.0071	0.0069	0.0068	0.0066	0.0064
−2.3	0.0107	0.0104	0.0102	0.0099	0.0096	0.0094	0.0091	0.0089	0.0087	0.0084
−2.2	0.0139	0.0136	0.0132	0.0129	0.0125	0.0122	0.0119	0.0116	0.0113	0.0110
−2.1	0.0179	0.0174	0.0170	0.0166	0.0162	0.0158	0.0154	0.0150	0.0146	0.0143
−2.0	0.0228	0.0222	0.0217	0.0212	0.0207	0.0202	0.0197	0.0192	0.0188	0.0183
−1.9	0.0287	0.0281	0.0274	0.0268	0.0262	0.0256	0.0250	0.0244	0.0239	0.0233
−1.8	0.0359	0.0352	0.0344	0.0336	0.0329	0.0322	0.0314	0.0307	0.0301	0.0294
−1.7	0.0446	0.0436	0.0427	0.0418	0.0409	0.0401	0.0392	0.0384	0.0375	0.0367
−1.6	0.0548	0.0537	0.0526	0.0516	0.0505	0.0495	0.0485	0.0475	0.0465	0.0455
−1.5	0.0668	0.0655	0.0643	0.0630	0.0618	0.0606	0.0594	0.0582	0.0571	0.0559
−1.4	0.0808	0.0793	0.0778	0.0764	0.0749	0.0735	0.0722	0.0708	0.0694	0.0681
−1.3	0.0968	0.0951	0.0934	0.0918	0.0901	0.0885	0.0869	0.0853	0.0838	0.0823
−1.2	0.1151	0.1131	0.1112	0.1093	0.1075	0.1056	0.1038	0.1020	0.1003	0.0985
−1.1	0.1357	0.1335	0.1314	0.1292	0.1271	0.1251	0.1230	0.1210	0.1190	0.1170
−1.0	0.1587	0.1562	0.1539	0.1515	0.1492	0.1469	0.1446	0.1423	0.1401	0.1379
−0.9	0.1841	0.1814	0.1788	0.1762	0.1736	0.1711	0.1685	0.1660	0.1635	0.1611
−0.8	0.2119	0.2090	0.2061	0.2033	0.2005	0.1977	0.1949	0.1922	0.1894	0.1867
−0.7	0.2420	0.2389	0.2358	0.2327	0.2296	0.2266	0.2236	0.2206	0.2177	0.2148
−0.6	0.2743	0.2709	0.2676	0.2643	0.2611	0.2578	0.2546	0.2514	0.2483	0.2451
−0.5	0.3085	0.3050	0.3015	0.2981	0.2946	0.2912	0.2877	0.2843	0.2810	0.2776
−0.4	0.3446	0.3409	0.3372	0.3336	0.3300	0.3264	0.3228	0.3192	0.3156	0.3121
−0.3	0.3821	0.3783	0.3745	0.3707	0.3669	0.3632	0.3594	0.3557	0.3520	0.3483
−0.2	0.4207	0.4168	0.4129	0.4090	0.4052	0.4013	0.3974	0.3936	0.3897	0.3859
−0.1	0.4602	0.4562	0.4522	0.4483	0.4443	0.4404	0.4364	0.4325	0.4286	0.4247
−0.0	0.5000	0.4960	0.4920	0.4880	0.4840	0.4801	0.4761	0.4721	0.4681	0.4641
0.0	0.5000	0.5040	0.5080	0.5120	0.5160	0.5199	0.5239	0.5279	0.5319	0.5359
0.1	0.5398	0.5438	0.5478	0.5517	0.5557	0.5596	0.5636	0.5675	0.5714	0.5753
0.2	0.5793	0.5832	0.5871	0.5910	0.5948	0.5987	0.6026	0.6064	0.6103	0.6141
0.3	0.6179	0.6217	0.6255	0.6293	0.6331	0.6368	0.6406	0.6443	0.6480	0.6517
0.4	0.6554	0.6591	0.6628	0.6664	0.6700	0.6736	0.6772	0.6808	0.6844	0.6879
0.5	0.6915	0.6950	0.6985	0.7019	0.7054	0.7088	0.7123	0.7157	0.7190	0.7224
0.6	0.7257	0.7291	0.7324	0.7357	0.7389	0.7422	0.7454	0.7486	0.7517	0.7549
0.7	0.7580	0.7611	0.7642	0.7673	0.7704	0.7734	0.7764	0.7794	0.7823	0.7852
0.8	0.7881	0.7910	0.7939	0.7967	0.7995	0.8023	0.8051	0.8078	0.8106	0.8133
0.9	0.8159	0.8186	0.8212	0.8238	0.8264	0.8289	0.8315	0.8340	0.8365	0.8389

Table A.3 Standard Normal Curve Areas *(cont.)* $\Phi(z) = P(Z \le z)$

z	0.01	0.02	0.03	0.04	0.05	0.06	0.07	0.08	0.09
1.0	0.8413	0.8438	0.8461	0.8485	0.8508	0.8531	0.8554	0.8577	0.8599	0.8621
1.1	0.8643	0.8665	0.8686	0.8708	0.8729	0.8749	0.8770	0.8790	0.8810	0.8830
1.2	0.8849	0.8869	0.8888	0.8907	0.8925	0.8944	0.8962	0.8980	0.8997	0.9015
1.3	0.9032	0.9049	0.9066	0.9082	0.9099	0.9115	0.9131	0.9147	0.9162	0.9177
1.4	0.9192	0.9207	0.9222	0.9236	0.9251	0.9265	0.9278	0.9292	0.9306	0.9319
1.5	0.9332	0.9345	0.9357	0.9370	0.9382	0.9394	0.9406	0.9418	0.9429	0.9441
1.6	0.9452	0.9463	0.9474	0.9484	0.9495	0.9505	0.9515	0.9525	0.9535	0.9545
1.7	0.9554	0.9564	0.9573	0.9582	0.9591	0.9599	0.9608	0.9616	0.9625	0.9633
1.8	0.9641	0.9649	0.9656	0.9664	0.9671	0.9678	0.9686	0.9693	0.9699	0.9706
1.9	0.9713	0.9719	0.9726	0.9732	0.9738	0.9744	0.9750	0.9756	0.9761	0.9767
2.0	0.9772	0.9778	0.9783	0.9788	0.9793	0.9798	0.9803	0.9808	0.9812	0.9817
2.1	0.9821	0.9826	0.9830	0.9834	0.9838	0.9842	0.9846	0.9850	0.9854	0.9857
2.2	0.9861	0.9864	0.9868	0.9871	0.9875	0.9878	0.9881	0.9884	0.9887	0.9890
2.3	0.9893	0.9896	0.9898	0.9901	0.9904	0.9906	0.9909	0.9911	0.9913	0.9916
2.4	0.9918	0.9920	0.9922	0.9925	0.9927	0.9929	0.9931	0.9932	0.9934	0.9936
2.5	0.9938	0.9940	0.9941	0.9943	0.9945	0.9946	0.9948	0.9949	0.9951	0.9952
2.6	0.9953	0.9955	0.9956	0.9957	0.9959	0.9960	0.9961	0.9962	0.9963	0.9964
2.7	0.9965	0.9966	0.9967	0.9968	0.9969	0.9970	0.9971	0.9972	0.9973	0.9974
2.8	0.9974	0.9975	0.9976	0.9977	0.9977	0.9978	0.9979	0.9979	0.9980	0.9981
2.9	0.9981	0.9982	0.9982	0.9983	0.9984	0.9984	0.9985	0.9985	0.9986	0.9986
3.0	0.9987	0.9987	0.9987	0.9988	0.9988	0.9989	0.9989	0.9989	0.9990	0.9990
3.1	0.9990	0.9991	0.9991	0.9991	0.9992	0.9992	0.9992	0.9992	0.9993	0.9993
3.2	0.9993	0.9993	0.9994	0.9994	0.9994	0.9994	0.9994	0.9995	0.9995	0.9995
3.3	0.9995	0.9995	0.9995	0.9996	0.9996	0.9996	0.9996	0.9996	0.9996	0.9997
3.4	0.9997	0.9997	0.9997	0.9997	0.9997	0.9997	0.9997	0.9997	0.9997	0.9998

Table A-4 The Incomplete Gamma Function $F(x; \alpha) = \int_0^x \frac{1}{\Gamma(\alpha)} y^{\alpha-1} e^{-y}\, dy$

x \ α	1	2	3	4	5	6	7	8	9	10
1	.632	.264	.080	.019	.004	.001	.000	.000	.000	.000
2	.865	.594	.323	.143	.053	.017	.005	.001	.000	.000
3	.950	.801	.577	.353	.185	.084	.034	.012	.004	.001
4	.982	.908	.762	.567	.371	.215	.111	.051	.021	.008
5	.993	.960	.875	.735	.560	.384	.238	.133	.068	.032
6	.998	.983	.938	.849	.715	.554	.398	.256	.153	.084
7	.999	.993	.970	.918	.827	.699	.550	.401	.271	.170
8	1.000	.997	.986	.958	.900	.809	.687	.547	.407	.283
9		.999	.994	.979	.945	.884	.793	.676	.544	.413
10		1.000	.997	.990	.971	.933	.870	.780	.667	.542
11			.999	.995	.985	.962	.921	.857	.768	.659
12			1.000	.998	.992	.980	.954	.911	.845	.758
13				.999	.996	.989	.974	.946	.900	.834
14				1.000	.998	.994	.986	.968	.938	.891
15					.999	.997	.992	.982	.963	.930

Table A.5 Critical Values $t_{\alpha,\nu}$ for the t Distribution

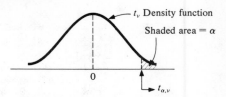

t_ν Density function

Shaded area $= \alpha$

0

$t_{\alpha,\nu}$

ν				α			
	.10	.05	.025	.01	.005	.001	.0005
1	3.078	6.314	12.706	31.821	63.657	318.31	636.62
2	1.886	2.920	4.303	6.965	9.925	22.326	31.598
3	1.638	2.353	3.182	4.541	5.841	10.213	12.924
4	1.533	2.132	2.776	3.747	4.604	7.173	8.610
5	1.476	2.015	2.571	3.365	4.032	5.893	6.869
6	1.440	1.943	2.447	3.143	3.707	5.208	5.959
7	1.415	1.895	2.365	2.998	3.499	4.785	5.408
8	1.397	1.860	2.306	2.896	3.355	4.501	5.041
9	1.383	1.833	2.262	2.821	3.250	4.297	4.781
10	1.372	1.812	2.228	2.764	3.169	4.144	4.587
11	1.363	1.796	2.201	2.718	3.106	4.025	4.437
12	1.356	1.782	2.179	2.681	3.055	3.930	4.318
13	1.350	1.771	2.160	2.650	3.012	3.852	4.221
14	1.345	1.761	2.145	2.624	2.977	3.787	4.140
15	1.341	1.753	2.131	2.602	2.947	3.733	4.073
16	1.337	1.746	2.120	2.583	2.921	3.686	4.015
17	1.333	1.740	2.110	2.567	2.898	3.646	3.965
18	1.330	1.734	2.101	2.552	2.878	3.610	3.922
19	1.328	1.729	2.093	2.539	2.861	3.579	3.883
20	1.325	1.725	2.086	2.528	2.845	3.552	3.850
21	1.323	1.721	2.080	2.518	2.831	3.527	3.819
22	1.321	1.717	2.074	2.508	2.819	3.505	3.792
23	1.319	1.714	2.069	2.500	2.807	3.485	3.767
24	1.318	1.711	2.064	2.492	2.797	3.467	3.745
25	1.316	1.708	2.060	2.485	2.787	3.450	3.725
26	1.315	1.706	2.056	2.479	2.779	3.435	3.707
27	1.314	1.703	2.052	2.473	2.771	3.421	3.690
28	1.313	1.701	2.048	2.467	2.763	3.408	3.674
29	1.311	1.699	2.045	2.462	2.756	3.396	3.659
30	1.310	1.697	2.042	2.457	2.750	3.385	3.646
40	1.303	1.684	2.021	2.423	2.704	3.307	3.551
60	1.296	1.671	2.000	2.390	2.660	3.232	3.460
120	1.289	1.658	1.980	2.358	2.617	3.160	3.373
∞	1.282	1.645	1.960	2.326	2.576	3.090	3.291

Source: This table is reproduced with the kind permission of the Trustees of Biometrica from E. S. Pearson and H. O. Hartley (eds.), *The Biometrica Tables for Statisticians,* vol. 1, 3rd ed., *Biometrica,* 1966.

Table A.6 Critical Values $X^2_{\alpha,\nu}$ for the Chi-Squared Distribution

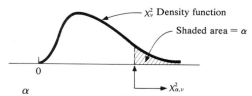

χ^2_ν Density function

Shaded area $= \alpha$

α

$X^2_{\alpha,\nu}$

ν	.995	.99	.975	.95	.90	.10	.05	.025	.01	.005
1	0.000	0.000	0.001	0.004	0.016	2.706	3.843	5.025	6.637	7.882
2	0.010	0.020	0.051	0.103	0.211	4.605	5.992	7.378	9.210	10.597
3	0.072	0.115	0.216	0.352	0.584	6.251	7.815	9.348	11.344	12.837
4	0.207	0.297	0.484	0.711	1.064	7.779	9.488	11.143	13.277	14.860
5	0.412	0.554	0.831	1.145	1.610	9.236	11.070	12.832	15.085	16.748
6	0.676	0.872	1.237	1.635	2.204	10.645	12.592	14.440	16.812	18.548
7	0.989	1.239	1.690	2.167	2.833	12.017	14.067	16.012	18.474	20.276
8	1.344	1.646	2.180	2.733	3.490	13.362	15.507	17.534	20.090	21.954
9	1.735	2.088	2.700	3.325	4.168	14.684	16.919	19.022	21.665	23.587
10	2.156	2.558	3.247	3.940	4.865	15.987	18.307	20.483	23.209	25.188
11	2.603	3.053	3.816	4.575	5.578	17.275	19.675	21.920	24.724	26.755
12	3.074	3.571	4.404	5.226	6.304	18.549	21.026	23.337	26.217	28.300
13	3.565	4.107	5.009	5.892	7.041	19.812	22.362	24.735	27.687	29.817
14	4.075	4.660	5.629	6.571	7.790	21.064	23.685	26.119	29.141	31.319
15	4.600	5.229	6.262	7.261	8.547	22.307	24.996	27.488	30.577	32.799
16	5.142	5.812	6.908	7.962	9.312	23.542	26.296	28.845	32.000	34.267
17	5.697	6.407	7.564	8.682	10.085	24.769	27.587	30.190	33.408	35.716
18	6.265	7.015	8.231	9.390	10.865	25.989	28.869	31.526	34.805	37.156
19	6.843	7.632	8.906	10.117	11.651	27.203	30.143	32.852	36.190	38.580
20	7.434	8.260	9.591	10.851	12.443	28.412	31.410	34.170	37.566	39.997
21	8.033	8.897	10.283	11.591	13.240	29.615	32.670	35.478	38.930	41.399
22	8.643	9.542	10.982	12.338	14.042	30.813	33.924	36.781	40.289	42.796
23	9.260	10.195	11.688	13.090	14.848	32.007	35.172	38.075	41.637	44.179
24	9.886	10.856	12.401	13.848	15.659	33.196	36.415	39.364	42.980	45.558
25	10.519	11.523	13.120	14.611	16.473	34.381	37.652	40.646	44.313	46.925
26	11.160	12.198	13.844	15.379	17.292	35.563	38.885	41.923	45.642	48.290
27	11.807	12.878	14.573	16.151	18.114	36.741	40.113	43.194	46.962	49.642
28	12.461	13.565	15.308	16.928	18.939	37.916	41.337	44.461	48.278	50.993
29	13.120	14.256	16.147	17.708	19.768	39.087	42.557	45.772	49.586	52.333
30	13.787	14.954	16.791	18.493	20.599	40.256	43.773	46.979	50.892	53.672
31	14.457	15.655	17.538	19.280	21.433	41.422	44.985	48.231	52.190	55.000
32	15.134	16.362	18.291	20.072	22.271	42.585	46.194	49.480	53.486	56.328
33	15.814	17.073	19.046	20.866	23.110	43.745	47.400	50.724	54.774	57.646
34	16.501	17.789	19.806	21.664	23.952	44.903	48.602	51.966	56.061	58.964
35	17.191	18.508	20.569	22.465	24.796	46.059	49.802	53.203	57.340	60.272
36	17.887	19.233	21.336	23.269	25.643	47.212	50.998	54.437	58.619	61.581
37	18.584	19.960	22.105	24.075	26.492	48.363	52.192	55.667	59.891	62.880
38	19.289	20.691	22.878	24.884	27.343	49.513	53.384	56.896	61.162	64.181
39	19.994	21.425	23.654	25.695	28.196	50.660	54.572	58.119	62.426	65.473
40	20.706	22.164	24.433	26.509	29.050	51.805	55.758	59.342	63.691	66.766

For $\nu > 40$, $x^2_{\alpha,\nu} \doteq \nu \left(1 - \dfrac{2}{9\nu} + z_\alpha \sqrt{\dfrac{2}{9\nu}} \right)^3$

Source: This table is reproduced with the kind permission of the Trustees of Biometrica from E. S. Pearson and H. O. Hartley (eds.), *The Biometrica Tables for Statisticians*, vol. 1, 3rd ed., *Biometrica*, 1966.

Table A.7 Critical Values F_{α, ν_1, ν_2} for the F Distribution

$\alpha = .05$

ν_2 \ ν_1	1	2	3	4	5	6	7	8	9	10	12	15	20	24	30	40	60	120	∞
1	161.4	199.5	215.7	224.6	230.2	234.0	236.8	238.9	240.5	241.9	243.9	245.9	248.0	249.1	250.1	251.1	252.2	253.3	254.3
2	18.51	19.00	19.16	19.25	19.30	19.33	19.35	19.37	19.38	19.40	19.41	19.43	19.45	19.45	19.46	19.47	19.48	19.49	19.50
3	10.13	9.55	9.28	9.12	9.01	8.94	8.89	8.85	8.81	8.79	8.74	8.70	8.66	8.64	8.62	8.59	8.57	8.55	8.53
4	7.71	6.94	6.59	6.39	6.26	6.16	6.09	6.04	6.00	5.96	5.91	5.86	5.80	5.77	5.75	5.72	5.69	5.66	5.63
5	6.61	5.79	5.41	5.19	5.05	4.95	4.88	4.82	4.77	4.74	4.68	4.62	4.56	4.53	4.50	4.46	4.43	4.40	4.36
6	5.99	5.14	4.76	4.53	4.39	4.28	4.21	4.15	4.10	4.06	4.00	3.94	3.87	3.84	3.81	3.77	3.74	3.70	3.67
7	5.59	4.74	4.35	4.12	3.97	3.87	3.79	3.73	3.68	3.64	3.57	3.51	3.44	3.41	3.38	3.34	3.30	3.27	3.23
8	5.32	4.46	4.07	3.84	3.69	3.58	3.50	3.44	3.39	3.35	3.28	3.22	3.15	3.12	3.08	3.04	3.01	2.97	2.93
9	5.12	4.26	3.86	3.63	3.48	3.37	3.29	3.23	3.18	3.14	3.07	3.01	2.94	2.90	2.86	2.83	2.79	2.75	2.71
10	4.96	4.10	3.71	3.48	3.33	3.22	3.14	3.07	3.02	2.98	2.91	2.85	2.77	2.74	2.70	2.66	2.62	2.58	2.54
11	4.84	3.98	3.59	3.36	3.20	3.09	3.01	2.95	2.90	2.85	2.79	2.72	2.65	2.61	2.57	2.53	2.49	2.45	2.40
12	4.75	3.89	3.49	3.26	3.11	3.00	2.91	2.85	2.80	2.75	2.69	2.62	2.54	2.51	2.47	2.43	2.38	2.34	2.30
13	4.67	3.81	3.41	3.18	3.03	2.92	2.83	2.77	2.71	2.67	2.60	2.53	2.46	2.42	2.38	2.34	2.30	2.25	2.21
14	4.60	3.74	3.34	3.11	2.96	2.85	2.76	2.70	2.65	2.60	2.53	2.46	2.39	2.35	2.31	2.27	2.22	2.18	2.13
15	4.54	3.68	3.29	3.06	2.90	2.79	2.71	2.64	2.59	2.54	2.48	2.40	2.33	2.29	2.25	2.20	2.16	2.11	2.07
16	4.49	3.63	3.24	3.01	2.85	2.74	2.66	2.59	2.54	2.49	2.42	2.35	2.28	2.24	2.19	2.15	2.11	2.06	2.01
17	4.45	3.59	3.20	2.96	2.81	2.70	2.61	2.55	2.49	2.45	2.38	2.31	2.23	2.19	2.15	2.10	2.06	2.01	1.96
18	4.41	3.55	3.16	2.93	2.77	2.66	2.58	2.51	2.46	2.41	2.34	2.27	2.19	2.15	2.11	2.06	2.02	1.97	1.92
19	4.38	3.52	3.13	2.90	2.74	2.63	2.54	2.48	2.42	2.38	2.31	2.23	2.16	2.11	2.07	2.03	1.98	1.93	1.88
20	4.35	3.49	3.10	2.87	2.71	2.60	2.51	2.45	2.39	2.35	2.28	2.20	2.12	2.08	2.04	1.99	1.95	1.90	1.84
21	4.32	3.47	3.07	2.84	2.68	2.57	2.49	2.42	2.37	2.32	2.25	2.18	2.10	2.05	2.01	1.96	1.92	1.87	1.81
22	4.30	3.44	3.05	2.82	2.66	2.55	2.46	2.40	2.34	2.30	2.23	2.15	2.07	2.03	1.98	1.94	1.89	1.84	1.78
23	4.28	3.42	3.03	2.80	2.64	2.53	2.44	2.37	2.32	2.27	2.20	2.13	2.05	2.01	1.96	1.91	1.86	1.81	1.76
24	4.26	3.40	3.01	2.78	2.62	2.51	2.42	2.36	2.30	2.25	2.18	2.11	2.03	1.98	1.94	1.89	1.84	1.79	1.73
25	4.24	3.39	2.99	2.76	2.60	2.49	2.40	2.34	2.28	2.24	2.16	2.09	2.01	1.96	1.92	1.87	1.82	1.77	1.71
26	4.23	3.37	2.98	2.74	2.59	2.47	2.39	2.32	2.27	2.22	2.15	2.07	1.99	1.95	1.90	1.85	1.80	1.75	1.69
27	4.21	3.35	2.96	2.73	2.57	2.46	2.37	2.31	2.25	2.20	2.13	2.06	1.97	1.93	1.88	1.84	1.79	1.73	1.67
28	4.20	3.34	2.95	2.71	2.56	2.45	2.36	2.29	2.24	2.19	2.12	2.04	1.96	1.91	1.87	1.82	1.77	1.71	1.65
29	4.18	3.33	2.93	2.70	2.55	2.43	2.35	2.28	2.22	2.18	2.10	2.03	1.94	1.90	1.85	1.81	1.75	1.70	1.64
30	4.17	3.32	2.92	2.69	2.53	2.42	2.33	2.27	2.21	2.16	2.09	2.01	1.93	1.89	1.84	1.79	1.74	1.68	1.62
40	4.08	3.23	2.84	2.61	2.45	2.34	2.25	2.18	2.12	2.08	2.00	1.92	1.84	1.79	1.74	1.69	1.64	1.58	1.51
60	4.00	3.15	2.76	2.53	2.37	2.25	2.17	2.10	2.04	1.99	1.92	1.84	1.75	1.70	1.65	1.59	1.53	1.47	1.39
120	3.92	3.07	2.68	2.45	2.29	2.17	2.09	2.02	1.96	1.91	1.83	1.75	1.66	1.61	1.55	1.50	1.43	1.35	1.25
∞	3.84	3.00	2.60	2.37	2.21	2.10	2.01	1.94	1.88	1.83	1.75	1.67	1.57	1.52	1.46	1.39	1.32	1.22	1.00

Table A.7 Critical Values F_{α,ν_1,ν_2} for the *F* Distribution (*cont.*)

$\alpha = .01$

ν_2＼ν_1	1	2	3	4	5	6	7	8	9	10	12	15	20	24	30	40	60	120	∞
1	4052	4999.5	5403	5625	5764	5859	5928	5981	6022	6056	6106	6157	6209	6235	6261	6287	6313	6339	6366
2	98.50	99.00	99.17	99.25	99.30	99.33	99.36	99.37	99.39	99.40	99.42	99.43	99.45	99.46	99.47	99.47	99.48	99.49	99.50
3	34.12	30.82	29.46	28.71	28.24	27.91	27.67	27.49	27.35	27.23	27.05	26.87	26.69	26.60	26.50	26.41	26.32	26.22	26.13
4	21.20	18.00	16.69	15.98	15.52	15.21	14.98	14.80	14.66	14.55	14.37	14.20	14.02	13.93	13.84	13.75	13.65	13.56	13.46
5	16.26	13.27	12.06	11.39	10.97	10.67	10.46	10.29	10.16	10.05	9.89	9.72	9.55	9.47	9.38	9.29	9.20	9.11	9.02
6	13.75	10.92	9.78	9.15	8.75	8.47	8.26	8.10	7.98	7.87	7.72	7.56	7.40	7.31	7.23	7.14	7.06	6.97	6.88
7	12.25	9.55	8.45	7.85	7.46	7.19	6.99	6.84	6.72	6.62	6.47	6.31	6.16	6.07	5.99	5.91	5.82	5.74	5.65
8	11.26	8.65	7.59	7.01	6.63	6.37	6.18	6.03	5.91	5.81	5.67	5.52	5.36	5.28	5.20	5.12	5.03	4.95	4.86
9	10.56	8.02	6.99	6.42	6.06	5.80	5.61	5.47	5.35	5.26	5.11	4.96	4.81	4.73	4.65	4.57	4.48	4.40	4.31
10	10.04	7.56	6.55	5.99	5.64	5.39	5.20	5.06	4.94	4.85	4.71	4.56	4.41	4.33	4.25	4.17	4.08	4.00	3.91
11	9.65	7.21	6.22	5.67	5.32	5.07	4.89	4.74	4.63	4.54	4.40	4.25	4.10	4.02	3.94	3.86	3.78	3.69	3.60
12	9.33	6.93	5.95	5.41	5.06	4.82	4.64	4.50	4.39	4.30	4.16	4.01	3.86	3.78	3.70	3.62	3.54	3.45	3.36
13	9.07	6.70	5.74	5.21	4.86	4.62	4.44	4.30	4.19	4.10	3.96	3.82	3.66	3.59	3.51	3.43	3.34	3.25	3.17
14	8.86	6.51	5.56	5.04	4.69	4.46	4.28	4.14	4.03	3.94	3.80	3.66	3.51	3.43	3.35	3.27	3.18	3.09	3.00
15	8.68	6.36	5.42	4.89	4.56	4.32	4.14	4.00	3.89	3.80	3.67	3.52	3.37	3.29	3.21	3.13	3.05	2.96	2.87
16	8.53	6.23	5.29	4.77	4.44	4.20	4.03	3.89	3.78	3.69	3.55	3.41	3.26	3.18	3.10	3.02	2.93	2.84	2.75
17	8.40	6.11	5.18	4.67	4.34	4.10	3.93	3.79	3.68	3.59	3.46	3.31	3.16	3.08	3.00	2.92	2.83	2.75	2.65
18	8.29	6.01	5.09	4.58	4.25	4.01	3.84	3.71	3.60	3.51	3.37	3.23	3.08	3.00	2.92	2.84	2.75	2.66	2.57
19	8.18	5.93	5.01	4.50	4.17	3.94	3.77	3.63	3.52	3.43	3.30	3.15	3.00	2.92	2.84	2.76	2.67	2.58	2.49
20	8.10	5.85	4.94	4.43	4.10	3.87	3.70	3.56	3.46	3.37	3.23	3.09	2.94	2.86	2.78	2.69	2.61	2.52	2.42
21	8.02	5.78	4.87	4.37	4.04	3.81	3.64	3.51	3.40	3.31	3.17	3.03	2.88	2.80	2.72	2.64	2.55	2.46	2.36
22	7.95	5.72	4.82	4.31	3.99	3.76	3.59	3.45	3.35	3.26	3.12	2.98	2.83	2.75	2.67	2.58	2.50	2.40	2.31
23	7.88	5.66	4.76	4.26	3.94	3.71	3.54	3.41	3.30	3.21	3.07	2.93	2.78	2.70	2.62	2.54	2.45	2.35	2.26
24	7.82	5.61	4.72	4.22	3.90	3.67	3.50	3.36	3.26	3.17	3.03	2.89	2.74	2.66	2.58	2.49	2.40	2.31	2.21
25	7.77	5.57	4.68	4.18	3.85	3.63	3.46	3.32	3.22	3.13	2.99	2.85	2.70	2.62	2.54	2.45	2.36	2.27	2.17
26	7.72	5.53	4.64	4.14	3.82	3.59	3.42	3.29	3.18	3.09	2.96	2.81	2.66	2.58	2.50	2.42	2.33	2.23	2.13
27	7.68	5.49	4.60	4.11	3.78	3.56	3.39	3.26	3.15	3.06	2.93	2.78	2.63	2.55	2.47	2.38	2.29	2.20	2.10
28	7.64	5.45	4.57	4.07	3.75	3.53	3.36	3.23	3.12	3.03	2.90	2.75	2.60	2.52	2.44	2.35	2.26	2.17	2.06
29	7.60	5.42	4.54	4.04	3.73	3.50	3.33	3.20	3.09	3.00	2.87	2.73	2.57	2.49	2.41	2.33	2.23	2.14	2.03
30	7.56	5.39	4.51	4.02	3.70	3.47	3.30	3.17	3.07	2.98	2.84	2.70	2.55	2.47	2.39	2.30	2.21	2.11	2.01
40	7.31	5.18	4.31	3.83	3.51	3.29	3.12	2.99	2.89	2.80	2.66	2.52	2.37	2.29	2.20	2.11	2.02	1.92	1.80
60	7.08	4.98	4.13	3.65	3.34	3.12	2.95	2.82	2.72	2.63	2.50	2.35	2.20	2.12	2.03	1.94	1.84	1.73	1.60
120	6.85	4.79	3.95	3.48	3.17	2.96	2.79	2.66	2.56	2.47	2.34	2.19	2.03	1.95	1.86	1.76	1.66	1.53	1.38
∞	6.63	4.61	3.78	3.32	3.02	2.80	2.64	2.51	2.41	2.32	2.18	2.04	1.88	1.79	1.70	1.59	1.47	1.32	1.00

Source: This table is reproduced with the kind permission of the Trustees of Biometrika from E. S. Pearson and H. O. Hartley (eds.), *The Biometrika Tables for Statisticians*, vol. 1, 3rd ed., Biometrika, 1966.

Table A.8 Critical Values $Q_{\alpha, m, \nu}$ for the Studentized Range Distribution

ν	α	2	3	4	5	6	7	8	9	10	11
5	.05	3.64	4.60	5.22	5.67	6.03	6.33	6.58	6.80	6.99	7.17
	.01	5.70	6.98	7.80	8.42	8.91	9.32	9.67	9.97	10.24	10.48
6	.05	3.46	4.34	4.90	5.30	5.63	5.90	6.12	6.32	6.49	6.65
	.01	5.24	6.33	7.03	7.56	7.97	8.32	8.61	8.87	9.10	9.30
7	.05	3.34	4.16	4.68	5.06	5.36	5.61	5.82	6.00	6.16	6.30
	.01	4.95	5.92	6.54	7.01	7.37	7.68	7.94	8.17	8.37	8.55
8	.05	3.26	4.04	4.53	4.89	5.17	5.40	5.60	5.77	5.92	6.05
	.01	4.75	5.64	6.20	6.62	6.96	7.24	7.47	7.68	7.86	8.03
9	.05	3.20	3.95	4.41	4.76	5.02	5.24	5.43	5.59	5.74	5.87
	.01	4.60	5.43	5.96	6.35	6.66	6.91	7.13	7.33	7.49	7.65
10	.05	3.15	3.88	4.33	4.65	4.91	5.12	5.30	5.46	5.60	5.72
	.01	4.48	5.27	5.77	6.14	6.43	6.67	6.87	7.05	7.21	7.36
11	.05	3.11	3.82	4.26	4.57	4.82	5.03	5.20	5.35	5.49	5.61
	.01	4.39	5.15	5.62	5.97	6.25	6.48	6.67	6.84	6.99	7.13
12	.05	3.08	3.77	4.20	4.51	4.75	4.95	5.12	5.27	5.39	5.51
	.01	4.32	5.05	5.50	5.84	6.10	6.32	6.51	6.67	6.81	6.94
13	.05	3.06	3.73	4.15	4.45	4.69	4.88	5.05	5.19	5.32	5.43
	.01	4.26	4.96	5.40	5.73	5.98	6.19	6.37	6.53	6.67	6.79
14	.05	3.03	3.70	4.11	4.41	4.64	4.83	4.99	5.13	5.25	5.36
	.01	4.21	4.89	5.32	5.63	5.88	6.08	6.26	6.41	6.54	6.66
15	.05	3.01	3.67	4.08	4.37	4.59	4.78	4.94	5.08	5.20	5.31
	.01	4.17	4.84	5.25	5.56	5.80	5.99	6.16	6.31	6.44	6.55
16	.05	3.00	3.65	4.05	4.33	4.56	4.74	4.90	5.03	5.15	5.26
	.01	4.13	4.79	5.19	5.49	5.72	5.92	6.08	6.22	6.35	6.46
17	.05	2.98	3.63	4.02	4.30	4.52	4.70	4.86	4.99	5.11	5.21
	.01	4.10	4.74	5.14	5.43	5.66	5.85	6.01	6.15	6.27	6.38
18	.05	2.97	3.61	4.00	4.28	4.49	4.67	4.82	4.96	5.07	5.17
	.01	4.07	4.70	5.09	5.38	5.60	5.79	5.94	6.08	6.20	6.31
19	.05	2.96	3.59	3.98	4.25	4.47	4.65	4.79	4.92	5.04	5.14
	.01	4.05	4.67	5.05	5.33	5.55	5.73	5.89	6.02	6.14	6.25
20	.05	2.95	3.58	3.96	4.23	4.45	4.62	4.77	4.90	5.01	5.11
	.01	4.02	4.64	5.02	5.29	5.51	5.69	5.84	5.97	6.09	6.19
24	.05	2.92	3.53	3.90	4.17	4.37	4.54	4.68	4.81	4.92	5.01
	.01	3.96	4.55	4.91	5.17	5.37	5.54	5.69	5.81	5.92	6.02
30	.05	2.89	3.49	3.85	4.10	4.30	4.46	4.60	4.72	4.82	4.92
	.01	3.89	4.45	4.80	5.05	5.24	5.40	5.54	5.65	5.76	5.85
40	.05	2.86	3.44	3.79	4.04	4.23	4.39	4.52	4.63	4.73	4.82
	.01	3.82	4.37	4.70	4.93	5.11	5.26	5.39	5.50	5.60	5.69
60	.05	2.83	3.40	3.74	3.98	4.16	4.31	4.44	4.55	4.65	4.73
	.01	3.76	4.28	4.59	4.82	4.99	5.13	5.25	5.36	5.45	5.53
120	.05	2.80	3.36	3.68	3.92	4.10	4.24	4.36	4.47	4.56	4.64
	.01	3.70	4.20	4.50	4.71	4.87	5.01	5.12	5.21	5.30	5.37
∞	.05	2.77	3.31	3.63	3.86	4.03	4.17	4.29	4.39	4.47	4.55
	.01	3.64	4.12	4.40	4.60	4.76	4.88	4.99	5.08	5.16	5.23

Table A.8 Critical Values $Q_{\alpha,m,\nu}$ for the Studentized Range Distribution (*cont.*)

					m						
12	13	14	15	16	17	18	19	20	α	ν
7.32	7.47	7.60	7.72	7.83	7.93	8.03	8.12	8.21	.05	5
10.70	10.89	11.08	11.24	11.40	11.55	11.68	11.81	11.93	.01	
6.79	6.92	7.03	7.14	7.24	7.34	7.43	7.51	7.59	.05	6
9.48	9.65	9.81	9.95	10.08	10.21	10.32	10.43	10.54	.01	
6.43	6.55	6.66	6.76	6.85	6.94	7.02	7.10	7.17	.05	7
8.71	8.86	9.00	9.12	9.24	9.35	9.46	9.55	9.65	.01	
6.18	6.29	6.39	6.48	6.57	6.65	6.73	6.80	6.87	.05	8
8.18	8.31	8.44	8.55	8.66	8.76	8.85	8.94	9.03	.01	
5.98	6.09	6.19	6.28	6.36	6.44	6.51	6.58	6.64	.05	9
7.78	7.91	8.03	8.13	8.23	8.33	8.41	8.49	8.57	.01	
5.83	5.93	6.03	6.11	6.19	6.27	6.34	6.40	6.47	.05	10
7.49	7.60	7.71	7.81	7.91	7.99	8.08	8.15	8.23	.01	
5.71	5.81	5.90	5.98	6.06	6.13	6.20	6.27	6.33	.05	11
7.25	7.36	7.46	7.56	7.65	7.73	7.81	7.88	7.95	.01	
5.61	5.71	5.80	5.88	5.95	6.02	6.09	6.15	6.21	.05	12
7.06	7.17	7.26	7.36	7.44	7.52	7.59	7.66	7.73	.01	
5.53	5.63	5.71	5.79	5.86	5.93	5.99	6.05	6.11	.05	13
6.90	7.01	7.10	7.19	7.27	7.35	7.42	7.48	7.55	.01	
5.46	5.55	5.64	5.71	5.79	5.85	5.91	5.97	6.03	.05	14
6.77	6.87	6.96	7.05	7.13	7.20	7.27	7.33	7.39	.01	
5.40	5.49	5.57	5.65	5.72	5.78	5.85	5.90	5.96	.05	15
6.66	6.76	6.84	6.93	7.00	7.07	7.14	7.20	7.26	.01	
5.35	5.44	5.52	5.59	5.66	5.73	5.79	5.84	5.90	.05	16
6.56	6.66	6.74	6.82	6.90	6.97	7.03	7.09	7.15	.01	
5.31	5.39	5.47	5.54	5.61	5.67	5.73	5.79	5.84	.05	17
6.48	6.57	6.66	6.73	6.81	6.87	6.94	7.00	7.05	.01	
5.27	5.35	5.43	5.50	5.57	5.63	5.69	5.74	5.79	.05	18
6.41	6.50	6.58	6.65	6.73	6.79	6.85	6.91	6.97	.01	
5.23	5.31	5.39	5.46	5.53	5.59	5.65	5.70	5.75	.05	19
6.34	6.43	6.51	6.58	6.65	6.72	6.78	6.84	6.89	.01	
5.20	5.28	5.36	5.43	5.49	5.55	5.61	5.66	5.71	.05	20
6.28	6.37	6.45	6.52	6.59	6.65	6.71	6.77	6.82	.01	
5.10	5.18	5.25	5.32	5.38	5.44	5.49	5.55	5.59	.05	24
6.11	6.19	6.26	6.33	6.39	6.45	6.51	6.56	6.61	.01	
5.00	5.08	5.15	5.21	5.27	5.33	5.38	5.43	5.47	.05	30
5.93	6.01	6.08	6.14	6.20	6.26	6.31	6.36	6.41	.01	
4.90	4.98	5.04	5.11	5.16	5.22	5.27	5.31	5.36	.05	40
5.76	5.83	5.90	5.96	6.02	6.07	6.12	6.16	6.21	.01	
4.81	4.88	4.94	5.00	5.06	5.11	5.15	5.20	5.24	.05	60
5.60	5.67	5.73	5.78	5.84	5.89	5.93	5.97	6.01	.01	
4.71	4.78	4.84	4.90	4.95	5.00	5.04	5.09	5.13	.05	120
5.44	5.50	5.56	5.61	5.66	5.71	5.75	5.79	5.83	.01	
4.62	4.68	4.74	4.80	4.85	4.89	4.93	4.97	5.01	.05	∞
5.29	5.35	5.40	5.45	5.49	5.54	5.57	5.61	5.65	.01	

Table A.9 Upper-Tail Critical Values and Probabilities for the Null Distribution of the Wilcoxon Signed-Rank Statistic S_+

$$P_0(S_+ \geq c_1) = P(S_+ \geq c_1 \text{ when } H_0 \text{ is true})$$

n	c_1	$P_0(S_+ \geq c_1)$	n	c_1	$P_0(S_+ \geq c_1)$
3	6	.125		78	.011
4	9	.125		79	.009
	10	.062		81	.005
5	13	.094	14	73	.108
	14	.062		74	.097
	15	.031		79	.052
6	17	.109		84	.025
	19	.047		89	.010
	20	.031		92	.005
	21	.016	15	83	.104
7	22	.109		84	.094
	24	.055		89	.053
	26	.023		90	.047
	28	.008		95	.024
8	28	.098		100	.011
	30	.055		101	.009
	32	.027		104	.005
	34	.012	16	93	.106
	35	.008		94	.096
	36	.004		100	.052
9	34	.102		106	.025
	37	.049		112	.011
	39	.027		113	.009
	42	.010		116	.005
	44	.004	17	104	.103
10	41	.097		105	.095
	44	.053		112	.049
	47	.024		118	.025
	50	.010		125	.010
	52	.005		129	.005
11	48	.103	18	116	.098
	52	.051		124	.049
	55	.027		131	.024
	59	.009		138	.010
	61	.005		143	.005
12	56	.102	19	128	.098
	60	.055		136	.052
	61	.046		137	.048
	64	.026		144	.025
	68	.010		152	.010
	71	.005		157	.005
13	64	.108	20	140	.101
	65	.095		150	.049
	69	.055		158	.024
	70	.047		167	.010
	74	.024		172	.005

Source: Adapted from W. J. Dixon and F. J. Massey, Jr., *Introduction to Statistical Analysis*, (3rd ed.), New York: McGraw-Hill Book Company, Copyright © 1969. Reprinted by permission.

Table A.10 Upper-Tail Critical Values and Probabilities for the Null Distribution of the Wilcoxon Rank-Sum Statistic W

$$P_0(W \geq c) = P(W \geq c \text{ when } H_0 \text{ is true})$$

m	n	c	$P_0(W \geq c)$	m	n	c	$P_0(W \geq c)$
3	3	15	.05			40	.004
	4	17	.057		6	40	.041
		18	.029			41	.026
	5	20	.036			43	.009
		21	.018			44	.004
	6	22	.048		7	43	.053
		23	.024			45	.024
		24	.012			47	.009
	7	24	.058			48	.005
		26	.017		8	47	.047
		27	.008			49	.023
	8	27	.042			51	.009
		28	.024			52	.005
		29	.012	6	6	50	.047
		30	.006			52	.021
4	4	24	.057			54	.008
		25	.029			55	.004
		26	.014		7	54	.051
	5	27	.056			56	.026
		28	.032			58	.011
		29	.016			60	.004
		30	.008		8	58	.054
	6	30	.057			61	.021
		32	.019			63	.01
		33	.010			65	.004
		34	.005	7	7	66	.049
	7	33	.055			68	.027
		35	.021			71	.009
		36	.012			72	.006
		37	.006		8	71	.047
	8	36	.055			73	.027
		38	.024			76	.01
		40	.008			78	.005
		41	.004	8	8	84	.052
5	5	36	.048			87	.025
		37	.028			90	.01
		39	.008			92	.005

Source: Adapted from W. J. Dixon and F. J. Massey, Jr., *Introduction to Statistical Analysis,* (3rd ed.), New York: McGraw-Hill Book Company, Copyright © 1969. Reprinted by permission.

Table A.11 Critical Constant c for the Wilcoxon Signed-Rank Interval

$$(\overline{x}_{(n(n+1)/2 \,-\, c+1)}, \ \overline{x}_{(c)})$$

n	Confidence level (%)	c	n	Confidence level (%)	c	n	Confidence level (%)	c
5	93.8	15	13	99.0	81	20	99.1	173
	87.5	14		95.2	74		95.2	158
6	96.9	21		90.6	70		90.3	150
	93.7	20	14	99.1	93	21	99.0	188
	90.6	19		95.1	84		95.0	172
7	98.4	28		89.6	79		89.7	163
	95.3	26	15	99.0	104	22	99.0	204
	89.1	24		95.2	95		95.0	187
8	99.2	36		90.5	90		90.2	178
	94.5	32	16	99.1	117	23	99.0	221
	89.1	30		94.9	106		95.2	203
9	99.2	44		89.5	100		90.2	193
	94.5	39	17	99.1	130	24	99.0	239
	90.2	37		94.9	118		95.1	219
10	99.0	52		90.2	112		89.9	208
	95.1	47	18	99.0	143	25	99.0	257
	89.5	44		95.2	131		95.2	236
11	99.0	61		90.1	124		89.9	224
	94.6	55	19	99.1	158			
	89.8	52		95.1	144			
12	99.1	71		90.4	137			
	94.8	64						
	90.8	61						

Source: Derived from W. J. Dixon and F. J. Massey, Jr., *Introduction to Statistical Analysis,* (3rd ed.), New York: McGraw-Hill Book Company, Copyright ©1969. Reprinted by permission.

Table A.12 Critical Constant *c* for the Wilcoxon Rank-Sum Interval

$$(d_{ij(mn-c+1)}, d_{ij(c)})$$

Larger Sample Size	Smaller Sample Size							
	5		6		7		8	
	Confidence Level (%)	*c*	Confidence Level (%)	*c*	Confidence Level (%)	*c*	Confidence Level (%)	*c*
5	99.2	25						
	94.4	22						
	90.5	21						
6	99.1	29	99.1	34				
	94.8	26	95.9	31				
	91.8	25	90.7	29				
7	99.0	33	99.2	39	98.9	44		
	95.2	30	94.9	35	94.7	40		
	89.4	28	89.9	33	90.3	38		
8	98.9	37	99.2	44	99.1	50	99.0	56
	95.5	34	95.7	40	94.6	45	95.0	51
	90.7	32	89.2	37	90.6	43	89.5	48
9	98.8	41	99.2	49	99.2	56	98.9	62
	95.8	38	95.0	44	94.5	50	95.4	57
	88.8	35	91.2	42	90.9	48	90.7	54
10	99.2	46	98.9	53	99.0	61	99.1	69
	94.5	41	94.4	48	94.5	55	94.5	62
	90.1	39	90.7	46	89.1	52	89.9	59
11	99.1	50	99.0	58	98.9	66	99.1	75
	94.8	45	95.2	53	95.6	61	94.9	68
	91.0	43	90.2	50	89.6	57	90.9	65
12	99.1	54	99.0	63	99.0	72	99.0	81
	95.2	49	94.7	57	95.5	66	95.3	74
	89.6	46	89.8	54	90.0	62	90.2	70

Larger Sample Size	Smaller Sample Size							
	9		10		11		12	
	Confidence Level (%)	*c*	Confidence Level (%)	*c*	Confidence Level (%)	*c*	Confidence Level (%)	*c*
9	98.9	69						
	95.0	63						
	90.6	60						
10	99.0	76	99.1	84				
	94.7	69	94.8	76				
	90.5	66	89.5	72				
11	99.0	83	99.0	91	98.9	99		
	95.4	76	94.9	83	95.3	91		
	90.5	72	90.1	79	89.9	86		
12	99.1	90	99.1	99	99.1	108	99.0	116
	95.1	82	95.0	90	94.9	98	94.8	106
	90.5	78	90.7	86	89.6	93	89.9	101

Source: Derived from W. J. Dixon and F. J. Massey, Jr., *Introduction to Statistical Analysis*, (3rd ed.), New York: McGraw-Hill Book Company, Copyright © 1969. Reprinted by permission.

Index